Nanobiotechnology

Jameel M. Al-Khayri · Mohammad Israil Ansari ·
Akhilesh Kumar Singh
Editors

Nanobiotechnology

Mitigation of Abiotic Stress in Plants

Springer

Editors
Jameel M. Al-Khayri
Department of Agricultural Biotechnology
King Faisal University
Al-Ahsa, Saudi Arabia

Mohammad Israil Ansari
Department of Botany
University of Lucknow
Lucknow, Uttar Pradesh, India

Akhilesh Kumar Singh
Department of Biotechnology
School of Life Sciences
Mahatma Gandhi Central University
Motihari, Bihar, India

ISBN 978-3-030-73605-7 ISBN 978-3-030-73606-4 (eBook)
https://doi.org/10.1007/978-3-030-73606-4

This Springer imprint is published by the registered company Springer Nature Switzerland AG
The registered company address is: Gewerbestrasse 11, 6330 Cham, Switzerland

Preface

The emergence of nanotechnology have opened up exciting opportunities for novel applications in agriculture, food, medicine, and biotechnology industries. Nanotechnology has the potential to modernize agricultural research and practice, although it has gained momentum in the agriculture sector over the last decade. Abiotic stresses are important constraints that adversely affect the production of agricultural crops. Nanobiotechnology may be a boon for the mitigation of plant abiotic stress impact.

This book provides up-to-date knowledge of the promising field of nanobiotechnology with emphasis on the mitigation approaches to combat plant abiotic stress factors including drought, salinity, waterlog, temperature extremes, mineral nutrients, and heavy metals. These factors adversely affect the growth as well as yield of crop plants worldwide especially under the global climate change. The book consists of 24 chapters discussing the status and prospects of this cutting-edge technology in relation to the mitigation of the adverse impact of the abovementioned stress factors. Moreover, it highlights contemporary knowledge of tolerance mechanisms and the role of signaling molecules and enzyme regulation as well as the applications of nanobiotechnology in agriculture.

The book is perceived as an important reference source for plant scientists and breeders interested in understanding the mechanisms of abiotic stress in pursue of developing stress-tolerant crops to support agricultural sustainability and food security. It is valuable for professional researchers as well as advance graduate students interested in nanotechnology fundamentals and utilization.

The chapters are contributed by 61 internationally reputable scientists from 10 countries and subjected to review process to assure quality presentation and scientific accuracy.The chapters start with an introduction covering related backgrounds and provide in-depth discussion of the subject supported with a total 95 of high-quality color illustrations and relevant 31 data tables. The chapters conclude with recommendations for future research directions and a comprehensive list of up-to-date pertinent references to facilitate further reading. The editors convey their appreciation to all the contributors for their delegacy and to Springer for the opportunity to publish this work.

Al-Ahsa, Saudi Arabia — Jameel M. Al-Khayri
Lucknow, India — Mohammad Israil Ansari
Motihari, India — Akhilesh Kumar Singh

Contents

Editors and Contributors

About the Editors

Prof. Jameel M. Al-Khayri is Professor of plant biotechnology affiliated with the Department of Agricultural Biotechnology, King Faisal University, Saudi Arabia. He received B.S. in Biology in 1984 from the University of Toledo, M.S. in Agronomy in 1988, and Ph.D. in Plant Science in 1991 from the University of Arkansas. He is member the International Society for Horticultural Science and Society for In Vitro Biology as well as the National Correspondent of the International Association of Plant Tissue Culture and Biotechnology. For the last three decades, he dedicated his research efforts to date palm biotechnology. He has authored over seventy research articles in referred international journals, thirty chapters, and edited several journal special issues. In addition, edited twenty reference books on biotechnology, genetic resources and advances in plant breeding strategies. He has been involved in organizing international scientific conferences and contributed to numerous research presentations. In addition to teaching, students advising, and research, he held administrative responsibilities as the Assistant Director of Date Palm Research Center, Head of Department of Plant Biotechnology, and Vice Dean for Development and Quality Assurance. Dr. Al-Khayri served as Member of Majlis Ash-Shura (Saudi Legislative Council) for the 2009–2012 term. Currently, he is maintaining an active research program on date palm focusing in vitro culture, secondary metabolites production, genetic engineering and mutagenesis to enhance tolerance to abiotic and biotic stress.

Dr. Mohammad Israil Ansari is biologist with expertise in thrust areas viz. Plant Stress Physiology, Plant Molecular Biology, Nanotechnology, Post-harvest and Agricultural Biotechnology. Before joining University of Lucknow as Faculty, Dr. Ansari worked as Professor of Biotechnology at Amity Institute of Biotechnology, Amity University, Lucknow Campus, Lucknow, India (2007–2016). From 2000 to 2007 Dr. Ansari underwent international learning as Fellow of Academia Sinica and National Science Council Taiwan. Dr. Ansari has been associated with Institute of Botany, Academia Sinica, Taipei, Taiwan, Graduate Institute of Plant Biology, National Taiwan University, Taiwan, and Institute of Plant and Microbial Biology, Academia Sinica, Taipei, Taiwan. Dr. Ansari graduated and post-graduated from Department of Botany, Aligarh Muslim University, Aligarh, India, and doctoral degree from Dr. Ram Manohar Lohia Avadh University, Faizabad, India. Dr. Ansari has 75 research papers in international journals of high repute, several books (authored and edited Springer, Elsevier), book chapters, review articles, Post-doctoral, Ph.Ds. to his credit. Dr. Ansari has completed several research projects funded by Department of Science and Technology, Govt. of India. Dr. Ansari is editorial board member, reviewer of several international journals as well as member of several Academic Board/Faculty Board/Board of Studies/Research and Development bodies. Dr. Ansari has collaborated with several international groups and delivered a number of keynote, invited lecture at national, international level.

Dr. Akhilesh Kumar Singh is currently an Assistant Professor in the Department of Biotechnology, Mahatma Gandhi Central University (MGCU), Motihari, Bihar, India. Before joining MGCU, Dr. Singh worked as faculty at Amity Institute of Biotechnology, Amity University Uttar Pradesh, Lucknow Campus, Lucknow, India (September, 2008 to October, 2019). He has more than 12 years of teaching and multidisciplinary research experience in microbial physiology and industrial biotechnology. His research focus is on polyhydroxyalkanoate bioplastic production, metallic nanobiotechnology, nitric oxide signaling, and biofuel production. Dr. Singh holds a Ph.D. in agricultural and food engineering and an M.Tech. in agricultural biotechnology from prestigious Indian Institute of Technology (IIT), Kharagpur, India, with a master's degree in biotechnology from another prestigious institute, i.e., IIT, Roorkee, India. He has qualified Graduate Aptitude Test in Engineering (GATE) in Life Science conducted by IIT Kanpur on behalf of the Ministry of Human Resources Development, Government of India. He was awarded the prestigious Council of Scientific and Industrial Research (CSIR)-University Grants Commission (UGC) Junior Research Fellowship (CSIR-UGC JRF) by Council of Scientific and Industrial Research, India. In addition, Dr. Singh has also qualified Indian Council of Agriculture Research (ICAR)-National Eligibility Test (NET), i.e., ICAR-NET in "Agricultural Biotechnology", conducted by ICAR, India. He has been credited with 20 research papers and 10 review articles in international journals of high repute, 19 chapters in reputed books and has contributed to the book on "*Pseudomonas aeruginosa* MTCC 7925: Biofactory for Novel SCL-LCL-PHA". Dr. Singh has been on the "editorial board" as well as the "panel of reviewers" of several reputed journals of Elsevier, Springer Nature, John Wiley & Sons and so on. Dr. Singh has also supervised Ph.D., M.Tech., M.Sc. and B.Tech. students.

Contributors

Fadwa W. Abdulqahar Department of Food Sciences, University of Anbar, Ramadi, Iraq

Ausaf Ahmad Amity Institute of Biotechnology, Amity University Uttar Pradesh, Lucknow, India

Shamshad Ahmad Department of Environmental Science, School for Environmental Sciences, Babasaheb Bhimrao Ambedkar University, Lucknow, UP, India

Abdulkareem A. Alanezi Department of Pharmaceutics, College of Pharmacy, University of Hafr Al Batin, Hafr Al Batin, Saudi Arabia

Fawaz Alasmari Department of Pharmacology and Toxicology, College of Pharmacy, King Saud University, Riyadh, Saudi Arabia

Ayoob O. Alfalahi Department of Plant Protection, College of Agriculture, University of Anbar, Ramadi, Iraq

Qasim Alhadidi Department of Pharmacy, Salah Al-Din Health Directorate, Salah Al-Din, Iraq

Jameel M. Al-Khayri Department of Agricultural Biotechnology, College of Agriculture and Food Sciences, King Faisal University, Al-Ahsa, Saudi Arabia

Maysaa T. Alloosh Department of Biochemistry and Microbiology, Faculty of Pharmacy, Al-Baath University, Homs, Syria;
Fungal of Biochemistry and Microbiology, Faculty of Pharmacy, Al-Baath University, Homs, Syria;
Faculty of Pharmacy, Department of Biochemistry and Microbiology, Al-Baath University, Homs, Syria

Abdulsalam K. Almuhammady Arab Center for Nanotechnology, Cairo, Egypt

Lina M. Alnaddaf Biotechnology and Molecular Biology, Faculty of Agricultural, Al-Baath University, Homs, Syria

Faleh Alqahtani Department of Pharmacology and Toxicology, College of Pharmacy, King Saud University, Riyadh, Saudi Arabia

Saima Amjad Department of Zoology, University of Lucknow, Lucknow, UP, India

Mohammad Israil Ansari Department of Botany, University of Lucknow, Lucknow, Uttar Pradesh, India

Priyanka Bhagat School of Biotechnology, Shri Mata Vaishno Devi University, Katra J&K, India

Omer I. Fantoukh Department of Pharmacognosy, College of Pharmacy, King Saud University, Riyadh, Saudi Arabia

Garima Gupta Amity Institute of Biotechnology, Amity University Uttar Pradesh, Lucknow campus, Lucknow, India

Ziyaul Haque Amity Institute of Biotechnology, Amity University Uttar Pradesh, Lucknow, India

Mohammed Shariq Iqbal Amity Institute of Biotechnology, Amity University Uttar Pradesh, Lucknow campus, Lucknow, India

Zahra Iqbal Department of Biochemistry, Faculty of Science, Chulalongkorn University, Bangkok, Thailand

Syed Uzma Jalil Amity Institute of Biotechnology, Amity University Uttar Pradesh, Lucknow Campus, Lucknow, India

Ratna Kalita Department of Agricultural Biotechnology, Assam Agricultural University, Jorhat, Assam, India

K. Kisku Department of Botany, Ravenshaw University, Cuttack, Odisha, India

Mojtaba Kordrostami Nuclear Agriculture Research School, Nuclear Science and Technology Research Institute (NSTRI), Karaj, Iran

Charu Lata CSIR-National Institute of Science Communication and Information Resources, New Delhi, India

Mohammad Mafakheri Department of Horticulture Sciences, Faculty of Agricultural Sciences, University of Guilan, Rasht, Iran

Rakshit Manhas School of Biotechnology, Shri Mata Vaishno Devi University, Katra J&K, India

Anjuman Gul Memon Department of Biochemistry, College of Medicine, Qassim University, Buraydah, Kingdom of Saudi Arabia

Umesh C. Naik Department of Botany, Ravenshaw University, Cuttack, Odisha, India

Brijesh Pandey Department of Biotechnology, School of Life Sciences, Mahatma Gandhi Central University, Motihari, 845401 Bihar, India

Atin Kumar Pathak School of Energy Management, Faculty of Engineering, Shri Mata Vaishno Devi University, Katra J&K, India

Munmi Phukon Cruickshank Building Institute of Biotechnology & Environmental Science, University of Aberdeen, Aberdeen, UK

Paras Porwal Amity Institute of Biotechnology, Amity University Uttar Pradesh, Lucknow Campus, Lucknow, Uttar Pradesh, India

Vivek Prasad Department of Botany, University of Lucknow, Lucknow, India

Nasrin Rahman Department of Agricultural Biotechnology, Assam Agricultural University, Jorhat, Assam, India

Neelu Raina School of Biotechnology, Shri Mata Vaishno Devi University, Katra J&K, India

Vinaik Raina School of Biotechnology, Shri Mata Vaishno Devi University, Katra J&K, 182320 India

Nedhu Rani School of Biotechnology, Shri Mata Vaishno Devi University, Katra J&K, India

Saurabh Singh Rathore Department of Biotechnology, School of Life Sciences, Mahatma Gandhi Central University, Motihari, 845401 Bihar, India

Swati Sachdev Department of Environmental Science, School for Environmental Sciences, Babasaheb Bhimrao Ambedkar University, Lucknow, UP, India

Oliva Saha Department of Agricultural Biotechnology, Assam Agricultural University, Jorhat, Assam, India

Maysoun M. Saleh Genetic Resources Department, General Commission for Scientific Agricultural Research (GCSAR), Damascus, Syria

Khaled F.M. Salem Department of Biology, College of Science and Humanitarian Studies, Shaqra University, Qwaieah, Saudi Arabia

Mohammad Serajuddin Department of Zoology, University of Lucknow, Lucknow, UP, India

Laxuman Sharma Department of Horticulture, Sikkim University, Gangtok, India

Preeti Sharma School of Biotechnology, Shri Mata Vaishno Devi University, Katra J&K, India

Sakshi Sharma School of Biotechnology, Shri Mata Vaishno Devi University, Katra J&K, India

Akhilesh Kumar Singh Department of Biotechnology, School of Life Sciences, Mahatma Gandhi Central University, Motihari, 845401 Bihar, India

Rishi Kumar Singh School of Biotechnology, Mangalmay Institute of Management and Technology, Greater Noida, Uttar Pradesh, India

Sujeet Pratap Singh Amity Institute of Biotechnology, Amity University Uttar Pradesh, Lucknow Campus, Lucknow, Uttar Pradesh, India

Parvez Singh Slathia School of Biotechnology, Shri Mata Vaishno Devi University, Katra J&K, India

Garima Slathia School of Biotechnology, Shri Mata Vaishno Devi University, Katra J&K, India

Sashi Sonkar Department of Botany, Bankim Sardar College, Tangrakhali, South 24 Parganas, 743329 West Bengal, India

Asad Syed Department of Botany and Microbiology, College of Science, King Saud University, Riyadh, Saudi Arabia

Rimpi Thakur School of Biotechnology, Shri Mata Vaishno Devi University, Katra J&K, India

Shalini Tiwari Council of Scientific and Industrial Research-National Botanical Research Institute (CSIR-NBRI), Microbial Technologies Division, Lucknow, India

1.1 Introduction

Plants as sessile organisms began their evolution in a fundamentally hostile terrestrial environment approximately 700 million ago, and gradually made the land hospitable to be colonized by other organisms (Hotton et al. 2001; Selden and Edwards 1989). Since ever, plants have successfully developed a large array of adaptation mechanisms that enable them to response properly to environmental stressors (i.e., water stress: flood and drought, high salinity, extreme temperatures: cold and heat, heavy metal toxicity) (Bray 2000; Wani et al. 2016).

Of the most prevalent type of abiotic stresses, drought will severely affect nearly 45% of arable lands in the world by 2100 (Field et al. 2014). Water is the most essential component that if water would be available, every possible ecological niche regardless of how extreme it could be colonized by organisms (Wood 2005). Drought is a prolonged period with the absence of rainfall or irrigation and mainly expected in arid and semi-arid regions. The major water consumption is in the agriculture sector which accounts for over 70% of harvesting underground water resources, chiefly in underdeveloped nations. Around 90% of arable under cultivation lands worldwide directly depend on rainfall. By the end of the twenty-first century, drought will severely affect nearly 45% of arable lands in the world. Salinity is another common place for important biotic stress known for its notorious multidimensional effects on plant performance (Burke et al. 2006; Dai 2011; Vibha 2016). Salinization of arable land has increasingly become a limiting factor in agriculture in particular with the gradual increase in global temperature by roughly 1 °C over the last century (Fig. 1.1). This increase has exacerbated the situation through intensifying the evaporation rate from soil (Nouman et al. 2018; Zhao et al. 2017), thereby disturbing the hydrological paradigms that again is a major source of stress for the agriculture sector. Salinization occurs either naturally or anthropologically through mismanagement of water resources and soil degradation with intense agricultural practice. A large portion of arable land (i.e., ~1 million hectares) is experiencing negative impacts of salinity in addition to the fact that the superiority of irrigated lands over rain-fed in terms of yield facilitates the situation in the favor of salinization (Colla et al. 2010; Munns and Gilliham 2015). By the appearance of climate change-driven impacts, the incident of abiotic stresses is on the rise particularly for high temperature and heat waves, which intensifies the severity of other stresses, in which the only 1 °C rising in global temperature causes a massive reduction in crop productivity (Iizumi et al. 2017; Zhao et al. 2017). The occurrence of cold stress as another extreme weather events similarly affected by climate change. Even though scholars have mainly zeroed on high-temperature stress, low-temperature stress is threatening plant productivity in a large scale owing to variability in climatic phenomena in the recent decades (Budhathoki and Zander 2019; Thakur and Nayyar 2013).

Another dimension of climate change manifested itself in meteorological turmoil that causes unpredictability in terms of time and intensity with significant localization of rainfalls that have catastrophic floods in agricultural lands as aftermath

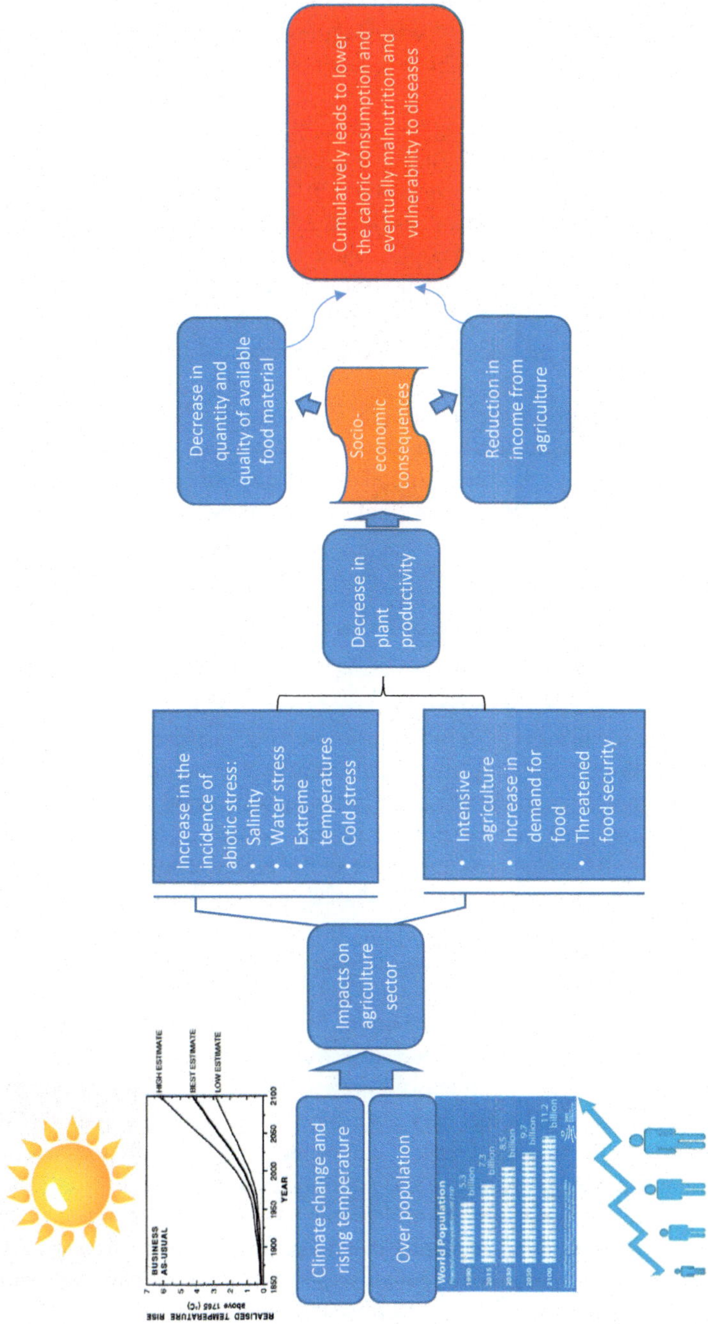

Fig. 1.1 Indirect socio-economic consequences of climate change and increasing population through negatively affecting agriculture sector

(Bailey-Serres and Voesenek 2010; Onyekachi et al. 2019). Concerning the toxicity of heavy metals in the rhizosphere, the accumulation of these elements can be attributed to various sources namely waterlogging (e.g., manganese, iron), erosion of bedrocks (e.g., nickel, cobalt, cadmium, lead), soil acidification (e.g., manganese, zinc) and anthropogenic activity (zinc, nickel, cobalt, copper, cadmium, molybdenum, chromium and lead) (Mengel et al. 2001; White et al. 2013; White and Pongrac 2017), which thanks to climate change, the sources are expanding and can considerably affect crops yield (Fageria et al. 2010). More severely, the occurrence of one stress facilitates the circumstances for other stresses, particularly high temperature and drought or salinity and drought, which often occur simultaneously (Sahin et al. 2018; Shah and Paulsen 2003). To make the matter worse, global climate change as a human-made phenomenon is dangerously jeopardizing the food production by imposing and increasing the incident of co-occurrence of the abiotic stresses at a dramatic rate.

This chapter summarizes the impacts of environmental stresses on social and economic status worldwide and give an updated perspective using most recent populations. Additionally, the morphological and physiological responses as well as tolerance mechanisms developed in plants against these stresses are discussed.

1.2 Socio-Economic Consequences of Abiotic Stress on Crop Production

A dramatic increase in average temperature over the last century has been enormously effective in orchestrating the circumstances for salinizing the arable land through increasing evaporation rate, instability in soil water content by floods or drought, and fluctuation in precipitation paradigms, which entirely severely affecting global food security. Human-caused increase in the global temperature reached 1 °C in 2017, which given the recent estimation that temperature will continue to rise ~0.2 °C per decade (Allen et al. 2018). So, conclusion that can be made is the exacerbation of the abiotic stress effects of crops and jeopardize the food security livelihood of a significant portion of the people on the earth. However, increasing the yield of some crops such as maize estimated to benefit from the rising global temperature in some areas, since a higher CO_2 concentration in the atmosphere as well as a higher temperature accelerates growth and development and biomass production. By 2025, complete water shortage will affect over 1.8 billion of the world population (Fig. 1.1), additionally, 65% of the population may face water stress (Lal 2018). Extreme fluctuation in climatic events generating socio-economic burdens worldwide specifically in developing nations, where agricultural products serve as an important source of financial income for families besides its role in providing food directly (Fig. 1.1). Among abiotic stresses, drought is probably the most economically disasters one, in a combination of drought and extreme high temperature stress during a 3-decade timeframe (1980–2010), a crop loss of worth about US $2 billion projected. Over half of rice cultivated farms as substantial food commodity in Asia is estimated to negatively influenced by water deficit stress (Bouman et al. 2005; Singh et al. 2018), whereas 95% of the world rice production is consumed in Asia, which indicates the scale of the threat that two-third world population will be faced (Dey et al. 2018). Lately, Ray et al. (2019) conducted a comprehensive assessment on climate change impact on the production of top ten crops (i.e., barley, cassava, maize, oil palm, rapeseed, rice, sorghum, soybean, sugarcane, and wheat: composing 85% of global consumable calories). Further, with a data collection from 1974 to 2013 at approximately 20,000 political units, it was discovered that roughly 1% decrease in crop production in these main crops accounted for abiotic stresses imposed by climate change worldwide, which means that effects of global warming are already in place. Of the three major cereal crops, rice, a major production loss was observed in India and Vietnam with approximately 2.2 million tons (Aggarwal and Mall 2002) and 1 million mt (Peng et al. 2004), respectively, wheat production similarly affected especially in Turkey with 0.8 million mt. Analogously, in South and North America changes in climate negatively influenced the production of three top kinds of cereal, whereas the changes seemed to be in favor of maize, sugarcane, oil palm and soybean production (Mourtzinis et al. 2015; Tack et al. 2015). During the abovementioned period, 3% reduction in consumable calories from the top ten crops of Australian people observed (Hochman et al. 2017). In the case of sub-Saharan Africa, this decrease was up to 1.8% despite the increase in Cassava production or some country-specific increases in the crops of interest. European country suffered the most with the highest production loss in top ten consumable calories owing to climate change generated negative effects in France 24%, Germany 11%, Hungary 10%, Romania 7%, Italy 7%, Spain 4% and Ireland 3% (Ray et al. 2019). The production losses worth billions of dollar, which renders vulnerable financial capability and food security; thus, putting a large portion of the world population in a greater risk. Additionally, half of the countries that have ongoing food insecurity issues, to make the matter worse, experience significant production losses.

Health-associated impacts of climate change on society can be reflected in the malnourishment in children that is projected to rise from 8.5 to 10.5% in a base case scenario. Interestingly, climate change-driven effects may be positive in temperate zones, however reduces the yield in tropical regions. Due to an increase in inputs required in the production process of agricultural products, the price in the favor of producers will rise, but affect the net consumers of agricultural products reside in urban or rural areas (Al et al. 2008; Budhathoki and Zander 2019).

Besides the substantial socio-economic impact of abiotic stress, these stresses cumulatively impose deleterious impacts on crop productivity through generating osmotic pressure, ionic toxicity, oxidative damage, and finally inadequacy in nutrient elements. As mentioned earlier, plants have evolved a countless number of adaptive tolerance mechanisms that can greatly contribute to stress tolerance (Bohnert et al. 2006; Waqas et al. 2019). Obtaining a profound understanding of how crops respond, develop, and employ tolerance mechanisms under stress is critical in having a clear picture to address the increasing impact of abiotic stress in the climate-changing era.

1.3 Crops Response to Abiotic Stress

Unlike animals, plants are immobile and cannot escape detrimental conditions that directly aim their overall function, hence crop plant responses to these situations enable them to cope with new changes. The abiotic stresses imposed are interlinked and may multidimensionally depress growth and yield formation of the crops through osmotic stress, oxidative stress and disruption of ion distribution, water relations, and plant cell homeostasis. These conditions can provoke the tolerance mechanisms that counteract the abiotic stress by a long list of morphological, physiological, and molecular modifications to exercise damage control (Bray 2000; Wani et al. 2016).

Responses to abiotic stresses involve changes that morphologically includes: leaf area reduction, increase in wax content and decrease in stem size, damage the productivity and reproduction processes under water stress and salinity, physiologically: disrupting water relations, stomatal conductivity, and transpiration, biochemically: increase in antioxidant and non-enzymes and osmoprotectants and finally molecularly: increase in biosynthesis of phytohormones in particular abscisic acid (ABA), specific proteins and transcription factors (DREB, ZIP, and WRKY) (Conesa et al. 2016; Ding et al. 2016; Lim et al. 2013; Singh et al. 2015). The abiotic stress trigger complex processes in crop plants that the precise interpretation and deciphering the network are rather difficult. Here we attempted to provide the responses and tolerance mechanisms in crop plants to cope with abiotic stresses.

1.3.1 Growth and Productivity

A determinative factor in the profitability of farming crops is a specific level of density that affects every agricultural practice up to harvest. Taking into consideration that intensity of abiotic stress on plant growth has a significant growth stage-dependency, the responses of crops could be specific. For example, if crops are in germination stage the consequences of abiotic stress would be detrimental by killing off a large percentage of seedlings that reduces the profitability of the whole farm (Okçu et al. 2005; Wang et al. 2009).

1.3.2 Germination and Early Seedling Stages

The occurrence of drought at early phases of germination can be harmful to the germination percentage owing to a deficiency in water uptake (Jain et al. 2019), reduction in water potential then improper enzymatic functions (Ahmad et al. 2009). Analogously high salinity in this stage prevents seed germination and emergence, since not only increases the necessity of water uptake due to high osmotic pressure but also limits the cell expansion and emergence of primary roots by a decline in water availability round the seed and in-between (Panuccio et al. 2014; Rauf et al. 2007). The crops responses to drought as well as salinity at early stages have been studied extensively, where recent observation on soybean showed the transferability of the drought stress effects from parents who experienced stress to progenies, which in this case manifested low germination rate and vigor (Wijewardana et al. 2019). The study conducted by Jovović et al. (2018) indicated a considerable variation in responses of various wheat cultivars to salinity during germination, which beside decline in germination rate and related features, delay in germination was observed. As mentioned earlier the accumulation of salt and decrease in osmotic potential could be responsible in deceleration or inhibition of water absorption vital for mobilizing nutrient components in the course of germination and/or sodicity damaged to the embryo. Similar results on crop plant responses to drought and salinity such as maize, soybean, barley, and sorghum have been reported. Germination of seeds additionally highly depended on optimum soil temperature and is vary from one crop to another. Whereas proper germination can start in wheat, as for temperate crops, around 4 °C with optimal temperature from 12 to 37 °C, the threshold in chilling sensitive crops such as rice is 20–35 °C, which similarly maize and rice have the same minimum critical temperature, 10 °C (Hasanuzzaman et al. 2013). Cold stress often defined under two terms: chilling (less than 20 °C) and freezing (less than 0 °C). Considering fluctuation in temperature pattern due to climate change, extreme temperatures also can be significantly destructive in particular in tropical and temperate regions, where the main part of grain crops (e.g., maize, wheat, rice and soybean) are produced (Beck et al. 2004; Savitch et al. 2011; Srinivasan et al. 1999; Yan et al. 2019). Cold stress can hamper the germination and root development mainly through simulating physiological effect similar to drought, for instance, in *Brassica napus* L. seeds exposed to chilling stress of 2 °C, only 50% germination was observed after almost two weeks, while in 3 days period under 8 °C the same germination rate recorded. In a recent study so as to monitor the responses of rice cultivars as germination index, coleoptile length, and radicle length to two weeks chilling stress (13 °C), Cong Dien and Yamakawa (2019) reported germination index of zero or no germination in 55 of 181 cultivars and germination index of 50% in solely 13 cultivars. The length of coleoptile under chilling stress downed by averagely 97.72% (2.7 mm), in the same manner, radicle length declined by 96.73% with 12.7 mm as the longest. Mainly, the reduction in water conductivity under chilling stress observed to be responsible for postponing the germination and emergence. Imbibition is considered to be the most sensitive phase of seed germination to abiotic stress, which cold stress specifically has the highest negative impact on germination rate in this phase. Mostly, because cold stress damages plasma membrane, which facilitates the situation of solutes (e.g., amino acids and carbohydrates) to leave the seeds, the condition is so-called 'chilling imbibition' (Lyons 1973). In a study where tomato seeds exposed to 4 °C symptoms of electrolyte leakage was observed (Bae et al. 2016). Further, undesirable effects of extremely high temperatures on the germination rate of crops have been investigated extensively, the germination rate of wheat seeds dramatically decreased in 45 °C, obviously owing to eventual drying up the water content of embryo and cell death in the course of early stages of germination. The combined effect of high (30 °C) and low (10 °C) temperatures, and salt-induced osmotic stress (−0.3 MPa) on *Triticum aestivum* seeds led to delay and inhibition of germination (Hampson and Simpson 1990).

Another major source of soil abiotic stress is the toxic level of heavy metals, which affects germination potential through reducing generating plenty of anomalies in seeds through toxicity and oxidative damage to vital biological membranes that disrupt the biosynthesis of carbohydrate and proteins. Among them, the toxic influence of Cd on seed germination has been investigated frequently, where effects are dose-dependent. However delayed germination, membrane leakage (Bae et al. 2016; Smiri et al. 2011), impediment of the process to mobilize the stored resources in seed by dysfunctioning the essential enzymes such as alpha-amylase and invertases (Sfaxi-Bousbih et al. 2010), hampering the production of amino acids and ultimately uncontrollable peroxidation of lipids have been frequently reported (Ahsan et al. 2007), analogously, Cu aims alpha-amylase and invertases, which inhibits the mobilization and finally production of energy to start the germination (Pena et al. 2011; Sfaxi-Bousbih et al. 2010). Similar responses in crop seeds during germination to some extend apply to other known heavy metal ions, for example, Ni in addition to the above-described reaction also impairs the activity of amylase, protease, and ribonuclease which again leads to arresting the digestion of reserved food in albedo (Ahmad and Ashraf 2012; Ashraf et al. 2011). Also, Pb majorly targets the energy production process in the cell by disrupting the absorption of nutrient elements (Fe and Mg) required for the function of enzymes participate in Calvin cycle, consequently inhibiting the germination process or root elongation (Mohamed 2011; Sethy and Ghosh 2013), (for review see Sethy and Ghosh (2013); Bae et al. (2016)).

1.3.3 Vegetative and Reproductive Stages

Overall, abiotic stress affect crops from early stages of growth up to the maturation, however germination and its quality is the key pillar of crop production with high vulnerability to abiotic stress. Abiotic stress aim at disrupting the energy production through imposing low turgor pressure, inhibiting enzymatic activity, which means even if the incidence of stress is after early stages of establishment, is will arrest the growth and development to production phases. The negative effects of drought stress with respect to growth and productivity can be properly observed in the study conducted by Colla et al. (2010) in which responses of hybrid lines of maize under drought resulted in a tremendous reduction in dry matter produced in shoot and root. Consequently, the yield reduced by 2–3-fold in comparison with control under normal condition. Reduction in growth vary within the plant organs, increasing the root:shoot ratio has been reported in maize lines responses to water deficit stress (Rahul et al. 2019), which is possibly due to lower sensitivity in roots toward low water potential (Wu and Cosgrove 2000). A ubiquitous response to drought stress is decreasing the leaf surface by folding, which is adaptation mechanisms leads to a reduction in light absorption and lessening the necessary component to maintain the ongoing biological processes, of course, such changes decreases the photosynthetic pigments as well that reflects in reducing the yield (Flagella et al. 2002; Hajibabaee et al. 2012). The important part is the involvement of the additive effects of other biotic stress in particular high temperature. Water shortage in soil and plants leads to rising temperature in the plant that triggers decreasing the leaf area in response or structural and functional modification in leaves such as minimizing the stomatal conductivity to improve water use efficiency (WUE), which ends up with a reduction in net photosynthesis. The concomitant of drought, heat, and salinity has been observed more often than not, however, owing to the difficulty in its assessments scholars tend to individual evaluation. Salinity responses of crops often compose of diminishing in shoot development and stunting by preventing the formation of internodes, as well as acceleration leaf shedding (Kozlowski and Pallardy 2002; Lacerda et al. 2003). Arresting growth and development can be attributed to aggregation of toxic ions leads to the removal of leaves (Hatfield and Prueger 2015; Lacerda et al. 2003). In pistachio rootstocks subjected to salinity necrosis symptoms in leaf were exhibited that had a high correlation with Na^+ and Cl^- accumulation (Rahneshan et al. 2018). In general, either low carbon fixation rate owing to the reduction in stomatal function as a result of decreased water potential (Hajiboland et al. 2014) and damage to photosynthetic pigments (Ashraf 2003), or direct preventative influence of accumulation of toxic ions (and unbalancing uptake of an essential ion such as K^+) (Munns 2002; Rahneshan et al. 2018) on cell division and elongation can be accountable for a decline in growth and biomass production under salinity. The incidence of chilling stress during growth may cause, as often have been reported, in damaging photosynthetic activity. Of course, mainly chilling is transient, and the intensity of its damage depends on the moment of occurrence whether the stomata are open or close. In watermelon plants subjected to 2 °C reduction in the activity of photosynthesis apparatus (Korkmaz and Dufault 2001) possibly owing to damage to oxidation production chain bridging two photosystems (I and II) in opened stomata that could not have a successful recovery after the course of stress was reported (Markhart III 1986). The arid and semi-arid region is prone to stimulate the combined effect of abiotic stress such as high light intensity and high temperature, or the latter one vs. salinity. In some case, combined effects of strictly regional with a superb instance is water deficit accompanied by low temperature stress in vineyards of north of China that happened to negatively influence the productivity considerably (Su et al. 2015). In another example, Mediterranean areas that environmental conditions facilitates the occurrence of combined effect of low temperature vs. high light stress (Loreto and Bongi 1989). Seasonal variation in atmospheric gases also sometimes contribute in make crops more sensitive to abiotic stress (Xu et al. 2007), in case of the point elevated O_3 concentration in winter increases the damage of low temperature stress in winter wheat (Barnes and Davison 1988) and/or O_3 vs. salinity exacerbated the reduction in productivity of *Oryza sativa* and *Cicer arietinum* (Welfare et al. 2002).

Productive phases of crops are susceptible to abiotic stress the most, the level of economic damage that stresses can cause is even much higher, since maintaining a farm in a region capable of severing abiotic stress requires a great deal of capital. Thus damages in critical stages of flowering, fertilization or filling in grain crops can be financially catastrophic. The responses of grain crops to abiotic stress during reproduction phases have been well-documented. Pollen formation and development as Achilles' heel of crop productivity in wheat is highly vulnerable to water deficit stress (Ashtox 1948; Ji et al. 2010), encourages out-crossing as often found linked to low grain set rate (Bingham 1966). Likewise, between the pollen mother cell and leptonema in sorghum showed the highest vulnerability to low-temperature stress (Brooking 1976). Extreme temperatures singly or in combination with drought and salinity can be terminal in pollen germination as heat stress in cereals overall led to a dramatic reduction in grain-filling time (Jagadish et al. 2007; Wardlaw and Wrigley 1994). This unfavorable conditions affecting the functionality of starch production enzymes subsequently incomplete grain-filling and low yield in cereals (Zahedi et al. 2003). However, the main vulnerability accountable for the reduction in grain number is before the appearance of ear and panicle out of leaf sheath. Even after meiosis as the most sensitive stage toward stress in rice and wheat, water deficiency and low-temperature stress caused a significant degree of infertility (Ji et al. 2010; Oliver et al. 2005). Male sterility additionally enhanced under drought stress (Saini 1997). The influence of temperature could be sometimes very specific, as the low temperature in rice enhances the number of grains but notably reduces the fertility of pollens (Dolferus et al. 2011; Okada et al. 2018). From an evolutionary perspective, the size of grain in undomesticated plants is more important since the fecundity of larger seeds is higher, while in grain crops this is actually number of grain that is determining the yield, a component of productivity which affects by biotic stress the most (Bingham 1966). Interestingly, while ovary is relatively stress-tolerant and reported to still be fertile, stress-simulated pollen sterility can be occurred in early stages of microspore (Hayashi et al. 2000).

Of the physiological responses linked to an increase in sterility is induction and concentration of ABA under stress conditions. The evidence such as a decrease in ABA content of anther of transgenic rice lines and their higher tolerance to low-temperature stress indicates the key role of ABA in the sterility of crops under stress. Similarly, a high level of male sterility in tomato under high-temperature stress and increased ABA is providing proof of its effects. The flowering and milk grain stages are observed to be the most vulnerable to drought in Chenopodium quinoa Willd (Blum 2011) which is coincidence with increasing the concentration of ABA in plant organs (Jacobsen et al. 2009; Razzaghi et al. 2011; Yang et al. 2016). Even the transient cold stress depressed the pollen germination in chickpea (Srinivasan et al. 1999). That is possibly owing to low energy that's frequently linked to inhibition influence of ABA on sugar and amino acid synthesis and supply through lessening the turgor pressure as an essential part of energy production that ultimately leads to hindering the growth of pollen tube, fertilization, and formation of seed (Clarke and Siddique 2004; Shivanna 1985; Thakur et al. 2010). Involvement of ABA in regulating anther sink strength recently attracted the attention of scholars as an important marker in screening germplasms for potential lines (Dolferus et al. 2011). Also using distinguishing phenotypic features such as fertility or sterility of pollen can provide remarkable breakthroughs that end up in exploring underlying molecular mechanisms.

1.4 Crop Water Relations

Understanding how crop plants behave regarding water relationships in critical periods of dealing with continuous or transient abiotic stress requires having reliable indices that truly convey the ongoing responses of the plant to the conditions (Passioura 2010). Abiotic stress often targets disturbing water relations in the aboveground and underground organs of the plant since growth and development to a large degree linked to a stable water relation. The most cited useful indicators of water relation in plants under stress are relative water contents (RWC), leaf water potential, osmotic potential, pressure potential, and transpiration rate (Kirkham 2005; Lazar et al. 2003; Okçu et al. 2005). Additionally, canopy temperature reported that can appropriately reflect the plant water potential status under heat, salinity, and drought stress because increasing in water potential means enhancement of photosynthetic activity which automatically lessens the canopy temperature (Ehrler et al. 1978; Siddique et al. 2000). Water relations is a delicate matter that defines the faith of plants dealing with long-term or short-term consequences of transient or permanent abiotic stress numerous processes involved in responses of plants to the stress-driven impacts on water status, nonetheless, they are mainly similar among various abiotic stresses. Owing to its predominant effect on productivity, tolerant genotypes and the application of comprehensive programs for their screening in germplasms can boost breeding programs (Chavarria and dos Santos 2012; Kirkham 2005).

1.4.1 Water Stress

Although each abiotic stress follows a specific damage mechanism their effects on water relation related-characteristics are similar to a large extend. Similar to extreme temperature stress responses of plant, imposing water deficit stress on soybean genotypes changed water relation through decreasing water potential in leaves, RWC, intensified exudation and expectedly enhanced temperature in the canopy, reduction in such features was delayed or not occurred in tolerance genotypes (Ouvrard et al. 1996). Likewise, sunflower manifested reduction in water relation associated features as RWC, leaf water potential when exposed to drought (Tezara et al. 2002). Stomatal closure, reduction in transpiration rate, and osmotic stress can be responsible for changing water relations in roots and shoots of crops under drought stress. A hydraulic gradient created by transpiration in plants that enables a constant flow of water from roots to leaves (Chavarria and dos Santos 2012). This connection depends on the availability of water in rhizosphere which by the increasing resistance in plant-soil relation transpiration leads to depletion of water content if stomata don't close down consequently reduction in water leaf potential and dehydration. The latter one is vary based on numerous factors such growth stage, atmosphere condition, the microclimate of aerial parts and water regime (Acosta-Motos et al. 2017). Commonly, response to the duration of drought stress is various, however, RWC, water potential, and osmotic pressure rise as the intensity of drought continues (de Campos et al. 2011). To save water content plants often tend to close the stomata which improved WUE while the rate of net photosynthesis decreased. WUE of genotypes and crops varies under drought (Abebe et al. 2003; Subramanian et al. 2006). Growth stage-dependency in WUE also has been reported in sunflower under water deficit stress, which is during reproduction phases, WUE markedly reduced in comparison to vegetative stages (Hussain et al. 2009). The absence of transpiration is accompanied by increase in respiration which means wasting stored resources that eventually recovery would be highly difficult or unlikely after irrigation (Franco et al. 2011; Sánchez-Blanco et al. 2004).

1.4.2 Extreme Temperatures

The influence of temperature on the water status of crop plants can be at multiple levels. That means changes in temperature beyond the optimal affects the enzymatic function directly through increased temperature or indirectly by imposing oxidative stresses that damage the activity of vital enzymes or causing osmotic stress which all lead to disruption the water relations (Bloom et al. 2004; Chavarria and dos Santos 2012; Ehrler et al. 1978; Kirkham 2005). Aerial parts of tomato (*Lycopersicon esculentum* L. cv. T5) that the roots exposed to low-temperature stress (5 °C) indicated the signs of low water potential and wilting. While another species of *Lycopersicon* (*L. hirsutum* LA 1778) known for its cold stress tolerance showed a higher level of water potential under the similar condition. Assessing the hydraulic conductance in either species proven to be similar, whereas stomatal behavior was a distinguishable difference. Further, stomatal closure in cold-tolerant species occurred in contrary to the sensitive one which stomata kept open until the temperature in root system dropped to 5 °C that resulted in sever wilt and injury. Interestingly, using grafting technique, the aerial part of one grafted to the roots of another, the response of stomata changed (Bloom et al. 2004). The stomatal behavior is a significant cold tolerance strategy in crops, which similar to the above-detailed study maize as tropical species vulnerable to cold failed to maintain water pressure that caused excessive transpiration under the cold condition of the soil. However, not due to reduction in root water hydraulic movement (de Juan Javier et al. 1997; Enders et al. 2019). Mainly, cold stress effects on crops are either individually through changing turgor or by formation ice that intracellularly stimulates a drought stress-like condition and drains water from cell to reach balance (Beck et al. 2004; Hansen and Beck 1988). Heat stress responses of crops are species-specific and duration of high-temperature stress is important, for instance, affecting water status in crops has extensively been reported, but heatwave in olive trees mainly lowered the CO_2 assimilation by damaging the photosystems and stomatal closure (Fahad et al. 2017; Haworth et al. 2018). The coincidence of high temperature and drought stress under field condition is common (Machado and Paulsen 2001; Velikova et al. 2009), water shortage in aerial part, leaves in particular during heat stress majorly attributed to intensification of transpiration rate (i.e., in the day time) and absence of equal response form roots. That all lead to depletion of water in leaves and drought stress while water is available in the soil. The unbalanced water potential in crops under high-temperature stress has been recorded in tomato (Morales et al. 2003), sugarcane (Wahid and Close 2007), and potato (Naz et al. 2018).

1.4.3 Salinity

Plant–water relations explain the behavior is a true reflection of plant responses to dehydration and ion toxicity caused by salinity (Passioura 2010). The salinity of soil and water resources, especially in arid and semi-arid regions can drastically reduce crop growth and yield. The level of intensity in effect on plant vary depending on species, season, tolerance threshold, duration of exposure to salinity stress, rainfall pattern during the growing season, intensity and type of salinity and soil physical and chemical properties. The salinity caused by sodium chloride is significantly higher than other salts and affects plant tissues in a higher rate, salts have a negative on water potential, water uptake, transpiration rate, stem water potential, osmotic potential and stomatal conductivity (Kirkham 2005; Munns and Gilliham 2015; Razzaghi et al. 2011). Salinity disturbs a plant's water relations owing to reduced availability of water from the soil solution as a result of negative water potential initiated by the toxic effects of the sodium and chloride ions (Munns 2005). This response has been observed in several species such as *Euonymus japonica* L., *Phlomis purpurea* L., and *Rosmarinus officinalis* L. (Alarcón et al. 2006; Álvarez et al. 2012; Gómez-Bellot et al. 2013).

The short-term responses of crops to salinity are highly analogs to water deficit specifically concerning osmotic stress (Navarro et al. 2008). Significant reduction in RWC, turgor pressure, and stomatal conductance of wheat genotypes subjected to a 4-week salinity (150 mM) during vegetative stages observed. However, while RWC affected by salt stress with no further modification during the experiment, the stomatal function considerably changed (Rivelli et al. 2002) which indicates the influence of hormonal regulation emerging from root system (Kaur et al. 2016; Passioura 1988). In general, crops (wheat and maize) responses related to water relations to high salt concentration in soil is either osmotic or aggregation of toxic ions in aerial parts (Azevedo Neto et al. 2004; Azizian and Sepaskhah 2014; Fortmeier and Schubert 1995). As mentioned earlier, responses often are highly situation-specific and depend on growth stage reaction might be different (Alarcón et al. 1999; Sánchez-Blanco et al. 2004). Fall in the number of water channels (or aquaporins) is probably responsible for the reduction of turgor and water conductivity (Kaldenhoff et al. 2008). Salt-treated *E. angustifolia* as a salt intolerance *and L. barbarum* with a higher ability to tolerate salinity showed a distinct difference in WUE as the former WUE decreased dramatically, whereas in latter one the water status was more stable (Acosta-Motos et al. 2017).

1.4.4 Heavy Metal

The negative effects of heavy metal stress on water relation frequently found to be linked with a change in aquaporins as the primary pass for water flow from roots that caused a reduction in hydraulic conductivity of water. This diminishment of aquaporins by heavy metal has been supported by experimental studies on *Alium cepa* L. and *Lupinus luteus* L. subjected to Pb (Przedpelska-Wasowicz and Wierzbicka 2011). The literature suggests that heavy metal stress mainly aim at the flow of water internally, which in this case notable reduction in transfer of water by xylem in *Ace saccharinum* L. under Cd stress was reported. A possible explanation can be justified with the decrease in xylem tissues capable of transferring water as well as shrinking in the size of vessels and clogging of xylem by fractions of gums or cell remnants (Lamoreaux and Chaney 1977). Disruption of water relations in heavy metal-treated plants sometimes caused by the accumulation of metal ions to a lethal level in root cells that led to cell death and limiting functional cells to uptake water, which simulating drought in aerial parts was the aftermath. Further, RWC is a sensitive indicator of changing in the water status of crops, however, its value under heavy metal stress reported to be stable, that is possibly due to the specific phenomenon known as vacuolization in various growth points in the plant (Gzyl et al. 1997; Przymusiński and Woźny 1985). This as a normal tolerance response in root cells helps to maintain RWC under heavy metal stress and mitigate water fluctuation in root cells subjected toxic ions. This development has been observed in meristematic cells of *Festuca rubra* (Davies and Zhang 1991) and maize (Doncheva 1998) exposed to a high level of Zn and Cu, respectively in addition to root epidermis and cortex cells of Ni-stressed *Psidium guajava* (Bazihizina et al. 2015). Induction of vacuolization in *L. luteus* received concentrations of Pb was similarly observed (Przedpelska-Wasowicz and Wierzbicka 2011). Increase in the number of stomata and reduction in their size clearly due to turgor pressure decrease in the heavy metal-treated plant also have been reported including *H. annuus* exposed to various levels of Pb, Cd, Cu and Zn (Kastori et al. 1992) and Cd-treated *Beta vulgaris* (Greger and Johansson 1992). However, contrary results on *S. bicolor* and *B. vulgaris* subjected to concentrations of Cd and Cu (Kasim 2006), and Zn (Sagardoy et al. 2010), respectively, indicated reduction in the number of stomata. Seemingly, responses have dose-and-species-dependency may to some extend explain the variation in results (Bazihizina et al. 2015; Doncheva 1998).

1.5 The Effect of Abiotic Stressors on Photosynthesis Pigments and Apparatus

An incredible ability of plants is to transform light energy into chemical energy through a delicate complicated chain of chemical reaction with H_2O_2 and CO_2. The process initiated by light breaking down water molecules into O_2 and hydrogen, the former one discharge out of the leaf, while the latter is rich in energy, runs respiration process to generate adenosine triphosphate (ATP) and nicotinamide adenine dinucleotide phosphate (NADPH). These two energy-carrier molecules used in the biosynthesis of carbohydrates as food for the plant. The assimilation of CO_2 in an efficient way is pivotal for plant growth and development. This elaborate apparatus as an energy generator of the plant affected severely by environmental stresses and respond differently to the exposure of various abiotic stresses (Colla et al. 2010; Maricle and Maricle 2012; Rahbarian et al. 2011; Sagardoy et al. 2010; Xu et al. 2007).

1.5.1 Water Stress

Fluctuation in the water content of cells affects photosynthetic machinery since the operation of mechanisms involved in CO_2 fixation largely influenced by the water potential of cells, by the reduction in turgor the stomatal closure occurs, which limits the accessibility of cells to CO_2. Decreased water potential generates unfavorable consequences in particular for photosynthetic and protective pigments (i.e. chlorophyll a/b and carotenoid), their enzymatic reactions, and thylakoid membranes, thus diminishing the growth and productivity (Haworth et al. 2018; Wahid and Close 2007). Water deficit-induced decrease in chlorophyll content has been previously reported in wheat (Xu et al. 2007), canola (Din et al. 2011) and chickpea (Talebi et al. 2013), whereas contradictory results also exist, the quantity of chlorophyll reduced in black gram (*Vigna mungo* L.) subjected to water stress (Ashraf and Karim 1991). The intensity of drought effects depends on growth stage and concentrations of chlorophylls may vary. For instance, Rahbarian et al. (2011) observed seedling and flowering stages where the highest reduction in chlorophyll and carotenoid content were recorded, interestingly the decrease in chlorophyll *a* was higher than chlorophyll *b* similar to Jain et al. (2010) results. This uneven pattern of changing in chlorophylls content may have been the result of a difference in enzymatic functions linked to the production of chlorophyll in each species (Fahad et al. 2017). While the photosynthesis process under water deficit may reduce or completely shut down, but in case of respiration, the fluctuation is possible however never fully disabled, which costs the plant consuming the assimilated materials. Therefore, the impedance of CO_2 entrance owing to stomatal closure leads the plant to switch to complete respiration to produce energy if even infinitesimal to continue, which gets more problematic as water shortage persisted (Franco et al. 2011; Lawlor and Tezara 2009). Another side of water stress is an excessive level of water that crop plants submerged as a result of flood, which has most of its response shared with drought. Stomatal behavior similar to drought under waterlogging attempt to close that leads to limitation in CO_2 and consequently reduction in carbon assimilation. As earlier discussed in influence of flood on nutrient uptake, flood-tolerant crops such as rice produce new adventitious root or develop aerenchyma. Reduction in the capacity of Rubisco observed to be responsible in lessening the assimilation of CO_2. The aerial parts, leaves in particular, are quick in sensing waterlogging that diminishing in chlorophyll content and CO_2 fixation, and altering respiration have been witnessed (Caudle and Maricle 2012; Maricle and Maricle 2012; Zhang et al. 2018).

1.5.2 Extreme Temperatures

Heat stress affects many cellular processes as photosynthesis is one of the most sensitive ones, that has a substantial decreasing effect on photosynthesis that eventually leads to a decline in growth and yield (Bahar et al. 2008; Fahad et al. 2017). The photochemical reaction in thylakoid membranes and carbon metabolism in chloroplast stroma are the first sites of high-temperature damage (Lamaoui et al. 2018). PS II has been recognized as the most vulnerable component of the photosynthesis apparatus to high-temperature stress and quiet numerous reasons in supporting this claim. In some experiments, the decrease in photosynthesis under heat stress was attributed to an increase in photorespiration. At high temperatures, due to the ability of Rubisco to act as oxygenase and decreased solubility of CO_2 compared with photorespiration increases and photosynthesis decreases (Fahad et al. 2017; Rahbarian et al. 2011). However, a decreased carbon fixation at high temperatures has been observed in both conditions of the presence and absence of photorespiration (Perdomo et al. 2017). This indicates that the reduction of photosynthesis can only partially be justified by photorespiration and reduction in Rubisco in enzymatic activity has a major role in photosynthesis decline when exposed to heat stress (Shah and Paulsen 2003). A decreased Rubisco activity under moderate heat stress is associated with a reduction in net photosynthesis that had an increase of ribulose 1,5-bisphosphate and a decrease of 3-phosphoglycerate (Demirevska-Kepova et al. 2005). Excessive high temperature can indirectly through reducing photosynthesis and increasing respiration and consequently, enhancing the concentration of carbon dioxide in the lower side of leaf facilitate stomatal closure. With the occurrence of water limitations stomatal closure can happen active or inactive as well as water-dependently (Chavarria and dos Santos 2012; Ehrler et al. 1978). However, under heat stress stomatal conductivity reduction is proportionately lower, in terms of time, than other photosynthetic processes, by comparison (Pirasteh-Anosheh et al. 2016). Research has shown that moderate heat stress conditions of (37–30 °C), inhibiting Rubisco activity, indirectly leading to a severe reduction in carbon assimilation. And under extreme heat stress (above 37 °C), decreased oxygen-evolving complex (OEC) activity, inhibition the electron transfer from QA to QB and overall damage to photosynthetic reaction centers were discovered two main factors in impairing photosynthesis apparatus (Heckathorn et al. 1998; Lu and Zhang 2000). Chlorophyll fluorescence value reflects the stability of thylakoid membranes and the relative efficiency of electron transfer from PSII to PS1 (Heckathorn et al. 1998).

Through altering the membrane properties, cold stress leads to an unbalance metabolism balance and by generating toxic metabolites that cause secondary injuries in the plant (Cai et al. 2019; Oliver et al. 2005). At low temperatures, the efficiency of the energy transferring rate to the center of PSII is reduced (Su et al. 2015). These all leads to the formation of ROS. Since photosynthetic processes are slow under low temperature, the existence of light and absence of balance between the absorbed light and photosynthesis pave the way to (van Buer et al. 2019). In this case, the induction of seedlings has been reported (Savitch et al. 2011). In cold-tolerant plants, the ROS formation is controlled and modified by enzymatic or non-enzymatic anti-radicals that eradicate and detoxify ROS, without regulation a variety of ROS can negatively change lipids, pigments, proteins, and nucleic acids and thylakoid membrane, which ends up in causing serious cellular damages (Korkmaz and Dufault 2001; Nakashima and Yamaguchi Shinozaki 2006; Srinivasan et al. 1999; van Buer et al. 2019). The decline in growth rate in plants exposed to low temperatures has been observed in maize with chlorosis symptoms on leaves (Dolstra et al. 1988) which manifested disruption in the biosynthesis of chlorophyll and also limited activity of enzymes, including particularly Rubisco (Rr and van Huystee 2011). The content of chlorophyll ab and total in sold-stressed rapeseed, which is most likely due to damage caused by free radicals (Yan et al. 2019). Also with decreasing temperature various phenotypic symptoms such as growth and leaf area decline, wilt, chlorosis, and necrosis have been observed (Yan et al. 2019). Also, often at low temperatures, water stress-like symptoms such as and leaf turgor are found in cold-sensitive plants, which are known as cold stress-induced signs (Miura and Tada 2014).

1.5.3 Salinity

In crops, besides negatively affecting yield and yield components, salinity also affects myriad processes involved in the growth and development of plants (Hajiboland et al. 2014; Hasana and Miyake 2017). Crops differ in their response to salinity, resources indicate that rapeseed, barley, cane, and cotton are classified as saline resistant crops, whereas canola is more resistant then wheat, therefore, may perform better in saline soils. Although based a worldwide field survey canola considered as a moderately salt resistant plant (Hasana and Miyake 2017; Rivelli et al. 2002; Shah et al. 2017; Welfare et al. 2002). In canola, salinity decreases root growth, leaf emergence, and early formation of nodes, also, to decrease plant height, pod number, and seed pod number in late growth stages (Ashraf and Ali 2008). Photosynthesis and cellular growth are processes that are rapidly affected by salinity stress, which associated with the reduction of CO_2 assimilation as an aftermath of limited stomatal conductance and chlorophyll degradation (Aziz and Khan 2001; Azizian and Sepaskhah 2014). Salinity-induced damages to chloroplasts structure and instability of pigment-protein compounds similarly have been reported. Carotenoids as protectors of the photosynthetic system against photooxidation experienced an induction in inhibition capability under salinity stress (Zhang et al. 2012). Lu and Vonshak (2002) stated that the quantum yield parameter of PS II as a reliable characteristic affected by salinity. When plants were exposed to environmental stresses, with increased salinity, non-photochemical fluorescence quenching was no longer able to remove excess electron energy. Absence of excitation resulting in the oxygen molecule acting as an alternative acceptor for the electron (Stepien and Johnson 2009). Sheng et al. (2008) reported that under salinity stress maximum photoconductive quantum efficiency, electron transfer rate, gas exchange and carbon assimilation reduced whereas non-photochemical fluorescence quenching increased by 40%, indicating an induction in the thermal loss of PS II. Zhao et al. (2007) noted that salinity stress reduced leaf area, dry matter content, photosynthesis rate, and stomatal conductance of the leaves. Salinity increases the amount of energy needed to maintain the cell's natural conditions, resulting in less energy left for plant growth (Fricke 2020). The salinity-induced decrease in photosynthesis is not owing to carbon fixation inefficiency per unite area but rather due to the decrease in the photosynthetic area (Acosta-Motos et al. 2017).

1.5.4 Heavy Metals

Among the processes affected by heavy metal stress, photosynthesis and photosynthetic pigments are the most common (Przedpelska-Wasowicz and Wierzbicka 2011; Schat et al. 1997). Photosynthesis is one of the most sensitive metabolic processes related to lead toxicity and further studies of photosynthetic inhibition in various plant species have been reported (Schat et al. 1997; Schutzendubel and Polle 2002; Yu et al. 2019). Pb may reduce photosynthesis by preventing stomatal closure facilitating damage to chloroplast ultrastructural, induce changes in photosynthetic metabolites, the substitution of ions such as Mg and Mn with Pb in chloroplasts and prevent the synthesis or degradation of photosynthetic pigments (Kopittke et al. 2007; Salt et al. 1995). Pb toxicity has also been implicated in oxidative stress through the generation of excessive concentration of ROS including superoxide radicals (O^{-}_2), hydroxyl radicals (OH), and hydrogen peroxide (H_2O_2) (Hossain et al. 2012; Yu et al. 2019). In a comparative study on the effects of Ni, Pb, Cu, Cd, and Zn in maize (Lu et al. 2015) and wheat (Hough et al. 2003) reduced net photosynthesis in Pb-treated seedlings was the highest. Total chlorophyll content in Ni-treated *P. vulgaris* experienced a significant decrease (Campanharo et al. 2010), a similar result reported in Ni-treated cabbage (Molas 2002). A reduced chlorophyll content Riccia sp. plants affected by cadmium also reported by Prasad et al. (2005). Accumulation and allocation of Cd in a plant vary according to species, cultivar, growth stage, and presence of other elements (Hossain et al. 2012). The toxicity of Cd in the plant is due to the reaction of this element with the sulfhydryl group present in the structure of enzymes and proteins. Cd negatively affects plant physiological and metabolic processes such as respiration, photosynthesis, plant water relationships, and gas exchange in the stomata. It also disrupts the pathway of chlorophyll biosynthesis, Calvin cycle, and photosynthetic (Greger and Johansson 1992; Kasim 2006; Khan et al. 2009; Mediouni et al. 2006). Toxic level of Cu directly inhibits photosynthetic electron transport as well as enzymatic activities during the Calvin cycle or net CO_2 assimilation. Additionally, by reducing the content of photosynthetic pigments, damaging the photosynthetic apparatus and chloroplast structure, altering the protein and lipid composition of the thylakoid membrane can affect plant growth (Doncheva 1998; Georgiadou et al. 2018; Kastori et al. 1992).

1.6 Conclusion and Future Prospects

Having a large array of approaches to mitigate abiotic stresses made plants thriving habitats on land. Their universal defense mechanisms confer the capability to endure environments impose multi stress, the systems that understanding them is still in great demand. Mainly to make applicable advancement in insuring food security, which has been notably jeopardized by an increase in global temperature (e.g., 3–5 °C increase over the next 100 years) in the era of climate change. An ongoing phenomenon that has triggered increment the occurrence of abiotic stresses. Markedly increase in the incidence of combinations of salinity, drought, and extreme temperatures are alarming and require urgent attention of breeders to develop crop cultivars to cope with unfavorable circumstances. The socio-economic consequences of the negative influence of abiotic stresses on crop productivity could be disastrous, albeit they are not the only culprit in threatening food security but of course the major one. Taking into consideration the various worldwide challenges and crises the world have experienced over the course of last decades especially the current one, Coronavirus (Covid-19) outbreak, once again reminded us how countries with agricultural system prone to abiotic stresses are fragile against other crisis as well and can have a significantly higher death rate owing to food shortage and malnutrition (Schellekens and Sourrouille 2020). Therefore, comprehensive scientific attempts to gain a reliable insight into tolerance mechanisms and developing stress tolerance crops are in prime importance.

References

Abebe T, Guenzi AC, Martin B et al (2003) Tolerance of mannitol-accumulating transgenic wheat to water stress and salinity. Plant Physiol 131:1748–1755

Acosta-Motos JR, Ortuño MF, Bernal-Vicente A et al (2017) Plant responses to salt stress: adaptive mechanisms. Agronomy 7:18. https://doi.org/10.3390/agronomy7010018

Aggarwal PK, Mall R (2002) Climate change and rice yields in diverse agro-environments of India. II. Effect of uncertainties in scenarios and crop models on impact assessment. Clim Change 52:331–343

Ahmad MSA, Ashraf M (2012) Essential roles and hazardous effects of nickel in plants. In: Whitacre D (ed) Reviews of environmental contamination and toxicology. Springer, New York, pp 125–167

Ahmad S, Ahmad R, Ashraf MY et al (2009) Sunflower (*Helianthus annuus* L.) response to drought stress at germination and seedling growth stages. Pak J Bot 41:647–654

Ahsan N, Lee SH, Lee DG et al (2007) Physiological and protein profiles alternation of germinating rice seedlings exposed to acute cadmium toxicity. CR Biol 330:735–746

Al W, Orking G, Clima O (2008) Climate change and food security: a framework document. FAO Rome
Alarcón J, Morales M, Ferrández T et al (2006) Effects of water and salt stresses on growth, water relations and gas exchange in *Rosmarinus officinalis*. J Hortic Sci Biotech 81:845–853
Alarcón J, Morales M, Torrecillas A et al (1999) Growth, water relations and accumulation of organic and inorganic solutes in the halophyte *Limonium latifolium* cv. Avignon and its interspecific hybrid *Limonium* caspia x *Limonium latifolium* cv. Beltlaard during salt stress. J Plant Physiol 154:795–801
Allen M, Dube O, Solecki W et al (2018) Framing and context. Global warming of 1:49–91
Álvarez S, Gómez-Bellot MJ, Castillo M et al (2012) Osmotic and saline effect on growth, water relations, and ion uptake and translocation in *Phlomis purpurea* plants. Environ Exp Bot 78:138–145
Ashraf M (2003) Relationships between leaf gas exchange characteristics and growth of differently adapted populations of Blue panicgrass (*Panicum antidotale* Retz.) under salinity or waterlogging. Plant Sci 165:69–75
Ashraf M, Ali Q (2008) Relative membrane permeability and activities of some antioxidant enzymes as the key determinants of salt tolerance in canola (*Brassica napus* L.). Environ Exp Bot 63:266–273
Ashraf M, Karim F (1991) Screening of some cultivars/lines of black gram (*Vigna mungo* L. Hepper) for resistance to water stress. Trop Agric 68:57–62
Ashraf MY, Sadiq R, Hussain M et al (2011) Toxic effect of nickel (Ni) on growth and metabolism in germinating seeds of sunflower (*Helianthus annuus* L.). Biol Trace Elem Res 143:1695–1703
Ashtox T (1948) Technique of breeding for drought resistance in crops. Technique of breeding for drought resistance in crops. Cambridge, London
Azevedo Neto ADd, Prisco JT, Enéas-Filho J et al (2004) Effects of salt stress on plant growth, stomatal response and solute accumulation of different maize genotypes. Braz J Plant Physiol 16:31–38
Aziz I, Khan MA (2001) Experimental assessment of salinity tolerance of *Ceriops tagal* seedlings and saplings from the Indus delta, Pakistan. Aquat Bot 70:259–268
Azizian A, Sepaskhah A (2014) Maize response to different water, salinity and nitrogen levels: agronomic behavior. Int J Plant Prod 8:107–130
Bae J, Benoit DL, Watson AK (2016) Effect of heavy metals on seed germination and seedling growth of common ragweed and roadside ground cover legumes. Environ Pollut 213:112–118
Bahar B, Yildirim M, Barutcular C et al (2008) Effect of canopy temperature depression on grain yield and yield components in bread and durum wheat. Not Bot Horti Agrobo 36:34–37
Bailey-Serres J, Voesenek LA (2010) Life in the balance: a signaling network controlling survival of flooding. Curr Opin Plant Biol 13:489–494
Barnes J, Davison A (1988) The influence of ozone on the winter hardiness of Norway spruce [*Picea abies* (L.) Karst.]. New Phytol 108:159–166
Bazihizina N, Colzi I, Giorni E et al (2015) Photosynthesizing on metal excess: copper differently induced changes in various photosynthetic parameters in copper tolerant and sensitive *Silene paradoxa* L. populations. Plant Sci 232:67–76
Beck EH, Heim R, Hansen J (2004) Plant resistance to cold stress: mechanisms and environmental signals triggering frost hardening and dehardening. J Biosci 29:449–459
Bingham J (1966) Varietal response in wheat to water supply in the field, and male sterility caused by a period of drought in a glasshouse experiment. Ann Appl Biol 57:365–377
Bloom A, Zwieniecki M, Passioura J et al (2004) Water relations under root chilling in a sensitive and tolerant tomato species. Plant, Cell Environ 27:971–979
Blum A (2011) Drought resistance–is it really a complex trait? Funct Plant Biol 38:753–757
Bohnert HJ, Gong Q, Li P et al (2006) Unraveling abiotic stress tolerance mechanisms–getting genomics going. Curr Opin Plant Biol 9:180–188
Bouman B, Peng S, Castaneda A et al (2005) Yield and water use of irrigated tropical aerobic rice systems. Agri Water Manag 74:87–105

Bray EA (2000) Response to abiotic stress. Biochem Mol Biol Plant 1158–1203
Brooking I (1976) Male sterility in *Sorghum bicolor* (L.) Moench induced by low night temperature. I. Timing of the stage of sensitivity. Funct Plant Biol 3:589–596
Budhathoki NK, Zander KK (2019) Socio-economic impact of and adaptation to extreme heat and cold of farmers in the Food Bowl of Nepal. Int J Environ Res Public Health 16:1578
Burke EJ, Brown SJ, Christidis N (2006) Modeling the recent evolution of global drought and projections for the twenty-first century with the Hadley Centre climate model. J Hydrometeorol 7:1113–1125
Cai X, Magwanga RO, Xu Y et al (2019) Comparative transcriptome, physiological and biochemical analyses reveal response mechanism mediated by CBF4 and ICE2 in enhancing cold stress tolerance in *Gossypium thurberi*. AoB Plant 11:plz045
Campanharo M, Monnerat P, Espindula M et al (2010) Toxicity symptoms of nickel in common bean. Rev Cienc Agron 41:490–494
Caudle KL, Maricle BR (2012) Effects of flooding on photosynthesis, chlorophyll fluorescence, and oxygen stress in plants of varying flooding tolerance. Trans Kans Acad Sci 115:5–18
Chavarria G, dos Santos HP (2012) Plant water relations: absorption, transport and control mechanisms. In: Montanaro G (ed) Advances in selected plant physiology aspects. Intechopen. Available via DIALOG. https://www.intechopen.com/books/advances-in-selected-plant-physiology-aspects/plant-water-relations-absorption-transport-and-control-mechanisms. Accessed 7 July 2020
Clarke H, Siddique K (2004) Response of chickpea genotypes to low temperature stress during reproductive development. Field Crops Res 90:323–334
Colla G, Rouphael Y, Leonardi C et al (2010) Role of grafting in vegetable crops grown under saline conditions. Sci Hortic 127:147–155
Conesa M, De La Rosa J, Domingo R et al (2016) Changes induced by water stress on water relations, stomatal behaviour and morphology of table grapes (cv. Crimson seedless) grown in pots. Sci Hortic 202:9–16
Cong Dien D, Yamakawa T (2019) Phenotypic variation and selection for cold-tolerant rice (*Oryza sativa* L.) at germination and seedling stages. Agriculture 9:162. https://doi.org/10.3390/agriculture9080162
Dai A (2011) Characteristics and trends in various forms of the palmer drought severity index during 1900–2008. J Geophys Res Atmos 116. https://doi.org/10.1029/2010jd015541
Davies WJ, Zhang J (1991) Root signals and the regulation of growth and development of plants in drying soil. Ann Rev Plant Biol 42:55–76
de Campos MKF, de Carvalho K, de Souza FS et al (2011) Drought tolerance and antioxidant enzymatic activity in transgenic 'Swingle' citrumelo plants over-accumulating proline. Environ Exp Bot 72:242–250
de Juan Javier P, José IJ, Manuel SD (1997) Chilling of drought-hardened and non-hardened plants of different chilling-sensitive maize lines changes in water relations and ABA contents. Plant Sci 122:71–79
Demirevska-Kepova K, Holzer R, Simova-Stoilova L et al (2005) Heat stress effects on ribulose-1, 5-bisphosphate carboxylase/oxygenase, Rubisco binding protein and Rubisco activase in wheat leaves. Biol Plant 49:521–525
Dey S, Ram K, Chhabra A et al (2018) Aerobic rice: smart technology of rice cultivation. Int J Curr Microbiol App Sci 7:1799–1804
Din J, Khan S, Ali I et al (2011) Physiological and agronomic response of canola varieties to drought stress. J Anim Plant Sci 21:78–82
Ding W, Fang W, Shi S et al (2016) Wheat WRKY type transcription factor gene TaWRKY1 is essential in mediating drought tolerance associated with an ABA-dependent pathway. Plant Mol Biol Rep 34:1111–1126
Dolferus R, Ji X, Richards RA (2011) Abiotic stress and control of grain number in cereals. Plant Sci 181:331–341

Dolstra O, Jongmans MA, de Jong K (1988) Improvement and significance of resistance to low-temperature damage in maize (*Zea mays* L.) I. Chlorosis resistance. Euphytica 39:117–123
Doncheva S (1998) Copper-induced alterations in structure and proliferation of maize root meristem cells. J Plant Physiol 153:482–487
Ehrler W, Idso S, Jackson RD et al (1978) Wheat canopy temperature: relation to plant water potential 1. Agron J 70:251–256
Enders TA, St. Dennis S, Oakland J et al (2019) Classifying cold-stress responses of inbred maize seedlings using RGB imaging. Plant Direct 3:e00104
Fageria NK, Baligar VC, Jones CA (2010) Growth and mineral nutrition of field crops. CRC Press, Florida
Fahad S, Bajwa AA, Nazir U et al (2017) Crop production under drought and heat stress: plant responses and management options. Front Plant Sci 8:1147. https://doi.org/10.3389/fpls.2017.01147
Field CB, Barros VR, Mastrandrea MD et al (2014) Summary for policymakers. Climate change 2014: impacts, adaptation, and vulnerability Part A: global and sectoral aspects contribution of working group II to the fifth assessment report of the intergovernmental panel on climate change. Cambridge University Press, London, pp 1–32
Flagella Z, Rotunno T, Tarantino E et al (2002) Changes in seed yield and oil fatty acid composition of high oleic sunflower (*Helianthus annuus* L.) hybrids in relation to the sowing date and the water regime. Eur J Agron 17:221–230
Fortmeier R, Schubert S (1995) Salt tolerance of maize (*Zea mays* L.): the role of sodium exclusion. Plant, Cell Environ 18:1041–1047
Franco J, Bañón S, Vicente M et al (2011) Root development in horticultural plants grown under abiotic stress conditions–a review. J Hortic Sci Biotech 86:543–556
Fricke W (2020) Energy costs of salinity tolerance in crop plants: night-time transpiration and growth. New Phytol 225:1152–1165
Georgiadou EC, Kowalska E, Patla K et al (2018) Influence of heavy metals (ni, cu, and zn) on nitro-oxidative stress responses, proteome regulation and allergen production in basil (*Ocimum basilicum* L.) Plants. Front Plant Sci 9:862. https://doi.org/10.3389/fpls.2018.00862
Gómez-Bellot MJ, Alvarez S, Castillo M et al (2013) Water relations, nutrient content and developmental responses of euonymus plants irrigated with water of different degrees of salinity and quality. J Plant Res 126:567–576
Greger M, Johansson M (1992) Cadmium effects on leaf transpiration of sugar beet, *Beta vulgaris*. Physiol Plant 86:465–473
Gzyl J, Przymusiński R, Woźny A (1997) Organospecific reactions of yellow lupin seedlings to lead. Acta Soc Bot Pol 66:61–66
Hajibabaee M, Azizi F, Zargari K (2012) Effect of drought stress on some morphological, physiological and agronomic traits in various foliage corn hybrids. Am Eurasian J Agric Environ Sci 12:890–896
Hajiboland R, Norouzi F, Poschenrieder C (2014) Growth, physiological, biochemical and ionic responses of pistachio seedlings to mild and high salinity. Trees 28:1065–1078
Hampson CR, Simpson G (1990) Effects of temperature, salt, and osmotic potential on early growth of wheat (*Triticum aestivum*) I. Germination. Can J Bot 68:524–528
Hansen J, Beck E (1988) Evidence for ideal and non-ideal equilibrium freezing of leaf water in frosthardy ivy (Hedera helix) and winter barley (Hordeum vulgare). Bot Acta 101:76–82
Hasana R, Miyake H (2017) Salinity stress alters nutrient uptake and causes the damage of root and leaf anatomy in maize. KnE Life Sci 3:219–225
Hasanuzzaman M, Nahar K, Fujita M (2013) Extreme temperature responses, oxidative stress and antioxidant defense in plants. In: Vahdati K, Leslie Ch (eds) Abiotic stress-plant responses and applications in agriculture. Intechopen. Available via DIALOG. https://www.intechopen.com/books/abiotic-stress-plant-responses-and-applications-in-agriculture/extreme-temperature-responses-oxidative-stress-and-antioxidant-defense-in-plants. Accessed 7 July 2020
Hatfield JL, Prueger JH (2015) Temperature extremes: effect on plant growth and development. Weather Clim Extremes 10:4–10

Haworth M, Marino G, Brunetti C et al (2018) The impact of heat stress and water deficit on the photosynthetic and stomatal physiology of olive (*Olea europaea* L.)—A case study of the 2017 heat wave. Plants 7:76

Hayashi T, Kashiwabara K, Yamaguchi T et al (2000) Effects of high nitrogen supply on the susceptibility to coolness at the young microspore stage in rice (*Oryza sativa* L.). Plant Prod Sci 3:323–327

Heckathorn SA, Coleman J, Hallberg R (1998) Recovery of net CO2 assimilation after heat stress is correlated with recovery of oxygen-evolving-complex proteins in *Zea mays* L. Photosynthetica 34:13–20

Hochman Z, Gobbett DL, Horan H (2017) Climate trends account for stalled wheat yields in Australia since 1990. Glob Chang Biol 23:2071–2081

Hossain MA, Piyatida P, da Silva JAT et al (2012) Molecular mechanism of heavy metal toxicity and tolerance in plants: central role of glutathione in detoxification of reactive oxygen species and methylglyoxal and in heavy metal chelation. J Bot. https://doi.org/10.1155/2012/872875

Hotton C, Hueber F, Griffing D et al (2001) Early terrestrial plant environments: an example from the Emsian of Gaspé, Canada. In: Gensel PG, Edwards D (eds) Plants invade the land: evolutionary and environmental perspective. Columbia University Press, New York, pp 179–212

Hough R, Young S, Crout N (2003) Modelling of Cd, Cu, Ni, Pb and Zn uptake, by winter wheat and forage maize, from a sewage disposal farm. Soil Use Manageme 19:19–27

Hussain M, Malik M, Farooq M et al (2009) Exogenous glycinebetaine and salicylic acid application improves water relations, allometry and quality of hybrid sunflower under water deficit conditions. J Agron Crop Sci 195:98–109

Iizumi T, Furuya J, Shen Z et al (2017) Responses of crop yield growth to global temperature and socioeconomic changes. Sci Rep 7:1–10

Jacobsen SE, Liu F, Jensen CR (2009) Does root-sourced ABA play a role for regulation of stomata under drought in quinoa (*Chenopodium quinoa* Willd). Sci Hortic 122:281–287

Jagadish S, Craufurd P, Wheeler T (2007) High temperature stress and spikelet fertility in rice (*Oryza sativa* L.). J Exp Bot 58:1627–1635

Jain M, Kataria S, Hirve M et al (2019) Water deficit stress effects and responses in maize. In: Hasanuzzaman M, Hakeem K, Nahar K, Alharby H (eds) plant abiotic stress tolerance. Springer, Cham, pp 129–151

Jain M, Tiwary S, Gadre R (2010) Sorbitol-induced changes in various growth and biochemici parameters in maize. Plant Soil Environ 56:263–267

Ji X, Shiran B, Wan J et al (2010) Importance of pre-anthesis anther sink strength for maintenance of grain number during reproductive stage water stress in wheat. Plant, Cell Environ 33:926–942

Jovović M, Tunguz V, Mirosavljević M et al (2018) Effect of salinity and drought stress on germination and early seedlings growth of bread wheat (*Triticum aestivum* L.). Genetika 50:285–298

Kaldenhoff R, Ribas-Carbo M, Sans JF et al (2008) Aquaporins and plant water balance. Plant, Cell Environ 31:658–666

Kasim WA (2006) Changes induced by copper and cadmium stress in the anatomy and grain yield of *Sorghum bicolor* (L.) Moench. Int J Agri Biol 8:123–128

Kastori R, Petrović M, Petrović N (1992) Effect of excess lead, cadmium, copper, and zinc on water relations in sunflower. J Plant Nutr 15:2427–2439

Kaur R, Bhardwaj R, Sharma R et al (2016) Hormonal regulation of drought stress in plants. In: Ahmad P (ed) Water stress and crop plants: a sustainable approach. Wiley, New York, pp 2:582–599

Khan N, Anjum N, Nazar R et al (2009) Increased activity of ATP–sulfurylase and increased contents of cysteine and glutathione reduce high cadmium-induced oxidative stress in mustard cultivar with high photosynthetic potential. Russ J Plant Physl 56:670–677

Kirkham M (2005) Principles of soil and plant water relations. Elsevier, Massachusetts

Kopittke PM, Asher CJ, Kopittke RA et al (2007) Toxic effects of Pb2+ on growth of cowpea (*Vigna unguiculata*). Environ Pollut 150:280–287

Rahneshan Z, Nasibi F, Moghadam AA (2018) Effects of salinity stress on some growth, physiological, biochemical parameters and nutrients in two pistachio (*Pistacia vera* L.) rootstocks. J Plant Interact 13:73–82

Rahul VD, Panda RK, Lenka D et al (2019) A study on the root characters of maize hybrid germplasm lines under moisture deficit stress. Int J Curr Microbiol App Sci 8:2836–2845

Rauf M, Munir M, ul Hassan M et al (2007) Performance of wheat genotypes under osmotic stress at germination and early seedling growth stage. Afri J Biotechnol 6:971–975

Ray DK, West PC, Clark M et al (2019) Climate change has likely already affected global food production. Plos One 14. https://doi.org/10.1371/journal.pone.0217148

Razzaghi F, Ahmadi SH, Adolf VI et al (2011) Water relations and transpiration of quinoa (*Chenopodium quinoa* Willd.) under salinity and soil drying. J Agron Crop Sci 197:348–360

Rivelli AR, James RA, Munns R et al (2002) Effect of salinity on water relations and growth of wheat genotypes with contrasting sodium uptake. Funct Plant Biol 29:1065–1074

Rr H, van Huystee R (2011) Chilling-induced chlorosis in maize (*Zea mays*). Can J Bot 63:711–715

Sagardoy R, Vázquez S, Florez-Sarasa I et al (2010) Stomatal and mesophyll conductances to CO2 are the main limitations to photosynthesis in sugar beet (*Beta vulgaris*) plants grown with excess zinc. New Phytol 187:145–158

Sahin U, Ekinci M, Ors S et al (2018) Effects of individual and combined effects of salinity and drought on physiological, nutritional and biochemical properties of cabbage (*Brassica oleracea* var. *capitata*). Sci Hortic 240:196–204

Saini HS (1997) Effects of water stress on male gametophyte development in plants. Sex Plant Reprod 10:67–73

Salt DE, Blaylock M, Kumar NP et al (1995) Phytoremediation: a novel strategy for the removal of toxic metals from the environment using plants. Bio/Technol 13:468–474

Sánchez-Blanco MJ, Rodríguez P, Olmos E et al (2004) Differences in the effects of simulated sea aerosol on water relations, salt content, and leaf ultrastructure of rock-rose plants. J Environ Qual 33:1369–1375

Savitch LV, Ivanov AG, Gudynaite-Savitch L et al (2011) Cold stress effects on PSI photochemistry in *Zea mays*: differential increase of FQR-dependent cyclic electron flow and functional implications. Plant Cell Physiol 52:1042–1054

Schat H, Sharma SS, Vooijs R (1997) Heavy metal-induced accumulation of free proline in a metal-tolerant and a nontolerant ecotype of Silene vulgaris. Physiol Plant 101:477–482

Schellekens P, Sourrouille D (2020, May 5) The unreal Dichotomy in COVID-19 mortality between high-income and developing countries. Retrieved from: https://www.brookings.edu/

Schutzendubel A, Polle A (2002) Plant responses to abiotic stresses: heavy metal-induced oxidative stress and protection by mycorrhization. J Exp Bot 53:1351–1365

Selden PA, Edwards D (1989) Colonisation of the land. In: Keith A, Briggs DEG (eds) Evolution and the fossil record. Smithsonian Institution Press, Washington, pp 122–152

Sethy SK, Ghosh S (2013) Effect of heavy metals on germination of seeds. J Nat Sci Biol Med 4:272–275

Sfaxi-Bousbih A, Chaoui A, El Ferjani E (2010) Cadmium impairs mineral and carbohydrate mobilization during the germination of bean seeds. Ecotoxicol Environ Saf 73:1123–1129

Shah N, Paulsen G (2003) Interaction of drought and high temperature on photosynthesis and grain-filling of wheat. Plant Soil 257:219–226

Shah ZH, Rehman HM, Akhtar T et al (2017) Redox and ionic homeostasis regulations against oxidative, salinity and drought stress in wheat (a systems biology approach). Front Genet 8:141. https://doi.org/10.3389/fgene.2017.00141

Sheng M, Tang M, Chen H et al (2008) Influence of arbuscular mycorrhizae on photosynthesis and water status of maize plants under salt stress. Mycorrhiza 18:287–296

Shivanna K (1985) The role of the pistil in screening compatible pollen. Theor Appl Genet 70:684–686

Siddique M, Hamid A, Islam M (2000) Drought stress effects on water relations of wheat. Bot Bull Acad Sinica 41

Singh M, Kumar J, Singh S et al (2015) Roles of osmoprotectants in improving salinity and drought tolerance in plants: a review. Rev Environ Sci Bio/Technol 14:407–426
Singh S, Prasad S, Yadav V et al (2018) Effect of drought stress on yield and yield components of rice (*Oryza sativa* L.) genotypes. Int J Curr Microbiol Appl Sci 7:2752–2759
Smiri M, Chaoui A, Rouhier N et al (2011) Cadmium affects the glutathione/glutaredoxin system in germinating pea seeds. Biol Trace Elem Res 142:93–105
Srinivasan A, Saxena N, Johansen C (1999) Cold tolerance during early reproductive growth of chickpea (*Cicer arietinum* L.): genetic variation in gamete development and function. Field Crops Res 60:209–222
Stepien P, Johnson GN (2009) Contrasting responses of photosynthesis to salt stress in the glycophyte Arabidopsis and the halophyte thellungiella: role of the plastid terminal oxidase as an alternative electron sink. Plant Physiol 149:1154–1165
Su L, Dai Z, Li S et al (2015) A novel system for evaluating drought–cold tolerance of grapevines using chlorophyll fluorescence. BMC Plant Biol 15:82
Subramanian K, Santhanakrishnan P, Balasubramanian P (2006) Responses of field grown tomato plants to arbuscular mycorrhizal fungal colonization under varying intensities of drought stress. Sci Hortic 107:245–253
Tack J, Barkley A, Nalley LL (2015) Effect of warming temperatures on US wheat yields. Proc Natl Acad Sci 112:6931–6936
Talebi R, Ensafi M, Baghebani N et al (2013) Physiological responses of chickpea (*Cicer arietinum*) genotypes to drought stress. Environ Exp Biol 11:9–15
Tezara W, Mitchell V, Driscoll S et al (2002) Effects of water deficit and its interaction with CO2 supply on the biochemistry and physiology of photosynthesis in sunflower. J Exp Bot 53:1781–1791
Thakur P, Kumar S, Malik JA et al (2010) Cold stress effects on reproductive development in grain crops: an overview. Environ Exp Bot 67:429–443
Thakur P, Nayyar H (2013) Facing the cold stress by plants in the changing environment: sensing, signaling, and defending mechanisms. In: Tuteja N, Singh Gill S (eds) Plant acclimation to environmental stress. Springer, New York, pp 29–69
van Buer J, Prescher A, Baier M (2019) Cold-priming of chloroplast ROS signalling is developmentally regulated and is locally controlled at the thylakoid membrane. Sci Rep 9:3022. https://doi.org/10.1038/s41598-019-39838-3
Velikova V, Tsonev T, Barta C et al (2009) BVOC emissions, photosynthetic characteristics and changes in chloroplast ultrastructure of *Platanus orientalis* L. exposed to elevated CO2 and high temperature. Environ Pollut 157:2629–2637
Vibha (2016) *Macrophomina phaseolina*: The most destructive soybean fungal pathogen of global concern. In: Kumar P, Gupta VK, Tiwari AK, Kamle M (eds) Current trends in plant disease diagnostics and management practices. Springer, Cham, pp 193–205
Wahid A, Close T (2007) Expression of dehydrins under heat stress and their relationship with water relations of sugarcane leaves. Biol Plant 51:104–109
Wang WB, Kim YH, Lee HS et al (2009) Analysis of antioxidant enzyme activity during germination of alfalfa under salt and drought stresses. Plant Physiol Biochem 47:570–577
Wani S, Kumar V, Shriram V et al (2016) Phytohormones and their metabolic engineering for abiotic stress tolerance in crop plants. Crop J 4:162–176
Waqas MA, Kaya C, Riaz A et al (2019) Potential mechanisms of abiotic stress tolerance in crop plants induced by thiourea. Front Plant Sci 10:1336. https://doi.org/10.3389/fpls.2019.01336
Wardlaw I, Wrigley C (1994) Heat tolerance in temperate cereals: an overview. Funct Plant Biol 21:695–703
Welfare K, Yeo AR, Flowers TJ (2002) Effects of salinity and ozone, individually and in combination, on the growth and ion contents of two chickpea (*Cicer arietinum* L.) varieties. Environ Pollut 120:397–403
White PJ, George TS, Dupuy LX et al (2013) Root traits for infertile soils. Front Plant Sci 4:193. https://doi.org/10.3389/fpls.2013.00193

White PJ, Pongrac P (2017) 12 heavy-metal Toxicity in Plants. In: Shabala S (ed) Plant stress Physiology, 2nd edn. pp 301–328

Wijewardana C, Reddy KR, Krutz LJ et al (2019) Drought stress has transgenerational effects on soybean seed germination and seedling vigor. PloS one 14. https://doi.org/10.1371/journal.pone.0214977

Wood AJ (2005) Eco-physiological adaptations to limited water environments. In: Jenks MA, Hasegawa PM (eds) Plant abiotic stress. Wiley, New York, pp 1–10

Wu Y, Cosgrove DJ (2000) Adaptation of roots to low water potentials by changes in cell wall extensibility and cell wall proteins. J Exp Bot 51:1543–1553

Xu H, Biswas D, Li WD et al (2007) Photosynthesis and yield responses of ozone-polluted winter wheat to drought. Photosynthetica 45:582–588

Yan L, Shah T, Cheng Y et al (2019) Physiological and molecular responses to cold stress in rapeseed (*Brassica napus* L.). J Integ Agric 18:2742–2752

Yang A, Akhtar S, Amjad M et al (2016) Growth and physiological responses of quinoa to drought and temperature stress. J Agron Crop Sci 202:445–453

Yu G, Ma, J, Jiang P, Li J et al (2019) The mechanism of plant resistance to heavy metal. In: IOP conference series: earth and environmental science international conference on the improvement of environmental quality (ICIEQ), Bogor, Indonesia 2019 August, vol. 310. IOP Publishing, p 052004

Zahedi M, Sharma R, Jenner CF (2003) Effects of high temperature on grain growth and on the metabolites and enzymes in the starch-synthesis pathway in the grains of two wheat cultivars differing in their responses to temperature. Funct Plant Biol 30:291–300

Zhang H, Feng P, Yang W et al (2018) Effects of flooding stress on the photosynthetic apparatus of leaves of two Physocarpus cultivars. J Forest Res 29:1049–1059

Zhang QY, Wang LY, Kong FY et al (2012) Constitutive accumulation of zeaxanthin in tomato alleviates salt stress-induced photoinhibition and photooxidation. Physiol Plant 146:363–373

Zhao C, Liu B, Piao S et al (2017) Temperature increase reduces global yields of major crops in four independent estimates. Proc Natl Acad Sci 114:9326–9331

Zhao G, Ma B, Ren C (2007) Growth, gas exchange, chlorophyll fluorescence, and ion content of naked oat in response to salinity. Crop Sci 47:123–131

Chapter 2
Plant Abiotic Stress Tolerance Mechanisms

Mohammad Mafakheri, Mojtaba Kordrostami, and Jameel M. Al-Khayri

Abstract In their life cycle, plants face a range of environmental stresses such as heat, cold, drought, salinity, etc., which greatly affects the performance of the plant and is one of the key factors in the distribution of plant species. Plants use special mechanisms to withstand these stresses. When plants are exposed to stress, the information is transmitted by the signal transduction pathway, and eventually, the response to these signals leads to physiological and biochemical changes in the plants. Usually, one type of stress is accompanied or followed by other stresses. For example, heat stress is followed by drought stress due to physical loss of water, and cold stress is followed by drought stress due to the physiological inaccessibility of water. Due to the large number of environmental hazards that the plant faces at a particular time, abiotic stress signaling is a very complex phenomenon. Plants have tools to avoid and deal with these stressors. On the one hand, they can produce inductive and appropriate responses that lead to a specific desired change for which special stress conditions are specialized. On the other hand, there is a significant overlap between the components of abiotic stress signaling and the starting points in which the pathways for stress signaling are coordinated. In this chapter summarized, how abiotic stresses in plants can be tolerated.

Keywords Adaptation · Antioxidant · Crop productivity · Drought · High temperature · Osomolytes · Transcriptome analysis

M. Mafakheri
Department of Horticultural Sciences, Faculty of Agricultural Sciences, University of Guilan, Rasht, Iran

M. Kordrostami (✉)
Nuclear Agriculture Research School, Nuclear Science and Technology Research Institute (NSTRI), Karaj, Iran
e-mail: mkordrostmi@aeoi.org.ir

J. M. Al-Khayri
Department of Agricultural Biotechnology, King Faisal University, Al Hofuf, Al-Ahsa, Saudi Arabia
e-mail: jkhayri@kfu.edu.sa

J. M. Al-Khayri et al. (eds.), *Nanobiotechnology*,
https://doi.org/10.1007/978-3-030-73606-4_2

2.1 Introduction

Growth and productive capacity of plants are adversely affected by abiotic stress worldwide. If environmental stresses did not occur, the actual yields would have to be equal to the potential yields of the plants. While in many crops the average yield of plants is 20–30% of their actual potential. Given their sessile nature and prevalence of abiotic stress, plants have evolved a plethora of tolerance mechanisms at macro and micro levels. They are often diverse from morphological to physiological and molecular changes that subjection to each abiotic stress initiates a cascade of resistance mechanisms across various levels (Grift et al. 2011; Sinclair and Muchow 2001).

This chapter deals with major abiotic stresses in plants, explores ways in which plants can cope with these stressors, and summarized how abiotic stresses in plants can be tolerated.

2.1.1 Morphological Flexibility Conferring Abiotic Stress Tolerance

2.1.1.1 Water Stress

Ample of survival strategies enable the plant to complete their life cycle by escape or avoid the harsh environment, which the former strategy encourages the plant to shorten the vegetative growth and enter the reproduction stage to yield seeds to take advantage of the current water resources that would not available by the onset of the drought period. Therefore, by producing progenies as primary goals before the arrival of fatal drought several plant species mostly early maturing inhibited in desert environment opt to escape while having complete reproduction (Fahad et al. 2017; Farooq et al. 2012). Whereas some other species choose to develop more sophisticated techniques to avoid dehydration with structural modification in favor of drought tolerability. These changes either minimize the rate of water loss from aerial parts or maximize the capability of the root system to uptake water that in both scenarios the main motive is to maintain a high water status so the growth and development can continue (Blum 2005). Reduction in biomass production under drought stress has been observed in various species with significant variation between aboveground and underground parts; also, decrease in the biomass of vegetative parts does not necessarily mean the absence of change or increase in root system biomass. As Xu et al. (2015) observed in rice that increases in root/shoot ratio was because of higher reduction rate in shoots not increasing the biomass of roots. The correlation between the root and shoot biomass was positive in drought-tolerant chickpea cultivars as higher biomass in aboveground was accompanied by high biomass production in root system which is a successful adaptation strategy to absorb water and nutrient through an extensive root system (Rahbarian et al. 2011). Changes in root architecture are essential to tackle the crisis caused by drought, which is normally involved in increasing the number of main roots and fine roots, the rapid expansion of root system to the upper layer of soil and enhancing root diameter and tissue density. Most importantly, deep roots greatly contribute to improving the water relation on drought-affected plants as their positive contribution in various crop plants exposed to drought have been observed such as rice (Manschadi et al. 2010; Wasson et al. 2012), maize (Hussain et al. 2019), and wheat (Krishnamurthy et al. 1999). Improvement in root length density up to 30 cm in the function area of root resulted in increased water and mineral nutrient absorption in chickpea subjected to drought stress (Sadok and Sinclair 2011). One the other hand, an extensive root system should not be considered advantageous constantly because a larger a root is the greater portion of allocated assimilated material for its maintenance will be, which can partially responsible for a decrease in aboveground biomass (Bramley et al. 2009; Li et al. 2010). Significant structural modification in leaves to avoid water deficit takes place in drought-stressed crop plants including reduction in volume of leaf as a prominent source of losing water. This modification has been known as an effective approach to resist water deficit by minimalizing water loss through transpiration; however, it penalize the productivity with low carbon assimilation efficiency (Lei et al. 2006; Sinclair and Muchow 2001). Additionally, enhancement in cuticle thickness, stomatal resistance, stomatal size and the last but not the least inclination of leaves toward vertical angles or rolling are other ways to lower the light interception and water consumption under drought stress (Sinclair and Muchow 2001).

A range of morpho-anatomical adaptation in plants to survive under flood stress have been developed to mitigate limitations caused by flood in particular reduction in CO_2 and exchange of gases overall. Several effective strategies exist that among them formation of novel roots with gas-filled spaces (aerenchyma) as result of cell separation or programmed cell death, is a characteristic common in numerous plant species provide oxygen for organs in plants submerged underwater as rice that creation of aerenchyma is by the termination of specific cells. Such adaptation enhanced the oxygen transport in rice effectively where submerging underwater prevents oxygen transfer (Cardoso et al. 2013; Colmer 2002; Drew et al. 1979). The capability of the root system to leakage oxygen to the anaerobic soil oxidizing the rhizosphere, therefore, reducing possible damages from reduced toxins like Fe^{2+} (Colmer 2002). In some crops such as maize, creation of aerenchyma can be induced with a reduction in oxygen and increase of ethylene which leads to initiating programmed cell death for cells in root cortex (Drew et al. 1979). Additionally, root architecture that confers adaptability to flooded plants exhibits less-branched main roots, large root diameter and deep roots (Aguilar et al. 2003, 2008; Visser et al. 1997). While in woody species the adaptations are to some extent different and it seems there is a positive correlation between drought tolerant with flood-tolerant as Sena Gomes and Kozlowski (1980) observed in *Eucalyptus* spp. In maize, if the shoot base submerged in water, the formation of aerenchyma in outer leaves may occur (Matsukura et al. 2000). The number of aerenchyma formed in flood stressed wheat affected by the level of tolerance in genotypes in which resistance genotypes had a higher number of aerenchyma (Johnson and Huang 1996). Obviously, owing to the organizational interconnection between the root system and aerial parts, the flood also affects the photosynthesis apparatus through depleting root area from oxygen and limiting the function of roots to uptake water, thus creates a drought-like condition, wilting leaves and consequently damages carbon assimilation. Further, N deficiency is common under flood stress but to tackle this problem, at least for a short-term, plants attempt to relocate N from older leaves to young leaves in order to continue growth (Drew and Sisworo 1977). Sometimes changes in leaves to resist water loss in crops under flood involve downward movement of leaves (epinasty) to lower the interception with light, this behaviour has been observed in tomato (Ellsworth and Freebairn 1969).

2.1.1.2 Extreme Temperature

A simple definition of thermos-tolerability of crops plant is the potential to endure heat stress, grow, develop and provide an economically reasonable yield, which candidate crop plants possibly morphologically enabled to stabilize its internal temperature just below the irreversible point where damages are terminal under heat stress. Such mechanism is achievable through changes similar to adaptation detailed earlier for drought stress as leaf rolling, change in leaf orientation to reduce the light absorption, reflecting solar radiation and regulating stomatal behavior as well as a promoted cuticle and wax layer on the surface of leaves to enhance stomatal resistance (Grift et al. 2011). Additionally, the capability to maintain photosynthesis activity, distribution of the fixed carbon and absorption of necessary minerals are also characteristics of tolerant crop cultivars (Cui et al. 2006; Grift et al. 2011). In wheat crops, rolling of flag leaf is to lower the temperature and consequently transpiration to improve the yield of photosynthetic activity (Perkons et al. 2014). The implication of escape-like strategies in plants inhabited areas with extreme fluctuation in temperature has been observed, these species commonly tend to early maturation and seed production before the arrival of heat stress period which is unavoidably has a drought in the company as well (Trachsel et al. 2009). To enhance the radiation use efficiency as well as saving the limited water resource some plants incline to form thinner blade leaves with reduced weight and prostrate growth habit (Richards 1996). A usual procedure in the plant to reduce the temperature and prevent heat stress damage is to promote transpiration. Employ this strategy requires deep root system to support the water use lowering the temperature in the canopy by transpiration proven to be effective as Chauhan et al. (2009) stated that tolerance of heat stress in wheat genotypes observed in those with lower temperature canopies when compared to genotypes with higher temperature canopies. Likewise, the positive features reported in tetraploid wheat had a higher yield (Bahar et al. 2008). A potent root system with branched yet deep main roots can uptake water and mineral nutrient that are highly in demand when the plant is challenged by heat or drought stress. On the adaptation strategy in crop plants against cold stress, we did not find any relevant information or published research indicating specific morphological adaptation in crop plant under low temperature possibly because cold stress resistance mechanisms in crops chiefly rely on physiological and molecular approaches as a quick relief to survive the suboptimal temperature.

2.1.1.3 Salinity

Salinity exerts multiple negative impacts on plants fitness. Plants have evolved with numerous morpho-anatomical adaptation strategies in their root system and aerial parts. The changes regarding morphological features in this section mainly concerns salt tolerance plant species that are not considered crop plants since these adaptations at this level mostly but not always have been seen in plant species inhabited in saline waters or areas with high salt content and dry soils. Halophytes (salt-tolerant plants) and glycophyte (e.g., are halophytes that only tolerate a low level of salinity) are mainly monocotyledon families and in some cases belong to dicotyledonous (Rozema 1991; Sharma et al. 2016). Since salinity is accompanied by extreme water deficit in soil root system architecture is more similar to the species adapted to drought stress and are deep, extensive, high length density and diameter as well as higher root weight. Enhanced root weight was recorded in grasses subjected to salinity (Seregin and Ivanov 2001). Maintaining biomass biosynthesis rate is associated with salt resistance, comparing salt-tolerant plant species *Leptochloa fusca* L. and *Puccinellia distans* with *Pennisetum divisum* L. as salt intolerant indicated a significant reduction in biomass of the latter species (Ashraf and Yasmin 1997). More, increasing, or maintaining the mesophyll area per leaf area in the tolerant species such as *Atriplex patula* L. provide the possible capability to protect photosynthesis apparatus against salt-induced terminal effects, unlike two glycophyte species *P. vulgaris* and *Gossypium hirsutum* that the ratio of mesophyll area significantly reduced. Longstreth and Nobel (1979) developing salt glands to secrete salt ions is probably the most common approach in halophytes which as an adaptive mechanism their density intensified in *Zoysia* spp. that improved salt tolerance and continued with the removal of those glands (Marcum et al. 1998). Another approach that some plant species enabled themselves to tolerate salinity under drought conditions is to collect water from the scarce resources into thick halo-succulent leaves to sustain the water status. Densely covered leaves with pubescence are specific to xerophytes and can help halophytes to lower the water loss and sustain growth and development in drylands with high salinity (Marcum et al. 1998).

2.1.1.4 Heavy Metal

The extensive array of adaptation mechanisms either avoidance or tolerant heavy metals have evolved, the studied survival mechanisms implemented by plants to protect the leaves or roots from lethal effects of heavy metals are mainly biochemical and molecular, and to this day a little attention has been paid to morpho-anatomical adaptations in plants against heavy metals (Shahid et al. 2017). However, the importance of some modification and organs in plants explored that found to be associated with heavy metal tolerance, for instance, leaf cuticular structure where heavy metal ion can penetrate and accumulate (Vu et al. 2013). Or symbiosis relationship with mycorrhizal fungi which as frequently has been observed serves the roots of hosts with limiting the mobility of metal ions while assisting the host with the uptake of essential nutrients to form soil matrix (Jentschke and Godbold 2000), in inoculated maize roots with a strain of Glomus sp. The concentration of heavy metal ions was significantly reduced by comparison to non-inoculated seedlings (Kaldorf et al. 1999). The preventative effect of fungi strains reported to vary from one species to another and several mechanisms have been suggested on the influence of symbiotic relationship with mycorrhiza with a reduction in transportation of heavy metal ions to aerial parts (Hall 2002), perhaps a combination of different mechanisms are involved.

2.2 Positive Physiological Modification to Tackle Abiotic Stress

Continuum of growth and development to reach productive stage is the prime intention of plans experiencing abiotic stresses. In addition to morphological changes, they resort to physiological modification enable them to maintain the vital water status by implicating osmotic regulation besides controlling and detoxifying the consequences of ROS-generated oxidative damage (Fig. 2.1) (Hasanuzzaman et al. 2013).

2.2.1 Antioxidants

The imbalanced water status leads to a reduction in photosynthesis activity owing to stomatal closure. An approach to prevent water loss through transpiration which by limiting carbon influx carbon reduction is in Calvin cycle declines that as result the electron acceptor in the photosynthesis process, oxidized $NADP^+$ is decreased (Blokhina et al. 2003; Wilson et al. 2014b). By excessive decline in iron-sulfur proteins in the course of electron transfer through a reaction process known as Water-Water Cycle or Mehler reaction harmful ROS are created such as $O2^{\bullet-}$ and H_2O_2 in the chloroplast, peroxisome, and mitochondria. This process starts a detrimental knock-on-effect leads to oxidative stress and peroxidation of lipids in the membrane, structural damage to proteins and large molecules, also more potent ROS like single oxygen ($^1O_2$) and $^{\bullet}OH$. Despite the detrimentally of ROS and their overproduction under stress condition, plant armed with effective enzymatic and non-enzymatic antioxidants to quench these lethal ROS, antioxidant defense system suggested as the main mechanism in plant tolerance to abiotic stress (Blokhina et al. 2003; Sairam et al. 2009; Wilson et al. 2014b; Xie et al. 2019; Zhang et al. 2007). Commonly, antioxidants experience an elevation in plant cells not only for

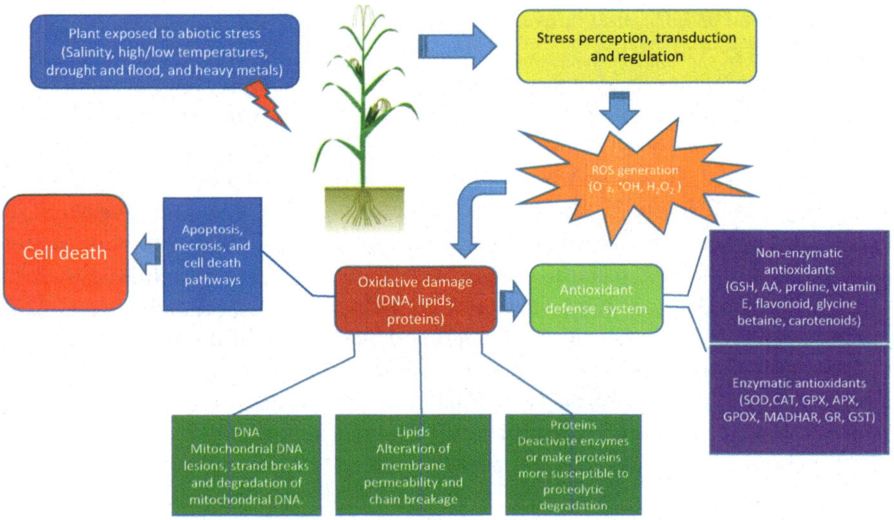

Fig. 2.1 Imposed oxidative stress as a result of abiotic stress (ROS, single oxygens; O_2^-, superoxide anion; H_2O_2, hydrogen peroxide; $^{\bullet}OH$, hydroxyl radical). Oxidative damage leads to either cell death eventually or with involvement of antioxidant defense system (enzymatic or non-enzymatic) the ROS detoxified. GSH, glutathione; AA, ascorbic acid; SOD, superoxide dismutases; CAT, catalase; GPX, glutathione peroxidase; APX, ascorbate peroxidase; MDHAR, monodehydroascorbate reductase; GR, glutathione reductase; GST, glutathione S-transferase

drought stress but in most stresses are superoxide dismutase (SOD), catalase (CAT), peroxidase (POX), glutathione reductase (GR), and ascorbate peroxidase (APX) and glutathione peroxidase (GPX) as enzymatic and, ascorbic acid (ASC), a-tocopherol, reduced glutathione (GSH), b-carotene, compatible solutes such as proline as non-enzymatic show increased concentration in crop plants in general (Polidoros et al. 2005; Xie et al. 2019).

The ability of antioxidants to alleviate the ROS and range of activity in terms of an organ is well-documented. Of the antioxidants that almost has no organizational limitations, SOD that is known to be the forefront of defense against radicals and can be found in all subcellular sections majorly catalyzing $O2^{\bullet-}$ reduction to H_2O_2 and finally to water possibly by GR and APX (Almeselmani et al. 2006; Lu et al. 2010). Moreover, ROS signaling-mediated by POD in various cellular organs leads to the accumulation of SOD, CAT, and APX (Mittler and Blumwald 2010; Mittler et al. 2011). Probably the highest efficiency to convert oxygen species into the water and/or oxygen is CAT with an estimated transformation rate of 6,000,000 H_2O_2 per molecule of CAT per min particularly in peroxisome where photorespiration takes place owing to the existence of tetra-ham in its chemical structure (Garg and Manchanda 2009). An essential combination of antioxidants with a high detoxification rate is the GR-AsA cycle that is greatly influenced by GR which is mainly chloroplast-located. Another H_2O_2 potent scavenger is APX with various isoforms each one specific to a cellular organ or GPX that converts H_2O_2 into the water using glutathione (Noctor et al. 2002; Xie et al. 2019).

In the case of non-enzymatic antioxidant AsA with high water solubility and potency against ROS, is an important participant of the cyclic pathway of enzymatic detoxification of ROS that can be found in high concentration in photosynthetically active cells and leaves with high chlorophyll content (Almeselmani et al. 2006; Foyer and Noctor 2011). This molecule, along with other components of the antioxidant system, protects plant cells against oxidative damage caused by aerobic respiratory metabolism, respiration, and even pollutions also are essential for alpha-tocopherol reproduction. It also acts as a secondary antioxidant in the recycling of alpha-tocopherol and other lipophilic antioxidants, similarly, alpha-tocopherol is a capable antioxidant owing to possessing methyl which makes it superior in comparison to its other three isoforms (Blokhina et al. 2003; Hasanuzzaman et al. 2013). Among photosynthetic pigments carotenoids play a key role as they are preserving photosynthetic apparatus, PS I in particular, against damaging surplus light radiation, scavenging ROS affecting thylakoid membranes and photoreceptors (Niyogi et al. 2001). Effect of essential versatile metabolites such as GSH is also important due to its role in detoxifying ROS and participate in various pivotal processes like signal transmission throughout the plant (Foyer and Noctor 2011; Wojciechowska et al. 2013). As an expected osmolyte in most of the abiotic stresses, proline also is capable of providing ROS scavenging to lessen the damage imposed by oxygen species (de Campos et al. 2011). Plant secondary metabolites especially phenolic compounds and flavonoids can be detected inclusively in cellular organs with a major concentration in vacuoles, possess a high antioxidant activity that their increment in plants under abiotic stress is common (Blokhina et al. 2003).

2.2.1.1 Water Stress

Water deficit with inhibiting the stomatal conductance and consequently photosynthesis activity prepares the ground for oxidative damage as a result accumulation of ROS which in turn cellular mechanisms provoke the production of an array of anti-radicals (Fig. 2.1). Ample of factors can influence the quality and quantity of antioxidants in water-stressed affected plants, Lin et al. (2006) in a comparative study subjected three cultivars of *Ipomoea batatas* L. plants to waterlogging and drought stress. The outcome indicated a high variation in SOD, phenol, flavonoid, and carotenoid content, but in general, reduction in these biochemical compounds were higher in flooded plants except for carotenoids that remained unaffected by both stressors. The content of nutrients in drought-stressed and growth stage observed to influence the concentration of antioxidants aim to improve the tolerance of crops. Case in point, canola seedlings at early stages subjected to water shortage and N deficiency had the highest CAT content while an increase in N application increased polyphenol oxidase and PO. However, the latter reduced in the following stage, or proline content that enhanced the most in the moderately water-stressed seedlings and incremented indiscriminately across at stages (Ahmadi et al. 2015). Reduction in antioxidants as a substantial contributor to plant adaptation by ontological advancement has been observed in maize (Zhang et al. 2007) which is possibly owing to their replacement with other mechanisms such as morphological adaptations to prevent the generation of ROS in the first place. Acquiring a reliable level of tolerance to water deficit is possible by having a proper level of antioxidant induction (Wang et al. 2009). Recent studies on detecting responsible genes for antioxidants yielded several important breakthroughs such as OsLG3 in rice which its overexpression associated with induction of numerous ROS scavenger-linked genes confer tolerance to drought-stressed rice (Lu et al. 2010). Similarly, rice plants expressing MnSOD (originally transferred from *Pisum sativum* L.) exhibited a lower level of peroxidation of lipids when compared with wild type, which implies the enhancement in antioxidants controlling ROS. Induced expression of APX and Cu/ZnSOD in *I. batatas* led to an improvement in drought adaptation with incremented photosynthetic activity (Lu et al. 2010). Accumulation of antioxidant in oxygen-deprived plant exposed to flood stress in the studied species is highly inconsistent which partially can be explained by the effect of species and experimental conditions (Blokhina et al. 2003). However, is one of the comprehensive assessments on eleven species composed of anoxia-sensitive and tolerance, the evaluation of the functionality of monodehydroascorbate (MDHAR) dehydroascorbate (DHAR) and quantity of AsA and GSH unveiled a notable increase in MDHAR and DHAR whereas the sensitive species the indices did not change or decreased. The content of AsA after GSH after anoxia treatments were incremented and decremented, respectively (Wollenweber-Ratzer and Crawford 1994). Studies also suggested a lower aggregation of ROS ($O2^{\bullet-}$ and H_2O_2) in flood stressed plants, for instance, stressed *V. radiata* L. the content of $O2^{\bullet-}$ and H_2O_2 reduced in sensitive genotypes while the ROS status did not change in the tolerant genotypes but APX and GR experienced an increase (Sairam et al. 2011). The exact similar pattern reported in *Cajanus cajan* L. treated with the flood (Sairam et al. 2009). The fluctuation in content and activity of antioxidants in hypoxia or anoxia treated plants as mentioned earlier is extremely circumstantial with significant dependency on duration possibly due to exacerbation of oxidative stress and the change in need for specific antioxidants.

2.2.1.2 Extreme Temperature

Plant responses and adaptation to abiotic stress mainly share a great deal of similarity specifically in respect of physiological and biochemical approaches to mitigate the terminal effects of oxidative stress caused by limiting CO_2 conductance and assimilation and photorespiration. Analogs to other abiotic stress, extreme temperatures (heat or cold) impose the generation of ROS in which antioxidants have been reported to have a key role in their regulation in plants under high/low temperatures. The best-case scenario to understand the behavioral paradigm in the effect of high/low temperature on specific antioxidants is to subject a given plant species to high/low temperature and evaluate the quality and quantity of antioxidants of interest. Soengas et al. (2018) conducted the similar study on two cold-season vegetable crop,

Kale (*Brassica oleracea acephala* group) and Cabbage (*B. oleracea capitate* group) which a general enhancement in CAT, SOD, GR and glutathione in crops of both groups exposed to high/low observed with no significant species effect. Contrasting reaction to high/low-temperature stress was observed in canola seedlings, under chilling stress, the activity of SOD reduced while CAT was unaffected but in heat-stressed seedlings, CAT indicated an increment and SOD decremented (Alscher et al. 1997; Blokhina et al. 2003; de Campos et al. 2011; Soengas et al. 2018). It's well-documented that accumulation of ROS generates damage to all cellular organs chloroplasts in particular. In a study screening the induction of antioxidants in five heat-stressed wheat genotypes, the stressed late-planted seedlings across the entire vegetative and reproductive stages experienced elevated content of SOD, CAT, and APX plus reduction in GR and POX. However, two genotypes showed proportionately higher values for all the evaluated antioxidants and also degradation in photosynthetic pigments and membrane was the least which is a vivid manifestation of essentiality of antioxidants in conferring heat stress tolerance (Almeselmani et al. 2006). Similarly, the heat-tolerance Kentucky bluegrass cultivar 'Eagleton' under high temperature exhibited an increase in chlorophyll content and enhanced level of antioxidants CAT, SOD and APX (He and Huang 2010). Moreover, in heat tolerance citrus cultivar, Carrizo, a significant elevation in antioxidants (GR, CAT, APX, and SOD) discovered to be responsible for its tolerability to high temperature while in the heat-sensitive cultivar, Cleopatra the reduction of antioxidant or lack of response observed, but proline content remained statistically unaffected in both cultivars (Zandalinas et al. 2017). The overexpression of AtGRXS17 in tomato seedlings enabled the genetically modified tomato to resist low-temperature stress with experiencing no growth-related negative effects most likely due to the reduced ion leakage and efficient photosynthetic activity under low temperature when compared with wild-type (Foyer and Noctor 2011). The accumulation of phenols, flavonoids, and glutathione in cold/heat-stressed have been reported as well (Raseetha et al. 2013; Wilson et al. 2014b; Wojciechowska et al. 2013).

2.2.1.3 Salinity

The main source of adverse impacts in salinity-stressed plants is oxidative stress induced by limitation in stomatal conductance, uptake and accumulation of toxic ions and deficiency in mineral nutrients (Shah et al. 2017). Plants evolved a potent capability to turn the cellular milieu into a battlefront to scavenge ROS and alleviate the resulting damage by enhancing the quantity of enzymatic and non-enzymatic antioxidants efficiently collaborate to tackle the matter (Aziz and Khan 2001; Azizian and Sepaskhah 2014; Parida and Das 2005). In date, palm seedlings resistance to salinity ('Umsila') the uptake of Na^+ reduced and notable increment in SOD, CAT, and APX in addition to proline, glutathione and polyphenolic substances in leaves was observed as compared to sensitive one 'Zabad' (Al Kharusi et al. 2019). Similarly in two soybean cultivars with contrasting tolerance to salinity exposed to various level of salt, the biomass production and healthiness in the tolerant cultivar was affected non-significantly with a dramatic induction in SOD, CAT, APX, and GR, whereas the otherwise observed in the sensitive one (Khan et al. 2009). More, in osmotic stress tolerance wheat genotypes treated with salt, an accumulation in CAT, PO, and APX recorded while its absence observed in the sensitive genotype (Siddiqui et al. 2017). Sometimes antioxidants exhibit organ-specificity, for instance CAT and DHAR showed increase universally in the plant under salt stress while SOD, APX, GST, and GR enhanced particularly in the root system (AbdElgawad et al. 2016). Given the presence of physiological evidence as a difficult to reject data supporting the essentiality of an antioxidant system for salt tolerance, several genes associated with induction of antioxidants aiming tolerance to oxidative stress have been reported (Guan et al. 2015). For example, a putative gene (PutAPX) in *Puccinellia tenuiflora* L. detected and its induction under oxidative stress conferred tolerance to salinity stress, following its transformation to *Arabidopsis* exhibited decrement in lipid peroxidation and terminal effect of salinity in general (Blokhina et al. 2003).

2.2.1.4 Heavy Metals

Impairing the photosynthetic apparatus, automation, and Fenton reaction and ion disequilibrium resulting in the formation of ROS beyond a controllable level (Schutzendubel and Polle 2002). We should keep that in mind that not all heavy metals are toxic for plants since elements as Fe, Mn, Cu, Zn, and Ni are essential for the very fundamental processes, required in small amounts and increasing their concentration after a certain point lead them to be juxtaposed with non-essential highly detrimental ions such as Cd, Cr, Pb, Co, Hg and Ar (Salt et al. 1995). Heavy metal ion classified into two groups with and without redox ability, the former composed of essential heavy metal ions except for Ni, also Cr considered in the first group that by participation in cycling reactions generate an excessive level of $^{\bullet}OH$ with high damaging potential to biological cellular organs, while the latter group includes the non-essential elements plus Ni (Emamverdian et al. 2015; Pandey 2018). An important dimension in heavy metal stress is the exhibition of antagonistic and/or synergistic behavior in the absorption of essential elements that tremendously contribute to the negative effects of heavy metal stress. Accelerated antioxidant defense system has been referred to as a viable mechanism against heavy metal-induced oxidative stress (Mousavi et al. 2013), which its diversity and concentration vary considerably depending on growth stage, type of metal ion, species, cultivar and even organs of the plant (Mediouni et al. 2006). Significant enhancement in CAT, SOD, POD, GR, GSH, PO, APX, phenolic compounds, osmolytes have been reported in plants subjected to heavy metal ions. Also, the intensity of oxidative stress vary between heavy metals. For instance, the seedlings of Ocimum basilicum L. received treatments of Ni, Cu, and Zn, which the order of the oxidative damage the seedlings experienced from high to low was in this order Cu > Zn > Ni, which indicated the difference in toxicity of heavy metals (Georgiadou et al. 2018). In some cases, the absence of responsiveness from antioxidants in heavy metal-treated plants as Milone et al. (2003) reported that SOD was ineffective in the root system of two contrasting Cd-treated wheat cultivars while APX and CAT exhibited a reduction in the sensitive one. In another study on maize seedlings affected by Cr, increase in SOD, CAT, and POD in early stages and followed by a dramatic reduction in the content of SOD and POD in leaves (Zou et al. 2009). The results projected in literature are overwhelmingly diverse and contradictory which reflects the complexness of antioxidant defense machinery and underlying mechanisms in plants against heavy metal stress.

2.2.2 Osmotic Adjustment

Of course, the adaptation of various strategies in plants to mitigate the negative changes caused by abiotic stress is in many cases are specific but if there is one biologically universal reaction exists against abiotic stress that could be osmotic adjustment which is a simple term is the accumulation of osmolytes. This mechanism aims to further reduce or tranquilize the water potentially affected by the interference of abiotic stress simulate water deficit, on water relations. Under abiotic stress-prone environments, water deficit, in particular, plants with the physiological potential to accumulate osmolytes to adjust the water relations have tremendous survival superiority. Since there is a misuse of terms in this area, some clarification would help to convey scientifically correct information; the term 'osmotic adjustment' is correctly applied to the aggregation of novel solutes occurred not when the reduction in water potential caused by the current solutes owing to water loss (Babu et al. 1999). Another one is the misapplication of 'osmoregulatory' and 'osmoprotectants,' it's safe to say that all osmolytes contribute in improving cell turgor pressure, which is imperative for efficient carbon fixation. While not all these osmolytes possess the antioxidative capability and lightweight molecules such as proline, glycine betaine (GB) and polyamines are both osmoregulators and osmoprotectants. Whereas sugars, polyols and mineral ions including K^+, Na^+, and Cl^- are solely osmoregulators, also these compounds collectively known as compatible solutes mainly accumulate in vacuoles. Additionally, a difference might appear between the osmotic adjustment of leaves vs. roots, where most believe that osmotic adjustment is higher in roots OA could be greater in roots than in leaves. Plenty of studies have revealed the critical role of osmolytes involves in osmotic adjustment in plants exposed to abiotic stress, a summary on each abiotic stress given as follows (Sharp and Davies 1979; Westgate and Boyer 1985; Hsiao and Xu 2000).

2.2.2.1 Water Stress

The first deadly impact of water deficit is unbalancing the water status of the plant which initiate a series of damaging changes specifically in cellular water potential that threatens the carbon assimilation. Therefore crop plants mainly aim to increase the accumulation of new compounds to lessen the water potential without a reduction in existing water content which often found to be an effective remedy as osmoregulation. It's a wonderful achievement for plant and unlike many other adaptations, the accumulation of osmolytes is not a bitter pill to swallow since the energy consumption is significantly low and has no tangible negative consequences (Cechin et al. 2006; Kiani et al. 2007). When it comes to crop plants survival does not properly apply and owing to the economic importance of crops, survival for crop plants means the production of an economically reasonable yield for which scientific discussion of scholars in this case majorly considered the osmotic adjustment as a mechanism that only serves the survival purposes rather than productivity, an issue that addressed by Blum (2017) using the current data on the effect of osmotic adjustment on yield in twelve main crops from cereals, legumes and oil crops who found that high yield of the crops under water deficit condition significantly associated with greater osmotic adjustment (for review see Turner [2018]). Osmotic adjustment assist plant in two main approaches, I. enhancing the cell turgor, makes a proper carbon fixation possible, II. enables the root system for correctly operate and absorb water from soil matrix which otherwise would be impossible or at least difficult, a third (III) approach can be provided by the antioxidant function of osmoprotectants (Kiani et al. 2007). Similar to other adaptation mechanisms, the concentration and type osmolytes changed in different growth stages of drought-stressed Bentgrass cultivars, is that an improved water relation was observed. Mainly due to the accumulation of sugars by the beginning of drought stress which replaced by proline as the stress continued, moreover, the non-organic ions remained almost unaffected (DaCosta and Huang 2006). As earlier explained, possible improvement in water uptake and stomatal behavior contributed to osmotic stress tolerance. Likewise, an enhanced yield in drought-stressed wheat found in association with the accumulation of compatible solutes through improving CO_2 conductance and consequently higher carbon fixation (Živcák et al. 2009). Irrespective of contrary reports, vacuole localized K^+ accumulation is a prerequisite to have an electrochemical gradient required for stomatal movements (Zhu 2001). The positive impact of increased accumulation of polyamines, putrescine, spermidine, and spermine in plants under drought stress has been reported (Singh et al. 2015; Yamaguchi et al. 2007). The protective role of GB on components of the photosynthetic system also has been suggested (Xing and Rajashekar 1999). Considering the damage to the root system in flooded plants, therefore the collapse of transpiration and carbon fixation, flood simulates a water deficit which similar to drought in flooded plants. Accumulation of osmolytes in an attempt to prevent or restore the situation has been studied in several plants species. In two wheat genotypes exposed to hypoxia during waterlogging, tolerance and recovery from the stress observed to be strongly associated with accumulation carbohydrates in root in the virtue of balancing water relations (Huang and Johnson 1995), while in another study on alfalfa, absence of a positive relationship between accumulation of osmolytes and flood stress was discovered (Barta 1988). In waterlogged young seedlings of Caisin (*Brassica rapa* subsp. *parachinensis*) a higher RWC and thus CO_2 conductance was accompanied by an improved osmotic adjustment (Issarakraisila et al. 2007). The versatility of osmolytes in dealing with two main stress-generated effects, osmotic- and oxidative-stress through reducing water potential and antioxidative function is a unique capability which so far a little attention has been paid from breeders possibly owing to the difficulty in using osmotic adjustment as a biochemical marker.

2.2.2.2 Extreme Temperatures

In plants subjected to high/cold temperature stress depress of photosynthetic apparatus by stimulate water deficit and lowering the transpiration rate leading to a reduction in the assimilation of carbon. Therefore, unleashing ROS as well as osmotic stress, for which, similar to water stress plants resort to adjustment of osmose through the accumulation of solutes (Argentel-Martínez et al. 2019). The behavior sometimes may be contradictory, however similarity in accumulation of compatible solutes have been observed. The prime example is high/cold stressed ecotypes of Iranian tall fescue ecotypes (*Festuca arundinacea* L.) exhibited that intensity of high/cold stress tolerance linked to the accumulation of carbohydrate and proline and the order to the tolerance of ecotypes was to some extent the same. Nevertheless, except for FA 1 and FA 2, the content of compatible cytosolutes dramatically reduced after 24 days, in those two ecotypes the osmolytes remained constantly high (Sheikh-Mohamadi et al. 2018). Acclimation of strawberry seedlings to overwintering indicated the importance of osmolytes in an immediate reduction in water potential, and prioritized reallocation of the solutes to younger leaves in which only young leaves survived. Also, the conversion of stored starch into soluble sugars is pivotal for reducing water potential in particular under cold stress (O'Neill 1983). The major contribution of a balanced water status is regular transpiration which accumulation of GB and decrementing water potential in the root system, as well as aerial parts. This reaction enabled heat-stressed plants of Triticum durum L. to have high transpiration and an efficient carbon fixation throughout the growth and reproduction stages, however, it was not with no cost as delaying in phenophases observed (Argentel-Martínez et al. 2019). Enhancement of GB content in crop plants under high-temperature stress has been observed by other scholars such as Rienth et al. (2014) who found a strong correlation between GB and proline accumulation (66 and 58%, respectively) with heat tolerance in vines. An increase in compatible solutes in heat-stressed crops also have been reported (Tian et al. 2017; Wilson et al. 2014a). Commonly proline has been known to greatly contribute to stress tolerance through osmotic adjustment (Zandalinas et al. 2017), on the other hand, plant secondary metabolites particularly flavonoids reported to confer tolerance to heat stress in citrus plants (Zandalinas et al. 2016). For a review see Lamaoui et al. (2018).

2.2.2.3 Salinity

Salinity stress in short leads to imbalanced water relation and osmotic stress through reduce the availability of water in the soil for the root system and ion toxicity which all result in a wilting and radical decrease in the amount of assimilated carbon (Ashraf 2003; Azizian and Sepaskhah 2014; Rahneshan et al. 2018; Sahin et al. 2018). Analogous to other abiotic stresses plants forced to seek adaptation by accumulating osmolytes in roots and leaves to lessen the water potential in the virtue of avoiding disrupt of cellular membranes and detoxify ROS with the aim of antioxidant activity in some of the compatible solutes. Despite the effectiveness of this approach the required energy is rather high and suggested as possible causes of growth reduction (Greenway and Gibbs 2003). The excessive uptake of Na^+ and Cl^- is the major source of nutritional deficiency in particular crucial mineral nutrients such as K^+ which affected by the antagonistic effect of Na^+ absorption and lack of discriminability from transporter channels against Na^+ that uses K^+ channels (Hmidi et al. 2018; Lacerda et al. 2003). Given the multidimensionality of salt effect, a combination of different mechanisms may trigger to address the situation but surely osmotic regulation is the main one. The linear increase in amino acids, proline, and GB have been frequently reported in salinity stressed plants (Bohnert et al. 1995). Accumulation of solutes in salt-treated *Cakile maritima* L. plants was varied in respect to either type and organ, is that osmotic adjustment in the stem was mainly (60%) attributed to Na^+ accumulation while this feature for proline was 36% in roots. Interestingly, the activity of ornithine-δ-aminotransferase which involved in N distribution under non-stress condition found to be highly correlated with proline synthesis which it can be implied that ornithine pathway significantly engaged in the production of proline (Hmidi et al. 2018), previous studies confirmed the involvement of ornithine pathway in proline biosynthesis which have accompanied symptoms of nitrogen deficiency (Öztürk et al. 2006; Roosens et al. 2002).

2.2.2.4 Heavy Metals

The exposure of plants to heavy metals stress have known to generate intense ROS accumulation that interferes with vital biological processes at the cellular level. The toxicity of heavy metals in soil targets root system by causing necrosis which depends on the level of tolerance and dose leads to reduction in water uptake from soil in addition to disables the photosynthesis apparatus thus carbon fixation through decrementing cell turgor. That leads the reduction in accumulation of cytosolutes to decline the cell turgor and preventing further water loss as well as imbalanced ionic condition. Moreover, the antioxidant capability of solutes such proline, GB, and glutathione can serve as dual-purpose compounds (Emamverdian et al. 2015; Georgiadou et al. 2018; Schat et al. 1997; Schutzendubel and Polle 2002). Therefore, the increased content of solutes is a significant mechanism to detoxify heavy metals and improve heavy metal stress tolerance. In heavy metal stressed Salvinia natans seedlings with Cd, Cu, Ni, Zn, Pb, Fe, Mn, and Cr treatments an increased level of GB, carbohydrates, and polyols observed (Schutzendubel and Polle 2002). Exogenous enhancement in the concentration of GB likewise has been reported in wheat subjected to heavy metal stress to provide a decrease in oxidative stress and improving osmotic adjustment (Alscher et al. 1997). Despite some paradoxical behavior in proline accumulation under heavy metal stress, often increase in proline content has been reported for example in Lactuca sativa (Costa and Morel 1994) and soybean (Balestrasse et al. 2005). However, proline accumulation and its association with osmotic adjustment have been questioned by Saradhi (1991) who observed enhanced water potential in Pb-treated wheat seedlings regardless of proline accumulation to a high level whereas in control seedlings the osmotic potential was significantly lower. The increase in accumulation of proline as key element in osmotic adjustment and oxidative stress is rather controversial, observed increased proline tied to water deficit since proline increased in transpiration suppressed discs of leaf as mu as or even higher in heavy metal stressed plants (Schat et al. 1997), following (Kastori et al. 1992) challenged their result by arguing that leaf discs exposed to heavy metal ions under full turgor experienced a dramatic increase in proline, however, there was a possibility of disruption in cellular membranes owing to a high level of toxic metal used (Ric de Vos et al. 1993). Observation made on Cu-stressed segments of rice leave raised the possibility of ABA involvement (Chen et al. 2001). Soluble sugars in plants under Co stress conferred adaptation by reducing cell turgor (Yu et al. 2019). Ultimately, Sharma and Dietz (2006) suggested that accumulation of amino acids and soluble sugars in heavy metal-stressed plants are not owing to disruption of biosynthetic process but rather tolerance mechanisms provide protection against heavy metal stress.

2.2.3 Molecular Strategies

Alongside various adaptation strategies that provide the plant with survival capability or more precisely enable the crop plant to have an economically reasonable yield under abiotic stress, the molecular level that plant as a response to abiotic stress many genes are triggered and make the occurrence of tolerance at the higher level possible, therefore the quantity of the expressed genes initiates a series of processes that acquiring an in-depth understanding is a necessity to reliably go forward on developing stress-tolerant crop cultivars which thanks to genomics and proteomics technologies an appreciable body of knowledge on model plants in particular and in many cases on crop plants exist that can be exploited for this purpose (Hossain et al. 2012; Shi et al. 2018; Yang et al. 2005).

2.2.3.1 Water Stress

Investigations regarding the identification of underlying genes responsible for conferring tolerance to high/low temperatures mainly in model plants and some crops using transcriptome and proteome analysis have enlightened our perspective toward the evolution of adaptive mechanisms. Notable experiments majorly those studied contrasting genotypes have yielded critical findings. In general, the flood- and drought-induced responsive genes by causing aggregation of specific proteins or hormones play their roles (Kavar et al. 2007). In most cases, the interplay of hormones and gene expression determines the tolerance to water stress as ABA a key hormone that it is enhanced biosynthesis in drought-stressed plants leads to tolerance-related regulation specifically the expression of associated genes (Yamaguchi-Shinozaki and Shinozaki 2006) which transferring of ABA-related *GhCBF3* from cotton to *Arabidopsis* conferred tolerance to water deficit through improving water relations and osmotic adjustment, and carbon assimilation (Ma et al. 2016). The expression of some genes can occur under different stresses, case in point SUB1A gene responsible for waterlogging tolerance in rice observed to bestow water-deficit resistance by enhancing antioxidants and improving the responsivity of ABA (Fukao et al. 2011). Considering the importance of circadian clock that induces the expression of a larger number of genes to obtain a level of tranquility in energy consumption and production in plants subjected to drought or flood stress. Two related genes, PRR7 (water deficit) and TOC1 (flood stress), which the latter provoked an increase in ABA content and linked with fluctuation in sugars during the day which manifests the dominant regulatory role of ABA in determining the energy homeostasis in drought and flood stressed-soybean plants (Syed et al. 2015). Water channels (aquaporins) are integral member proteins that are essential in the transformation of water across the membrane passively. That can amplify the water transfer by 10 to 20 times across membranes (Tyerman et al. 2002). Further, several transcription factors controlling the expression of drought-related genes have been identified including myeloblastosis protein (MYB), myelocytomatosis (MYC), DREB/CBF (drought-responsive *cis*-element binding protein/C-repeat-binding factor), ABF/AREB, NAC (NAM, ATAF, and CUC), and WRKY (Ishida et al. 2012; Nakashima et al. 2009; Sakuma et al. 2006; Tran et al. 2007). In *TaWRKY2*-transgenic wheat seedlings with limited water loss, enhanced content compatible solutes and higher survival rate proven to be drought resistance (Gao et al. 2018). Expression of *AtMYB60* and *AtMYB61* in rice localized in guard cells and found to be critical in controlling stomatal behavior under water deficit (Liou et al. 2005). Expression of genes involved in the induction of GB and proline biosynthesis evoked in under drought stress in several species including a number dicotyledons (Weretilnyk et al. 1989), Amaranthus hypochondriacus L. (Vojtěchová et al. 1997), barely (Jagendorf and Takabe 2001), cotton (Parida et al. 2008) and maize (Zenda et al. 2019). The significant up-regulation of xyloglucan endotransglycosylase (XET) gene family encoding enzymes involved in cell elongation observed in waterlogging-resistance maize genotype, HKI 1105, whereas it down-regulation occurred in flood-intolerant line, V 372 (van Veen et al. 2013). More, the resistant rice line indicated a higher induction level of IAA3, IAA14, and IAA16 genes (Wang et al. 2017). The interplay between ethylene and IAA in flood-tolerant maize is possibly linked to development of adventitious roots. Two of the regulatory energy-sensing protein kinases in carbon use under oxygen deprivation in *Arabidopsis*, KIN10 and KIN11, the former one encoding genes actively engage in carbohydrate and protein breakdown (Baena-González et al. 2007; Cho et al. 2012), additionally one of the two protein, KIN10/11, is responsible for EXORDIUM-LIKE1, an HUP critical for managing carbon in hypoxia circumstances (Schröder et al. 2011). The flood-stressed soybean seedlings the protomics analysis revealed an enhanced activity of ADH and delta-1-pyrroline-5-carboxylate synthase as well as presentation of seventeen various proteins such as β-glucosidase 31 and β-amylase 5 in both roots and leaves, which probably β-amylase 5 expression may have been linked to reallocation of carbohydrates to aerial part specially leaves (Wang et al. 2017).

2.2.3.2 Extreme Temperature

The main determinate of plant growth and productivity is temperature due to the substantial influence of temperature on biological processes which evidenced that high/low-temperature stress can severely affect crop plants. Plants are capable of triggering organizational tolerance response to extreme temperatures. These responses often leads to the expression of numerous genes initiating osmotic adjustment, antioxidants or regulating water status of under stress plants by modifying in the transcriptome, proteome, and metabolome or even in some cases by evoking programmed cell death to detach a specific organ or eliminating the whole plant (Lamaoui et al. 2018; Qi et al. 2011; Shi et al. 2018). Of the critical commonplace proteins such as heat shock proteins (HSPs) and heat stress transcription factors (HSfs) reported that their expression genotypes are associated with thermos-tolerance (Cheng et al. 2016; Sanghera et al. 2011), of course, a large array of metabolites as osmolytes and antioxidants involved in tolerance mechanisms which primarily influenced by the nature of genotype (Challinor et al. 2007; Kaya et al. 2001). Studies covering the molecular mechanisms conferring heat tolerance are often conducted on model plant species and a small portion have dedicated to crop plants which we try to mainly focus on the latter. The effects of HSPs majorly concern their role as molecular chaperones prevent misfolding and enhancing and stabilizing the structure of the protein so they function properly under heat stress, however knocking out this protein had a marginal negative influence of thermos-tolerance (Schramm et al. 2008; Yoshida et al. 2011). The association of heat stress tolerance with the superfamily of HSP70 has been discovered in several crop plants including, rice (Sarkar et al. 2013), maize (Rochester et al. 1986), and wheat (Duan et al. 2011). Isolated HSP proteins (HSP70 and HPS90) from Sitodiplosis mosellana as a harmful pest insect of wheat has obligatory diapause to avoid extreme temperatures in larval stages, and their transformation to wheat plants indicated contrasting capability in these two genes in which HSP70 and HPS90 conferred thermo-tolerance and cold-stress resistance, respectively (Cheng et al. 2016). 32 HSP70 has been identified rice and the involvement of HSP70 also have been proven (Sarkar et al. 2013). A high level of accumulation of HSP70 in heat-stress tolerance grape genotypes also reported (Zhang et al. 2005). Whereas HSP101 observed to not be inducible by heat stress but is essential for growth in root in maize, nevertheless, the heat-stressed maize experienced amount in the transcript of HSP101 in developing tassel, ear, silks, endosperm, and embryo but it was without an increment in HSP101 protein (Nieto-Sotelo et al. 2002; Young et al. 2001). The otherwise was identified for HSP101 in *Arabidopsis* which was critical in expression of antioxidants to lessen the damage caused by heat stress (Queitsch et al. 2000). More, HSP100 aim at the chloroplast intermembrane space and stroma to facilitate protein import of nuclear-encoded proteins (Nielsen et al. 1997). Comparing contrasting genotypes in rice using proteomic analysis revealed that high-temperature tolerance genotype had a significantly higher accumulation of HSPs a direct correlation between thermos-tolerability in rice genotypes and level of HSPs observed (Jagadish et al. 2009). Of the important most studied transcription factors responsive to cold stress, dehydration-responsive element-binding (DREB) protein/C-repeat binding factors (CBFs) or DREBs/CBFs bind to promotor cis-element CRT/DRE to regulate the expression of these cold-responsive genes. The overexpressed CBF1/DREB1b and CBF3/DREB1a observed to induce cold regulated genes (Gilmour et al. 2000; Kasuga et al. 1999), additionally, overexpression of DREB1A conferred tolerance to low temperature and water deficit in wheat and peanut (Kasuga et al. 2004; Pellegrineschi et al. 2004). Using comparative transcriptome Cai et al. (2019) discovered low-temperature sensitive genes, CBF4, ICE2, and several other ABA-associated ones. Successive gene silencing indicated the effective role of CBF4, ICE2 in the reduction of lipid oxidation in membranes as well as increasing SOD and proline to assist the adaptation to cold stress in *Gossypium thurberi* L. CBFs similarly identified and isolated in several crop plants such as rice, tomato, wheat, barley and maize (Shi et al. 2018). The expression of CBFs negatively affected by MYB15 by direct binding to conserved MYB motif in their promoters (Agarwal et al. 2006), moreover, Su et al. (2010) observed the inhibition of the expression of cold-sensitive OsDREB1B by OsMYBS3 in rice.

2.2.3.3 Salinity

Similar to other abiotic stresses addressed above, plants molecular mechanism to tackle the negative influence of salinity which majorly is osmotic- and oxidative-stress includes a vast range salinity-specific gene families and general stress-sensitive genes that express upon the occurrence of abiotic stress. In both cases transcriptomic analysis provided a comprehensive image of underlying genes conferring tolerance to plants by triggering the biosynthesis of antioxidants or ABA production that each one initiates sets of modification (Fujita et al. 2013; Johnson et al. 2002). Wu et al. (2019) in salt-stressed flax plants identified several TFs namely bZIP, HD-ZIP, NAC, MYB, GATA, CAMTA, and B3 which are renowned as stress-responsive TFs for expression of compatible solutes, hormones and signal transduction, in particular bZIPs that regulates effective tolerance mechanisms in plants under salinity. Moreover, salt specific molecular mechanisms are expressed in salinity-affected plants. Salt Overly Sensitive (SOS) signaling pathway is a classical signal pathway involved in salinity stress resistance, SOS composed of three proteins regulating vital processes such as Na^+/H^+ antiporter in the plasma membrane, transport of Na^+ from the root system to aerial organs, activating Ca^+ signals under salinity and signaling stress (Liu et al. 2000; Ma et al. 2019; Shi et al. 2000). Similar to SOS especially SOS3, the function of SNF1-related protein kinases (SnRKs) is long-distance signaling and regulation of ion translocation from roots to shoots. By its engagement with SnRK2.4 and SnRK2.10 in regulating numerous antioxidants to balance the ROS production in salt-stressed *Arabidopsis* has been reported (Szymańska et al. 2019). Limited ion toxicity and ROS aggregation in salt-treated *Arabidopsis* seedlings associated with $Ca2^+$-dependent protein kinases (CDPKs), since its silencing led to a significant salt-stress vulnerability (Zhang et al. 2018). In respect of ionic stress and related tolerance mechanisms, high-affinity potassium-type transporter 1 (HKT1) is crucial to maintain cytoplasmic K+/Na+ homeostasis under salinity stress which is outmost essential for salinity tolerance (Ali et al. 2019). Regarding transcription factors, the inner ability of cultivars can sometimes affect their level of expression, as an example, upregulation of bZIP in salt-intolerant wheat genotypes occurred while in salt-tolerant genotype down-regulated (Johnson et al. 2002). Overexpressed OsNAC6, a member of NAC TFs, in rice, regulated multiple stress tolerance including drought, salinity, and blast disease. Additionally, using microarray analysis it was discovered that OsNAC6 induced two other genes associated with stress tolerance (Nakashima et al. 2007). From other TFs, regulation of salt-tolerance in rice found be strongly controlled byDREB1/CBF, DREB2, and AREB/ABF (Fujita et al. 2013; RoyChoudhury et al. 2008). One of the latest discovered salt-stress related TFs in rice was SALT-RESPONSIVE ERF1 (SERF1) with root-localized induction under salinity stress which its silencing defected the upregulation of salt-induced mitogen-activated protein kinase (MAPK) related genes, following studies indicated the H2O2 responsiveness of SERF1-dependent genes with specific promoters (Schmidt et al. 2013).

2.2.3.4 Heavy Metals

A vast range of molecular mechanisms involves heavy metal tolerance which possibly indicates the evolution of tolerance to the edaphic limitation in these heavy metals before other abiotic stressors. Similar to the previously discussed general TFs expressed under abiotic stresses, in heavy metal stressed plants the expression of TFs has been reported frequently. For instance, HSPs, molecular chaperones responsible for repairmen of damaged proteins and protecting the proteins from denaturation. The inducibility of HSPs have been shown in plants stressed with heavy metal ions (Zn, Cu, Cd, Hg, Al and Cr) (Gupta 2010; Sarry et al. 2006; Zhen et al. 2007), from which HSP70 a potent protein and known for conferring significant tolerance to plants under high temperature, it's the accumulation in Cd-treated plants was reported (Neumann et al. 1994; Sarry et al. 2006). Similarly, the aggregation of HSP70 in seaweed and several other freshwater plants under Cd-stress have been observed (Ireland et al. 2004). The sea thrift (*Armeria maritima* L.) grown in Cu-contaminated soils experienced the induction of small HSP (sHSP) in the root system (Neumann et al. 1995). Likewise, HSP25 accumulated in soybean as a reaction to Al toxicity (Zhen et al. 2007). Interestingly, a transit heat-stress exposure in the plant caused Cd-tolerance which can be suggested the heat-responsive HSPs such as HSP70/90 induced by the heat stress that led to tolerance against Cd. Considering the destructive effect of heavy metal toxicity, HSPs' preservation of proteins could be a viable reason for the induction of HPS under heavy metal stress, nonetheless, the actual influence of HPSs in heavy metal stressed plants yet to be clarified (Hossain et al. 2012). The metal responsive TFs known as MTF-1 has a significant effect on heavy metal tolerance through initiating the expression of genes account for absorption, transport, and detoxification (Fusco et al. 2005; van de Mortel et al. 2008). Several MTF-1 have been reported for instance WRKY, bZIP, ethylene-responsive factor (ERF), and MYB, which regulate the induction of Cd-tolerance genes (Yang et al. 2005). For the bZIP TFs family, an important OsbZIP39 was identified which its induction led to the endoplasmic reticulum responds to the aggregation of unfolded proteins (Takahashi et al. 2012). The very specific metal stress proteins are metallothioneins (MTs), a family of small, conserved metal-binding proteins critical for the toxicity of heavy metals, which induced by phytohormones, cytotoxic agents, and heavy metals such as Cd, Zn, Hg, Cu, Au, Ag, Co, Ni, and bismuth (Bi) (Kägi 1991; Yang et al. 2005). The tissue specificity in MT genes regarding the growth stage and the effect of different heavy metal ions (Castiglione et al. 2007). (Ahn et al. 2012)reported the dependent-expression of 3 MT genes (BrMT1, BrMT2, and BrMT3) to heavy metal ions Mn, Zn, Fe, and Cu. Upregulation of related genes to MTs by TFs such as heat shock transcription factor A4a (HsfA4a), have been reported to improve the Cu-resistance in rice (Shim et al. 2009). Al tolerance incremented in rice seedlings with the induction of ASR5 that suggested acting as a TF which express genes protect cells against Al toxicity. Similarly, another TF, C2H2, induce ART1 protein localized in root that expresses genes associated with Al-tolerance (Arenhart et al. 2013). Nakashima and Yamaguchi-Shinozaki (2006) detected the DREB TFs down-regulated in heavy metal-stressed rice seedlings, which they introduced the possibility of DREB assistance in osmotic adjustment to decrement the uptake of heavy metals.

2.3 Conclusion and Future Perspective

Along with the pace of change in edaphic and atmospheric properties on earth the evolution enabled the primary forms of plants to survive and propagate. In the modern era, the definition of surviving in domesticated plants changed to the capability of the plants of interest to yield agricultural products under various conditions, a potential that is now jeopardized more than ever by a dramatic increase in the incidence of abiotic stresses. This resulted from the accelerated rhythm of negative changes caused by climate change majorly driven from exacerbating anthropological activity. During the last half-century, statistically speaking, the probability of severe water deficit, salinization of soil and water resources, and extreme temperatures has dramatically incremented in addition to the risk of simultaneous occurrence of these stressors. This all comes down to the fact that from the few limited choices to improve the food security of a world that population is increasing, improving the knowledge and understanding of crop plants response and tolerance mechanisms as a critical input for breeding programs can be perhaps the most viable approach if not only. Using phenotypic screening of germplasms responses to abiotic stresses under open-field conditions since lab trials may not be realistic. Probing for a pattern in response to abiotic stresses could be significantly helpful, of course, abiotic stress tolerance in crops controlled by multiple genes. However, commonality of dehydration due to damage to root system or unavailability of water as well as failure in proper stomatal reaction to the stress needs be considered together with the physiological level, where the general response is the disruption of water relations and/or reduction in carbon assimilation. Generating osmotic stress and oxidative stress is the lethal approach of abiotic stresses that force the plant to compulsorily initiate universal and sometimes small scale costly (i.e., from both the percentage of energy budget and the prospective reduction in ultimate yield) approaches as decreasing the carbon fixation. This often happens by the obligatory morphological changes in leaves to lower the light interception and biosynthesis of compatible solutes (i.e., GB, proline, and glutathione) and antioxidants (i.e., SOD, POD, CAT, APX, GR, GSH, and PO, etc.,) by expression of stress-inducible TFs (DREB, bZIP, CBFs, NAC and BrMT1, etc.). A body of knowledge that's a compilation of traditional and modern approaches is a valuable possession that can broaden our perspective to develop crop plants resilient for the climate-changing era.

References

AbdElgawad H, Zinta G, Hegab MM et al (2016) High salinity induces different oxidative stress and antioxidant responses in maize seedlings organs. Front Plant Sci 7:276. https://doi.org/10.3389/fpls.2016.00276

Agarwal M, Hao Y, Kapoor A et al (2006) A R2R3 type MYB transcription factor is involved in the cold regulation of CBF genes and in acquired freezing tolerance. J Biol Chem 281:37636–37645

Aguilar E, Turner D, Gibbs D et al (2003) Oxygen distribution and movement, respiration and nutrient loading in banana roots (*Musa* spp. L.) subjected to aerated and oxygen-depleted environments. Plant Soil 253:91–102

Aguilar EA, Santos PJA, Tamisin L Jr (2008) Root characteristics of five local banana cultivars (*Musa* spp.) under waterlogged conditions. Philipp J Crop Sci 33:14–23

Ahmadi SAK, Ebadi A, Jahanbakhsh S et al (2015) Changes in enzymatic and nonenzymatic antioxidant defense mechanisms of canola seedlings at different drought stress and nitrogen levels. Turk J Agric Forest 39:601–612

Ahn YO, Kim SH, Lee J et al (2012) Three *Brassica rapa* metallothionein genes are differentially regulated under various stress conditions. Mol Biol Rep 39:2059–2067

Al Kharusi L, Al Yahyai R, Yaish MW (2019) Antioxidant response to salinity in salt-tolerant and salt-susceptible cultivars of date palm. Agriculture 9:8. https://doi.org/10.3390/agriculture9010008

Ali A, Maggio A, Bressan RA et al (2019) Role and functional differences of HKT1-type transporters in plants under salt stress. Int J Mol Sci 20:1059. https://doi.org/10.3390/ijms20051059

Almeselmani M, Deshmukh P, Sairam R et al (2006) Protective role of antioxidant enzymes under high temperature stress. Plant Sci 171:382–388

Alscher RG, Donahue JL, Cramer CL (1997) Reactive oxygen species and antioxidants: relationships in green cells. Physiol Plant 100:224–233

Arenhart RA, De Lima JC, Pedron M et al (2013) Involvement of ASR genes in aluminium tolerance mechanisms in rice. Plant, Cell Environ 36:52–67

Argentel-Martínez L, Garatuza-Payan J, Yepez EA et al (2019) Water regime and osmotic adjustment under warming conditions on wheat in the Yaqui Valley. Mexico. PeerJ 7: https://doi.org/10.7717/peerj.7029

Ashraf M (2003) Relationships between leaf gas exchange characteristics and growth of differently adapted populations of Blue panicgrass (*Panicum antidotale* Retz.) under salinity or waterlogging. Plant Sci 165:69–75

Ashraf M, Yasmin N (1997) Response of some arid zone grasses to brackish water. J Agric Trop Subtrop 98:3–12

Aziz I, Khan MA (2001) Experimental assessment of salinity tolerance of Ceriops tagal seedlings and saplings from the Indus delta, Pakistan. Aqua Botan 70:259–268

Azizian A, Sepaskhah A (2014) Maize response to different water, salinity and nitrogen levels: agronomic behavior. Int J Plant Prod 8:107–130

Babu RC, Pathan MS, Blum A et al (1999) Comparison of measurement methods of osmotic adjustment in rice cultivars. Crop Sci 39:150–158

Baena-González E, Rolland F, Thevelein JM et al (2007) A central integrator of transcription networks in plant stress and energy signalling. Nature 448:938–942

Bahar B, Yildirim M, Barutcular C et al (2008) Effect of canopy temperature depression on grain yield and yield components in bread and durum wheat. Not Bot Horti Agrobo 36:34–37

Balestrasse KB, Gallego SM, Benavides MP et al (2005) Polyamines and proline are affected by cadmium stress in nodules and roots of soybean plants. Plant Soil 270:343–353

Barta AL (1988) Response of field grown alfalfa to root waterlogging and shoot removal. I. Plant injury and carbohydrate and mineral content of roots. Agron J 80:889–892

Blokhina O, Virolainen E, Fagerstedt KV (2003) Antioxidants, oxidative damage and oxygen deprivation stress: a review. Ann Bot 91:179–194

Blum A (2005) Drought resistance, water-use efficiency, and yield potential—are they compatible, dissonant, or mutually exclusive? Aust J Agric Res 56:1159–1168

Blum A (2017) Osmotic adjustment is a prime drought stress adaptive engine in support of plant production. Plant, Cell Environ 40:4–10

Bohnert HJ, Nelson DE, Jensen RG (1995) Adapt environ stress. Plant Cell 7:1099–1111

Bramley H, Turner NC, Turner DW et al (2009) Roles of morphology, anatomy, and aquaporins in determining contrasting hydraulic behavior of roots. Plant Physiol 150:348–364

Cai X, Magwanga RO, Xu Y et al (2019) Comparative transcriptome, physiological and biochemical analyses reveal response mechanism mediated by CBF4 and ICE2 in enhancing cold stress tolerance in Gossypium thurberi. AoB Plants 11:plz045. https://doi.org/10.1093/aobpla/plz045

Cardoso JA, Rincón J, JdlC Jiménez et al (2013) Morpho–anatomical adaptations to waterlogging by germplasm accessions in a tropical forage grass. AoB Plants 5:plt047. https://doi.org/10.1093/aobpla/plt047

Castiglione S, Franchin C, Fossati T et al (2007) High zinc concentrations reduce rooting capacity and alter metallothionein gene expression in white poplar (*Populus alba* L. cv. Villafranca). Chemosphere 67:1117–1126

Challinor AJ, Wheeler TR, Craufurd PQ et al (2007) Adaptation of crops to climate change through genotypic responses to mean and extreme temperatures. Agric Ecosyst Environ 119:190–204
Chauhan S, Srivalli S, Nautiyal A et al (2009) Wheat cultivars differing in heat tolerance show a differential response to monocarpic senescence under high-temperature stress and the involvement of serine proteases. Photosynthetica 47:536–547
Chen CT, Chen LM, Lin CC et al (2001) Regulation of proline accumulation in detached rice leaves exposed to excess copper. Plant Sci 160:283–290
Cheng W, Li D, Wang Y et al (2016) Cloning of heat shock protein genes (hsp70, hsc70 and hsp90) and their expression in response to larval diapause and thermal stress in the wheat blossom midge, *Sitodiplosis mosellana*. J Insect Physiol 95:66–77
Cho YH, Hong JW, Kim EC et al (2012) Regulatory functions of SnRK1 in stress-responsive gene expression and in plant growth and development. Plant Physiol 158:1955–1964
Cechin I, Rossi S, Oliveira V et al (2006) Photosynthetic responses and proline content of mature and young leaves of sunflower plants under water deficit. Photosynthetica 44:143–146
Colmer T (2002) Aerenchyma and an inducible barrier to radial oxygen loss facilitate root aeration in upland, paddy and deep-water rice (*Oryza sativa L.*). Ann Bot 91:301–309
Costa G, Morel JL (1994) Water relations, gas exchange and amino acid content in Cd-treated lettuce. Plant Physiol Biochem 32:561–570
Cui L, Cao R, Li J et al (2006) High temperature effects on ammonium assimilation in leaves of two Festuca arundinacea cultivars with different heat susceptibility. Plant Growth Regul 49:127–136
DaCosta M, Huang B (2006) Osmotic adjustment associated with variation in bentgrass tolerance to drought stress. J Am Soc Hortic Sci 131:338–344
de Campos MKF, de Carvalho K, de Souza FS et al (2011) Drought tolerance and antioxidant enzymatic activity in transgenic 'Swingle' citrumelo plants over-accumulating proline. Environ Exp Bot 72:242–250
Drew M, Jackson M, Giffard S (1979) Ethylene-promoted adventitious rooting and development of cortical air spaces (aerenchyma) in roots may be adaptive responses to flooding in *Zea mays* L. Planta 147:83–88
Drew M, Sisworo E (1977) Early effects of flooding on nitrogen deficiency and leaf chlorosis in barley. New Phytol 79:567–571
Duan YH, Guo J, Ding K et al (2011) Characterization of a wheat HSP70 gene and its expression in response to stripe rust infection and abiotic stresses. Mol Biol Rep 38:301–307
Ellsworth RS, Freebairn HT (1969) Ethylene, seed germination, and epinasty. Plant Physiol 44:955–958
Emamverdian A, Ding Y, Mokhberdoran F et al (2015) Heavy metal stress and some mechanisms of plant defense response. Sci World J. https://doi.org/10.1155/2015/756120
Fahad S, Bajwa AA, Nazir U et al (2017) Crop production under drought and heat stress: plant responses and management options. Front Plant Sci 8:1147. https://doi.org/10.3389/fpls.2017.01147
Farooq M, Hussain M, Wahid A et al (2012) Drought stress in plants: an overview. In: Aroca R (ed) Plant responses to drought stress. Springer, Switzerland, pp 1–33
Foyer CH, Noctor G (2011) Ascorbate and glutathione: the heart of the redox hub. Plant Physiol 155:2–18
Fujita Y, Yoshida T, Yamaguchi-Shinozaki K (2013) Pivotal role of the AREB/ABF-SnRK2 pathway in ABRE-mediated transcription in response to osmotic stress in plants. Physiol Plant 147:15–27
Fukao T, Yeung E, Bailey-Serres J (2011) The submergence tolerance regulator SUB1A mediates crosstalk between submergence and drought tolerance in rice. Plant Cell 23:412–427
Fusco N, Micheletto L, Dal Corso G et al (2005) Identification of cadmium-regulated genes by cDNA-AFLP in the heavy metal accumulator *Brassica juncea* L. J Exp Bot 56:3017–3027
Gao H, Wang Y, Xu P et al (2018) Overexpression of a WRKY transcription factor TaWRKY2 enhances drought stress tolerance in transgenic wheat. Front Plant Sci 9:997. https://doi.org/10.3389/fpls.2018.00997
Garg N, Manchanda G (2009) ROS generation in plants: boon or bane? Plant Biosys 143:81–96

Georgiadou EC, Kowalska E, Patla K et al (2018) Influence of heavy metals (Ni, Cu, and Zn) on nitro-oxidative stress responses, proteome regulation and allergen production in Basil (*Ocimum basilicum* L.) plants. Front Plant Sci 9:862

Gilmour SJ, Sebolt AM, Salazar MP et al (2000) Overexpression of the *Arabidopsis* CBF3 transcriptional activator mimics multiple biochemical changes associated with cold acclimation. Plant Physiol 124:1854–1865

Greenway H, Gibbs J (2003) Mechanisms of anoxia tolerance in plants. II. Energy requirements for maintenance and energy distribution to essential processes. Funct Plant Biol 30:999–1036

Grift TE, Novais J, Bohn M (2011) High-throughput phenotyping technology for maize roots. Biosys Eng 110:40–48

Guan Q, Wang Z, Wang X et al (2015) A peroxisomal APX from Puccinellia tenuiflora improves the abiotic stress tolerance of transgenic *Arabidopsis* thaliana through decreasing of H2O2 accumulation. J Plant Physiol 175:183–191

Gupta SD (2010) Metal toxicity, oxidative stress and antioxidative defense system in plants. Reactive oxygen species and antioxidants in higher plants. CRC Press, Boca Raton, FL, pp 193–220

Hall J (2002) Cellular mechanisms for heavy metal detoxification and tolerance. J Exp Bot 53:1–11

Hasanuzzaman M, Nahar K, Fujita M (2013) Extreme temperature responses, oxidative stress and antioxidant defense in plants. In: Vahdati K, Leslie Ch (eds) Abiotic stress-plant responses and applications in agriculture. Intechopen. Available via DIALOG. https://www.intechopen.com/books/abiotic-stress-plant-responses-and-applications-in-agriculture/extreme-temperature-responses-oxidative-stress-and-antioxidant-defense-in-plants. Accessed 7 July 2020

He Y, Huang B (2010) Differential responses to heat stress in activities and isozymes of four antioxidant enzymes for two cultivars of Kentucky bluegrass contrasting in heat tolerance. J Am Soc Hortic Sci 135:116–124

Hmidi D, Abdelly C, Athar HUR et al (2018) Effect of salinity on osmotic adjustment, proline accumulation and possible role of ornithine-δ-aminotransferase in proline biosynthesis in *Cakile maritima*. Physiol Mol Biol Plants 24:1017–1033

Hossain MA, Piyatida P, da Silva JAT et al (2012) Molecular mechanism of heavy metal toxicity and tolerance in plants: central role of glutathione in detoxification of reactive oxygen species and methylglyoxal and in heavy metal chelation. J Bot. https://doi.org/10.1155/2012/872875

Hsiao TC, Xu LK (2000) Sensitivity of growth of roots versus leaves to water stress: biophysical analysis and relation to water transport. J Exp Bot 51:1595–1616

Huang B, Johnson JW (1995) Root respiration and carbohydrate status of two wheat genotypes in response to hypoxia. Ann Bot 75:427–432

Hussain HA, Men S, Hussain S et al (2019) Interactive effects of drought and heat stresses on morpho-physiological attributes, yield, nutrient uptake and oxidative status in maize hybrids. Sci Rep 9:1–12

Ireland HE, Harding SJ, Bonwick GA et al (2004) Evaluation of heat shock protein 70 as a biomarker of environmental stress in *Fucus serratus* and *Lemna minor*. Biomarkers 9:139–155

Ishida A, Ohta N, Suzuki Y et al (2012) Expression of pendrin periostin in allergic rhinitis chronic rhinosinusitis. Allergol Int 61:589–595

Issarakraisila M, Ma Q, Turner DW (2007) Photosynthetic and growth responses of juvenile Chinese kale (*Brassica oleracea* var. *alboglabra*) and Caisin (*Brassica rapa* subsp. *parachinensis*) to waterlogging and water deficit. Sci Hortic 111:107–113

Jagadish SVK, Muthurajan R, Oane R et al (2009) Physiological and proteomic approaches to address heat tolerance during anthesis in rice (*Oryza sativa* L.). J Exp Bot 61:143–156

Jagendorf AT, Takabe T (2001) Inducers of glycinebetaine synthesis in barley. Plant Physiol 127:1827–1835

Jentschke G, Godbold D (2000) Metal toxicity and ectomycorrhizas. Physiol Plant 109:107–116

Johnson JW, Huang B (1996) Responses of triticale and wheat to hypoxia. In Triticale: Today and Tomorrow, Springer, Dordrecht, pp. 489–495

Johnson RR, Wagner RL, Verhey SD et al (2002) The abscisic acid-responsive kinase PKABA1 interacts with a seed-specific abscisic acid response element-binding factor, TaABF, and phosphorylates TaABF peptide sequences. Plant Physiol 130:837–846

Kägi JR (1991) Overview of metallothionein. Methods Enzymol 205:613–626

Kaldorf M, Kuhn A, Schröder W et al (1999) Selective element deposits in maize colonized by a heavy metal tolerance conferring arbuscular mycorrhizal fungus. J Plant Physiol 154:718–728

Kastori R, Petrović M, Petrović N (1992) Effect of excess lead, cadmium, copper, and zinc on water relations in sunflower. J Plant Nutr 15:2427–2439

Kasuga M, Liu Q, Miura S et al (1999) Improving plant drought, salt, and freezing tolerance by gene transfer of a single stress-inducible transcription factor. Nature Biotechnol 17:287–291

Kasuga M, Miura S, Shinozaki K et al (2004) A combination of the *Arabidopsis* DREB1A gene and stress-inducible rd29A promoter improved drought-and low-temperature stress tolerance in tobacco by gene transfer. Plant Cell Physiol 45:346–350

Kavar T, Meglič V, Rozman L (2007) Diversity of Slovenian maize (*Zea mays*) populations by Hbr (MITE) markers and morphological traits. Russ J Genet 43:989–995

Kaya H, Shibahara KI, Taoka KI et al (2001) FASCIATA genes for chromatin assembly factor-1 in *Arabidopsis* maintain the cellular organization of apical meristems. Cell 104:131–142

Khan F, Siddiqi TO, Mahmooduzzafar M et al (2009) Morphological changes and antioxidant defence systems in soybean genotypes as affected by salt stress. J Plant Interact 4:295–306

Kiani SP, Grieu P, Maury P et al (2007) Genetic variability for physiological traits under drought conditions and differential expression of water stress-associated genes in sunflower (*Helianthus annuus* L.). Theor Appl Genet 114:193–207

Krishnamurthy L, Johansen C, Sethi S (1999) Investigation of factors determining genotypic differences in seed yield of non-irrigated and irrigated chickpeas using a physiological model of yield determination. J Agron Crop Sci 183:9–17

Lacerda CFd, Cambraia J, Oliva MA et al (2003) Osmotic adjustment in roots and leaves of two sorghum genotypes under NaCl stress. Braz J Plant Physiol 15:113–118

Lamaoui M, Jemo M, Datla R et al (2018) Heat and drought stresses in crops and approaches for their mitigation. Front Chem 6:26

Lei C, Shin Y, Magnuson JK et al (2006) Characterization of functionalized nanoporous supports for protein confinement. Nanotechnology 17:5531

Li W, Zhang S, Ding S et al (2010) Root morphological variation and water use in alfalfa under drought stress. Acta Ecol Sin 30:5140–5150

Lin KH, Chao PY, Yang CM et al (2006) The effects of flooding and drought stresses on the antioxidant constituents in sweet potato leaves. Bot Stud 47:417–426

Liou J, Kim ML, Do Heo W et al (2005) STIM is a Ca2+ sensor essential for Ca2+-store-depletion-triggered Ca2+influx. Curr Biol 15:1235–1241

Liu J, Ishitani M, Halfter U et al (2000) The *Arabidopsis thaliana* SOS2 gene encodes a protein kinase that is required for salt tolerance. Proc Natl Acad Sci 97:3730–3734

Longstreth DJ, Nobel PS (1979) Salinity effects on leaf anatomy: consequences for photosynthesis. Plant Physiol 63:700–703

Lu YY, Deng XP, Kwak SS (2010) Over expression of CuZn superoxide dismutase (CuZn SOD) and ascorbate peroxidase (APX) in transgenic sweet potato enhances tolerance and recovery from drought stress. Afr J Biotechnol 9:8378–8391

Ma LF, Li Y, Chen Y et al (2016) Improved drought and salt tolerance of Arabidopsis thaliana by ectopic expression of a cotton (*Gossypium hirsutum*) CBF gene. Plant Cell Tissue and Organ Cult (PCTOC) 124:583–598

Ma Y, Wang L, Wang J et al (2019) Isolation and expression analysis of Salt Overly Sensitive gene family in grapevine (*Vitis vinifera*) in response to salt and PEG stress. PLoS ONE 14: https://doi.org/10.1371/journal.pone.0212666

Manschadi A, Christopher J, Hammer G et al (2010) Experimental and modelling studies of drought-adaptive root architectural traits in wheat (Triticum aestivum L.). Plant Biosyst 144:458–462

Marcum KB, Anderson SJ, Engelke M (1998) Salt gland ion secretion: a salinity tolerance mechanism among five zoysiagrass species. Crop Sci 38:806–810
Matsukura C, Kawai M, Toyofuku K et al (2000) Transverse vein differentiation associated with gas space formation—fate of the middle cell layer in leaf sheath development of rice. Ann Bot 85:19–27
Mediouni C, Benzarti O, Tray B et al (2006) Cadmium and copper toxicity for tomato seedlings. Agron J 26:227–232
Milone MT, Sgherri C, Clijsters H et al (2003) Antioxidative responses of wheat treated with realistic concentration of cadmium. Environ Exp Bot 50:265–276
Mittler R, Blumwald E (2010) Genetic engineering for modern agriculture: challenges and perspectives. Annu Rev Plant Biol 61:443–462
Mittler R, Vanderauwera S, Suzuki N et al (2011) ROS signaling: the new wave? Trends Plant Sci 16:300–309
Mousavi SR, Galavi M, Rezaei M (2013) Zinc (Zn) importance for crop production—a review. Int J Agron Plant Prod 4:64–68
Nakashima K, Ito Y, Yamaguchi-Shinozaki K (2009) transcriptional regulatory networks in response to abiotic stresses in *Arabidopsis* and grasses. Plant Physiol 149:88–95
Nakashima K, Tran LSP, Van Nguyen D et al (2007) Functional analysis of a NAC-type transcription factor OsNAC6 involved in abiotic and biotic stress-responsive gene expression in rice. Plant J 51:617–630
Nakashima K, Yamaguchi-Shinozaki K (2006) Regulons involved in osmotic stress-responsive and cold stress-responsive gene expression in plants. Physiol Plant 126:62–71
Neumann D, Lichtenberger O, Günther D et al (1994) Heat-shock proteins induce heavy-metal tolerance in higher plants. Planta 194:360–367
Neumann D, Zur Nieden U, Lichtenberger O et al (1995) How does *Armeria maritima* tolerate high heavy metal concentrations? J Plant Physiol 146:704–717
Nielsen E, Akita M, Davila-Aponte J et al (1997) Stable association of chloroplastic precursors with protein translocation complexes that contain proteins from both envelope membranes and a stromal Hsp100 molecular chaperone. EMBO J 16:935–946
Nieto-Sotelo J, Martínez LM, Ponce G et al (2002) Maize HSP101 plays important roles in both induced and basal thermotolerance and primary root growth. Plant Cell 14:1621–1633
Niyogi KK, Shih C, Chow WS et al (2001) Photoprotection in a zeaxanthin-and lutein-deficient double mutant of *Arabidopsis*. Photosynth Res 67:139–145
Noctor G, Gomez L, Vanacker H et al (2002) Interactions between biosynthesis, compartmentation and transport in the control of glutathione homeostasis and signalling. J Exp Bot 53:1283–1304
O'Neill SD (1983) Osmotic adjustment and the development of freezing resistance in *Fragaria virginiana*. Plant Physiol 72:938–944
Öztürk M, Waisel Y, Khan MA et al (2006) Biosaline agriculture and salinity tolerance in plants. Birkhäuser, Basel
Pandey N (2018) Antioxidant defense system in plants exposed to metal toxicity. In: Hasanuzzaman M, Nahar K, Fujita M (eds) Plants under metal and metalloid stress. Springer, Singapore, pp 107–148
Parida AK, Dagaonkar VS, Phalak MS et al (2008) Differential responses of the enzymes involved in proline biosynthesis and degradation in drought tolerant and sensitive cotton genotypes during drought stress and recovery. Acta Physiol Plant 30:619–627
Parida AK, Das AB (2005) Salt tolerance and salinity effects on plants: a review. Ecotoxicol Environ Saf 60:324–349
Pellegrineschi A, Reynolds M, Pacheco M et al (2004) Stress-induced expression in wheat of the *Arabidopsis thaliana* DREB1A gene delays water stress symptoms under greenhouse conditions. Genome 47:493–500
Perkons U, Kautz T, Uteau D et al (2014) Root-length densities of various annual crops following crops with contrasting root systems. Soil Till Res 137:50–57

Polidoros AN, Mylona PV, Pasentsis K et al (2005) The maize alternative oxidase 1a (Aox1a) gene is regulated by signals related to oxidative stress. Redox Rep 10:71–78

Qi Y, Wang H, Zou Y et al (2011) Over-expression of mitochondrial heat shock protein 70 suppresses programmed cell death in rice. FEBS Lett 585:231–239

Queitsch C, Hong SW, Vierling E et al (2000) Heat shock protein 101 plays a crucial role in thermotolerance in *Arabidopsis*. Plant Cell 12:479–492

Rahbarian R, Khavari-Nejad R, Ganjeali A et al (2011) Drought stress effects on photosynthesis, chlorophyll fluorescence and water relations in tolerant and susceptible chickpea (*Cicer arietinum* L.) genotypes. Acta Biol Cracoviensia Seri Bot 53:47–56

Rahneshan Z, Nasibi F, Moghadam AA (2018) Effects of salinity stress on some growth, physiological, biochemical parameters and nutrients in two pistachio (*Pistacia vera* L.) rootstocks. J Plant Interac 13:73–82

Raseetha S, Leong SY, Burritt DJ et al (2013) Understanding the degradation of ascorbic acid and glutathione in relation to the levels of oxidative stress biomarkers in broccoli (*Brassica oleracea* L. *italica* cv. Bellstar) during storage and mechanical processing. Food Chem 138:1360–1369

Ric de Vos C, Ten Bookum W, Vooijs R et al (1993) Effect of copper on fatty acid composition and peroxidation of lipids in the roots of copper tolerant and sensitive *Silene cucubalus*. Plant Physiol Biochem (Paris) 31:151–158

Richards R (1996) Defining selection criteria to improve yield under drought. Plant Growth Regul 20:157–166

Rienth M, Romieu C, Gregan R et al (2014) Validation and application of an improved method for the rapid determination of proline in grape berries. J Agric Food Chem 62:3384–3389

Rochester DE, Winer JA, Shah DM (1986) The structure and expression of maize genes encoding the major heat shock protein, hsp70. EMBO J 5:451–458

Roosens NH, Al Bitar F, Loenders K et al (2002) Overexpression of ornithine-δ-aminotransferase increases proline biosynthesis and confers osmotolerance in transgenic plants. Mol Breed 9:73–80

RoyChoudhury A, Gupta B, Sengupta DN (2008) Trans-acting factor designated OSBZ8 interacts with both typical abscisic acid responsive elements as well as abscisic acid responsive element-like sequences in the vegetative tissues of indica rice cultivars. Plant Cell Rep 27:779–794

Rozema J (1991) Growth, water and ion relationships of halophytic monocotyledonae and dicotyledonae; a unified concept. Aquat Bot 39:17–33

Sadok W, Sinclair TR (2011) Crops yield increase under water-limited conditions: review of recent physiological advances for soybean genetic improvement. In: Sparks DL (ed) Advances in agronomy, vol 113. Elsevier, Cambridge, MA, pp 325–345

Sahin U, Ekinci M, Ors S et al (2018) Effects of individual and combined effects of salinity and drought on physiological, nutritional and biochemical properties of cabbage (*Brassica oleracea* var. *capitata*). Sci Hortic 240:196–204

Sairam R, Kumutha D, Ezhilmathi K et al (2009) Waterlogging induced oxidative stress and antioxidant enzyme activities in pigeon pea. Biol Plan 53:493–504

Sairam RK, Dharmar K, Lekshmy S et al (2011) Expression of antioxidant defense genes in mung bean (*Vigna radiata* L.) roots under water-logging is associated with hypoxia tolerance. Acta Physiol Plan 33:735–744

Sakuma Y, Maruyama K, Osakabe Y et al (2006) Functional analysis of an Arabidopsis transcription factor, DREB2A, involved in drought-responsive gene expression. Plant Cell 18:1292–1309

Salt DE, Blaylock M, Kumar NP et al (1995) Phytoremediation: a novel strategy for the removal of toxic metals from the environment using plants. Bio/technol 13:468–474

Sanghera GS, Wani SH, Hussain W et al (2011) Engineering cold stress tolerance in crop plants. Curr Genomics 12:30–43

Saradhi PP (1991) Proline accumulation under heavy metal stress. J Plant Physiol 138:554–558

Sarkar NK, Kundnani P, Grover A (2013) Functional analysis of Hsp70 superfamily proteins of rice (Oryza sativa). Cell Stress Chaperones 18:427–437

Sarry JE, Kuhn L, Ducruix C et al (2006) The early responses of *Arabidopsis thaliana* cells to cadmium exposure explored by protein and metabolite profiling analyses. Proteomics 6:2180–2198

Schat H, Sharma SS, Vooijs R (1997) Heavy metal-induced accumulation of free proline in a metal-tolerant and a nontolerant ecotype of *Silene vulgaris*. Physiol Plant 101:477–482

Schmidt R, Mieulet D, Hubberten HM et al (2013) SALT-RESPONSIVE ERF1 regulates reactive oxygen species–dependent signaling during the initial response to salt stress in rice. Plant Cell 25:2115–2131

Schramm F, Larkindale J, Kiehlmann E et al (2008) A cascade of transcription factor DREB2A and heat stress transcription factor HsfA3 regulates the heat stress response of Arabidopsis. Plant J 53:264–274

Schröder F, Lisso J, Müssig C (2011) EXORDIUM-LIKE1 promotes growth during low carbon availability in Arabidopsis. Plant Physiol 156:1620–1630

Schutzendubel A, Polle A (2002) Plant responses to abiotic stresses: heavy metal-induced oxidative stress and protection by mycorrhization. J Exp Bot 53:1351–1365

Sena Gomes A, Kozlowski T (1980) Effects of flooding on *Eucalyptus camaldulensis* and *Eucalyptus globulus* seedlings. Oecologia 46:139–142

Seregin I, Ivanov V (2001) Physiological aspects of cadmium and lead toxic effects on higher plants. Russ J Plant Physiol 48:523–544

Shah ZH, Rehman HM, Akhtar T et al (2017) Redox and ionic homeostasis regulations against oxidative, salinity and drought stress in wheat (a systems biology approach). Front Genet 8:141. https://doi.org/10.3389/fgene.2017.00141

Shahid M, Dumat C, Khalid S et al (2017) Foliar heavy metal uptake, toxicity and detoxification in plants: a comparison of foliar and root metal uptake. J Hazard Mater 325:36–58

Sharma R, Wungrampha S, Singh V et al (2016) Halophytes as bioenergy crops. Front Plant Sci 7:1372. https://doi.org/10.3389/fpls.2016.01372

Sharma SS, Dietz KJ (2006) The significance of amino acids and amino acid-derived molecules in plant responses and adaptation to heavy metal stress. J Exp Bot 57:711–726

Sharp RE, Davies WJ (1979) Solute regulation and growth by roots and shoots of water-stressed maize plants. Planta 147:43–49

Sheikh-Mohamadi MH, Etemadi N, Arab M (2018) Correlation of heat and cold tolerance in Iranian tall fescue ecotypes with reactive oxygen species scavenging and osmotic adjustment. HortScience 53:1062–1068

Shi H, Ishitani M, Kim C et al (2000) The *Arabidopsis thaliana* salt tolerance gene SOS1 encodes a putative Na+/H+ antiporter. Proc Natl Acad Sci 97:6896–6901

Shi Y, Ding Y, Yang S (2018) Molecular regulation of CBF signaling in cold acclimation. Trends Plant Sci 23:623–637

Shim D, Hwang JU, Lee J et al (2009) Orthologs of the class A4 heat shock transcription factor HsfA4a confer cadmium tolerance in wheat and rice. Plant Cell 21:4031–4043

Siddiqui MN, Mostofa MG, Akter MM et al (2017) Impact of salt-induced toxicity on growth and yield-potential of local wheat cultivars: oxidative stress and ion toxicity are among the major determinants of salt-tolerant capacity. Chemosphere 187:385–394

Sinclair TR, Muchow RC (2001) System analysis of plant traits to increase grain yield on limited water supplies. Agron J 93:263–270

Singh M, Kumar J, Singh S et al (2015) Roles of osmoprotectants in improving salinity and drought tolerance in plants: a review. Rev Environ Sci Bio/Technol 14:407–426

Soengas P, Rodríguez VM, Velasco P et al (2018) Effect of temperature stress on antioxidant defenses in *Brassica oleracea*. ACS Omega 3:5237–5243

Su CF, Wang YC, Hsieh TH et al (2010) A novel MYBS3-dependent pathway confers cold tolerance in rice. Plant Physiol 153:145–158

Syed NH, Prince SJ, Mutava RN et al (2015) Core clock, SUB1, and ABAR genes mediate flooding and drought responses via alternative splicing in soybean. J Exp Bot 66:7129–7149

Szymańska KP, Polkowska-Kowalczyk L, Lichocka M et al (2019) SNF1-related protein kinases SnRK2. 4 and SnRK2. 10 modulate ROS homeostasis in plant response to salt stress. Int J Mol Sci 20:143. https://doi.org/10.3390/ijms20010143
Vu DT, Huang L, Nguyen AV et al (2013) Quantitative methods for estimating foliar uptake of zinc from suspension-based Zn chemicals. J Plant Nutr Soil Sci 176:764–775
Takahashi H, Kawakatsu T, Wakasa Y et al (2012) A rice transmembrane bZIP transcription factor, OsbZIP39, regulates the endoplasmic reticulum stress response. Plant Cell Physiol 53:144–153
Tian F, Wang W, Liang C et al (2017) Overaccumulation of glycine betaine makes the function of the thylakoid membrane better in wheat under salt stress. Crop J 5:73–82
Trachsel S, Messmer R, Stamp P et al (2009) Mapping of QTLs for lateral and axile root growth of tropical maize. Theor Appl Genet 119:1413–1424
Tran LSP, Nakashima K, Shinozaki K et al (2007) Plant gene networks in osmotic stress response: from genes to regulatory networks. In: Häussinger D, Sies H (eds) Methods in Enzymology, vol 428. Academic Press, Cambridge, MA, pp 109–128
Turner NC (2018) Turgor maintenance by osmotic adjustment: 40 years of progress. J Exp Bot 69:3223–3233
Tyerman S, Niemietz C, Bramley H (2002) Plant aquaporins: multifunctional water and solute channels with expanding roles. Plant, Cell Environ 25:173–194
van de Mortel JE, Schat H, Moerland PD et al (2008) Expression differences for genes involved in lignin, glutathione and sulphate metabolism in response to cadmium in Arabidopsis thaliana and the related Zn/Cd-hyperaccumulator Thlaspi caerulescens. Plant, Cell Environ 31:301–324
van Veen H, Mustroph A, Barding GA et al (2013) Two Rumex species from contrasting hydrological niches regulate flooding tolerance through distinct mechanisms. Plant Cell 25:4691–4707
Visser E, Nabben R, Blom C et al (1997) Elongation by primary lateral roots and adventitious roots during conditions of hypoxia and high ethylene concentrations. Plant, Cell Environ 20:647–653
Vojtěchová M, Hanson AD, Muñoz-Clares RA (1997) Betaine-aldehyde dehydrogenase from amaranth leaves efficiently catalyzes the nad-dependent oxidation of dimethylsulfoniopropionaldehyde to dimethylsulfoniopropionate. Arch Biochem Biophys 337:81–88
Wang WB, Kim YH, Lee HS et al (2009) Analysis of antioxidant enzyme activity during germination of alfalfa under salt and drought stresses. Plant Physiol Biochem 47:570–577
Wang X, Khodadadi E, Fakheri B et al (2017) Organ-specific proteomics of soybean seedlings under flooding and drought stresses. J Proteom 162:62–72
Wasson AP, Richards R, Chatrath R et al (2012) Traits and selection strategies to improve root systems and water uptake in water-limited wheat crops. J Exp Bot 63:3485–3498
Weretilnyk EA, Bednarek S, McCue KF et al (1989) Comparative biochemical and immunological studies of the glycine betaine synthesis pathway in diverse families of dicotyledons. Planta 178:342–352
Westgate ME, Boyer JS (1985) Osmotic adjustment and the inhibition of leaf, root, stem and silk growth at low water potentials in maize. Planta 164:540–549
Wilson ME, Basu MR, Bhaskara GB et al (2014a) Plastid osmotic stress activates cellular stress responses in Arabidopsis. Plant Physiol 165:119–128
Wilson RA, Sangha M, Banga S et al (2014b) Heat stress tolerance in relation to oxidative stress and antioxidants in *Brassica juncea*. J Environ Biol 35:383
Wojciechowska R, Hanus-Fajerska EJ, Kolton A et al (2013) The effect of seedling chilling on glutathione content, catalase and peroxidase activity in *Brassica oleracea* L. var. *italica*. Acta Soc Bot Pol 82. https://doi.org/10.5586/asbp.2013.020
Wollenweber-Ratzer B, Crawford R (1994) Enzymatic defence against post-anoxic injury in higher plants. P Roy Soc Edinb B Bio 102:381–390
Wu J, Zhao Q, Wu G et al (2019) Comprehensive analysis of differentially expressed unigenes under nacl stress in flax (*Linum usitatissimum* L.) using RNA-Seq. Int J Mol Sci 20:369
Xie X, He Z, Chen N et al (2019) The roles of environmental factors in regulation of oxidative stress in plant. Biomed Res Int. https://doi.org/10.1155/2019/9732325

Xing W, Rajashekar CB (1999) Alleviation of water stress in beans by exogenous glycine betaine. Plant Sci 148:185–192
Xu W, Cui K, Xu A et al (2015) Drought stress condition increases root to shoot ratio via alteration of carbohydrate partitioning and enzymatic activity in rice seedlings. Acta Physiol Plant 37:9
Yamaguchi-Shinozaki K, Shinozaki K (2006) Transcriptional regulatory networks in cellular responses and tolerance to dehydration and cold stresses. Annu Rev Plant Biol 57:781–803
Yamaguchi K, Takahashi Y, Berberich T et al (2007) A protective role for the polyamine spermine against drought stress in Arabidopsis. Biochem Biophys Res Commun 352:486–490
Yang XE, Jin XF, Feng Y et al (2005) Molecular mechanisms and genetic basis of heavy metal tolerance/hyperaccumulation in plants. J Integ Plant Biol 47:1025–1035
Yoshida T, Ohama N, Nakajima J et al (2011) Arabidopsis HsfA1 transcription factors function as the main positive regulators in heat shock-responsive gene expression. Mol Genet Genomics 286:321–332
Young TE, Ling J, Geisler-Lee CJ et al (2001) Developmental and thermal regulation of the maize heat shock protein, HSP101. Plant Physiol 127:777–791
Yu G, Ma, J, Jiang P, Li J et al (2019). The mechanism of plant resistance to heavy metal. In: IOP conference series: earth and environmental science. International conference on the improvement of environmental quality (ICIEQ), Bogor, Indonesia 2019 August, vol 310. IOP Publishing, pp 052004
Zandalinas SI, Balfagón D, Arbona V et al (2017) Modulation of antioxidant defense system is associated with combined drought and heat stress tolerance in citrus. Front Plant Sci 8:953. https://doi.org/10.3389/fpls.2017.00953
Zandalinas SI, Sales C, Beltran J et al (2016) Activation of secondary metabolism in citrus plants is associated to sensitivity to combined drought and high temperatures. Front Plant Sci 7:1954. https://doi.org/10.3389/fpls.2016.01954
Zenda T, Liu S, Wang X et al (2019) Key maize drought-responsive genes and pathways revealed by comparative transcriptome and physiological analyses of contrasting inbred lines. Int J Mol Sci 20:1268. https://doi.org/10.3390/ijms20061268
Zhang H, Zhang Y, Deng C et al (2018) The arabidopsis Ca2+—dependent protein kinase CPK12 is involved in plant response to salt stress. Int J Mol Sci 19:4062. https://doi.org/10.3390/ijms19124062
Zhang J, Huang W, Pan Q et al (2005) Improvement of chilling tolerance and accumulation of heat shock proteins in grape berries (*Vitis vinifera* cv. *Jingxiu*) by heat pretreatment. Postharvest Biol Tec 38:80–90
Zhang LX, Li SX, Zhang H et al (2007) Nitrogen rates and water stress effects on production, lipid peroxidation and antioxidative enzyme activities in two maize (*Zea mays* L.) genotypes. J Agron Crop Sci 193:387–397
Zhen Y, Qi JL, Wang SS et al (2007) Comparative proteome analysis of differentially expressed proteins induced by Al toxicity in soybean. Physiol Plant 131:542–554
Zhu JK (2001) Plant salt tolerance. Trends Plant Sci 6:66–71
Živčák M, Repkova J, Olšovská K, Brestič M (2009) Osmotic adjustment in winter wheat varieties and its importance as a mechanism of drought tolerance. Cereal Res Commun 37:569–572
Zou J, Yu K, Zhang Z et al (2009) Antioxidant response system and chlorophyll fluorescence in chromium (VI)—treated *Zea Mays* L. seedlings. Acta Biol Cracov Bot 51:23–33

Chapter 3
Biotechnology Strategies to Combat Plant Abiotic Stress

Syed Uzma Jalil and Mohammad Israil Ansari

Abstract Adverse environmental conditions cause major challenge to crop production and have significant decreases in crop yields worldwide. Developing stress tolerance varieties against wide range of abiotic stresses is a widely supported approach that allows the environment to adapt to these methods. The reactive oxygen species (ROS) in plants are generated due to abiotic stresses that causes lipid peroxidation, inactivation of enzymes, DNA damage in plant cells. Biotechnological approaches propose numerous applications in crop improvement including stress resistance and quality enhancement. Identification and functional characterization of various target genes involves in signaling, transcription, antioxidant defense system for understanding the molecular mechanism of abiotic stress tolerance has been employed to developed stress resistant plants by biotechnological techniques. Employing genetic engineering approaches, tissue culture techniques, functional validation of genes and transcription factors and genome editing approaches for example Zinc Finger Nucleases (ZFNs), transcription activator-like effector nucleases (TALENs) as well as advanced molecular tool CRISPR-Cas9 systems which provides simplicity and precision of targeted gene editing methods. These biotechnological approaches engage in different processes to enhance abiotic stress resistant in different plants. Present chapter provide inclusive outline to draw the consideration of investigators with advances in biotechnological techniques to improve the tolerance of abiotic stresses in various plants to increase plant productivity.

Keywords Abiotic stresses · Genetic engineering · Genome editing · Metabolic engineering · Stress tolerance

S. U. Jalil
Amity Institute of Biotechnology, Amity University Uttar Pradesh,
Lucknow Campus, Lucknow 226 028, India

M. I. Ansari (✉)
Department of Botany, University of Lucknow, Lucknow 226 007, India
e-mail: ansari_mi@lkouniv.ac.in

J. M. Al-Khayri et al. (eds.), *Nanobiotechnology*,
https://doi.org/10.1007/978-3-030-73606-4_3

3.1 Introduction

In the current era, plant biotechnology faces an integral challenge to develop tolerance in plants for combating hostile environmental conditions particularly under the continuously changing climatic conditions. Abiotic stresses negatively influence the growth and development of plants causing severe reduction in crop yield as shown in Fig. 3.1 (Jalil and Ansari 2020a). Different techniques have been discovered to elevate plant stress resistance to ensure plant survival and enhance crop productivity (Chikkaputtaiah and Marwein 2019). Conventional breeding strategies utilize existing hereditary variations that occur in different germplasms or induced through intergeneric and interspecific hybridization, induced mutation and somaclonal variation. Despite their success, these strategies have numerous limitations as they resulted in developing limited number of new stress resistant varieties capable of field survival (Lin et al. 2013). In addition, conventional techniques are constrained with limited genetic variation of plants that can thrive in adverse environmental conditions, the intricacy of stress resistance characters and in proficient selection approaches (Lin et al. 2010).

Recent strategies to enhance resistance against abiotic stresses in plants have received momentous accomplishments. Genetic engineering methods for tolerating abiotic stress on plants depend on genetic expression that influences stress management as well as signaling mechanism (Ohama et al. 2017), stress responsive genes (Ansari et al. 2011; Ma et al. 2018), and enzymes involved in metabolites production (Li et al. 2018; Liu et al. 2018). The plant genetic engineering strategies to improve stress tolerance comprises of stimulating endogenous systems via interceding at various responses, degrees of reactions, regulatory components, transcription factors and antioxidants Fig. 3.2 (Jacob et al. 2017). Moreover, the significant achievements

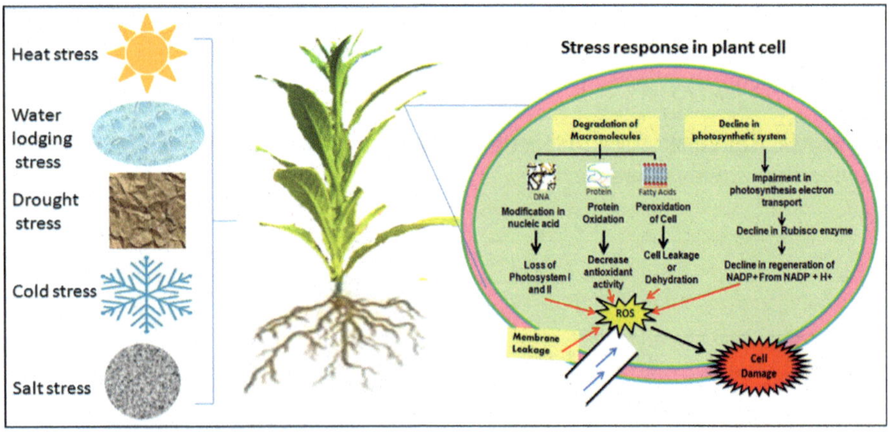

Fig. 3.1 Schematic representation of abiotic stress responses in plant cell. High accumulation of ROS in plants are the result of abiotic stress that causes lipid peroxidation, inactivation of enzymes, DNA damage in plant cells that leads to cell death

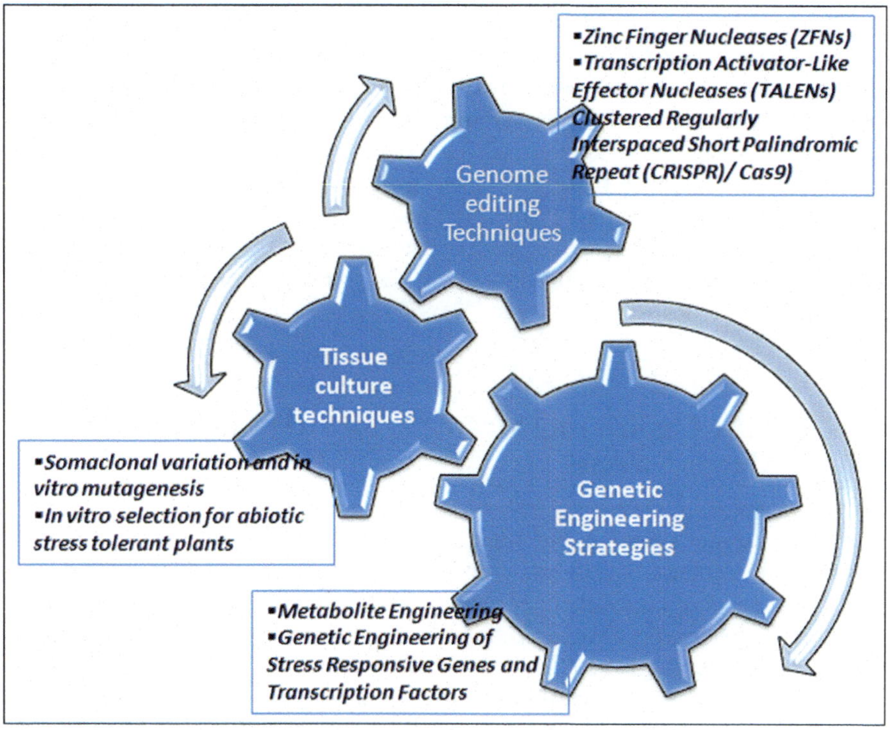

Fig. 3.2 Biotechnological approaches for mitigating abiotic stresses in plants

in the utilization of molecular techniques, which possibly improve the abiotic stress resistance in plants.

Recently, gene editing technologies pave a new way to manipulate primary and specific metabolisms. It engaged in the modification of the identified sequences in a genomic data, targeting only limited nucleotide bases, which make possible the functional study of phenotypic traits responsive genes (Mahfouz et al. 2014). Gene editing tools has been generally used for improving the crop quality and production, as well as improved plant resistance against abiotic stresses Fig. 3.2 (Abdallah et al. 2016). These are also considered as non-genetically modified strategies while it does not involve in manipulation of huge sequences and transformation of plants with foreign genes (Abdallah et al. 2016). These biotechnological approaches regulated the transcriptional and post-transcriptional factors as well as improved the resistance of plants through prompting to the fabrication of different metabolites as well as bio molecules and these techniques provides further perceptive of the mechanism of signaling compounds involved in stress tolerance.

Advancement in biotechnological approaches have transformed our proficiency for gene detection, studies of tissue-specific promoters, characterization of genes, and the development of proficient methods for genetic engineering and gene editing in plants (Chikkaputtaiah and Marwein 2019).This biotechnological tool offers the prospect to attain stress resistance, enhanced crop production, and improved nutritional qualities. the study of plant molecular mechanisms involved in resistance to abiotic pressure, the formation of signaling pathways for hormones, cell interactions, and crosstalk between pathways can improved the strategies to develop stress tolerant plant varieties that will be beneficial for increasing crop production. This chapter presents advancement in biotechnological techniques to study the plant molecular mechanisms involved in abiotic stress resistance for development of stress resistant plants or for improving abiotic stress tolerance in plants by molecular regulation of genes, genetic engineering and very recently the genome editing tools for increasing productivity of crops.

3.2 Genetic Engineering Strategies for Resistance to Abiotic Stresses

As per the current situation, the population around the world increases drastically therefore the demand of increase crop productivity is very high in limited arable land and under continuous changes in the environmental conditions. To conquer this issue, various endeavors have been applied to improve breeding and biotechnological approaches for developing stress resistant crops with high yield and enhanced nutritional value. Conversely, it is still the biggest challenge in agricultural sector (Joshi et al. 2020). The traditional breeding strategies have very limited success because of the intricacy of abiotic stress resistance traits, whereas advanced biotechnological approaches such as genetic engineering strategies has been in the spotlight for the researchers as these modern approaches includes introduction of new foreign genes or modification of endogenous stress related gene expressions to transform plants to improving stress resistance ability of plants. Development of stress resistance transgenic plants required crucial knowledge on the molecular mechanisms that involved in the transduction of stress signals to cellular compartment to activate versatile responses which required for the identification of important stress related genes and pathways (Sanchez et al. 2011). Several studies have been done to illuminate the molecular mechanism of stresses responsive genes using advanced sequencing and functional genomics approaches (Chikkaputtaiah et al. 2017).

3.2.1 Metabolite Engineering for Improving Abiotic Stress Tolerance

Osmoprotectant are the ammonium compounds include betaine, polyols as well as sugars, and few amino acids (glycine, proline) which accumulated in plants under adverse environmental conditions (Ashraf and Harris 2004). These osmoprotectants involves in stabilizing proteins and membranes for managing the osmotic adjustment of the cells during stress conditions (Zandalinas et al. 2018). Various crops have synthesizing very limited amount of osmoprotectants and deficiency of these osmoprotactant is the rationale in using metabolic engineering approaches to accumulate the synthesis of osmoprotectants in such plants for enhancing their tolerance in response to different abiotic stress conditions (McNeil et al. 1999). Proline are accumulated in several plants under abiotic stresses, in a study it has been revealed that the activities of proline biosynthesis enzyme, 1-pyrroline-5-carboxylate synthetase increased in *Ailanthus altissima* (Miller) during salinity and dehydration stress (Filippou et al. 2014). Moreover, high accumulation of proline observed in *Zea mays* plant under stress conditions to combat with abiotic stress responses (Huang et al. 2018).

The accumulation of osmoprotactants occurs during osmotic adjustment under adverse environmental condition have been used to develop stress-resistant genetically modified crops (Per et al. 2018). Furthermore, salinity stress resistance was shown in *Triticum aestivum* by transferring *betA* gene for glycine betaine biosynthesis (He et al. 2010). *Zea mays* as well as *Glycine max* plants shows increased glycine betaine 1 content in transgenic plants as compared to control plants due to the expression level of GB1 gene (Castiglioni et al. 2018). Bacterial choline dehydrogenase gene introduced genetically modified *Oryza sativa* plants showed high level of glycine betaine accumulation for improving resistance in response to drought and cold stresses (Quan et al. 2004). Moreover, the *Triticum aestivum* introduced with pyrroline T carboxylate synthetase (P5CS) gene of Vignaa conitifolia reveal resistance against drought conditions. Furthermore, *O. sativa* seedlings expressing P5CS gene illustrate enhance level of proline (Sawahel and Hassan 2002). The overexpression of Ornithine delta-aminotransferase gene (OsOAT) of *O. sativa* showed enhanced level of δ-OAT and proline content in genetically modified rice plants for combating different stress conditions (You et al. 2012). Improving trehalose biosynthesis enhanced yield potential in transgenic *O. sativa* under drought and salt stress (Joshi et al. 2020). GABA shunt pathway regulating the GABA metabolism during hostile conditions, GABA play important role as osmolyte that reduced the negative effect in plants during adverse environmental conditions (Jalil et al. 2019; Jalil and Ansari 2020b). Isoprenoids are the terpenes that involved in defence mechanism of plants against abiotic stresses and oxidizing conditions of the environment. Isoprenoids synthesized in plants via two different biosynthetic pathways. Genetic modification of these pathway genes helps in isoprenoids overexpression that responses against abiotic stress by modification of ROS generation with direct reactions, indirect reactions with ROS, which developed stress tolerance ability in plants for combating abiotic stresses (Jalil and Ansari 2020c).

3.2.2 Genetic Engineering of Stress Responsive Genes and Transcription Factors

Introduction of diversify stress regulatory genes includes molecular chaperons, heat shock proteins (HSPs) encoded genes, regulation of histones, modification of helicases and micro RNA and various transcription factors (TFs) are utilized for enhancing stress resistance in plants using genetic engineering approach. Molecular chaperones are responsible for various cell activities such as protein folding, translocation, and degradation and it observed that various proteins showed chaperone like activities. Conversely, various chaperones are stress responsive proteins and characterized as HSPs (Wang et al. 2004). Furthermore, overexpression of Hsp90 genes of *Glycine max* in *A. thaliana* showed reduced stress induced responses and improved yield and phynotypic traits during heat stress (Xu et al. 2013). Moreover, over expressing OsHSP18.6 in genetically modified rice seedlings showed improved resistance in response to abiotic stresses. The exogenous expression of the HSP70 gene in transformed rice seedlings showed increased resistance against salt stress (Hoang et al. 2015). Moreover, sugarcane plants induced with EaHSP70 shows resistance against drought and salt stress (Augustine et al. 2015). In another study, heat-induced HsfA2 gene (transcription factor) from *Z. mays*, ZmHsf04, was overexpressed in *A. thaliana* showed tolerance in response to heat and salt stress (Jiang et al. 2017).

Modification in N terminal region of histones was done by methylation, acetylation, phosphorylation, and ubiquitination, these modifications activated or silenced stress responsive genes (Asensi-Fabado et al. 2017; Kim et al. 2015). Moreover, the hyperacetylation of histones H3K9, H3K14ac, and H3K27ac activates OsDREBb gene of *O. sativa* showed improved resistance against salt and chilling stresses (Scott et al. 2014). Furthermore, histone modification in the MYB, bZIP, and AP2/DREB TFs of *G. max* shows increased resistance in response to salt stress (Ci et al. 2015). Histone deacetylase gene HDA6 is essential for plants to survived in low temperature (To et al. 2011), and respond in salinity condition (Chen and Wu 2010). Moreover, HD2 proteins show improved resistance against extreme salinity condition (Luo et al. 2012).

Several researches have reported the beneficiary functions of helicases in countering the effect of adverse environmental conditions (Macovei et al. 2012; Tuteja et al. 2012; Liu and Imai 2018; Raikwar et al. 2015). It has been reported that transgenic *Nicotiana tabacum* with overexpressing pea PDH45 gene shows resistance against salt stress (Sanan-Mishra et al. 2005). Furthermore, OsABP (*O. sativa* ATP binding protein) showed response against abiotic stresses (Macovei et al. 2012). Moreover, overexpression of pea DNA helicase PDH45 in transgenic chilli involved in alleviating various abiotic stress conditions by elevating the activities of antioxidants and increasing the expression of stress genes (Shivakumara et al. 2017). Additionally, Overexpressing Pea p68 genein genetically modified *Nicotiana* plants shows accumulation of low Na1 and high K1 than wild type plants that showed late foliar senescence in the transgenic plants in different stresses (Tuteja et al. 2014).

In another study, groundnut plants expressing PgeIF4A gene improved resistance against drought and salinity stress conditions (Rao et al. 2017).

MicroRNAs are the non-coding RNAs which negatively regulates the posttranscriptional expressions of stress related genes and any modifications can influence expression confer adaptation benefits, therefore characterizing these objectives may be helpful in elucidating the regulatory role of miRNA in response to adverse environmental conditions (Chen et al. 2017). Characterization of these RNAs by means of genomic and proteomic approaches can facilitate in developing stress tolerant transgenic plants (Shriram et al. 2016). Furthermore, overexpression of OsSPL2 and OsSPL14 genes via miR529a in *O. sativa* showed improved tolerance against oxidative stress (Yue et al. 2017). Furthermore, overexpression of miR408 in *Cicer arietinum* showed enhanced resistance against drought stress (Hajyzadeh et al. 2015). It has been observed that overexpression of soybean miRNA172c and MIR394a in transgenic *A. thaliana* shows improved tolerance against salinity and dehydration stress (Li et al. 2016; Ni et al. 2012). The overexpressed *O. sativa* seedlings with miR319a and miR319b showed tolerance in response to chilling stress (Yang et al. 2013). Moreover, it has been studied that suppression ESK1 with siRNA gene expressions along with overexpression of CBF gene in *A. thaliana* showed resistance against drought stress (Xu et al. 2014).

3.3 Tissue Culture Techniques

In vitro plant tissue culture (PTC) technique is a promising tool and vital technology that has huge demand in agriculture. It prompts the advancement of biotechnological tools utilized for crop improvement. PTC produce disease-free/stress tolerant crop plants that will be require for increasing global population (Chatenet et al. 2001). PTC technology is used for the development of genetic engineered plants through introgression of stress responsive genes and selection of stress-resistant plants via *in vitro* selection (Pérez-Clemente and Gómez-Cadenas 2012). PTC techniques are widely applied in breeding as well as biotechnological techniques; this is an efficient method obligatory for the validation and utilization of data produced by these influential molecular tools. Execution of vigorous procedure for regeneration is consequently essential for genetic engineering and other tissue-culture derived methods to produce abiotic stress tolerant plants.

3.3.1 Somaclonal Variation and In Vitro Mutagenesis

PTC produces genetic variations in plants that included in plant breeding programmes (Jain 2001). It is notable that somaclonal variation concerning callus development and somatic embryogenesis has the ability to produce genetic variation (Larkin and Scowcroft 1981). The probability of developing somaclones by tissue culture approaches has been studied in *Pisum sativum* (Griga et al. 1995) and *Cajanus cajan* (Chintapalli et al. 1997). These variations are not enviable for certain purposes for example transgenic approach and enormous micropropagation, however can be helpful for breeding. These approaches, independently or with mutagenesis, develops diversity that are main breeding objective for enhancing the abiotic stress resistant in crop plants.

In vitro mutagenesis approaches have been utilized in crop breeding for abiotic stress tolerance in crops (Fuller and Eed 2003). These approaches have been affected by the recalcitrant plants to regenerate and the minimal efficiency of screening the desired phenotypes. However, the progress of regeneration methods of various plants and the activity of induced mutant crops showed that *in vitro* mutagenesis significantly involved crop breeding (Khan et al. 2001). In fact, associating mutagenesis approaches with other advance biotechnological techniques will make mutagenesis more progressive and appropriate for crop improvement. The main complexity with these approaches is the requirement of large number of individuals for screening the required trait. However, by utilizing *in vitro* selection approach this drawback can be reduced.

3.3.2 In Vitro Selection for Abiotic Stress Tolerant Plants

In vitro selection has been utilized for stress tolerant plants. Salt stress is the foremost abiotic stress that has been deal with this approach (Flowers 2004), however, has been reported for other stresses also (Samantaray et al. 1999). Furthermore, these approaches are significantly coordinated with conventional breeding practices (Svabova and Lebeda 2005). *In vitro* selection applied in *Medicago sativa* for screening against *Colletotrichum* (Cucuzza and Kao 1986), *Fusarium* (Cvikrova et al. 1992) and *Verticillium* species (Koike and Nanbu 1997). These reports revealed the possibility of *in vitro* selection, even though no resistant plants were documented. Moreover, in another study, *in vitro* screening was done in Guava transformed with endochitinase gene against *Fusarium oxysporum* (Mishra et al. 2016) and in this study resistant line against wilt disease were reported (Mishra et al. 2014). This approach can also be joined with other approaches along with somaclonal variation. Classical breeding as well as genetic engineering methods utilized for developing stress resistant plants which can be monitored by *in vitro* selection approach. These are especially appealing for adverse environmental conditions, where proper screening strategies are inaccessible or less proficient.

3.4 Gene Editing Tools for Improving Stress Resistance in Plants

Genome editing approaches by nuclease-specific sequences have emerged as powerful tools for genetic and plant development studies. The availability of high-quality genome editing tools raises the possibility of improving the quality of plants for efficient and directed growth of plant traits, especially enhanced tolerance to abiotic stress conditions. The invention of the engineered nucleases develops a double-stranded break that alters the biology of cells by opening a new way of genetic engineering of genes of interest. Genome editing by ZFNs (Kim et al. 1996) and TALENs (Christian et al. 2010), but now days it turn emphasized by the innovation of clustered regularly interspaced short palindromic repeats CRISPR/Cas systems (Jinek et al. 2012) that offers the flexibility as well as simplicity of targeted genome editing.

3.4.1 Zinc Finger Nucleases (ZFNs)

Zinc Finger Nucleases (ZFNs) is a genome editing method that uses well-designed nucleases formed after exposure to localization of Cys2-His2 zinc finger (ZF) (Kim et al. 1996; Pabo et al. 2001; Palpant and Dudzinski 2013). Though the primary study on ZFs, successfully used a few organisms together with plants (Gaj et al. 2013). ZFNs focus on the unfunctional of endogenous genes in various plants (Zhang et al. 2010; Shukla et al. 2009; Townsend et al. 2009). Site-specific mutagenesis and base substitution are important for genetic engineering of plants (Osakabe et al. 2010). Consequently, ZFNs are important tool used in several crops to improve abiotic stress resistance. ZFNs was developed specially formulated alongside HSPs to modify the AP2/ERF family definition, ABA-INSENSITIVE 4, which is involved in abiotic stresses (Osakabe et al. 2010). In *Arabidopsis*, ZFNs are designed to target DNA to introduce insertion and deletion mutations (dePater et al. 2009; Hou et al. 2014).

3.4.2 Transcription Activator-Like Effector Nucleases (TALENs)

TALENs protein nucleases produced and regulated the targeted DNA by the help of specific protein activator-like effectors (TALE) protein (Jankele and Svoboda 2014). Proteins contains a space that focuses on DNA binding, cellular signaling, and the filling site as a gene translation tool (Schornack et al. 2006). Therefore, the DNA inhibition potential of these proteins, and after few years, scientists decodes the identification code of the targeting DNA pattern containing TALE proteins (Boch et al. 2009). TALENs promote site-based mutagenesis in the target gene (Hou et al.

2014; Mahfouz et al. 2011). TALENs effectively transfer to various plants including *A. thaliana, N. tabacum* and *O. sativa* (Curtin et al. 2012; Shan et al. 2013). In addition, TALEN has created various mutations in *O. sativa* (Zhang et al. 2016). Furthermore, the introduction of TALEN in potatoes has shown cold storage and processing properties (Clasen et al. 2016).

3.4.3 Clustered Regularly Interspaced Short Palindromic Repeat (CRISPR)/Cas9)

Advancement in genome editing tools provide new opportunities for genetic editing using targeted genes for specific plant traits. CRISPR/CRISPR-related protein-9, the Cas9 received an astonishing speculation by researchers due to its tangible benefits over other genome editing tools such as ZFN and TALEN (Mao et al. 2013). The nucleases that use protein-targeted proteins, CRISPR-Cas9 relies on the approval of RNA-DNA to form a double strand. The various desirable conditions of CRISPR-Cas9 in addition to these nucleases are specific to the target system, the efficiency of introducing mutation by formulating Cas9 and guide RNA, as well as the easiness of multiplexing intensive on deviations in several genetic expressions (Zafar et al. 2019).

In addition to mutagenesis, this method can be used to activate (CRISPR or CRISPRa implementation) or to deactivate (CRISPR impedance or CRISPRi) quality integration through merging a chemically active Cas9 with a transcriptional activator or repressor (Bortesi and Fischer 2015). Regardless, reports regarding the concentration of abiotic stress resilencing are inadequate for such studies. This method interfered in the heat-resistant genetic modification achieved via directing stress genes such as SIAGAMOUS-LIKE 6 (SIAGL6) on tomatoes, which increased plant resistance under heat stress (Klap et al. 2017). Consequently, a focus on multiple attributes in a single living form using CRISPR-Cas9 has also been successfully developed in various plants (Char et al. 2017; Gao et al. 2017; Miao et al. 2013; Wang et al. 2016).

Thus, this genome-editing method have the potential to develop plants that are more tolerant of stress through concentrating on a several stress related genes in the favorable but critical crop yields. In addition, tolerant genes, can be overexpressed using CRISPR which is commonly used for genetic mutations. Recently, the use of this method in genome editing has continued to increase the use of this method in comprehensive genome studies to improve quality of crops (Mahas and Mahfouz 2018; Rodríguez-Leal et al. 2017). It could be another way of growing traditional plants which depends on finding plants communities including enough genetic diversity to carry attractive traits to plants. This approach may introduce novel allelic differences in plants, so novel alleles associated with a particular attractive phenotype will be recognizable in the sequence of the guide RNA (Eid et al. 2018). So, this tool has played a major role for developing stress-resistant varieties.

3.5 Conclusion and Prospects

In view of the enormous loss of crop productivity because of adverse environment conditions, there is an imperative requirement to focus our research on developing plants with improved abiotic stress tolerance along with high yields by means of biotechnological strategies to alleviate the negative impacts of environmental changes on plants. Biotechnological approaches have also been widely used to study the plant molecular mechanisms associated with different abiotic stress resistance for development of stress resistant crop plant by molecular regulation of genes, genetic engineering and very recently the genome editing tools specifically CRISPR-Cas9. Presently, genome editing tools are the momentous innovation of agricultural biotechnology. Generally considered as a non-genetically modified technique, CRISPR/Cas tool has appeared as a gene editing approaches that can be a possible turning point for intensifying our researches on producing stress tolerant crops without changing nutritional quality of plants for the future.

References

Abdallah NA, Prakash CS, McHughen AG (2016) Genome editing for crop improvement: challenges and opportunities. GM Crops Food 6:183–205

Ansari MI, Lin TP, Liu MS (2011) Molecular characterization of TLP 18.3 gene of *Arabidopsis thaliana*. Int J Integr Biol 11:44–51

Asensi-Fabado M-A, Amtmann A, Perrella G (2017) Plant responses to abiotic stress: the chromatin context of transcriptional regulation. Biochem Biophys Acta (BBA) 1860:106–122

Ashraf M, Harris PJC (2004) Potential biochemical indicators of salinity tolerance in plants. Plant Sci 166:3–16

Augustine SM, Narayan JA, Syamaladevi DP et al (2015) *Erianthus arundinaceus* HSP70 (EaHSP70) overexpression increases drought and salinity tolerance in sugarcane (*Saccharum* spp. hybrid). Plant Sci 232:23–34

Boch J, Scholze H, Schornack S et al (2009) Breaking the code of DNA binding specificity of TAL-type III effectors. Science 326:1509–1512

Bortesi L, Fischer R (2015) The CRISPR/Cas9 system for plant genome editing and beyond. Biotechnol Adv 33:41–52

Castiglioni P, Bell E, Lund A, Rosenberg AF et al (2018) Identification of GB1, a gene whose constitutive overexpression increases glycine betaine content in maize and soybean. Plant Direct 2:1–7

Char SN, Neelakandan AK, Nahampun H et al (2017) An Agrobacterium-delivered CRISPR/Cas9 system for high-frequency targeted mutagenesis in maize. Plant Biotechnol Journal 15:257–268

Chatenet M, Delage C, Ripolles M, Irey M, Lockhart BLE, Rott P (2001) Detection of sugarcane yellow leaf virus in quarantine and production of virus-free sugarcane by apical meristem culture. Plant Dis 85(11):1177–1180

Chen LT, Wu K (2010) Role of histone deacetylases HDA6 and HDA19 in ABA and abiotic stress response. Plant Signal Behav 5:1318–1320

Chen XY, Yang Y, Ran LP et al (2017) Novel insights into miRNA regulation of storage protein biosynthesis during wheat caryopsis development under drought stress. Front Plant Sci 8

Chikkaputtaiah C, Debbarma J, Baruah I et al (2017) Molecular genetics and functional genomics of abiotic stress-responsive genes in oilseed rape (*Brassica napus* L.): a review of recent advances and future. Plant Biotechnol Rep 11:365–384

Chikkaputtaiah C, Marwein R (2019) Genetic engineering/Genome editing approaches to modulate signaling processes in abiotic stress tolerance In: Iqbal M, Khan R (eds) Plant signalling molecules: role and regulation under stressfull environment, 1st edn. Elsevier, pp 64–77

Chintapalli PL, Moss JP, Sharma KK, Bhalla JK (1997) *In vitro* culture provides additional variation for pigeonpea *Cajanus cajan* (L) Millsp crop improvement. Vitro Cell Dev Biol-Plant 33:30–37

Christian M, Cermak T, Doyle EL et al (2010) Targeting DNA double-strand breaks with TAL effector nucleases. Genetics 186:757–761

Ci D, Song Y, Tian M, Zhang D (2015) Methylation of miRNA genes in the response to temperature stress in *Populus simonii*. Front Plant Sci 6

Clasen BM, Stoddard TJ, Luo S et al (2016) Improving cold storage and processing traits in potato through targeted gene knockout. Plant Biotechnol J 14:169–176

Cucuzza JD, Kao J (1986) Invitro assay of excised cotyledons of alfalfa (*Medicago sativa*) to screen for resistance to *Colletotrichum trifolii*. Plant Dis 70:111–115

Curtin SJ, Voytas DF, Stupar RM (2012) Genome engineering of crops with designer nucleases. Plant Genome J 5:42

Cvikrova M, Binarova P, Eder J, Nedelnik J (1992) Accumulation of phenolic-acids in filtrate-treated alfalfa cell-cultures derived from genotypes with different susceptibility to *Fusarium oxysporum*. J Plant Physiol 140:21–27

de Pater S, Neuteboom LW, Pinas JE, Hooykaas PJJ, van der Zaal BJ (2009) ZFN-induced mutagenesis and gene-targeting in *Arabidopsis* through Agrobacterium-mediated floral dip transformation. Plant Biotechnol J 7:821–835

Eid A, Alshareef S, Mahfouz MM (2018) CRISPR base editors: genome editing without double-stranded breaks. Biochem J 475:1955–1964

Filippou P, Bouchagier P, Skotti E, Fotopoulos V (2014) Proline and reactive oxygen/nitrogen species metabolism is involved in the tolerant response of the invasive plant species *Ailanthus altissima* to drought and salinity. Environ Exp Bot 97:1–10

Flowers TJ (2004) Improving crop salt tolerance. J Exp Bot 55:307–319

Fuller MP, Eed MHI (2003) The development of multiple stress-resistant cauliflower using mutagenesis in conjunction with a microshoot tissue culture technique. In XXVI International Horticultural Congress: Environmental stress and horticulture crops. Toronto, Canada, 2003. Ed I.A. Horticulturae

Gaj T, Gersbach CA, Barbas CF III (2013) ZFN, TALEN, and CRISPR/Cas-based methods for genome engineering. Trends Biotechnol 31:397–405

Gao W, Long L, Tian X et al (2017) Genome editing in cotton with the CRISPR/Cas9 system. Front Plant Sci 8:1364

Griga M, Stejskal J, Beber K (1995) Analysis of tissue culture-derived variation in pea (*Pisum sativum* L). Preliminary Results Euphytica 85:335–339

Hajyzadeh M, Turktas M, Khawar KM, Unver T (2015) miR408 overexpression causes increased drought tolerance in chickpea. Gene 555:186–193

He C, Yang A, Zhang W et al (2010) Improved salt tolerance of transgenic wheat by introducing beta gene for glycine betaine synthesis. Plant Cell Tiss Organ Cult 101:65–78

Hoang TML, Moghaddam L, Williams B et al (2015) Development of salinity tolerance in rice by constitutive-overexpression of genes involved in the regulation of programmed cell death. Front Plant Sci 6:175

Hou H, Atlihan N, Lu ZX (2014) New biotechnology enhances the application of cis genesis in plant breeding. Front Plant Sci 5

Huang Q, Wang M, Xia Z (2018) The SULTR gene family in maize (*Zea mays* L.): gene cloning and expression analyses under sulphate starvation and abiotic stress. J Plant Physiol 220:24–33

Jacob P, Hirt H, Bendahmane A (2017) The heat- shock protein/chaperone network and multiple stress resistance. Plant Biotechnol J 15:405–414

Jain SM (2001) Tissue culture-derived variation in crop improvement. Euphytica 118:153–166
Jalil SU, Ansari MI (2020a) Stress implication and crop productivity (ed 1) plant ecophysiology and adaptation under climate change: mechanism and prespectives. Springer Nature, Singapore, pp 73–86
Jalil SU, Ansari MI (2020b) Physiological role of Gamma-Aminobutyric Acid in salt stress tolerance (ed) salt and drought stress tolerance in plants, signaling and communication in plants. Springer Nature, Switzerland AG, pp 337–350
Jalil SU, Ansari MI (2020c) Isoprenoids in plant protection against abiotic stress (ed) protective chemical agents in the amelioration of plant abiotic stress: biochemical and molecular perspectives. Wiley, pp 424–436
Jalil SU, Khan MAA, Ansari MI (2019) Tremendous role of GABA shunt metabolic pathway in plant system. J Biol Chem Res 36(1):188–196
Jankele R, Svoboda P (2014) TAL effectors: tools for DNA targeting. Brief Funct Genomics 13:409–419
Jiang Y, Zheng Q, Chen L, Liang Y, Wu J (2017) Ectopic overexpression of maize heat shock transcription factor gene ZmHsf04 confers increased thermo and salt-stress tolerance in transgenic *Arabidopsis*. Acta Physiol Plant 40:9
Jinek M, Chylinski K, Fonfara I et al (2012) A programmable dual-RNA-guided DNA endonuclease in adaptive bacterial immunity. Science 337:816–821
Joshi R, Sahoo KK, Anil Kumar Singh AK (2020) Enhancing trehalose biosynthesis improves yield potential in marker-free transgenic rice under drought, saline, and sodic conditions. J Exp Botany 71:653–668
Khan AJ, Hassan S, Tariq M, Khan T (2001) Haploidy breeding and mutagenesis for drought tolerance in wheat. Euphytica 120:409–414
Kim H, Kim ST, Kim SG et al (2015) Targeted genome editing for crop improvement. Plant Breed Biotech 3:283–290
Kim YG, Cha J, Chandrasegaran S (1996) Hybrid restriction enzymes: zinc finger fusions to Fok I cleavage domain. Proceedings of the National Academy of Sciences of the United States ofAmerica 93:1156–1160
Klap C, Yeshayahou E, Bolger AM et al (2017) Tomato facultative parthenocarpy results from SlAGAMOUS-LIKE 6 loss of function. Plant Biotech J 15:634–647
Koike M, Nanbu K (1997) Phenylalanine ammonia-lyase activity in alfalfa suspension cultures treated with conidia and elicitors of *Verticillium alboatrum*. Biol Plant 39:349–353
Larkin PJ, Scowcroft WR (1981) Somaclonal variation—a novel source of variability from cell-cultures for plant improvement. Theor Appl Genet 60:197–214
Li M, Li Y, Zhao J et al (2016) GpDSR7, am novel E3 ubiquitin ligase gene in *grimmiapilifera* is involved in tolerance to drought stress in *Arabidopsis*. PLoS One 11
Li Q, Yu HM, Meng XF et al (2018) Ectopic expression of glycosyltransferase UGT76E11 increases flavonoid accumulation and enhances abiotic stress tolerance in *Arabidopsis*. Plant Biol 20:10–19
Lin KH, Huang HC, Lin CY (2010) Cloning, expression and physiological analysis of broccoli catalase gene and Chinese cabbage ascorbate peroxidase gene under heat stress. Plant Cell Rep 29:575–593
Lin KH, Kuo WS, Chiang CM, Hsiung TC, Chiang MC, Lo HF (2013) Study of sponge gourd ascorbate peroxidase and winter squash superoxide dismutase under respective flooding and chilling stresses. Sci Hortic 162:333–340
Liu S, Jia Y, Zhu Y et al (2018) Soybean matrix metalloproteinase Gm2-MMP relates to growth and development and confers enhanced tolerance to high temperature and humidity stress in transgenic *Arabidopsis*. Plant Mol Biol Rep 36:94–106
Liu Y, Imai R (2018) Function of plant DExD/H-Box RNA helicases associated with ribosomal RNA biogenesis. Front Plant Sci 9:125
Luo M, Wang YY, Liu X et al (2012) HD2C interacts with HDA6 and is involved in ABA and salt stress response in *Arabidopsis*. J Exp Bot 63:3297–3306

Ma X, Chen C, Yang M et al (2018) Cold-regulated protein (SlCOR413IM1) confers chilling stress tolerance in tomato plants. Plant Physiol Biochem 124:29–39
Macovei A, Vaid N, Tula S, Tuteja N (2012) A new DEAD-box helicase ATP-binding protein (OsABP) from rice is responsive to abiotic stress. Plant Signal Behav 7:1138–1143
Mahas A, Mahfouz M (2018) Engineering virus resistance via CRISPR–Cas systems. Curr Opin Virol 32:1–8
Mahfouz MM, Li L, Shamimuzzaman M, Wibowo A, Fang X, Zhu JK (2011) De novo-engineered transcription activator-like effector (TALE) hybrid nuclease with novel DNA binding specificity creates double-strand breaks. Proc Natl Acad Sci USA 108:2623–2628
Mahfouz MM, Piatek A, Stewart CN (2014) Genome engineering via TALENs and CRISPR/Cas9 systems: challenges and perspectives. Plant Biotechnol J 12:1006–1014
Mao Y, Zhang H, Xu N et al (2013) Application of the CRISPR–Cas system for efficient genome engineering in plants. Mol Plant 6:2008–2011
McNeil SD, Nuccio ML, Hanson AD (1999) Betaines and related osmoprotectants. Targets for metabolic engineering of stress resistance. Plant Physiol 120:945–949
Miao J, Guo D, Zhang J et al (2013) Targeted mutagenesis in rice using CRISPR-Cas system. Cell Res 23:1233–1236
Mishra M, Jalil SU, Mishra RK, Kumari S, Pandey BK (2016) *In vitro* screening of guava plantlets transformed with endochitinase gene against *Fusarium oxysporum*f.sp. psidii. Czech J Gen Plant Breed 52(1):6–13
Mishra M, Jalil SU, Sharma N, Hudedwani U (2014) An *Agrobacterium* mediated transformation system of guava *(Psidium guajava L.)* with *endochitinase* gene. Crop Breed App Biotechnol 14:232–237
Ni Z, Hu Z, Jiang Q, Zhang H (2012) Overexpression of gma- MIR394a confers tolerance to drought in transgenic *Arabidopsis thaliana.* Biochem Biophys Res Commun 427:330–335
Ohama N, Sato H, Shinozaki K, Yamaguchi-Shinozaki K (2017) Transcriptional regulatory network of plant heat stress response. Trends Plant Sci 22:53–65
Osakabe K, Osakabe Y, Toki S (2010) Site-directed mutagenesis in *Arabidopsis* using custom-designed zinc finger nucleases. Proc Natl Acad Sci USA 107:12034–12039
Pabo CO, Peisach E, Grant RA (2001) Design and selection of novel Cys_2His_2 zinc finger proteins. Annu Rev Biochem 70:313–340
Palpant NJ, Dudzinski D (2013) Zinc finger nucleases: looking toward translation. Gene Ther 20:121–127
Per TS, Khan MIR, Anjum NA et al (2018) Jasmonates in plants under abiotic stresses: crosstalk with other phytohormones matters. Environ Exp Bot 145:104–120
Pérez-Clemente RM, Gómez-Cadenas A (2012) In vitro tissue culture, a tool for the study and breeding of plants subjected to abiotic stress conditions, recent advances in plant in vitro culture, Annarita Leva and Laura M. R. Rinaldi, Intech Open. https://doi.org/10.5772/50671
Quan R, Shang M, Zhang H et al (2004) Engineering of enhanced glycine betaine synthesis improves drought tolerance in maize. Plant Biotechnol J 2:477–486
Raikwar S, Srivastava VK, Gill SS et al (2015) Emerging importance of helicases in plant stress tolerance: characterization of *Oryza sativa* repair helicase XPB2 promoter and its functional validation in tobacco under multiple stresses. Front Plant Sci 6:1094
Rao SRB, Vijaya NJ, Sudhakar RP et al (2017) Expression of pennisetumglaucum eukaryotic translational initiation factor 4A (PgeIF4A) confers improved drought, salinity, and oxidative stress tolerance in groundnut. Front Plant Sci 8:453
Rodríguez-Leal D, Lemmon ZH, Man J et al (2017) Engineering quantitative trait variation for crop improvement by genome editing. Cell 171:470–480
Samantaray S, Rout GR, Das P (1999) *In vitro* selection and regeneration of zinc tolerant calli from *Setaria italica* L. Plant Sci 143:201–209
Sanan-Mishra N, Pham XH, Sopory SK, Tuteja N (2005) Pea DNA helicase 45 overexpression in tobacco confers high salinity tolerance without affecting yield. Proc Natl Acad Sci USA 102:509

Sanchez DH, Pieckenstain FL, Szymanski J et al (2011) Comparative functional genomics of salt stress in related model and cultivated plants identifies and overcomes limitations to translational genomics. PLoS ONE 6:e17094

Sawahel WA, Hassan AH (2002) Generation of transgenic wheat plants producing high levels of the osmoprotectant proline. Biotechnol Lett 24:721–725

Schornack S, Meyer A, R¨omer P et al (2006) Gene-for-gene-mediated recognition of nuclear-targeted AvrBs3-like bacterial effector proteins. J Plant Physiol 163:256–272

Scott KC, Roy D, Paul A, Roy A, Ghosh R, Ganguly P et al (2014) Differential acetylation of histone H3 at the regulatory region of OsDREB1b promoter facilitates chromatin remodelling and transcription activation during cold stress. PLoS ONE 9:e100343

Shan Q, Wang Y, Chen K et al (2013) Rapid and efficient gene modification in rice and brachypodium using TALENs. Mol Plant 6:1365–1368

Shivakumara TN, Sreevathsa R, Dash PK et al (2017) Overexpression of Pea DNA Helicase 45 (PDH45) imparts tolerance to multiple abiotic stresses in chili (*Capsicum annuum* L.). Sci Rep 7:2760

Shriram V, Kumar V, Devarumath RM et al (2016) MicroRNAs as potential targets for abiotic stress tolerance in plants. Front Plant Sci 7

Shukla VK, Doyon Y, Miller JC et al (2009) Precise genome modification in the crop species *Zea mays* using zinc-finger nucleases. Nature 459:437–441

Svabova L, Lebeda A (2005) *In vitro s*election for improved plant resistance to toxin producing pathogens. J Phytopathol 153:52–64

To TK, Kim JM, Matsui A et al (2011) *Arabidopsis* HDA6 regulates locus-directed heterochromatin silencing in cooperation with MET1. PLoS Genet 7:28

Townsend JA, Wright DA, Winfrey RJ (2009) High frequency modification of plant genes using engineered zinc finger nucleases. Nature 459:442–445

Tuteja N, Banu MS, Huda KM et al (2014) Peap68, a DEAD-box helicase, provides salinity stress tolerance in transgenic tobacco by reducing oxidative stress and improving photosynthesis machinery. PLoS One 9

Tuteja N, Singh S, Tuteja R (2012) Helicases in improving abiotic stress tolerance in crop plants: improving crop resistance to abiotic stress. Wiley-VCH Verlag GmbH & Co. KGaA, Weinheim, pp 435–449

Wang W, Akhunova A, Chao S, Akhunov E (2016) Optimizing multiplex CRISPR/Cas9-based genome editing for wheat. bioRxiv 051342

Wang W, Vinocur B, Shoseyov O, Altman A (2004) Role of plant heat-shock proteins and molecular chaperones in the abiotic stress response. Trends Plant Sci 9:244–252

Xu F, Liu Z, Xie H et al (2014) Increased drought tolerance through the suppression of ESKMO1 gene and overexpression of CBF-related genes in *Arabidopsis*. PLoS One 9

Xu J, Xue C, Xue D et al (2013) Overexpression of GmHsp90s, a heat shock protein 90 (Hsp90) gene family cloning from soybean, decrease damage of abiotic stresses in *Arabidopsis thaliana*. PLoS One 8

Yang C, Li D, Mao D et al (2013) Overexpression of microRNA319 impacts leaf morphogenesis and leads to enhanced cold tolerance in rice (*Oryza sativa* L.). Plant, Cell Environ 36:2207–2218

You J, Hu H, Xiong L (2012) An ornithine delta-aminotransferase gene OsOAT confers drought and oxidative stress tolerance in rice. Plant Sci 197:59–69

Yue E, Liu Z, Li C et al (2017) Overexpression of miR529a confers enhanced resistance to oxidative stress in rice (*Oryza sativa* L.). Plant Cell Rep 36:1171–1182

Zafar SA, Zaidi SS, Gaba Y et al (2019) Engineering abiotic stress tolerance via CRISPR/Cas-mediated genome editing. J Exp Botany 71:470–479

Zandalinas SI, Mittler R, Balfagon D et al (2018) Plant adaptations to the combination of drought and high temperatures. Physiol Plant 162:2–12

Zhang F, Maeder ML, Unger-Wallaced E et al (2010) High frequency targeted mutagenesis in *Arabidopsis thaliana* using zinc finger nucleases. Proc Natl Acad Sci USA 107:12028–12033

Zhang H, Gou F, Zhang J, Liu W, Li Q, Mao Y et al (2016) TALEN-mediated targeted mutagenesis produces a large variety of heritable mutations in rice. Plant Biotechnol J 14:186–194

Chapter 4
Nanomaterials Fundamentals: Classification, Synthesis and Characterization

Abdulsalam K. Almuhammady, Khaled F. M. Salem, Maysaa T. Alloosh, Maysoun M. Saleh, Lina M. Alnaddaf, and Jameel M. Al-Khayri

Abstract Nanotechnology is an exciting recent-day technology. Different metals can be prepared as nanomaterials (NM) that can be used in various fields. Therefore, ongoing research focuses on developing many methods for synthesizing nanoparticles (NPs). Nanoparticles have unique physicochemical, structural and morphological characteristics that are important for a wide range of applications in conjunction with the fields of electronic, optoelectronic, optical, electrochemical, environmental

A. K. Almuhammady (✉)
Arab Center for Nanotechnology, Cairo, Egypt

K. F. M. Salem
Department of Plant Biotechnology, Genetic Engineering and Biotechnology Research Institute (GEBRI), University of Sadat City, Sadat City, Egypt

Department of Biology, College of Science and Humanitarian Studies, Shaqra University, Qwaieah, Saudi Arabia

K. F. M. Salem
e-mail: khaled.salem@gebri.usc.edu.eg; khaledfathi@su.edu.sa

M. T. Alloosh
Department of Biochemistry and Microbiology, Faculty of Pharmacy, Al-Baath University, Homs, Syria
e-mail: malloosh@albaath-univ.edu.sy

M. M. Saleh
Genetic Resources Department, General Commission for Scientific Agricultural Research (GCSAR), Damascus, Syria
e-mail: mzainsamasaleh@gcsar.gov.sy; mzainsamasaleh@gmail.com

L. M. Alnaddaf
Biotechnology and Molecular Biology, Faculty of Agricultural-Al-Baath University-Homs, Homs, Syria
e-mail: lalnaddaf@albaath-univ.edu.sy

J. M. Al-Khayri
Department of Agricultural Biotechnology, College of Agriculture and Food Sciences, King Faisal University, Al-Ahsa, Saudi Arabia
e-mail: jkhayri@kfu.edu.sa

J. M. Al-Khayri et al. (eds.), *Nanobiotechnology*,
https://doi.org/10.1007/978-3-030-73606-4_4

and biomedical. This chapter presents an insight on the fundamental concepts, properties and classifications of NPs including their preparation methods as well as importance in obtaining suitable NPs for several uses. In addition, the present chapter also explains the characterization tools of NPs such as X-ray diffraction, transmission electron microscopy, scanning electron microscopy, energy dispersion spectroscopy, Fourier transform infrared and ultraviolet-visible spectroscopy.

Keywords Characterization · Classification · Nanotechnology · Nanoparticles · Properties of nanoparticles · Quantum effects

4.1 Introduction

The term "nano" is derived from the Greek word "Nanos" which means dwarf. Mathematically, the nano is a unit of measurement, which is equal to one part of billion, i.e., nanometer is 10^{-9} m. The nanomaterials (NM) are defined as a set of substances where at least one dimension is less than approximately 100 nm (Wu et al. 2020). For example, in order to imagine the small nanometer, the thickness of a single human hair is 50 micrometers or 50,000 nm. The smallest things a person can see with the naked eye displays about 10,000 nm wide. Ten hydrogen atoms placed in a line that touches each other will measure one nanometer. The science that deals with these materials and studying their properties is nanoscience (Madkour 2019). The application and engineering of these sciences to produce useful inventions is nanotechnology, which deals with materials by dimensions not exceeding 100 nm (Wu et al. 2020). The unique thing about the nano or "Nano Scale" is that most of the basic properties of materials and machines such as thermal or electrical conductivity, hardness, melting point and physical properties, depend on size like no other in any scale larger than nano, for example, the nano scale wire or conductor does not necessarily follow Ohm's law whose equation links current and voltage (Dhand et al. 2015). The resistance value depends on the principle of electrons flowing into the wire. When the width of the wire is one atom, the passing of electrons through it will be very difficult. Also, the color of the material in nano scale changes, because of the difference in the absorption coefficient of light, since the wavelength that is reflected from the material in the nano scale will change and it determines the color of the material, for example, in gold nano particles will be diagonal to red on the abnormal, which is yellow (Ealias and Saravanakumar 2017).

The chemical properties will also change greatly due to the large surface area compared to the volume, as well as the difference of the ionization factor due to the increase in the number of free electrons, and, therefore the chemical stability will be less than of bulk materials (Viswanathan 2014). The physical properties will differ for the same reasons in addition to the change in heat capacity and heat exchange. The change of the properties in the case of nano structure materials is the key to understanding broad nano science (Ealias and Saravanakumar 2017). Therefore, knowledge of the benefits gained and the side effects of this new technological revolution and the changes that will bring about our lives to reap its benefits and avoid or reduce its harms and be sure that it will not poison us at the same time (Madkour 2019; Yaqoob et al. 2020). For a more understanding of the state of nanostructures, one nano linearly contains three carbon atoms. In this case, the square of the length 1 nm contains 9 carbon atoms. The cube with the length of 1 nm contains 27 carbon atoms (in a cube 1 m × 1 m × 1 m thus the number of atoms are 2.7×10^{28}C atoms) (Madkour 2019). This chapter presents brief overview on nanotechnology as well as NM together with their properties, methods for preparation and the tools of characterization.

4.2 Nanomaterials

The NM are very small dimension materials usually in the range of 1–100 nm (Wu et al. 2020). In these dimensions, the various properties such as optical, magnetic, electrical, mechanical, chemical, thermal and physical can be altered or improved (Vollath 2008). NM have the potential for great applications in electronics, industry, medical, agriculture, energy, environment, water treatment, space and all engineering applications (Madkour 2019). Some NM may exist naturally without interference, however it is especially important that NM are already designed to be used in applications that serve humanity in general (Horikoshi and Serpone 2013).

4.3 Classification of Nanomaterials

The NM are classified into four kinds based on their shapes and dimensions (Madkour 2019). The first kind called, zero dimension: Materials have three nanoscale dimensions and their are so small and close to zero, for example, nano dots (Viswanathan 2014). The second kind is one dimension. It has one length (x, y or z) in the nano scale and the other two dimensions are out rang of nano, for example, very thin surface coatings or a single sheet of graphite (graphene oxide) (Pokropivny 2007). The third kind is two dimensions, the material has two dimensions in the range of nanometer and its third dimension out the range of nano, like nano carbon tube. The last kind of NM is three dimension, all three dimensions are in nano scale. It differs from the zero dimensions in that its dimensions are slightly larger and not neglected (often greater than 10 nm) (Madkour 2019; Shao et al. 2020).

4.4 Quantum Effects

Classical mechanics in the case of NM become unable to illustrate the phenomena that occur in these models because their small dimensions. Quantum mechanics managed to explain phenomena, which classical physics cannot explain (Madkour 2019). For instance changing the color of gold when it turns to nano. Another example is the principle of uncertainty in determining the position of an electron, as classical physics fail to explain this phenomenon (Pokropivny 2007). Through the use of quantum mechanics, many things can be described in small limits of sizes. Classical system, such as Newton's law of motion may not be able to explain this, especially when the dimensions of materials reach less than 10 nm. In this case, the Earth's gravitational force is neglected because it does not effect on objects in very small dimensions (Vollath 2008).

Types of electrons confinement was shown in (Fig. 4.1). The type one is quantum well" or two dimensional system. This means that the electron can move in two directions and confinement in the other. It occurs in one dimensional NM (Madkour 2019). Whereas, if the electron was free in one direction and became confinement in two directions, the two dimensions are not sufficient to give the electron freedom of movement. This is type two which called the quantum wire or one-dimensional system. This case found in two-dimensional NM (Pokropivny 2007). While the third

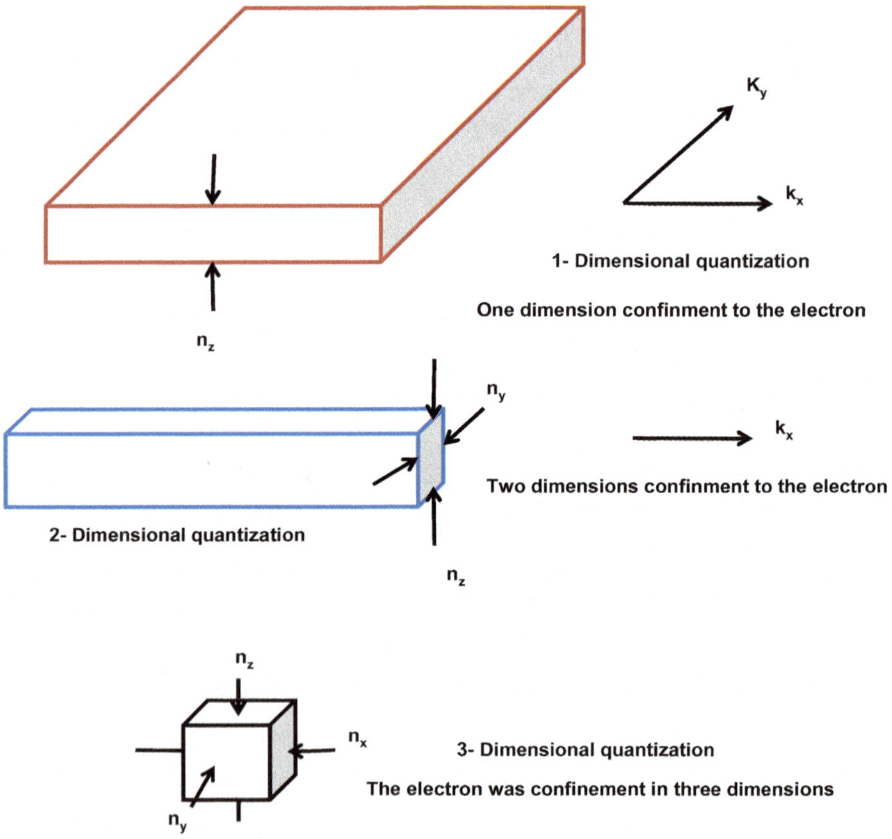

Fig. 4.1 Types of electrons confinement (Figure constructed by A. K. Almuhammady)

type is a zero dimensional system or quantum dot. The electron is not able to move in the three directions because there is not a sufficient dimension to move it. This occurs in zero dimensional NM (Sethi et al. 2020).

4.5 Unique Properties of Nanomaterials

NM differ from materials of their natural dimensions in properties (Ealias and Saravanakumar 2017). The change in characteristics happens for the many reasons, the main one is: The NM are distinguished having a very large surface area when compared to size. Let consider a sphere of radius "r", its surface area $= 4\pi r^2$, its volume $= 4/3\pi r^3$, surface area to volume ratio $= 3/r$. In this case, if assume that the value of (r) in the equation is 1 nm, then the ratio of area to the volume will be $3/r = 3/10^{-9} = 3 \times 10^9$. It is a very large amount and to clarify the situation more (Horikoshi and Serpone 2013).

If a cubic length of 1 m, then the surface area is 6 m^2. If the cube divided into two equal parts, the area is 12 m^2. If it's divide into three equal parts, then the area will be 18 m^2. As a result, if subdivided a bulk material into an ensemble of individual NM, the total volume remains the same, but the collective surface area is greatly increased (Fig. 4.2) (Kulkarni 2015). In addition, the NM may reach very small dimensions so the quantum effect becomes dominant in the behavior of matter at the nano scale (Pokropivny 2007). Quantum mechanics will describe motion and energy instead of the classical mechanics model. When the three dimensions of NM become very small (less than 10 nm), then the gravitational forces become very small and neglected within certain limits because it depends on the factors of distance and mass

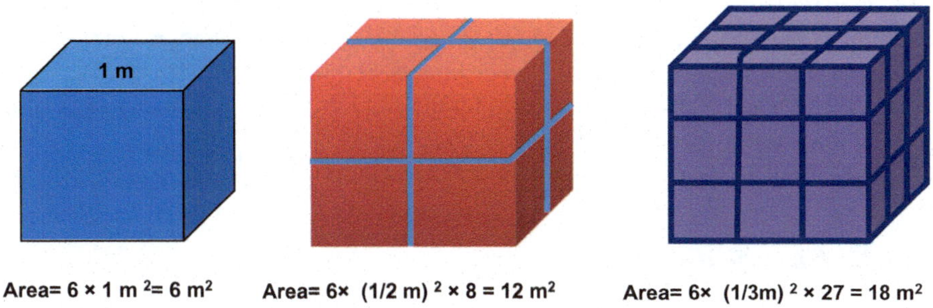

Fig. 4.2 The relationship between size and surface area (Figure constructed by A. K. Almuhammady)

that are very small and have no effect (Viswanathan 2014). In nanoparticles (NPs), the electromagnetic forces, in this case, are very high. Another reason to change the properties of NM is the band gap and lattice parameters are depending on the size of particles (Vollath 2008).

4.5.1 Physical Properties

The NM crystal structure is identical to bulk materials. However, it differs in the lattice parameter (Kulkarni 2015). Wonderful specific properties that may vary significantly from the physical properties of bulk materials. Some of these unusual characteristics are known, but many are still being discovered (Madkour 2019). The physical properties of NM for different origins are identical for the large atomic distribution on the surface of the material and what happens because of this distribution from the interactions with the external environment, large surface energy, due to the increased surface area and spatial confinement of the electron spatial because of the small dimensions, which does not give the electron freedom of movement (Mageswari et al. 2016; Viswanathan 2014).

4.5.2 Optical Properties

One of the most important characteristics of NM, optical properties because many applications depend on it (Juh 2007). These applications include an optical detector, laser, sensor, imaging, solar cells, photo catalysis, photochemistry and biomedicine (Madkour 2019). The optical properties of NM rely on criteria such as molecule size, shape, surface properties and other variables including increased the activity of interaction with the surrounding environment or other nanostructures (Tshabalala et al. 2020). The color of NM changes by the particle size due to the change of the optical absorption coefficient (Zhang 2009).

4.5.3 Chemical Properties

The electronic structure of nanoscale materials depends on their size which mainly affects chemical stability and reaction (Bunaciu et al. 2015). NM have a high surface area, that increases the possibility of interaction with the external environment, Then it has low chemical stability (Madkour 2019). In other words, NM tend to interact more than bulk materials (Mageswari et al. 2016). Inaddition, among the reasons that can lead to less chemical stability are the changes in an electronic structure and the relatively high ionization factor (Viswanathan 2014).

4.5.4 Electrical Properties

The intensity of the energy states in the conduction range of the NM varies differently from the bulk due to the difference of the electronic structure between it (Kulkarni 2015). When the spacing energy between two levels is more than the $k_B T$ (k or k_B) is the product of Boltzmann constant, and T the temperature), an energy gap is created. Relevant nanoscale parts will produce different sizes on different electronic structures and different energy level separators (Madkour 2019). Ionizing possibilities in NPs are higher than in bulk materials. The most reasons that lead to a large difference in the electrical properties of NM are quantum confinement of electrons with their freedom in movement, quantum size effect, the energy bands between valance and conduction band which lead to determine the electronic transition from level to upper and charge quantization (Viswanathan 2014).

In quantum wire, two dimensions are reduced and one dimension remains large. Therefore, the electrical resistivity of quantum wire can be calculated using the following conventional formula: R = ρ L/A; where ρ: is the resistivity of the conductor, L: is the length of the conductor, and A: is the cross sectional area (Madkour 2019; Viswanathan 2014). Therefore, to solve this equation and calculate the resistance value in quantum wire, it needs to determine the cross section area of the wire (A), which needs two dimension and one of them reaches to zero, the resistance is infinite, this meaning that it is impossible to cross any electron and electron confinement was obtained (Madkour 2019). Figure 4.3 shows the transitions between the valence beam to the conduction band in semiconductors.

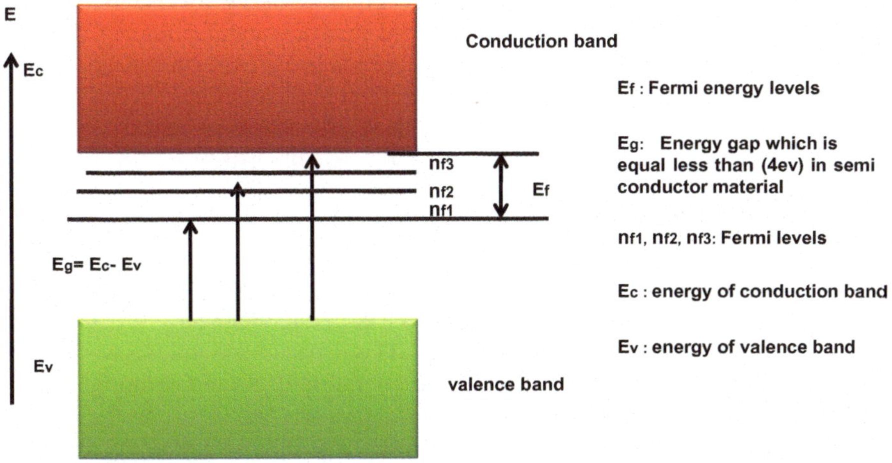

Fig. 4.3 Transitions between valence band to conduction band in semiconduction (Figure constructed by A. K. Almuhammady)

4.5.5 Magnetic Properties

Magnetic properties of NPs are dominated by two main features: finite-size effects (single-domain, multi-domain structures and quantum confinement) and surface effects, which results from the symmetry breaking of the crystal structure at the surface of the particle, oxidation dangling bonds, surface stain (Katz 2020). Surface effects become significant as the particle size decreases because of the ratio of the number of surface atoms to the core atoms increases (Kulkarni 2015). It is well established that several magnetic properties such as magnetic anisotropy, the magnetic moment per atom, curie temperature, and the coercivity field of NPs can be different than those of bulk material (Madkour 2019). The magnetization in the surface decreases faster with increasing temperature than the magnetization in the interior of a particle. The low symmetry around surface atoms can result in a large contribution to the magnetic anisotropy of NPs (Viswanathan 2014). Moreover, the magnetic structure in the surface and around defects in the interior may be influenced by a reduced number of magnetic neighbor atoms, and this can lead to non-collinear spin structures in ferrimagnetic particles (Kulkarni 2015).

Large surface area to volume ratio in magnetic materials develops a substantial proportion of atoms having a different magnetic coupling with neighboring atoms, leading to differing magnetic properties (Madkour 2019). Superparamagnetism is observed in magnetic NPs by which the magnetizations of the particles are randomly oriented and aligned only under an applied magnetic field and the alignment disappears once the external field is withdrawn. This is due to the presence of only one domain in magnetic NPs as compared with the multiple domains of bulk Magnetic (Kulkarni 2015). The charge localized at the particle surface gives rise to ferromagnetic like behavior. This observation indicated that the modifications of the band structure by chemical bonding can develop ferromagnetic like behavior in metallic clusters (Huh et al. 2020; Viswanathan 2014).

4.5.6 Mechanical Properties

Compared with bulk materials, the mechanical properties of NM change greatly. Due to an increase in the number of atoms on the surface, interatomic distance decreases which causes an increase in interatomic force. This rise in interatomic force increases the shearing strength of NM increases (Wu et al. 2020). As the shearing strength increases the young modulus {it is a mechanical property that measures the stiffness of a solid material and can be defined as the relationship between stress (force per unit area) and strain (proportional deformation) in a material in the linear elasticity regime of auniaxial deformation} of nanosolids also increases (Bunaciu et al. 2015). Also, the NPs have the biggest tensile property, less plastic deformation, more fragile and fewer surface defects compared to bulk materials. These causes and changes give better mechanical properties (Kulkarni 2015; Madkour 2019; Mageswari et al. 2016; Viswanathan 2014; Vollath 2008).

4.6 Synthesis Methods of Nanomaterials

4.6.1 Physical Methods for Synthesis Nanomaterials

Physical methods for preparing NM are among the oldest methods used, especially those that rely on manual grinding (Viswanathan 2014). Physical methods apply mechanical pressure, high-energy radiation, thermal energy or electrical energy to cause material corrosion, smelting, evaporation or condensation to generate NPs (Madkour 2019; Mageswari et al. 2016). These physical methods work on a top-down system and beneficial due to free of solvent contamination and produce standardized mono-NPs. Economically and operationally, physical methods are cheap (Madkour 2019). There are several methods including high energy ball milling (Dhand et al. 2015), electron beam lithography (Madkour 2019), inert gas condensation synthesis method (Dhand et al. 2015; Mageswari et al. 2016), physical vapor deposition method (Viswanathan 2014) and laser pyrolysis method (Kulkarni 2015).

4.6.2 Chemical Methods for Synthesis Nanomaterials

Chemical methods depend mainly on chemical reactions that lead to mechanical or thermal forces capable of forming molecules or collecting atoms to produce materials with nanoscale sizes (Omar et al. 2019). The primary chemicals are a mixture of chlorides, oxides and minerals, all of which react through a grinding or heat treatment process to produce a powder in which the ultra-pure particles are dispersed within a stable salt matrix (Mageswari et al. 2016). Using appropriate solvents for each washed compound, to recover these particles from removal selectivity of the matrix, the most important chemical methods are sol-gel method (Boutamart et al. 2020), hydrothermal synthesis (Dhand et al. 2015), polyol synthesis (Kulkarni 2015), micro emulsion technique (Viswanathan 2014) and microwave assisted synthesis (Madkour 2019) (Fig. 4.4).

4.6.3 Green Methods for Synthesis Nanomaterials

The green synthesis of NPs has risen due to the rise of costs and toxicity of physical and chemical methods (Thunugunta and Reddy 2015). Hence, researchers have started using biological molecules in search of cheaper alternatives that act

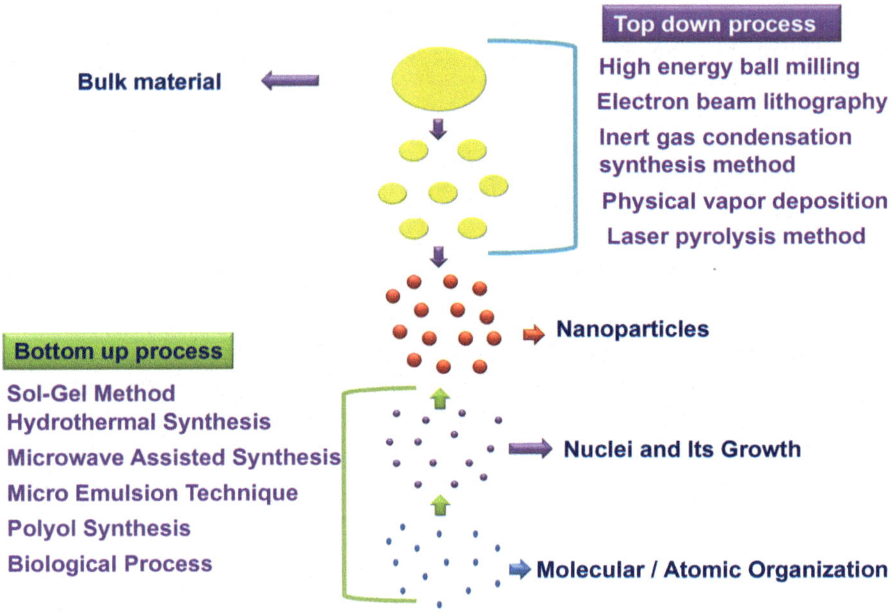

Fig. 4.4 Synthesis methods of NM (physical (top down process), chemical and biological (bottom up process) (Figure constructed by A. K. Almuhammady)

as reducing agents, including microorganisms (Roy et al. 2019), biomolecules (Tsekhmistrenko et al. 2020), plants and plant extracts (Bartolucci et al. 2020). Biomolecules are typically responsible for reducing metals into their respective NPs (Ahmad et al. 2019). The biosynthesis method follows the bottom-up approach and includes either a reduction or oxidation. The success of green synthesis relies on the solvent medium, the environmentally friendly reducing agent and the material used to stabilize non-toxically (Samaira et al. 2020).

4.7 Characterization of Nanoparticles

4.7.1 X-Ray Diffraction (XRD)

X-Ray diffraction (XRD) is a form of non-destructive crystalline material characterization. It provides information on the structure of crystals, phase, preferred direction of crystals (texture) and other structural parameters such as average grain size, crystal, stress and crystal defects (Gour and Jain 2019). X-ray diffraction peaks are created by a constructive overlap of a monochrome X-ray beam. That is reflected at specific angles from each group of capillary planes in a sample. The intensity peak is determined by the distribution of the atoms in a lattice (Andreeva et al. 2011). Consequently, these are expressed from each community of capillary planes in a sample at different angles. Atom distribution in the lattice determines the peak strength (Murray et al. 2010). The XRD pattern thus represents a fingerprint of the atomic arrangements within a material evolved gas analysis (EAG). Multiple X-ray diffraction systems are equipped with interchangeable optical units, according to the requirements of the analysis, without affecting the positioning accuracy (Andreeva et al. 2011). The change is easy between the focus on lines and point of the X-ray source, allowing simple switching from a regular XRD configuration to high-resolution XRD configuration as needed. Various combinations of optical units enable the analysis of powders, coatings, thin films, panels, fabricated parts and top films. The evolved gas analysis also has accurate refract meters with 2D region detectors for small spot XR ($< 50\ \mu m$), providing a good signal noise even with small X-ray beam sizes (Murray et al. 2010). The main applications of XRD analysis (a) Quantification of the crystalline phase. It Measures the average size of crystals, strain, or effects of a partial strain on loose materials and thin films and (b) Determination of the crystalline ratio of amorphous materials in materials and thin film. Quantification of preferred texture in thin films, multilayer piles and fabricated parts (Bunaciu et al. 2015) (Fig. 4.5).

For deflect of the electromagnetic radiation, the spacing must be in the grates must be of order as the wavelength. In crystals the typical inter atomic spacing ~ 2-3 Å so the suitable radiation is X-rays hence, X-rays can be used for the study of crystal structures, neutrons and electrons are also used for diffraction studies from materials, and neutron diffraction is especially useful for studying the magnetic ordering in materials (Horikoshi and Serpone 2013). The diffraction peak position is a product of inter planar spacing, as calculated by Bragg's law: $n\ \lambda = 2d\sin\theta$ when n: Is an integer, λ: The wavelength of incident light, d: Is the inter planar spacing of the crystal and θ: Is the angle of incidence (Goldstein et al. 2003; Kulkarni 2015) (Fig. 4.6).

4.7.2 Transmission Electron Microscopy (TEM)

Transmission electron microscopy (TEM) uses a beam of electrons to examine and test samples, and when the scanning electron microscope examines the surfaces of the samples and characterizes their properties (Gour and Jain 2019). The penetrating microscope is characterized by its ability to penetrate the sample. It is placed in the path of the electron flame coming from the electronic radiation generation source (Fig. 4.7) (Egerton et al. 2004; Goodhew 2011). Electrons are produced by thermal emission, through heating a wick made of mostly tungsten, where an accelerating voltage is applied to this filament ranging between 60–100 kilovolt (KV) (Goodhew 2011). The accelerated electrons have energy controlled by the user as required. The electron beam passes through the vacuum microscope column. This beam is focused on utilizing a group of electromagnetic lenses along this column (Egerton et al. 2004). The control vents along this column also control the width of the electron

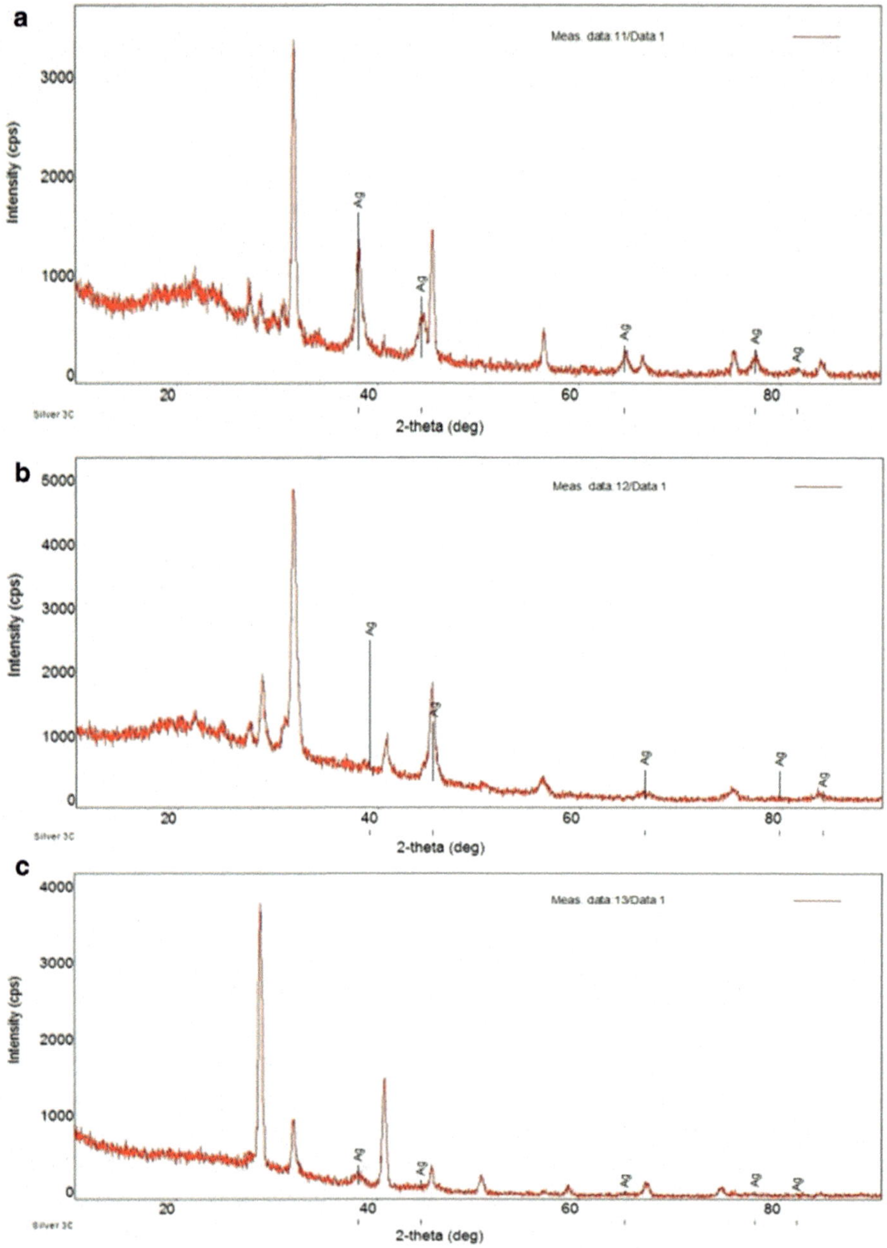

Fig. 4.5 The XRD pattern for a Silver NPs with water hyacinth extract, b Water hyacinth extract only, c Silver NPs with coontail extract and d Coontail extract only. The results showed that the biosynthesis process of conversion of the silver ions to NM turned them partly into a noncrystalline material which gave them an amorphous state mainly (Photo by A. K. Almuhammady and F. Abdulqahar)

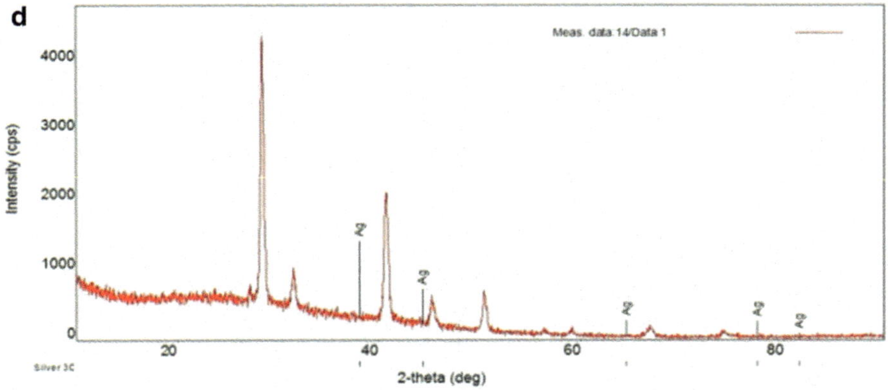

Fig. 4.5 (continued)

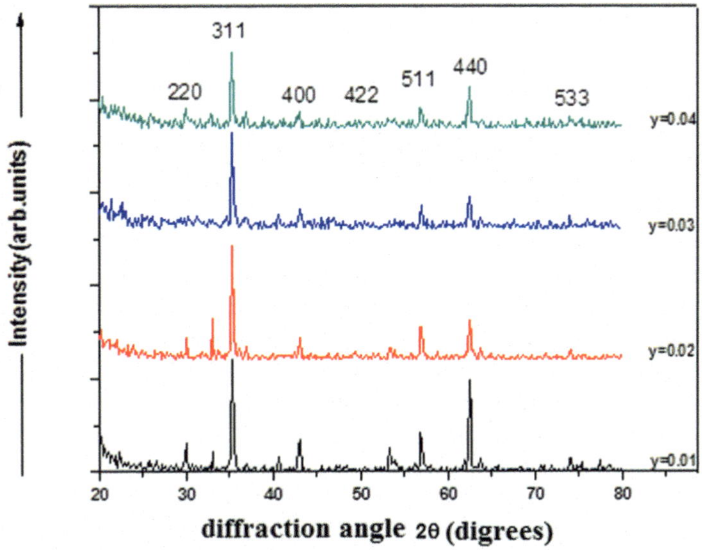

Fig. 4.6 Shows the X-ray diffraction patterns for ($Co_{1+x}Ti_xFe_{2-2x}O_4$) samples where $0 \leq x \leq 0.7$. The patterns show the existence of a single phase cubic spinel ferrites with small secondary phase exist in case of $x = 0.7$. The figure shows strong diffraction from the planes 220, 311, 400, 511 and 440 as well as a weak diffraction from the planes 222, 422, 533 and 620. The data also show that, all planes are characterized by the spinel ferrite and the peak intensity depends on the concentration of magnetic ions in the lattice. The comparison shows that the conversion in the system takes place where a small secondary phase was expected to appear at $x = 0.7$. This means that at this high concentration Ti ion scan not dissolved in the structure completely (Photo by A. K. Almuhammady)

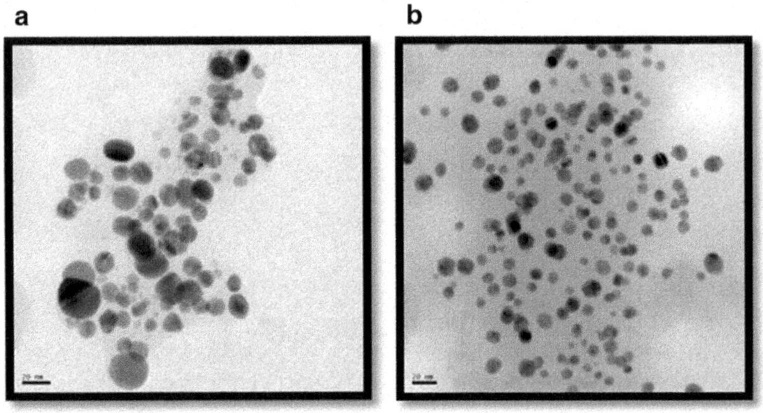

Fig. 4.7 Transmission electron microscopy images. **a** Silver NPs with water hyacinth extract, **b** Silver NPs with coontail extract (Photos by A. K. Almuhammady and F. Abdulqahar)

beam, by blocking the separated electrons. Then electron beam reaches the sample and this results in an interaction of these electrons with the surface of the sample, where a portion of the falling beam called the effective beam is executed, which is a valid electronic beam without deviation, and electronic and dispersed and deviated electronic bundles of the sample atoms and molecules (Brydson 2011). Electrons are produced by thermal emission, which leads to heat a fuse made mostly of tungsten, where an acceleration voltage is applied to this filament ranging between 60–100 kV (Horikoshi and Serpone 2013). Electromagnetic lenses and fluorescent screen control as the picture. The image contains dark and bright areas depending on the type of sample (Goldstein et al. 2003). The dark areas indicate that the electrons did not reach the screen from these regions and this occurs as a result of their absorption from the atoms of these regions or the large scattering, which indicates that the electrons did not suffer from any absorption or significant dispersion of the atoms of these regions. This mean that the sample in these seemingly light regions is contained elements of light atoms (small atomic numbers) (Goodhew 2011).

The amount and scale of the information which can be extracted by TEM depend critically on four parameters: The resolving power of the microscope (usually smaller than 0.3 nm) (Juh 2007), the energy spread of the electron beam (often several eV) (Kulkarni 2015), the thickness of the specimen (almost always significantly less than 1 μm) and the composition and stability of the specimen (Vollath 2008). The first and second of these depend largely on the depth of pocket so the more spend then the better of microscope parameters. The third is usually determined by experimental skill while the last depends on the choice of a suitable experimental system (Juh 2007; Vollath 2008).

4.7.3 Scanning Electron Microscope (SEM)

It is used in analyzing and specifying the properties of the thick or thin sample surfaces of the material, knowing its shape and determining its external dimensions. Its magnification strength reaches half a million times with good accuracy. This microscope can determine the elements involved in the sample composition and its relation (Fig. 4.8) (Chen et al. 2004). The scanning electron microscope works with the production of the following steps of electrons using thermal emission. This is done using a heating primer usually made of tungsten and an acceleration voltage of varying value 0.1–30 kV is applied to them. Then the electron beam passes through the vacuum microscope column (Orłowska et al. 2020). This package is focused on a set of lenses electromagnetism along the column. The width of the electron beam is controlled by the holes located along the microscope column, where the distracted and deviated electrons are trapped from the path of the beam. The sample is placed inside the microscope room, which is a completely closed and empty space (Goldstein et al. 2003). The electron beam collides with the sample. These interactions are translated to signals. The most important is the secondary electron emission signal (SE) and the emission of the back-dispersed electrons (BSE). The signals are analyzed, processed and shown as X-ray images and signals that are translated into an analytical spectrum

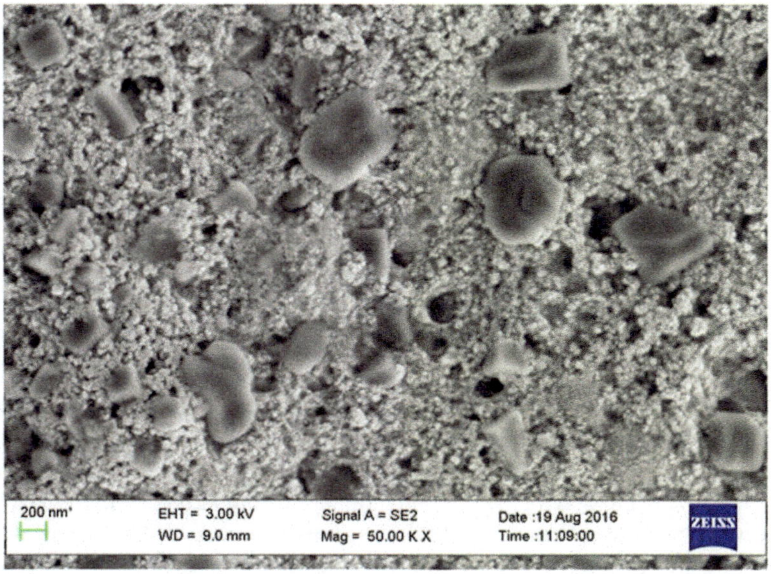

Fig. 4.8 Scanning electron microscope image for silver NPs biosynthesis using Iraq aquatic weeds extract water hyacinth. This figure shows a nano-silver compound prepared in biological method and it is clear that the compound is not homogeneous, as the nano forms the largest part, but some particles appeared in larger dimensions. This may be due to the presence of other compounds that are not silver that belong to the plant compounds itself (Photo by A. K. Almuhammady and F. Abdulqahar)

(Chen et al. 2004; Liu et al. 2005). The microscope of the scan is suitable tool for understanding and analyzing the morphology of micro-structure and characterizing the chemical composition. The naked eye distinguishes objects that curve around 1/60° optical angle, which correspond to an accuracy of ~ 0.1 mm (at the optimal viewing distance of cm) (Horikoshi and Serpone 2013). The optical microscope has an accuracy limit of ~ 2000 Å by magnification using a magnifying optical lens. Light microscopy is popular and still prefer in the scientific research. Practical experiments demonstrated the ability to deviate from the magnet field. For development of the electronic microscope it happens by replaced an electronic high power kit in the light source (Kulkarni 2015; Zhou et al. 2006).

4.7.4 Energy Dispersion Spectroscopy

The X-ray energy dispersion spectroscopy (EDS or energy-dispersion X-ray spectroscopy) is an analytical technique used to analyze the sample chemical properties and is a type of X-ray spectroscopy (Andreeva et al. 2011). The principle of this technique is based on the fact that X-rays, which result from the mutual effect between charged particles such as a beam of electrons with the sample material are distinct to the corresponding elements in the sample, thus the structure can be known. In other words, since each element has its distinct atomic structure. It has a set of distinct peaks in the X-ray spectrum (Goldstein et al. 2003).To obtain the distinctive X-rays of matter, the atoms must first become irritated. This happens by throwing matter with a beam of electrons, as in a scanning electron microscope or with a beam of X-rays, as in the X-ray brilliance. As a result, an electron is released from the internal atomic orbitals. Then the excitation and instability occur as a result of an electronic vacancy. Which is filled from higher atomic orbitals (Huang et al. 2020). When electrons travel from the highest atomic orbitals to the lowest. They release X-rays that have energy corresponding to the energy difference between the atomic orbitals. The difference in energy is characteristic of every chemical element. Each element has several permissible transitions between orbitals (Goldstein et al. 2003). These transitions, which are described as quantum transfers (from quantum chemistry) are denoted by ($K\alpha$, $K\beta$, $L\alpha$) (Andreeva et al. 2011).The detector measures the energy of the resulting X-ray photons. When the photon detector absorbs within the sensitive region. This results in a proportional number of electrons. An amplification of which occurs so that we obtain a quantum standard (Günther et al. 2019). This uses an index to give a value that appears as an energy value on the x-axis in the resulting spectrum. The detector's precision value is between 120–140 electron volts. The detector consists of many types of semiconductors, for example, silicon or germanium (Goldstein et al. 2003).

4.7.5 Fourier Transform Infrared (FTIR) Spectrometer

Fourier transform infrared (FTIR) spectrometry is a technique used to obtain an infrared spectrum to absorb or emit a solid, liquid or gaseous substance. At the same time, the Fourier spectrum gathers high-resolution spectral data over a wide spectral range. This provides a major advantage over the spectrum dispersion scale, which measures the intensity over a narrow range of wavelengths simultaneously (Gour and Jain 2019).The term FTIR comes from the fact that in order to convert the primary data into the actual spectrum, the mathematical process of the Fourier transform must be performed (Breton 2001). The principle underlying spectroscopy (UV-VIS, FTIR), it is the determination of the amount of absorption of a specific wavelength by the sample. The most obvious way of doing this, the technique of scattering spectroscopy, is to project a monochromatic beam of light onto a sample, measure the amount of light absorption and repeat each different wavelength (Kumar 2006).

Fourier spectroscopy is a less easy way to obtain the same information. Instead of shining a beam of monochromatic light (a beam consisting of only one wavelength) in the sample, this technique brightens a beam that has many light frequencies simultaneously and measures the amount of absorption that the beam absorbs from the sample. The beam is then adjusted to contain a different set of frequencies, which gives a second data point. This is repeated many times. Next, the computer takes all this data and calculations can be made to calculate absorption at each wavelength (Chu et al. 2004). The FTIR is an effective analytical technique to quickly determine the chemical composition to determine accurately and quickly, the chemical family of substance. Usually, organic and polymeric compounds (and to a lesser extent inorganic compounds) are produced the fingerprint infrared spectrum, which can be compared to the EAG comprehensive reference database and can identify the chemical family of the unknown component or actual identity (Tsekhmistrenko et al. 2020). Fingerprint infrared spectrum, which can be compared to the EAG comprehensive reference database and can identify the chemical family of the unknown component or actual identity. Fourier infrared spectrometer measures the absorption of infrared light by a sample and generates a spectrum dependent on the functional groups of the material (Gour and Jain 2019). In addition to typical sample preparation methods (such as micro-extraction, dilution, potassium bromide (KBr) packages and reflection methods), EAG also uses several accessories total reflection (ATR), which allow examination of insoluble or multilayer samples directly. The objective of any absorption spectrometer FTIR is to measure the extent to which the sample absorbs light at each wavelength. The most obvious way to do this, the scatter spectroscopy technique, is to project a monochrome light beam onto a sample, measure the amount of light absorption and repeat each different wavelength (Chu et al. 2004).

This spectroscopy is concerned with electronic transitions from the bottom state to the excited state and thus prepares as a complement to the fluorescent spectroscopy, which studies the fluorescence resulting from the transition from the excited state to the bottom state (Breton 2001). In the spectroscopy of visible and ultraviolet rays, molecules are exposed to electromagnetic radiation in the visible and ultraviolet fields, which leads to irritation and excitation of valence electrons (such as p or d electrons in outer orbits), that is, they acquire energy and electron transport occurs within the energy levels of the molecule. In this transition, the energy difference between the levels at which the transition was made must be consistent with the amount of energy. The photon absorbed as a result of the transition (Breton 2001; Chu et al. 2004) (Figs. 4.9, 4.10 and 4.11).

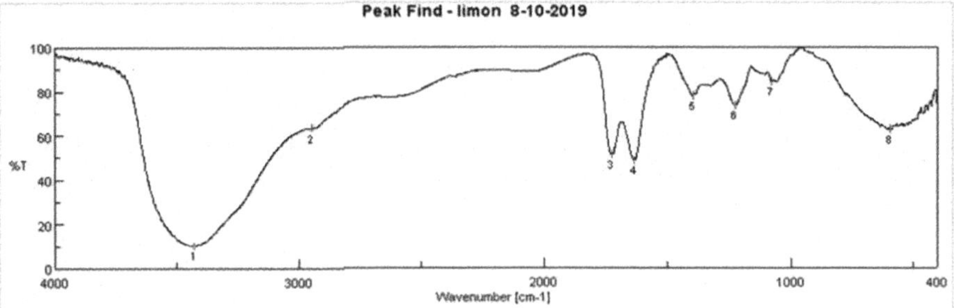

[Result of Peak Picking]

No.	Position	Intensity	No.	Position	Intensity	No.	Position	Intensity	No.	Position	Intensity
1	3428.81	9.9	2	2948.63	63.3	3	1725.01	50.9	4	1636.3	48.3
5	1399.1	77.5	6	1227.47	73.3	7	1078.98	84.3	8	592.039	62.5

Fig. 4.9 Fourier infrared spectrometer for lemon before added silver NPs. The peak picking at the number (3,4) at a position of 1725 cm^{-1} has existed only in lemon (Photo by L. M. Alnaddaf)

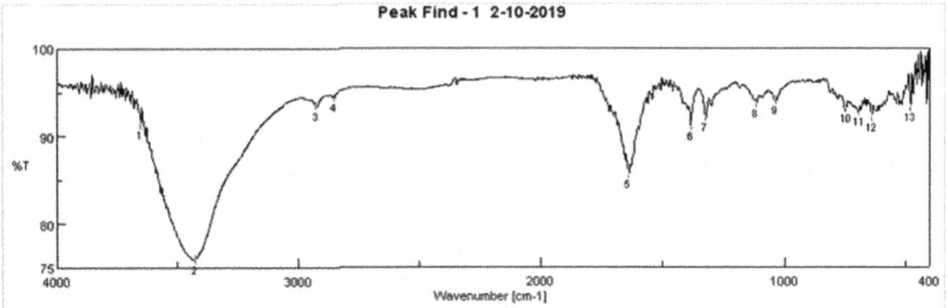

[Result of Peak Picking]

No.	Position	Intensity	No.	Position	Intensity	No.	Position	Intensity	No.	Position	Intensity
1	3650.59	91.3	2	3424.96	75.9	3	2926.45	93.4	4	2854.13	94.4
5	1637.27	85.5	6	1384.64	91.0	7	1324.86	92.0	8	1115.62	93.6
9	1035.59	94.0	10	743.424	93.0	11	691.355	92.7	12	641.215	92.1
13	476.331	93.2									

Fig. 4.10 Fourier infrared spectrometer of silver NPs obtained on (4:1) lemon juice: silver nitrate (10 Mm $AgNO_3$).The FTIR analysis showed the presence of bands due to O–H stretching (around 3434 cm^{-1}), CH stretching (around 2930 cm^{-1}) and C–O stretch (around 1125 cm^{-1}). The peaks at 1384 and 1324 cm^{-1} were corresponding to various functional groups (Photo by L. M. Alnaddaf)

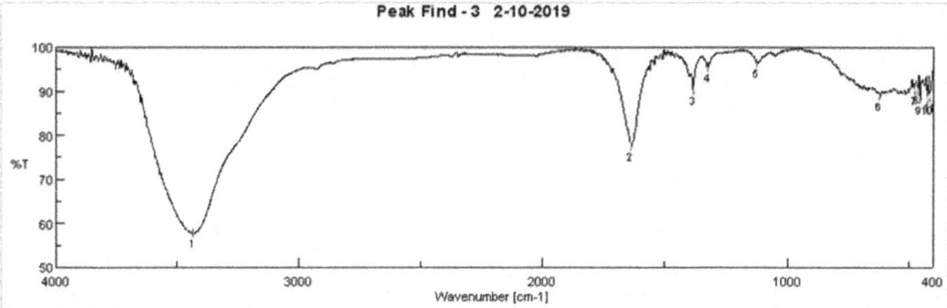

[Result of Peak Picking]

No.	Position	Intensity	No.	Position	Intensity	No.	Position	Intensity	No.	Position	Intensity
1	3434.6	57.7	2	1637.27	77.3	3	1384.64	90.1	4	1325.82	95.2
5	1124.3	96.2	6	619.038	88.8	7	474.403	90.2	8	463.796	90.2
9	453.19	87.9	10	421.37	87.9	11	412.692	88.6			

Fig. 4.11 Fourier infrared spectrometer of silver NPs obtained on 1:4 lemon juice: silver nitrate (10 Mm $AgNO_3$). The FTIR analysis showed the presence of the C = C group (around 1600 cm^{-1}) (Photo by L. M. Alnaddaf)

4.7.6 Ultraviolet-Visible (UV/Vis) Spectroscopy

UV/Vis spectroscopy is a type of spectroscopy categorized under absorption spectroscopy that occurs both in the UV spectrum and in the spectrum visible. In other words, this method of study uses light in a broad range, beginning from ultraviolet radiation in the visible region to areas of the near-infrared spectrum (NIR) (Gour and Jain 2019). The absorption or reflection in the visible field influences the sense of color seen in chemicals, where electronic changes occur as a result of the effect of electromagnetic radiation (Fig. 4.12). UV/Vis spectroscopy has become a common tool used every day in many laboratories in the life sciences (Mohammed 2018). This is mainly due to its simplicity, it does not require complicated sample planning, easy to implement and in seconds results are obtained. Typical measurement requires only a small amount of the sample (Gour and Jain 2019). As it is considered a non-destructive method, samples may be used for the following analyses the life sciences field applies UV/Vis spectroscopy to nucleic acids, proteins and bacterial cell culture. The life sciences field applies UV/Vis spectroscopy to nucleic acids, proteins and bacterial cell culture (Mohammed 2018). The UV/Vis rays of life sciences deepen in our brief guide to the most popular ones: (a) Measurement of the concentration of nucleic acids (DNA and RNA) and (b) Measuring the concentration of proteins through direct measurements or color assays, studying enzymatic reactions and controlling growth curves in bacterial cell emulsions (Kumar 2006).

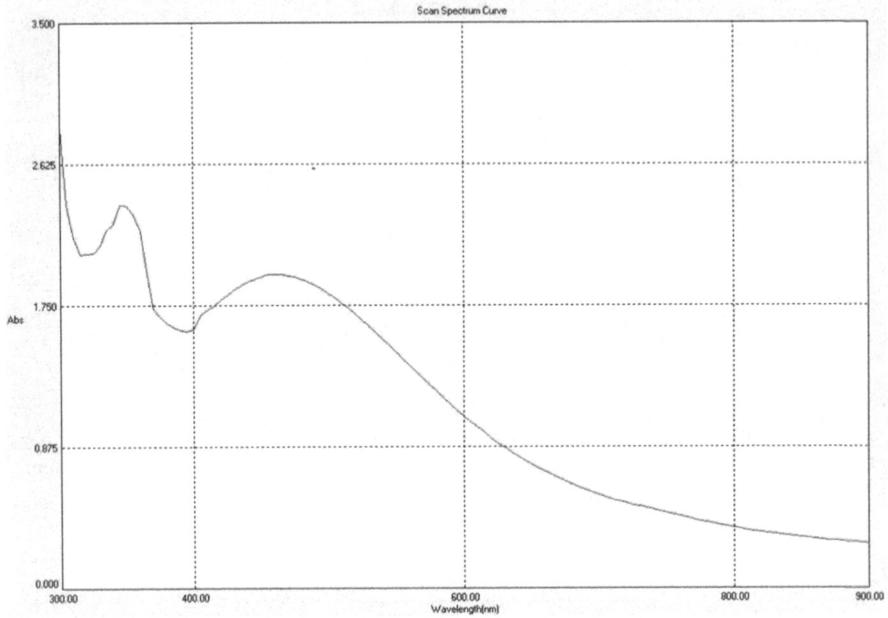

Fig. 4.12 Ultraviolet-visible spectroscopy of silver NPs obtained on 4:1 lemon juice: silver nitrate (10 Mm $AgNO_3$) absorption peak 460 nm (Photo by L. M. Alnaddaf)

4.8 Conclusion and Prospects

This chapter introduced various methods for synthesizing and manufacturing NPs with different properties and explained several tools to determine the shapes and sizes of NPs used in many applications. The size and forms of NPs can be expected according to the preparation factors, reaction conditions, materials used and their concentration. So more research and studies still require to arrive at the size, shape, quality and the actual cost of production, type of product and its specifications in terms of structure. The synthesis needs to be further explored and optimized which can help to develop economically viable technologies. Chemical and physical methods of NPs synthesis cannot be neglected as each method has its importance both in determining the type and shape of NM to the appropriate application for it. The green synthesis of NPs is one of the modern fields in preparing NPs to use organic molecules as reducing agents. Recent studies tend to define strategies for the scalable production of NPs either from plants or using microorganisms that have multiple effects and can be used in agriculture, water waste treatment, engineering, medicine and food industries. Considerable results have been achieved in some NM studies. Some NM were applied in many fields. However, related work on the molding mechanism and process reinforcement of NM microstructure are still relatively little, and many areas still need to be searched. Nanomaterial,s unique properties endow them with broad prospects for application and enormous potential future value. Therefore, it must continue to investigate NM and deepen our understanding of their molding mechanism, process reinforcement, and modification methods to improve their properties. Despite the fact that NM show different positive results in their use, their toxicity remains a concern. This toxicity depends on different aspects, mainly on the concentration of NM and their size and form. However, the exact mechanism of NM interaction is not clear yet. Thus, these interactions could be a prospect for future research.

References

Ahmad S, Munir S, Zeb N et al (2019) Green nanotechnology: a review on green synthesis of silver nanoparticles-an ecofriendly approach. Inter J Nanomedicine 14:5087–5107

Andreeva P, Stoilov V, Petrov O (2011) Application of X-Ray diffraction analysis for sedimentological investigation of Middle Devonian dolomites from Northeastern Bulgaria. GeologicaBalc 40(1–3):31–38

Bartolucci C, Antonacci A, Arduini F et al (2020) Green nanomaterials fostering agrifood sustainability. TrAC Trend Anal Chem 125: https://doi.org/10.1016/j.trac.2020.115840

Boutamart M, Briche S, Nouneh K et al (2020) Transparent and self-cleaning surfaces based on nanocomposite sol-gel coatings. ChemistrySelect 5(28):8522–8531. https://doi.org/10.1002/slct.202002027

Breton J (2001) Fourier transform infrared spectroscopy of primary electron donors in type I photosynthetic reaction centers. Biochim et BiophysicaActa (BBA)-Bioenerg 1507:180–193. https://doi.org/10.1016/s0005-2728(01)00206-7

Brydson R (2011) Aberration-corrected analytical transmission electron microscopy. Wiley, p 280 https://doi.org/10.1002/9781119978848

Bunaciu AA, UdriştioiuEg, Aboul-Enein HY (2015) X-Ray diffraction: instrumentation and applications, critical reviews in analytical chemistry, 45(4):289–299. https://doi.org/10.1080/10408347.2014.949616

Chen Y, Campbell L, Zhou W (2004) Self-catalytic branch growth of SnO2 nanowire junctions. MRS Proceedings Library Archive 818. https://doi.org/10.1557/PROC-818-M5.36.1

Chu HA, Hillier W, Debus RJ (2004) Evidence that the C-terminus of the D1 polypeptide of photosystem II is ligated to the manganese ion that undergoes oxidation during the S1 to S2 transition: an isotope-edited FTIR study. Biochem 43:3152–3166

Dhand C, Dwivedi N, Jun Loh X et al (2015) Methods and strategies for the synthesis of diverse nanoparticles and their applications: a comprehensive overview. RSC Adv 5:105003–105037. https://doi.org/10.1039/c5ra19388e

Ealias AM, Saravanakumar MP (2017) A review on the classification, characterisation, synthesis of nanoparticles and their application. IOP Conf Ser Mater Sci Eng 263: https://doi.org/10.1088/1757-899X/263/3/032019

Egerton RF, Li P, Malac M (2004) Radiation damage in the TEM and SEM. Micron 35:399–409. https://doi.org/10.1016/j.micron.2004.02.003

Goldstein J, Newbury D, Joy D et al (2003) Scanning electron microscopy and X-ray microanalysis. Springer, pp 1–270. ISBN 978-1-4615-0215-9

Goodhew P (2011) General introduction to transmission electron microscopy (TEM). In: Brydson R (ed) Aberration corrected analytical transmission electron microscopy, First edition, Wiley, pp 1–19. https://doi.org/10.1002/9781119978848.ch1

Gour A, Jain NK (2019) Advances in green synthesis of nanoparticles. Artif Cells Nanomed Biotechnol 47:844–851. https://doi.org/10.1080/21691401.2019.1577878

Günther B, Dierolf M, Achterhold K, Pfeiffer F (2019) Device for source position stabilization and beam parameter monitoring at inverse Compton x-ray sources. J. Synchrotron Rad 26(5):1546–1553. ISSN 1600-5775

Horikoshi S, Serpone N (2013) Microwaves in nanoparticle synthesis: fundamentals and applications. Wiley-VCH Verlag GmbH & Co. KGaA, Düsseldorf, Germany, pp 352 ISBN: 978-3-527-33197-0 352

Huang J, Günther B, Achterhold K et al (2020) Energy-dispersive X-ray absorption spectroscopy with an inverse compton source. Sci Rep 10:8772. https://doi.org/10.1038/s41598-020-65225-4

Huh S, Song KH, Yang HJ et al (2020) Solvothermal synthesis and interfacial magnetic interaction of β-fese/srtio3-x nanocomposites. ChemSel 5(30):9517–9522. https://doi.org/10.1002/slct.202000871

Juh TL (2007) Physical properties of nanomaterials. Encyclopedia of nanoscience and nanotechnology, pp 1–46. ISBN: 1-58883-058-6/$35.00

Katz E (2020) Magnetic nanoparticles. Magnetochemistry, pp 1–6 https://doi.org/10.3390/magnetochemistry6010006

Kulkarni SK (2015) Nanotechnology: principles and practices. Springer Nature Switzerland AG, pp 379 https://doi.org/10.1007/978-3-319-09171-6

Kumar S (2006) Spectroscopy of organic compounds. New Age International Pvt Ltd Publishers, New Delhi

Liu JJ, Yu MH, Zhou WL (2005) Well-aligned Mn-doped ZnO nanowires synthesized by a chemical vapor deposition method. Appl Phys Lett 87(17): https://doi.org/10.1063/1.2084321

Madkour LH (2019) Nanoelectronic materials fundamentals and applications. Advanced structured materials. Springer, Cham. https://doi.org/10.1007/978-3-03021621-4-10

Mageswari A, Srinivasan R, Subramanian P et al (2016) Nanomaterials: classification, biological synthesis and characterization. Nanoscience in food and agriculture, vol 3, Springer, pp 31–71 https://doi.org/10.1007/978-3-319-48009-12

Mohammed AM (2018) UV-visible spectrophotometric method and validation of organic compounds. EJERS 3(3):8–11. https://doi.org/10.24018/ejers.2018.3.3.622

Murray CE, Ying AJ, Polvino SM et al (2010) Nanoscale strain characterization in microelectronic materials using X-ray diffraction. Powder Diff 25(2):108–113

Omar R, Afreen S, Talreja N et al (2019) Impact of nanomaterials on the microbial system. In: Prasad R (ed) Microbial nanobionics: Nanotechnology in the Life Sciences. Springer, Nature, Switzerland AG, pp 141–158

Orłowska K, Mognaschi ME, Kwoka K et al (2020) A method of magnetic field measurement in a scanning electron microscope using a microcantilever magnetometer. MetrolMeasSyst 27:141–149

Pokropivny V (2007) Introduction to nanomaterials and nanotechnology. Tartu University, Tallinn University, frantsevich institute for problems of materials science of NASU, pp 191. ISBN: 978–9949–11–741–3

Roy A, Bulut O, Some S et al (2019) Green synthesis of silver nanoparticles: biomolecule-nanoparticle organizations targeting antimicrobial activity. RSC Adv 9:2673–2702

Samaira Y, Shazia N, Haq NB et al (2020) Green synthesis, characterization and photocatalytic applications of silver nanoparticles using Diospyros lotus. Green Process Synth 9:87–96. https://doi.org/10.1515/gps-2020-0010

Sethi SK, Kadian S, Goel A et al (2020) Fabrication and analysis of zno quantum dots based easy clean coating: a combined theoretical and experimental investigation. ChemSel 5(29):8942–8950. https://doi.org/10.1002/slct.202001092

Shao S, Wang P, Gong Z et al (2020) Cobalt oxide grown on biomass carbon as a three-dimensional self-supporting negative electrode with high area specific capacit. ChemSel 5(29):8881–9312. https://doi.org/10.1002/slct.202001680

Thunugunta T, Reddy Reddy L (2015) Green synthesis of nanoparticles: current prospectus. Nanotechnol Rev 4(4):303–323

Tsekhmistrenko S, Bityutskii V, Tsekhmistrenko O et al (2020) Bacterial synthesis of nanoparticles: a green approach. Biosyst Divers 28(1):9–17. https://doi.org/10.15421/012002

Tshabalala ZP, Oosthuizen DN, Swart HC, Motaung DE (2020) Tools and techniques for characterization and evaluation of nanosensors. In: Han B, Tomer V, Nguyen TA, Farmani A, Singh PK (eds) Nanosensors for smart cities. Micro and Nano Tech, pp 85–110. ISBN: 9780128198704

Viswanathan B (2014) Nanomaterials. Narosa Publishing House, pp 250. ISBN: 978-81-7319-936-3

Vollath D (2008) Nanomaterials an introduction to synthesis, properties and application. Environ EngManag J 7(6):865–870

Wu Q, Ws Miao, Zhang Yd et al (2020) Mechanical properties of nanomaterials: a review. Nanotechnol Rev 9:259–273. https://doi.org/10.1515/ntrev-2020-0021

Yaqoob AA, Parveen T, Umar K, Ibrahim MNM (2020) Role of nanomaterials in the treatment of wastewater: a review. Mdpi water 495:1–30. https://doi.org/10.3390/w12020495

Zhang JZ (2009) Optical properties and spectroscopy of nanomaterials. World Scientific, pp 400. https://doi.org/10.1142/7093

Zhou W, Apkarian R, Wang ZL, Joy D (2006) Fundamentals of scanning electron microscopy (SEM). In: Zhou W, Wang ZL (eds) Scanning microscopy for nanotechnology. Springer, New York, NY. https://doi.org/10.1007/978-0-387-39620-01

Chapter 5
Nanotechnology in Agriculture

Ratna Kalita, Oliva Saha, Nasrin Rahman, Shalini Tiwari, and Munmi Phukon

Abstract Nanotechnology has a very significant role in the field of agriculture. The global requirement of food is increasing whereas traditional farming techniques have failed to increase the productivity and are unable to repair ecosystems damage caused by existing farming techniques. Thus, nanotechnology has been a boon to the society with broad range of opportunities and advantages in agriculture and in our daily life. Nanotechnology can be implemented in agriculture through the use of nano-fertilizers for increasing efficiency of nutrient uptake, and nano-pesticides for controlling pest and pathogen. This chapter provides information on the recent advancements in nanoscience research in agriculture, application of nanoformulations in controlling plant diseases, and microorganisms-based biosynthesis of nanoparticles. Present chapter also provides a brief idea of nano-sensors types, different nano-based smart delivery systems, use of nanoparticles in recycling of agricultural waste, use of nanotechnology in crop biotechnology, and use of nanotech for development in agricultural sector.

Keywords Agriculture · Nano-fertilizers · Nanoparticles · Nanotechnology

5.1 Introduction

Agriculture is the major occupational pillar of most of the developing countries. It has been predicted that our earth has to shelter around 9 billion people by 2050. This speedy population growth will cause a serious impact in water, food and energy

R. Kalita (✉) · O. Saha · N. Rahman
Department of Agricultural Biotechnology, Assam Agricultural University, Jorhat 13, Assam, India
e-mail: ratna.kalita@aau.ac.in

S. Tiwari
CSIR-National Botanical Research Institute, Rana Pratap Marg, Lucknow 226001, India

M. Phukon
Cruickshank Building Institute of Biotechnology & Environmental Science, University of Aberdeen, Aberdeen AB24 3UU, UK

J. M. Al-Khayri et al. (eds.), *Nanobiotechnology*,
https://doi.org/10.1007/978-3-030-73606-4_5

supply (Marchiol 2018). Therefore, there is an immediate need of sustainable intensification to increase crop production with an eco-friendly manner in existing cultivable land. Exploitation of limited natural resources such as depletion of land, water and soil, with the rapid increase in global population, demands development in agriculture to be economically feasible and environment-friendly. Therefore, novel strategies need to be taken for agricultural development as a whole (Singh et al. 2015; Prasad et al. 2014). Recently, promising applications of nanotechnology has been proposed in agriculture as potential technology of the twenty-first century that paved new lanes of research in academics as well as in industries (Dasgupta et al. 2015). Nanotechnology in agriculture has emerged as a multidisciplinary science that holds quite significance in agricultural practices in areas namely disease detection, management, nutrient delivery system and tissue engineering (Srilatha 2011). It deals with the minute possible materials, which facilitate improving agricultural productivity by enhancing management as well as conservation of agricultural inputs (Abobatta 2018). Nanomaterials are natural or artificially synthesized particles, with the size range of approximately 1–100 nm, including nanoparticles (NPs), nanocapsules and nanocrystals which has the potential to modernize agricultural practices. It increases bioavailability of pesticides, insecticides and fertilizers, thus increasing crop productivity. Therefore, implementation of nanotechnology in agriculture will sustain an outstanding effect in the areas of disease diagnosis, disease management, nutrient delivery, methodologies in design and development of product, and instrumentation of food safety and bio-security. Thus, this technology will assist to modernize agricultural practices at global scale by offering various techniques that will reduce crop production cost, increase crop productivity and maintain ecological balance. Moreover, the uniqueness in the properties of nanomaterials makes them highly desirable for designing and development of novel nano-technological tools in favor of agriculture sustainability.

This chapter addresses how NPs are synthesized, their classifications and applications in agriculture. The constructive and destructive approaches for synthesis of different classes of NPs. The use of microorganisms for preparation of NPs is also described. Nanotechnology has proved many implications in crop improvement research namely, for their use in seed germination, and as nano-fertilizers, nano-pesticides, nano-herbicide and nano-biosensors. Different smart delivery systems for the nanoparticles such as nanoformulation, nanoemulsion and nano-encapsulation which are also discussed.

5.2 Approaches for Synthesis of NPs

NPs are different from that of the bulk particles in its physico-chemical and biological characteristics. These novel properties of nanoparticle, thus act as a link between the bulk and atomic or molecular structures (Khandelwal and Joshi 2018). NPs could be synthesized with two approaches (Fig. 5.1), (a) Bottom-up or constructive approach: builds up material from atom to clusters of nano-size particles by

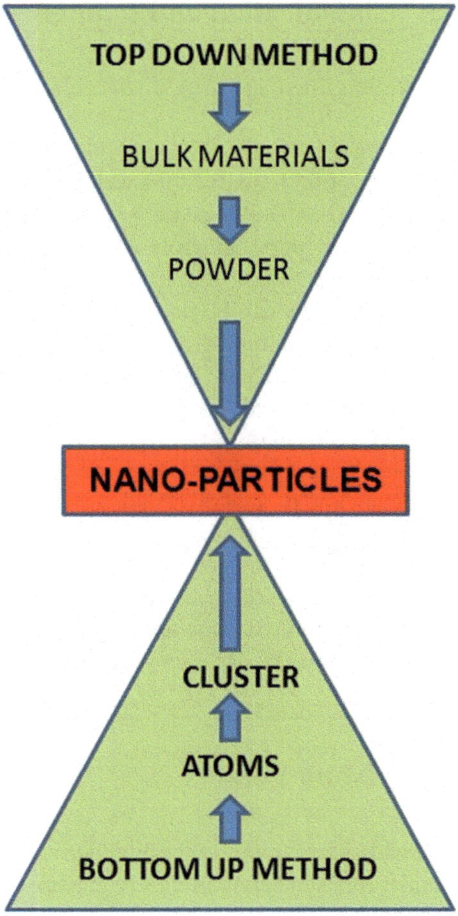

Fig. 5.1 Approaches for synthesis of NPs (Figure constructed by Ratna Kalita)

spinning, chemical vapour deposition (CVD), sol-gel, pyrolysis and biosynthesis; and (b) Top-down or destructive approach: reduces bulk material to powder and then to nano-scale particles by mechanical milling, laser ablation, nanolithography, sputtering and thermal decomposition (Ealias and Saravanakumar 2017; Khandelwal and Joshi 2018). However, in terms of the objects size, both methods are approximately similar as both approaches tend to converge in similar object size range. Though, the former approach, tends to be more extensively used due to the material quality, varieties of design, and nanometric control, whereas the latter approach put more emphasis on the procurement of materials, and control might not be as strong.

5.3 Classification and Examples of NPs

NPs are classified into three distinct classes on the basis of their source and synthesis, (a) Organic-based NPs (frequently biodegradable), (b) Inorganic-based NPs (non-biodegradable), and (c) Hybrid NPs. The most widely used nanoparticles in agriculture are organic-based nanoparticles. This group comprises of starch, lignin, lactalbumin, chitosan, cellulose derivatives, lecithin, phospholipids, alginates, propylene-glycol, polylactides, and polysorbate. The active ingredients are encapsulated in the organic-based NPs which facilitates controlled release of the nanocarriers. Natural and organic NPs can be produced by the host such as microorganisms, plants and animals or from natural processes like forest fires and volcanic eruptions. On the other hand, inorganic NPs are non-toxic, biocompatible, and are more stable than organic NPs are mainly used in pharmaceutical industries. Metallic inorganic nanoparticles including gold, silver, iron, zinc and silica nanoparticles are widely used nowadays. However, several studies also reported their role in the field of agricultural crop production and protection (Pandey et al. 2019). Hybrid nanoparticles are constructed via combining both organic and inorganic nanoparticles to overcome the limits of both the types of NPs and to achieve novel multifunctional properties. Hybrid nanostructures such as yolk-shell, core shell and dot-in-nonotubes are extremely important in pushing their promising applications forward (Ma 2019). Following sub-headings are the examples of commonly known nanoparticles used in agricultural system.

5.3.1 Nano Silver

Its broad-spectrum antimicrobial activity makes it one of the most researched and extensively used nanoparticles to prevent various plants diseases. Antifungal activity of colloidal nano silver solution (average 1.5 nm in diameter) has proved to be highly effective in prevention of powdery mildew of rose caused by infection of *Sphaerotheca pannosa*. In addition to its antifungal property, it is a potent growth regulator for crops, and extends the post-harvest longevity or shelf life of cut flowers and different ornamental foliages (Byczyńska 2017).

5.3.2 Nano Alumino-Silicate

One of the efficient formulations of pesticide at nano scale are the alumino-silicate nano tubes. When nano alumino-silicate tubes containing active ingredients are sprinkled on surfaces of leaves, these can be easily pick up by insects. The insects eat up pesticide-filled nano tubes by grooming actively on them. These pesticides, by nature, are biologically active and environment friendly (Sharon et al. 2010).

5.3.3 *Titanium Dioxide NPs (nTio2)*

One of the mostly used metal-based nanomaterials (MBN) is TiO_2. TiO_2 photocatalyst technique is largely used in protection of plants due to its non-toxic and non-lethal properties and it has great disinfection efficiency for pathogen (Yao et al. 2009). Under normal and stressed soil conditions, $nTiO_2$ enhances plant growth and accumulate photosynthetic pigments (Latef et al. 2017). In barley, $nTiO_2$ treatment greatly affects the concentration of amino acids, and also it has influences on the food chain (Mattiello and Marchiol 2017).

5.3.4 *Carbon NPs*

Carbon NPs are a form of carbon resembling two-dimensional graphene sheet rolling into a tube, forming single-walled (SWCNTs) and multi-walled carbon nanotubes (MWCNTs). SWCNTs possess a notable position among various engineered nanomaterials in various nano-biotechnology applications. Carbon Nanotubes (CNTs) function as a vehicle in delivering the necessary molecules into targeted seeds during germination to protect them against various plant diseases. They are expected to be non-toxic since they are mere growth promoting particles (Gandhi et al. 2010).

5.3.5 *Magnetic NPs*

This type of NPs is employed for site-targeted drug delivery. Thus, it can be used to treat site-specific disease in a plant through site-targeted delivery system. Tracking of internalized magnetic NPs would make it possible to relocate them to specific area from where the chemicals are designed to be released (Jurgons et al. 2006).

5.4 Biogenic/Green NPs

Microorganisms such as virus, bacteria and fungi are used to synthesize nanoparticles (also called green NPs). Use of fungi to prepare NPs has become prevalent due to easy recovery and purification, easy function and their potentiality to secrete immense amount of enzymes. Fungi like *Verticillium* sp., *Phanerochaete chrysoparium, Fusarium oxysporum, Aspergillus furnigatus* and *Aspergillus flavus* were found to be very efficient to synthesize metal and metal sulphide containing NPs (Shang et al. 2019). The separation of gold and silver NPs from *Trichoderma*, a rhizosphere fungus, has been used for bioremediation, antimicrobial efficacy, and against biotic stress (Kumari et al. 2017). Several plants associated and beneficial microbes are known to be globally used as biofertilizers or for bioremediation to promote growth and yield of crops (Tiwari et al. 2016, 2017; Tiwari and Lata 2018). These bacteria have received utmost attention for synthesis of nanoparticle. Bacteria such as *Pseudomonas*, *Klebsiella*, *Clostridium* and *Desulfovibrio* are used for the synthesis of silicon, cadmium sulphide, zinc sulphide and gold NPs, respectively (Shang et al. 2019). The unique life cycle of plant viruses i.e. its ability in infecting and delivering its nucleic acid genome followed by host cell lyses have made them an important candidate to be used in green nanotechnology (Young et al. 2008).

5.5 Application of Nanotechnology in Agriculture

Novel research in agricultural nanotechnology is believed to ease out and frame new phase in advancement of genetically modified crops, inputs for animal husbandry, chemical pesticides, fertilizers, and techniques in precision farming. Recent development and advances in chemistry and nanoscience have prompted excellence in nanoparticle technology, with vast implications in agriculture (Singh et al. 2015). Presently, rapid shift towards green nanotechnology has been witnessed to reduce agricultural waste and greenhouse gas emissions (Prasad et al. 2014). The uptake of nanoparticles adsorbed by plant surface depends largely on nanoparticles size and its surface properties. The small sized nanoparticles are taken up through cuticles and large sized nanoparticles can penetrate through hydathodes, stigma of flowers and stomata. NPs pass into the plant cell wall before entering the cell membrane of protoplasts. Anjum et al. (2016) reported that only NPs with < 5 nm diameter will be able to efficiently navigate the cell wall of undamaged cell. Mukhopadhyay (2014), reported that the nanotechnology can be used to treat acidic soil. The use of nanozeolites provide a better soil environment for crop growth, helps in lowering the future cost of importing farm-technologies, and helps in maintaining a sustainable agriculture. Thus, implementation of nanotechnology in agriculture has an outstanding effect in the field of agriculture that has been illustrated in the Fig. 5.2 and Table 5.1.

5.5.1 Role of Nanotechnology in Seed Germination

Nearly 60% of agricultural land in India is under rainfed area. Most of the seeds resist germinating in rainfed condition due to lack of moisture. Therefore, various groups of researcher have come up with new nano-technological approaches to enhance seed germination in these areas (Yadav and Yadav 2016). Carbon nanotubes and metal oxide NPs have shown immense potential in seed germination. The silicon dioxide (SiO2) NPs are reported to have improved tomato seed germination and other favouring germination factors (Siddiqui and Al-Whaibi 2014). *Boswellia ovalifoliolata*, commonly known as Indian Olibanum, is an endangered medicinal plant. It has shown improved seed germination and seedling growth when treated with

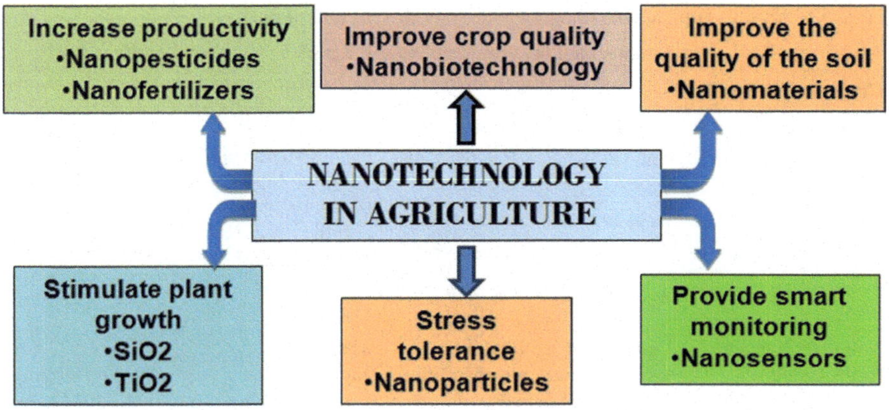

Fig. 5.2 Various applications of nanotechnology in agriculture (Figure constructed by Ratna Kalita)

silver nanoparticles. Silver NPs help nutrients and water uptake through seed coat and enhances germination of the seed (Savithramma et al. 2012).

5.5.2 Nano-Fertilizers for Better Crop Production

Eutrophication, imbalanced fertilization, depletion of soil organic matter, and low fertilizer use efficiency are some of the common problems in farming due to rigorous use of fertilizers. Recently, in order to solve these problems, nano-based fertilizers have gained the attention of researchers and agronomist. Nano-fertilizers regulate the release of nutrient based on need of the crop (Liu et al. 2006). Application of slow-release fertilizers (SRF) is a recent concept of nano-fertilizers. SRF releases their nutrients gradually and fulfill the plants nutrient requirement. These fertilizers are prepared by covering the conventional fertilizer granules with NPs for the controlled release of fertilizers in the soil. The pattern of coating of the NPs determines the rate of release and water solubility of the fertilizers.

5.5.3 Nano-Pesticides and Nano-Herbicide for Crop Protection

Application of conventional herbicides and pesticides often gets lost in a given environment and never reaches the targeted sites needed for active pest control. Excessive use of conventional pesticides is expensive and harmful to the environment (Nuruzzaman et al. 2016; Shang et al. 2019). Therefore, a new plant protection technology has been formulated in the agricultural field, which has modernized the use

Table 5.1 Various applications of nanotechnology in agricultural field

Area	Nanotechnology Applications	Examples	References
Crop production			
Plant protection products	Nanocapsules, NPs, and nanoemulsions for disease and pest control	Thymol nanoemulsion as antimicrobial; Anise oil emulsions as antimicrobial; Jojoba seed oil emulsions as insecticidal	Kumari et al. (2018); Topuz et al. (2016); Sh et al. (2015)
Nano-fertilizers and Naopesticides	Nanocapsules, NPs for enhanced nutrients absorption and targeted delivery of nutrients; use of nanoherbicide, nanoinsecticide, nanofungicide	Zinc Oxide Nanoparticles Iron and magnesium NPs	Milani et al. (2015); Delfani et al. (2014)
Soil improvement			
Water/liquid retention	Zeolites and nano-clays, for water or liquid agrochemicals retention in the soil	Zeolite based nanofertlizer	Manikandan and Subramanian (2016); Pulimi and Subramanian (2016)
Diagnostic			
Nanosensors and diagnostic devices	Carbon nanotubes, nanofibers and fullerenes, as bio-sensors to monitor environmental conditions, soil environments, plant health and growth. Also for precise application of fertilizers and pesticides	Liposome-based nano-biosensor	Vamvakaki and Chaniotakis (2007); Kaushal and Wani (2017)
Plant breeding			
Plant genetic modification	NPs carrying DNA or RNA to be delivered to plant cells for their genetic transformation	Mesoporous silica NPs; iron oxide NPs; calcium phosphate NPs	Torney et al. (2007); Zhao et al. (2017); Naqvi et al. (2012)

of herbicides and pesticides. Nano-herbicides and nano-pesticides are promising in tackling the increasing demand of agricultural products to obtain better and higher crop yields (He et al. 2019). Nano-pesticides promise a number of benefits to agriculture (Fig. 5.3). Some of the benefits of using nano-pesticides in agriculture over conventionally used pesticides are, increase in the water solubility of insoluble active ingredients, toxic organic solvents elimination, faster decomposition in soil and/or plants, controlled release of the active ingredients, improves stability of formulation

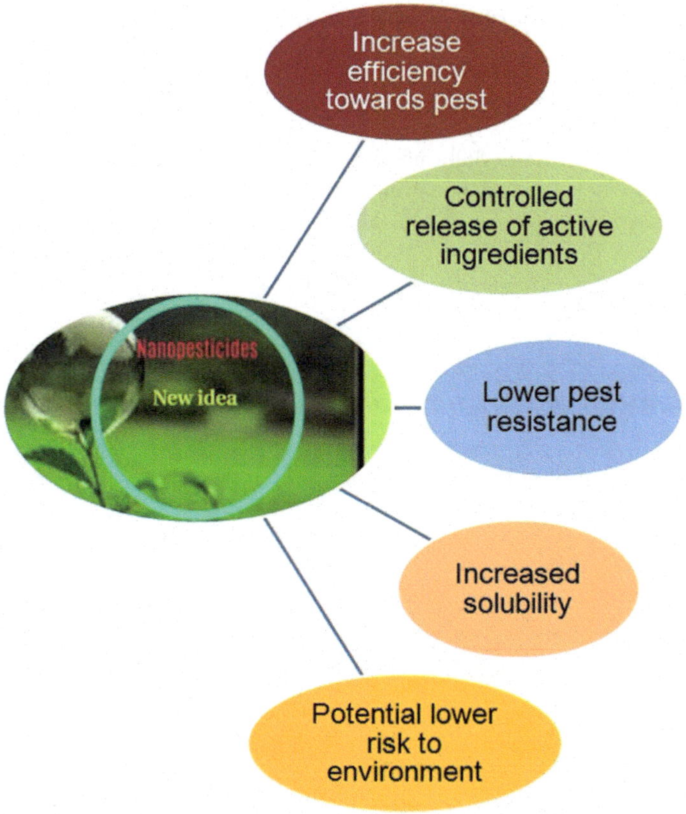

Fig. 5.3 Benefits of nano-pesticides applications in agriculture (Figure constructed by Ratna Kalita)

to prevent their early degradation, improves mobility and uptake. A number of nano-pesticides have been examined for their effectiveness against different economically important pests (Zahir et al. 2012). For example, normal glycerol is used to control pests, but when nano-glycerin is used, it is required in a very small quantity and proved highly efficient against most of the pests within a span of 6 h. It has altered the conventional method of requiring a heavy amount and taking longer duration of about 24 h (Fawzy et al. 2018). Nano-herbicides are very small, thus are easily mixed with the soil. They eliminate weeds without depleting the soil and prevent growth of those weed species that evolved resistance against conventional herbicides (Prasad et al. 2014). The parasitic weed control by using nano-herbicides reduces the degree of herbicide phytotoxicity in the plant (Perea-de-Lugue and Rubiales 2009).

5.5.4 Nanotechnology in Plant Disease Detection

An utmost need for early detection of plant disease will help to safeguard tons of food and can protect the possible outbreak of disease. An early diagnosis using nano-based technique is an immediate need to protect the food and agriculture against microorganisms such as virus, bacteria, and fungi. Since it involves less time, simple and easy to carry and operate, and accurate, this can be operated even by a farmer. The nano-sensors linked with the GPS system can be used for real-time monitoring of the crop and soil conditions (Singh et al. 2015).

5.5.5 Nano-Biosensors for Monitoring Agricultural Field

These techniques employ materials in nanoscale range to serve as diagnostic tools and deliver the nano-based particles to targeted sites in a governed manner. The nano-biosensor is described as a compact analytical device that incorporates a biological or biologically-derived element linked to a physico-chemical transducer (Turner 2000). Application of nano-biosensors in the cultivated field emerges as a diagnostic tool to revolutionize the agro-industry. It helps to promote sustainable agriculture, for soil quality and disease assessment, analysis in food products, detection of contaminants, pests, plant nutrients and impact of plant abiotic stresses such as drought, extreme temperature, and salinity (Rai et al. 2012). Nano-biosensors and nano-based smart delivery technologies enable precision farming which include systems for geographic information, devices for remote sensing and systems for satellites positioning. This smart technique has also helped the scientists to study plant's hormone regulation. The nano-sensor reacts with auxin level of the plant and thus help scientists to understand the plant root adaptation mechanism (McLamore et al. 2010). Similarly, nano-based systems are effectively used to sense insecticides, pesticides, fertilizers, herbicide, pathogens, and their precise and controlled use can enhance crop productivity and support sustainable agriculture (Sekhon 2014; González-Melendi et al. 2008).

5.5.6 Nanotechnology in Recycling and Elimination of Agricultural Wastes

Nanotechnology is also copiously applied to prevent agricultural waste. The agricultural waste so obtained can be converted into useful end product with the help of nanotechnology. For example, 25% of cellulose is eliminated while processing of cotton into garments or fabrics. Nano-techniques and solvents electrospinning produces 100 nm diameter fibers and are used as fertilizer or pesticide absorbent (Lang 2003). Similarly, during burning of rice husks, massive amount of high-quality nano-silica is produced and these could be used to make glass and concrete materials through nanotechnology. Thus, nanotechnology can be broadly used in managing agricultural wastage (Liou and Wu 2010).

5.5.7 *Role of Nanotechnology in Plant Genome Manipulation*

Nanoparticles have proved to be highly efficient magic bullets. Nanotechnology has a role in nanoparticle-mediated gene transfer and in development of genetically modified (GM) crops. It delivers DNA and other necessary agro-chemicals like nano-pesticides and nano-fertilizers into target site of the plant to prevent the plant from harmful insect pests (Torney 2009; Kamle et al. 2020). The bullet carries gene of interest or desired chemicals to a targeted plant part and releases their content gradually. Nano-capsules or bullets allow effective and easy penetration of the gene of interest or desired chemicals through plant cuticles and other tissues. The effective penetration is followed by slow and steady release of the DNA, RNA, siRNA or desired chemicals that then gradually integrates into the host genome (Perea-de-Lugue and Rubiales 2009). It has been found that 3 nm mesoporous silica nanoparticle (MSN) coating can successfully deliver genetic materials and chemicals into cells of tobacco and corn plants. The coating helps the plant to carry the particles across the cell walls, up to the area of genes insertion and activation, without any negative effects (Torney et al. 2007). Nanotechnology has also showed a great achievement in tissue engineering and nanomaterial-based smart delivery system in genome editing for development of genetically modified crops (Shang et al. 2019).

5.6 Nano-Based Smart Delivery Systems for Nano-Fertilizers and Nano-Pesticides

Technology advancements also paves the ways for "smart delivery systems" for nano-fertilizers and nano-pesticides to improve fertilizer formulation by increasing nutrient uptake and minimizing nutrient loss in plant cell and by supplying targeted and controlled nano-pesticides concentration (Solanki et al. 2015). Following are the smart delivery systems that have been used for improved delivery of nutrient and pesticides.

5.6.1 Nanoformulation

Fertilizers and chemicals used for plant protection are conventionally applied to crop fields by spraying. However, only a very low concentration of the applied chemical reaches the targeted site of the crop due to different prevailing situations especially through chemical leaching, photolysis, hydrolysis and degradation by microbes. Hence, nanoformulation goal is to increase the activity of bioactive agents as well as helps in their targeted delivery.

5.6.2 Nanoemulsion

This is a complex colloidal solution system consisting of oil phase, water and the surfactant, with optically transparent and kinetically stable. The size of the dispersed droplet ranges between 20 and 200 nm. It is a common observation that nanoparticle suspensions present instability in physical and chemical properties during their storage. Hence to overcome these challenges elimination of water from the aqueous phase to convert them in a dry solid form was performed. Nanoemulsion facilitates encapsulation of active ingredients within their dispersed droplets that enables reduction in chemical degradation (Salim et al. 2011).

5.6.3 Nano-Encapsulation

This method helps to reduce environmental pollution by reducing leaching and evaporation of toxic substances. Solid, liquid or gaseous nanoparticles are used to encapsulate traditional fertilizers onto a matrix. Nano-encapsulation helps in slow release of fertilizers in soil, thus making judicious use of fertilizers. Carrageenan and chitosan are commonly used secondary nano-materials used in encapsulating fertilizers, pesticides or herbicides (Duhan et al. 2017).

5.6.4 Mode of Administration

Ndlovu et al. (2020) reported the three distinct classification of nano-fertilizers on the basis of their mode and dilvery of nanofertilizer: (a) Nanoscale fertilizers: are nanomaterials containing nutrients. Zeolites and nanoclays are examples of nanoscale fertilizers. Zeolites are natural mineral that has a crystal structure similar to a honeycomb. When essential macro and micro nutrients are loaded onto zeolites, they ensure slow and continuous release of nutrients throughout the crop cycle. This helps in increasing the nutrient use efficiency of crops (Joshi et al. 2019); (b) Nanoscale coating: nanoparticles coated on traditional fertilizers. This is one of the most important strategies used for nanonutrient supplementation. Coating of traditional fertilizers with nanoparticles ensures stability of fertilizers in soil. Coating increases life span of fertilizers in soil and enhances its availability for uptake by plants (Fawzy et al. 2018). Nanotubes and slow-release fertilizers are common examples of nanocoating; and (c) Nanoscale additives: NPs when mixed with traditional fertilizers. Layered silicates are one of the most widely used nanoadditives, particularly montmorillonite, a natural mineral found in bentonite deposits (Ray and Okamoto 2003).

5.7 Conclusions and Prospects

Nanotechnology acts as a bridge mainly for developing countries to attain food security and food safety by ameliorating issues related to sanitary conditions, water scarcity, poor input use efficiency, and other related problems experienced by poor nations. At present, nanotechnology is thought to be a sustainable solution for the challenges faced by food and agricultural sector. But, nanotechnology still has to face issues regarding its safety on human health, environment, biodiversity and ecosystem. Nanotechnology is considered as a novel key to unlock solutions to numerous agricultural issues via developing nanoparticles, nano-capsules and nano-crystals for disease detection, management, delivery system, tissue engineering practices for plant growth monitoring, plant protection, and generation of improved crop varieties. NPs like nano-silver, alumino-silicate, titanium dioxide are largely used due to its broad spectrum antimicrobial activities, while nano-formulation of insecticides, pesticides and fertilizers help to reach the target site more accurately, thus giving a better crop yield. In present scenario, biosynthesis of green nanoparticles is gaining momentum due to its non-hazardous nature and for being economically advantageous. Therefore, such brief study showcases the importance of nanotechnology in the realm of agriculture.

References

Abobatta WF (2018) Nanotechnology application in agriculture. Acta Sci Agric 2(6):99–102

Anjum NA, Rodrigo MAM, Moulick A et al (2016) Transport phenomena of nanoparticles in plants and animals/humans. Env Res 151:233–243

Byczyńska A (2017) Nano-silver as a potential biostimulant for plant—a review. WSN 86(3):180–192

Dasgupta N, Ranjan S, Mundekkad D et al (2015) Nanotechnology in agro-food: from field to plate. Food Res Int 69:381–400

Delfani M, Firouzabadi MB, Farrokhi N, Makarian H (2014) Some physiological responses of black-eyed pea to iron and magnesium nanofertilizers. Commun Soil Sci Plant Anal 45:530–540

Duhan JS, Kumar R, Kumar N et al (2017) Nanotechnology: the new perspective in precision agriculture. Biotechnol Rep 15:11–23

Ealias AM, Saravanakumar MP (2017) IOP Conf Series Mater Sci Eng 263:032019

Fawzy ZF, Yunsheng L, Shedeed SI, El-Bassiony AM (2018) Nanotechnology in agriculture—current and future situation. Res Rev J Agricult Allied Sci 7(2):73–76

Gandhi G, Girgila PS, Aggarwal RK, Buttar BS (2010) Propensity for DNA damage in psoriasis patients genotyped for two candidate genes. J Carcinogene Mutagene 1:112

González-Melendi P, Fernández-Pacheco R, Coronado MJ et al (2008) Nanoparticles as smart treatment delivery systems in plants: assessment of different techniques of microscopy for their visualisation in plant tissues. Ann Bot 101:187–195

He X, Deng H, Hwang H-M (2019) The current application of nanotechnology in food and agriculture. J Food Drug Anal 27(1):1–21

Joshi H, Choudhary P, Mundra S (2019) Future prospects of nanotechnology in agriculture. Int J Chem Stud 7:957–963

Jurgons R, Seliger C, Hilpert A et al (2006) Drug loaded magnetic nanoparticles for cancer therapy. J Phys: Condens Matter 18:S2893–S2902

Kamle M, Mahato DK, Devi S et al (2020) Nanotechnological interventions for plant health improvement and sustainable agriculture. 3 Biotech 10:168

Kaushal M, Wani SP (2017) Nanosensors: Frontiers in precision agriculture. In: Prasad R, Kumar M, Kumar V (eds) Nanotechnology. Springer, Singapore, pp 279–291

Khandelwal A, Joshi R (2018) Synthesis of nanoparticles and their application in agriculture. Acta Sci Agri 2(3):10–13

Kumari M, Pandey S, Bhattacharya A et al (2017) Protective role of biosynthesized silver nanoparticles against early blight disease in *Solanum lycopersicum*. Plant Physiol Biochem 121:216–225

Kumari S, Kumaraswamy RV, Choudhary RC et al (2018) Thymol nanoemulsion exhibits potential antibacterial activity against bacterial pustule disease and growth promotory effect on soybean. Sci Rep 8(1):1–12

Lang SS (2003) Waste fiber can be recycled into valuable products using new technique of electrospinning. Cornell Chronicle. Available at https://news.cornell.edu/stories/2003/09/electrospinning-cellulose-waste-fiberelectrospinning-cellulose-waste-fiber. Accessed on 30 May 2020

Latef A, Srivastava A, El-sadek M et al (2017) Titanium dioxide nanoparticles improve growth and enhance tolerance of broad bean plants under saline soil conditions. Land Degrad Dev 29(4):1065–1073

Liou TH, Wu SJ (2010) Kinetics study and characteristics of silica nanoparticles produced from biomass-based material. Indus Eng Chem Res 49(18):8379–8387

Liu X, Feng Z, Zhang S et al (2006) Preparation and testing of cementing nano-subnano composites of slow- or controlled release of fertilizers. Sci Agri Sinica 39:1598–1604

Ma D (2019) Hybrid nanoparticles: an introduction. In: Mohapatra S, Nguyen TA, Nguyen-Tri P (eds) Noble metal-metal oxide hybrid nanoparticles, 1st edn. Woodhead Publishing, pp 3–6

Manikandan A, Subramanian KS (2016) Evaluation of zeolite based nitrogen nano-fertilizers on maize growth, yield and quality on inceptisols and alfisols. Int J Plant Soil Sci 9(4):1–9

Marchiol L (2018) Nanotechnology in agriculture: new opportunities and perspectives. In: Çelik Ö (ed) New vision in plant science. Intech, London, UK, pp 121–141

Mattiello A, Marchiol L (2017) Application of nanotechnology in agriculture: assessment of TiO_2 nanoparticle effects on barley. In: Janus M (ed) Application of titanium dioxide. InTech, London, UK, pp 23–39

McLamore ES, Diggs A, Marzal PC et al (2010) Non-invasive quantication of endogenous root auxin transport using an integrated flux microsensor technique. Plant J 63(6):1004–1016

Milani N, Hettiarachchi GM, Kirby JK et al (2015) Fate of zinc oxide nanoparticles coated onto macronutrient fertilizers in an alkaline calcareous soil. PLoS ONE 10(5):

Mukhopadhyay SS (2014) Nanotechnology in agriculture: prospects and constraints. Nanotechnol Sci Appln 7:63–71

Naqvi S, Maitra AN, Abdin MZ et al (2012) Calcium phosphate nanoparticle mediated genetic transformation in plants. J Mater Chem 22(8):3500–3507

Ndlovu N, Mayaya T, Muitire C, Munyengwa N (2020) Nanotechnology applications in crop production and food systems. Int J Plant Breed 7(1):624–634

Nuruzzaman M, Rahman MM, Liu YJ, Naidu R (2016) Nanoencapsulation, nano-guard for pesticides a new window for safe application. J Agric Food Chem 64:1447–1483

Pandey S, Mishra A, Giri VP, Kumari M, Soni S (2019) A Green nano-synthesis to explore the plant microbe interactions. In: Verma JP, Macdonald C, Gupta V, Podile AR (eds) New and future developments in microbial biotechnology and bioengineering, 1st edn. Elsevier, pp 85–105

Perea-de-Lugue A, Rubiales D (2009) Nanotechnology for parasitic plant control. Pest Manag Sci J 5(65):540–545

Prasad R, Kumar V, Prasad KS (2014) Nanotechnology in sustainable agriculture: present concerns and future aspects. Afr J of Biotechnol 13(6):705–713

Pulimi M, Subramanian S (2016) Nanomaterials for soil fertilisation and contaminant removal. In: Ranjan S, Dasgupta N, Lichtfouse E (eds) Nanoscience in Food and Agriculture 1. Springer, Cham, pp 229–246

Rai V, Acharya S, Dey N (2012) Implications of nanobiosensors in agriculture. J Biomat Nanobiotechnol 3:315–324

Ray SS, Okamoto M (2003) Polymer/layered silicate nanocomposites: a review from preparation to processing. Progress in Polymer Sci 28:1539–1641

Salim N, Basri M, Rahman MBA et al (2011) Phase behaviour, formation and characterization of palm-based esters nanoemulsion formulation containing ibuprofen. J Nanomedic Nanotechnol 2(4):1000113

Savithramma N, Ankanna S, Bhumi G (2012) Effect of nanoparticles on seed germination and seedling growth of Boswellia ovalifoliolata an endemic and endangered medicinal tree taxon. Nano Vis 2(1):2

Sekhon BS (2014) Nanotechnology in agri-food production: an overview. Nanotechnol Sci Appln 7:31–53

Sh A, Abdelrazeik AB, Rakha OM (2015) Nanoemulsion of jojoba oil, preparation, characterization and insecticidal activity against *Sitophilus oryzae* (coleoptera: Curculionidae) on wheat. Int J Agri Innov Res 4(1):72–75

Shang Y, Hasan M, Ahammed GJ et al (2019) Applications of nanotechnology in plant growth and crop protection: a review. Mol 24(14):2558

Sharon M, Choudhary AK, Kumar R (2010) Nanotechnology in agricultural diseases and food safety. J Phytol 2(4):83–92

Siddiqui MH, Al-Whaibi MH (2014) Role of nano-SiO2 in germination of tomato (*Lycopersicum esculentum* seeds Mill). Saudi J Biol Sci 21(1):13–17

Singh S, Singh BK, Yadav SM, Gupta AK (2015) Applications of nanotechnology in agricultural and their role in disease management. Res J Nanosci Nanotechnol 5(1):1–5

Solanki P, Bhargava A, Chhipa H, Jain N, Panwar J (2015) Nano-fertilizers and their smart delivery system. In: Rai M, Ribeiro C, Mattoso L, Duran N (eds) Nanotechnologies in food and agriculture. Springer, cham, pp 81–101

Srilatha B (2011) Nanotechnology in agriculture. J Nanomedic Nanotechnol 2:123

The European Commission. Commission Recommendation of 18 October 2011 on the Definition of Nanomaterial. Brussels, Belgium: Official Journal of the European Union. http://eur-lex.europa.eu/LexUriServ/LexUriServ.do?uri=OJ:L:2011:275:0038:0040:EN:PDF

Tiwari S, Lata C (2018) Heavy metal stress, signaling, and tolerance due to plant-associated microbes: an overview. Front Plant Sci 9:452

Tiwari S, Lata C, Chauhan PS et al (2016) *Pseudomonas putida* attunes morphophysiological, biochemical and molecular responses in *Cicer arietinum* L. during drought stress and recovery. Plant Physiol Biochem 99:108–117

Tiwari S, Prasad V, Chauhan PS et al (2017) *Bacillus amyloliquefaciens* confers tolerance to various abiotic stresses and modulates plant response to phytohormones through osmoprotection and gene expression regulation in rice. Front Plant Sci 8:1510

Topuz OK, Özvural EB, Zhao Q et al (2016) Physical and antimicrobial properties of anise oil loaded nanoemulsions on the survival of foodborne pathogens. Food Chem 203:117–123

Torney F (2009) Nanoparticle mediated plant transformation. Emerging technologies in plant science research. Interdepartmental Plant Physiology Major Fall Seminar Series Phys 696

Torney F, Trewyn B, Lin V, Wang K (2007a) Mesoporous silica nanoparticles deliver DNA and chemicals into plants. Nat Nanotech 2(5):295–300

Torney F, Trewyn BG, Lin VSY, Wang K (2007b) Mesoporous silica nanoparticles deliver DNA and chemicals into plants. Nat Nanotechnol 2(5):295–300

Turner AP (2000) Biosensors-sense and sensitivity. Science 5495(290):1315–1317

Vamvakaki V, Chaniotakis NA (2007) Pesticide detection with a liposome-based nano-biosensor. Biosens Bioelectron 22(12):2848–2853

Yadav S, Yadav KM (2016) Review on applications of nanotechnology in agriculture. Int J Dev Res 6(11):9942–9945

Yao KS, Li SJ, Tzeng KC et al (2009) Fluorescence silica nanoprobe as a biomarker for rapid detection of plant pathogens. Adv Mater Res 79–82:513–516

Young M, Willits D, Uchida M, Douglas T (2008) Plant viruses as biotemplates for materials and their use in nanotechnology. Annu Rev Phytopathol 46:361–384

Zahir AA, Bagavan A, Kamaraj C, Elango G, Rahuman AA (2012) Efficacy of plant-mediated synthesized silver nanoparticles against Sitophilus oryzae. J Biopest 5:95

Zhao X, Meng Z, Wang Y et al (2017) Pollen magnetofection for genetic modification with magnetic nanoparticles as gene carriers. Nat plants 3(12):956–964

Chapter 6
Contributions of Nano Biosensors in Managing Environmental Plant Stress Under Climatic Changing Era

Mojtaba Kordrostami, Mohammad Mafakheri, and Jameel M. Al-Khayri

Abstract The necessity of novel technologies to address the long-standing and worsened status of food security and sustainable agriculture is out of the question. Of the most recent promising approaches to alleviate constraints of the increasing occurrence of biotic and abiotic stressors on crop plants in the climate change era is nanobiotechnology involving potent methodologies with a large spectrum of applications covering both biotechnology and agriculture. Versatility of nanotechnology has made possible the establishment of quite a few biosensors that allow not only plant signaling biomolecules to communicate and actuate visually electronic monitoring devices to assess health status in real-time but also facilitates the allocation of resources. These include water and agrochemicals efficiency prior to the occurrence of water-deficit, salinity, extreme temperatures as well as phytopathogen-associated stress in order to maximize crop productivity and, consequently economical gains. Being new to the field of agriculture, achievements may not be commonly accessible worldwide. However, significant steps have been taken and reports are accumulating in relation to the development of nano biosensors for real-time monitoring. Particularly, using genetically encoded sensors that facilitate and accelerate the progress in this field by markedly improving current understanding of communications in stressed plants at the cellular level, the deciphering of which has been the purpose of plenty of recent biological studies. Nevertheless, a significant amount of research is required to optimize the application of nanomaterials to enhance feasibility of sustainable agriculture. This chapter aimed to summarize the most recent approaches

M. Kordrostami (✉)
Nuclear Agriculture Research School, Nuclear Science and Technology Research Institute (NSTRI), Karaj, Iran
e-mail: mkordrostmi@aeoi.org.ir

M. Mafakheri
Department of Horticultural Sciences, Faculty of Agricultural Sciences, University of Guilan, Rasht, Iran

J. M. Al-Khayri
Department of Agricultural Biotechnology, King Faisal University, Al-Hofuf Al-Ahsa, Saudi Arabia
e-mail: jkhayri@kfu.edu.sa

J. M. Al-Khayri et al. (eds.), *Nanobiotechnology*,
https://doi.org/10.1007/978-3-030-73606-4_6

to develop nano biosensors to bring worldwide food security into reality as the consequences of climate change become more tangible.

Keywords Abiotic stress · Nanomaterials · Real-time monitoring · Resource allocation · Signaling molecules

6.1 Introduction

Over the past few decades, the agriculture sector has been experiencing an unprecedented challenge driven by changes in climate which are accompanied by constant increase in human population that projected to reach 9.1 billion by 2050 (Mafakheri and Kordrostami 2020). This projected population growth will occur mostly in the developing countries, consequently intensifies the demand for food and energy (Das and Das 2019; Jalil and Ansari 2019; Rockström et al. 2017). Taking into account the current status, the exploitation of scarce natural resources for expanding food production, agriculture sector has the booming urbanization as a significant rival that imposes higher pressure on water and land, which itself is related to demographic changes in rural areas, where a large portion of the population have to migrate to cities (Alexandratos 2009). A universal swift shift in diet resulting from urbanization particularly in highly populous countries with growth of middle-class will bring diversity in consumption of food products. It is expected that the ratio of cereals reduces, while the portion of vegetables and fruits rises, a trend that is already in action in developing nations. Nevertheless, still over one billion tons of increase in cereals is necessary to meet the population demand of 2050. Such burdensome demands that require to be met by the mid-twenty-first century is accompanied by a challenged food production system. That already has over one billion people majorly hosted in hunger hotspots in developing countries, where obtaining the basic daily calories in addition to threatened ecological and natural resources. Incremented global demand for biofuels to lower carbon emission is another source of stress on land and water (Alexandratos 2009; Bala 2016; Spiess 2016; Tilman et al. 2011; Wang et al. 2016). Regarding water resources, a large percentage of agricultural production (~50%) comes from irrigated land that is composed of one-fifth of arable lands. Whereas several countries will face water shortage in the forthcoming decades. Of the irrigated land, 19.5% is salt-affected worldwide and the rest of arable lands account for semi-arid and arid agriculture, which 2.1% affected by salinity (Alexandratos 2009; Dagar et al. 2016; Ghassemi et al. 1995). An increase in global temperature is additionally putting further pressure on the ongoing trembling situation, since this rise will intensify the severity of other stresses in which the only 1 °C rising in global temperature causes a massive reduction in crop productivity (Iizumi et al. 2017; Zhao et al. 2017). The change in meteorological paradigm beyond the common threshold would also promote the possibility of a sudden drop in temperature in tropical and subtropical zones, where a large portion of crops are produced (Budhathoki and Zander 2019; Thakur and Nayyar 2013). The ecological upheaval caused by climate change will also impair the condition in favor of pest insects, phytopathogens (i.e., viral, microbial and fungal) through decreasing the population of their natural enemies, which provides the ground for devastating breakouts impose significant damages to crop yield (Luck et al. 2011; Post 2013). All the above-mentioned stressors that their incidence is foreseen to rise exponentially are major contributors to crop loss beyond the economic threshold, which unbalances the supply vs. demand and will bring the achievability of projected food products required to a halt.

Possible economically logical approaches to address this challenge is the establishment of management system to meticulously monitor the crop plants in real-time. Monitoring is necessary to effectively allocate the scarce resources to maximize the productivity and minimize the waste in resources (Chhipa 2019). This can be achieved mostly by remote sensing techniques, a revolutionary approach to improve crop productivity through managing the biotic and abiotic stressors, which nanotechnology could be a prime choice (Giraldo et al. 2019). Although conventional methods for diagnosis and differentiation of various types of plant responses to pathogens or deficiencies can be sensitive and inexpensive, biosensors allow for immediate analysis. Nanotechnology plays a key role in the development of biosensors, where the detection limits are improved to the nanoscale (Afsharinejad et al. 2015; Kwak et al. 2017). Nanomaterials are matrices that one of their dimensions is 1 to 100 nm, thus provide a high surface-to-volume ratio and exhibit a set of unique physical and chemical properties. The biosensor is described as a dense analytical tools, which use biological substances (for example, enzymes, antibodies, receptors, tissues, and nucleic acids) commonly called "analyte" and are measurable to be displayed as electrical, chemical or physical signals (Bakhori et al. 2013; Siddiquee et al. 2014; Walia et al. 2018). The basis of the diagnosis is on the specificity of the contact of the analyte with the bioassay element. Converters used in biosensors include optical, electrochemical, piezoelectric and thermometers. Biosensors can be categorized according to the type of analyte, how the transducer operates and its applications. Various nanostructures have been studied to evaluate their applicability in biosensors from which carbon nanotubes have received significant attention (Giraldo et al. 2019; Kwak et al. 2017). Specific bonding results in changes in one or more physical and chemical properties (such as pH, electron transfer, heat transfer, adsorption or release of volatile compounds), and may be measured with a converter. The main contribution is the generation of an electronic signal proportional to the magnitude and frequency of the analyte concentration that binds to the biosensor element. Biosensors can be divided into different groups: optical, magnetic, electrochemical, piezoelectric, and thermometer (Jianrong et al. 2004; Kumar and Arora 2020).

Nanomaterials have represented a long list of unique optical, electronic, physical, catalytic or mechanical features that have actively contributed in every discipline in science with no limitation in form or shape to provide the application of interest. In general, the structural forms of nanomaterials composes of nanotubes, dendrimers, and quantum dots (QD), nanoparticles, nanowires and fullerenes (Jeevanandam et al. 2018; Liu et al. 2012). The critical features of biosensors including sensitivity, flexibility, repeatability, precision, and accuracy have been notably boosted by incorporating nanomaterials (Jianrong et al. 2004; Kumar and Arora 2020; Kwak et al.

2017). The application of nanomaterial have already been used for numerous detection targets such as pathogens, viruses, bacteria, fungi, food quality control, and environmental monitoring. However, as mentioned earlier their applications for plants sciences is not as developed as for other fields in particular food safety (Farber and Kurouski 2018; Giraldo et al. 2019; Khiyami et al. 2014; Lin et al. 2017; Werres et al. 2001). The effectiveness of utilizing nano biosensor systems and their capabilities as viable monitoring approaches in the most recent plant-based applications are comprehensively described in this chapter.

6.2 Nanosensors for Plant Health Status Monitoring

The momentum for sustainable agriculture is needed now more than ever, the agriculture section requires taking a more sustainable strategy by meticulously monitoring the allocation of agricultural inputs. In the last decade nanosensors have manifested their strong capability for constant evaluation of the plant health in terms of nutrient deficiencies (phosphate, nitrogen and potassium), pathogens, salinity, drought, cold stress and soil organic matter (Álvarez et al. 2016; Antonacci et al. 2018; Ghaffar et al. 2020; Salouti and Derakhshan 2020). Various types of nanosensors have been explored in plants such as plasmonic nanosensors, fluorescence resonance energy transfer (FRET)-based nanosensors, carbon-based electrochemical nanosensors, nanowire nanosensors and antibody nanosensors. Additionally, various molecular methods such as polymerase chain reaction (PCR), real-time PCR, Raman spectroscopy, fluorescence spectroscopy, infrared spectroscopy and surface-enhanced Raman spectroscopy (SERS) (Farber and Kurouski 2018; Li et al. 2018b; Lin et al. 2014; Wang et al. 2017) exist. Albeit, the application of nanosensors in plants is lagging behind (Rai et al. 2012), promising experiments have implicated genetically encoded nanosensors or FRET-based nanosensors for improving the resource allocation efficiency for pathogens or resource deficiency through early identification and magnification. The recent most important findings regarding abiotic and biotic stress detection are discussed below.

6.2.1 Abiotic and Biotic Stress

Given the sever impacts of environmental stressors on crops and the importance of measures to decrease the crop loss, time-saving monitoring technologies are required to efficiently follow the responses of plants at physiological, biochemical, and morphological level (Fiorani and Schurr 2013). The importance of such a system is to visualize or predict the onset of insensible water deficit, low temperature, UV radiation, and elevated ozone or any other possible deficiency in nutrient elements before the occurrence of normal symptoms (Jansen et al. 1998). The current monitoring remote sensing systems use spectroscopy (Gitelson and Merzlyak

1996), chlorophyll fluorescence (Kalaji et al. 2011) and imaging (Zia et al. 2013). They utilize these methods to measure stress-related responses such as chlorophyll pigments ratios/contents (Gitelson and Merzlyak 1996; Kalaji et al. 2011), changes in leaf area (Born et al. 2014), or water relations of plants (Zia et al. 2013), which provide a wealth of information on health status of plants. However, these methods are not viable option for early identification of negative changes driven by abiotic stresses. Moreover, they lack specificity in detection of plant stresses, besides, to be time-consuming and costly for monitoring plants individually (Li et al. 2014; Mahlein 2016; Zarco-Tejada et al. 2012). The reduced quantity of photosynthetic pigments and active leaf area are manifestations of stress-induced responses. These effects interfere with the plant growth process. However, the decrements in chlorophyll fluorescence may not be the most reliable indicator because it is not the earliest response to abiotic stress nor has stress-specificity. Chlorophyll fluorescence reduction can occur in response to different stress factors such as water deficit, high salinity or phytopathogens (Al-Tamimi et al. 2016; Cohen et al. 2005; Li et al. 2014).

Of the customary phenotyping systems, Raman and infrared spectroscopy, although generate essential data on chemical interaction and cellular compositions but known to have low signal/noise ratios and laborious analytical process (Dong and Zhao 2017). Further, the application of nanobiotechnology-based approaches have reported frequently to enhance the sensitivity and reliability of current remote sensing systems and efficiently. These novel methods cover the drawbacks through securing an avenue to transduce the stress-specific precise signaling biochemicals into visual forms recordable by existing phenotyping tools (Kim et al. 2010; Lee et al. 2016). The plant health status can be reliably monitored using key signaling molecules such as reactive oxygen species (ROS) (Fig. 6.1), phytohormones including indole butyric acid (IBA), abscisic acid (ABA), jasmonic acid, methyl salicylate and ethylene, inorganic elements (calcium), primary metabolites (glucose, sucrose) and gaseous ROS, nitric oxide (NO) (for review see Giraldo et al. 2019). These molecules are in the frontline signaling network to evoke proper adaptation responses in plants, particularly ROS and calcium are systematically acknowledged secondary messengers involve in responses to a wide spectrum of stresses (Kiegle et al. 2000; Mazars et al. 2010; Mittler 2017; Suzuki et al. 2013). Hormonal signaling initiates the reactions to limit stomatal conductance or enables root extension under resource deficiencies often linked to ABA, IBA, and ethylene. The involvement of jasmonate signaling has been reported in tolerance of plants exposed to water stress, salinity, frost or physical damage (Howe et al. 2018). Plant pathogen-related resistance responses often associated with ethylene, NO and methyl salicylate. Carbohydrates such as glucose and sucrose also numerously found to be associated with conferring tolerance to abiotic-stress subjected plants (Delledonne et al. 1998; Lin et al. 2017; van Loon et al. 2006).

A combined application of nanomaterials and the recognized key signal molecules (Fig. 6.1) can indeed transform our perspectives by improving the performance of current monitoring devices and facilitate the understanding of the crosstalk between the key molecules that determine the tolerance mechanisms (Yoshida et al. 2014; Zhu et al. 2017). The in capabilities and challenges in Raman spectroscopy application to

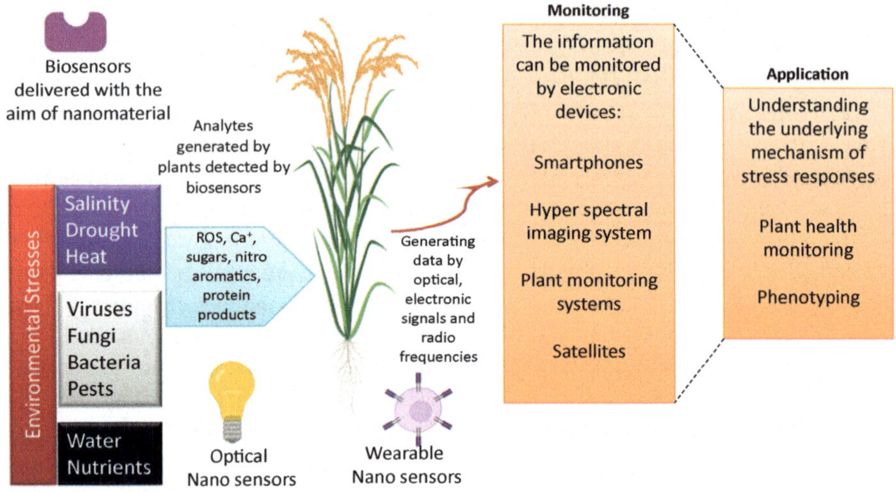

Fig. 6.1 Nanobiotechnological methodologies confer real-time communication between crop plants and monitoring systems by relying on precise detection of analytes. The generation several of plant signaling molecules are universal when plant expose to abiotic and biotic stresses or nutrient deficiency. Among them, plant responses, ROS (e.g., H_2O_2), Ca, sugars (e.g., glucose), nitroaromatics and protein products have been utilized as signaling molecules to develop anomaterial-mediated delivery of genetically encoded sensors. This accomplishment can help to gain an insight into underlying mechanisms and cross-talks involved in plant stress responses which can make designing commercial types of these sensors possible. Nanotechnology-based optical and wearable sensors by transferring biochemical analytes into radio waves and electric signals bring real-time monitoring of plant health status into reality. The introduction of nanobiosensors into agriculture is fairly recent, however, the application of these nanosensors can changes the borders of knowledge in this filed for ever and allow smart nano-based biosensors control and optimize the environment for crop plants. This schematic is adapted from Giraldo et al. (2019) and Kwak et al. (2017)

detect plant responses to abiotic stresses have been majorly addressed by using SERS. Relying on metal nanoparticles, SERS turned into a potent highly sensitive detection tool (Dong and Zhao 2017; Li et al. 2018b). Wang et al. (2017) managed the transformation of IBA into Raman-inactive and resonant biomolecule using Ehrlich reaction p-(dimethylamino) benzaldehyde (PDAB) combined with gold nanoparticles (Au NPs) hotspots to enable selective identification of phytohormones with indole ring by SERS. The multispectral overlap between Au NPs, IBA-PDAB and exciting laser onsets visible resonances to produce extremely high Raman scattering with significantly low limitation to IBA (2 nM). This system was latter applied to detection and quantification of IBA in seedlings of legumes (*Pisum sativum* L., *Vigna radiata* L., *Glycine max* L. and *Phaseolus vulgaris* L.). Among various methods exploited recently to develop nano biosensors, genetically encoded nanosensors stand out as highly capable approach with a promising future.

6.2.2 Nanoscale Sensors to Monitor Abiotic Stress in Plants

Application of NPs with genetically encoded nanoscale to screen key stress-responsive biomolecules is the most recent approach on designing smart biosensors in which delivers genetically encoded sensors that have the ability of real-time reporting of possible fluctuation in level or activity of the biomolecules of interest (Gjetting et al. 2013; Walia et al. 2018). Genetically encoded sensors provide a new insight of physiological-related sensitivities with temporal resolutions capable to identify dynamic key signaling biomolecules and be recorded and observed on the scale of seconds. Nonetheless, utilizing this approach in crop plants is restricted to the availability of efficient DNA transformation procedures for delivering DNA cassettes or plasmids as well as rather slowly optimization in each crop plant (Giraldo et al. 2019; Toyota et al. 2018; Walia et al. 2018). Thus, so far, the application of genetically encoded sensors has genetic amenability as a prerequisite for transformation methodologies as *Agrobacterium*-mediated delivery of plasmid or gene gun particle bombardment (Humplík et al. 2015). Moreover, transformation of genetically encoded nanoscale sensors using DNA cassettes or plasmids to expression in undomesticated plant species may pave the way to it for utilization beyond genetically amenable species. For instance, DNA-coated single-walled carbon nanotubes (SWCNT) are capable of inactive and spontaneous translocation, which confer them successful penetration into phospholipid bilayer. This process indicates the unique capability of SWCNTs as a delivery frameworks to transfer DNA into plant cells (Giraldo et al. 2014; Wong et al. 2016). Through creating a bonded cassette-SWCNTs and its administration with pneumatic injection to the epipodium, DNA-transfected plants may be possible without the need to use *Agrobacterium*-mediation and particle bombardment transformation (Demirer et al. 2019; Kwak et al. 2019). Exploiting this method, delivering a foreign plasmid DNA to chloroplasts of several plant species, the expression of yellow fluorescent protein (YFP) made possible without using gene gun, where its expression was further evaluated with confocal laser scanning microscopy (Kwak et al. 2019). Considering the tendency of biosensors cytosolic localization, an exceptional opportunity provided by the plastid genome to develop biosensors that are not restricted to the nuclear genome.

The genetically encoded nanosensing methods in plants mainly depend on biochemical interactions as protein-protein or change in chemical bonds. These parameters influential on fluorescence intensity or wavelength in the visible range of the electromagnetic radiation spectrum (Giraldo et al. 2019). Lately, using key singling molecules and fluorescent protein (roGFP2-Orp1), Nietzel et al. (2019) have been developed biosensors that identify H_2O_2 in *Arabidopsis* and largely clarifies the dynamic of H_2O_2 with no sensitivity to pH fluctuation during elicitor induced oxidative bursts with 15 s temporal resolution. In another study, Exposito-Rodriguez et al. (2017) took advantage of chloroplast-nuclei communication through H_2O_2 signaling, when cells experience fluctuation in light intensity. They developed a genetically encoded fluorescent H_2O_2 sensor, HyPer2, proved the involvement of direct transformation of H_2O_2 from chloroplast to nuclei to induce possible expression of responsible genes. The most recent application of key signaling biomolecules, H_2O_2, was reported by Wu et al. (2020). They assessed the stressed leaves of *Arabidopsis* with abiotic stressors (i.e., high light and UV-B) and biotic stress simulated by flg22, for induced H_2O_2, utilizing near-infrared fluorescent SWCNTs, which manifested high sensitivity (10–100 μM, within the physiological range of H_2O_2) and discriminability. However, no optical monitoring response was observed for mechanically wounded leaves.

Further, utilizing genetically encoded sensors combined with motorized fluorescence stereomicroscope, Toyota et al. (2018) revealed presentation of glutamate as a wound signal in attacked plants by caterpillar. Where *GLUTAMATE RECEPTOR-LIKE* family as actual sensors to transforms the signal leads to an intracellular incrementing level of calcium throughout the plant organs that then induces proper defensive responses. The main supplier of energy in plants is glucose, also known as a sugar-signaling molecule as well. Constant spatial or temporal monitoring of its intercellular concentration could be substantial step in deciphering signaling pattern of glucose in abiotic or biotic stressed plants. The FRET-based nanosensors, FLIPglu–2 μΔ13 and FLIPglu–600 μΔ13, utilize signaling pattern of glucose to measure cellular glucose dynamic in cytosol of rice transgenic lines exposed to abiotic (drought, salinity and cold/high temperatures) and biotic (flg22 and chitin) stresses with a high temporal resolution (10 s) (Zhu et al. 2017). Similarly, the potentiality of glucose signaling in leaves and roots has been reported in several studies on *Arabidopsis* with analogous temporal resolution (Chaudhuri et al. 2008, 2011; Deuschle et al. 2006).

Application of FRET-based genetically encoded calcium sensor to detect and understand the underlying signaling network has extensively investigated for crop plant viral pathogens (Keinath et al. 2015; Krebs et al. 2012; Loro et al. 2016). Other key signaling biomolecules, ROS, and salicylic acid are key are highly responsive to stressors and their accumulation takes place in chloroplasts. Therefore, they could be a reliable indicator of plant health status that their detection and monitoring may be possible by using genetically encoded nanosensors with generalized methodologies (Giraldo et al. 2019). The above-described recent groundbreaking studies revealed advancement towards designing a platform for an efficient phenotyping system using fluorescence imaging.

Of the other fundamentally important signaling molecules, lipid phosphatidic acid (PA) with a large array of involvement in different plant biological processes that they mainly remained unclear owing to the absence of reliable precise monitoring system for PA. Recently, Li et al. (2019) aimed to create a biosensor with enough sensitivity to monitor and visualize the concentration of PA at subcellular level, when induced by an increased ABA level and salt. They developed a FRET-based PA-specific optogenetic biosensor that monitors and reports the cellular concentration of PA. In addition, its bioimaging indicated the present of high tissue specificity in salt-induced AP accumulation. Comparative study on wild type and knockoff mutants of *Arabidopsis* (pldα1) lacking phospholipase Dα1 (PLDα1) required for PA biosynthesis showed that the latter had a low and delayed aggregation of PA. Additionally, their comparison showed the interplay of PA with the proteins in target as well as PLD/PA all regulated by pH. It could be asserted from these results that the response in plant to salinity may orchestrated by PA signaling and cellular pH. Exploiting such potent biosensors as PAleon can unmask the underlying mechanisms and dynamics of PA. As it was mentioned earlier, various platforms have been developed to employ certain metabolites to create nanobiosensors. In this case, glucose is a main products generated by photosynthesis apparatus and a critical energy component to run cellular processes in plants. That inspired developing an optical in vivo glucose sensor in undomesticated plants exploiting QD ratiometric method. Li et al. (2018a) fabricated an optical probe with a coupled QDs (thioglycolic acid-capped) used as glucose insensitive internal indicator (control) and boronic acid (BA)-conjugated QDs (BA-QD) to reduce their fluorescence intensity in response to glucose, The QD probe fluorescence was in the spectrum of the visible light. The probe manifested a high discrimination ability against other sugar compounds that exist in plant with detection limit of 500 μM in the physiological range. The potentials of this probe in in vivo quantification and reporting of glucose by Raspberry Pi camera and confocal laser scanning microscopy in a single chloroplast and intact algal cells (*Chara zeylanica* Willd) to leaf tissues of *Arabidopsis* were demonstrated. This nanoprobe has a significant capability in in vivo screening of glucose in photoautotrophs organisms.

6.2.3 Detection of Toxic Elements in Water and Soil

The contamination of natural resources primarily water and soil by heavy metals can initiate a dangerous chain reaction that may jeopardize the very being of organisms especially humans. Therefore, cost-effective, easy-to-use and efficient analyses approaches required, hopefully, to precisely monitor the content of toxic mineral elements in rhizosphere and water various nano-inspired biosensors have been developed (Ansari et al. 2020; Rai et al. 2012; Salouti and Derakhshan 2020). So as to understand the unique capability of carbon nanotubes in conferring and improving the electrochemical function of specific biomolecules as well as enhancing proteins involved in electron transfer reactions (Sagadevan and Periasamy 2014), Shi et al. (2017) developed a SWCNT-based biosensor to detect mercury (ll; Hg^{2+}) in water environment and serum samples with high sensitivity (0.84 pM). To design real-time and high precision nanodevices to detect heavy metal ions, FRET and QD as potent platforms also have been utilized (Ejeian et al. 2018). For example, Wu et al. (2010) developed a nanobiosensor-based QD-DNAzyme using FRET. This structure was a conjugation of carboxyl-silanized QDs onto quencher-labeled DNAzymes in which by the present of heavy metal ions as lead and copper that causes DNAzyme cleavage and then restoration of emission. Having a single laser excitation source the detection process can be completed within 25 min. With a distinctive discrimination ability, QD can help to differentiate between different metal ions by reflecting ion-specific fluorescence signals (Ejeian et al. 2018). The application of SERS for detection of heavy metal ions such as Hg^{2+} also has been successful; Xu et al. (2015) designed a single-stranded DNA (ssDNA)/AuNPs complex from a double helix DNA by T–Hg^{2+}–T base pairs. A significantly high sensitivity was acquired in detection of Hg^{2+} (0.45 pg mL^{-1}), and further assessments proved in applicability to detect Hg^{2+} content in real samples. Additionally, examining the discriminatory of this nanosensor for other heavy metal cations (Zinc; Zn^{2+}, Cu^{2+}, Nickle; Ni^{2+}, Pb^{2+}, and chromium; Cr^{2+}) revealed the absence of significant response to these ions at the concentrations up to 20 μM. In another study, simple and highly accurate nanobiosensor to detect Hg^{2+} (1 pM and 100 nM limitation) was developed based on SERS employing ssDNA and SWCNT conjugation which observed to have a considerable capability in detecting Hg^{2+} (Yang et al. 2017). As a notorious heavy metal ion, high precision monitoring of Pb^{2+} also has received significant attention from scholars. Using modified glassy carbon electrode (GCE) with self-doping polyaniline nanofiber mesoporous carbon nitride and bismuth was proposed to determine the heavy metal ions Cd^{2+} and Pb^{2+} by anodic stripping mechanism under voltammetry. This approach involves the use of a calibration curve from 5 to 80 nM for Cd^{2+} and Pb^{2+} and limits of detection (LOD) of 0.7 nM for Cd^{2+} and 0.2 nM for Pb^{2+} (Zhang et al. 2016). Exploiting DNAzyme graphene QDs and AuNPs, an optical nanobiosensor fabricated detection of Pb^{2+} by (Niu et al. 2018), which exhibited a notably wide detection range, 50 nM to 4 μM, for Pb^{2+} and16.7 nM as detection limit.

In an innovative and promising approach, Wong et al. (2017) fabricated a CoPhMoRe-based Bombolitin II nanosensors (BSWNTs) and after its internal placement into leaf tissues of wild-type spinach plants, which turned the plants into active nanodevices. This plant acted as independent autosamplers and preconcentrators of signaling molecules in underground water, to identify analytes present there, and interestingly an IR communication system transfers these data to be displayed via user's smartphone. Moreover, they took advantage of two NIR fluorescent nanosensors implanted inside leaf mesophyll of spinach plants. The first one was created utilizing a conjugation of peptide Bombolitin II with SWNTs to detect nitroaromatics via emission of IR fluorescent at 1,100 nm and temporal resolution of 5–15 min following the introduction of picric acid (PA, 2,4,6-trinitrophenol; 400 μM) to the roots. The latter IR was a PVA–activated SWNT (P-SWNT) that plays as a fixed control signal. When roots uptake of nitroaromatic pollutants in water solution go into leaf tissues as stem then they aggregated in the mesophyll where the SWNT nanosensors are implanted. This outcome potently projects the significant potential of nanomaterials-living wild-type plants interface to develop a robust plant-based chemical monitoring system for underground water with real-time communication.

The content of nitrates in water must strictly follows a certain standard and demands constant precise monitoring, so as to achieve, which Mura et al. (2015) designed a efficient and easy to use colorimetric test utilizing cysteamine-modified AuNPs for non-mediated nitrates detection in water. Citrate-stabilized AuNPs altered with cysteamine with a high affinity to nitrates, and comprehensive assessment and quantification of its capability to trap nitrates were carried out. More, Ali et al. (2017) developed a microfluidic impedimetric sensor utilizing nanosheets of graphene oxide (GO) and poly (3,4 ethylenedioxythiophene) nanofibers. In GO oxygen-containing functional groups enables an enhanced resistance in charge transfer of the electrochemical electrode. The level of sensitivity that sensor delivers is 61.15 Ω/(mg/L)/cm2 nitrate ions concentrations from 0.44 to 442 mg/L in farmland soils.

Urea is a main nitrogen source for agriculture; however, it high susceptibility to hydrolysis to baker's ammonia (ammonium carbonate) makes hazardous as it can inflects seedlings in germinating stage or even young plants and negative effect on nitrite. Thus, having a convenient approach for quantification urea content in soils is in high demand (Antonacci et al. 2018). A monitoring system for H_2O_2 based on AuNP-catalyzed 3,3′,5,5′-tetramethylbenzidine (peroxidase sensitive dye) utilized as a highly sensitive pH indicator introduced (Deng et al. 2016). A framework of this nanosensor and HRP enzyme exploited for the detection of urea, urease, and urease inhibitor with the sensitivity of 5 μM and 1.8 U/L for urea and urease. Nano-inspired biosensors based on Aptamer seem to be reliable devices for evaluation the source and reaction of metabolites in plant root system produced by living cells and also for studying investigating the controlled-release agricultural fertilizers (Salouti and Derakhshan 2020).

6.2.4 Pests and Pathogen-Related Stresses

An effective and comprehensive plant disease and pest management strategy can be possible with the aim of novel technologies for accurate evaluation of intensity, prediction and firsthand diagnosis. To diagnose phytopathogens a large array of molecular methods have been introduced, which although they are highly precise but laborious and costly. On the other hand, nanobiosensors far more superior than conventional techniques in terms of cost-saving, efficiency, nondestructive and accuracy (Pandita 2020; Singh et al. 2009; Vikram Singh and Sitti 2016). Detection of viral, fungal or bacterial pathogens in plants have been performed utilizing current technologies. For example, quantitative polymerase chain reaction (Q-PCR) (Khiyami et al. 2014), enzyme-linked immunosorbent assay (ELISA) and reverse transcription-polymerase chain reaction (RT-PCR) (Lin et al. 2014). Despite the relatively high sensitivity and in many cases selectivity, these methodologies are generally significantly complicated, laborious, time-consuming and cost-intensive. Additionally, in order to obtain a precise analysis, a considerable quantity of the target tissue is required (Galvão and Fankhauser 2015; Khater et al. 2017; Kwak et al. 2017; Neethirajan et al. 2018; Shang et al. 2019).

Progress has been reported using nano biosensors with ample of advantages for real-time assessment of plant pathogens in crop plants before appearance of their symptoms. This makes the application of agrochemicals tremendously efficient. In the case of insect pests, early detection means less devoted resources for an effective quick control (Chhipa 2019; Singh et al. 2010, 2015).

By fabricating a nonstructural layer using copper oxide nanoparticles (CuO NPs) detection of *Aspergillus niger*, commonly known as black mold is one the most common fungal diseases mainly infect fruit and vegetable crops (for example, apricot, peanut, apricot, and onion) made possible. Electrical resistance assessed to evaluate the biosensing features (Etefagh et al. 2013; Sagadevan and Periasamy 2014). Exploiting antibody-conjugated-fluorescent silica nanoparticles (FSNP) Yao et al. (2009) were capable to detect *Xanthomonas axonopodis* pv. *Vesicatoria,* a gram-negative bacteria that causes bacterial leaf spots in *Capsicum* sp. and *Solanum lycopersicum* L. a devastating pathogen particularly in the warm season. They used circular dye, tris-2, 2'-bipyridyl dichlororuthenium (II) hexahydrate (Rubpy)-doped SNPs with diameters of 50 ($\pm$4.2) nm manifested a significant photostability. When underwent collisional quenching fluorescence test, their fluorescence intensity by exposing to the bacteria of interest notably enhanced in comparison to non-conjugated antibody. Being multitalented, gold nanoparticles (AuNPs) have a unique fluorescence-quenching capability, the synthesis of DNA oligos labeled with 2 nm AuNPs fluorescent at 3' and 5' can be carried out, which is suitable when DNA profiling is compulsory or screening phytoplasma disease such as *flavescence doree* in grapevine is the case (Firrao et al. 2005). An AuNRs-based fiber optic particle plasmon resonance (FOPPR) immunosensor was designed to aim to identify *Cymbidium mosaic* virus (CymMV) and *Odontoglossum ringspot* virus (ORSV), common virus disease in orchids, utilizing AuNRs. An nIR created, the developed immunosensor observed to be capable to detect the viruses in leaf sap matrix with high detection limit as low as 48 and 42 pg/mL, respectively. This was achieved at a temporal resolution of 10 min, which is highly superior in terms of efficiency and sensitivity when compared to ELISA (Lin et al. 2014). Moreover, the hybridization of oligos with the DNA in the target may intensify the fluorescence signals probably resulting in its application as nano-transducer in hybridized DNA. Raman spectroscopy which recently applied to discriminate healthy maize kernels from contaminated ones with fungal agents with 100% accuracy in a non-destructive manner (Farber and Kurouski 2018).

Nanotechnology in some cases strikingly enhanced the capability of method such as Raman. It provides an improved tool as SERS utilized for monitoring pathogenic agents jeopardizing the landscape (Yüksel et al. 2015). A prime example could be *Phytophthora ramorum* found to be related with symptoms on *Rhododendron* and *Viburnum*, which its detection process majorly rely on microbiological or PCR-based approaches (Bilodeau et al. 2007; Lamour 2013; Schena et al. 2008; Werres et al. 2001). The existence of adenine with potent spectral characteristics conveys the availability of hybridization with complementary target DNA, exhibited the identification of *P. ramorum* target DNA with high sensitivity (30 nucleotides) excreted from infected tissue using SERS (Werres et al. 2001).

Cutting-edge techniques to detect the presents of pathogens in advance have been developed. For instance utilizing FRET-based nanosensors for real-time assessment of the signaling molecules such as glucose-induced by biotic stress stimuli namely pathogen-derived conserved peptides, flg22, or derivatives of glucose main components in fungi cell walls, chitin with significantly high temporal resolution (Zhu et al. 2017). Another exploitation of key signaling biomolecules is a study conducted by Wu et al. (2020) to monitor plant health with nIR fluorescent SWCNTs aim to report induction of H_2O_2 in leaves of Arabidopsis stressed with stimuli flg22 with considerable high sensitivity (10–100 μM). Development of nanosensors has also expanded to detect pathogens in soil milieu as Siddiquee et al. (2014) designed a nano biosensor to by incorporation of ZnO NPs, chitosan nanocomposite and bare gold electrode capable to identify *Trichoderma harzianum*, a useful fungi with fungicide capability against several species of fungus.

Application of nanosensors for insect pest detection in croplands has been limited to couple of publications, such as Afsharinejad et al. (2015). They developed wireless nanosensors for continuous monitoring of plants for insect attack by relying on organic volatile compounds release from plants in response to pest attack with high discriminatory power among various host plant species. More, to detect an insect *Polymyxa betae* vector of a destructive viral disease in sugar beet, Rhizomania, that beet necrotic yellow vein virus is responsible for. Safarpour et al. (2012) developed a quantum dot-FRET-based nano biosensor that successfully with 100% stability detected the vector. Similarly, using a quantum dot-FRET-based system to detect *Ganoderma boninense*-related oligos was designed. Modification of QDs with single-stranded DNA and DNA probes and their conjugation with the targeted DNA made the G. *boninense gene* sequences detectable by FRET signals with notable sensitivity (Bakhori et al. 2013).

6.3 Optical Nanobiosensors for in Vivo Sensing

The very exceptional properties of nanomaterials can facilitate the possibility of real-time screening of signalling molecules in vivo. Having significantly low photobleaching has enabled them to fluorescence in transparent-setting of living tissue which make the identification of signaling molecules in target possible with a distinctively high quality resolution (Guo et al. 2014; Hong et al. 2015). To construct frameworks for fluorophore based nanobiosensor in mammalian systems, SWCNT has been frequently employed (Cui et al. 2010; Khazi-Syed et al. 2019). Its application in plants also has been reported (Giraldo et al. 2014; Giraldo et al. 2015; Kwak et al. 2017; Wong et al. 2017). By interaction of organic component of SWCNT with signaling molecules in the context, modulation of SWCNT fluorescence with the nIR transparency window for plants can occur. As mentioned earlier in this chapter, the high sensitivity and flexibility of SWCNT based biosensors has led to detect a great variety of analytes, namely oxidants, calcium, sugars (glucose), dopamine, nitroaromatics, and protein products (Giraldo et al. 2014; Kruss et al. 2014; Kruss et al. 2017; Son et al. 2011; Wong et al. 2017; Zhang et al. 2013). To deliver optically active nano-inspired biosensors various high-throughput lab-based phenotyping approaches or easier novel procedures such as needle free injection via the leaf lamina, topical delivery system and Agroinfiltration. By placement of SWCNT based biosensors in leaf tissue provides the opportunity for instantaneous monitoring of temporary analytes including NO and ROS (Deuschle et al. 2006; Giraldo et al. 2015; Giraldo et al. 2019). The intensity of SWCNT fluorescence is sensitive to fluctuation in H_2O_2 concentration and transfers the change to monitor in the matter of seconds.

Additionally, nanoparticles have extensively been applied for optical sensing which amongst them QDs are common. Of the critical stress-associated signaling molecules, glucose, can be a reliable index for evaluating plant health status which recently has been exploited to develop highly precise and selective QD-based optical probes (Li et al. 2018a). As nanoscale semiconductors with optical characteristics, QDs are able to generate tunable light by photoluminescence covering visible spectrums to the nIR in addition to their easily manipulatable either structure or chemical characteristics (Borovaya et al. 2015; Jiang et al. 2020; Kruss et al. 2014; Ornes 2016). Using boronic acids to activate enables them bond with glucose to optically identify the analyte in plant tissues and report it to the monitoring system about a couple of minutes (Hansen and Christensen 2013; Sun and James 2015). As for now, the optical systems detection mechanism rely on nanobiosensors placed inside plant tissue with low signal-to-noise ratios compared with ideal conditions and in the absence of specific plant tissue, or cellular components. To address the shortcomings in this approach, AuNPs have been suggested to intensify signal (Hong et al. 2015).

Another proposed method to tackle the abovementioned issue is to thinly coat plant viruses with a layer of gold to confer an improved field while avoiding jeopardizing the virus targeting potential. However, the applicability of this novel approach for imaging and spectroscopy still has not been confirmed, but it could be viable method to enhance the optically communicating nano-inspired biosensor against the natural setting in the field (Giraldo et al. 2019; Kwak et al. 2017). To tackle the challenges involve delivering nanobiosensors to the plant tissues in target another approach is synthesizing the sensors under in situ that in this case metal–organic frameworks (MOFs) as 3D platforms with a significant capacity constructed by ions and organic ligands (Kruss et al. 2017). Richardson and Liang (2018) attempted to develop in situ synthesized fluorescent MOFs in plants that enabled them to record and report the concentration of acetone in atmosphere by changing fluorescence when contact with these unstable volatile substances. Further, nanobiosensors coated with chemical moieties of interest namely peptides can enable the organ-specific localization (Yu et al. 2012). The highly responsive analytes to environmental stressors like calcium and ROS, often fluctuate swiftly in the cellular compartments, single cells or in plant overall which make them a potent candidate for monitoring plant health. However, a stable level of ROS, for instance, is unlikely to provide useful information whereas its spatiotemporal fluctuation does (Chaudhuri et al. 2008; Giraldo et al. 2019; Kwak et al. 2017; Mazars et al. 2010)

Optically active nanobiosensors, as discussed earlier, have been extensively explored and designed in mammals. Application issues such as localization of emitting signals from cellular organs involved in this technology in regard of monitoring analytes in either mammalian or plant systems are notably analogous (Alfinito et al. 2010; Butnariu and Butu 2019; Kruss et al. 2017; Liu et al. 2009).

Briefly, kinetic-based sensors are capable to demonstrate whether detection of signaling molecules in plant can address the occurred stress or not. To closely investigate the expression pattern of analytes in time and space under various biological condition and, observe and prognosticate the fluorescence response, these sensors can provide scholars with this prime opportunity (Giraldo et al. 2019; Meyer et al. 2017). Using such predictions and optical observations for quantification and reliably discriminate chemical signaling from cell or organelle. Consequently, with aim of these outcomes a solid background to develop well-structured and optimized biosensors while avoiding laborious experimental methods. In addition, a correlation between plant signaling paradigm with the stressor(s) and/or nutrient shortage can be made utilizing these novel technologies.

Considering the current advancement in nanobiosensors, in the upcoming years, the communication between sensors and analytes can be analyzed to determine their discriminatory power against different signaling molecules. Therefore, their accuracy and selectivity of the analyte of interest and precise translation of plant responses under various environmental conditions will be possible which ultimately can bring the commercially acceptable nanobiosensors into reality to improve the agricultural productivity.

6.4 Conclusions and Prospects

Interdisciplinary approaches are the prerequisite of developing effective sustainable strategies to narrow the gap between supplies versus demands for agricultural commodities. Preventative measures to reduce the yield loss caused by environmental stressors, where the possibility of their co-occurrence is ever high owing to climate change, should be taken. Nanotechnology is a potent candidate for real-time monitoring offering high spatial and temporal resolution sensors relying on key signaling biomolecules with high sensitivity and accuracy. It also has the potential of delivering platforms for genetically encoded biosensors. Such approaches need to be comprehensively investigated to produce smart plant sensors that accurately communicate with machines. Informing the system for the presence of pathogens or the probability of resource deficiency in advance through translating biochemical signaling into wireless, electrical and optical signals, nano biosensors hugely contribute to impeding the reduction in plant growth. Therefore, enhancing yield with well-adjusting the scarce resources. The monitoring technologies developed based on nanotechnology pave the way to their utilization such as high-throughput screening of chemical phenotypes that is a capable system for breeding projects to identify the potential candidates. Although arrival of nanotechnology to plant science is relatively recent, significant accomplishments have been realized. For instance, developing nanoscale sensors exploiting the abiotic- and biotic-induced signaling molecules. Moreover, genetically encoded sensors are true manifestation of the capability of nanotechnology approach that has revolutionized the design of nano biosensors through eliminating species-specificity by acting as genetic agent, or methodological-associated constraints or laboriousness. Detection of the deficiency of mineral elements is a serious growing threat to the production of agricultural products due to taking intense agriculture methods and soil degradation globally. Thus, nano biosensors can open a new window to an efficient allocation of fertilizers in a real-time manner. Nanotechnology is a promising approach that holds plenty of potentials that will soon emerge. It is perhaps the sole technology with boundless applications in various scientific fields driven by humankind imagination in peruse of enriched wellbeing.

References

Afsharinejad A, Davy A, Jennings B et al (2015) Performance analysis of plant monitoring nanosensor networks at THz frequencies. IEEE Internet Things J 3:59–69

Al-Tamimi N, Brien C, Oakey H et al (2016) Salinity tolerance loci revealed in rice using high-throughput non-invasive phenotyping. Nat Commun 7:13342. https://doi.org/10.1038/ncomms13342

Alexandratos N (2009) Highlights and views from MID-2009: paper for the expert meeting on. In: World food and agriculture to 2030/50, Rome, 24 June 2009

Alfinito E, Pennetta C, Reggiani LJS et al (2010) Olfactory receptor-based smell nanobiosensors: an overview of theoretical and experimental results. Sensor Actuat B-Chem 146:554–558

Ali MA, Jiang H, Mahal NK et al (2017) Microfluidic impedimetric sensor for soil nitrate detection using graphene oxide and conductive nanofibers enabled sensing interface. Sensor Actuat B-Chem 239:1289–1299

Álvarez SP, López NEL, Lozano JM et al (2016) Plant fungal disease management using nanobiotechnology as a tool. In: Parsad R (ed) Advances and applications through fungal nanobiotechnology. Springer, Cham, pp 169–192

Ansari MH, Lavhale S, Kalunke RM et al (2020) Recent advances in plant nanobionics and nanobiosensors for toxicology applications. Curr Nanosci 16:27–41

Antonacci A, Arduini F, Moscone D et al (2018) Nanostructured (Bio)sensors for smart agriculture. TrAC Trend in Anal Chem 98:95–103

Bakhori NM, Yusof NA, Abdullah AH et al (2013) Development of a fluorescence resonance energy transfer (FRET)-based DNA biosensor for detection of synthetic oligonucleotide of *Ganoderma boninense*. Biosensors 3:419–428

Bala BK (2016) Principles of drying. In: Drying and storage of cereal grains. pp 1–3

Bilodeau G, Lévesque C, de Cock AWAM et al (2007) Molecular detection of *Phytophthora ramorum* by real-time polymerase chain reaction using TaqMan, SYBR Green, and molecular beacons. Phytopathology 97:632–642

Born N, Behringer D, Liepelt S et al (2014) Monitoring plant drought stress response using terahertz time-domain spectroscopy. Plant Physiol 164:1571–1577

Borovaya M, Burlaka O, Yemets A et al (2015) Biosynthesis of quantum dots and their potential applications in biology and biomedicine. In: Fesenko O, Yatsenko L (eds) Nanoplasmonics, nano–optics, nanocomposites, and surface studies, vol 167. Springer, Cham, pp 339–362

Budhathoki NK, Zander KK (2019) Socio-economic impact of and adaptation to extreme heat and cold of farmers in the food bowl of Nepal. Int J Env Res Pub He 16:1578. https://doi.org/10.3390/ijerph16091578

Butnariu M, Butu A (2019) Plant nanobionics: application of nanobiosensors in plant biology. In: Parsad R (ed) Plant nanobionics. Springer, Cham, pp 337–376

Chaudhuri B, Hörmann F, Frommer WB (2011) Dynamic imaging of glucose flux impedance using FRET sensors in wild-type Arabidopsis plants. J Exp Bot 62:2411–2417

Chaudhuri B, Hörmann F, Lalonde S et al (2008) Protonophore-and pH-insensitive glucose and sucrose accumulation detected by FRET nanosensors in Arabidopsis root tips. Plant J 56:948–962

Chhipa H (2019) Applications of nanotechnology in agriculture. In: Methods in microbiology. Academic Press, pp 115–142

Cohen Y, Alchanatis V, Meron M et al (2005) Estimation of leaf water potential by thermal imagery and spatial analysis. J Exp Bot 56:1843–1852

Cui HF, Vashist SK, Al-Rubeaan K et al (2010) Interfacing carbon nanotubes with living mammalian cells and cytotoxicity issues. Chem Res Toxicol 23:1131–1147

Dagar J, Sharma P, Chaudhari S et al (2016) Climate change vis-a-vis saline agriculture: impact and adaptation strategies. In: Dagar JC, Chander Sharma P, Sharma D, Singh AK (eds) Innovative saline agriculture. Springer, Delhi, pp 5–53

Das A, Das B (2019) Nanotechnology a potential tool to mitigate abiotic stress in crop plants. In: Abiotic and biotic stress in plants. Intech Open. Available via DIALOG. https://www.intechopen.com/books/abiotic-and-biotic-stress-in-plants/nanotechnology-a-potential-tool-to-mitigate-abiotic-stress-in-crop-plants. Accessed 7 Jul 2020

Delledonne M, Xia Y, Dixon RA et al (1998) Nitric oxide functions as a signal in plant disease resistance. Nature 394:585–588

Demirer GS, Zhang H, Matos JL et al (2019) High aspect ratio nanomaterials enable delivery of functional genetic material without DNA integration in mature plants. Nat Nanotechnol 14:456–464

Deng HH, Hong GL, Lin FL et al (2016) Colorimetric detection of urea, urease, and urease inhibitor based on the peroxidase-like activity of gold nanoparticles. Anal Chim Acta 915:74–80

Deuschle K, Chaudhuri B, Okumoto S et al (2006) Rapid metabolism of glucose detected with FRET glucose nanosensors in epidermal cells and intact roots of *Arabidopsis* RNA-silencing mutants. Plant Cell 18:2314–2325

Dong D, Zhao C (2017) Limitations and challenges of using Raman spectroscopy to detect the abiotic plant stress response. Proc Natl Acad Sci 114. https://doi.org/10.1073/pnas.1707408114

Ejeian F, Etedali P, Mansouri-Tehrani H-A et al (2018) Biosensors for wastewater monitoring: A review. Biosensors Bioelectron 118:66–79

Etefagh R, Azhir E, Shahtahmasebi N (2013) Synthesis of CuO nanoparticles and fabrication of nanostructural layer biosensors for detecting *Aspergillus niger* fungi. Sci Iranica 20:1055–1058

Exposito-Rodriguez M, Laissue PP, Yvon-Durocher G et al (2017) Photosynthesis-dependent H2O2 transfer from chloroplasts to nuclei provides a high-light signalling mechanism. Nat Commun 8:49. https://doi.org/10.1038/s41467-017-00074-w

Farber C, Kurouski D (2018) Detection and identification of plant pathogens on maize kernels with a hand-held Raman spectrometer. Anal Chem 90:3009–3012

Fiorani F, Schurr U (2013) Future scenarios for plant phenotyping. Annu Rev Plant Biol 64:267–291

Firrao G, Moretti M, Rosquete MR et al (2005) Nanobiotransducer for detecting flavescence dorée phytoplasma. J Plant Pathol 87:101–107

Galvão VC, Fankhauser C (2015) Sensing the light environment in plants: photoreceptors and early signaling steps. Curr Opin Neurobiol 34:46–53

Ghaffar N, Farrukh MA, Naz S (2020) Applications of nanobiosensors in agriculture. In: Javad S (ed) Nanoagronomy. Springer, Cham, pp 179–196

Ghassemi F, Jakeman AJ, Nix HA (1995) Salinisation of land and water resources: human causes, extent, management and case studies. CAB international

Giraldo JP, Landry MP, Faltermeier SM et al (2014) Plant nanobionics approach to augment photosynthesis and biochemical sensing. Nat Mater 13:400–408

Giraldo JP, Landry MP, Kwak SY et al (2015) A ratiometric sensor using single chirality near-infrared fluorescent carbon nanotubes: application to in vivo monitoring. Small 11:3973–3984

Giraldo JP, Wu H, Newkirk GM et al (2019) Nanobiotechnology approaches for engineering smart plant sensors. Nat Nanotechnol 14:541–553

Gitelson AA, Merzlyak MN (1996) Signature analysis of leaf reflectance spectra: algorithm development for remote sensing of chlorophyll. J Plant Physiol 148:494–500

Gjetting SK, Schulz A, Fuglsang AT (2013) Perspectives for using genetically encoded fluorescent biosensors in plants. Front Plant Sci 4:234

Guo Z, Park S, Yoon J et al (2014) Recent progress in the development of near-infrared fluorescent probes for bioimaging applications. Chem Soc Rev 43:16–29

Hansen JS, Christensen JB (2013) Recent advances in fluorescent arylboronic acids for glucose sensing. Biosensors 3:400–418

Hong G, Diao S, Antaris AL, Dai H (2015) Carbon nanomaterials for biological imaging and nanomedicinal therapy. Chem Rev 115:10816–10906

Howe GA, Major IT, Koo AJ (2018) Modularity in jasmonate signaling for multistress resilience. Annu Rev Plant Biol 69:387–415

Humplík JF, Lazár D, Husičková A et al (2015) Automated phenotyping of plant shoots using imaging methods for analysis of plant stress responses—a review. Plant Methods 11:29. https://doi.org/10.1186/s13007-015-0072-8

Iizumi T, Furuya J, Shen Z et al (2017) Responses of crop yield growth to global temperature and socioeconomic changes. Sci Rep 7. https://doi.org/10.1038/s41598-017-08214-4

Jalil SU, Ansari MI (2019) Nanoparticles and abiotic stress tolerance in plants: synthesis, action, and signaling mechanisms. In: Khan N, Khan MIR, Ferrante A, Reddy PS (eds) Plant signaling molecules. Elsevier, Cambridge, pp 549–561

Jansen MA, Gaba V, Greenberg BM (1998) Higher plants and UV-B radiation: balancing damage, repair and acclimation. Trends Plant Sci 3:131–135

Jeevanandam J, Barhoum A, Chan YS et al (2018) Review on nanoparticles and nanostructured materials: history, sources, toxicity and regulations. Beilstein J Nanotechnol 9:1050–1074

Jiang X, Jin H, Gui RJB (2020) Visual bio-detection and versatile bio-imaging of zinc-ion-coordinated black phosphorus quantum dots with improved stability and bright fluorescence. Biosensors Bioelectron 165. https://doi.org/10.1016/j.bios.2020.112390

Jianrong C, Yuqing M, Nongyue H et al (2004) Nanotechnology and biosensors. Biotechnol Advan 22:505–518

Kalaji HM, Bosa K, Kościelniak J et al (2011) Chlorophyll a fluorescence—a useful tool for the early detection of temperature stress in spring barley (*Hordeum vulgare* L.). OMICS: J Integrative Biol 15:925–934

Keinath NF, Waadt R, Brugman R et al (2015) Live cell imaging with R-GECO1 sheds light on flg22-and chitin-induced transient [Ca2+] cyt patterns in Arabidopsis. Mol Plant 8:1188–1200

Khater M, de la Escosura-Muñiz A, Merkoçi A (2017) Biosensors for plant pathogen detection. Biosensors Bioelectron 93:72–86

Khazi-Syed A, Hasan MT, Campbell E et al (2019) Single-walled carbon nanotube-assisted antibiotic delivery and imaging in s. epidermidis strains addressing antibiotic resistance. Nanomaterials (Basel) 9. https://doi.org/10.3390/nano9121685

Khiyami MA, Almoammar H, Awad YM et al (2014) Plant pathogen nanodiagnostic techniques: forthcoming changes? Biotechnol Biotechnol Equip 28:775–785

Kiegle E, Moore CA, Haseloff J et al (2000) Cell-type-specific calcium responses to drought, salt and cold in the Arabidopsis root. Plant J 23:267–278

Kim TH, Böhmer M, Hu H et al (2010) Guard cell signal transduction network: advances in understanding abscisic acid, CO2, and Ca2+ signaling. Annu Rev Plant Biol 61:561–591

Krebs M, Held K, Binder A et al (2012) FRET-based genetically encoded sensors allow high-resolution live cell imaging of Ca(2)(+) dynamics. Plant J 69:181–192

Kruss S, Landry MP, Vander Ende E et al (2014) Neurotransmitter detection using corona phase molecular recognition on fluorescent single-walled carbon nanotube sensors. Am Chem Soc 136:713–724

Kruss S, Salem DP, Vuković L et al (2017) High-resolution imaging of cellular dopamine efflux using a fluorescent nanosensor array. Proc Natl Acad Sci 114:1789–1794

Kumar V, Arora K (2020) Trends in nano-inspired biosensors for plants. Mater Sci Ener Technol 3:255–273

Kwak S-Y, Lew TTS, Sweeney CJ et al (2019) Chloroplast-selective gene delivery and expression in planta using chitosan-complexed single-walled carbon nanotube carriers. Nat Nanotechnol 14:447–455

Kwak S-Y, Wong MH, Lew TTS et al (2017) Nanosensor technology applied to living plant systems. An Rev Anal Chem 10:113–140

Lamour K (2013) Characterizing Phytophthora populations. In: Lamour K (ed) *Phytophthora*: a global perspective. CABI, pp 28–36
Lee H, Choi TK, Lee YB et al (2016) A graphene-based electrochemical device with thermoresponsive microneedles for diabetes monitoring and therapy. Nat Nanotechnol 11:566. https://doi.org/10.1038/nnano.2016.38
Li J, Wu H, Santana I et al (2018a) Standoff optical glucose sensing in photosynthetic organisms by a quantum dot fluorescent probe. ACS Appl Mater Interfaces 10:28279–28289
Li L, Zhang Q, Huang D (2014) A review of imaging techniques for plant phenotyping. Sensors 14:20078–20111
Li W, Song T, Wallrad L et al (2019) Tissue-specific accumulation of pH-sensing phosphatidic acid determines plant stress tolerance. Nat Plant 5:1012–1021
Li Y, Li Q, Wang Y et al (2018b) A reagent-assisted method in SERS detection of methyl salicylate. Spectrochim Acta A 195:172–175
Lin HY, Huang C-H, Lu SH et al (2014) Direct detection of orchid viruses using nanorod-based fiber optic particle plasmon resonance immunosensor. Biosensors Bioelectron 51:371–378
Lin Y, Qasim M, Hussain M et al (2017) The herbivore-induced plant volatiles methyl salicylate and menthol positively affect growth and pathogenicity of entomopathogenic fungi. Sci Rep 7:1–11
Liu Y, Tan J, Thomas A et al (2012) The shape of things to come: importance of design in nanotechnology for drug delivery. Ther Deliv 3:181–194
Liu Z, Tabakman S, Welsher K et al (2009) Carbon nanotubes in biology and medicine: in vitro and *in vivo* detection, imaging and drug delivery. Nano Res 2:85–120
Loro G, Wagner S, Doccula FG et al (2016) Chloroplast-specific *in vivo* Ca^{2+} imaging using Yellow Cameleon fluorescent protein sensors reveals organelle-autonomous Ca^{2+} signatures in the stroma. Plant Physiol 171:2317–2330
Luck J, Spackman M, Freeman A et al (2011) Climate change and diseases of food crops. Plant Pathol 60:113–121
Mafakheri M, Kordrostami M (2020) Role of molecular tools and biotechnology in climate–resilient agriculture. In: Hasanuzzaman M (ed) Plant ecophysiology and adaptation under climate change: mechanisms and perspectives II. Springer, Singapore, pp 491–529
Mahlein A-K (2016) Plant disease detection by imaging sensors–parallels and specific demands for precision agriculture and plant phenotyping. Plant Dis 100:241–251
Mazars C, Thuleau P, Lamotte O et al (2010) Cross-talk between ROS and calcium in regulation of nuclear activities. Mol Plant 3:706–718
Meyer D, Hagemann A, Kruss S (2017) Kinetic requirements for spatiotemporal chemical imaging with fluorescent nanosensors. ACS Nano 11:4017–4027
Mittler R (2017) ROS are good. Trends Plant Sci 22:11–19
Mura S, Greppi G, Roggero PP et al (2015) Functionalized gold nanoparticles for the detection of nitrates in water. Int J Environ Sci Technol 12:1021–1028
Neethirajan S, Ragavan K, Weng X (2018) Agro-defense: Biosensors for food from healthy crops and animals. Trends Food Sci Technol 73:25–44
Nietzel T, Elsasser M, Ruberti C et al (2019) The fluorescent protein sensor roGFP2-Orp1 monitors *in vivo* H2O2 and thiol redox integration and elucidates intracellular H2O2 dynamics during elicitor-induced oxidative burst in *Arabidopsis*. New Phytol 221:1649–1664
Niu X, Zhong Y, Chen R et al (2018) A "turn-on" fluorescence sensor for Pb2+ detection based on graphene quantum dots and gold nanoparticles. Sensors Actuators B: Chem 255:1577–1581
Ornes S (2016) Core concept: quantum dots. Proc Natl Acad Sci 113:2796–2797
Pandita D (2020) Nano-enabled agriculture can sustain "farm to fork" chain. In: Hakeem KR, Pirzadah TB (eds) Nanobiotechnology in agriculture. Springer, Cham, pp 35–61
Post E (2013) Erosion of community diversity and stability by herbivore removal under warming. P Roy Soc B-Biol Sci 280. https://doi.org/10.1098/rspb.2012.2722
Rai V, Acharya S, Dey N (2012) Implications of nanobiosensors in agriculture. J Biomater Nanobiotechnol 3:315–324

Richardson JJ, Liang KJS (2018) Nano biohybrids: *in vivo* synthesis of metal–organic frameworks inside living plants. Small 14. https://doi.org/10.1002/smll.201702958

Rockström J, Williams J, Daily G et al (2017) Sustainable intensification of agriculture for human prosperity and global sustainability. Ambio 46:4–17

Safarpour H, Safarnejad MR, Tabatabaei M et al (2012) Development of a quantum dots FRET-based biosensor for efficient detection of *Polymyxa betae*. Can J Plant Pathol 34:507–515

Sagadevan S, Periasamy M (2014) Recent trends in nanobiosensors and their applications—a review. Rev Adv Mater Sci 36:62–69

Salouti M, Derakhshan FK (2020) Biosensors and nanobiosensors in environmental applications. In: Ghorbanpour M, Bhargava P, Varma A, Choudhary DK (eds) Biogenic nano-particles and their use in agro-ecosystems. Springer, Singapore, pp 515–591

Schena L, Duncan J, Cooke D (2008) Development and application of a PCR-based 'molecular tool box' for the identification of Phytophthora species damaging forests and natural ecosystems. Plant Pathol 57:64–75

Shang Y, Hasan M, Ahammed GJ et al (2019) Applications of nanotechnology in plant growth and crop protection: a review. Molecules 24. https://doi.org/10.3390/molecules24142558

Shi L, Wang Y, Chu Z et al (2017) A highly sensitive and reusable electrochemical mercury biosensor based on tunable vertical single-walled carbon nanotubes and a target recycling strategy. J Mater Chem B 5:1073–1080

Siddiquee S, Rovina K, Yusof NA et al (2014) Nanoparticle-enhanced electrochemical biosensor with DNA immobilization and hybridization of *Trichoderma harzianum* gene. Sens Biosensing Res 2:16–22

Singh A, Patil R, Kasture M et al (2009) Synthesis of Ag–Pt alloy nanoparticles in aqueous bovine serum albumin foam and their cytocompatibility against human gingival fibroblasts. Colloid Surface B 69:239–245

Singh AV, Raymond M, Pace F et al (2015) Astrocytes increase ATP exocytosis mediated calcium signaling in response to microgroove structures. Sci Rep 5:1–8

Son D, Park SY, Kim B et al (2011) Nanoneedle transistor-based sensors for the selective detection of intracellular calcium ions. ACS Nano 5:3888–3895

Spiess WE (2016) Challenges to food security in a changing world. In: Jackson P, Spiess WE, Sultana F (eds) Eating, drinking: surviving. Springer, Cham, pp 57–73

Sun X, James DJ (2015) Glucose sensing in supramolecular chemistry. Chem Rev 115:8001–8037

Suzuki N, Miller G, Salazar C et al (2013) Temporal-spatial interaction between reactive oxygen species and abscisic acid regulates rapid systemic acclimation in plants. Plant Cell 25:3553–3569

Thakur P, Nayyar H (2013) Facing the cold stress by plants in the changing environment: sensing, signaling, and defending mechanisms. In: Tuteja N, Gill SS (eds) Plant acclimation to environmental stress. Springer, New York, pp 29–69

Tilman D, Balzer C, Hill J et al (2011) Global food demand and the sustainable intensification of agriculture. Proc Natl Acad Sci 108:20260–20264

Toyota M, Spencer D, Sawai-Toyota S et al (2018) Glutamate triggers long-distance, calcium-based plant defense signaling. Science 361:1112–1115

V Singh A, AS A, N Gade W et al (2010) Nanomaterials: new generation therapeutics in wound healing and tissue repair. Curr Nanosci 6:577–586

van Loon LC, Geraats BP, Linthorst HJ (2006) Ethylene as a modulator of disease resistance in plants. Trends Plant Sci 11:184–191

Vikram Singh A, Sitti M (2016) Targeted drug delivery and imaging using mobile milli/microrobots: a promising future towards theranostic pharmaceutical design. Curr Pharm Des 22:1418–1428

Walia A, Waadt R, Jones AM (2018) Genetically encoded biosensors in plants: pathways to discovery. Ann Rev Plant Biol 69:497–524

Wang F, Gu X, Zheng C et al (2017) Ehrlich reaction evoked multiple spectral resonances and gold nanoparticle hotspots for raman detection of plant hormone. Anal Chem 89:8836–8843

Wang X-j, Zhang J-y, Shahid S et al (2016) Adaptation to climate change impacts on water demand. Mitig Adapt Strat Gl 21:81–99

Werres S, Marwitz R, In't Veld WAM et al (2001) *Phytophthora ramorum* sp. nov., a new pathogen on Rhododendron and Viburnum. Mycol Res 105:1155–1165

Wong MH, Giraldo JP, Kwak SY et al (2017) Nitroaromatic detection and infrared communication from wild-type plants using plant nanobionics. Nat Mater 16:264–272

Wong MH, Misra RP, Giraldo JP et al (2016) Lipid exchange envelope penetration (LEEP) of nanoparticles for plant engineering: a universal localization mechanism. Nano Lett 16:1161–1172

Wu CS, Khaing Oo MK, Fan X (2010) Highly sensitive multiplexed heavy metal detection using quantum-dot-labeled DNAzymes. ACS Nano 4:5897–5904

Wu H, Nißler R, Morris V et al (2020) Monitoring Plant health with near infrared fluorescent h2o2 nanosensors. Nano Lett 20:2432–2442

Xu L, Yin H, Ma W et al (2015) Ultrasensitive SERS detection of mercury based on the assembled gold nanochains. Biosensors Bioelectron 67:472–476

Yang X, He Y, Wang X, Yuan R (2017) A SERS biosensor with magnetic substrate CoFe2O4@ Ag for sensitive detection of Hg2+. Appl Surf Sci 416:581–586

Yao KS, Li S, Tzeng K et al (2009) Fluorescence silica nanoprobe as a biomarker for rapid detection of plant pathogens. Adv Mater Res 79:513–516

Yoshida T, Mogami J, Yamaguchi-Shinozaki K (2014) ABA-dependent and ABA-independent signaling in response to osmotic stress in plants. Curr Opin Plant Biol 21:133–139

Yu MK, Park J, Jon SJT (2012) Targeting strategies for multifunctional nanoparticles in cancer imaging and therapy. Theranostics 2:3–44

Yüksel S, Schwenkbier L, Pollok S et al (2015) Label-free detection of *Phytophthora ramorum* using surface-enhanced Raman spectroscopy. Analyst 140:7254–7262

Zarco-Tejada PJ, González-Dugo V, Berni JA (2012) Fluorescence, temperature and narrow-band indices acquired from a UAV platform for water stress detection using a micro-hyperspectral imager and a thermal camera. Remote Sens Environ 117:322–337

Zhang C, Zhou Y, Tang L et al (2016) Determination of Cd2+ and Pb2+ based on mesoporous carbon nitride/self-doped polyaniline nanofibers and square wave anodic stripping voltammetry. Nanomaterials 6. https://doi.org/10.3390/nano6010007

Zhang J, Landry MP, Barone PW et al (2013) Molecular recognition using corona phase complexes made of synthetic polymers adsorbed on carbon nanotubes. Nat Nanotechnol 8:959–968

Zhao C, Liu B, Piao S et al (2017) Temperature increase reduces global yields of major crops in four independent estimates. Proc Natl Acad Sci 114:9326–9331

Zhu Q, Wang L, Dong Q et al (2017) FRET-based glucose imaging identifies glucose signalling in response to biotic and abiotic stresses in rice roots. J Plant Physiol 215:65–72

Zia S, Romano G, Spreer W et al (2013) Infrared thermal imaging as a rapid tool for identifying water-stress tolerant maize genotypes of different phenology. J Agron Crop Sci 199:75–84

Chapter 7
Nanobiotechnology: A Process to Combat Abiotic Stress in Crop Plants

K. Kisku and Umesh C. Naik

Abstract In recent years, the dire climatic change has increased the exposure of the crop plants to regular but various types of biotic and abiotic stresses. Reports on abiotic stresses imposing potential adverse effects on crop productivity worldwide are more than biotic stresses. Abiotic stresses mainly drought, salinity, flooding, metal toxicity, and rising temperature due to global warming disrupts the ionic and osmotic balance of the plant cell. As a result, there is restriction of diverse crop farming declining agricultural production over large areas. The declining crop production leads to negative and inevitable effects on the livelihoods of the farmers and mankind for their survival. According to a report, the maximum yield associated with abiotic stress factors is estimated to vary between 54 and 82%. Not only these stresses adversely affect the sustainability of the agricultural industry, but it also threatens the national economy and food security. Therefore, the major challenge is to manage the abiotic stress to improve crop production under abiotic stress. In the changing environmental scenario, nanobiotechnology has gained greater importance to mitigate the constraints associated with environmental stresses and is considered as a promising solution for improving crop production. The present chapter reviews the responses of the crop plants to different abiotic stresses and the potential roles of nanotechnology towards modulating the stress factors in order to secure the future of sustainable agriculture worldwide.

Keywords Abiotic stress · Biotic stress · Drought · Metal toxicity · Nanobiotechnology · Salinity · Sustainable agriculture

---

K. Kisku · U. C. Naik (✉)
Department of Botany, Ravenshaw University, Cuttack, Odisha 753003, India
e-mail: unaik.bot@ravenshawuniversity.ac.in

K. Kisku
e-mail: kanikakisku.bot@ravenshawuniversity.ac.in

J. M. Al-Khayri et al. (eds.), *Nanobiotechnology*,
https://doi.org/10.1007/978-3-030-73606-4_7

7.1 Introduction

The discouraging conditions of the environment including different types of biotic and abiotic stresses are long been known to impose vulnerable impacts on the plants. They define themselves to be a significant threat to the sustainability of our crop production. Abiotic stresses are the major concern of today's research than those of biotic because of their unavoidable and recurring nature. These stresses mainly comprise of drought, flood, salinity, heat and cold, and heavy metal toxicity. Plants are also frequently exposed to UV radiation which is responsible for dropping the optimum crop growth and productivity. These stresses result in the average reduction of yield by more than 50% (Bray et al. 2000). Water stress has been chalked out as the most prevalent abiotic stress worldwide, in which drought and salinity being its literal members. The stresses are gaining a constant position in reducing the survivability and growth of the plants and thereby limiting their productivity (Boyer 1982). An estimation by Ashraf (1994) states that by the next 25 years, drought and salinity could cause serious salinization of all the arable lands up to 30% followed by detrimental effects globally. A land diminution up to 50% is expected by 2050. Several morphological, physiological, biochemical, and molecular changes become obvious due to long exposure to various types of abiotic stress. The demands of crop products continue to increase amidist persistent impact of abiotic stress on agricultural productivity (Zhao et al. 2017). A very common situation in Germany is that summer is much dryer due to rising termperature beyond 30°C in most of its days. The precipitation is getting elevated by about 11% during the winter season for the past 100 years favoring the environmental conditions for the growth of various types of pathogens affecting crop fertility and productivity.

The world population is expected to reach around 9.8 billion by 2050 and 11.2 billion by 2100 (United Nations 2017). To address the rising demand of the growing population, significant research and technological advancements are been made in the field of agriculture (Dwivedi et al. 2016; Kou et al. 2018; Xiao et al. 2013). Agriculture supports almost entire world population and more than 60% of the livelihood depends on it. Though agriculture is regarded the backbone of the developing countries, there are handful of challenges facing the agricultural sector. The climatic conditions, global warming, overexploitation of resources, usage of chemical fertilizers and various biotic, and abiotic stresses are some of the major challenges that need immediate response (Raliya et al. 2017). Abiotic stresses are more detrimental than biotic and therefore hold much attention for scientific advancement. Abiotic stress faced by the crop plants has stimulated research interests since the last decade because of it is unavoidable and recurring in nature. Considering these inevitable occurring, nanotechnologies have gained immense attention in the recent years because of its critical role in increasing and improving the quality of the crop production. The growing applications of biotechnologies and nanotechnologies in the field of material energy, medicine, medical drug and catalysis have gained good scientific attention (Ghidan and Antary 2019). Nanotechnology is an interdisciplinary field that combines the application of biology, physics, chemistry, medicine and engineering (Abd-Elrahman and Mostafa 2015). Advancement in nanotechnological research has unraveled various strategies to enhance the tolerance capacity of the crop plants against abiotic stresses. This book chapter intends to highlight the mitigation of diverse and hostile effects of abiotic stresses on crop plants using emerging nanobiotechnology.

7.1.1 Climate Change

Climate change has now become a major determinant for the agricultural sector. It has worsened drastically since the last few years. Humans have been utilizing the mother earth for its own benefit from decades thereby creating constant challenges for the agricultural sector to meet the demand of the ever rising population. The time is not very far when we will put a question mark on the sustainability of the planet. There is a significant rise in the temperature of the planet by 1.4 °C in the preceding century and has been estimated to increase up to 11.5 °C in the next century, thereby putting the climatic change at peak priority (IPCC 2014). Earth's carbon concentration, heat waves, acidification of oceans, highly varying temperature and rise in sea levels have seen steady escalation since the 1950s, whereas the untimely rainfall has imposed catastrophic impact on the agricultural productivity (Calanca 2017; Chen et al. 2017). The disturbance in climate is caused more by human activities leading to global warming, than by natural climate itself, rendering it difficult for the plants to adapt in these continuously changing environments. The results of these actions have become an eminent project for researchers and the response mechanism pathways of the crop plants to the overlapping stresses are under investigation.

7.1.2 Stress Types

Stress holds varying definitions based on its general, physical and biological applications. However, stress according to agricultural terms is described as adverse forces that reduce the biomass and productivity of the crop plants. Biological definition of the biotic and abiotic stresses are shown in Fig. 7.1. Stress symptoms include measurable injuries that are not visible through naked eyes like alteration in enzyme activities and membrane structure, growth and development arrest, plant injuries like necrosis, chlorosis, and discoloration, changes in the composition of microorganisms, and their interference and alternation in the genomic structures of the species (Cramer et al. 2011; Dresselhaus and Huckelhoven 2018).

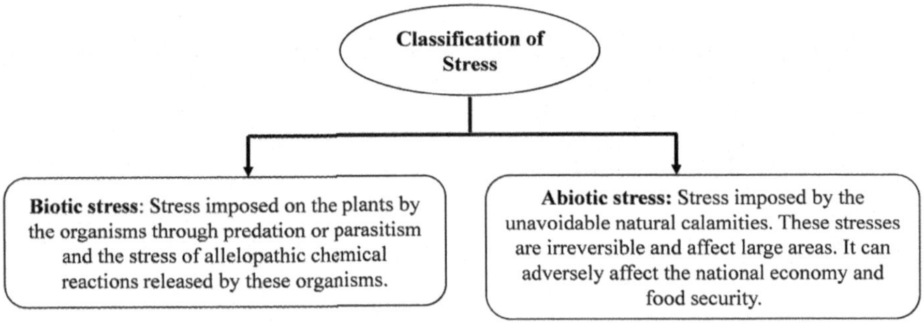

Fig. 7.1 Classification of different types of stresses (Figure constructed by K.Kisku)

7.1.2.1 Biotic Stress

The effect of biotic stresses also holds equal importance as abiotic stresses but the difference that lies is very narrow; the adverse impact on crop plants by biotic stresses can be avoided to too much extent by being more cautious during farming, whereas abiotic stresses cannot be controlled directly. The crop plants elicit different types of responses towards the various environmental strains by adapting and bringing changes in their response mechanism (Bartels and Salamini 2001). Microbial communities in association with the plants that constitute a complex environment function as holobionts rather than functioning as individual genotypes and can aid in bringing new resistance pathways to the pathogenic microbial infections. Continuous exposure to these tensions by plant species alter their gene patterns in order to withstand these stress conditions. One such secondary metabolite from grasses that show chemical defense against biotic strains is Benzoxazinoids (BXs). A study that was conducted by Niculaes et al. (2018) demonstrated the biosynthesis of BXs as well as its further biological activities.

7.1.2.2 Abiotic Stress

The stresses that occur naturally but unexpectedly and affect the crop plants indirectly through the physical environment, fall under abiotic stresses. This impose a negative impact on the crop yield, survivability, total biomass and quality production (Grover et al. 2001; Calanca 2017). As food is one of the prime factors of human existence, agriculture and agricultural land have always been taken care of. But due to population overflow, agricultural lands are being dissolved into non-arable lands to accommodate the exploding human number. Because of unconcerned human activities, there is misbalanced in the climate timings and have raised the global temperature to a great extent. Plants growing in stress conditions have come up with better-surviving strategies and adaptability than the plants growing in normal and optimum growth conditions. The stress responses of the former are much more complex. Since the last century, earth's atmospheric temperature has risen drastically, the plant productivity is continuously facing limitations making it difficult to meet the global needs. It has become a paramount concern because of its frequent exposure (Cramer et al. 2011; Kumar et al. 2018). The responses of the plants to these overlapping abiotic stresses represent very dynamic and sophisticated mechanisms The plant community is constantly facing exploitation due to climate change and environmental stresses. The advancement of the techniques and technologies has helped to reduce these exploitations to a great amount. Combating abiotic stresses is a constant challenge in the present scenario. Though the progress in the molecular and physiological biology that have improved our understanding of stress tolerance, researches demand insight into the mechanism and gene structure using various bio and nanotechnology tools (Table 7.1).

7.2 Plant Adaptation to Abiotic Stresses

The plants on frequent exposures to the stresses from various biotic and abiotic sources have evolved some advanced mechanisms to cope with the environmental conditions. Different plants adapt to relatively different responses and tolerance properties and vary widely among each other. Development and expansion of the bio-technologies and omics-technologies have answered many of the challenges in unraveling the mechanism of tolerance and resistance against these stresses both via genomic and proteomic analyses. Highly efficient omics tools deliver throughput information in gene discovery and aid to understand their genomic function. Being constantly exposed to stresses at short intervals, plants have undergone an evolution of diverse growth patterns and surviving habits besides changing their mechanism of the stress response. Plant genes have undergone different transcriptional and translational changes producing protein products that are specific to the stresses (Cushman and Bohnert 2000). Response to these strains varies widely among species and even among genotypes. Adaptation to tolerance by the crop plants has been divided into four groups as Fig. 7.2.

Osmolytes, for instance, mannitol, glycine, proline, betaine, antioxidants like catalase, ascorbate, peroxidase, stress-induced proteins such as LEA proteins, chaperons, antifreeze proteins, heat shock proteins, protein kinase and trans-acting factors like DREB1/CBF, AP2/ERF, DREB2, NAC, basic leucine zipper proteins and zinc-finger are the substantial plant molecules that are highly targeted for the genomic modification (Tayal et al. 2004; Sangam et al. 2005). Plants alter their genes that consequently produce specific byproducts, which serves as the stress response mitigating adaptation. Functional genes responsible for stress resistance in plants fall under the category of signaling factors that contain signal transduction proteins, and transcriptional factors that function in maintaining the ion homeostasis and integrity of the cell. However, in addition to these tolerance factors, there is another important factor known as functional protein. Functional proteins regulate the biosynthesis of

Table 7.1 Different types of abiotic stresses, its impact on plants and their studied mitigation steps

Types	Definition	Stress impacts	Mitigation steps	References
Drought	The prolonged period of no rainfall or very low rainfall in a region which leads to water stress. Types a) Meteorological drought: significant deficiency in the amount of rainfall. b) Hydrological drought: decreased flow rate of river and streams and below level groundwater c) Agricultural drought: a long dry period that results in crop stress and productivity	• Poor vegetative and reproductive growth. • Compromised seed germination and seedling development • Suppressed plant height and reduced leaf area. • Significant reduction in photosynthesis and in leaf weight. • Decreased total dry mass and reduced stomatal conductance	• For flowering and formation of grains: 2% Diammonium phosphate (DAP) + 1% Potassium chloride (KCl) foliar spray • For moisture stress management: 3% Kaoline spray • For extreme water stress management: 500 ppm Cycocel foliar spray	Vijayalakshmi (2018); Aslam et al. (2017)
Flood/ Waterlogging	Caused due to the overflow of water from the river, sudden increase in the water level due to heavy precipitation Types a) Flash flood/ Short duration flood: Sudden and uncontrollable flood that lasts a few weeks and imposes less impact as compared to deep flood. Here, the water level can reach up to 50 cm b) Deep waterflood: Uncontrollable heavy rain that lasts for a prolonged period time. It can largely affect the crop yield, even damage the whole productivity	• Leaves start to die, and decay at a huge rate. Wilting, abscission, epinasty and lenticels formation are the other stress impacts	• For stagnant water: proper drainage system can be provided to drain excess water and use it for the root system • For promoting growth: 500 ppm Cycocel spray. It has growth retarding properties that arrest apical dominance and promotes lateral growth • For flowering and formation of grains: 2% DAP + 1% KCl foliar spray (MOP) • For increasing photosynthetic activity: 0.5 ppm brassinolide foliar spray • For moisture stress management: 100 ppm salicylic acid foliar spray is used to enhance stem reserve utilization • For extreme stress management: 0.3% Boric acid + 0.5% $ZnSO_4$ + 0.5% $FeSO_4$ + 1.0% urea foliar spray is used. Sufficient amount of potassium (K) must be provided	Hattori et al. (2011); Vijayalakshmi (2018)

(continued)

Table 7.1 (continued)

Types	Definition	Stress impacts	Mitigation steps	References
Salinity	Water and soil containing excessive amounts of soluble salts that affect the normal functioning of the crop plants There are two main sources of salinity: a) Primary/ Natural sources Resulting from weathering of minerals and the soils developed/derived from saline parent rocks b) Secondary salinization Caused by human factors such as irrigation, deforestation, overgrazing, or intensive cropping	• Salt puts two types of primary effects on plants: osmotic stress and ionic toxicity • It changes the chemical properties of soil • It alters the physical properties like a decrease in hydraulic conductivity and reduced speed of irrigation water • Salinity of soil also affects the residing microflora	• Seeds can be hardened with 10 mM NaCl and 50% Gypsum may be added in the soil before planting • For increasing photosynthetic activity: 0.5 ppm brassinolide foliar spray may be used • For extreme stress management: 2% DAP + 1% KCl (MOP). • For preventing pre-mature fall of flowers/ buds/ fruits: 40 ppm of NAA and 100 ppm salicylic acid spray • Application of N and K fertilizers in excess is recommended • For enhancing the crop quality: Ascorbic acid spray can increase the number of leaves and leaf areas and can increase the total biomass and height of the plant if combined with $ZnSO_4$ • Application of plant growth hormones exogenously can diminish the adverse effects of salt stress, thereby promoting germination, growth, development, quality seed yields • Abscisic acid (ABA) application can reduce leaf abscission in plants in salt stress conditions • A combination of 4 mM ascorbic and 4 mM gibberellin could alleviate the adverse effects of salinity on plants while Jasmonic Acid can ameliorate salt stress post-application • K/Na ratio should be maintained	Ashraf and Wu (1994); Lal and Khanna (1994); Vijayalakshmi (2018); Flowers (2004)

(continued)

Table 7.1 (continued)

Types	Definition	Stress impacts	Mitigation steps	References
Temperature	Temperature above or below the normal threshold level that cause a reduction in growth, biological and physiological activities, and morphological development Direct exposure to high temperatures can even induce tissue injury, and genetic mutation Types: a) High temperature/ Heat b) Low temperature/ Cold	• Reduction in plant growth, development, and total biomass Untimely drying of leaves and seeds • Photosynthesis and pollen development are hampered, often cause sterility • Yield quality is highly affected	Mitigating heat stress: • Shade cultivation of plants is recommended • Gibberellic Acid (GA) stimulates the α-Amylase synthesis thereby promoting seed germination • BAP can be used to reduce the leaf senescence and lipid peroxidation • Application of salicylic acid enhances the thermotolerance capacity and glycine betaine can help reduce the ion leakage • Overhead irrigation can avoid plant sunburn • Seed germination can be enhanced by the application of Ethylene Mitigating cold stress: • 0.15% Ammonium molybdate foliar spray and cryoprotectants can reduce the cold stress effect too much extent • GA3 and proline can enhance seed germination • Ion leakage can be minimized by using 50 ppm uniconzole • Paclobutrazol can increase the activity of scavenging enzymes • Application of ABA can increase the freeze tolerance of the crop plants	Greaves (1996); Levitt (1980); Vijayalakshmi (2018)

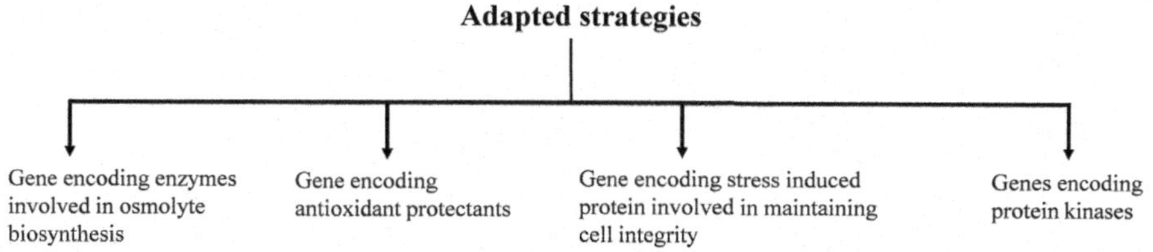

Fig. 7.2 Strategies adapted by crop plants. (Figure constructed by K. Kisku based on Tayal et al. [2004] and Sangam et al. [2005])

scavenging proteins, antioxidant molecules, and growth hormones like abscisic acid (ABA), various stress-induced proteins, and protectants.

The introduction of high throughput omics technology has enhanced our understanding of the newly evolved response mechanisms and pathways adapted by the plants. It has broadened our vision into the gene regulations, thereby regulating the synthesis of specific functional proteins. Omics tools like genomics (involves the analysis of gene with mutations and regulatory functions), proteomics (offers insight into the regulatory pathways and structural arrangements of the proteins), metabolomics (analysis of different metabolites or molecules involved in the metabolic activities) and transcriptomics (used for profiling diverse coding and non-coding genes and the expression of their respective mRNA's) has wide application in the study of stress tolerance proteins (Vij and Tyagi 2007) bioinfor. Besides, bioinformatics has also put its footmark by providing information about the resistance gene similarity between the species and its expression. The collection of all the information data through various omics tools provides better knowledge of the stress-induced plants. Factors involved in the environmental stress response and arrangements of various factor proteins can also be checked by structural and system biology. Various researches have been done taking individuals into concern like drought, heat, and flood. Studies report that these three stresses have a major impact on the plants and impose notable damage to the growth and developmental stages both physically and biochemically. Plants give physiological responses to maintain its rate of transpiration, photosynthesis, respiration, and osmotic balance, however, though protecting membrane breakage is a top priority. Heat stress brings deleterious effect on the leaf parts including its reproductive organs, which makes it difficult for the plants to reproduce and this might contribute up to 40% yield losses. Plant responses against heat stress involve the induction of nitric oxide (NO), reactive oxygen species (ROS), heat stress factors (HSF), and several other tolerant proteins and signaling factors (Nadeem et al. 2018). Drought on the other hand induces osmotic disbalance and the photosynthesis rate is also hampered. Studies have shown the need to assess all the wild species of plants and filter the molecular techniques that would help in elucidating the underlying mechanisms of stress responses. Application of transgenic plants to withstand the stressful environmental conditions and reduce the loss of production is now being recommended by most of the researchers. Using water efficiently by plants could increase their survivability during drought stress by limiting the transpiration and respiration rates. Some C_3 plants like *Arabidopsis* sp. naturally have this capability of using water efficiently than C_4 plants, however, this ability has its drawbacks too. It reduces the uptake of carbon dioxide, thereby affecting the photosynthesis which ultimately reduces the crop yield. Growth and development of such plants are also halted at intervals that result in low productivity. Stress tolerance genes identification can be done using QTLs analysis that helps to identify the specific tolerant genes. The data could be of use to understand their mechanisms to generate new modified crop plants.

7.3 Existing Biotechnological Strategies for Abiotic Stress Tolerance

Crop plants are constantly exposed to various types of climatic and environmental fluctuation. Although plants have developed several internal mechanisms to adapt to the physical environment, not all the plants are able to do it efficiently. Also, plants lack many of the strategies available to the animals to protect themselves from the environmental challenges. Plants bring about changes in their metabolism, genes and proteins to combat with the fluctuation of the external environment. Abiotic stresses, for example, salinity brings detrimental effects on the plants and can cause potential damage. Variation in the environmental solute concentration can impose osmotic stress to the crop plants at the cellular level and weaken the turgor pressure. This will lead to water unavailability eventually causing growth hinderance (Kumar et al. 2018). To enhance the nutrient uptake capacity and identify the specific genes involved in the stress response, novel transgenic techniques and nanoparticles have now evolved (Ashraf and Wu 1994) (Fig. 7.3).

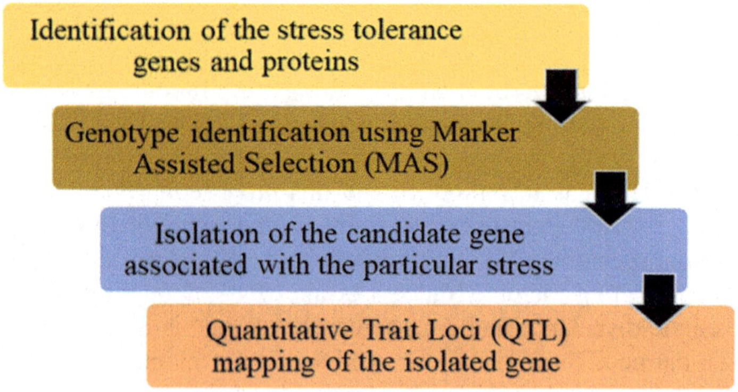

Fig. 7.3 Biotechnological strategies to combat abiotic stress (Kumar et al. 2018) (Figure constructed by K. Kisku)

The hereditary investigation have revealed multiple loci that offer abiotic stress tolerance. There are many genes yet to be recognized who carry crucial features. Identification of such agronomically important crop plants and their stress tolerating genes would create novel variety of highly stress resistant crop species. Genome studies have identified many stress tolerance genes as well as proteins followed by subsequent studies. Marker assisted selection (MAS) is utilized to target the specific gene of interest rather than collecting the entire gene. MAS commonly utilizes restriction fragment length polymorphisms (RFLP) markers. PCR-based markers like random amplified polymorphic DNA (RAPD), Amplified fragment length polymorphism (AFLP), Simple sequence repeats (SSR) and sequence tagged sites (STS) are current new markers utilized in the study of genetic map (Gupta and Rustogi 2004; Sehgal et al. 2008). Identification and isolation of candidate genes associated with the stress show molecular polymorphisms of the trait involved. It is hereditarily related with quantitative trait locus (QTL), which is represented by polygenes. Candidate genes markers are used in identifying, separating and cloning the stress resistant genes and provides an insight into the molecular pathways and mechanisms involved. Polygenes representing QTLs exhibit minimum effect on the traits of interest. It is highly influenced by the environment. Thus the application of molecular markers provide quantitative assessment of the inherited traits responsible for the stress tolerance and help to distinguish between the genetic maps of loci accurately. Moreover, these markers brings many unanswered questions to light by its outstanding contribution in the agricultural sector (Broman and Speed 1999).

The production of genetically modified crops, are also on the go (Ran et al. 2017; Kim et al. 2012). Studies have provided breakthrough results in genetically engineered crop plants, tissue engineering, and in the production of nanoparticles. One of the preeminent accomplishments so far is the targeted delivery of CRISPR (clustered regularly interspaced short palindromic repeats)/ Cas mRNA (CRISPR-associated protein) and sgRNA via nanoparticles. Besides, nanotechnology has the requisite answers to the existing agricultural problems with the addition of the application of nanosensors to enhance the defense ability of the crop plants against stress and disease (Afsharinejad et al. 2016; Kwak et al. 2017).

7.4 Transgenic Plants as Alternative

Since the importance of protecting to the crop plants to fulfill the demands of the growing population is touching its peak, the emerging idea of transgenic plants has helped us achieve it to some extent. Genetic engineering has confirmed its application in gene regulation and the assessment of the biochemical process, besides increasing our understanding of using the engineering tools to improve the crop plants. Development of the plant is hampered by the adverse effects of the diverse biotic and abiotic stress, therefore making it necessary to stabilize the transgene expression by the application of various promoters such as tissue-specific promoters at different stages. From the basic research to the development of better quality plant materials and its economic availability, there is rising need of variety of promoters for transgene expression and so do their isolation from broad range of organisms. Now, the question arises which types of promoters are meant to be isolated from the organisms. Stress inducible promoters including tissue-specific, constitutive, viral, and synthetic promoters (Potenza et al. 2004) are highly recommended by various studies. Different stress-inducible promoters vary in their transgene expression pattern that holds influence on the tolerance and resistance of the crop plants to abiotic stresses, thereby making the selection of inducible promoters an important job. Preferring stable and legitimate stress-inducible promoter incubates successful transgene expression via proper gene transfer that accounts for their tolerance activity against abiotic stresses at different developmental stages. Stress tolerance capacity can be increased by using stress-inducible promoters (having relatively less expression) with their specific transgenes (Bhatnagar-Mathur et al. 2008). Promoters are the DNA sequence that lies upstream in the coding region (Buchanan et al. 2001). This coding region of the gene is recognized by various proteins and their expression is regulated during transcription.

The expansion of genetic engineering in plants has unraveled diverse information about the activity of promoters. These promoters possess the ability to get induced when they encounter abiotic stresses, like salinity or heat stress. One such promoter is a constitutive promoter. Their expression neither depends on endogenous factors nor on the environmental factors. Constitutive promoters are expressed in all tissue and are commonly found active in almost all species. One of such overexpressed promoter is found in plant virus-like CaMV 35S promoter (Cauliflower Mosaic virus) that causes a rise in the expression of their specific transgene both in dicots and monocots (Benfey et al. 1990). Promoters from a diverse group of endogenous plants, especially those derived from actin (a basic cytoskeletal component) and ubiquitin (mainly known for its transforming property), are used to regulate the transgene expression (Gupta et al. 2001; Dhankher et al. 2002).

7.5 Emerging Field of Nanotechnology

There is undoubtedly no wonder that the world's population is indeed increasing drastically with an estimation of approximately 9 billion by the year 2050 (United Nations 2017). With the magnification of the human population, the resource demands are also on its way to touching heights. However, the existing economy and human-technology interface is not sufficient to fulfill the never-ending needs of society. Therefore, up-gradation and validation of the technical methods and protocol are imperative (Godfray et al. 2010). Plant genetic engineering has worked as a boon in making the crop plants more resistant and tolerant against the adverse effects of the distinct abiotic stresses. Plants face frequent episodes of environmental stresses mainly abiotic stress. Heat, salinity, and drought are the major limitations to the crop plants lowering the productivity to a remarkable extent. To survive these conditions, there is a constant endeavor by various concerns to combat the stress effects and enhance the tolerating potential of the plants. The world agriculture crop production challenge now is to reach a 70% increment in the crop production for the upcoming additional population estimated to be reached by 2050 (United Nations 2017). This has bought scientists together from all over the world to develop environment friendly and efficient technology to overcome the crop yield challenges. Nanobiotechnology is a rising technological field that involves the application of the biotechnologies at the nano-level (i.e., nanoparticles) to unveil the complex biological systems to win the battle against the biotic and abiotic stresses (Banerjee and Kole 2016; Cheng et al. 2016). It is proving its efficiency in multi-disciplinary areas as well and is emerging as a promising method to reduce the crop damage by various environmental constraints. Nanoparticles are synthesized by various methods from metal, metal oxides, or from plants. Using physical, biological, or chemical methods to derive nanoparticles out of metals has been carried out for decades which sometimes prove to be very expensive. Researchers have now come up with the idea of green nanoparticles that can be synthesized from plants and is cost-effective (Iravani 2011; Sharma et al. 2009). The potentiality and efficiency of plant synthesized nanoparticles are under investigation to evaluate their ability in protecting the crop plants from stresses, limiting the losses, increasing the growth and development of the plants and, thereby enhancing the crop productivity. An overview of the research works done until 2020 on agriculture nanotechnology are depicted in Fig. 7.4. The data presented in Fig. 7.4, obtained from PubMed and ScienceDirect, shows that USA is most concerned about its food security.

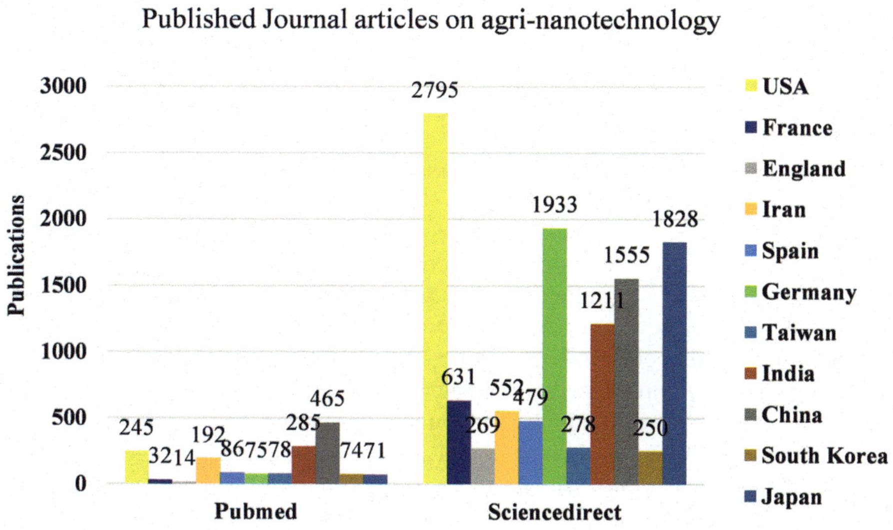

Fig. 7.4 List of publications on agri-nanotechnology from 2010 to 20 (Figure constructed by K. Kisku)

Nanotechnology-based application requires nanoformulations of agrochemicals for field application. Nanosensors and nanodevices are widely used for the identification of plant diseases and agrochemical resideue. Nontheless, nanotechnology has the potential to diagnose and treat several plant and animal diseases besides minimizing the side effects (Ghidan and Al-Antary 2019). The available types of nanoparticles and their use in various crop plants is shown in Table 7.2.

7.5.1 *Types of Nanoparticles*

The emergence of nanotechnology has paved the way for the broad spectrum application in the multidisciplinary field. Nanoparticles (NPs) exists naturally and can be synthesized by various natural calamities like volcanic eruptions, simple soil erosion, forest fires, and photochemical reactions. NPs can also be produced from plants, animals, and microorganisms (Love et al. 2005; Dahoumane et al. 2017). The conventional method of nanoparticle synthesis from metals (e.g., Au, Ag, Pd) or metal oxides (e.g., ZnO, SiO_2, TiO_2) includes physical, chemical, and biological methods (Singh et al. 2016). Green synthesis is another alternative method for the production of some metallic NPs that utilizes the phytochemicals and the enzymes present in the plant parts or their extracts. The reduction of metal salts to their respective nanoparticles is assisted by the phytochemical compounds of the plant that act as a reducing agent during the synthesis. This ecofriendly synthesized nanoparticle imprints its usage in enhancing the plant growth, its defense capacity, and total yield by reaching the high surface to volume ratio. These compounds act as reducing and stabilizing agent during synthesizing of metal nanoparticles from the metal salts that help in finding most promising and eco-friendly nanoparticle synthesis solutions, which provides a controlled synthesis with well-defined size and shape but also prevent the atmosphere pollution (Kumar and Yadav 2009; Sharma et al. 2009; Siddiqui et al. 2014). Nanoparticles (NPs) attain high surface to volume ratio which enhances their bioavailability, bioactivity, and other biochemical activities (Dubchak et al. 2010). Therefore, with increased advances made by applying tools of nanobiotechnology in the agricultural sector, it is assumed that it will help to augment plant growth, development and productivity, and biotic and abiotic stress tolerance. It was also observed that under certain conditions plants are capable of producing naturally mineralized nano-materials (NMs) necessary to their growth. It is also expected that as the understanding of nanotechnology deepens, it will help to exploit nanotechnology to become a major economic driving force that will benefit consumers as well as farmers with no adverse effect on humans and the environment. The different types of nanoparticles commonly used today is shown is Fig. 7.5.

Table 7.2 List of commonly used nanoparticles in agriculture

Particle name	Chemical formula	Preferred particle size	Doses	Particle description	Stress	Commonly used crop plants	Role in plants	References
Silicon dioxide nanoparticle	SiO_2 NPs	< 7.0 nm	2.5 mM/L	It is the second most abundant found in the earth's crust after O2. It is green and eco-friendly alternative	Drought, Salinity, Cold, Heavy metal toxicity	*Oryza sativa*	• Promotes growth and development of plant • Maintains xylum humidity and balances turger pressure • Improved water translocation and water uptake capacity • Enhanced stress tolerance capacity • Ameliorated tolerance against heavy metal toxicity	Liang et al. (2007); Dantnoff et al.; Ma (2004); Rastogi et al. (2019); Wang et al. (2016)
Titanium dioxide nanoparticle	TiO_2 NPs	< 0.1 µm	0.25% suspension	It has photocatalytic nature and antimicrobial and antioxidant properties	Salinity	*Spinacia oleracea*	• Ameliorated plant growth • Increased protein, chlorophyll and nitrogen content • Increased total biomass • Enhanced antioxidant stress tolerance by stimulating antioxidant enzymes	Hong et al. (2005); Gohari et al. (2020); Yang et al. (2007); Janmohammadi et al. (2015)
Zinc oxide nanoparticles	ZnO NPs	<25 nm	10 mg/L, 20 mg/L, 0.2 µM, 1 µM	Zn is non-toxic, eco-friendly and bio-friendly material. It can be found commonly in the soil. It is also used as green reagent having antimicrobial properties	Drought, Salinity	*Coffea Arabica, Cyamopsis tetragonoloba Triticum aestivum, Nicotiana tabacum*	• Improved metabolic activities in plants • Enhanced plant growth, nutrient content, photosynthesis rate • Increased biomasss accumulation and enzymatic activities • Reduced toxic heavy metal uptake • Provides important micronutrients to plants	Pandey et al. (2010); Sabir et al. (2014); Das et al. (2019); Rossi et al. (2019); Raliya and Tarafdar (2013); Du et al. (2019); Tirani et al. (2019)

(continued)

Table 7.2 (continued)

Particle name	Chemical formula	Preferred particle size	Doses	Particle description	Stress	Commonly used crop plants	Role in plants	References
Silver nanoparticle	AgNPs	1–100 nm	50 mg/L, 75 mg/L, 50–100 μg/mL	Synthetic/engineered nanomaterials have industrial application	Heat stress, Water stress	*Triticum aestivum, Vigna sinensis, Vigna unguiculata*	• Effective pest manager • It has antifungal properties • Increases soil microbial communities and biomass accumulation • Enhances activity of antioxidant enzymes • Promotes growth and stimulate root nodulation • Could inhibit growth of *Xanthomonas axonopodis pv. malvacearum* and *Xanthomonas campestris pv. campestris* in vitro	Davies (2008); Anand and Bhagat (2019); Iqbal et al. (2019); Pallavi Mehta et al. (2016); Vanti et al. (2019)
Aluminium oxide nanoparticles	Al_2O_3 NPs	2–10 nm	400 mg/L	Mainly used in industries, have good thermal conductivity, has high strength and stiffness	Temperature	*Solanum lycopersicum*	• Direct exposure leads to phytotoxicity • Imposes negative impact on growth and development of plants • Could control root rot caused by fungus	Burklew et al. (2012); Vardar and Yanik (2015); Shenashen et al. (2017)

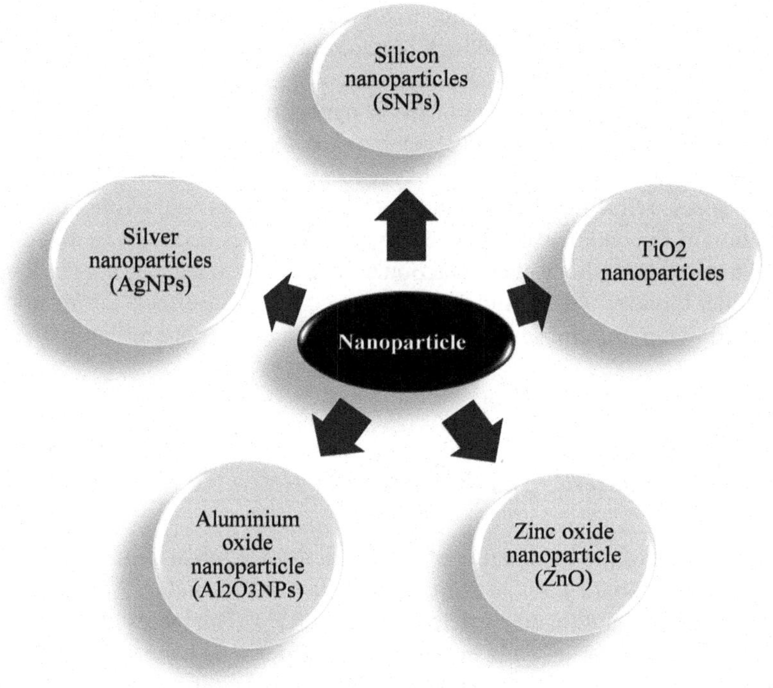

Fig. 7.5 Different types of nanoparticles in use (Figure constructed by K. Kisku)

7.5.2 Role of Nanoparticles on Plant

Engineering nanoparticles in agricultural development is gaining momentum recently as one of the new and updated technological aspects. Because nanoparticles are very small molecules and come in different sizes ranging from 1–100 nm, they could be of great interest not only in the field of agriculture but also in the field of pharmacy, food processing, environmental science, and industrial production (ASTM E2456-06 2006). The characteristics features of nanoparticles like improved fertilizer delivery, targeted release of biofertilizer, and eco-friendly nature makes it a reliable method (Abobatta 2018). Besides nanotechnology applications have helped mitigate some of the novel agricultural challenges and have offered sustainable agriculture to great extent. However, the detailed information on the mechanism of nanoparticle-based application is still under progressive study. In recent years, 'precision farming' has been globally introduced that inspects over the site-specific agricultural practices covering horticulture crops to field crops through wireless network and sensors (Gouma et al. 2016; Chhipa and Joshi 2016). Nanoparticles (NPs) derived from silver, gold, or hybrid nanomaterials are reported to have a response to foreign stimuli which results in the balanced release of macromolecules from the NPs. Since they have unique and distinct physicochemical properties such as varying particle morphology, enhanced reactivity due to high surface to volume ratio, wide pore size range, reduced surface area, and high surface energy, they are employed as vehicles to transfer agrochemicals or other molecules of interest to different plants.

Though the intelligent application of the nanoparticles in various fields of agriculture has been accepted worldwide, they may, on the other hand, display toxicity by the accumulation of toxic molecules like reactive oxygen species (ROS) and reactive nitrogen species (RNS) at various concentrations. Studies on the role of NPs in toxic production at the cellular and molecular level have not been covered yet. Furthermore, knowledge on their mode of action and response to the varying environmental stress either biotic or abiotic and their advantages and side effects during application in the field of agricultural science are scanty and require more research to unveil the cover.

7.5.3 Development of Green Nanoparticles (GNPs)

With the advent of the application of nanoparticles for target delivery, derived from metals and metal oxides, there is a constant demand for natural, eco-friendly, and economically available nanoparticles (Iravani 2011). This greased the way for the imminent nano-technology and hence the idea of deriving nanoparticles from plant-microbes interaction came into existence. The more simple approach, i.e., the green synthesized nanoparticles was explored and accepted widely. The microorganisms commonly used for the synthesis include bacteria like *Lactobacillus sporogens* (ZnO2-NPs synthesis), diatoms (Si-NPs synthesis), *Bacillus megaterium* (Ag/Pb/Cd-NPs synthesis), *Pseudomonas stutzeri* (Ag-NPs synthesis), *Desulfovibrio desulfuricans* (Pd-NPs synthesis) and many more (Chokriwal et al. 2014). The green nanoparticle is mostly derived from plants and microorganisms and have now become a matter of great attention. Because of the availability and eco-friendly nature of the green nanoparticles, they are undergoing various research investigations. More emphasis is given on the different methods to produce them cost-effectively and to obtain them in large quantities. It does not exploit mineral ores and has no side effects as compared to metal nanoparticles. Some of the microorganisms that are considered to be of great source of green nanoparticles are *Aspergillus*, *Verticillium* sp, and *Fusarium oxysporum* (Kwak et al. 2017).

7.6 Application of Nanobiotechnology in Agronomy

Nanotechnology has become the answer to every industrial and agricultural question. With its new tools, nanotechnology has the capacity to address the obstacles laid in the way of machine, medicine and agriculture. It can now execute rapid detection and treatment of plant disease using sensors and smart delivery system (Abd-Elrahman and Mostafa 2015). Nanoparticles can be a promising approach in plant science and its potency can be effectively utilized to counter the calumnious impact of the various biotic and abiotic stresses besides improving the plant growth quality (Panpatte et al. 2016). It can be used directly in the agricultural fields as nano-fertilizers (whose influential entry can release required contents at the targeted sites of the plants) as well as growth promoters. Routine use of the bulk quantity of fertilizer was common in practice to increase crop productivity, which does not meet the requirements of the plants due to its uneven absorption because of which most of their parts stay unused. This may be due to many factors such as hydrolysis, decomposition, soil leaching, and photolysis (Singhal et al. 2015). Application of nanofertilizers has potential benefits in crop improvement, increment in biomass, and productivity. Because of its small size, it is very useful for target delivery. Nanoparticles can be targeted to the specific gene of interest that could help enhance the resistivity of the plants against the environmental constraints. 'Precision farming' can be achieved by using engineered nanosensors and nano pesticides that not only contribute to crop protection but also enhance the soil quality.

7.6.1 Application of Nanofertilizers

Crop plants are traditionally cultivated using chemical fertilizers to enhance the yield, but they do not absorb all the chemical nutrients provided to them which consequently lead to the accumulation of a large amount of chemical fertilizer in the soil and this ultimately renders the soil infertile. Therefore, the nano fertilizers are now recommended over chemical and biofertilizers as they have the potential to provide a requisite amount of nutrients as per need in just a small quantity. These nanofertilizers provide efficacy to the crop plants to tolerate environmental constraints as well as to reduce its toxicity. In one of the studies, it has been reported that vegetative features of the plants can be improved physiologically and morphologically via the application of Si-nanoparticles and Si-fertilizer under stress conditions (salinity stress) (Janmohammadi et al. 2015).

Water being the imperative source of life can greatly hamper agriculture if not present in a sufficient amount or if present in excessive amount. One such condition that is commonly faced by the crops of the arid region is drought. This environmental stress reduces the productivity to nearly half of the bulk resulting in the nutritionless growth of the crops. Nanoparticles have become the solutions to such problems and Si-NPs particularly have made the task easier. Zn-NPs have also shown promising results (Cakmak et al. 1996). Application of nano-fertilizers on various crop species have been shown in Table 7.2.

7.6.2 *Mode of Application*

Delivery of the nanomaterials to the crop plants can be done through their roots, shoots, or leaves. From the leaves, they can be absorbed via micro openings like stomata, hydathodes, and bark pores (Eichert et al. 2008). Nanomaterials used in nano fertilizers are now available in various shapes and sizes for example 2D forms (i.e., sheets, disks, rod, tubes), nanoclays, and nanoemulsions. Delivery methods have an impact on the uptake efficiency as reported by some of the studies. Once these NPs are absorbed by the different organs after crossing the external multi-membrane layers (containing cellulose or hemicellulose, various other proteins), they are transferred passively to the targeted sites. Cuts or wounds on the plant parts caused somehow also act as a route to enter into the plant system. Understanding the physiology of nanomaterials-plant interface is always recommended by various studies to better the knowledge of their dynamics.

7.7 Conclusions and Prospects

To meet the demands of the growing population, paramount advancement has been achieved not only in the technological field but also in the field of plant science. So far we have seen improvement in crop productivity by engaging transgenic tools and genetically engineered crops in the agricultural area. The understanding of the stress biology has guided us to manipulate the stress responses by characterizing each stress gene and measuring their efficacy in environmental stress tolerance. Transgenic and genetically engineered approaches have provided a better understanding of the metabolic pathways and molecular mechanisms of the stress-tolerant plants, thereby facilitating the stress resistance or stress tolerance potentiality of crop plants. The implication of distinct nanoparticles of varying shapes and sizes in the crop field has shown remarkable enhancement in the nutrient uptake, thereby increasing crop productivity. Scientific reports have shown that targeted delivery by the nanomaterials assists better protection against diverse biotic and abiotic stresses whilst augmenting the quality crop yield. Furthermore, the emergence of nanosensors in 'precision farming' has bridge the gap between soil health and plant productivity to a major extent. It evaluates the crop yield, quality of biomass produce, soil health, monitors the nutrient uptake efficacy, and promotes protection against disease. Therefore, the application of nanotechnology offers breakthrough advantages over conventional methods of agricultural farming and quality crop produce. It minimizes the misconception related to the usage of nanomaterials. Besides, it also lay out strategies to elevate the crop tolerance to abiotic constraints (mainly heat, drought, and salinity).

World agriculture has seen revolutionary alternations in the method of farming with the advantages of nano-fertilizers, and nanoparticles, besides the application of transgenic crops. The development of nanotechnology in the field of agriculture is gaining interest widely, yet its applications are still in the lag phase. There lies a wide gap in the understanding of its potential in reducing the risks of farming like nutrient uptake capacity, adaptation mechanism of withstanding stress, and undesirable toxic effects imposed by the nanoparticles. Further research is needed to extrapolate the untouched areas of nanotechnology especially in the field of agriculture. A number of scientific investigations have been done on the potential benefits of nanoparticles in the field of pharmaceutical, industrial and medicine. But very few studies have been carried out on the side- effects of nanoparticles, accumulation of its toxic materials and nanoparticle-cell interactions. There are handful of challenging areas in the agri-nanotechnology sectors. Genetic crop engineering by the incorporation of novel stress-tolerant characters also leaves some of the questions unanswered. With the increase in industrial utilization of nanoparticles as nanobiosensors for secondary metabolites detection or as nanofertilizers in farming, there is a need to optimize the size and dose of NPs before application. Understanding the interaction of the plant cells with the nanomaterials, transfer of DNA, and proteins to the competent cell would also prove to be new research aspects in the future.

References

Abd-Elrahman SH, Mostafa MAM (2015) Applications of nanotechnology in agriculture: an overview. Egypt J Soil Sci *55*(2):197–214

Abobatta WF (2018) Nanotechnology application in agriculture. Acta Scientific Agric 2(6). (ISSN: 2581–365X)

Afsharinejad A, Davy A, Jennings B, Brennan C (2016) Performance analysis of plant monitoring nanosensor networks at THz frequencies. IEEE Internet Things J 3:59–69

Anand R, Bhagat M (2019) Silver nanoparticles (AgNPs): as nanopesticides and nanofertilizers. MOJ Biol Med 4(1):19–20

Ashraf M (1994) Breeding for salinity tolerance in plants. Crit Rev Plant Sci 13:17–42

Ashraf M, Wu L (1994) Breeding for salinity tolerance in plants. Crit Rev Plant Sci 13(1):17–42

Aslam Z, Khattak JZK, Ahmed M (2017) Drought tolerance in cereal grain crops under changing climate. In: Ahmed A, Stockle CO (eds) Quantification of climate variability, adaptation and mitigation for agricultural sustainability. Springer, Cham, pp 18–209

ASTM E2456-06 (2006) Standard terminology relating to nanotechnology, pp 5–6. https://doi.org/10.1520/e2456-06r12

Banerjee J, Kole C (2016) Plant nanotechnology: an overview on concepts, strategies and tools. In: Kole C et al (eds) Plant nanotechnology. Springer, Cham. pp 1–14. https://doi.org/10.1007/978-3-319-42154-4_1

Bartels D, Salamini F (2001) Desiccation tolerance in the resurrection plant Craterstigma plantagineum. A contribution to the study of drought tolerance at the molecular level. Plant Physiol 127(4):1346–1353

Benfey PN, Ren L, Chua NH (1990) Tissue-specific expression from CaMV 35S enhancer subdomains in early stages of plant development. EMBOJ 9:1677–1684

Bhatnagar-Mathur P, Devi MJ, Reddy DS et al (2008) Stress-inducible expression of at DREB1A in transgenic peanut (Arachis hypogaea L.) increases transpiration efficiency under water-limiting conditions. Plant Cell Rep 26:2071–2082

Boyer JS (1982) Plant productivity and environment. Sci 218(4571):443–448

Bray EA, Bailey-Serres J, Weretilnyk E (2000) Responses to abiotic stresses. In: Buchanan BB, Gruissem W, Jones RL (eds) Biochemistry and molecular biology of plants. ASPB, Rockville, MD, pp 1158–1203

Broman KW, Speed TP (1999) A review of methods for identifying QTLs in experimental crosses. In: Seiller-Moiseiwitsch F (ed) Statistics in molecular biology and genetics, IMS lecture notes monograph series 33, pp 114–142
Buchanan BB, Gruissem W, Jones RL (2001) Biochemistry and molecular biology of plants. Plant Growth Reg 35:105–106
Burklew CE, Ashlock J, Winfrey WB, Zhang B (2012) Effects of aluminium oxide nanoparticles on the growth, development, and microRNA expression of tobacco (Nicotiana tabacum). PLoS ONE 7(5):e34783
Cakmak I, Yilmaz A, Kalayci M, Ekiz H, Torun B, Ereno B, Braun HJ (1996) Zinc deficiency as a critical problem in wheat production in Central Anatolia. Plant and soil 180(2):165–172
Calanca PP (2017) Effects of abiotic stress in crop production. In: Ahmed M, Stockle CO (eds) Quantification of climate variability, adaptation and mitigation for agricultural sustainability. Springer, Cham, pp 165–180
Chen WY, Suzuki T, Lackner M (eds) (2017) Handbook of climate change mitigation and adaptation. Springer International Publishing. https://doi.org/10.1007/978-3-319-14409-2
Cheng HN, Klasson KT, Asakura T, Wu Q (2016) Nanotechnology in agriculture. In: Cheng HN, Doemeny L, Geraci CL, Schmidt DG (eds) Nanotechnology: delivering on the promise, vol 2. ACS, Washington, DC, pp 233–242
Chhipa H, Joshi P (2016) Nanofertilisers, nanopesticides and nanosensors in agriculture. In: Ranjan S, Dasgupta N, Lichtfouse E (eds) Nanoscience in food and agriculture 1, Sustainable Agriculture Reviews, vol 20. Springer, Cham. pp 247–282. https://doi.org/10.1007/978-3-319-39303-2
Chokriwal A, Sharma MM, Singh A (2014) Biological synthesis of nanoparticles using bacteria and their applications. Am J Pharm Tech Res 4(6):38–61
Cramer GR, Urano K, Delrot S et al (2011) Effect of abiotic stress on plants: a systems biology perspective. BMC Plant Biol 11:163. https://doi.org/10.1186/1471-2229-11-163
Cushman JC, Bohnert HJ (2000) Genomic approaches to plant stress tolerance. Curr Opin Plant Biol 3(2):117–124
Dahoumane SA, Jeffryes C, Mechouet M, Agathos SN (2017) Biosynthesis of inorganic nanoparticles: a fresh look at the control of shape, size and composition. Bioeng 4(1):14
Das S, Yadav A, Debnath N (2019) Entomotoxic efficacy of aluminium oxide, titanium dioxide and zinc oxide nanoparticles against Sitophilus oryzae (L.): a comparative analysis. J Stored Prod Res 83:92–96
Davies JC (2008) Nanotechnology oversight: an agenda for the new administration. Project on Emerging Nanotechnologies. Woodrow Wilson International Center for Scholars, Washington, DC
Dhankher OP, Li Y, Rosen BP et al (2002) Engineering tolerance and hyperaccumulation of arsenic in plants by combining arsenate reductase and γ-glutamylcysteine synthetase expression. Nat Biotech 20(11):1140–1145
Dresselhaus T, Huckelhoven R (2018) Biotic and abiotic stress responses in crop plants. Agron 8(11):267. https://doi.org/10.3390/agronomy8110267
Du W, Yang J, Peng Q et al (2019) Comparison study of zinc nanoparticles and zinc sulphate on wheat growth: from toxicity and zinc biofortification. Chemosphere 227:09–116
Dubchak S, Ogar A, Mietelski JW, Turnau K (2010) Influence of silver and titanium nanoparticles on arbuscular mycorrhiza colonization and accumulation of radiocaesium in Helianthus annuus. Span J Agric Res 8:s103–S108
Dwivedi S, Saquib Q, Al-Khedhairy AA, Musarrat J (2016) Understanding the role of nanomaterials in agriculture. In: Singh DP, Singh HB, Prabha R (eds) Microbial inoculants in sustainable agricultural productivity. Springer, New Delhi, India, pp 271–288
Eichert T, Kurtz A, Steiner U, Goldbach HE (2008) Size exclusion limits and lateral heterogeneity of the stomatal foliar uptake pathway for aqueous solutes and water-suspended nanoparticles. Physiol Plant 134(1):151–160. https://doi.org/10.1111/j.1399-3054.2008.01135.x
Flowers TJ (2004) Improving crop salt tolerance. J Exp Bot 55(396):307–319

Ghidan AY and Al Antary TM (2019) Applications of nanotechnology in agriculture. In: Applications of nanobiotechnology. IntechOpen

Godfray HCJ, Beddington JR, Crute IR et al (2010) Food security: the challenge of feeding 9 billion people. Science 327(5967):812–818

Gohari G, Mohammadi A, Akbari A et al (2020) Titanium dioxide nanoparticle (TiO_2 NPs) promote growth and ameliorate salinity stress effects on essential oil profile and biochemical attributes of Dracocephalum moldavica. Sci Rep 10:912. https://doi.org/10.1038/s41598-020-57794-1

Gouma PI, Simon SR, Stanacevic M (2016) Nano-sensing and catalysis technologies for managing food-water-energy (FEW) resources in farming. Mater Today Chem 1(2):40–45

Greaves JA (1996) Improving suboptimal temperature tolerance in maize- the search for variation. J Exp Bot 47:307–323

Grover A, Kapoor A, Lakshmi OS et al (2001) Understanding molecular alphabets of the plant abiotic stress responses. Curr Sci (Special Section: Plant Mol Biol) 80:206–216

Gupta P, Raghuvanshi S, Tyagi AK (2001) Assessment of the efficiency of various gene promoters via biolistics in leaf and regenerating seed callus of millets, Eleusine coracana and Echinochloa crusgalli. Plant Biotechnol 18(4):275–282

Gupta PK, Rustogi S (2004) Molecular markers from the transcribed/expressed region of the genome in higher plants. Func Integr Genomics 4:139–162

Hattori YK, Nagai M, Ashikari (2011) Rice growth adapting to deep water. Curr Opin Plant Biol 14:100–105

Hong F, Yang F, Liu C et al (2005) Influence of nano-TiO_2 on the chloroplast aging of spinach under light. Biol Trace Elem Res 104:249–260

IPCC Climate Change (2014) Synthesis report. Available from: http://www.ipcc.ch/report/ar5/syr/

Iqbal M, Raja NI, Hussain M et al (2019) Effect of silver nanoparticles on growth of wheat under heat stress. IJST A Sci 43:387–395

Iravani S (2011) Green synthesis of metal nanoparticles using plants. Green Chem 13:2638–2650

Janmohammadi M, Sabaghnia N, Ahadnezhad A (2015) Impact of silicon dioxide nanoparticles on seedling early growth of lentil (Lens culinaris Medik.) genotypes with various origins. Poljoprivreda i Sumarstvo 61(3):19

Khan MR, Rizvi TF (2014) Nanotechnology: scope and application in plant disease management. Plant Pathol J 13(3):214–231

Kim SW, Jung JH, Lamsal K et al (2012) Antifungal effects of silver nanoparticles (AgNPs) against various plant pathogenic fungi. Mycobiol 40(1):53–58

Kou TJ, Yu WW, Lam SK et al (2018) Differential root responses in two cultivars of winter wheat (Triticum aestivum L.) to elevated ozone concentration under fully open-air field conditions. J Agron Crop Sci 204:325–332

Kumar A, Sengar RS, Singh A, Dixit R, Singh R (2018) Biotechnological tools for enhancing abiotic stress tolerance in plant. In: Sengar R, Singh A (eds) Eco-friendly agro-biological techniques for enhancing crop productivity. Springer, Singapore

Kumar V, Yadav SK (2009) Plant-mediated synthesis of silver and gold nanoparticles and their applications. J Chem Technol Biotechnol 84(2):151–157

Kwak SY, Wong MH, Lew TTS et al (2017) Nanosensor technology applied to living plant systems. Annu Rev Anal Chem (Palo Alto Calif) 10(1):113–140

Lal B, Khanna S (1994) Selection of salt-tolerant Rhizobium isolates of Acacia nilotica. World J Microbiol Biotechnol 10(6):637–639

Levitt J (1980) Responses of Plants to Environmental Stress, Volume 1: Chilling, Freezing, and High Temperature Stresses. Academic Press

Liang Y, Sun W, Zhu YG, Christie P (2007) Mechanisms of silicon-mediated alleviation of abiotic stresses in higher plants: a review. Environ Pollut 147(2):422–428. https://doi.org/10.1016/j.envpol.2006.06.008

Love JC, Estroff LA, Kriebel JK et al (2005) Self-assembled monolayers of thiolates on metals as a form of nanotechnology. Chem Rev 105(4):1103–1169. https://doi.org/10.1021/cr0300789

Ma JF (2004) Role of silicon in enhancing the resistance of plants to biotic and abiotic stresses. Soil Sci Plant Nutr 50:11–18
Nadeem MA, Nawaz MA, Shahid MQ, Doğan Y, Comertpay G, Yıldız M, Baloch FS (2018) DNA molecular markers in plant breeding: current status and recent advancements in genomic selection and genome editing. Biotechnol Biotechnol Equip 32(2):261–285
Niculaes C, Abramov A, Hannemann L, Frey M (2018) Plant protection by benzoxazinoids—recent insights into biosynthesis and function. Agronomy 8(8):143
Pallavi Mehta CM, Srivastava R, Arora S, Sharma AK (2016) Impact assessment of silver nanoparticles on plant growth and soil bacterial diversity. 3 Biotech 6:254
Pandey AC, Sanjay SS, Yadav RS (2010) Application of ZnO nanoparticles in influencing the growth rate of Cicer arietinum. J Exp Nanosci 488–497
Panpatte DG, Jhala YK, Shelat HN, Vyas RV (2016) Nanoparticles: the next generation technology for sustainable agriculture. In: Sing D, Sing H, Prava R (eds) Microbial inoculants in sustainable agricultural productivity. Springer, New Delhi, India, pp 289–300. https://doi.org/10.1007/978-81-322-2644-4_18
Potenza C, Aleman L, Sengupta-Gopalan C (2004) Targeting transgene expression in research, agricultural, and environmental applications: promoters used in plant transformation. In Vitro Cell Dev Biol-Plant 40(1):1–22
Raliya R, Saharan V, Dimkpa C, Biswas P (2017) Nanofertilizer for precision and sustainable agriculture: current state and future perspectives. J Agric Food Chem 66(26):6487–6503
Raliya R, Tarafdar JC (2013) ZnO nanoparticle biosynthesis and its effect on phosphorous-mobilizing enzyme secretion and gum contents in Clusterbean (Cyamopsis tetragonoloba L.). Agric Res 2:48–57
Ran Y, Liang Z, Gao C (2017) Current and future editing reagent delivery systems for plant genome editing. Sci China Life Sci 60(5):490–505
Rastogi A, Tripathi DK, Yadav S et al (2019) Application of silicon nanoparticles in agriculture, vol 3. Biotech 9(3):90
Rossi L, Fedenia LN, Sharifan H et al (2019) Effects of foliar application of zinc sulfate and zinc nanoparticles in coffee (Coffea arabica L.) plants. Plant Physiol Biochem 135:160–166
Sabir S, Arshad M, Chaudhari SK (2014) Zinc oxide nanoparticle for revolutionizing agriculture: synthesis and applications. Sci World J Article ID 925494, 8 pages
Sangam S, Jayasree D, Reddy KJ, Chari PVB, Sreenivasulu N, Kishor PB (2005) Salt tolerance in plants-transgenic approaches. J Plant Biotechnol 7(1):1–15
Sehgal D, Bhat V, Raina SN (2008) Applicability of DNA markers for genome diagnostics of grain legumes. In: Kirti PB (ed) Handbook of new technology for genetic improvement of grain legumes. CRC Press, New York, pp 497–557
Sharma VK, Yngard RA, Lin Y (2009) Silver nanoparticles: green synthesis and their antimicrobial activities. Adv Colloid Interface Sci 145(1–2):83–96
Shenashen M, Derbalah A, Hamza A et al (2017) Antifungal activity of fabricated mesoporous alumina nanoparticles against root rot disease of tomato caused by Fusarium oxysporium. Pest Manag Sci 73:1121–1126
Siddiqui MH, Al-Whaibi MH, Faisal M, Al Sahli AA (2014) Nano-silicon dioxide mitigates the adverse effects of salt stress on Cucurbitapepo L. Environ Toxicol Chem 33(11):2429–2437. https://doi.org/10.1002/etc.2697
Singh A, Singh S, Prasad SM (2016) Scope of nanotechnology in crop science: profit or loss. Res Rev J Bot Sci 5(1):1–4
Singhal P, Jan AT, Azam M, Haq QMR (2015) Plant abiotic stress: a prospective strategy of exploiting promoters as alternative to overcome the escalating burden. Front Life Sci 9(1):52–63
Tayal D, Srivastava PS, Bansal KC (2004) Transgenic crops for abiotic stress tolerance. In: Plant biotechnology and molecular markers. Kluwer Academic Publishers, Springer, Dordrecht, pp 346–365

Tirani MM, Haghjou MM, Ismaili A (2019) Hydroponic grown tobacco plants respond to zinc oxide nanoparticles and bulk exposures by morphological, physiological and anatomical adjustments. Funct Plant Biol 46:360–375

United Nations (2017) Department of Economic and Social Affairs, Population Division. World population prospects: the 2017 revision, key findings and advance tables. Working Paper No. ESA/P/WP/248 (ed) New York, United Nations

Vanti GL, Nargund VB, Basavesha KN et al (2019) Synthesis of Gossypium hirsutum-derived silver nanoparticles and their antibacterial efficacy against plant pathogens. Appl Organomet Chem 33:e4630

Vardar F, Yanik F (2015) Toxic effects of Aluminium oxide (Al_2O_3) nanoparticles on root growth and development in Triticum aestivum. Water Air Soil Pollut 226:296

Vij S, Tyagi AK (2007) Emerging trends in the functional genomics of the abiotic stress response in crop plants. Plant Biotechnol J 5(3):361–380. https://doi.org/10.1111/j.1467-7652.2007.00239.x

Vijayalakshmi D (2018) Abiotic stress and its management in agriculture. TNAU Agritech, Coimbatore, pp 361–387

Wang SH, Wang FY, Gao SC, Wang XG (2016) Heavy metal accumulation in different rice cultivars as influenced by foliar application of nano-silicon. Water Air Soil Pollut 227:228

Xiao M, Song F, Jiao J et al (2013) Identification of the gene Pm47 on chromosome 7BS conferring resistance to powdery mildew in the Chinese wheat landrace Hongyanglazi. Theor Appl Genet 126:1397–1403

Yang F, Liu C, Gao F et al (2007) The improvement of spinach growth by nano-anatase TiO_2 treatment is related to nitrogen photoreduction. Biol Trace Elem Res 119:77–88

Zhao C, Liu B, Piao S et al (2017) Temperature increase reduces global yields of major crops in four independent estimates. Proc Natl Acad Sci USA 114(35):9326–9331. https://doi.org/10.1073/pnas.1701762114

Chapter 8
Green Synthesis of Nanoparticles Using Different Plant Extracts and Their Characterizations

Lina M. Alnaddaf, Abdulsalam K. Almuhammady, Khaled F. M. Salem, Maysaa T. Alloosh, Maysoun M. Saleh, and Jameel M. Al-Khayri

Abstract The green nanoparticles synthesis is a modern field that currently resonates compared to other preparation methods due to its characteristics that make it used in all fields. This chapter briefly explained traditional and biological methods for preparing nanomaterials and mentioned the advantage and disadvantage to these methods, then explained in more detail the phytofabrication of nanoparticles from different parts of the plant, which are considered a good source for biological molecules that act as reducing agents and modifies metal ions into nanoparticles that

L. M. Alnaddaf (✉)
Biotechnology and Molecular Biology, Faculty of Agriculture, Al-Baath University, Homs, Syria
e-mail: lalnaddaf@albaath-univ.edu.sy

A. K. Almuhammady
Arab Center for Nanotechnology, Cairo, Egypt

K. F. M. Salem
Department of Plant Biotechnology, Genetic Engineering and Biotechnology Research Institute (GEBRI), University of Sadat City, Sadat City, Egypt

Present Address:
Department of Biology, College of Science and Humanitarian Studies, Shaqra University, Qwaieah, Saudi Arabia

K. F. M. Salem
e-mail: khaled.salem@gebri.usc.edu.eg; Khaledfathi@su.edu.sa

M. T. Alloosh
Faculty of Pharmacy, Department of Biochemistry and Microbiology, Al-Baath University, Homs, Syria
e-mail: malloosh@albaath-univ.edu.sy

M. M. Saleh
Genetic Resources Department, General Commission for Scientific Agricultural Research (GCSAR), Damascus, Syria
e-mail: mzainsamasaleh@gcsar.gov.sy

J. M. Al-Khayri
Department of Agricultural Biotechnology, College of Agriculture and Food Sciences, King Faisal University, Al-Ahsa, Saudi Arabia
e-mail: jkhayri@kfu.edu.sa

J. M. Al-Khayri et al. (eds.), *Nanobiotechnology*,
https://doi.org/10.1007/978-3-030-73606-4_8

have unique properties. It also illustrates the green methods for preparing nanoparticles such as silver, zinc oxide and copper in some detail and their reaction conditions which influence the size, shape and structure of NPs. In addition to mechanisms of their formation and the different biomolecules that contribute to its synthesis.

Keywords Biological methods · Green synthesis · Nanoparticles characterization · Nanoparticles mechanism · Plant extracts · Traditional methods

8.1 Introduction

Green synthesis is a modern technology used to prepare nanomaterials (NM) that lead to a new era that reveals the plant potential in synthesizing stable nanoparticles (NPs) and increases the life of NPs (Ahmad et al. 2019). Constraints on chemical and physical methods are also overcome, as the first depends primarily on chemical reactions with its risks and side effects. While physical methods are rather complicated and economically expensive (Thunugunta et al. 2015). NPs possess various applications in different scientific and technological fields and this leads to high demand for produced nanoparticles (Dhand et al. 2015). To search for safe, cheap and environmentally friendly methods for synthesizing NPs, several manufacturing methods have been chosen. Further, it is necessary to understand, biochemical and molecular mechanisms of NPs production. Secondary metabolites from natural product extract act as reducers, stabilizers and capping agents in the process of nanomaterial synthesis (Bartolucci et al. 2020). These agents are present in biological entities and act as terminators of growth and inhibit agglomeration processes, thus enhancing NPs stability and persistence (Dhand et al. 2015). This chapter includes brief content about nanotechnology, nanomaterials, its properties, methods for its preparing and the tools of characterization. This chapter presents the importance of biological methods over traditional methods concerning NPs formation, factors affecting their preparation together with the biochemical mechanisms to produce NPs.

8.1.1 Traditional Methods

The two unique methodologies for combining NPs are top-down and bottom-up. Top-down using various techniques such as grinding, milling and sputtering or thermal/laser ablation. The suitable bulk material is broken down into smaller fine particles by size reduction (Singh et al. 2018). While, at bottom-to-top NPs are formatting by chemical and biological methods through the self-assembly of atoms into new nuclei that grow into particles of nanosize. These include electrochemical and chemical reduction methods (Gour and Jain 2019). NPs are obtained using diverse techniques including physical, chemical and green methods (Rawat et al.

2018). Physical methods for the synthesis of NPs involve physical vapor deposition (PVD), thermolysis, microwave-assisted synthesis, pulsed laser method, high energy ball milling, laser ablation, melt mixing, ion implantation, deposition of the electric arc and deposition of the sputter (Thunugunta et al. 2015). Each of these mechanisms converts one of the physical factors, for example, thermolysis changes temperature, ball friction pressure, ion implantation pH and laser ablation radiation changes. The desired size and shape of the NPs can be obtained through the optimization and maintenance of optimized parameters. The main disadvantages of physical methods contain time-consuming procedures, elevated parameters high equipment costs, Pollution from milling media and/or atmosphere, consolidate the powder product without the nanocrystalline microstructure being, Recycling and disposal. That no hard-and-fast safe policies on nanomaterial disposal have evolved. Furthermore, these NPs are not environmentally friendly owing to associated toxicity (Alagarasi 2011; Singh et al. 2018). Chemical synthesis is producing large quantities of NPs in a short span. Such as the co-precipitation, sol-gel process, radiofrequency plasma method and solvothermal synthesis and chemical vapor deposition. These methods employ strong reducing agents for nanoparticle reduction and capping agents to control the size and stabilize synthesized NPs such as oleic acid, thioglycerol and triethanolamine. This method utilizes toxic chemicals that may be dangerous to the living (Park et al. 2012; Singh et al. 2018; Thunugunta et al. 2015) (Fig. 8.1).

8.1.2 Biological Methods

The synthesis process is started in the aqueous solution of metal ions by adding extracts obtained from plant parts such as leaves, roots and fruits. With the compounds existent in the plant extract, such as protein, sugar, enzyme, flavonoid, organic acid and polymer, effective as a reducing agent, the bioinduction of metal ions into NPs occurs (Bartolucci et al. 2020). These chemical groups able to bind the particle surface, then providing stabilization and reduced toxicity (Gour and Jain 2019). For example, polyphenol of tea and coffee extracts were employed in the silver and palladium NPs synthesis (Nadagouda and Varma 2008), agricultural residuum waste (red grape pomace from winery) in produce various metal NPs, such as gold (Au) and silver (Ag) (Baruwati and Varma 2009). The success of NPs biosynthesis depends on bioavailable molecules in the green synthesis and this attaches to seasonality, organic growth conditions and extraction methods that influence concentrations of antioxidant chemical species and generate varied characteristics nanostructures (Gonnelli et al. 2015). In addition to plant extract, living plants can perform green NPs synthesis and maintain NPs in their tissue. The metal synthesis mechanism of NPs in a living plant depends on three main steps, the first step is the activation phase where ions are reduced and NPs form nuclei. In the second step, the growth phase, small NPs become interconnected to form larger particles. NPs then join together with the growth development to form an assortment of morphologies such as cubes, spheres, triangles, hexagons, pentagons, rods and wires. The growth stage results

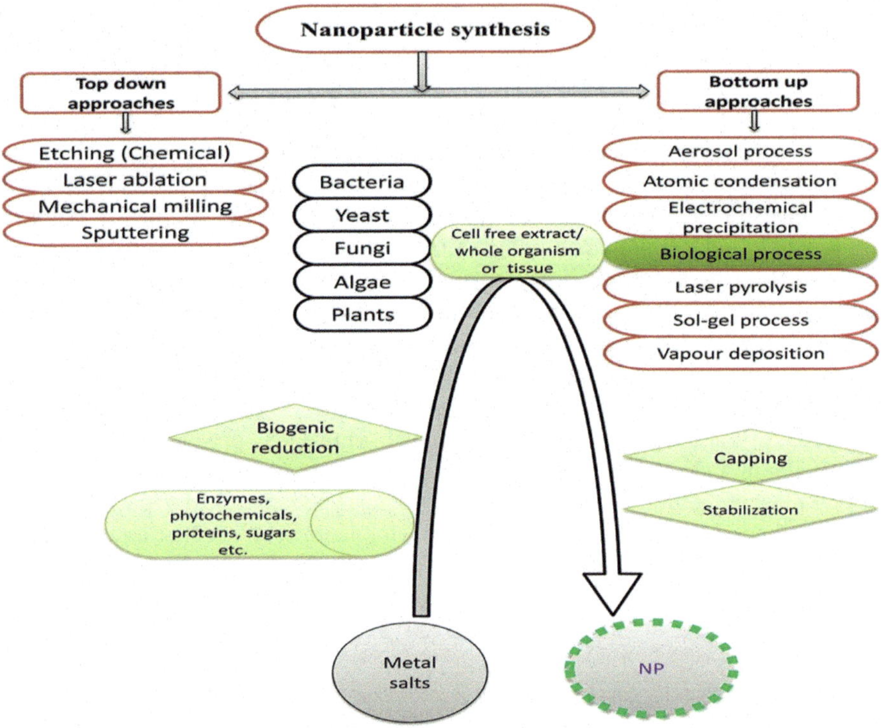

Fig. 8.1 NPs synthesis methodologies (*Source* Hussain et al. 2016)

in increased thermodynamic durability of NPs. In the third phase, NPs have stable thermodynamic effects in their final shape (Taghizadeh et al. 2018). Recent studies hypothesized that big-reducing sugars and fructose in chloroplasts were responsible to convert metal salts into NPs (Gan and Li 2012; Marchiol et al. 2014). In the living plants of *Medicago sativa*, *Helianthus annus* and *Brassica juncea*, NPs of zinc (Zn), nickel (Ni), silver (Ag), cobalt (Co) and copper (Cu) were synthesized. Response to the presence of high NPs concentrations, the cells alter their subcellular organization, then due to sulfhydryl enzyme inhibition, the membrane permeability changes in cell membranes that causes damage to cells. The major drawbacks to the industrial applications of NPs that produce in a living plant extract involve purification of synthesized NPs, heterogeneity of the form and size, low efficiency and high cost (Rawat et al. 2018) (Fig. 8.2).

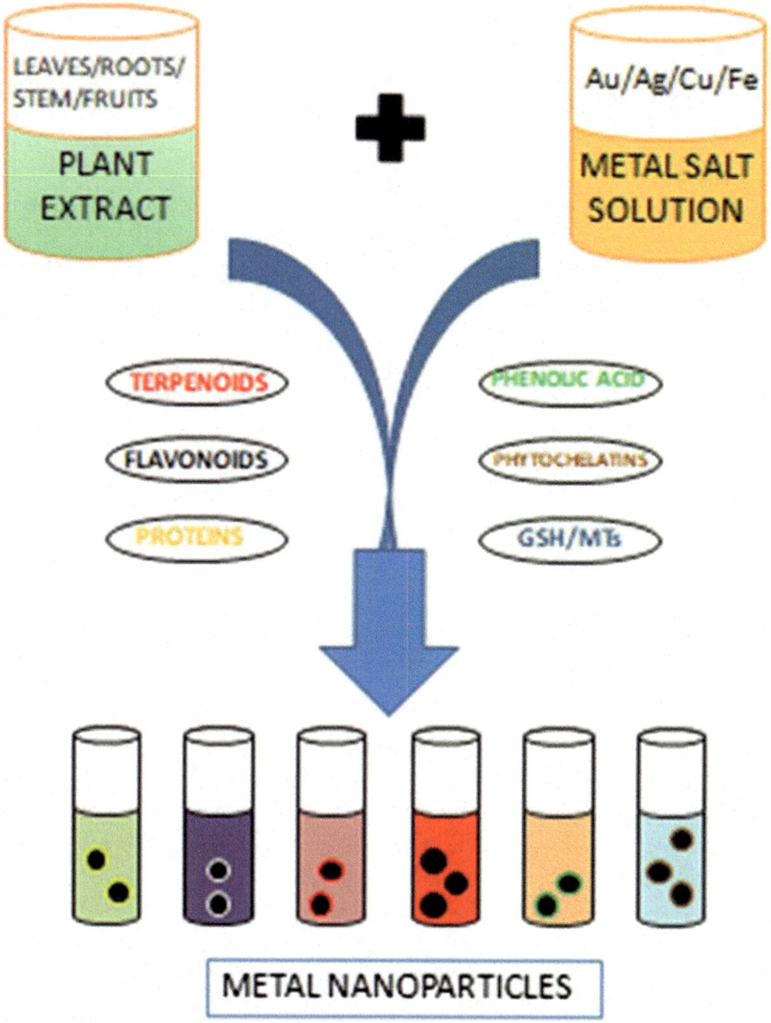

Fig. 8.2 Schematic representation of a collective mechanistic approach for different metal nanoparticle synthesis using the plant (*Source* Oza et al. 2020)

8.2 Green Synthesized NPs Using Plant Extract

8.2.1 Plant Material

8.2.1.1 Roots

The root of plants was widely used in the synthesis of metal NPs like silver (Ag), gold (Au), zinc oxide (ZnO), nickel oxide (NiO), titanium dioxide (TiO_2) and ferric oxide (Fe_3O_4) NPs (Table 8.1). The majority of these plants are medicinal herbs belonging to different families. The aqueous extract was used in most of the research (easy and fast extraction method). In a few cases, an alcoholic extract (ethanol) was prepared. The roots demonstrated a high ability to manufacture NPs in a very short duration (1 min) like synthesis Ag NPs using ginger root (Vijaya et al. 2017). Mostly, the completion of the reaction needs about 1–5 h and in the worst cases 48 h under normal conditions (room temperature). Sometimes NPs manufacture need heating with stirring like biosynthesis of NiO NPs from garlic and ginger by heating the mixture of root extract with NaOH to 90 °C for two hours (Haider et al. 2020) (Fig. 8.3).

8.2.1.2 Shoots

The researchers used the shoots, bark and stem on a relatively small scale compared to the rest of the plant parts, such as roots, leaves, fruits and seeds. In aforementioned context, the Ag and ZnO NPs are biosynthesized by stem extract in the last decade, where most of the time, the particular were synthesized at room temperature. Also, it was observed that the manufactured NPs were sometimes large in size (Mohammed 2016; Wang et al. 2020) (Table 8.2). Similarly, Tanase et al. (2020) found that the size of the Ag NPs manufactured using spruce (*Picea abies* L.) bark extract was in the range of 100–500 nm with 226 nm diameter on average, but the synthesis was fast and completed within three hours (Fig. 8.4).

8.2.1.3 Leaves

Several (mostly medicinal) plants have recently been used in the manufacture of silver, gold, copper oxide, zinc oxide, titanium oxide and sulfide. Spherical NPs were obtained mostly in relatively small sizes ranging from 2 to 45 nm (Table 8.3). Vijayan et al. (2019) used the microwave in synthesized Ag and Au NPs. The reaction mixture of silver nitrate and *Bauhinia purpurea* leaf extract solutions changed its color from colorless to yellowish-brown after 90 s of microwave irradiation. The auric chloride is light yellow and its mixture with the leaf extract changes to dark violet color after 30 s of microwave irradiation due to gold NPs (Fig. 8.5). Also, Dash et al. (2020)

Table 8.1 Various plant root extracts used to produce metallic NPs

Common name	Scientific name	Family	Extracted materials	Compound	Type of metal	Applications	Shape[a]	Size (nm)[a]	Synthesize condition	Reference
Palmyra palm	*Borassus aethiopum* Mart	Arecaceae	Aqueous	Silver	Ag	Nanolarvicidal on mosquito	Ui[b]	Ui	80 °C, 30 min	Danbature et al. (2020)
Ginger	*Zingiber officinale* Roscoe	Zingiberaceae	Aqueous	Silver	Ag	Degradation of environmental pollutants	Spherical, triangular, hexagonal	20.4	60 °C, 3.5 h	Barman et al. (2020)
Beet	*Beta vulgaris* L.	Amaranthaceae	Aqueous	Silver	Ag	Oxidative stress in normal and cancerous human hepatic cells in vitro	Spherical	5–100	Room temperature, 25 min	Bin-jumah et al. (2020)
Common barberry	*Berberis vulgaris* L.	Berberidaceae	Aqueous	Silver	Ag	Antibacterial activity	Spherical	30–70	Room temperature, 1 h	Behravan et al. (2019)
Sickle senna	*Senna tora* (L.) Roxb.	Fabaceae	Aqueous	Silver	Ag	Antimicrobial activity	Spherical	20–30	Room temperature, sunlight, 3 h	Shaikh et al. (2019)
peanut	*Arachis hypogaea* L.	Apiaceae	Aqueous	Silver	Ag	Antibacterial and clinical applications	Spherical, irregular	30	48 h	Sankaranarayanan et al. (2017)
Gouji in Chinese	*Cibotium barometz* (L.) J.Sm.	Cibotiaceae	Aqueous	Silver	Ag	Antimicrobial activity	Spherical	23	80 °C, 25 min	Wang et al. (2017)
Ebony	*Diospyros sylvatica* Roxb	Ebenaceae	Methanol	Silver	Ag	Antimicrobial activity	Spherical	8–10	Room temperature	Pethakamsetty et al. (2017)
Mountain soursop	*Annona muricate* L.	Annonaceae	Aqueous	Silver	Ag	Antimicrobial activity	Spherical	22±2	25 °C in the dark, 20 min	Ezealisiji et al. (2017)
Ginger	*Zingiber officinale* Roscoe	Zingiberaceae	Aqueous	Silver	Ag	Antibacterial	Spherical	10	60 s in the microwave Oven (1200 w, 50 Hz)	Vijaya et al. (2017)

(continued)

Table 8.1 (continued)

Common name	Scientific name	Family	Extracted materials	Compound	Type of metal	Applications	Shape[a]	Size (nm)[a]	Synthesize condition	Reference
Chinese rhubarb	*Rheum palmatum* L.	Polygonaceae	Aqueous	Silver	Ag	Antimicrobial activity	Hexagon, spherical	10–90	50–250 °C, 55 min	Arokiyaraj et al. (2017)
Ebony	*Diospyros paniculate* Willd	Ebenaceae	Methanol	Silver	Ag	Antibacterial activity	Spherical	14–28	Room temperature, 2 h	Rao et al. (2016)
Mamey	*Mammea suriga* (Buch.-Ham. ex Roxb.) Kosterm.	Calophyllaceae	Aqueous	Silver	Ag	Antibacterial activity	Square-shaped	50	80 °C, pH 10	Poojary et al. (2016)
Ginseng	*Panax ginseng* C.A. Mey.	Araliaceae	Aqueous	Silver	Ag	Antimicrobial activity	Spherical	10–30	80 °C, 2 h	Singh et al. (2016)
greater celandine,	*Chelidonium majus* L.	Papaveraceae	Aqueous	Silver	Ag	Antibacterial activity	Spherical	15	Room temperature	Alishah et al. (2016)
Malabar nut	*Justicia adhatoda* L.	Acanthaceae	Aqueous	Silver	Ag	Ui	Spherical	25	60 °C, 20 min	Ponvel et al. (2015)
Vinca rosea	*Catharanthus roseus* (L.) G.Don	Apocynaceae	Aqueous	Silver	Ag	Larvicidal effects	Spherical	35–55	60 °C	Rajagopal et al. (2015)
Fischer Euphorbia	*Euphorbia fischeriana* Steud	Euphorbiaceae	-	Gold	Au	Myocardial infarction	Spherical	20–60	Dark, room temperature, few hours	Zhang et al. (2020)
Penawar jambi	*Cibotium barometz* (L.) J.Sm.	Cibotiaceae	Aqueous	Gold	Au	Antimicrobial activity	Spherical	6	25 min	Wang et al. (2017)
Chinese ginseng	*Panax ginseng* C.A. Mey.	Araliaceae	Aqueous	Gold	Au	Antimicrobial activity	Spherical	10–40	80 °C, 5 min	Singh et al. (2016)
Surangi	*Mammea suriga* (Buch.-Ham. ex Roxb.) Kosterm.	Calophyllaceae	Aqueous	Gold	Au	Antibacterial activity	Square-shaped, spherical	22	80 °C, pH 8	Poojary et al. (2016)

(continued)

Table 8.1 (continued)

Common name	Scientific name	Family	Extracted materials	Compound	Type of metal	Applications	Shape[a]	Size (nm)[a]	Synthesize condition	Reference
Creeping-oxeye	*Sphagneticola trilobata* (L.) Pruski	Asteraceae	Aqueous	Zinc oxide	ZnO	Toxic metal removal, sowing germination and fostering of plant growth	Irregular	65–80	60 °C, 2 h	Shaik et al. (2020)
Radish	*Raphanus sativus* L.	Brassicaceae	Aqueous	Zinc oxide	ZnO	Effective wound dressing agents for diabetic foot ulcers	Hexagonal	15–25	80 °C, pH 5	Liu et al. (2020)
Milkworts	*Polygala tenuifolia* Willd	Polygalaceae	Aqueous	Zinc oxide	ZnO	Antioxidant and anti-inflammatory activities	Spherical	33.03–73.48	150 °C, 5 h	Nagajyothi et al. (2015)
Garlic	*Allium sativum* L.	Liliaceae	Aqueous	Nickel oxide	NiO	Bactericidal and catalytic	Spherical	11–59	90 °C, pH 12, 2 h	Haider et al. (2020)
Ginger	*Zingiber officinale* L.	Zingiberaceae	Aqueous	Nickel oxide	NiO	Bactericidal and catalytic	Spherical	16–52	90 °C, pH 12, 2 h	Haider et al. (2020)
Jack in the bush	*Chromolaena odorata* (L.) R.M. King & H. Rob	Asteraceae	Aqueous	Ferric oxide	Fe_3O_4	Ui	Spherical	5.6–16.8	70 °C, pH 13, 1 h	Nnadozie and Ajibade (2020)

[a]by transmission electron microscopy (TEM)
Ui[b]: un identify

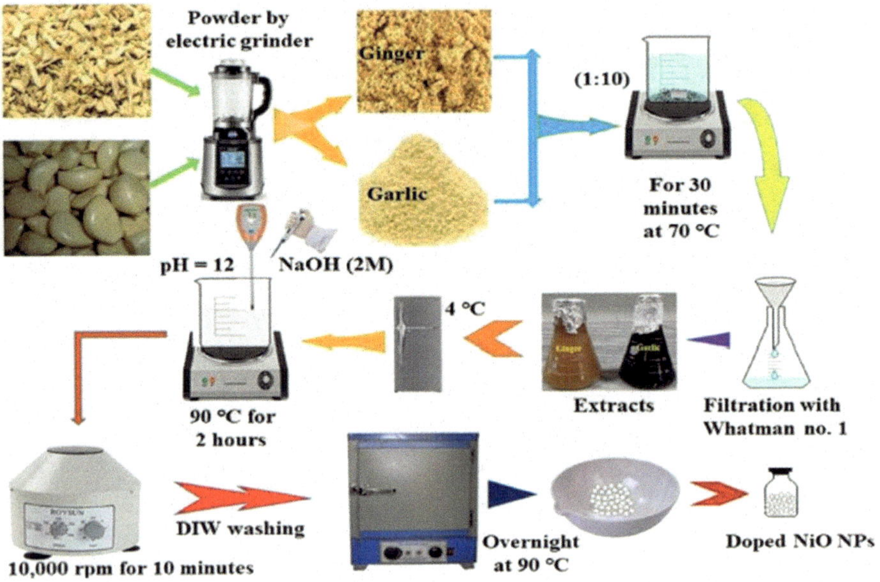

Fig. 8.3 Green synthesis of nickel oxide (NiO) NPs scheme using ginger and garlic root extracts (*Source* Haider et al. 2020)

used *Cinnamomum tamala* leaf extract to synthesis Ag NPs within 30 min and the reaction was completed after 2 h (Fig. 8.6).

8.2.1.4 Flowers

Recent studies indicate the use of flower extract in the biosynthesis of NPs especially Ag NPs. Medicinal plant flower extracts showed speediness and efficacy in manufacturing NPs under simple conditions (Table 8.4). For instance, Varadavenkatesan et al. (2019) used *Ipomoea digitata* flower extract for synthesized Ag NPs within 10 min (Fig. 8.7).

8.2.1.5 Fruits

Recent studies indicated the use of an aqueous extract of some fruits in the production of silver, gold, iron oxide, zinc oxide and copper NPs at different conditions (Table 8.5). Copper oxide (CuO) NPs were synthesized by heated the mixture of oak fruit extract with copper acetate to 500 °C (Sorbiun et al. 2018). However, others used autoclave in the same synthesis method that involved heating the mixture to 190 °C (Rostamizadeh et al. 2020). While, Jayaprakash et al. (2017) used a microwave oven to synthesis Ag NPs within 3 min. In contrast, some treatments

Table 8.2 Different plant shoot and stem extracts used to generate metallic NPs

Common name	Scientific name	Family	Extracted materials	Compound	Metal	Applications	Shape[a]	Size[a] (nm)	Synthesize condition	Reference
coco yam	*Colocasia esculenta* (L.) Schott	Araceae	Aqueous	Silver	Ag	Larvicidal activity	Spherical	13–50	70 °C, 30 min	Mondal et al. (2019)
Cuckoo flower	*Chasmanthera dependens*	Manispermaceae	Aqueous	Silver	Ag	Antimicrobial Antioxidant, anticoagulant, thrombolytic, and larvicidal	Cubically	24.53–92.38	Room temperature, 2 h	Aina et al. (2019)
Jasmine	*Jasminum auriculatum* Vahl	Oleaceae	Aqueous	Silver	Ag	Antibacterial activity	Spherical[b]	10–20[b]	Room temperature, 2 h	Balasubramanian et al. (2019)
Cowpea	*Vigna unguiculata* (L.) Walp.	Fabaceae	Aqueous	Silver	Ag	Ui[d]	Not determint	~25	Room temperature, pH 9, 5 h	Dawodu et al. (2019)
Bamboo	*Bambusa* sp	Poaceae	Chloroform, aqueous ethanol	Silver	Ag	Ui	Spherical[b]	53.2[c]	Dark condition, 15–20 min	Rashmi et al. (2018)
Arta	*Calligonum comosum* L'Her.	Polygonaceae	Aqueous	Silver	Ag	Antimycotic potential	Spherical	105.8	80 °C, 10 min, room temperature	Mohammed (2016)
Mustard	*Salvadora persica* L.	Salvadoraceae	Aqueous	Silver	Ag	Photocatalytic activity	Spherical	1–6	120 °C	Tahir et al. (2015)
Basota	*Premna barbata* Wall. ex Schauer	Lamiaceae	Aqueous	Zinc oxide	ZnO	Antibacterial property	Spherical	81.5	60 °C, 10 min	Prachi and Negi (2019)

[a]transmission electron microscopy (TEM)
[b]scanning electron microscopy (SEM)
[c]dynamic light scattering (DLS)
Ui[d]: un identify

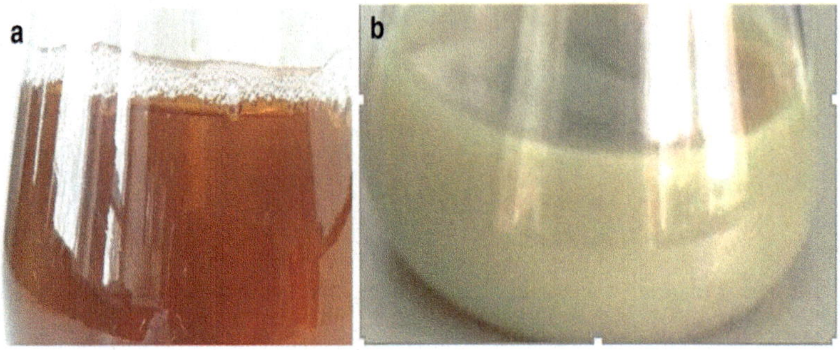

Fig. 8.4 Color modification of silver-green synthesizes NPs using spruce (*Picea abies*) bark extract. **a** *Picea abies* extract with silver nitrate ($AgNO_3$) solution at the beginning (0 min), **b** *Picea abies* extract with silver nitrate ($AgNO_3$) solution after three hours (*Source* Tanase et al. 2020)

require long time, for instance, the synthesis Ag NPs using *Cleome viscous* fruit extract require up to 24 h (Lakshmanan et al. 2018) (Fig. 8.8).

8.2.1.6 Other Plant Tissues

Numerous studies have examined the use of different plant parts (fruit peels, seeds, seed coat, bulb, latex) in the biosynthesis of NPs like Ag, TiO_2, Mn_3O_4, Fe_3O_4, CuO, MgO and ZnO NPs (Table 8.6).

8.2.2 Synthesis of NPs

8.2.2.1 Biosynthesis of Zinc Oxide NPs Using Root Extract of Ginger (*Zingier Officinal*)

Preparation of the Root Extract

Fresh ginger (*Zingier officinal*) was washed well with clean water and then soaked in distend water in order to remove the contaminants present in the skin. The moisture content removed from roots by dried it completely. Dried ginger was taken and the outer skin was peeled off, which were then weighed about (5 g). Then, it was cut into pieces and kept in a hot air oven at 50 °C for about an hour. The dried roots are taken and crushed in a mortar and pestle by slowly adding 25 ml of deionized water. The extract was filtered using Whatman filter paper No.1. For future use, the extract could preserve at about 4 °C (Gnanasangeetha and Thambavani 2013).

Table 8.3 Diverse plant leaf extracts used to synthesize metallic NPs

Common name	Scientific name	Family	Extracted materials	Compound	Metal	Applications	Shape[a]	Size (nm)[a]	Synthesize condition	Reference
Tej patta	*Cinnamomum tamala* (Buch.-Ham.) T.Nees & C.H. Eberm.	Lauraceae	Aqueous	Silver	Ag	Antimicrobial activity	Spherical	10–12	70 °C, 30 min	Dash et al. (2020)
Chinaberry	*Melia azedarach* L.	Meliaceae	Aqueous	Silver	Ag	Antifungal activities	Spherical	18–30	10 min	Jebril and Dridi (2020)
Rhubarb	*Rheum ribes* L.	Polygonaceae	Ethanolic	Silver	Ag	Anticarcinogenic and antimicrobial potential	Spherical	3.32±0.58	75 °C, 2 days	Aygün et al. (2020)
Ferula	*Ferula latisecta* L.	Apiaceae	Aqueous	Silver	Ag	Antibacterial activity	Spherical	5–30	70 °C	Mohammadi et al. (2019)
Nimba	*Azadirachta indica* A.Juss.	Meliaceae	Aqueous	Silver	Ag	Antibacterial activities	Spherical	34	Room temperature, dark, 15 min	Ahmed et al. (2016)
Camel's foot tree	*Bauhinia purpurea* L.	Fabaceae	Aqueous	Silver	Ag	Anticancer, antimicrobial, antioxidant and catalytic activities	Spherical	Ui	Microwave	Vijayan et al. (2019)
				Gold	Au		Hexagonal, triangular, nanorods			
Arjuna	*Terminalia arjuna* (Roxb.) Wight & Arn	Combretaceae	Aqueous	Gold	Au	Antibacterial activity	Spherical	15–30	Room temperature, 15 min	Dudhane et al. (2019)
Ailanthus	*Ailanthus altissima* (Mill.) Swingle	Simaroubaceae	Aqueous	Copper oxide	CuO	Antibacterial activity	Spherical	20	Room temperature, 4 h	Awwad and Amer (2020)
Madagascar periwinkle	*Catharanthus roseus* (L.) G.Don	Apocynaceae	Aqueous	Copper oxide	CuO	Photocatalytic activity	Rods	Deferent size	UV light	Mari et al. (2020)
Curryleaf tree	*Murraya koenigii* (L.) Spreng.	Rutaceae	Aqueous	Copper oxide	CuO	Catalytic activity	Spherical	8.4	pH 11, 60 min	Nordin and Shamsuddin (2019)

(continued)

Table 8.3 (continued)

Common name	Scientific name	Family	Extracted materials	Compound	Metal	Applications	Shape[a]	Size (nm)[a]	Synthesize condition	Reference
Khat	*Catha edulis* (Vahl) Forssk. ex Endl.	Celastraceae	Aqueous	Copper oxide	CuO	Antibacterial Activity	Ui[d]	Ui	50–60 °C, 6 min, pH 11	Gebremedhn et al. (2019)
Indian jujube	*Ziziphus mauritiana* Lam.	Rhamnaceae	Aqueous	Copper oxide	CuO	Ui	Spherical	20–45	Ui	Pansambal et al. (2017)
Yellow guava	*Psidium guajava* L.	Myrtaceae	Aqueous	Copper oxide	CuO	Photocatalytic degradation	Spherical	2–6	60 °C, 4 h, pH 7.5	Singh et al. (2019)
Pomegranate	*Punica granatum* L.	Lythraceae		Copper oxide	CuO	Ui	Spherical	20.33	Ui	Vidovix et al. (2019)
Fig	*Ficus carica* L.	Moraceae	Aqueous	Copper oxide	CuO	Ui	Spherical	7.5–35[b]	Room temperature	Hassan et al. (2018)
Basil	*Ocimum basilicum* L.	Lamiaceae	Aqueous	Zinc oxide	ZnO	Antimicrobial activity	Hexagonal	30–40	25 °C	Irshad et al. (2020)
Java plum	*Syzygium cumini* (L.) Skeels.	Myrtaceae	Aqueous	Zinc oxide	ZnO	Seed germination and wastewater purification	Spherical	64–78[b]	60 °C, 2 h	Rafique et al. (2020)
Garden sage	*Salvia officinalis* L.	Lamiaceae	Aqueous	Zinc oxide	ZnO	Antibacterial activity	Spherical[b]	21.19[c]	pH 13, 1 h	Shreema et al. (2020)
Basota	*Premna barbata* Wall. ex Schauer	Lamiaceae	Aqueous	Zinc oxide	ZnO	Antibacterial	Spherical	22.1	60 °C, 10 min	Prachi and Negi (2019)
Rosemary	*Rosmarinus officinalis* L.	Lamiaceae	Aqueous	Sulfur	S	Nematicidal activity	Spherical	5–80	Room temperature	Al Banna et al. (2020)
Indian mulberry	*Morinda citrifolia* L.	Rubiaceae	Ethanol	Titanium oxide	TiO_2	Antimicrobial properties	Quasi-spherical	15–19[b]	120 °C, 8 h	Sundrarajan et al. (2017)

[a]by Transmission Electron Microscopy TEM
[b]by Scanning Electron Microscopy (SEM)
[c]by X-ray
Ui[d]: un identify

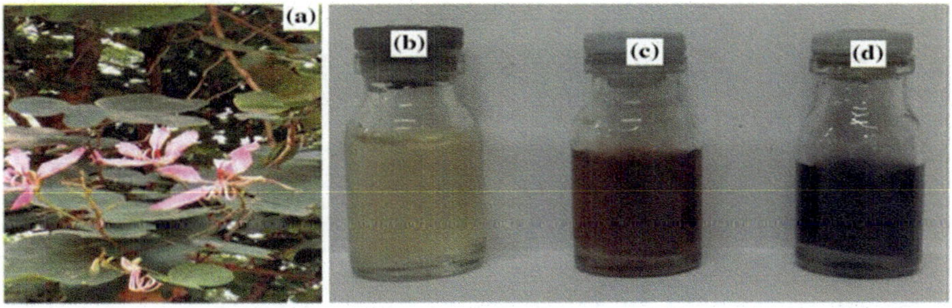

Fig. 8.5 Green synthesis of silver (Ag) and gold (Au) NPs using *Bauhinia purpurea* leaf extract. **a** Photographs of a *Bauhinia purpurea* plant, **b** *Bauhinia purpurea* leaf extract, **c** Biological silver (Ag) NPs, **d** Biological gold (Au) NPs (*Source* Vijayan et al. 2019)

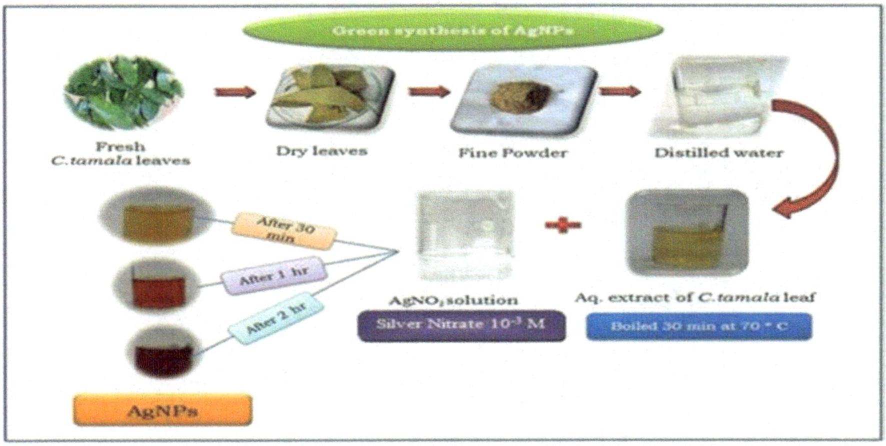

Fig. 8.6 Green synthesis of silver (Ag) NPs using bay leaf extract (*Source* Dash et al. 2020)

Preparation of Zinc Oxide NPs

Fifty ml of zinc acetate dihydrate was dissolved in deionized distilled water by stirring vigorously and pH was adjusted to 12 with sodium hydroxide (NaOH). The solution was placed in the stirrer for two hours until a white precipitate was observed. Then, the mixture was centrifuged at 10,000 rpm for 10 min. Deionized water washed the granules and then dried for about 12 h in a hot air oven at 100 °C. The resultant white powder was carefully extracted for characterization (Gnanasangeetha and Thambavani 2013) (Fig. 8.9).

Table 8.4 Various plant flower extracts used to produce metallic NPs

Common name	Scientific name	Family	Extracted materials	Compound	Metal	Applications	Shape[a]	Size(nm)[a]	Synthesize condition	Reference
Chavil-e-Roshan ball[b]	*Ferulago macrocarpa*	Apiaceae	Aqueous	Silver	Ag	Antibacterial, Antifungal	Spherical	14–25	80 °C, pH 11, 2.5 h	Azarbani and Shiravand (2020)
Angel hair	*Cascuta reflexa* Roxb	Convolvulaceae	Aqueous	Silver	Ag	Antioxidant and antibacterial properties	Spherical	20–50	Sunlight, 4 h	Shaikh et al. (2020)
Horseradish tree	*Moringa oleifera* Lam.	Moringaceae	Aqueous	Silver	Ag	Antimicrobial and sensing properties	Spherical	8	30 min	Bindhu et al. (2020)
Mountain knotgrass	*Aerva lanata* (L.) Juss. ex Schult.	Amaranthaceae	Ui	Silver	Ag	Environmental applications	Spherical	90	30 min	Kanniah et al. (2020)
Morning glory	*Ipomoea digitate* Jacq.	Convolvulaceae	Aqueous	Silver	Ag	Antibacterial activity	Spherical	111	80 °C, 10 min	Varadavenkatesan et al. (2019)
Angel's trumpet	*Datura inoxia* Mill.	Solanaceae	Aqueous	Silver	Ag	Cytotoxic activity	Polygonal	15–73	37 °C, 60 min	Gajendran et al. (2019)
Fritillaria	*Fritillaria* sp. Tourn. ex L.	Liliaceae	Aqueous	Silver	Ag	Antibacterial activity	Spherical	10	2 h	Hemmati et al. (2019)
Rush broom	*Spartium junceum* L.	Fabaceae	Aqueous	Silver	Ag	Ui[b]	Spherical	15–25	80 °C, 20 min	Nasseri et al. (2019)
Mowra	*Madhuca longifolia* (J.Konig) J.F.Macbr.	Sapotaceae	Aqueous	Silver	Ag	Antibacterial potential	Spherical and oval	30–50	40 °C, 20 min	Patil et al. (2018)
Camel's foot tree	*Bauhinia purpurea* L.	Fabaceae	Aqueous	Silver	Ag	Antibacterial activity against clinical pathogens	Spherical	Av 20	pH 7, 24 h	Chinnappan et al. (2017)

(continued)

Table 8.4 (continued)

Common name	Scientific name	Family	Extracted materials	Compound	Metal	Applications	Shape[a]	Size(nm)[a]	Synthesize condition	Reference
Dwarf banana	*Musa acuminate* Colla	Musaceae	Ethanolic	Silver	Ag	Antibacterial, anticancer activity	Spherical	10.1–15.6	Room temperature	Valsalam et al. (2019)
Betony	*Betonica officinalis* L.	Lamiaceae	Aqueous	Gold	Au	Phytotoxic effects	Ui	10	80 °C, 24 h	Dobrucka et al. (2020)
Grey mangrove	*Avicennia marina* (Forssk.) Vierh.	Acanthaceae	Aqueous	Zinc oxide	ZnO	Electrochemical activity	Non-uniform	30–100	Ui	Karpagavinayagam and Vedhi (2019)
Priyangu	*Aglaia elaeagnoidea* (A.Juss.) Benth.	Meliaceae	Aqueous	Copper oxide	CuO	Catalytic	Spherical	36–54	Room temperature	Manjari et al. (2017)

[a]by Transmission Electron Microscopy (TEM)
Ui[b]: un identify

Fig. 8.7 Green synthesis of silver (Ag) NPs using *Ipomoea digitata* (ID) flower extract. **a** Picture of *Ipomoea digitata* Linn. Flower, **b** Color change of *Ipomoea digitata* flower extract (reddish-brown, left) to ID-SNP (golden brown, right) (*Source* Varadavenkatesan et al. 2019)

8.2.2.2 Biosynthesis of Silver NPs Using Turmeric (*Curcuma Longa*) Tuber Powder

Preparation of Extract The turmeric *C. longa* tubers were washed to remove the adhering mud particles and potential impurities and dried for a week under sunlight to eliminate any moisture. The tubers were cut into small pieces, powdered in a mixer and then sewn to obtain a uniform size range using a 20-mesh sieve. For extract output 0.1 g turmeric. Tumeric tuber powder was added to Erlenmeyer flask with 20 ml sterile distilled water and then mixed for 4 h at room temperature.

Synthesis of AgNPs using *C. longa* Emulsion In short, *C. longa* tubers (0.1 g) extract of water from *C. longa* was added to a 20 ml diluted deionized water with vigorous four-hour stirring. Silver nitrate ($AgNO_3$) (40 ml) was blended for 24 h at 25 °C. During the incubation period, Ag-NPs were gradually obtained. During the reduction process, the solution was kept at room temperature in the darkness to avoid any photochemical reactions. The solution component was clean out with nitrogen gas before use. Subsequently, reduction continued to remove oxygen in the presence of nitrogen. Biological silver (Ag/*C. longa*) colloidal suspensions obtained. Then, *C. longa* mixture was centrifuged at 15,000 rpm for 20 min and washed four times

Table 8.5 Diverse plants fruit extracts used to form metallic NPs

Common name	Scientific name	Family	Extracted materials	Compound	Metal	Applications	Shape[a]	Size(nm)[a]	Synthesize condition	Reference
Pincushion tree	*Nauclea latifolia* Smith	Rubiaceae	Aqueous	Silver	Ag	Antimicrobial and Antioxidant activities	Irregularly	<10 nm	24–72 h	Odeniyi et al. (2020)
Syrian mesquite	*Prosopis farcta* (Banks & Sol.) J.F.Macbr.	Fabaceae	Aqueous	Silver	Ag	Antioxidant and antibacterial activity	Spherical	10.26–14.65	20–45 min	Salari et al. (2019)
Chinese boxthorn	*Lycium chinense* Mill.	Solanaceae	Aqueous	Silver	Ag	Antibacterial activity	Spherical	50–200	80 °C in an oil bath, 25 min	Chokkalingam et al. (2019)
Pomace	*Kalipatti sapota* (Chiku)	Sapotaceae	Aqueous	Silver	Ag	Antimicrobial activity	Spherical	8–16	20 min	Vishwasrao et al. (2019)
Sabestan Plum.	*Cordia dichotoma* G.Forst	Boraginaceae	Aqueous	Silver	Ag	Antibacterial, antibiofilm and photocatalytic activity	Spherical	6–20	Room temperature (27±1 °C), a few minutes	Bharathi et al. (2018)
Tickweed	*Cleome viscosa* L.	Cleomaceae	Aqueous	Silver	Ag	Antibacterial and anticancer activity	Spherical	20–50	In dark, 50±2 °C, 24 h	Lakshmanan et al. (2018)
Tamarind	*Tamarindus indica* L.	Fabaceae	Aqueous	Silver	Ag	Antibacterial	Spherical	5–12	Microwave oven (input power 230 V, 50 Hz) for 180 s	Jayaprakash et al. (2017)
Chinese boxthorn	*Lycium chinense* Mill.	Solanaceae	Aqueous	Gold	Au	Antibacterial activity	Polydispersed	20–100	80 °C in an oil bath, 1 min 15 s	Chokkalingam et al. (2019)
Cannonball tree	*Couroupita guianensis* Aubl.	Lecythidaceae	Aqueous	Ferric oxide	Fe_3O_4	Antibacterial and cytotoxicity activities	Spherical	7–80	Room temperature for 30 min. then, the pH was adjusted to 10.5 and heated at 80 °C for 30 min	Sathishkumar et al. (2018)
Cornel cherry	*Cornus mas* Cornelian cherry	Cornaceae	Aqueous	Ferric oxide	Fe_2O_3	Growth-promoting in Barley	Spherical	20–40	190 °C for 14 h under pressure in autoclave	Rostamizadeh et al. (2020)

(continued)

Table 8.5 (continued)

Common name	Scientific name	Family	Extracted materials	Compound	Metal	Applications	Shape[a]	Size(nm)[a]	Synthesize condition	Reference
Tree of Heaven	*Ailanthus altissima* (Mill.) Swingle	Simaroubaceae	Aqueous	Zinc oxide	ZnO	Antibacterial activity	Spherical	5–18	70-80 °C, for 2 h	Awwad et al. (2020)
English oak	*Quercus robur* L.	Fagaceae	Aqueous	Zinc oxide	ZnO	Photocatalytic degradation	Uniform spherical	34[b]	60 °C, for 4 h then at room temperature overnight	Sorbiun et al. (2018)
			Aqueous	Copper oxide	CuO	Photocatalytic degradation	Quasi-spherical	40[b]	Boil then heated at 500 °C for 4 h	

[a]by Transmission Electron Microscopy TEM
[b]by Scanning Electron Microscopy (SEM)

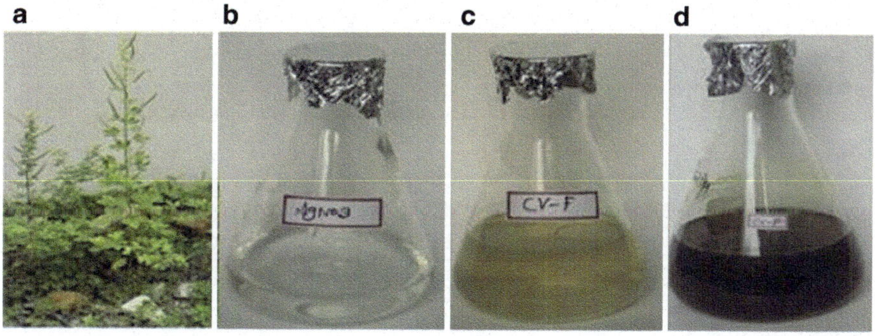

Fig. 8.8 **a** *Cleome viscosa* plant; **b** Silver nitrate ($AgNO_3$), **c** Aqueous fruit extract, **d** Synthesis of silver nitrate ($AgNO_3$) (*Source* Lakshmanan et al. 2018)

to remove silver ions from the precipitate NPs. Then, it was dried overnight at 30 °C under vacuum to obtain the Ag/*C. longa* (Shameli et al. 2012).

Biosynthesis of Silver NPs using Ginger Extract To prepare Ag NPs, 5 gm of ginger is cut into small pieces, then dissolved with 100 ml of 70% ethanol at 70 °C for two hours. The extract is filtered and stored at 4 °C. To manufacture nanoscale silver, ginger extract is mixed with $AgNO_3$ mmol/L by heating at 85 °C and the color of the solution will be observed within 20 min (Yang et al. 2017).

8.2.2.3 Biosynthesis of Copper NPs Using Sumac

Preparation of Sumac (*Rhus coriaria* L.) Extract The plant extract 2% (*w/v*) prepared by drying and well grinding 2 g of sumac. Twenty-minute grapes, boiling up to 100 ml of deionized water and finishing. Bio extract was kept at 4 °C for further experiments.

Copper NPs Synthesis using Sumac Extract To synthesize NPs using sumac as a stabilizer, 3 ml of plant extract were added to 1 ml of $CuSO_4.5H_2O$ (0.01 M), 2 ml of NaOH (0.01 M) and 1 ml of hydrazine 2% *v:v* at room temperature then complete the volume to 10 ml with deionized water. The reduction reaction could be expressed as:

$$2Cu^{+2} + 2\,N_2H_4 + 4OH^- \rightarrow 2Cu + N_2 + 4NH_4OH$$

The mixture was heated for 10 min in a water bath. The solutions color shifted from dark blue to light yellow and becomes deep red by heating (Ismail 2020).

Table 8.6 Various plant fruit shells, seeds and other explant extracts used to produce metallic NPs

Common name	Scientific name	Family	Part of plant	Compound	Metal	Extracted materials	Applications	Shape[a]	Size (nm)[a]	Synthesize condition	Reference
Tamarind	*Tamarindus indica* L.	Fabaceae	Fruit shell	Silver	Ag	Aqueous	Anticancer activity (breast cancer)	Spherical	20–52	45 °C, 2 h	Gomathi et al. (2020)
Jackfruit	*Artocarpus heterophyllus* Lam.	Moraceae	Seed	Silver	Ag	Aqueous	Antibacterial activity	Irregular	10.78	121 °C for 5 min	Jagtap and Bapat (2013)
Pomegranate	*Punica granatum* L.	Lythraceae	Seed	Silver	Ag	Aqueous	Antimicrobial activity	Spherical	19–54	Sunlight, room temperature, pH 7, 30 s	Mohseni et al. (2020)
Hirsute artocarpus	*Artocarpus hirsutus* Lam.	Moraceae	Seed	Silver	Ag	Ui[d]	Antibacterial activity	Ui	Ui	Ui	Shobana et al. (2020)
Indiantree spurge	*Euphorbia Tirucalli* L.	Euphorbiaceae	Latex	Silver	Ag	Aqueous	Management Meloidogyne incognita	Spherical and cubic[b]	20–30[b]	In the dark, 24 h	Kalaiselvi et al. (2019)
Grapevine	*Vitis vinifera* L.	Vitaceae	Fruit shell	Silver	Ag	Aqueous	Antibacterial activity	Spherical	3–14	90 °C, 20 min	Soto et al. (2019)
Sweet orange	*Citrus sinensis* (L.) Osbeck (pro sp.)	Rutaceae							5–50		
Kiwi	*Actinidia chinensis* Planch.	Actinidiaceae	Fruit shell	Silver	Ag	Aqueous	Antifungal activity	Spherical	18–35	Room temperature, 1 h	Al-Othman and Abeer (2019)
Pear	*Pyrus* sp. L.	Rosaceae							5–10		
Indian squill	*Ledebouria revoluta* (L.f.) Jessop	Asparagaceae	Bulb	Titanium oxide	TiO_2	Aqueous	Larvicidal, Histopathological, antibacterial and anticancer activity	Spherical	47[c]	50 °C, 4 h	Aswini et al. (2020)
Lipstick tree	*Bixa orellana* L.	Bixaceae	Seed	Titanium oxide	TiO_2	Ethanolic	Solar cells	Irregular spherical	16±2	7 °C, 2 h	Maurya et al. (2019)

(continued)

Table 8.6 (continued)

Common name	Scientific name	Family	Part of plant	Compound	Metal	Extracted materials	Applications	Shape[a]	Size (nm)[a]	Synthesize condition	Reference
Sunflower	*Helianthus Annuus* L.	Asteraceae	Seed	Manganese oxide	Mn_3O_4	Aqueous	Effect on *Vigna radiata* growth	Spherical	10–70	8 °C, dark room., 72 h	Ramesh et al. (2020)
Toddy palm	*Borassus flabellifer* L. [excluded]	Arecaceae	Seed coat	Ferric oxide	Fe_3O_4	Ethanolic	Antimicrobial activity	Hexagonal[b]	35[c]	Room Temperature, pH 10–11	Sandhya and Kalaiselvam (2020)
Pomegranate	*Punica granatum* L.	Lythraceae	Seed	Ferric oxide	Fe_2O_3	Aqueous	Photocatalytic activity	Semi spherical	25–55	70 °C, 15 min	Bibi et al. (2019)
Custard apple	*Annona squamosal* L.	Annonaceae	Seed	Magnesium oxide	MgO	Aqueous	Catalytic and antibacterial activity	Irregular	27–86	80 °C, 4 h	Sharma et al. (2020)
Bonduc	*Caesalpinia bonducella* L.	Fabaceae	Seed	Copper oxide	CuO	Aqueous	Antibacterial activity	Rice-grain[b]	13.07[c]	7 h	Sukumar et al. (2020)
Bread wheat	*Triticum aestivum* L.	Poaceae	Seed	Copper oxide	CuO	Aqueous	Catalytic activity	Spherical	22±1.5	25 °C, 1 h	Buazar et al. (2019)
Dwarf banana	*Musa acuminata* Colla	Musaceae	Fruit shell	Zinc oxide	ZnO	Aqueous	Photocatalytic activity	Different morphologies	30–80	60 °C, pH 12, 3 h	Abdullah et al. (2020)
Scarlet Spiral-Ginger	*Costus woodsonii* Maas	Costaceae	Bulb	Zinc oxide	ZnO	Aqueous	Ui	Spherical and hexagonal	20–25	60 °C and stirring With 500–600 rpm for 3 h	Khan et al. (2019)

[a]by
[b]by Scanning Electron Microscopy (SEM)
[c]by X-ray
Ui[d]: un identify

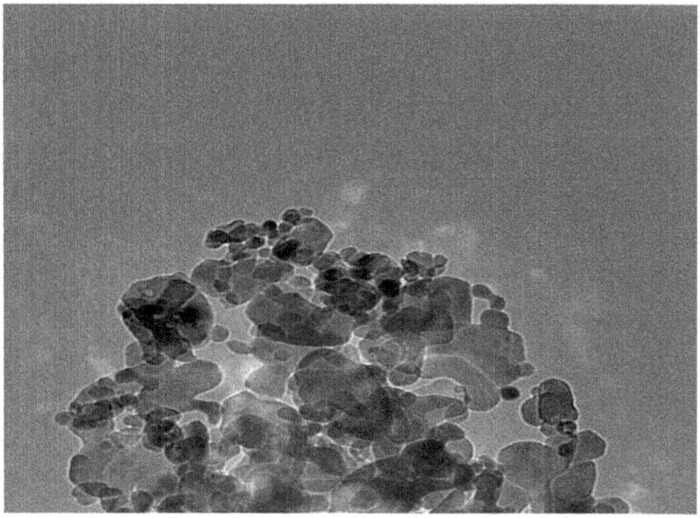

Fig. 8.9 Transmission electron microscopy (TEM) images for zinc oxide (ZnO) NPs (Photo by A.K. Almuhammady and F. Abdulqahar)

8.2.3 Reaction Conditions

Many factors occur in the synthesis, characterization and application of NPs. Numerous researches have indicated that the production of nanomaterials depends on the method of preparation, environment, temperature, the extract concentration to be implemented, reaction time and pH. It is important to know that the effect of these factors varies from one method to another and also varies with the substances and compounds to use. The methods of collecting and synthesizing NPs differ with the methods that are used to produce the chemical, physical, mechanical or biological materials and using various organic or organic materials and even microorganisms (Kharissova et al. 2013; Vadlapudi and Kaladhar 2014). The pH factor is an important factor affecting the synthesis of NPs by green technology. Research has confirmed that the pH of the solution medium affects the size and texture of complex NPs (Armendariz et al. 2004; Gamez et al. 2003). Therefore, the size of the NPs can be controlled by changing the pH of the solution media. Soni and Prakash (2011) found that pH has affected the shape and size of the prepared silver NPs. The nature of chemical reactions depends on the direct relationship between the pH of the bulk solution and the properties of the chemical surface (Ratchagar and Jagannathan 2016). Some studies on the preparation of NPs for yellow gold showed that metal reduction was more appropriate in the acid medium achieved with hydrochloric acid. This confirms that the corresponding absorption range for the remaining plasmon was due to the relatively low acidity. While the complexes of hydroxides could not be reduced at a high pH. Therefore, almost no gold molecules were formed and this process was accompanied by a major and significant change in the absorption spectrums (Mukha et al. 2016). Focusing on obtaining the final bimetallic Ag, Au particles, some experimental conditions have changed and the pH range of gold has been expanded (pH = 2, 4 and 6). Also, the use of nitric acid to control acidity reduced the effect of chlorine ions on the process of reducing silver except for the formation of low soluble chloride. The composition and stability of the Au and Ag NPs in the presence of tryptophan are strongly influenced by the acidity of the primary components. According to mass and absorption spectroscopy data, the tryptophan conversion in these systems passes through the kynurenine pathway. The highest persistence and least dispersion of nonmetric measurements occur in the case of metal reduction with amino acids in the case of anion present in the primary alkali medium (Mukha et al. 2016). Assuming that the transformation in the surface of the plasmon peak resonance indicates a change in the size of the Ag Nps and thus any change will lead to a decrease in the size of the particles. This leads to an increase in the pH resulting in the formation of smaller size NPs and vice versa (Alqadi et al. 2014). Temperature is one of the important factors that affect the structure of NPs in all methods used to prepare nanomaterials. Where the physical method requires a higher degree and may reach 350 °C or more spermatic. While chemical methods require a temperature much lower than this degree and, in most cases, as for biological methods or green technology, the synthesis of NPs does not require high temperatures. Since it is most likely to be less than 100 °C as the temperature is determined by the reaction medium. Thus, it determines the shape and size of the prepared particles (Rai et al. 2006). Both the hydrolysis and condensation reactions, like any other chemical reactions, depend heavily on the reaction temperatures. The reaction rate will increase dramatically when the temperature is high. We can prepare NPs, but this quick reaction makes them quick, in this case, it can lead to larger sizes (Matijevic 1977, 1985). In green technology, the consistency and form of complex NPs are greatly affected by the duration or shortening of time during which the reaction medium is incubated (Darroudi et al. 2011). Likewise, the change of the properties of compound NPs may change over time and be greatly influenced by the tuning process, exposure to light, storage conditions and different environmental conditions (Kuchibhatla et al. 2012; Mudunkotuwa et al. 2012). Time differences may affect several methods, such as collecting particles with several long storage periods and the particles may shrink or grow during long storage. The chemical materials may have a shelf life and so on, leading to a reduction in its effectiveness and impact (Baer 2011; Mudunkotuwa et al. 2012).

8.2.4 Mechanism of NPs Formation

The nano-sized nanomaterial scale allows control of different properties, mainly stability, size and shape (Martínez-Fernândez et al. 2016). These properties can be amended by modifying conditions of the reaction and chemical concentrations. The metals such as Fe, Co, Mg, Zn, Cu, Au, Ce, Ag, Ni, Mn and their oxides are the nanomaterials most frequently synthesized. Fe, Co, Mg, and Mn are widely

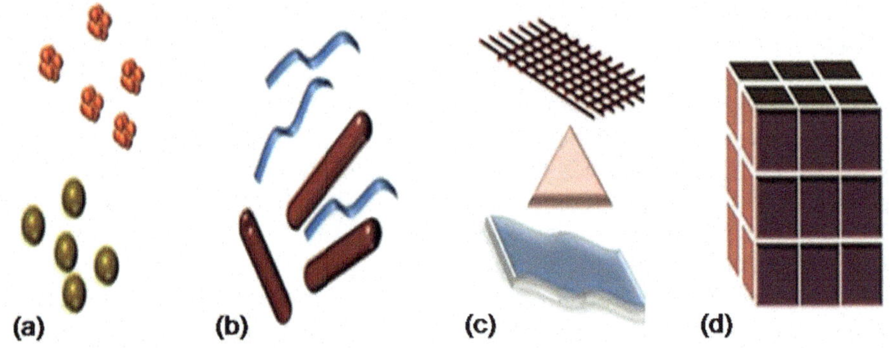

Fig. 8.10 Nanomaterial classification **a** 0 D spheres and clusters, **b** 1 D wires, nanofibres and rods, **c** 2 D plates, films and networks, **d** 3 D nanomaterials (*Source* Trotta and Mele 2019)

used for magnetic NP synthesis (Gupta et al. 2020). Plant extracts have different biomolecules, including polyphenols, proteins, enzymes, aldehydes, polysaccharides, flavones, amino acids, carboxylic acids, caffeine, ascorbic acids, ketones and terpenoids, which can reduce metal ions and stabilize NPs to the desired shapes and sizes (Pirtarighat et al. 2019). This explains the nanoparticle morphological diversity (triangular, hexagonal, spherical, cubic and pentagon) which connected by biomolecular reactions with metal ions (Gupta et al. 2020) (Fig. 8.10).

The aqueous extract of Artemisia (*Seriphidium quettense (Podlech)*) contains phenolic and Flavonoids. These considered as scave of free radicals and lipid peroxidation inhibition. Extracts of *Salvia* species have shown the presence of flavonoids and carnosic acid that have a community of carboxylate (Pirtarighat et al. 2019). Negatively charged groups presented in the plant extract such as carboxylate (COO^-) and polar groups such as CO and OH have a higher inclination to attach themselves to the Ag^+ surface. Thus, these groups share both the Ag ions reduction and stabilization (Ajitha et al. 2014). The use of plant extracts with acidic pH for the synthesis of NPs could increase the efficacy of NPs as catalysts for Fenton (Makarov et al. 2014). Because of the formation of OH radicals, which attack bonds in the dye molecules that may be in solution or adsorbed on the catalyst surface. Thus NPs can catalyze the degradation of bromothymol blue, methylene orange, methylene blue and monochlorobenzene (Makarov et al. 2014). It is important throughout the year to assess green synthesis efficiency according to natural extract characteristics related to seasonality (Santos et al. 2019). It should be careful to take the physicochemical characterizations for substances responsible for the green synthesis of metallic NPs. It can be observed that flavonoid and phenolic acids, mainly depend on ripening stages of fruit, so that there is a significant modification in antioxidant compound concentration that is essential for adjustment in green synthesis routes (Backx and Santana 2018; Blank et al. 2018). Also, Plant genetic and environmental conditions have significant efficacy in metallic nanoparticle synthesis. The use of an extract of

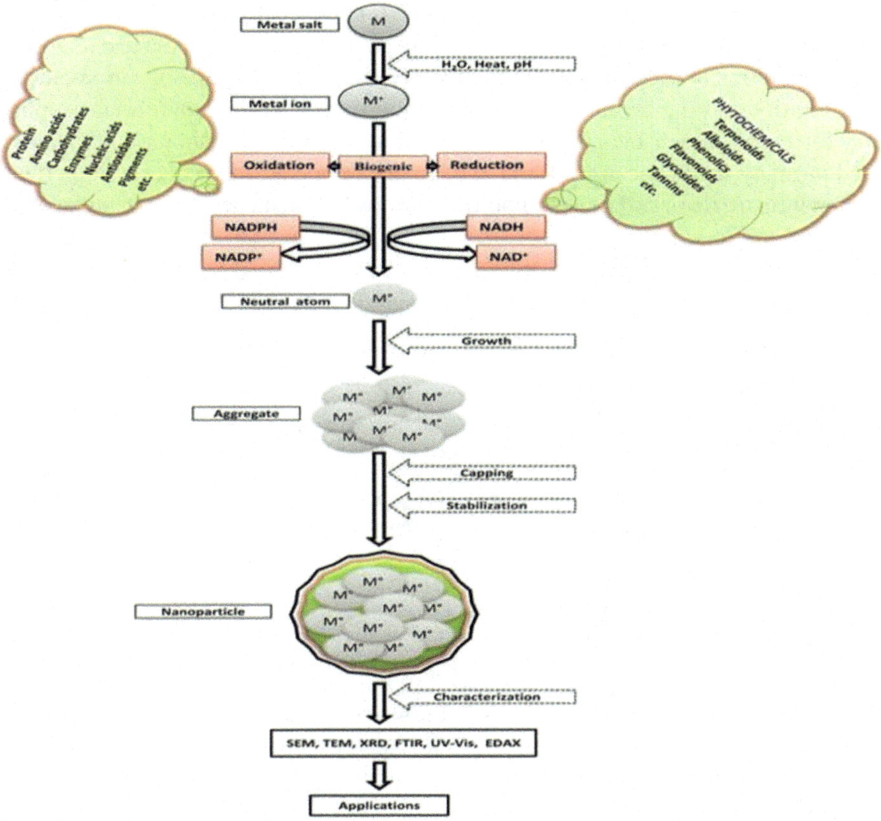

Fig. 8.11 Diagram for mechanism nanoparticle synthesis by different biomolecules (*Source* Hussain et al. 2016)

blackberry (*Morus nigra* L.) leaves leads to successful synthesis routes for nanoparticle formation. In contrast nanoparticle synthesis is not functional in the winter. This is related to seasonal characteristics because the morphological leaf and its size have decreased during this period (Biasiolo et al. 2004) (Fig. 8.11).

8.3 Conclusion and Prospects

Synthesis of NPs by green synthesis methods are considered important methods to manufacture NPs for some reasons, including that they are environmentally friendly and easy to apply, inexpensive, the size of the produced NPs can be expected according to the preparation factors, reaction conditions, materials used and their concentration. In addition, raw materials are generally available, but this method still requires more research and studies to reach mass production. At the same time, chemical and physical methods of NPs synthesis cannot be neglected as each method has its importance both in terms of production quantity, type of product and its specifications in terms of form, and structure. Finally, the type and shape of nanomaterials, size, quality and the actual cost of production determine the appropriate application for them. Green biosynthesis also produces stabilized nanoparticles that are used in many fields because of a lack of toxic chemicals and reduced side effects. Thus nanoparticles can be used in agriculture, water waste treatment, engineering, medicine and food industries.

References

Abdullah FH, Bakar NA, Bakar MA (2020) Low-temperature biosynthesis of crystalline zinc oxide nanoparticles from *Musa acuminata* peel extract for visible-light degradation of methylene blue. Optik 206:164279

Ahmed S, Saifullah Ahmad M et al (2016) Green synthesis of silver nanoparticles using *Azadirachta indica* aqueous leaf extract. J Radiat Res Appl Sci 9:1–7. https://doi.org/10.1016/j.jrras.2015.06.006

Ahmad S, Munir S, Zeb N et al (2019) Green nanotechnology: a review on green synthesis of silver nanoparticles-an ecofriendly approach. Inter J Nanomedicine 14:5087–5107

Aina DA, Owolo O, Lateef A et al (2019) Biomedical applications of *Chasmanthera dependens* stem extract mediated silver nanoparticles as antimicrobial, antioxidant, anticoagulant, thrombolytic and Larvicidal agents. Karbala Int J Mod Sci 5:70–80. https://doi.org/10.33640/2405-609X.1018

Ajitha B, Reddy YAK, Reddy PS (2014) Biogenic nano-scale silver particles by *Tephrosia purpurea* leaf extract and their inborn antimicrobial activity. Spectrochimica Acta A 121:164–172. https://doi.org/10.1016/j.saa.2013.10.077

Al Banna LS, Salem NM, Jaleel GA et al (2020) Green synthesis of sulfur nanoparticles using *Rosmarinus officinalis* leaves extract and nematicidal activity against *Meloidogyne javanica*. Chem Int 6:137–143

Alagarasi A (2011) Introduction to nanomaterials. In: Viswanathan B (ed) Nanomaterials. Narosa Publishing House pp 1–76

Alishah H, Seyedi SP, Ebrahimipour SY et al (2016) A green approach for the synthesis of silver nanoparticles using root extract of *Chelidonium majus*: characterization and antibacterial evaluation. J Clust Sci 27:421–429

Al-Othman MR, Abeer AE (2019) Effect of green synthesis silver nanoparticles from five fruits peel on protein capped and anti-fungal properties. Int J Adv Res Biol Sci 6(2):156–165

Alqadi MK, Abo Noqtah OA, Alzoubi FY et al (2014) pH effect on the aggregation of silver nanoparticles synthesized by chemical reduction. Mater Sci Pol 32:107–111. https://doi.org/10.2478/s13536-013-0166-9

Armendariz V, Herrera I, Peralta-Videa JR et al (2004) Size controlled gold nanoparticle formation by *Avena sativa* biomass: use of plants in nanobiotechnology. J Nanopart Res 6:377–382. https://doi.org/10.1007/s11051-004-0741-4

Arokiyaraj S, Vincent S, Saravanan M et al (2017) Green synthesis of silver nanoparticles using *Rheum palmatum* root extract and their antibacterial activity against *Staphylococcus aureus* and *Pseudomonas aeruginosa*. Artif Cell Nanomed Biotechnol 45:372–379

Aswini R, Murugesan S, Kannan K (2020) Bio-engineered TiO_2 nanoparticles using *Ledebouria revoluta* extract: Larvicidal, histopathological, antibacterial and anticancer activity. Int J Environ Anal Chem 1–11 https://doi.org/10.1080/03067319.2020.1718668

Awwad AM, Amer MW (2020) Biosynthesis of copper oxide nanoparticles using *Ailanthus altissima* leaf extract and antibacterial activity. Chem Int 6:210–217

Awwad AM, Amer MW, Salem NM et al (2020) Green synthesis of zinc oxide nanoparticles (ZnO-NPs) using *Ailanthus altissima* fruit extracts and antibacterial activity. Chem Int 6:151–159

Aygün A, Gülbağça F, Nas MS et al (2020) Biological synthesis of silver nanoparticles using *Rheum ribes* and evaluation of their anticarcinogenic and antimicrobial potential: a novel approach in phytonanotechnology. J Pharm Biomed Anal 179:1–24. https://doi.org/10.1016/j.jpba.2019.113012

Azarbani F, Shiravand S (2020) Green synthesis of silver nanoparticles by *Ferulago macrocarpa* flowers extract and their antibacterial, antifungal and toxic effects. Green Chem Lett Rev 13:41–49

Backx BP, Santana JCS (2018) Green synthesis of polymer blend impregnated with silver nanoparticles in *Euterpe Oleracea* dispersive medium. Inter J Green Herbal Chem 7:424–429. https://doi.org/10.24214/IJGHC/GC/7/2/42429

Balasubramanian S, Jeyapaul U, Kala SMJ (2019) Antibacterial activity of silver nanoparticles using *Jasminum auriculatum* stem extract. Int J Nanosci 18:1850011

Baer DR (2011) Surface characterization of nanoparticles: critical needs and significant challenges. J Surf Anal 17(3):163–169. https://doi.org/10.1384/jsa.17.163

Barman K, Chowdhury D, Baruah PK (2020) Bio-synthesized silver nanoparticles using *Zingiber officinale* rhizome extract as efficient catalyst for the degradation of environmental pollutants. Inorg Nano-Met Chem 50:57–65. https://doi.org/10.1080/24701556.1661468

Bartolucci C, Antonacci A, Arduini F et al (2020) Green nanomaterials fostering agrifood sustainability. TrAC Trend Anal Chem 125: https://doi.org/10.1016/j.trac.2020.115840

Baruwati B, Varma RS (2009) High value products from waste: grape pomace extract-a three-in-one package for the synthesis of metal nanoparticles. Chem Sus Chem 2:1041–1044

Behravan M, Panahi AH, Naghizadeh A et al (2019) Facile green synthesis of silver nanoparticles using *Berberis vulgaris* leaf and root aqueous extract and its antibacterial activity. Int J Biol Macromol 124:148–154

Bharathi D, Vasantharaj S, Bhuvaneshwari V (2018) Green synthesis of silver nanoparticles using *Cordia dichotoma* fruit extract and its enhanced antibacterial, anti-biofilm and photo catalytic activity. Mater Res Express 5:1–24

Biasiolo M, MTD C, Tornadore N (2004) Micromorphological characterization of ten mulberry cultivars (*Morus* spp). Econ Bot 58:639–646. https://doi.org/10.1663/00130001

Bibi I, Nazar N, Ata S et al (2019) Green synthesis of iron oxide nanoparticles using pomegranate seeds extract and photocatalytic activity evaluation for the degradation of textile dye. J Mater Res Technol 8:6115–6124

Bindhu MR, Umadevi M, Esmail GA et al (2020) Green synthesis and characterization of silver nanoparticles from *Moringa oleifera* flower and assessment of antimicrobial and sensing properties. J Photochem Photobiol B:1–29

Bin-Jumah M, Monera A-A, Albasher G et al (2020) Effects of green silver nanoparticles on apoptosis and oxidative stress in normal and cancerous human hepatic cells in vitro. Int J Nanomedicine 15:1537–1548

Blank DE, Justen D, Fraga S et al (2018) Chemical composition and antioxidant activity of *Bunchosia glandulifera* fruit at different ripening stages. Food Nutr Sci 9:1147–1159. https://doi.org/10.4236/fns.2018.910083

Buazar F, Sweidi S, Badri M et al (2019) Biofabrication of highly pure copper oxide nanoparticles using wheat seed extract and their catalytic activity: a mechanistic approach. Green Process Synth 8:691–702

Chinnappan S, Kandasamy S, Arumugam S et al (2017) Biomimetic synthesis of silver nanoparticles using flower extract of *Bauhinia purpurea* and its antibacterial activity against clinical pathogens. Environ Sci Pollut Res:1–7. https://doi.org/10.1007/s11356-017-0841-1

Chokkalingam M, Singh P, Huo Y et al (2019) Facile synthesis of Au and Ag nanoparticles using fruit extract of *Lycium chinense* and their anticancer activity. J Drug Deliv Sci Technol 49:308–315

Dhand C, Dwivedi N, Jun Loh X et al (2015) Methods and strategies for the synthesis of diverse nanoparticles and their applications: a comprehensive overview. RSC Adv 5:105003–105037. https://doi.org/10.1039/c5ra19388e

Danbature WL, Shehu Z, Yoro M, Adam MM (2020) Nanolarvicidal effect of green synthesized Ag-Co bimetallic nanoparticles on *Culex quinquefasciatus* Mosquito. Adv Biol Chem 10:16

Darroudi M, Ahmad MB, Zamiri R et al (2011) Time-dependent effect in green synthesis of silver nanoparticles. Int J Nanomed 6:677–681. https://doi.org/10.2147/IJN.S17669

Dash SS, Samanta S, Dey S et al (2020) Rapid green synthesis of biogenic silver nanoparticles using *Cinnamomum tamala* leaf extract and its potential antimicrobial application against clinically isolated multidrug-resistant bacterial strains. Biol Trace Elem Res. https://doi.org/10.1007/s12011-020-02107-w

Dawodu FA, Onuh CU, Akpomie KG et al (2019) Synthesis of silver nanoparticle from *Vigna unguiculata* stem as adsorbent for malachite green in a batch system. SN Appl Sci 1:346. https://doi.org/10.1007/s42452-019-0353-3

Dudhane AA, Waghmode SR, Dama LB et al (2019) Synthesis and characterization of gold nanoparticles using plant extract of Terminalia arjuna with antibacterial activity. Int J Nanosci Nanotechnol 15:75–82

Ezealisiji KM, Noundou XS, Ukwueze SE (2017) Green synthesis and characterization of monodispersed silver nanoparticles using root bark aqueous extract of *Annona muricata* Linn and their antimicrobial activity. Appl Nanosci 7:905–911

Gajendran B, Durai P, Varier KM et al (2019) Green synthesis of silver nanoparticle from *Datura inoxia* flower extract and its cytotoxic activity. BioNanoSci 9:564–572

Gamez G, Gardea-Torresdey JL, Tiemann KJ et al (2003) Recovery of gold (III) from multielemental solutions by alfalfa biomass. Adv Environ Res 7(2):563–571. https://doi.org/10.1016/S1093-0191(02)00021-7

Gan PP, Li SF (2012) Potential of plant as a biological factory to synthesize gold and silver nanoparticles and their applications. Rev Environ Sci Bio 11(2):169–206. https://doi.org/10.1007/s11157-012-9278-7

Gebremedhn K, Kahsay MH, Aklilu M (2019) Green synthesis of CuO nanoparticles using leaf extract of *Catha edulis* and its antibacterial activity. J Pharm Pharmacol 7:327–342. https://doi.org/10.17265/2328-2150/2019.06.007

Gnanasangeetha D, Thambavani DS (2013) Biogenic production of zinc oxide nanoparticles using *Acalypha indica*. J Chem Biol Phys Sci 4(1):238–246

Gomathi AC, Rajarathinam SX, Sadiq AM et al (2020) Anticancer activity of silver nanoparticles synthesized using aqueous fruit shell extract of *Tamarindus indica* on MCF-7 human breast cancer cell line. J Drug Deliv Sci Technol 55:101376

Gonnelli C, Cacioppo F, Giordano C et al (2015) *Cucurbita pepo* L. extracts as a versatile hydrotropic source for the synthesis of gold nanoparticles with different shapes. Green Chem Lett Rev 8:139–47

Gour A, Jain NK (2019) Advances in green synthesis of nanoparticles. Artif Cells Nanomed Biotechnol 47:844–851. https://doi.org/10.1080/21691401.2019.1577878

Gupta M, Tomar RS, Mishra RK (2020) Factors affecting biosynthesis of green nanoparticles. Our Heritage 68:10535–10555

Haider A, Ijaz M, Ali S et al (2020) Green synthesized phytochemically (*Zingiber officinale* and *Allium sativum*) reduced nickel oxide nanoparticles confirmed bactericidal and catalytic potential. Nanoscale Res Lett 15:1–11

Hassan KH, Saadi SK, Jarullah AA et al (2018) Green synthesis and structural characterisation of CuO nanoparticles prepared by using fig leaves extract. Pak J Sci Ind Res Ser Phys Sci 61:59–65

Hemmati S, Rashtiani A, Zangeneh MM et al (2019) Green synthesis and characterization of silver nanoparticles using fritillaria flower extract and their antibacterial activity against some human pathogens. Polyhedron 158:8–14

Hussain I, Singh NB, Singh A et al (2016) Green synthesis of nanoparticles and its potential application. Biotechnol Lett 38:545–560

Irshad S, Riaz M, Anjum AA et al (2020) Biosynthesis of ZnO nanoparticles using *Ocimum Basilicum* and determination of its antimicrobial activity. J Anim Plant Sci 30:185–191

Ismail MIM (2020) Green synthesis and characterizations of copper nanoparticles. Mater Chem Phys 240: https://doi.org/10.1016/j.matchemphys.2019.122283

Jagtap UB, Bapat VA (2013) Green synthesis of silver nanoparticles using *Artocarpus heterophyllus* Lam. seed extract and its antibacterial activity. Ind Crops Prod 46:132–137. https://doi.org/10.1016/j.indcrop.2013.01.019

Jayaprakash N, Vijaya JJ, Kaviyarasu K et al (2017) Green synthesis of Ag nanoparticles using Tamarind fruit extract for the antibacterial studies. J Photochem Photobiol, B 169:178–185

Jebril S, Dridi C (2020) Green synthesis of silver nanoparticles using *Melia azedarach* leaf extract and their antifungal activities: *in vitro* and *in vivo*. Mater Chem Phys 122898

Kalaiselvi D, Mohankumar A, Shanmugam G et al (2019) Green synthesis of silver nanoparticles using latex extract of *Euphorbia tirucalli*: a novel approach for the management of root-knot nematode, *Meloidogyne incognita*. Crop Prot 117:108–114

Kanniah P, Radhamani J, Chelliah P et al (2020) Green synthesis of multifaceted silver nanoparticles using the flower extract of *Aerva lanata* and evaluation of its biological and environmental applications. Chem Select 5:2322–2331

Karpagavinayagam P, Vedhi C (2019) Green synthesis of iron oxide nanoparticles using *Avicennia marina* flower extract. Vacuum 160:286–292

Khan MM, Saadah NH, Khan ME et al (2019) Phytogenic synthesis of band gap-narrowed ZnO nanoparticles using the bulb extract of *Costus woodsonii*. BioNanoSci 9:334–344

Kharissova OV, Dias HR, Kharisov BI et al (2013) The greener synthesis of nanoparticles. Trends Biotechnol 31(4):240–248

Kuchibhatla SV, Karakoti AS, Baer DR et al (2012) Influence of aging and environment on nanoparticle chemistry: implication to confinement effects in nanoceria. J Phys Chem C 116(26):14108–14114

Lakshmanan G, Sathiyaseelan A, Kalaichelvan PT et al (2018) Plant-mediated synthesis of silver nanoparticles using fruit extract of *Cleome viscosa* L.: assessment of their antibacterial and anticancer activity. Karbala Int J Mod Sci 4:61–68

Liu D, Liu L, Yao L et al (2020) Synthesis of ZnO nanoparticles using radish root extract for effective wound dressing agents for diabetic foot ulcers in nursing care. J Drug Deliv Sci Technol 55:

Makarov V, Love AJ, Makarova SS, Sinitsyna OV (2014) Biosynthesis of stable iron oxide nanoparticles in aqueous extracts of *Hordeum vulgare* and *Rumex acetosa* plants. Langmuir 30(20):5982–5988. https://doi.org/10.1021/la5011924

Matijevic E (1977) The role of chemical complexing in the formation and stability of colloidal dispersions. J Colloid Interf Sci 58(2):374–389

Matijevic E (1985) Production of monodispersed colloidal particles. Annu Rev Mater Sci 15(1):483–516

Manjari G, Saran S, Arun T et al (2017) Catalytic and recyclability properties of phytogenic copper oxide nanoparticles derived from *Aglaia elaeagnoidea* flower extract. J Saudi Chem Soc 21:610–618. https://doi.org/10.1016/j.jscs.2017.02.004

Marchiol L, Mattiello A, Pošćić F et al (2014) *In vivo* synthesis of nanomaterials in plants: location of silver nanoparticles and plant metabolism. Nanoscale Res Lett 9:101. https://doi.org/10.1186/1556-276x-9-101

Mari A, Vincent MV, Mookkaiah R et al (2020) *Catharanthus roseus* leaf extract mediated facile green synthesis of copper oxide nanoparticles and its photocatalytic activity. Chem Method 424–436

Martínez-Fernández D, Barroso D, Komárek M (2016) Root water transport of *Helianthus annuus* L. under iron oxide nanoparticle exposure. Environ Sci Pollut Res 23(2):1732–1741

Maurya IC, Singh S, Senapati S et al (2019) Green synthesis of TiO_2 nanoparticles using *Bixa orellana* seed extract and its application for solar cells. J Sol Energy 194:952–958

Mohammadi F, Yousefi M, Ghahremanzadeh R (2019) Green synthesis, characterization and antimicrobial activity of silver nanoparticles (AgNPs) using leaves and stems extract of some plants. Adv J Chem A 2:266–275

Mohammed AE (2016) Arta (*Calligonum Comosum*, L'Her.) shoot extracts: bio-mediator in silver nanoparticles formation and antimycotic potential. Nano Biomed Eng 8:128–135

Mohseni MS, Khalilzadeh MA, Mohseni M et al (2020) Green synthesis of Ag nanoparticles from pomegranate seeds extract and synthesis of Ag-Starch nanocomposite and characterization of mechanical properties of the films. Biocatal Agric Biotechnol 101569:1–31. https://doi.org/10.1016/j.bcab.2020.101569

Mondal A, Hajra A, Shaikh WA et al (2019) Synthesis of silver nanoparticle with *Colocasia esculenta* (L.) stem and its larvicidal activity against *Culex quinquefasciatus* and *Chironomus* sp. Asian Pac J Trop Biomed 9:510–517

Mudunkotuwa IA, Pettibone JM, Grassian VH (2012) Environmental implications of nanoparticle aging in the processing and fate of copper-based nanomaterials. Environ Sci Technol 46(13):7001–7010. https://doi.org/10.1021/es203851d

Mukha I, Vityuk N, Severynovska O et al (2016) The pH-dependent stucture and properties of Au and Ag nanoparticles produced by tryptophan reduction. Nanoscale Res Lett 11:101. https://doi.org/10.1186/s11671-016-1318-8

Nadagouda MN, Varma RS (2008) Green synthesis of silver and palladium nanoparticles at room temperature using coffee and tea extract. Green Chem 10:859–862

Nagajyothi PC, Cha SJ, Yang IJ et al (2015) Antioxidant and anti-inflammatory activities of zinc oxide nanoparticles synthesized using *Polygala tenuifolia* root extract. J Photochem Photobiol, B 146:10–17

Nasseri MA, Shahabi M, Allahresani A et al (2019) Eco-friendly biosynthesis of silver nanoparticles using aqueous solution of *Spartium junceum* flower extract. Asian J Green Chem 3:382–390

Nnadozie EC, Ajibade PA (2020) Green synthesis and characterization of magnetite (Fe_3O_4) nanoparticles using *Chromolaena odorata* root extract for smart nanocomposite. Mater Lett 263:127145

Nordin NR, Shamsuddin M (2019) Biosynthesis of copper (II) oxide nanoparticles using *Murayya koeniggi* aqueous leaf extract and its catalytic activity in 4-nitrophenol reduction. Malays J Fundam Appl Sci 15:218–224

Odeniyi MA, Okumah VC, Adebayo-Tayo BC et al (2020) Green synthesis and cream formulations of silver nanoparticles of *Nauclea latifolia* (African peach) fruit extracts and evaluation of antimicrobial and antioxidant activities. Sustain Chem Pharm 15:100197

Oza G, Reyes-Calderón A, Mewada A et al (2020) Plant-based metal and metal alloy nanoparticle synthesis: a comprehensive mechanistic approach. J Mater Sci 55:1309–1330. https://doi.org/10.1007/s10853-019-04121-3

Pansambal S, Gavande S, Ghotekar S et al (2017) Green synthesis of CuO nanoparticles using *Ziziphus Mauritiana* L. extract and its characterizations. Int J Sci Res Sci Tech 3:1388–1392

Park J, Kwon SG, Jun SW et al (2012) Large-scale synthesis of ultra-small-sized silver nanoparticles. Chem Phys Chem 13:2540–2543

Patil MP, Singh RD, Koli PB et al (2018) Antibacterial potential of silver nanoparticles synthesized using *Madhuca longifolia* flower extract as a green resource. Microb Pathog 121:184–189. https://doi.org/10.1016/j.micpath.2018.05.040

Pethakamsetty L, Kothapenta K, Nammi HR et al (2017) Green synthesis, characterization and antimicrobial activity of silver nanoparticles using methanolic root extracts of *Diospyros sylvatica*. J Environ Sci 55:157–163

Pirtarighat S, Ghannadnia M, Baghshahi S (2019) Green synthesis of silver nanoparticles using the plant extract of *Salvia spinosa* grown in vitro and their antibacterial activity assessment. J Nanostructure Chem 9:1–9. https://doi.org/10.1007/s40097-018-0291-4

Ponvel KM, Narayanaraja T, Prabakaran J (2015) Biosynthesis of silver nanoparticles using root extract of the medicinal plant *Justicia adhatoda*: characterization, electrochemical behavior and applications. Int J Nano Dimens 6(4):339–349

Poojary MM, Passamonti P, Adhikari AV (2016) Green synthesis of silver and gold nanoparticles using root bark extract of *Mammea suriga*: characterization, process optimization and their antibacterial activity. BioNanoSci 6:110–120

Prachi AN, Negi DS (2019) Plant mediated synthesis of zinc oxide nanoparticles using leaf and twig extract of *Premna Barbata*: a comparative approach on its physical characteristic and antibacterial property. Plant Arch 19:2469–2475

Soni N, Prakash NS (2011) Factors affecting the geometry of silver nanoparticles synthesis in *Chrysosporium tropicum* and *Fusarium oxysporum*. Am J Nanotechnol 2:112–121. https://doi.org/10.3844/ajnsp.2011.112.121

Rafique M, Tahir R, Gillani SSA et al (2020) Plant-mediated green synthesis of zinc oxide nanoparticles from *Syzygium Cumini* for seed germination and wastewater purification. Int J Environ Anal Chem. https://doi.org/10.1080/03067319.2020.1715379

Rai A, Singh A, Ahmad A, Sastry M (2006) Role of halide ions and temperature on the morphology of biologically synthesized gold nanotriangles. Langmuir 22(2):736–741

Rajagopal T, Jemimah IAA, Ponmanickam P et al (2015) Synthesis of silver nanoparticles using *Catharanthus roseus* root extract and its larvicidal effects. J Environ Biol 36:1283

Ramesh R, Catherine G, John Sundaram S et al (2020) Synthesis of Mn_3O_4 nano complex using aqueous extract of *Helianthus annuus* seed cake and its effect on biological growth of *Vigna radiata*. Mater Today Proc:8. https://doi.org/10.1016/j.matpr.2020.02.883

Rao NH, Lakshmidevi N, Pammi SVN et al (2016) Green synthesis of silver nanoparticles using methanolic root extracts of *Diospyros paniculata* and their antimicrobial activities. Mater Sci Eng, C 62:553–557

Rashmi TS, Divyashree K, Gagan KK (2018) Bioactive compound profiling and green synthesis of silver nanoparticle; its characterization using bamboo shoot extract. Int J Biol Med Res 9:6235–6244

Ratchagar V, Jagannathan K (2016) Effect of pH on magnetic, thermal and dielectric properties of SnO_2 nanomaterials. J Alloys Compd 689:1088–1095

Rawat A, Kumar R, Bhatt B, Ram P (2018) Nanotechnology in agriculture—a review. Int J Curr Microbiol App Sci 7(8):969–978

Rostamizadeh E, Iranbakhsh A, Majd A et al (2020) Green synthesis of Fe_2O_3 nanoparticles using fruit extract of *Cornus mas* L. and its growth-promoting roles in barley. J Nanostruct Chem 10:125–130. https://doi.org/10.1007/s40097-020-00335-z

Salari S, Bahabadi SE, Samzadeh-Kermani A et al (2019) *In-vitro* evaluation of antioxidant and antibacterial potential of green synthesized silver nanoparticles using *Prosopis farcta* fruit extract. Iran J Pharm Res 18(1):430–455

Sandhya J, Kalaiselvam S (2020) Biogenic synthesis of magnetic iron oxide nanoparticles using inedible borassus flabellifer seed coat: characterization, antimicrobial, antioxidant activity and *in vitro* cytotoxicity analysis. Mater Res Express 7:015045. https://doi.org/10.1088/2053-1591/ab6642

Sankaranarayanan A, Munivel G, Karunakaran G et al (2017) Green synthesis of silver nanoparticles using *Arachis hypogaea* (ground nut) root extract for antibacterial and clinical applications. J Clust Sci 28:995–1008

Sathishkumar G, Logeshwaran V, Sarathbabu S et al (2018) Green synthesis of magnetic Fe_3O_4 nanoparticles using *Couroupita guianensis* Aubl. fruit extract for their antibacterial and cytotoxicity activities. Artif Cells Nanomed Biotechnol 46:589–598

Shaik AM, Raju MD, Reddy DR (2020) Green synthesis of zinc oxide nanoparticles using aqueous root extract of *Sphagneticola trilobata* L. in and investigate its role in toxic metal removal, sowing germination and fostering of plant growth. Inorg Nano-Metal Chem 50:569–579. https://doi.org/10.1080/24701556.2020.1722694

Shaikh NS, Shaikh RS, Kashid S (2020) In vitro biosynthesis of silver nanoparticles using flower extract of parasitic plant *Cascuta reflexa* and evaluation of its biological properties. Asian J Nanosci Mater 3:121–130

Shaikh R, Zainuddin Syed I, Bhende P (2019) Green synthesis of silver nanoparticles using root extracts of *Cassia toral* L. and its antimicrobial activities. Asian J Green Chem 3:70–81

Shameli K, Ahmad M, Zamanian A et al (2012) Green biosynthesis of silver nanoparticles using *Curcuma longa* tuber powder. Int J Nanomedicine 7:5603–5610

Sharma SK, Khan AU, Khan M et al (2020) Biosynthesis of MgO nanoparticles using *Annona squamosa* seeds and its catalytic activity and antibacterial screening. Micro Nano Lett 15:30–34
Shobana S, Veena S, Sameer SSM et al (2020) Green synthesis of silver nanoparticles using *Artocarpus hirsutus* seed extract and its antibacterial activity. Curr Pharm Biotechnol. https://doi.org/10.2174/1389201021666200107115849
Shreema K, Priyadharshini K, Mathammal R et al (2020) Green synthesis of zinc oxide nanoparticles using leaf extract of *Salvia officinalis*. Stud Indian Place Names 40:1175–1187
Santos DSM, Santos DLAO, Filho AS et al (2019) Can green synthesis of nanoparticles be efficient all year long? Nanomater Chem Technol 1:32–36
Singh A, Singh NB, Hussain I et al (2016) Green synthesis of nano zinc oxide and evaluation of its impact on germination and metabolic activity of *Solanum lycopersicum*. J Biotechnol 233:84–94
Singh J, Dutta T, Kim KH et al (2018) Green synthesis of metals and their oxide nanoparticles: applications for environmental remediation. Nanobiotechnol 16:84. https://doi.org/10.1186/s12951-018-0408-4
Singh J, Kumar V, Kim KH et al (2019) Biogenic synthesis of copper oxide nanoparticles using plant extract and its prodigious potential for photocatalytic degradation of dyes. Environ Res 177:108569. https://doi.org/10.1016/j.envres.2019.108569
Sorbiun M, Mehr ES, Ramazani A et al (2018) Green synthesis of zinc oxide and copper oxide nanoparticles using aqueous extract of oak fruit hull (jaft) and comparing their photocatalytic degradation of basic violet 3. Int J Environ Res 12:29–37
Soto KM, Quezada-Cervantes CT, Hernández-Iturriaga M et al (2019) Fruit peels waste for the green synthesis of silver nanoparticles with antimicrobial activity against foodborne pathogens. LWT 103:293–300
Sukumar S, Rudrasenan A, Nambiar DP (2020) Green-synthesized rice-shaped copper oxide nanoparticles using *Caesalpinia bonducella* seed extract and their applications. ACS Omega 5:1040–1051. https://doi.org/10.1021/acsomega.9b02857
Sundrarajan M, Bama K, Bhavani M et al (2017) Obtaining titanium dioxide nanoparticles with spherical shape and antimicrobial properties using *M. citrifolia* leaves extract by hydrothermal method. J Photochem Photobiol, B 171:117–124
Taghizadeh S, Taghizadeh SM, Ghasemi Y et al (2018) Living plant-mediated synthesis of nanoparticles. J Adv Med Sci Appl Technol 4:1–6
Tahir K, Nazir S, Li B et al (2015) An efficient photo catalytic activity of green synthesized silver nanoparticles using *Salvadora persica* stem extract. Sep Purif Technol 150:316–324
Tanase C, Berta L, Mare A et al (2020) Biosynthesis of silver nanoparticles using aqueous bark extract of *Picea abies* L. and their antibacterial activity. Europ J Wood Prod 78:281–291
Thunugunta T, Reddy A, Reddy Reddy L (2015) Green synthesis of nanoparticles: current prospectus. Nanotechnol Rev 4(4):303–323
Trotta F, Mele A (2019) Nanomaterials: classification and properties. In: Trotta F, Mele A (eds) Nanosponges: synthesis and applications, 1st edn. Wiley-VCH Verlag GmbH & Co KGaA
Vadlapudi V, Kaladhar DSVGK (2014) Review: green synthesis of silver and gold nanoparticles. Middle East J Sci Res 19(6):834–842
Valsalam S, Agastian P, Esmail GA et al (2019) Biosynthesis of silver and gold nanoparticles using *Musa acuminata* colla flower and its pharmaceutical activity against bacteria and anticancer efficacy. J Photochem Photobiol, B 201:111670. https://doi.org/10.1016/j.jphotobiol.2019.111670
Varadavenkatesan T, Selvaraj R, Vinayagam R (2019) Dye degradation and antibacterial activity of green synthesized silver nanoparticles using *Ipomoea digitata* Linn. flower extract. Int J Environ Sci Technol 16:2395–2404
Vidovix TB, Quesada HB, Januário EFD et al (2019) Green synthesis of copper oxide nanoparticles using *Punica granatum* leaf extract applied to the removal of methylene blue. Mater Lett 257:126685. https://doi.org/10.1016/j.matlet.2019.126685

Vijaya JJ, Jayaprakash N, Kombaiah K et al (2017) Bioreduction potentials of dried root of *Zingiber officinale* for a simple green synthesis of silver nanoparticles: antibacterial studies. J Photochem Photobiol, B 177:62–68

Vijayan R, Joseph S, Mathew B (2019) Anticancer, antimicrobial, antioxidant, and catalytic activities of green-synthesized silver and gold nanoparticles using *Bauhinia purpurea* leaf extract. Bioprocess Biosyst Eng 42:305–319

Vishwasrao C, Momin B, Ananthanarayan L (2019) Green synthesis of silver nanoparticles using sapota fruit waste and evaluation of their antimicrobial activity. Waste Biomass Valorization 10:2353–2363

Wang D, Markus J, Wang C et al (2017) Green synthesis of gold and silver nanoparticles using aqueous extract of *Cibotium barometz* root. Artif Cells Nanomedicine Biotechnol 45:1548–1555

Wang Z, Que B, Gan J et al (2020) Zinc oxide nanoparticles synthesized from Fraxinus rhynchophylla extract by green route method attenuates the chemical and heat induced neurogenic and inflammatory pain models in mice. J Photochem Photobiol, B 202:111668

Yang N, Li F, Jian T et al (2017) Biogenic synthesis of silver nanoparticles using ginger (*Zingiber officinale*) extract and their antibacterial properties against aquatic pathogens. Acta Oceanol Sin 36:95–100. https://doi.org/10.1007/s13131-017-1099-7

Zhang T, Dang M, Zhang W, Lin X (2020) Gold nanoparticles synthesized from *Euphorbia fischeriana* root by green route method alleviates the isoprenaline hydrochloride induced myocardial infarction in rats. J Photochem Photobiol, B 202:111705. https://doi.org/10.1016/j.jphotobiol.2019.111705

Chapter 9
Applications of Plant-Derived Nanomaterials in Mitigation of Crop Abiotic Stress

Maysoun M. Saleh, Lina M. Alnaddaf, Abdulsalam K. Almuhammady, Khaled F.M. Salem, Maysaa T. Alloosh, and Jameel M. Al-Khayri

Abstract Nanotechnology is a great and promising future science for pressing global climate change solutions and increasing the global population through its various uses. This chapter conversed the implementation of nanotechnology in various environmental, medical and agricultural fields. In addition, it discussed the application of various nanoparticle biosynthesis as fertilizers in multiple forms including soaking, foliar fertilization and soil fertilization. Then explained in more detail the effectiveness of NPs such as (nano silicon, zinc oxide NPs, titanium dioxide NPs, silver NPs) on crop growth, phenological and physiological development under abiotic stress. Also, this chapter highlights the Mechanism of NPs uptake and accumulation in crops. Therefore, nanotechnology offers an effective

---

M. M. Saleh (✉)
Genetic Resources Department, General Commission for Scientific Agricultural Research (GCSAR), Damascus, Syria
e-mail: mzainsamasaleh@gcsar.gov.sy

L. M. Alnaddaf
Biotechnology and Molecular Biology, Faculty of Agricultural, Al-Baath University, Homs, Syria
e-mail: lalnaddaf@albaath-univ.edu.sy

A. K. Almuhammady
Arab Center for Nanotechnology, Cairo, Egypt
e-mail: lalnaddaf@albaath-univ.edu.sy

K. F.M. Salem
Department of Biology, College of Science and Humanitarian Studies, Shaqra University, Qwaieah, Saudi Arabia
e-mail: khaled.salem@gebri.usc.edu.eg; khaledfathi@su.edu.sa

M. T. Alloosh
Department of Biochemistry and Microbiology, Faculty of Pharmacy, Al-Baath University, Homs, Syria
e-mail: malloosh@albaath-univ.edu.sy

J. M. Al-Khayri
Department of Agricultural Biotechnology, College of Agriculture and Food Sciences, King Faisal University, Al-Ahsa, Saudi Arabia
e-mail: jkhayri@kfu.edu.sa

J. M. Al-Khayri et al. (eds.), *Nanobiotechnology*,
https://doi.org/10.1007/978-3-030-73606-4_9

application to reduce pollution, management of soil nutrients and achieve sustainable development.

Keywords Abiotic stress · Nanoparticles accumulation · Nano fertilization · Nanoparticles · Nanoparticles uptake · Sustainability development

9.1 Introduction

9.1.1 Applications of Nanoparticles

9.1.1.1 Environmental Remediation Processes

The synthesis of nanoparticles (NPs) could serve as a future direction for many life fields. Metal NPs that are biologically synthesized using plant extract have various applications, in pharmaceutical and diagnostic industries, industrial scale, used water waste treatment, engineering, agriculture industries and food industries (Seqqat et al. 2019). A large number of recent studies have shown that NPs remediate the environment such as nanofiltration, nano-adsorbents, nanobiocides and nanocatalysts currently being utilized for water and wastewater pollutant remediation to reduce the risks posed to human and environmental receptors by radiological and chemical contaminants (Bratovcic 2019). The benefits of applying nanomaterials for remediation would be faster or more cost-effective waste cleanup. Therefore, exploring a more reliable and sustainable process for nanomaterial synthesis is vitally important. Researchers continue to strive for the development of easy, effective and reliable green chemistry processes for nanomaterial production (Pérez et al. 2019). These may include actinomycetes, bacteria, yeast, fungi, viruses and plants to produce NPs that are well-functional and stable (Saif et al. 2016). Metal NPs are considered suitable water treatment filters due to very large surface area of NPs and their excellent electron relaying capacity and more efficient in the degradation of several organic dyes in wastewater than conventional methods (Fig. 9.1). When AgNPs flow-through systems, certain problems such as excessive pressure drops, low hydraulic conductivity, difficult separation from treated water and agglomeration of particles appeared (Kango and Kumar 2016). Therefore, some supports such as zeolite, fiberglass, polyurethane foam and sand must be covered with AgNPs for practical application (Hua et al. 2012). Farhana and Meera (2016) tried nanosilver-coated sand, which was synthesized with papaya fruit, neem leaf and bamboo leaf extracts. The highest percentage of silver coatings was obtained for the nanosilver-coated sand synthesized using papaya fruit extract. So, it could be used as an effective filter medium for microorganism removal in water/wastewater treatment. The stem of cowpea *Vigna unguiculata* L. was used alone in synthesis silver NPs. So, this was not sufficient due to a relatively low percentage removal (21.6% at 200 ppm) in the complete adsorption of malachite green dye (Dawodu et al. 2019). Many novel studies have pointed out the potential of Fe NPs to remediate of environment because the iron NPs had a

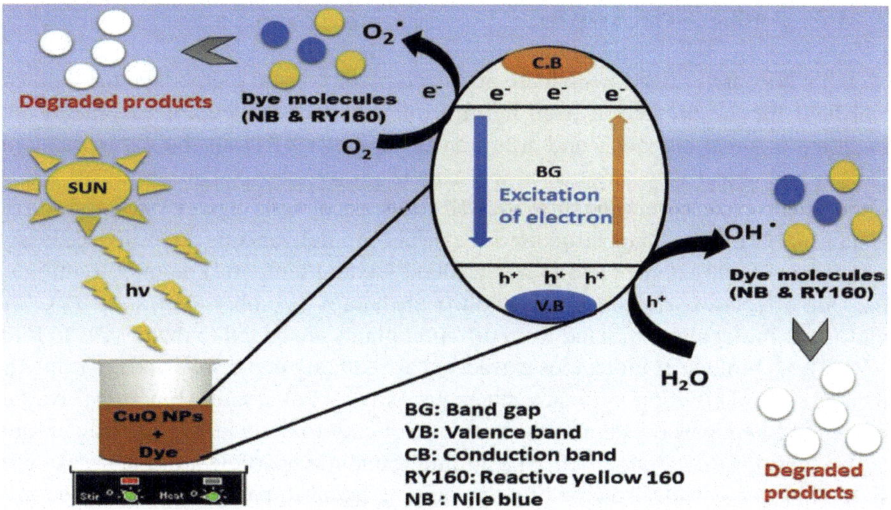

Fig. 9.1 Schematic of potential dye degradation process by CuO NPs (*Source* Singh et al. 2019a)

large surface area to volume ratio. Chrysochoou et al. (2012) reported that iron NPs synthesized with polyphenol enriched green tea solution using two granular media, refined silica sand, as well as aluminum hydroxide sand coated. As a result, the redox potential rose from 150 to 550 mV because effecting of polyphenols. He and Zhao (2005) utilized starch mediated bimetallic iron/lead (Fe/Pd) NPs for trichloroethylene (TCE) degradation. Results from this study showed that the NPs of starched Fe showed considerably less agglomeration, but higher dechlorination power than those produced without a stabilizer. Starch NPs at 0.1 g/L of 98% of TCE were degraded in water within one hour. Wang et al. (2014a) used eucalyptus leaf extracts to synthesize iron NPs for the treatment of eutrophic wastewater. After twenty-one days, the total percentage of phosphorus, nitrogen and chemical oxygen demand (COD) were taken away 30, 71.7 and 84.5%, respectively. The lack of precipitating agents like calcium, magnesium or aluminum was the reason why phosphorus was removed very low. Wang et al. (2014b) applied leaf extracts of eucalyptus and green tea separately to form Fe NPs and employed them to remove nitrate from wastewater. Eucalyptus and green tea Fe NPs were able to take off 41.4 and 59.7% of nitrate from wastewater, respectively. Njagi et al. (2011) investigated the phenolic compounds of various branches of sorghum to synthesize NPs made of metallic iron and its use as a catalyst for the degradation of blue bromothymol. Venkateswarlu et al. (2013) exercised waste plantain peel extract for the reduction of iron salt to form Fe_3O_4 NPs to remove toxic metals and dyes. Zinc oxide (ZnO) NPs can be used as environmental pollutant photocatalytic degradation materials. Zinc oxide NPs bulk and thin films have proven highly sensitive to many toxic gasses.

9.1.1.2 Antibacterial Activity

Metallic NPs are utilized, which are produced by biological methods in the biomedical field for the protection from harmful microorganisms, cancer treatment, bioimaging, medical diagnosis and drug transport. These NPs must be stable, biocompatible and selectively targeted at a particular body site (Nadaroğlu et al. 2017). This can be done by conjugating the NPs with acceptable ligands (Fahimirad and Hatami 2019). Smart nanostructured materials can deliver drugs at reduced dosage frequencies to target sites and in a controlled (spatial/temporal) manner to mitigate the side effects experienced with traditional therapy (Lombardo et al. 2019). The biological molecules extracted from different plants are added to metal salts to form NPs. These biological molecules extract act as reducing metal salts and covering the formed NPs. This capping is advantageous as it acts as a multi-functional way of preventing nanoparticulate agglomeration, reducing toxicity and improving antimicrobial action (Roy et al. 2019). The antimicrobial nanoparticle attitude bases on its size. This size is well preserved by the capping agent impact. Interestingly enough now, if the capping agents themselves have anti-microbial activity. It could offer enhanced antimicrobial action. Plants with antimicrobial action can thus be successfully used to develop antimicrobial action-enhanced NPs (Muniandy et al. 2019). The mechanism of antibacterial NPs includes affecting cell membrane permeability resulting from direct reactions between NPs and cell surfaces and these NPs induce oxidative stress in bacterial cells, afterward inhibits cell growth than its death (Sharma and Gothalwal 2019). Numerals researchers published several scientific papers on the synthesis of silver, zinc and copper oxide NPs using various plant extracts. There are many studied have been conducted towards the antibacterial performance of plant extracts such as squash (*Cucurbita pepo*) seed powder (Singh and Shrivastava 2018), lemon (*Citrus limon*) leaves (Dhinek and Vanitha 2016), leaf extracts of banana (*Musa balbisiana*), neem (*Azadirachta indica*) and black tulsi (*Ocimum tenuiflorum*) (Banerjee et al. 2014), onion (*Allium cepa*) extract (Saxena et al. 2010), leaves of rocket (*Eruca sativa*), spinach (*Spinacia oleracea*) and cheese weed (*Malva parviflora*) (Mohammad and Al-Jubouri 2019), aqueous extracted from sea holly (*Acanthus ilicifolius*) (Mohamad et al. 2019), leaf elephant yam (*Amorphophallus paeoniifolius*) (Gomathi et al. 2019), leaf banyan (*Ficus benghalensis*) (Saxena et al. 2012), cinnamon (*Cinnamomum zylinicum*) bark (Almalah et al. 2019), leaf artichoke (*Cynara scolymus*) (Erdogan et al. 2019), *Salvia spinosa* (Pirtarighat et al. 2019), *Lemon peel* (samreen et al. 2018), *Seripheidium quettense* (Nasara et al. 2019), *Eucalyptus* leaf oil (Heydari et al. 2017), *M. parviflora* (Farhan et al. 2017), leaf extract of *Catharanthus roseus* (Gupta et al. 2018) and *Tridax procumbens* (Gopalakrishnan et al. 2012) (Fig. 9.2).

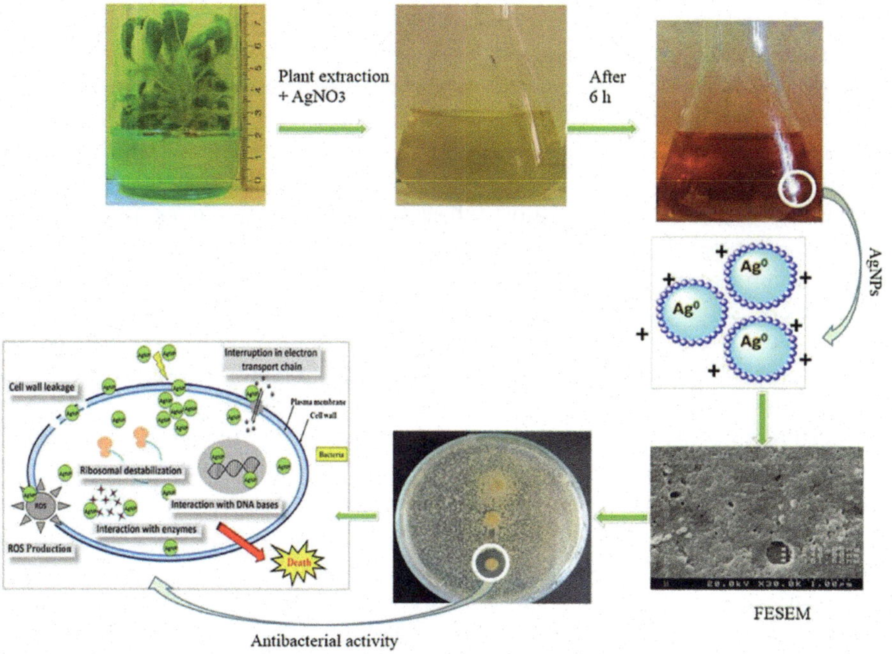

Fig. 9.2 Green synthesis of silver NPs using the *Salvia spinosa* plant extract (grown *in vitro*) and its antibacterial activity evaluation (*Source* Pirtarighat et al. 2019)

9.1.1.3 Application in Agriculture

Biotechnological advances and rapid and more accurate diagnostic tools using NPs have a great and promising future for modern farming practices such as precise nutrient and fertilizer delivery and early diagnosis of disease. NPs can be loaded with fertilizers, herbicides, nucleic acids, fungicides or nutrients and target specific plant tissues in order to release their charge to the demand part of the plant for the desired results (Duhan et al. 2017). Increased food production through excess nitrogen use accounts for 80% of the increase in atmospheric nitrous oxide (N_2O) (a greenhouse gas), which causes higher atmospheric temperatures and, therefore contributes to global warming. It is estimated that approximately 50–70% of potassium, 40–70% of nitrogen and 80–90% of phosphorus chemical fertilizers are going to the environment. Thus, plants are not able to absorb these and, therefore cause environmental pollution (Bartolucci et al. 2020). The use of nanocoated fertilizer reduces the dissolution rate of the fertilizer and allows a slow, sustained release of coated fertilizer that is more efficiently absorbed by plant roots and minimizes pollution of the environment. Polymer biocompatible NPs (chitosan) and kaolin have potential applications in fertilizer slow-release behavior (Corradini et al. 2010). Nanotechnology can supply micronutrients to plants by spraying or fertilization. The use of green nanotechnology is very important due to synthetic pesticides considered harmful to the environment.

Polyethylene glycol coated NPs increased the insecticidal activity of essential garlic oil against *Tribolium castaneum* (red flour beetle) (Duhan et al. 2017). With an efficiency of 80% due to the slow and permanent relief from NPs of active components. The green synthesis of (AgNPs) using rosemary *Rosmarinus officinalis* leaf extract and its influence on wheat and tomato plants have been studied. The results indicated that AgNPs have a noticeable stress effect on tomato plants as dry weight and lower chlorophyll a. In addition, wheat germination percentage, pigment fractions and dry weight have a non-significant impact AgNPs catalyze malondialdehyde (MDA) accumulation in wheat and tomato plants. There was an evident various effect of AgNPs on antioxidant enzymes as catalase and peroxidase and soluble proteins between these two plants (Farghaly and Nafady 2015). In contrast, copper (Cu) NPs were prepared from onion extract enhancing wheat growth as compared to control but their desired effect depends on their concentration so that treated 35 ppm Cu NPs produced remarkably higher root dry weight, shoot length, root length, chlorophyll content, germination percentage and fresh weight. The copper oxide (CuO) NPs with intermediate concentration about 0.025 mg/mL which it is prepared with biological synthesis by the *Adiantum lunulatum* extract and applied to *Lens culinaris* seeds. This concentration increased the root length, phenol and flavonoid levels and antioxidative enzymes. While all these parameters decreased at higher concentrations of CuO NPs. Thus, CuO NPs at an optimum concentration not only have the potential to affect the physiological condition but can also modulate the innate immune system of model plants like lentil *L. culinaris* (Sarkar et al. 2020). AgNPs can produce from leaf extracts of neem (*A. indica*), black tulsi (*O. tenuiflorum*) and banana (*M. balbisiana*). The positive effect of the pervious extracts on chickpea (*Cicer arietinum*) and moong bean (*Vigna radiata*) seeds treated with AgNP solutions was on oxidative stress enzymes activity and germination rates (Banerjee et al. 2014). The seeds of squash (*C. pepo*) were soaked in neem and saisban silver NPs solution along with distilled water as a control for 20 min. There was no influence on the percentage of growth while for all the several concentrations of silver NPs a remarkable change in seeding speed and length has been observed. Silver NPs treated with saisban increase the seeding speed of *C. pepo*. While seeds treated with silver NPs prepared using neem have been noted with the highest length of hypocotyls and semi root (Bamsaoud and Bahwirth 2017). Sabir et al. (2018) assessed the effect of silver (Ag) NPs that synthesized with the leaf extract drumstick tree (*Moringa oliefera*) on seed germination and seedling growth of bread wheat (*Triticum aestivum* L.). Several Ag NPs concentrations (25, 50, 75 and 100 ppm) were applied and tested against a control. The results detected that Ag NPs increased wheat germination. The considerable improvement was registered in root dry weight, root fresh weight, root elongation and root length at 100 ppm of Ag NPs. However, negative results in root length, shoot length, root fresh weight, shoot fresh weight, total protein and total chlorophyll content in White Lupin (*Lupinus termis*) seedlings with different concentrations of Ag NPs from coriander (*Coriandrum sativum*) leaf extract. So, further studies needed to determine the effectiveness, longevity and toxicity of Ag NPs towards photosynthetic systems and antioxidant parameters to improve researches (Alhuqail et al. 2018). Sehnal et al. (2019) examined the effect of green synthesized

AgNPs on germinated plants of maize such as the basic growth and physiological parameters of the plants. The following sequence control < AgNPs < Ag (I) ion was proven to be phytotoxic. Silver NPs exhibited a significant effect on photosynthetic pigments and repression growth above-ground of plant parts was 40%. Alnaddaf et al. (2019) reported the biosynthesis of silver NPs by mixing different ratios of Lemon (*C. limon*) and the concentration of silver nitrate. In addition to the effect of silver NPs of *C. limon* juice solution on seed germination as well as seedling growth of the Syrian durum wheat cultivar Sham 7. The variety Sham 7 seeds were immersed for 30, 60, 90 and 120 minin Ag NPs solution of *C. limon* juice in different mixing ratios (1:1, 1:4 and 4:1) along with distilled water and *C. limon* juice as control. Seedling length, seed germination, leaves number, root number, root length were measured after 14 days. Germination percent of soaked grains for 60 min have a noticeable effect with different mixing ratios. Whereas soaking wheat seeds for 120 minin silver NPs solution of *C. limon* juice with different mixing ratios have an inhibitory effect on germination and seedling growth of wheat (Fig. 9.3). Immersed seeds with 4:1 lemon juice to 15 Mm silver nitrate for 90 min have a significant positive influence on wheat leave numbers. However, another treatment with a ratio of 1:4 lemon juice to 10 Mm silver nitrate for 30 min has a significant positive influence on wheat root length (Fig. 9.4). Whereas, on shoot length, no noticeable efficacy was observed.

Gopinath et al. (2014) reported the green synthesis of gold NPs (Au NPs) from the Arjun Myrobalan (*Terminalia arjuna*) fruit extract, for gloriosa lily (*Gloriosa superba*) increased seed germination activity. Two different concentration 500 and 1000 lM of Au NPs were handled for *G. superba* seed. The concentration of 1000 lM has the most important influence on seed germination rate and *G. Superba* vegetative growth. Singh et al. (2016) made zinc oxide (ZnO) NPs from the Russian olive

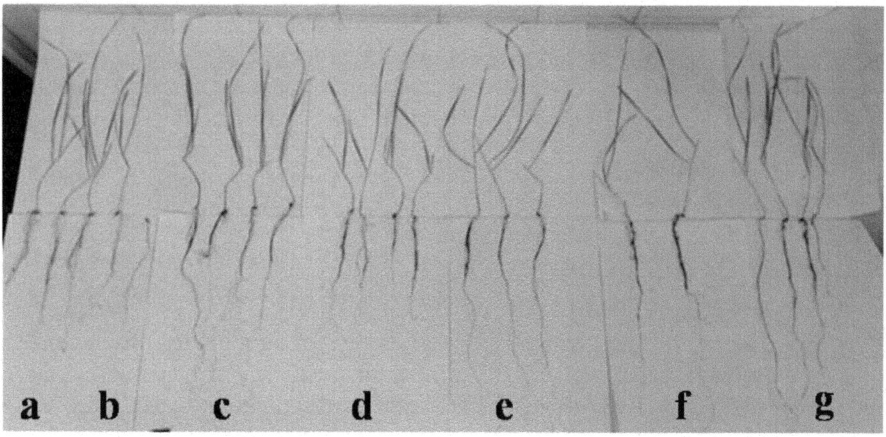

Fig. 9.3 Changes in wheat seedling length with different mixing ratios. **(a, b)** Control, **(c)** Lemon juice: silver nitrate (4:1, 10 Mm $AgNO_3$, 90 min), **(e, g)** Lemon juice: silver nitrate (4:1, 10 Mm, 15 Mm $AgNO_3$, 60 min), **(d, f)** Lemon juice: silver nitrate (4:1, 10 Mm, 15 Mm $AgNO_3$, 120 min) (*Source* Alnaddaf et al. 2019)

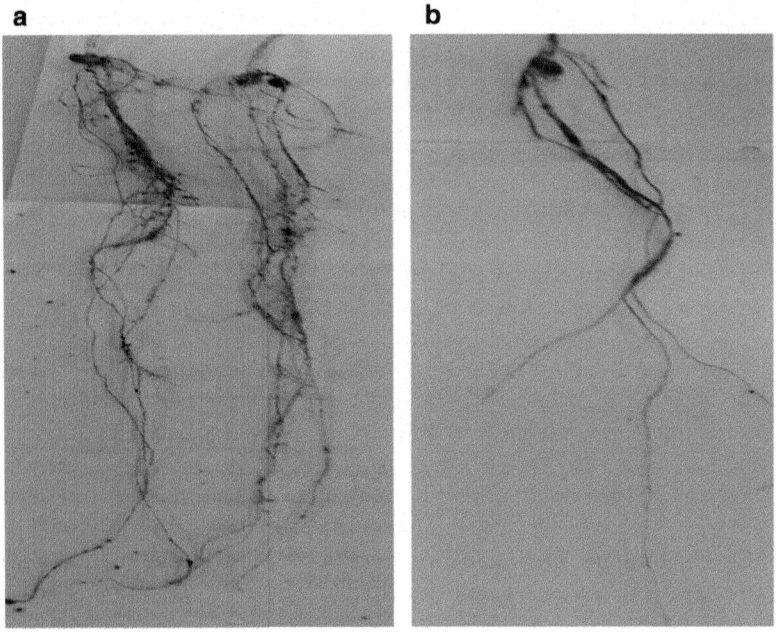

Fig. 9.4 Change in roots length and its numbers in durum wheat cultivar Shame 7. **a** Lemon juice: silver nitrate (1:4, 10 Mm $AgNO_3$) for 30 min; **b** Control (*Source* Alnaddaf et al. 2019)

(*Elaeagnus angustifolia*) flower extract and estimated its impact on chlorophyll, germination, seedling vigor, sugar and protein content as well as lipid peroxidation and antioxidant enzyme activity of tomato (*Solanum lycopersicum*). The highest reply was with a 6.1 mM concentration whereas the lowest response was with a 1.2 mM concentration. Thus, we can understand that the influence of NPs on plant growth varies with different plant extracts, NPs, concentrations and reaction conditions. Nanomaterial efficacy may be stimulation, inhibition or no effect on growth processes and plant development.

9.2 Nano Fertilization

Plant nutrition is the most vital factor for successful agricultural production and quality. NPs have been shown to exhibit both positive and negative effects on the growth of plants. Even though nanomaterials display many positive results in plants. However its phytotoxicity remains a problem. This relies on different factors primarily on the concentration of nanomaterials and their shape and size. Nano fertilization may be done using three methods seed priming, foliar application and soil incorporation. But various factors are responsible for the efficiency of each process. Including mode of absorption by different sections of plants, the process of application and environmental considerations (Goswami et al. 2019).

9.2.1 Soaking

Recently, Acharya et al. (2020) used AgNPs as nanopriming agents for watermelon (*Citrullus lanatus* L.) seeds. Transmission electron microscopy confirmed the internalization of nanomaterials. The rate of seedling emergence at fourteen days after sowing in AgNP treated triploid seeds were significantly higher compared with other treatments. Soluble sugar (fructose and glucose) contents were improved during germination in the AgNP-treated seeds at 96 h, higher yield compared to control. Also, Ramesh et al. (2020) studied the influence of manganese oxide complex NPs in germination and growth attributed to mung bean (*V. radiata* L.) in sandy loam soil. The results showed decreased protein content and rose in full chlorophyll content compared to untreated plants. The manganese deficiency increased the root surface zone due to increased consumption, reduced the quantity of manganese translocation caused reduced shoot growth and yielded an increased root ratio. Rafique et al. (2020) tested the effect of ZnO NPs on pearl millet (*Pennisetum glaucum* L.) seed germination. The seeds treated showed a significant increase in shoot length, root length and germination rate. Seeds treated with (1.5 g ZnO-NPs/30 ml for 2 h) were reported with the highest shoot length, seedling vigor index, and germination rate. Results indicate that ZnO-NPs at lower concentrations promoted vigor index, shoot length and germination rate, whereas at higher concentrations (3 g ZnO-NPs/30 ml) reduced vigor index, shoot length and germination rate. However increased root length. Abdel-Aziz (2019) investigated the effect of two different concentrations (0.05, 0.1%) of chitosan NPs (CsNPs 20 $\pm$ 2 nm) as priming solutions for 6 h of broad bean (*Vicia faba* L.) seeds. Both concentrations of chitosan NPs caused negative effects on germination and seedling growth compared to control (seed treatment with distilled water). On the other hand, the lower concentration of CsNPs (0.05%) improved the contented of total phenols and the antioxidant enzymes (catalase, ascorbate peroxidase, peroxidase and polyphenol oxidase) which improved the protection system of seeds. Another study showed that the French bean (*Phaseolus vulgaris* L.) plants seed rinsing process with nanochitosan (Cs) or carbon nanotubes (CNTs) caused considerable decreases in all parameters of growth compared with control (Abdel-Aziz et al. 2019). The effect of polyvinylpyrrolidone (PVP) stabilized platinum NPs (Pt: PVP) on seed germination and on growth efficiency of pea (*Pisum sativum* L,) was studied. The germination rate decreased by 45% when the seeds submerged in 1.0 mM Pt: PVP 3 h. Hydrosol concentration and the germination time raised from 2.5 days to 7 days. After 30 days of germination of saplings from treated seeds for 3 h. The root to shoot length was 20% lower than that of untreated seed saplings. Also, a significant reduction in rhizobial colonization was observed in samples treated with Pt-NPs compared to control plates, especially when the seeds were soaked for 3 h (Rahman et al. 2020). Boutchuen et al. (2019) used a new method of presoak seed using a drop of hematite NP fertilizer to dramatically increase leguminous plant growth by 230–830% depending on the crop. Although the growth pattern varied between different leguminous species (*Cicer arientinum*, *Vigna radiate* and *P. vulgaris*). The seeds treated with a high NP concentration (1.1 g/L Fe) generally showed the fastest growth compared to those soaked in low NP concentration (0.022 g/L Fe) and control deionized water. The NP treated plants exhibited double faster initial pod production, twice as many pods per plant and a longer life span overall compared to control. Itroutwar et al. (2019) primed rice (*Oryza sativa* L.) seeds with biosynthesis ZnO NPs at 10 mg/L for 12 h. This treatment showed an improvement of the seed germination (100%), shoot length, root length, seedling length, leaf length, sub-root number, seedling vigor and dry matter production compared to the control. Also, Raj and Chandrashekara (2019) checked the effect of seed treatment application with Zn on growth, yield and economics of cotton (*Gossypium hirsutum* L.). Among seed treatments, higher seed cotton yield, plant height, number of monopodial and sympodial branches, leaf area index, leaf area duration and chlorophyll meter values were recorded with ZnO NPs seed treatment (1 g/kg seeds) than seed priming with zinc NPs solution (1000 ppm) and chelated $ZnSO_4$ (4 g/kg seeds) treatment. Rizwan et al. (2019) investigated the effects of seed priming with (Fe NPs) and (ZnO NPs) on the wheat (*T. aestivum*) growing and cadmium (Cd) accumulation. The results showed that wheat photosynthesis was positively impacted by NPs relative to control. On the other hand, decreased Cd concentrations noticed in shoots, roots and grains which treatment with NPs. As well as the amount of Cd in the grains was less than the verge level of Cd for cereals when the seeds were touched with higher NPs. Generally, NPs play a vital role in increasing biomass, nutrients and decrease the toxicity of Cd in wheat. Xiao et al. (2019) investigated the toxicity of MgO and ZnO NPs for pomelo (*Citrus maxima* Merr.) by soaking seedlings in hydroponic systems containing 0 (control), 250, 500 and 1000 mg/L of MgO or ZnO NPs. Results showed that Mg^{2+} and MgO exposure at all concentrations indicated extreme toxicity and high oxidative stress. Zinc oxide NPs showed only mild toxicity. Whereas Zn^{2+} caused chlorosis of the vein in the leaf and heavy oxidative stress in plant shoots. Several researchers have also examined the inhibitory influence of NPs on seed germination of different plants. Silica, Pd, Au and Cu NPs have been found to have a significant adverse germination effect of seed in cucumber (*Cucumis sativus* L.) and lettuces lettuce (*Lactuca sativa* L.). In addition, silver, Au and Fe_3O_4 NPs showed a moderate influence on seed germination in cucumber and lettuce (Barrena et al. 2009). Shah and Belozerova (2009) studied the effects of NPs on seed germination. Copper (Cu) NPs had some beneficial effects on the germination of lettuce and mung beans. Yet, they were phytotoxic to growing seedlings.

9.2.2 Foliar Fertilization

Nanofertilizers NFs have very useful in improving the growth, yield and health of fruit crops. The application of NFs such as nitrogen (N), boron (B), Zn, ZnO, chelate, Fe and its compounds and chitosan on different plants such as pomegranate (*Punica granatum* L.), almond (*Prunus amygdalus* Batsch), grapes (*Vitis vinifera* L.), mango (*Mangifera indica* L.), date palm (*Phoenix dactylifera* L.), coffee (*Coffea canephora* Pierre ex A.Froehner) and strawberry (*Fragaria* sp L.) have positive results. When sprayed at very low concentrations, these compounds have a straight effect by improving the growth, final products and quality of fruits. Higher concentrations of NFs may have harmful effects and even toxicity (Zahedi et al. 2020). In order to food safety, researchers seem to be finding ways to increase fertilizer productivity without losing or polluting. Nanotechnology is a useful means of producing agricultural products, especially in fertilization programs. Because nanoferta is an effective alternative to traditional fertilizers. As it achieves many advantages due to its use of low chemicals, fast absorption by the factory and its high stability under different conditions. This increases the capacity it is stored for long-term use. Nanotechnology can also be used to detect and treat plant diseases, increase crop yields, improve their quality and ensure sustainable crops (Al-Hchami and Alrawi 2020). The requirements of macronutrients such as nitrogen (N), phosphorus (P), potassium (K) and to a lesser degree from calcium (Ca), sulfur (S) and magnesium (Mg). Plants require micronutrients such as copper (Cu), iron (Fe), zinc (Zn), boron (B) to ensure both crop productivity and high-quality. Such elements behavior is significantly influenced by very small changes in environmental factors such as pH, mechanical composition, and soil organic matter. So, Phan et al. (2019) developed a new fertilizer to provide nutrients more effectively (Ag, Zn, Cu, Co and Fe) with hydroxyapatite and studied its impacts on asparagus (*Asparagus officinalis* L.) germination. The ten-day-long seeds test proved a faster germination rate compared to that of normal treatment. As well as asparagus was used the micronutrient nanosystem had grown faster than unused. Also, the effect of nano-fertilization, mineral on growth and yield of sorghum (*Sorghum bicolor* L. Moench) by the foliar sprayed method was studied by Rehab et al. (2020). The study was about the weed control methods, mineral NPK, Nano NPK and their interaction affected the yield and its sorghum components. The results showed that the highest value of sorghum yield characters was achieved by applying hand hoeing once with fertilizing herbicide by 50% NPK mineral and 50% NPK NPs fertilization. As well as, these treatments minimizing the impact of weeds on the field. Wasaya et al. (2020) suggested foliar application by 20 ppm silver NPs with 6 ppm Zn NPs on mung bean under the arid area. This level of application increased the number of branches and pods per plant, chlorophyll amount and seed yield which increased by 26% of seed yield. Abbasifar et al. (2020) used Zn and Cu NPs as a foliar spray on basil (*Ocimum basilicum* L.). Nutrient treatments with 4000 ppm Zn NPs and 2000 ppm Cu NPs caused a significant increase in most morphological traits. The application of the Zn and Cu NPs has significantly affected the concentration of chlorophyll a, chlorophyll b and carotenoid in the leaves. The highest flavonoid and phenolic content were obtained for 4000 ppm Zn NPs and 2000 ppm Cu NPs treatment. Plants treated with 4000 ppm Zn NPs and 0 ppm Cu NPs showed the highest antioxidant activity. Also, Bala et al. (2019) studied the impact of ZnO NPs foliar spray on rice. The foliar application of ZnO NPs (5 g/L) significantly enhanced the growth and yield parameters. However, root characteristics achieved the maximum values at 1.0 g/L ZnO NPs. The soil microbial amounts and enzyme actions such as total viable cell and dehydrogenase activity were detected to be the maximum at 5.0 g/L ZnO NPs. Overall, ZnO NPs treatments successfully reverted the Zn-deficiency symptoms, besides, improve plant Zn contents. Though the response was concentration dependent. These results indicated that ZnO NPs can be successfully used for remediation and Zinc protection in rice cultivated under low soil Zn concentrations. Kahlel et al. (2020) studied the effect of spraying with some NPs fertilizer on the vegetative growth of broad bean (*V. faba* L.). The result indicated that 50 ppm zinc nanomaterials caused a significant increase in broad bean length, number of branches, number of leaves and the leaf area. Noaema et al. (2019) found that boron foliar spray by 10 mg.dm^{-3} increased the nitrogen, phosphorus, boron and chlorophyll content of faba bean leaves and plant dry matter compared to control. As well as Poornima and Koti (2019) studied the effects of Nano ZnO and bulk $ZnSO_4$ on sorghum growth, yield and grain Zn content by two application methods. Foliar spray of 500 ppm ZnO was found more effective than 1000 ppm $ZnSO_4$. Seed treatment with bulk $ZnSO_4$ gives high total dry matter and grain yield but grain zinc content was highest in nano ZnO treatments. Among the method of foliar spray application, it was much better than seed priming. The inhibitory effect was observed at nano ZnO concentration >1000 ppm, revealing the toxicity and need careful use of applications foliar NPs. Abdel-Aziz et al. (2019) showed that nanochitosan (Cs) foliar application or carbon nanotubes (CNTs) treatment on French bean increased all plant growth parameters significantly compared with control. Foliar usage reduced the days to harvest without reducing yield as compared with seed priming treatment. Of importance, Cs NPs acted to improve growth and yield parameters extra than CNTs in foliar application treatment. Elshamy et al. (2019) studied the effect of foliar treatment of chitosan CS NPs packed with nitrogen, phosphorus and potassium (NPK) on the development and yielding traits, chemical metabolites and nutritional content of potato (*Solanum tuberosum* L.) cultivar Spunta cultivated in sandy loam soil. It was applied to leaf faces escaping straight contact with soil systems. The results showed that NPs were taken up and transported within phloem tissues. Foliar use of nano CS-NPK levels (10, 50 and 100%) significantly increased all the growth and yield traits, photosynthetic pigments, chemical constituents of potato tuber at harvest and macronutrients in potato leaves and tubers as compared with the control. In this respect, the highest successful treatment was 10% Nano CS-NPK compared with the other two treatments. Abdel-Aziz et al. (2018) found that the foliar application of nano chitosan nitrogen, phosphorus and potassium (NPK) fertilizer decreased the life cycle of wheat plants with the ratio of 23.5%. Treatment of wheat with nano chitosan NPK fertilizer caused significant increases in all yield traits as compared with non-fertilized and normal fertilized NPK. Transmission electron microscopy showed that NPs were present in phloem tissues and transported through phloem way from leaves to stem then to roots. Hussein et al. (2019) used (SeNPs) to induce growth enhancements of groundnut (*Arachis hypogaea* L.) cultivars (Gregory and Giza 6, NC) during the vegetative stage, foliar application of SeNPs was applied in many concentrations (0–40 ppm). The effect of selenium NPs on growth depends on its concentration and cultivars used. Use of SeNPs improved Gerogory cultivar production and Giza 6, while the cultivar NC growth parameters affected negatively by SeNP treatments. The effects of SeNPs on groundnut cultivars growth were associated with biochemical and physiological traits. The changes in photosynthetic pigments, antioxidants enzymes (peroxidase, catalase and ascorbic acid peroxidase), lipid peroxidation, total flavonoids, total phenols and total soluble sugars. In general, nano selenium acts as a stimulant and/or stress or enhancing the antioxidant defense systems in groundnut cultivars tested leads to improve the stress tolerance under sandy soil conditions. Kheyri et al. (2019) proved that applied small amounts of Si and Zn (300 g/ha) nano-scale as foliar spray fertilizers on rice (*O. sativa* L.) provided a benefit that was similar or greater than large amounts of traditional fertilizers (9 kg/ha Zn and 392 kg/ha Si) which improved yield and element accumulation in rice grain.

9.2.3 Soil Fertilization

From the plant roots of the nanomaterials such as ZnO, TiO_2, CeO_2, Fe_3O_4, $Ni(OH)_2$, C70 fullerenes, Al, Cu, Ag and carbon nanotubes (CNT) are uptake and translocated to plant stem where partly are deposited (C70, Fe_3O_4,CeO_2, $Ni(OH)_2$) or partly are foliar deposited (Al, Ag, Cu, Zn, ZnO, CeO_2, Fe_3O_4, C70) (Predoi et al. 2020). A plant root cell has different absorption zones for various forms of nanomaterials. For example, Fe_3O_4 has absorption areas in epidermis, cortex and cambium, $Ni(OH)_2$ in the epidermis, cambium, cortex, and metaxylem, Ag in epidermis and cortex and Ag^{2+} in epidermis, cortex, endodermis and metaxylem (Predoi et al. 2020). Singh et al. (2019b) studied the effect of nano-zinc oxide (ZnO NPs) and zinc sulfate ($ZnSO_4$) which added in suspension and ionic form at various concentrations, respectively on rice. The application of the ZnO NPs improved seed germination, root and shoot growth, seedling vigor index, chlorophyll content, concentration and yield of grain zinc. Grain yield was also increased over control and $ZnSO_4$ by 8.84% and 3.89%, respectively. Right-dose delivery of Zn nutrient via ZnO NPs and of the right size could be efficient and beneficial in enhancing rice crop growth and yield traits. There is a possibility of reducing Zn dose with nanostructured fertilizer such as ZnO NPs. A field experiment conducted to identify the effect of 20: 20: 20 NPK NPs and mineral fertilizer adding methods and fertilizer levels on corn (*Zea mays* L.) growth and productivity. Results showed that treatment with 1.5 g/L + 7.5 kg/ha NPK NPs mixing with soil was significantly succeeded in vegetative growth and yield by giving the maximum mean in plant height, total number of leaves, leaves area index, total chlorophyll content and grain yield per plant. Moreover, spraying treatment with NPK NPs 1.5 g/L + 7.5 kg/ha mixing with soil achieved significantly increased and recorded the maximum means in root content of nutrients (N, P, K) (Al-Gym and Al-Asady 2020). Abdelsalam et al. (2019) studied the effect of a nanoparticular fertilizer compound on productivity and genotoxicity in two wheat cultivars compared to traditional mineral fertilizer in field conditions. Fertilization with NPK NPs caused an increase in yield. However, root-tip cells showed various types of chromosomal aberrations compared with control treatments that showed normal mitotic stages. The wheat root tip cells instantly internalize NPK NPs that could interfere with normal cell function. The use of nanotechnologies in agriculture is considered the best product for the plant resistance after the detection of different biotic or abiotic indicators known as induced resistance. El-Sherif et al. (2019) examined the effect of three mineral and nano-fertilizer particles (mineral-Zn; nano-Zn oxide; mineral-Fe; nano-Fe oxide; mineral Fe-Zn oxide; nano-Fe-Zn oxide) as a soil change in controlling root-knot nematode *Meloidogyne incognita* on tomato under greenhouse conditions. The result showed that the application of nano-fertilizers was more effective than mineral ones. Also, the Zn oxide nano fertilizer was the most effective among all the treatments. Kachel et al. (2019) evaluated the quality of virgin oil pressed from spring rape (*Brassica napus* L.) seeds which treatment with 0.01% colloidal nanosilver and 0.005% nanocopper solutions as soaking seeds for one hour and foliar fertilizer on florescence stage. The results showed increased the content of carotenoid pigments in the oil. Both the acid and peroxide numbers values were upper in the studied oils as compared to the control but did not out do the required acceptable levels. Also, Mansoor et al. (2019) studied the effects of (ZnO NPs) and ZnO bulk powder on seed germination and early growth parameters of bread wheat (*T. aestivum*). The resulte proved that low concentration of 300 ppm ZnO NPs can have stimulatory or more effect on wheat germination and seedling growth parameters (Shoot, root and seedling length). Salachna et al. (2019) studied the effects of various concentrations of AgNPs (0, 25, 50, 100, and 150 ppm) and their application methods (pre-planting bulb soaks, foliar sprays and substrate drenches) on the growth, flowering, morphological traits, leaf photosynthetic pigments content, basic macronutrients and complex biomolecules on lily (*Lilium* L.) cv. Mona Lisa. Soaking the bulbs in a AgNPs solution turned out to be the most successful method for growth and promotion of blooms. Silver NPs (100 ppm) stimulated plant growth, which was evidenced by increased leaf and bulb biomass accumulation and accelerated flowering. Plants treated with silver NPs also displayed a higher leaf greenness index, flowers and flowered longer.

9.3 Mechanism of NPs Uptake and Accumulation in Crops

Efficacy of NPs usage in plants relies on their uptake and accumulation, which stays to date under discussions. Usually, NPs concentrate on cells, or extracellular space (Husen and Siddiqi 2014).

9.3.1 Root System

The absorption of NPs from soil begins when the roots absorb water. Afterward, they transfer into the xylem tracheary elements which its structure controls the movement of water and NPs (Fig. 9.5) (Mishra et al. 2014). Researches relating the possibility of plant root uptake of engineered NPs (ENPs) indicated that soil type and chemistry play the main role in the availability of nanoparticle uptake via roots. Zhu et al. (2008) explained for the first-time details about plant root uptake of ENPs. Then the accurate methods dealing with root uptake via pores were investigated by many scientists (Feng et al. 2013; Judy and Bertsch 2014; Judy et al. 2015; Watts-Williams

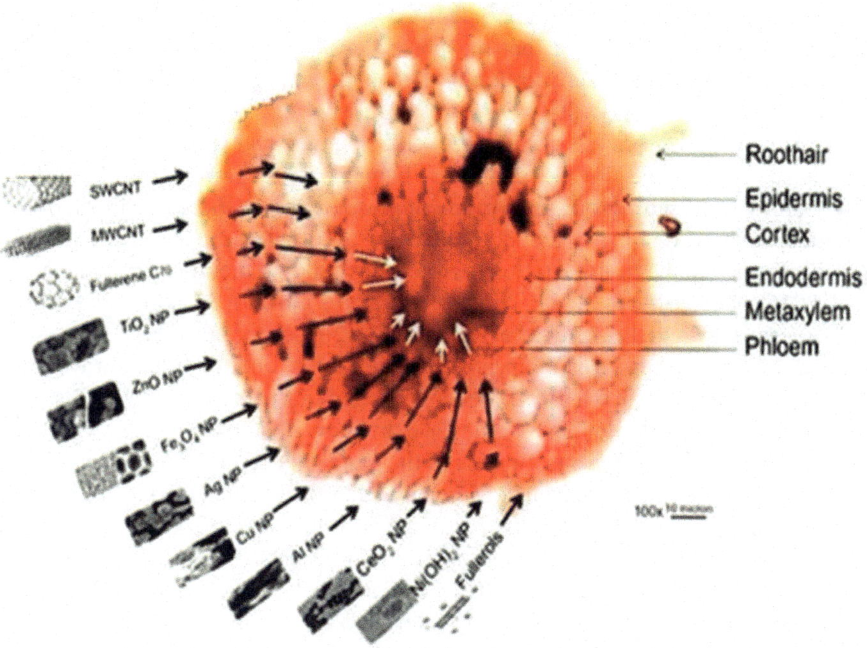

Fig. 9.5 Explains the root anatomical composition and the uptake of various NPs metal via plant root (*Source* Mishra et al. 2014)

et al. 2014). The diameter of almost all roots pores is smaller than 8 nm, whereas few pores described as mesopores with a diameter of 50 nm (Adani et al. 2011). The nature of plants regarding the mechanism of absorption and uptake activity should be well known before applying any type of nonmaterial. In addition, the level of accumulation differs according to time exposure, size of NPs and plant species. Many scientists evaluated the accessibility of several NPs to different species of plants (Miralles et al. 2012; Rico et al. 2011). Nanoparticle nature and type affect the roots ability to uptake. The absorbed metal oxide NPs via tomato roots remained in roots and did not accumulate in shoots. While the uptake of other metal NPs like Co, Ni and Ag revealed their accumulation in different parts of the plant (Antisari et al. 2014). As an example, the uptake, translocation and accumulation of Fe_3O_4 NPs differ between crops and even species and its accumulation occur in different plant tissues. As results of Van Aken (2015) revealed differentiation in the existence of Fe_3O_4 NPs in the xylem vessels between lima beans and pumpkin. While results of Corredor et al. (2010) illustrated that the presence of Fe_3O_4 NPs was distinguished in the cytoplasm of pumpkin plants as well as outside the cell membrane in the epidermis of the stem. After seven days of exposure, Fe_3O_4 NPs enhanced their accumulation in *Lemna gibba* plants (Barhoumi et al. 2015). Numerus scientific discussions reported how a plant absorbs and uptakes NPs. Some researchers declared that NPs described with the high surface area absorbed by plant tissue because they attached easily to the organic chemicals or carrier proteins (Xu et al. 2011). Rico et al. (2011) mentioned that the presence of ion transporters in the plasma membrane facilitates plant absorption of the metal NPs. Zhu et al. (2008) mentioned that the accumulation of Fe_3O_4 NPs by lima bean and pumpkin plants mostly occurred close to the roots and move towards the leaf tissues. Hong et al. (2005) found that the translocation of Fe_3O_4 NPs can easily happen in pumpkin plants. The NPs are absorbed by roots with water, then gradient pressure drives the water transportation to root apoplast. Whereas osmotic gradients, osmotic pressure and capillary action are all involved in the transportation between surrounding membrane routes (Patrick et al. 2015). According to Shankar et al. (2003), NPs move to the cortex after passing the epidermis, then to the plant stele. Patrick et al. (2015) reported that NPs sometimes make a way into plasma membranes of each cell in the endoderm and move toward the steel via the apoplast which is blocked by accumulated lignin in cell walls. While passage cells which do not have lignin ease the movement of NPs (Fig. 9.6).

The accumulation of NPs in all parts of the plant depends also on their transportation through the phloem. Gonzalez-Melendi et al. (2008) confirmed the existence of NPs in Cucurbita plants inside cells and spaces between cells. White (2012) explained about factors affecting the uptake of all compounds especially plant genotype and nature of the absorbed component. Then declared that the accumulation of mineral elements commonly found in cells. However, the findings of Lin et al. (2009) refereed to the existence of engineered NPs in the apoplast (plasma membrane and cell wall). Lin and Xing (2008) found NPs in the nuclei, apoplast and cytoplasm of ryegrass endodermal cells. NPs could move to the endodermis with no need to cross the cortical and epidermal cells borders, but in most cases, they finally accumulate in the endodermis (Larue et al. 2012; Patrick et al. 2015; Zhao et al. 2012b).

9.3.2 Vegetative System

Root uptake is not the only way that plants use to accumulate the applied NPs, other ways are available like foliar uptake via pores which occurs when NPs are applied on leaves surfaces (Lin et al. 2009). The up-taken of NPs through the vegetative system is conditioned by the relation between plants and the surrounding circumstances. In general, the method of traversing the pores by NPs needs more clarification and supportive scientific studies. However, researches afforded functional information about the mechanism of nanomaterials NMs uptake and accumulation via the vegetative system. Stomata or cuticle are two routes. Where NPs enter the leaf in the foliar application (Pérez-de-Luque 2017). Then distribute in the stem and move via the phloem to root cells (Deepa et al. 2014). The entrance of NPs is controlled by the cuticle according to the size of NPs. Concerning stomata, both symplastic and apoplastic routes are the only ways for cellular transfer of NPs which size is more than 10 nm (Pérez-de-Luque 2017). According to Tripathi et al. (2017) Transfer of NPs with size 50–200 nm takes place in between cells (through apoplast). While the transfer of smaller NPs 10–50 nm happens by symplastic road throughout the

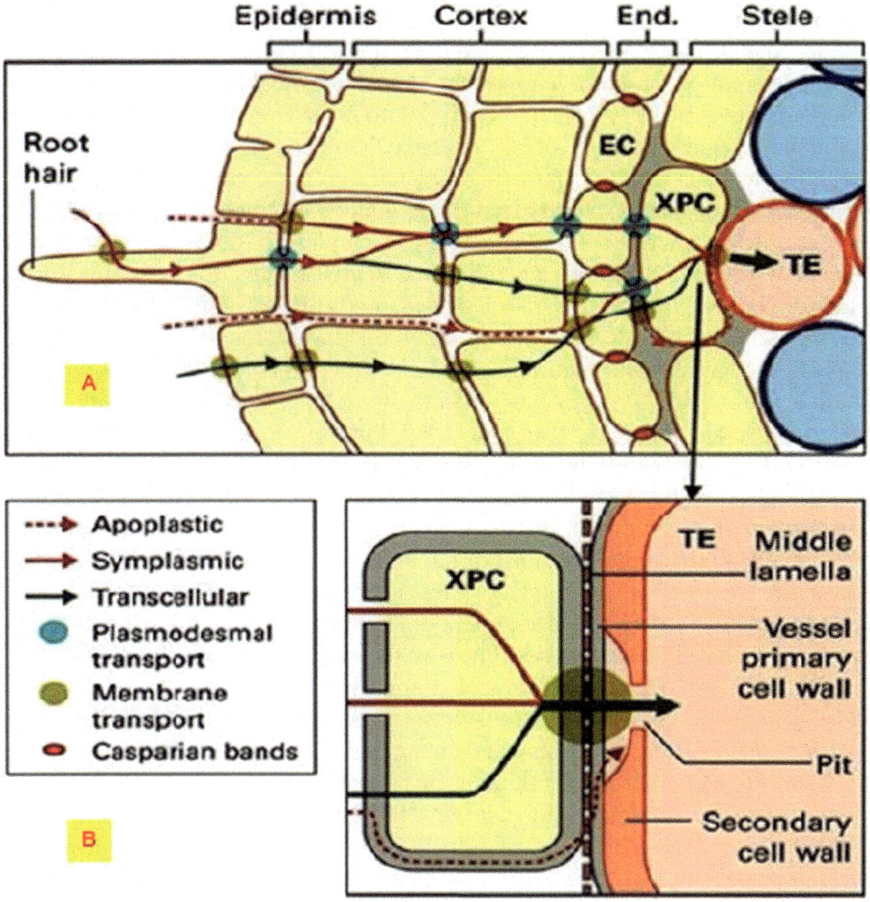

Fig. 9.6 Feasible paths of NPs uptake and transfer via xylem in a plant structure (containing symplasmic, apoplastic, Casparian band, membrane and plasmodesmal transport, and transcellular). **a** The illustration shows arranged root parts from hair to endoderm end with a stele. Transportation routes of water and nutrient are considered as the main ways of transfer via plasma membranes and plasmodesmata. Sometimes Casparian bands in cell walls consisted of lignin blocks the root apoplast. In the endoderm, some cells called passage cells since they do not have lignin. **b** Illustration indicates some thoughts about xylem parenchyma cells (XPCs) which loads tracheary elements in xylem (TE) assisted by particular membrane carriers (*Source* Patrick et al. 2015)

cytoplasm of closest cells, then interior NPs move with sugar transportation via the phloem and accumulates in the stem, roots, grains and other plant parts (Raliya et al. 2016). Limitations due to stomata size affect the method of cuticular diffusion, stomata size differs according to plant type from 0.6 nm in some plants to 4 nm in *Coffee arabica* and could be more or less in other plants. Thus, cuticular diffusion is considered by researchers as an exhausting way of entry due to variation of stomata size (Schönherr 2006; Schreiber 2005). Some results revealed that the stomata pore size must be more than 40 nm to permit the uptake of fluorescent polystyrene NPs of 43 nm size. While others showed that the usage of CeO_2 NPs with a size of 37 nm as spray or solution on maize leaves were remained by leaves and did not transfer to stem (Birbaum et al. 2010). Eichert et al. (2008) considered the size of 43 nm as the maximum expected limit size of NPs to make their entry available via leaves pores. According to their results from using a specific microscope in *V. faba*, the polymeric NPs whose size is 43 nm entered leaf via pores. Whereas the size of 1.1 μm was not able to enter pores and (Eichert and Goldbach 2008) agreed with them. Wang et al. (2013) refereed to a number of essential factors that affect the efficient uptake of NPs via vegetative parts such as methods of NPs application and concentration and size of NPs. Researches indicated that the ecological circumstances like temperature, sunlight and humidity could also affect the vegetative uptake of NPs (Sharma et al. 2015). As they have effects on the nanoparticle capability to entre plant via traversing the pores (Lin et al. 2009; Punshon et al. 2003; Schönherr 2006; Schreiber 2005). NPs trapping on leaves surfaces depends on some important factors concerning leaves like the existence of wax and exudates, chemical composition and leaves morphology (Larue et al. 2014; Schreck et al. 2012). Different researches were carried out to estimate the absorption, transportation and accumulation of NPs capacity throughout the foliar uptake. Taiz and Zeiger (2010) found that the uptake and accumulation via the foliar treatment of Fe, Mn and Cu nano-fertilizers through leaf spores could be more efficient than by the soil treatment since some of these elements are lost in the soil. Foliar application of some NPs like copper, TiO_2, Iron oxide and ZnO have been used (for their better accumulation) to fertilizing various species of plant such as rape, bean and cucumber (Verma et al. 2018; Saharan et al. 2016). In order to control the releasing of phosphorus, potassium and nitrogen in wheat, vegetative application of chitosan NPs (natural polymer) is used recently (Abdel-Aziz et al. 2016). Larue et al. (2014) reported that NPs can enter the plant via leaves pores. Hong et al. (2014) declared that the treated cucumber leaves of CeO_2 NPs resulted in the accumulation of Ce in the roots. Results of Wang et al. (2012) mentioned that copper oxide NPs accumulate in roots by phloem after applying it on maize vegetative system (shoot). Many scientists confirmed according to their researches the foliar application of NPs, for example, Wang et al. (2013) applied a spray of (ZnO, Fe_2O_3, MgO, TiO_2) NPs on watermelon plants, they concluded that NPs equal or less than 100 nm were up-taken by the leaves pores and transported and accumulated in both stems and roots. Taran et al. (2014) applied Zn, Mn, and Fe NPs solutions and examined the NPs concentration in wheat seedlings after germination and after growth. Their results insured the uptake of Zn and Mn NPs from leaves epidermis. Raliya et al. (2015) mentioned that NPs of TiO_2 and ZnO can penetrate tomato plants leaves only if their size is 25 nm ($\pm$3.5), whereas Adhikari et al. (2016) found that CuO with the size of <50 nm enters via *Z. mays* leaves. Whereas, Deepa et al. (2014) reported that calcium transfers from leaves to stem and roots via phloem in groundnut crop when foliar applied with CaO NPs.

9.4 Effect of NPs on Plant Growth Under Abiotic Stresses

Abiotic stress can be realized as the environmental condition which hinders the plants' normal functioning. The main abiotic stresses are salinity, drought, thermal, flooding and heavy metals. The duration and intensity of stress were determined its damage (Iyarin et al. 2019). Accumulation of heavy metals in soil due to continuous anthropogenic activities (mining, vehicle exhaust, sewage disposal) is of global concern consequent to their detrimental toxicity on plants with subsequent soil quality and fertility reduction (Azeez et al. 2019). It also constitutes burdens on human health when eventually transferred through plant uptake into the food chain (Lamhamdi et al. 2013; Liu et al. 2015; Azeez et al. 2019). Lead (Pb) and cadmium (Cd) are heavy metals that are not usually required for any essential in plant cell activity but are highly toxic. Various management systems for remediating contaminated heavy metal soil such as soil replacement, surface capping, vitrification, chemical immobilization. In contrast, encapsulation, phytoremediation, phytostabilization, soil flushing and bioremediation were applied (Azeez et al. 2019; Liu et al. 2018; Rizwan et al. 2018). These stress command to oxidative stress by forming reactive oxygen species (ROS) excessive ROS formation leads to cyto and genotoxicity. with consequent physiological and morphological changes affecting plant growth, germination, quantity, fruiting, quality and nutrient translocation (Azeez et al. 2019). Likewise, major effects of higher temperatures on plant include growth retardation by decreasing cell division and cell elongation resulting in dwarf plants, as well as lowering root growth, root number and root diameter (Iqbal et al. 2017). High temperatures are also responsible for significant pre- and post-harvest damage, including leaf and twig scorching, leaf sunburn, branches and stems, leaf senescence and abscission, root and shoot growth inhibition, fruit discoloration, and damage (Wahid 2007). Plants face these various environmental stresses through developing their defense at various levels by biochemical, modulating the molecular and physiological pathways like enzymes and antioxidants (Iqbal et al. 2017). One of the most pressing solutions is Nanomaterials that facilitate plant growth and potential plant tolerance to environmental stress. Nanomaterials may imitate the role of antioxidant enzymes such as catalase, peroxidase and superoxide dismutase (Iyarin et al. 2019). Due to its size larger surface area and more reactive areas (Kasim et al. 2017). NPs assisted in enzyme activity connected to tolerance stress. The negatively charged surfaces of metal NPs can help accumulate within plants and show relatively less capacity for translocation than nanomaterials based on carbon. Supporting plant growth and crop protection thereby carbon nanotubes (CNTs) and graphene may permeate and move from root to shoot and leaf into seed coat (Iyarin et al. 2019). NPs are appropriate adsorbents with improved efficacy due to their morphological characteristics, quick soil dispersal, ease of delivery, strong affinity and high target metal sorption capacity (Azeez et al. 2019). Moreover, their phytostimulatory, phytopathogenic and attributes regarding the promotion of seed germination, growth enhancement, physiological tolerance, nitrogen metabolism, photosynthesis and antifungal properties have increased their applications significantly (Azeez et al. 2017, 2019; Galdames et al. 2017; Gong et al. 2018; Li et al. 2017; Liu et al. 2015; Ochoa et al. 2018; Praveen et al. 2017). NPs such as multi-walled carbon nanotube, carbon nanotube, graphene oxide, fullerene, silver, titanium oxide, magnetite, iron phosphate, nickel oxide, magnesium oxide, zero-valent iron, silicon, copper oxide, aluminum oxide and zinc NPs were used as functional adsorbents and inactivating agents for the removal of phenanthrene, naphthalene, pesticides, Cd, As, Cr, Ni, Pb, Zn, Cu, Fe, Al, antibiotics and rhodamine B in water and soil (Azeez et al. 2019; Galdames et al. 2017; Gong et al. 2018; Liu et al. 2015; Ochoa et al. 2018; Praveen et al. 2017; Venkatachalam et al. 2016) (Fig. 9.7).

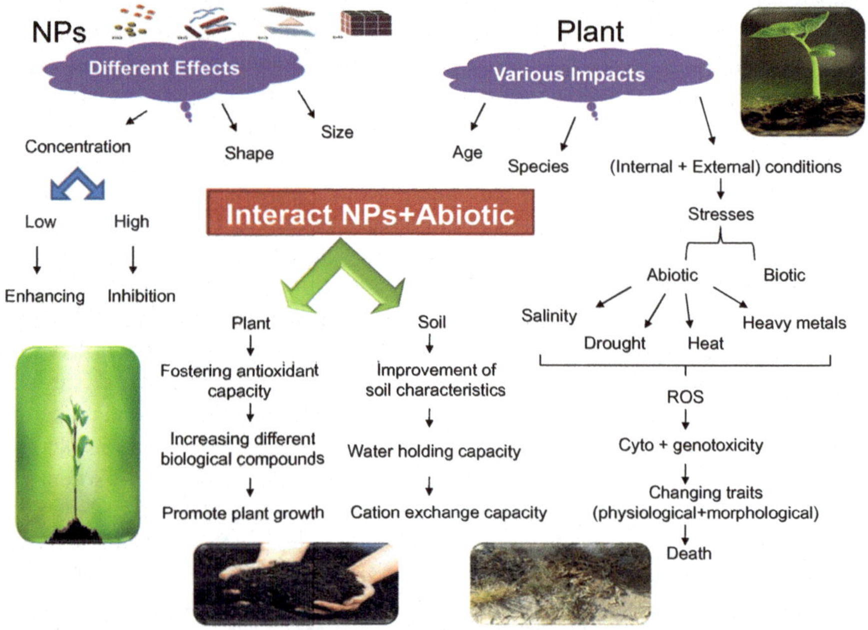

Fig. 9.7 The way nanoparticles (NPs) interact with the plants. NPs are different effectively with shape, size, and concentration. Additionally, the plants response to NPs treatment varies depending on species, age and external and internal conditions, some are inhibitory and others are growth activation. The impact of various stresses on plant growth. NPs interact with various stresses and how this reaction affects plants and soil. NPs reactive oxygen species (ROS) (Constructed by L.M. Alnaddaf)

9.4.1 Silicone NPs Nano-Sio_2

Silicone NPs promote plant growth despite environmental stress and the use of nano-SiO_2 enhances proline accumulation, chlorophyll, leaves weight, nutrients, amino acids and enzyme action (Siddiqui et al. 2014). Silicone becomes a vital plant protection factor against many biotic and abiotic stresses such as toxicity to diseases, pests, drought, salinity and heavy metals (Alsaeedi et al. 2017, 2018). Silicon has considerable potential improvements to soil characteristics such as soil texture, water holding capacity, soil erosion, soil organic matter stability and cation exchange capacity (Alsaeedi et al. 2019). The positive effect of silicon has been to induce cucumber plant growth under water deficit and salinity stresses. Attributed to alleviating oxidative stress by fostering antioxidant capacity and raising the concentration of nutrients in the cucumber shoot (Alsaeedi et al. 2019). Jafari et al. (2012) notified that silicone decreased H_2O_2, lipid peroxidation levels, proline and ion leakage in osmotic stressed cucumber plants. They also found silicon enhanced antioxidant capacity of cucumber plants by increasing non-enzymatic antioxidants, flavonoids, anthocyanins, total phenolic compounds and Ca^{2+}, Si, K in the shoot and phenylalanine lyase activity. Low silicone levels (1 g/kg added as SiO_2) improved wheat plant biomass production compared to control. However, higher silicon concentration gradually decreased the shooting part of biomass production. In addition, the grain yield increased significantly at a rate of 10 g/kg SiO_2 (Neu et al. 2017). Similarly, rice grain yields increased considerably when silicone was applied at a rate of 100–400 kg/ha (Cuong et al. 2017). Cucumber fruit yields were higher for lower silicon doses 200 mg/kg, while higher silicon levels decreased the fruit yield (Alsaeedi et al. 2019). In addition, enhancing the efficiency of nutrient use, silicon increases the rate of photosynthesis by modifying the position and orientation of plant leaves due to silicon precipitation in the cell wall, which makes leaves more erect and thereafter improves the characteristics of light interception. Many researchers found that silicon ameliorates the absorption of many nutrients such as phosphorus, nitrogen (Singh et al. 2005), potassium (Singh et al. 2005; Pati et al. 2016), zinc (Curie and Briat 2003) and iron (Mali and Aery 2009). Greger et al. (2018) indicated that silicon not only affected the availability and uptake of nutrients but also the moving of nutrients from root to shoot. Since Mg is the main element in chlorophyll structure, silicon increased the translocation of Mg into the shooting part improving the photosynthesis rate (Alsaeedi et al. 2019). Many reports have referred that the uptake and accumulation of potassium in plant tissues is strongly consistent with providing varying rates of silicon to plants. Increased potassium intake by cucumber (Alsaeedi et al. 2018; Hasanuzzaman et al. 2018), tomato (Al-Aghabary et al. 2004), barley (Liang et al. 2006), sugarcane (Ashraf 2009) resulting of silicon applied particularly under salinity stress. This is explained as silicon rises the efficiency of the proton-pump ATPase (H^+-ATPase) located in the plasma membrane through creating electrochemical gradients in the plasma membrane which operatives K^+ channels and carriers across the plasma membrane which increases cellular uptake of potassium (Liang et al. 2006). The mechanisms proposed to improve the absorption of water by plants treated with silicone are silicon improves aquaporin activity in the cell membrane by upregulating aquaporin genes in addition to scavenging the reactive oxygen species that inhibit aquaporin activity (Alsaeedi et al. 2019). Silicon increases root xylem sap osmosis by increasing osmo regulators such as amino acids, soluble sugars and potassium. Silicone increases the ratio of root to shoot due to root growth (Chen et al. 2018). Neu et al. (2017) observed that silicon is more deposited in old wheat plant leaves compared to younger ones. It is precipitated by phytoliths in leaf blades of old wheat plant leaves treated with silica NPs at a rate of 10 and 50 g/kg. Under salinity condition, SiO_2 nanofertilizer application can have a positive effect on plant growth and cucumber yield by improving nitrogen and phosphorus uptake and reducing the Na content (Alsaeedi et al. 2019). Zahedi et al. (2019) investigated the beneficial role of SeNPs in mitigating the adverse effects of soil-salinity on growth and yield of strawberry *Fragaria ananassa*. The foliar spray of 10 and 20 mg/L SeNPs enhanced the strawberry growth and yield traits grown on different saline soils and non-saline which was attributed to their ability to protect photosynthetic pigments. As well as SeNPs improved fruit quality and nutritional values (Fig. 9.8). Zurccani (2008) also established that Silicone's application counteracts the adverse effects of salinity in *P. vulgaris* as antioxidant enzyme activity rises and stomatal conductance decreases. The SiO_2 nanoparticle as a foliar application prevented the loss of leaching N and helped to accumulate more nitrogen in the leaf (Siddique 2014), and help to increase turgidity, strength and elasticity of the cell wall during growth extension (Yassen et al. 2017). The silica presence on the epidermis of the leaf raises the tolerance for ultraviolet (Goto et al. 2003) and reduces the damage caused by ultraviolet-B on the cell membrane (Shen et al. 2010). In the case of rice, it increases resistance to lodging by reinforcing the stems (Liang et al. 2013) resulting in reduced leaf heat load, which provides an effective cooling mechanism and thus improves plant tolerance at high temperatures. It is also important for plants growing under drought conditions, as a double layer of silica cuticular is formed below the epidermis of the leaf, which in turn reduces water loss due to cuticular transpiration (Snehal and Lohani 2018). Foliar application with 2.5 mM nano-silicon reduced Cd stress in rice seedlings as a result of increasing chlorophyll content, accessibility of nutrition for Mg, Zn and Fe and decrease accumulation and translocation of Cd from root to shoot (Wang et al. 2014b). However, those nano-Si handled plants had minimum malondialdehyde (MDA) but greater glutathione (GSH) content and varied antioxidant enzyme activities pointing a higher Cd tolerance in them (Shi et al. 2010; Zeng et al. 2011; Wang et al. 2011). Also, these plants are excreted phenolics such as quercetin and catechins that have strong chelating abilities. These phenolic compounds, on the other hand, Aluminum detoxify by forming hydroxyl-aluminum silicates in the apoplast (Wang et al. 2004). Liu et al. (2015) demonstrated that silicon NPs eased the phytotoxicity effects of Pb and increased growth and biomass of rice seedlings. Likewise, the extraction of water from deeper layers of soil increases, as silica promotes root elongation (Hattori et al. 2005; Snehal and Lohani 2018). The soluble silicate available in the soil becomes hydrolyzed and produces gelatinous metasilicic acid which has the property of retaining heavy metals (Gu et al. 2011; Snehal and Lohani 2018).

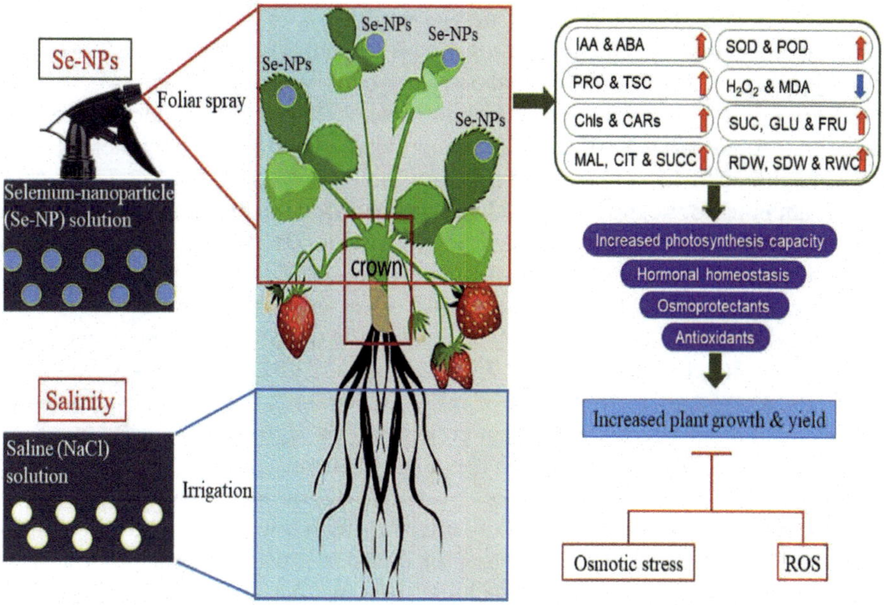

Fig. 9.8 The diagram explains the proposed salt stress tolerance mechanisms in strawberry plants by using (Se-NPs). Se-NPs can be used to improve growth and yield by improving photosynthetic capacity through protecting photosynthetic pigments, increased proline and total soluble carbohydrates for enhanced osmoprotectant, antioxidant system activated to maintain efficient reactive oxygen reaction homeostasis (ROS), improving the levels of indole-3-acetic acid (IAA) and abscisic acid (ABA) to enhance root biomass and maintain the proper osmotic status of cells, glucose, GLU, Chls, carotenoids, relative water content (RWC), chlorophylls (CARs), total soluble carbohydrates (TSC), hydrogen peroxide (H_2O_2), root dry weight (RDW), malondialdehyde (MDA), proline (PRO), sucrose (SUC), malic (MAL), succinic (SUCC), citric (CIT), shoot dry weight (SDW), superoxide dismutase (SOD), peroxidase (POD), fructose (FRU) (*Source* Zahedi et al. 2019)

9.4.2 Zinc Oxide NPs

The NPs simplify fertilizer absorption and improve the impact of Hoagland solution by raising the availability of Fe and Zn which are associated with the mechanism of salt tolerance. Hussein and Abou Baker (2018) reported nano Zn application increased root penetration and nutrient uptake, resulting in significant changes in fresh and dry rice weight (Upadhyaya et al. 2015), sunflower biomass production (Torabian et al. 2016), wheat grain yield under salt stress (Babaei et al. 2017) and maize yield under drought stress (Farnia et al. 2015). Soliman et al. (2015) confirmed that enhancing enzyme activity related to salt tolerance may alleviate the salt stress in moringa plants using foliar applications of ZnO and Fe_3O_4 NPs containing Hoagland solution. An important indicator of salt tolerance in plants is the accumulation of less Na. In addition, potassium (K) content reflects salt tolerance in plants that alleviates

NaCl adverse effects on nutrient uptake by improving root growth, preventing nutritional disorders and increasing root uptake of nutrients (El-Fouly et al. 2002). The foliar application of 200 ppm ZnO to the stressed cotton crop contributes to increasing cotton growth and yield (Hussein and Abou-Baker 2018). In maize under water stress conditions, foliar spraying of nano Zn increased yield and yield components (Amin and Mohammad 2015). Green synthesis of the ZnO NPs using *Sphagneticola trilobata* in aqueous root extract and evaluating their effect on germination, chromium reduction activity and fenugreek seed growth promotion. The nitrogen-containing phytochemical constituents are engaged in the creation of irregularly shaped NPs and the size ranges from 65–80 nm. The efficiency extraction of NPs chromium metal was detected to be more than 80% with an 8 h interaction time. The percentage reduction of chromium metal of 38.17, 53.33, 55.83 and 81.17%, respectively, was perceived at 0.10, 0.25, 0.5 and 1 g/L dosage of NPs. The treatment of zinc NPs showed an improvement in seed germination, root growth and plant growth compared to control and zinc sulfate (Shaik et al. 2020). The ameliorative role of ZnO which prepared from leaf extract of *C. sativum* and its effect on seed priming with different concentrations (10, 50,100 ppm) for different periods 3 and 6 h with 100 ppm ZnO for 6 h mediated the alleviation of Cu toxicity by increasing growth criteria, chlorophyll b (Chl b), chlorophyll a (Chl a), total soluble carbohydrates, carotenoids and protein (Kasim et al. 2017). This improvement may be due to that Zn plays an important role in chloroplast structure, photosynthetic electron transfer (Fathi et al. 2017), chlorophyll synthesis (Corredor et al. 2009) cell elongation and cell division, membrane stability by Zn directing to rise in fresh and dry weights (Cakmak 2000; Sedghi et al. 2013). Likewise, Venkatachalam et al. (2016) communicated that biomass and the plant growth tolerance index were promoted by ZnO nanoparticle under Cd and Pb stress. The most favorable, efficient and cost-effective method for the priming of *V. faba* seeds with ZnO which can be used to alleviate the inhibitory effects of Cu stress. The 150 mM $CuSO_4$ irrigation resulted in a remarkable reduction in fresh and dry root and shoot weights, root length, leaf area, shooting height and photosynthetic pigments (carotenoids, Chl a and Chl b). Whereas the ratio Chl a/b under treatment with Cu was enhanced total soluble carbohydrates and protein content were greatly depleted (Kasim et al. 2017). These reductions could evidence to inhibition of Cu stress in cell division and cell elongation that is eventually translated into the impaired shoot and root growth (Agami 2016). The chlorophyll reduction recorded could be attributed to the Cu-induced Fe deficiency or Cu substitution of the central Mg chlorophyll ion (Kasim et al. 2017).

9.4.3 Titanium Dioxide NPs

Using 0.02% of titanium dioxide TiO_2 NPs enhanced wheat crop growth. So, it is mentioned that titanium dioxide NPs must be applied under water deficit conditions (Iyarin et al. 2019). Titanium enhances the activity of rubisco and increases the metabolism of CO_2, increases photosynthesis and improves yield (Gao et al. 2006).

Gluten and starch content of wheat decreased under water stress conditions. So, the application of nano TiO_2 improves these contents due to the positive correlation between titanium application and photosynthesis rate (Jaberzadeh et al. 2013; Zhao et al. 2008). Titanium NPs activate progressive growth, development and productivity of *T. aestivum* under abiotic stress.

9.4.4 *Silver NPs AgNPs*

AgNPs that synthesized from cocoa pod extracts to demobilize, decontaminate and adsorb Cd and Pb in addition to their photostimulation effects on *M. oleifera* for heavy metal-induced toxicity attenuation. Silver NPs can accumulate heavy metals by diminishing their mobility and absorption. And to foster the growth of *M. oleifera* exposed to stress heavy metals of Pb and Cd. Through the ability to scavenge free radicals and inhibited physiological tolerance towards stress notable in their relative water contents, root and shoot lengths, antioxidant activities, growth tolerance index, photosynthetic, pigment contents and polyphenolic contents. Then the study has shown that the biosynthesized AgNPs can be valuable in agrosystems to mitigate the heavy metals deleterious effects in crop production. Moreover, may contribute to the bioremediation of environments contaminated with heavy metals by boosting the growth of remediating plants (Azeez et al. 2019). Silver NPs can be employed to find metal ions in contaminated water with the addition of even the smallest amount of 35 μl metal ion. Due to the variation in optical density (Kaur and Komal 2019). Iqbal et al. (2017) used *Moringa oleifera* plant extract for AgNPs synthesis and its effect on the regulation of wheat growth under heat stress. At the trifoliate stage, different concentrations of 25, 50, 75 and 100 mg/L AgNPs were used on wheat. The stress of heat was active in the range of 35–40 °C during 3 h/day for about three days. Silver NPs contributed to improving the wheat morphological traits (root number, length of root shoot, leaf area, leaf number, leaf fresh weight and dry weight). Effective results were noticed for 50 and 75 mg/L AgNPs under heat stress.

9.5 Application of Biosynthesis NPs in Agriculture for Sustainability Development

Modern agriculture relies on the application of nanotechnology (Parashuram et al. 2020). For sustainable development of agriculture, nanotechnology is widely used to improve crop production for the growing population demands (Grillo et al. 2016). NPs considered essential keys for sustainable development (Peralta-Videa et al. 2011). The green-based synthesized NPs are more suitable than other NPs (Vadlapudi and Kaladhar 2014). Nanotechnology applications are used in many fields to sustain the development of agriculture in different types such as nanopesticides,

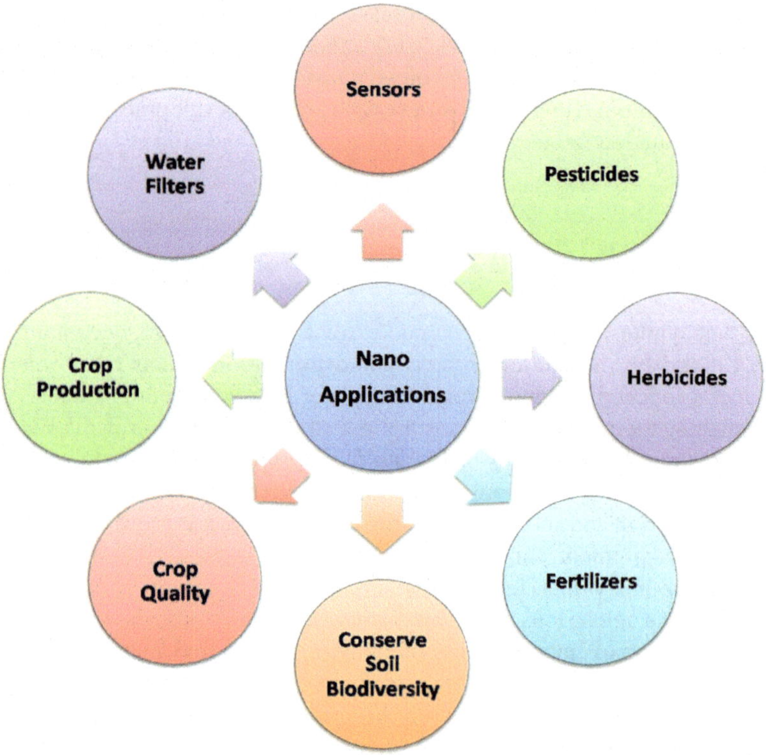

Fig. 9.9 Some applications of biosynthesis nanotechnology for sustainable agriculture (Constructed by M.M. Saleh)

nanoherbicides, nanofertilizers, nanobiosensors (Agrawal and Rathore 2014; Kah 2015; Servin et al. 2015; Tripathi et al. 2015, 2016) (Fig. 9.9). Nanosensors are one of the nanotechnology application which developed to identify and investigate results at the atomic level, used to detect chemicals like herbicides, pesticides, glucose and urea, pathogens, analysis of metabolic products and enzymes (Rai and Ingle 2012), even reduce pollution and management of soil nutrients (Ingale and Chaudhari 2013). Production of green synthesized NMs and appliances sustains agriculture (Bartolucci et al. 2020), such as the production of reduced graphene oxide Gro/gold (Au) NPs biosensor from *E. tereticornis* leave for detecting tryptophan (Nazarpour et al. 2020). Servin et al. (2015) reported that NPs were used to detect crop diseases caused by microorganisms. Also, each of the silicon and Au NPs were used for pathogen detection on plants (Rico et al. 2011). Brock et al. (2011) refereed to the possibility of using NPs in detecting plant viruses and fungi and in the analysis of soil nutrients concentration. Concerning plant protection, applying nanopesticides in agriculture avoid the reduction of soil biodiversity and birds' habitats and other dangers that could be caused by traditional pesticides (Ghormade et al. 2011). El-bendary and El-Helaly (2013) mentioned that applying nanosilica on tomato was useful to destruct the primary pest *Spodoptera littoralis* which is resistant to almost all pesticides. Rice pests and silkworm disease were successfully controlled by usage of different NPs such as titanium oxide, Ag and zinc oxide (Goswami et al. 2010). Nanoherbicides are target distinctive that can kill weeds without and damage or reduction in crop yield (Bickel and Killorn 2001), whereas the traditional herbicides may cause a reduction in crop yield (Deva and Kadiri 2016). The nano combination of silver-chitosan proved its antifungal ability to prevent the mycelium growth of seed fungal pathogen (Kaur et al. 2012). Usage of nanofertilizers is suggested by many scientists since the applying of conventional fertilizers is in most cases unsafe for plant and nature (Parashuram et al. 2020). For sustainable agriculture, it is better to use nanofertilizers for their effective role (El-Ramady 2014). The concept of using nanofertilizers is raising in some countries, although their production is only available by registered companies (Dimkpa and Bindraban 2017). Nanofertilizers are characterized as nature-friendly, with the ability to enhance the absorption of nutrients by plants when they applied in an efficient encapsulated type (Mazzaglia et al. 2017). Naderi et al. (2013) confirmed that nanofertilizers release nutrients in a controlled way and gradually into the soil which makes nutrients more effective than the conventional fertilizer and the application number of normal fertilizer, in addition to the toxicity level of the soil may decline by the usage of nanofertilizers.

The scarcity of water becomes the most problem threatened agriculture in the world accompanied by limited resources of pure water. So utilization of other water sources like rainwater harvest, recycling water plus reuse of filtered brackish water, seems to be a promising solution for irrigation (Ghermandi and Messalem 2009). Nanofilters were used successfully to eliminate irrigation of organic particles and turbidity (Mrayed et al. 2011; Riera et al. 2013) and to purify brackish and salty water effectively (Sotto et al. 2013) and in the same time using of nanofilters to ensure the preservation of essential ions for plant growth (Zhao et al. 2012a).

9.6 Conclusion and Prospects

Nanotechnology is a current revolution used in several sciences which started in the 1990s. It was incorporated into all the sectors of medicine, industry, agriculture, environment, water desalination and pharmaceutical. Except it has opponents like the same as any modern technology, but other people believe it as a hope for a better future. However, the fact is that if we use this technology well, it will transfer the whole world to a new and favorable stage, especially as scientific development has made this technology easy, simple and accessible to all. Several methods already exist to produce NPs, scientific responsibility is to develop the processes which are environment-friendly, more effective, affordable manufacture NPs. The world has started investing billions of countries in all areas of nanotechnology, especially in America, Germany, China, Russia and Japan. In addition, it may open broad horizons for space conquest if used in a scientifically. On our part, we believe that we must not neglect the safety and security side, by seeking the side effects neutrally, and publishing all research on safety and security based on correct scientific research. Nanomaterials have various characteristics, such as their surface charges, shape, size and chemical properties. NPs have influence plant growth through their ability to translocate within the plant increases crop yields. In addition, it is functional to assist plants to face different stresses and make it more tolerant by various mechanisms. however, the effects phytotoxicity on plant growth need much research to determine it. This depends on one of the factors such as nanomaterial size, concentration, and shape. Additionally, plants and NPs interact and their mechanism could offer future research prospects.

References

Abbasifar A, Shahrabadi F, ValizadehKaji B (2020) Effects of green synthesized zinc and copper nano-fertilizers on the morphological and biochemical attributes of basil plant. J Plant Nutr 43:1104–1118

Abdel-Aziz H (2019) Effect of priming with chitosan nanoparticles on germination, seedling growth and antioxidant enzymes of broad beans. Catrina 18:81–86

Abdel-Aziz HMM, Hasaneen MNA, Omer AM (2016) Nano chitosan-NPK fertilizer enhances the growth and productivity of wheat plants grown in sandy soil. Spanish J Agric Res 14:1–9. https://doi.org/10.5424/sjar/2016141-8205

Abdel-Aziz HMM, Hasaneen MNA-G, Omer AM (2018) Foliar application of nano chitosan NPK fertilizer improves the yield of wheat plants grown on two different soils. Egypt J Exp Biol Bot 14:63–72

Abdel-Aziz HMM, Hasaneen MNA, Omer AM (2019) Impact of engineered nanomaterials either alone or loaded with NPK on growth and productivity of French bean plants: Seed priming vs foliar application. South Afr J Bot 125:102–108. https://doi.org/10.1016/j.sajb.2019.07.005

Abdelsalam NR, Kandil EE, Al-Msari MAF et al (2019) Effect of foliar application of NPK nanoparticle fertilization on yield and genotoxicity in wheat (*Triticum aestivum* L.). Sci Total Environ 653:1128–1139. https://doi.org/10.1016/j.scitotenv.2018.11.023

Acharya P, Jayaprakasha GK, Crosby KM et al (2020) Nanoparticle-mediated seed priming improves germination, growth, yield, and quality of watermelons (*Citrullus lanatus*) at multi-locations in texas. Sci Rep 10:1–16

Adani F, Papa G, Schievano A et al (2011) Nanoscale structure of the cell wall protecting cellulose from enzyme attack. Environ Sci Technol 45:1107–1113

Adhikari T, Sarkar D, Mashayekhi H et al (2016) Growth and enzymatic activity of maize (*Zea mays* L.) plant: solution culture test for copper dioxide nanoparticles. J Plant Nutr 39:99–115

Agami RA (2016) Pre-soaking in indole-3-acetic acid or spermidine enhances copper tolerance in wheat seedlings. S Afr J Bot 104:167–174

Agrawal S, Rathore P (2014) Nanotechnology pros and cons to agriculture: a review. Int J Curr Microbiol Appl Sci 3(3):43–55

Al-Aghabary K, Zhu Z, Shi Q (2004) Influence of silicon supply on chlorophyll content, chlorophyll fluorescence and antioxidative enzyme activities in tomato plants under salt stress. J Plant Nutr 27(12):2101–2115

AL-Gym AJK, Al-Asady MHS (2020) Effect of the method and level of adding NPK nanoparticles and mineral fertilizers on the growth and yield of yellow corn and the content of mineral nutrient of some plant parts. Plant Arch 20:38–43

Al-Hchami SHJ, Alrawi TK (2020) Nano fertilizer, benefits and effects on fruit trees: a review. Plant Arch 20:1085–1088

Almalah HI, Alzahrani HA, Abdelkader HS (2019) Green synthesis of silver nanoparticles using *Cinnamomum Zylinicum* and their synergistic effect against multi-drug resistance bacteria. J Nanotechnol Res 1(3):95–107. https://doi.org/10.26502/jnr.2688-8521008

Alnaddaf L, Saleh M, Almuhammady A, Atiyeeh A (2019) Synthesis of silver nanoparticles from *Citrus limon* juice and its effect on the germination and growth of durum wheat seedlings. Challenges and innovations in nanotechnology workshop, HIAST, 18–19 November 2019, Damascus, Syria

Alsaeedi A, El-Ramady H, Alshaal T et al (2019) Silica nanoparticles boost growth and productivity of cucumber under water deficit and salinity stresses by balancing nutrients uptake. Plant Physiol Bioch 139:1–10

Alsaeedi A, El-Ramady H, Alshaal T, Almohsen M (2017) Enhancing seed germination and seedlings development of common bean (*Phaseolus vulgaris*) by SiO_2 nanoparticles. Egypt J Soil Sci 57(4):407–415

Alsaeedi A, El-Ramady H, Alshaal T et al (2018) Exogenous nanosilica improves germination and growth of cucumber by maintaining K^+/Na^+ ratio under elevated Na^+ stress. Plant Physiol Biochem 125:164–171

Amin F, Mohammad MO (2015) Effect of nano-zinc chelate and nano-biofertilizer on yield and yield components of maize (*Zea mays* L.), underwater stress condition. Indian J Nat Sci 5(29):4614–4620

Antisari LV, Carbone S, Gatti A et al (2014) Uptake and translocation of metals and nutrients in tomato grown in soil polluted with metal oxide (CeO2, Fe3O4, SnO2, TiO2) or metallic (Ag Co, Ni) engineered nanoparticles. Environ Sci Pollut Res 22(3):1–14. https://doi.org/10.1007/s11356-014-3509-0

Ashraf M (2009) Biotechnological approach of improving plant salt tolerance using antioxidants as markers. Biotechnol Adv 27:84–93

Azeez L, Adejumo AL, Lateef A et al (2019) Zero-valent silver nanoparticles attenuate Cd and Pb toxicities on *Moringa oleifera* via immobilization and induction of phytochemicals. Plant Physiol Bioch 139:283–292

Azeez L, Lateef A, Adebisi SA (2017) Silver nanoparticles (AgNPs) biosynthesized using pod extract of Cola nitida enhances antioxidant activity and phytochemical composition of *Amaranthus caudatus* Linn. Appl Nanosci 7(1–2):59–66

Babaei K, Sharifi RS, Pirzad A, Khalilzadeh R (2017) Effects of bio fertilizer and nano Zn-Fe oxide on physiological traits, antioxidant enzymes activity and yield of wheat (*Triticum aestivum* L.) under salinity stress. J Plant Interact 12:381–389

Bala R, Kalia A, Dhaliwal SS (2019) Evaluation of efficacy of ZnO nanoparticles as remedial zinc nanofertilizer for rice. J Soil Sci Plant Nutr 19:379–389. https://doi.org/10.1007/s42729-019-00040-z

Bamsaoud SF, Bahwirth MA (2017) The effect of biologically synthesized silver nanoparticles on the germination and growth of *Cucurbita pepo* seedlings. J Arab American Uni 3:34–47

Banerjee P, Satapathy M, Mukhopahayay A, Das P (2014) Leaf extract mediated green synthesis of silver nanoparticles from widely available Indian plants: synthesis, characterization, antimicrobial property and toxicity analysis. Bioresour Bioprocess 1:3. https://doi.org/10.1186/s40643-014-0003-y

Barhoumi L, Oukarroum A, Taher LB et al (2015) Effects of super paramagnetic iron oxide nanoparticles on photosynthesis and growth of the aquatic plant *Lemna gibba*. Arch Environ Contam Toxicol 68(3):510–520

Barrena R, Casals E, Colón J et al (2009) Evaluation of the ecotoxicity of model nanoparticles. Chemosphere 75:850–857. https://doi.org/10.1016/j.chemosphere.2009.01.078

Bartolucci C, Antonacci A, Arduini F et al (2020) Green nanomaterials fostering agrifood sustainability. TrAC Trend Anal Chem 125: https://doi.org/10.1016/j.trac.2020.115840

Bickel A, Killorn R (2001) Spatial response of corn to banded zinc sulfate fertilizer in Iowa. In: 31st Northcentral extension industry soil fertility conference, 14–15 November 2001, Potash and Phosphate Institute Avenue South Brookings, SD 57006, pp 144–155

Birbaum K, Brogioli R, Schellenberg M et al (2010) No evidence for cerium dioxide nanoparticle translocation in maize plants. Environ Sci Technol 44:8718–8723
Boutchuen A, Zimmerman D, Aich N et al (2019) Increased plant growth with hematite nanoparticle fertilizer drop and determining nanoparticle uptake in plants using multimodal approach. J Nanomaterial 2019: ID6890572. https://doi.org/10.1155/2019/6890572
Bratovcic A (2019) Different applications of nanomaterials and their impact on the environment. Inter J Mater Sci Eng 5:1–7
Brock DA, Douglas TE, Queller DC et al (2011) Primitive agriculture in a social amoeba. Nature 469:393–396
Cakmak I (2000) Role of zinc in protecting plant cells from reactive oxygen species. New Phytol 146:185–205
Chen D, Wang S, Yin L, Deng X (2018) How does silicon mediate plant water uptake and loss under water deficiency? Front Plant Sci 9:281. https://doi.org/10.3389/fpls.00281
Chrysochoou M, McGuirea M, Dahalb G (2012) Transport characteristics of green-tea nano-scale zero valent iron as a function of soil mineralogy. Chem Eng Trans 28:122–126
Corradini E, Moura MR, Mattoso LHC (2010) A preliminary study of the incorporation of NPK fertilizer into chitosan nanoparticles. eXPRESS Polym Lett 4:509–515
Corredor E, Risueno MC, Testillano PS (2010) Carbon iron magnetic nanoparticles for agronomic use in plants promising but still a long way to go. Plant Signal Behav 5:1295–1297
Corredor E, Testillano PS, Coronado MJ (2009) Nanoparticle penetration and transport in living pumpkin plants: in situ subcellular identification. BMC Plant Biol 9:45–54
Cuong TX, Ullah H, atta A, Hanh TC (2017) Effects of silicon-based fertilizer on growth, yield and nutrient uptake of rice in tropical zone of Vietnam. Rice Sci 24(5):283–290
Curie C, Briat JF (2003) Iron transport and signaling in plants. Annu Rev Plant Biol 54:183–206
Dawodu FA, Onuh CU, Akpomie KG et al (2019) Synthesis of silver nanoparticle from *vigna unguiculata* stem as adsorbent for malachite green in a batch system. SN Appl Sci 1:346. https://doi.org/10.1007/s42452-019-0353-3
Deepa K, Singha S, Panda T (2014) Doxorubicin nanoconjugates. J Nanosci Nanotechnol 14:892–904
Deva S, Kadiri L (2016) A review on impact of nanoparticles on plant growth and development. Adv Life Sci 5(1):16–21
Dhinek A, Vanitha S (2016) Synthesis and characterization of silver nanoparticle from *Citrus limon* leaves and its antibacterial activity. Inter J Sci Res 7:2319–7064
Dimkpa CO, Bindraban PS (2017) Nanofertilizers: new products for the industry? J Agric Food Chem 66(26):6462–6473. https://doi.org/10.1021/acs.jafc.7b02150
Duhan JS, Kumar R, Kumar N et al (2017) Nanotechnology: the new perspective in precision agriculture. Plant Physiol Biochem 15:11–23
Eichert T, Goldbach HE (2008) Equivalent pore radii of hydrophilic foliar uptake routes in stomatous and astomatous leaf surface–Further evidence for a stomatal pathway. Physiol Plant 132:491–502
Eichert T, Kurtz A, Steiner U et al (2008) Size exclusion limits and lateral heterogeneity of the stomatal foliar uptake pathway for aqueous solutes and water-suspended nanoparticles. Physiol Plan 134:151–160
El-bendary HM, El-Helaly AA (2013) First record nanotechnology in agricultural: silica nanoparticles a potential new insecticide for pest control. Appl Sci Rep 4(3):241–246
El-Fouly MM, Mobarak ZM, Salsma ZA (2002) Micronutrient foliar application increase salt tolerance of tomato seedlings. Proceedings of the symposium on Techniques to control salination for horticultural productivity. Acta Hortic 573:337–385
El-Ramady HR (2014) Integrated nutrient management and postharvest of crops. In: Lichtfouse E (ed) Sustainable Agriculture Reviews, Springer International Publishing Switzerland, 13:163–274
Elshamy MT, Husseiny SM, Farroh KY (2019) Application of nano-chitosan NPK fertilizer on growth and productivity of potato plant. J Sci Res Sci 36:424–441

El-Sherif AG, Gad SB, Megahed AA et al (2019) Induction of tomato plants resistance to *Meloidogyne incognita* infection by mineral and nano-fertilizer. J Entomol Nematol 11(2):1–26. https://doi.org/10.5897/jen2018.0224

Erdogan O, Abbak M, Demirbolat GM et al (2019) Green synthesis of silver nanoparticles via *Cynara scolymus* leaf extracts: the characterization, anticancer potential with photodynamic therapy in MCF7 cells. PLoS ONE 14(6): https://doi.org/10.1371/journal.pone.0216496

Fahimirad S, Hatami M (2019) Nanocarrier-based antimicrobial phytochemicals. Adv phytonanotechnology 12:299–314

Farghaly FA, Nafady NA (2015) Green synthesis of silver nanoparticles using leaf extract of *Rosmarinus officinalis* and its effect on tomato and wheat plants. J Agric Sci 7(11):277–287

Farhan AM, Jassim RA, Kadhim NJ et al (2017) Synthesis of silver nanoparticles from *Malva parviflora* extract and effect on Ecto-5'- Nucleotidase(5'-NT), ADA and AMPDA enzymes in sera of Patients with arthrosclerosis. Baghdad Sci J 14(4):742–750. https://doi.org/10.21123/bsj.2017.14.4.0742

Farhana MA, Meera VB (2016) Synthesis of nanosilver coated sand using plant extracts. Proc Technol 24:188–195

Farnia A, Omidi MM, Farnia A (2015) Effect of nano-zinc chelate and nano-biofertilizer on yield and yield components of maize (*Zea mays* L.) under water stress condition. Indian J Nat Sci 5:4614–4646

Fathi A, Zahedi M, Torabian S, Khoshgoftar A (2017) Response of wheat genotypes to foliar spray of ZnO and Fe_2O_3 nanoparticles under salt stress. J Plant Nut 29:1543–1566

Feng Y, Cui X, He S et al (2013) The role of metal nanoparticles in influencing arbuscular mycorrhizal fungi effects on plant growth. Environ Sci Technol 47:9496–9504

Galdames A, Mendoza A, Orueta M et al (2017) Development of new remediation technologies for contaminated soils based on the application of zero-valent iron nanoparticles and bioremediation with compost. Resour Effic Technol 3(2):166–176

Gao F, Hong F, Liu C et al (2006) Mechanism of nano-anatase TiO_2 on promoting photosynthetic carbon reaction of spinach. Biol Trace Elem Res 11(3):239–254

Ghermandi A, Messalem R (2009) Solar-driven desalination with reverse osmosis: the state of the art. Desalin Water Treat 7(1–3):285–296

Ghormade V, Deshpande MV, Paknikar KM (2011) Perspectives for nano-biotechnology enabled protection and nutrition of plants. Biotechnol Adv 29:792–803

Gonzalez-Melendi P, Fernandez Pacheco R, Coronado MJ et al (2008) Nanoparticles as smart treatment-delivery systems in plants: assessment of different techniques of microscopy for their visualization in plant tissues. Ann Bot 101:187–195

Gomathi M, Prakasam A, Rajkumar PV (2019) Green Synthesis, characterization and antibacterial activity of silver nanoparticles using *Amorphophallus paeoniifolius* leaf extract. J Cluster Sci. https://doi.org/10.1007/s10876-019-01559-y

Gong X, Huang D, Liu Y et al (2018) Remediation of contaminated soils by biotechnology with nanomaterials: bio-behavior, applications and perspectives. Crit Rev Biotechnol 38(3):455–468

Gopalakrishnan K, Ramesh C, Ragunathan V, Thamilselvana M (2012) Antibacterial activity of Cu_2O nanoparticles on *E. coli* synthesized from *Tridax procumbens* leaf extract and surface coating with polyaniline. Dig J Nanomater Bios 7:833–839

Gopinath K, Gowri S, Karthika V, Arumugam A (2014) Green synthesis of gold nanoparticles from fruit extract of *Terminalia arjuna*, for the enhanced seed germination activity of *Gloriosa superb*. J Nanostruct Chem 4:115. https://doi.org/10.1007/s40097-014-0115-0

Goswami A, Roy I, Sengupta S et al (2010) Novel applications of solid and liquid formulations of nanoparticles against insect pests and pathogens. Thin Solid Films 519:1252–1257

Goswami P, Yadav S, Mathur J (2019) Positive and negative effects of nanoparticles on plants and their applications in agriculture. Plant Science Today 6(2):232–242. https://doi.org/10.14719/pst.2019.6.2.502

Goto M, Ehara H, Karita S, Takabe K (2003) Protective effect of silicon on phenolic biosynthesis and ultraviolet spectral stress in rice crop. Plant Sci 164(3):349–356

Greger M, Landberg T, Vaculík M (2018) Silicon influences soil availability and accumulation of mineral nutrients in various plant species. Plants 7(2):41. https://doi.org/10.3390/plants7020041

Grillo R, Abhilash PC, Fraceto LF (2016) Nanotechnology applied to bio-encapsulation of pesticides. J Nanosci Nanotechnol 16:1231–1244

Gu HH, Qiu H, Tian T, Zhan SS et al (2011) Mitigation effects of silicon rich amendments on heavy metal accumulation in rice (*Oryza sativa* L.) planted on multi-metal contaminated acidic soil. Chemosphere 83:1234–1240. https://doi.org/10.1016/j.chemosphere.2011.03.014

Gupta M, Tomar RS, Kaushik S et al (2018) Effective antimicrobial activity of green ZnO nano particles of *Catharanthus roseus*. Front Microbiol 9:2030. https://doi.org/10.3389/fmicb.2018.02030

Hasanuzzaman M, Bhuyan MHMB, Nahar K et al (2018) Potassium: a vital regulator of plant responses and tolerance to abiotic stresses. Agron 8:31

Hattori T, Inanaga S, Araki H, An P et al (2005) Application of silicon enhanced drought tolerance in *Sorghum bicolor*. Physiol Plant 123:459–466. https://doi.org/10.1111/j.1399-3054.2005.00481.x

He F, Zhao D (2005) Preparation and characterization of a new class of starch-stabilized bimetallic nanoparticles for degradation of chlorinated hydrocarbons in water. Environ Sci Technol 39:3314–3320

Heydari MA, Mobini M, Salehi M (2017) The Synergic activity of eucalyptus leaf oil and silver nanoparticles against some pathogenic bacteria. Arch Pediatr Infect Dis. https://doi.org/10.5812/pedinfect.61654

Hong J, Peralta-Videa JR, Rico C et al (2014) Evidence of translocation and physiological impacts of foliar applied CeO_2 nanoparticles on cucumber (*Cucumis sativus*) plants. Environ Sci Technol 48(8):4376–4385. https://doi.org/10.1021/es404931g

Hua M, Zhang S, Pan B et al (2012) Heavy metal removal from water/wastewater by nanosized metal oxides: a review. J Hazardous Mater 212:317–331

Hong TK, Yang HS, Choi CJ (2005) Study of the enhanced thermal conductivity of Fe nanofluids. J Appl Phys 97:1–4

Husen A, Siddiqi KS (2014) Phytosynthesis of nanoparticles: concept, controversy and application. Nanoscale Res Lett 9:229

Hussein H-AA, Darwesh OM, Mekki BB (2019) Environmentally friendly nano-selenium to improve antioxidant system and growth of groundnut cultivars under sandy soil conditions. Biocatal Agric Biotechnol 18:

Hussein MM, Abou-Baker NH (2018) The contribution of nanozinc to alleviate salinity stress on cotton plants. R Soc Open Sci 5(8):

Ingale AG, Chaudhari AN (2013) Biogenic synthesis of nanoparticles and potential applications: an eco-friendly approach. J Nanomed Nanotechol 4(165):1–7

Iqbal M, Raja NI, Mashwani ZU et al (2017) Effect of silver nanoparticles on growth of wheat under heat stress. Iran J Sci Technol Trans Sci.

Itroutwar PD, Govindaraju K, Tamilselvan S et al (2019) Seaweed-based biogenic ZnO nanoparticles for improving agro-morphological characteristics of rice (*Oryza sativa* L.). J Plant Growth Regul. https://doi.org/10.1007/s00344-019-10012-3

Iyarin E, Mahil T, Kumar BNA (2019) Foliar application of nanofertilizers in agricultural crops-a review. J Farm Sci 32(3):239–249

Jaberzadeh A, Moaveni P, Moghadam HRT, Zahedi H (2013) Influence of bulk and nanoparticles titanium foliar application on some agronomic traits, seed gluten and starch contents of wheat subjected to water deficit stress. Not Bot Horti Agrobot 41:201–207

Jafari M, Haghighi J, Zare H (2012) Mulching impact on plant growth and production of rainfed fig orchards under drought conditions. J Food Agric Environ 10(1):428–433

Judy JD, Bertsch PM (2014) Bioavailability, toxicity and fate of manufactured nanomaterials in terrestrial ecosystems. In: Sparks D (ed) Advances in Agronomy. Elsevier, Amsterdam, Netherlands, pp 1–64

Judy JD, Kirby JK, Creamer C et al (2015) Effects of silver sulfide nanomaterials on mycorrhizal colonization of tomato plants and soil microbial communities in biosolid-amended soil. Environ Pollut 206:256–263

Kachel M, Matwijczuk A, Sujak A et al (2019) The Influence of copper and silver nanocolloids on the quality of pressed spring rapeseed oil. Agron 9:643. https://doi.org/10.3390/agronomy9100643

Kah M (2015) Nanopesticides and nanofertilizers: emerging contaminants or opportunities for risk mitigation. F Chem 3(64):1–6

Kahlel A, Ghidan A, Al-Antary TA et al (2020) Effects of nanotechnology liquid fertilizers on certain vegetative growth of broad bean (*Vicia faba* L.). Fresen Env Bull 29:4763–4768

Kango S, Kumar R (2016) Magnetite nanoparticles coated sand for arsenic removal from drinking water. Environ Earth Sci 75:381. https://doi.org/10.1007/s12665-016-5282-5

Kasim W A, Abokassem EM, Ragab GA (2017) Ameliorative effect of yeast extract, IAA and green-synthesized nano Zinc oxide on the growth of Cu-stressed *Vicia faba* seedlings. The 7th Inter Conference, Plant & Microbial Biotech & their Role in the Development of the Society, pp 1–16

Kaur P, Komal R (2019) Synthesis, analysis and application of noble metal nanoparticles by *Cucurbita pepo* using different solvents. Asian J Nano Mat 2:376–398

Kaur P, Thakur R, Choudhary A (2012) An in vitro study of the antifungal activity of silver/chitosan nanoformulations against important seed borne pathogens. Int J Sci Technol Res 1:83–86

Kheyri N, Norouzi HA, Mobasser HR et al (2019) Effects of silicon and zinc nanoparticles on growth, yield, and biochemical characteristics of rice. Agron J 111:3084–3090

Lamhamdi M, ElGaliou O, Bakrim A et al (2013) Effect of lead stress on mineral content and growth of wheat (*Triticum aestivum*) and spinach (*Spinacia oleracea*) seedlings. Saudi J Biol Sci 20:29–36

Larue C, Castillo-Michel H, Sobanska S et al (2014) Foliar exposure of the crop *Lactuca sativa* to silver nanoparticles: evidence for internalization and changes in Ag speciation. J Haz Mat 264:98–106. https://doi.org/10.1016/j.jhazmat2013.10.053

Larue C, Laurette J, Herlin-Boime N (2012) Accumulation, translocation and impact of TiO_2 nanoparticles in wheat *Triticum aestivum* spp.: influence of diameter and crystal phase. Sci Total Environ 431:197–208

Li X, Ke M, Zhang M et al (2017) The interactive effects of diclofop-methyl and silver nanoparticles on *Arabidopsis thaliana*: Growth, photosynthesis and antioxidant system. Environ Pollut 232:212–219

Liang SJ, Li ZQ, Li XJ et al (2013) Effects of stem structural characters and silicon content on lodging resistance in rice (*Oryza sativa* L.). Res Crops 14:621–636

Liang YC, Zhang WH, Chen Q (2006) Effect of exogenous silicon (Si) on H^+-ATPase activity, phospholipids and fluidity of plasma membrane in leaves of salt-stressed barley (*Hordeum vulgare* L.). Environ Exp Bot 57:212–219

Lin D, Xing B (2008) Root uptake and phytotoxicity of ZnO nanoparticles. Environ Sci Technol 42:5580–5585

Lin S, Reppert J, Hu Q et al (2009) Uptake, translocation, and transmission of carbon nanomaterials in rice plants. Small 5(10):1128–1132

Liu L, Li W, Song W, Guo M (2018) Remediation techniques for heavy metal contaminated soils: principles and applicability. Sci Total Environ 633:206–219

Liu W, Tian S, Zhao X et al (2015) Application of stabilized nanoparticles for in situ remediation of metal-contaminated soil and groundwater. Curr Pollut Rep 1:280–291. https://doi.org/10.1007/s40726-015-0017-x

Lombardo D, Kiselev MA, Caccamo MT (2019) Smart nanoparticles for drug delivery application: development of versatile nanocarrier platforms in biotechnology and nanomedicine. J Nanomater 3702518:1–26 p. https://doi.org/10.1155/2019/3702518

Mali M, Aery NC (2009) Effect of silicon on growth, biochemical constituents and mineral nutrition of cowpea. Commun Soil Sci Plant Anal 40:1041–1052

Mansoor N, Younus A, Jamil Y et al (2019) Impact of nanosized and bulk ZnO on germination and early growth response of *Triticum aestivum*. Pak J Agric Sci 56(4):879–884
Mazzaglia A, Fortunati E, Kenny J et al (2017) Nanomaterials in plant protection. In: Axelos MAV, Van de Voorde M (eds) Nanotechnology in agriculture and food science, pp 115–134
Miralles P, Church TL, Harris AT (2012) Toxicity, uptake, and translocation of engineered nanomaterials in vascular plants. Environ Sci Technol 46(17):9224–9239
Mishra V, Mishra RK, Dikshit A, Pandey AC (2014) Interactions of nanoparticles with plants: an emerging prospective in the agriculture industry. In: Ahmad P, Rasool S (eds) Emerging technologies and management of crop stress tolerance: biological techniques, vol 1, pp 159–180
Mohamad NM, Abdul Razak SR, Ahmed MA et al (2019) Characterization, antimicrobial activity of silver nanoparticles biosynthesized by *Acanthus ilicifolius*. American J Current App Res Microbiol 1(1):9–16
Mohammad DA, Al-Jubouri SHK (2019) Comparative antimicrobial activity of silver nanoparticles synthesized by *Corynebacterium glutamicum* and plant extracts. Baghdad Sci J 16(3):689–696. https://doi.org/10.21123/bsj.2019.16.3(Suppl.).0689
Mrayed SM, Sanciolo P, Zou I et al (2011) An alternative membrane treatment process to produce low-salt and high-nutrient recycled water suitable for irrigation purposes. Desalination 274(19):144–149
Muniandy SS, Sasidharan S, Lee LH (2019) Green synthesis of ag nanoparticles and their performance towards antimicrobial properties. Sains Malaysiana 48(4):851–860. https://doi.org/10.17576/jsm-2019-4804-17
Nadaroğlu H, Güngör AA, İnce S (2017) Synthesis of nanoparticles by green synthesis method. Inter J Innovative Res Rev 1(1):6–9
Naderi MR, Danesh-Shahraki A, Nair BG et al (2013) Nanofertilizers and their roles in sustainable agriculture. Int J Agric Crop Sci 5(19):2229–2232
Njagi EC, Huang H, Stafford L et al (2011) Biosynthesis of iron and silver nanoparticles at room temperature using aqueous sorghum bran extracts. Langmuir 27:264–271
Nasara MQ, Zohra T, Khalil AT et al (2019) Seripheidium quettense mediated green synthesis of biogenic silver nanoparticles and their theranostic applications. Green Chem Lett Rev 12(3):310–322. https://doi.org/10.1080/17518253.2019.1643929
Nazarpour S, Hajian R, Sabzvari MH et al (2020) A Novel nanocomposite electrochemical sensor based on green synthesis of reduced graphene oxide/gold nanoparticles modified screen printed electrode for determination of tryptophan using response surface methodology approach. Microchem J 15:10463. https://doi.org/10.1016/j.microc.2020.104634
Neu S, Schaller J, Dudel EG (2017) Silicon availability modifies nutrient use efficiency and content, C: N:P stoichiometry and productivity of winter wheat (*Triticum aestivum* L.). Sci Rep 7:40829. https://doi.org/10.1038/srep40829
Noaema AH, Alhasany ARk, Altai DSk et al (2019) Effect of nano-boron spraying on the concentration of some nutrients in leaves and dry matter of two *Vicia faba* L. (Partim) cultivars. Ann UMCS Sect E Agric 74:34–45. https://doi.org/10.24326/as.2019.4.2
Ochoa L, Zuverza-Mena N, Medina-Velo IA et al (2018) Copper oxide nanoparticles and bulk copper oxide, combined with indole-3-acetic acid, alter aluminum, boron and iron in *Pisum sativum* seeds. Sci Total Environ 634:1238–1245
Parashuram B, Dennis T, Muhammad SR et al (2020) How to cut down the gap between the Zn requirement and supply of food chain and crop growth: A critical review. Int J Plant Anim Environ Sci 10(1):1–26
Pati S, Pal B, Badole S et al (2016) Effect of silicon fertilization on growth, yield and nutrient uptake of rice. Commun Soil Sci Plant Anal 47:284–290
Patrick JW, Tyerman SD, Bel AJE (2015) Long-distance transport. In: Buchanan BB, Gruissem W, Jones RL (eds) Biochemistry and molecular biology of plants, 2nd edn. Wiley, West Sussex, pp 658–710
Peralta-Videa JR, Zhao L, Lopez-Moreno ML et al (2011) Nanomaterials and the environment: a review for the biennium 2008–2010. J Hazard Mater 186:1–15

Pérez-de-Luque A (2017) Interaction of Nanomaterials with Plants: What do we need for real applications in agriculture? Front Environ Sci 5:12

Pérez GM, Luqueño FF, Nuñez EV et al (2019) Remediating polluted soils using nanotechnologies: environmental benefits and risks. Pol J Environ Stud 28(3):1–18

Phan KS, Nguyen HT, Le TTH et al (2019) Fabrication and activity evaluation on *Asparagus officinalis* of hydroxyapatite based multimicronutrient nano systems. Adv Nat Sci Nanosci Nanotechnol 10:1–7. https://doi.org/10.1088/2043-6254/ab21cc

Pirtarighat S, Ghannadnia M, Baghshahi S (2019) Green synthesis of silver nanoparticles using the plant extract of *Salvia spinosa* grown in vitro and their antibacterial activity assessment. J Nanostructure Chem 9:1–9. https://doi.org/10.1007/s40097-018-0291-4

Poornima R, Koti RV (2019) Effect of nano zinc oxide on growth, yield and grain zinc content of sorghum (*Sorghum bicolor*). J Pharmacogn Phytochem 8:727–731

Praveen A, Khan E, Ngiimei S et al (2017) Iron oxide nanoparticles as nano-adsorbents: a possible way to reduce arsenic phytotoxicity in Indian mustard (*Brassica juncea*L.). J Plant Growth Regul 37(2):612–624 https://doi.org/10.1007/s00344-017-9760-0

Predoi D, Ghita RV, Iconaru SL et al (2020) Application of nanotechnology solutions in plants fertilization. In: Solankey SS, Akhtar S, Maldonado AIL (eds) Urban horticulture-necessity of the future, Intech Open. https://doi.org/10.5772/intechopen.91240

Punshon T, Gaines KG, Bertsch PM et al (2003) Bioavailability of uranium and nickel to vegetation in a contaminated riparian ecosystem. Environ Toxicol Chem 22(5):1146–1154

Rafique M, Tahir R, Gillani SSA et al (2020) Plant-mediated green synthesis of zinc oxide nanoparticles from *Syzygium cumini* for seed germination and wastewater purification. Int J Environ Anal Chem. https://doi.org/10.1080/03067319.2020.1715379

Rahman MS, Chakraborty A, Mazumdar S et al (2020) Effects of poly (vinylpyrrolidone) protected platinum nanoparticles on seed germination and growth performance of *Pisumsativum*. Nano-Structures Nano-Object 21:

Rai M, Ingle A (2012) Role of nanotechnology in agriculture with special reference to management of insect pests. Appl Microbiol Biotechnol 94(2):287–293

Raj NP, Chandrashekara CP (2019) Nano zinc seed treatment and foliar application on growth, yield and economics of Bt cotton (*Gossypium hirsutum* L.). Int J Curr Microbiol App Sci 8:1624–1630

Raliya R, Franke C, Chavalmane S et al (2016) Quantitative understanding of nanoparticle uptake in watermelon plants. Front Plant Sci 7:1228

Raliya R, Nair R, Chavalmane S et al (2015) Mechanistic evaluation of translocation and physiological impact of titanium dioxide and zinc oxide nanoparticles on the tomato (*Solanum lycopersicum* L.) plant. Metallomics 7:1584–1594

Ramesh R, Catherine G, John Sundaram S et al (2020) Synthesis of Mn_3O_4 nano complex using aqueous extract of *Helianthus annuus* seed cake and its effect on biological growth of *Vigna radiata*. Materialstoday Proc. https://doi.org/10.1016/j.matpr.2020.02.883

Rehab IF, Kordy AM, Salim B (2020) Assessment of sorghum (*Sorghum bicolor* L.) productivity under different weed control methods, mineral and nano fertilization. Egypt Acad J Biol Sci H Bot 11:1–11

Rico CM, Majumdar S, Duarte-Gardea M et al (2011) Interaction of nanoparticles with edible plants and their possible implications in the food chain. J Agric Food Chem 59(8):3485–3498

Riera FA, Suarez A, Muro C et al (2013) Nanofiltration of UHT flash cooler condensates from a dairy factory: characterization and water reuse potential. Desalination 309:52–63

Rizwan M, Ali S, Abbas T et al (2018) Residual effects of biochar on growth, photo-synthesis and cadmium uptake in rice (*Oryzasativa*L.) under Cd stress with different water conditions. J Environ Manag 206:676–683

Rizwan M, Ali S, Ali B et al (2019) Zinc and iron oxide nanoparticles improved the plant growth and reduced the oxidative stress and cadmium concentration in wheat. Chemosphere 214:269–277

Roy A, Bulut O, Some S et al (2019) Green synthesis of silver nanoparticles: biomolecule-nanoparticle organizations targeting antimicrobial activity. RSC Adv 9:2673–2702. https://doi.org/10.1039/c8ra08982e

Sabir S, Arshad M, Satti SH et al (2018) Effect of green synthesized silver nanoparticles on seed germination and seedling growth in wheat. Int J Agron Agric Res 12(4):1–7

Saharan V, Kumaraswamy RV, Choudhary RC et al (2016) Cu-chitosan nanoparticle mediated sustainable approach to enhance seedling growth in maize by mobilizing reserved food. J Agric Food Chem 64:6148–6155. https://doi.org/10.1021/acs.jafc.6b02239

Saif S, Tahir A, Chen Y (2016) Green synthesis of iron nanoparticles and their environmental applications and implications. Nanomater 6(11):209. https://doi.org/10.3390/nano6110209

Salachna P, Byczyńska A, Zawadzińska A et al (2019) Stimulatory effect of silver nanoparticles on the growth and flowering of potted oriental lilies. Agron 9:610. https://doi.org/10.3390/agronomy9100610

Samreen FG, Muzaffar R, Nawaz M et al (2018) Synthesis, characterization and anti-microbial activity of *Citrus limon* mediated nanoparticles. Preprints. https://doi.org/10.20944/preprints201 8110417.v1

Sarkar J, Chakraborty N, Chatterjee A et al (2020) Green synthesized copper oxide nanoparticles ameliorate defence and antioxidant enzymes in *Lens culinaris*. Nanomater 10:312. https://doi.org/10.3390/nano10020312

Saxena A, Tripathi RM, Singh RP (2010) Biological synthesis of silver nanoparticles by using onion (*Allium cepa*) extract and their antibacterial activity. Dig J Nanomater Bios 5:427–432

Saxena A, Tripathi RM, Zafar FB, Singh RP (2012) Green synthesis of silver nanoparticles using aqueous solution of Ficus benghalensis leaf extract and characterization of their antibacterial activity. Mater Lett 67:91–94. https://doi.org/10.1016/j.matlet.2011.09.038

Schönherr J (2006) Characterization of aqueous pores in plant cuticles and permeation of ionic solutes. J Exp Bot 57(11):2471–2491

Schreck E, Foucault Y, Sarret G et al (2012) Metal and metalloid foliar uptake by various plant species exposed to atmospheric industrial fallout: mechanisms involved for lead. Sci Total Environ 427:253–262

Schreiber L (2005) Polar paths of diffusion across plant cuticles: new evidence for an old hypothesis. Ann Bot 95:1069–1073

Sedghi M, Hadi M, Toluie SG (2013) Effect of nano zinc oxide on the germination of soybean seeds under drought stress. Ann West Uni Timişoara Ser Biol 2:73–78

Sehnal K, Hosnedlova B, Docekalova M et al (2019) An assessment of the effect of green synthesized silver nanoparticles using sage leaves (*Salvia officinalis* L.) on germinated plants of maize (*Zea mays* L.). Nanomater 9:1550 https://doi.org/10.3390/nano9111550

Seqqat R, Blaney L, Quesada D et al (2019) Nanoparticles for environment, engineering, and nanomedicine. Hindawi J Nanotechnol Volume 2019, Article ID 2850723, 2 pp. https://doi.org/10.1155/2019/2850723

Servin A, Elmer W, Mukherjee A et al (2015) A review of the use of engineered nanomaterials to suppress plant disease and enhance crop yield. J Nanopart Res 17:1–21

Shah V, Belozerova I (2009) Influence of metal nanoparticles on the soil microbial community and germination of lettuce seeds. Water Air Soil Pollut 197:143–148. https://doi.org/10.1007/s11270-008-9797-6

Shaik AM, David Raju M, Rama Sekhara Reddy D (2020) Green synthesis of zinc oxide nanoparticles using aqueous root extract of *Sphagneticola trilobata* L. in and investigate its role in toxic metal removal, sowing germination and fostering of plant growth. Inorg Nano-Met Chem 50(7):569–579 https://doi.org/10.1080/24701556.2020.1722694

Shankar SS, Ahmad A, Sastry M (2003) Geranium leaf assisted biosynthesis of silver nano particles. Biotechnol Prog 19(44):1627–1631

Sharma S, Gothalwal R (2019) Green biosynthesis of metallic nanoparticle for medical diagnostic. Res J Life Sci Bioinform Pharm Chem Sci 5(1):63–82. https://doi.org/10.26479/2019.0502.05

Sharma VK, Filip J, Zboril R et al (2015) Natural inorganic nanoparticles – formation, fate, and toxicity in the environment. Chem Soc Rev 44:8410–8423

Shen X, Zhou Y, Duan L et al (2010) Silicon effects on photosynthesis and antioxidant parameters of soybean seedlings under drought and ultraviolet-B radiation. J Plant Physiol 167:1248–1252

Shi G, Cai Q, Liu C, Wu L (2010) Silicon alleviates cadmium toxicity in peanut plants in relation to cadmium distribution and stimulation of antioxidative enzymes. Plant Growth Regul 61:45–52

Shukla PK, Misra P, Kolie C (2016) Uptake, translocation, acumulation, transformation, and generational transmission of nanoparticles in plants. In: Kole C, Kumar DS, Khodakovskaya MV (eds) Plant nanotechnolgy principles and practices, Springer International Publishing Switzerland, pp 183 – 218

Siddiqui MH, Al Whaibi MH, Faisal M, Al Sahli AA (2014) Nanosilicon dioxide mitigates the adverse effects of salt stress on *Cucurbita pepo* L. Environ Toxicol Chem 33(11):2429–2437. https://doi.org/10.1002/etc.2697

Singh AK, Singh R, Singh K (2005) Growth, yield, and economics of rice (*Oryza sativa*) as influenced by level and time of silicon application. Indian J Agron 50(3):190–193

Singh A, Singh NB, Hussain I et al (2016) Green synthesis of nano zinc oxide and evaluation of its impact on germination and metabolic activity of *Solanum lycopersicum*. J Biotechnol 233:84–94

Singh A, Shrivastava JN (2018) Antibacterial potency of green synthesized silver nanoparticles. Inter J Basic Appl Biol 5:109–113

Singh J, Kumar V, Kim K-H et al (2019a) Biogenic synthesis of copper oxide nanoparticles using plant extract and its prodigious potential for photocatalytic degradation of dyes. Environ Res 177: https://doi.org/10.1016/j.envres.2019.108569

Singh K, Madhusudanan M, Ramawat N (2019b) Synthesis and characterization of zinc oxide nano particles (ZnO NPs) and their effect on growth, Zn content and yield of rice (*Oryza sativa* L.). Synthesis 6(3):9750–9754

Snehal S, Lohani P (2018) Silica nanoparticles: Its green synthesis and importance in agriculture. J Pharmacogn Phytochem 7(5):3383–3393

Soliman A, El-feky S, Darwish E (2015) Alleviation of salt stress on *Moringa peregrine* using foliar application of nanofertilizers. J Hortic For 7(2):36–47

Sotto A, Arsuaga JM, VanderBruggen B (2013) Sorption of phenolic compounds on NF/RO membrane surfaces: influence on membrane performance. Desalination 309:64–73

Taiz L, Zeiger E (2010) Plant physiology, 5th edn. Sinauer Associates Inc, Massachusetts, p 781

Taran N, Batsmanova L, Konotop Y et al (2014) Redistribution of elements of metals in plant tissues under treatment by non-ionic colloidal solution of biogenic metal nanoparticles. Nanoscale Res Lett 9:354–357

Torabian S, Zahedi M, Khoshgoftarmanesh A (2016) Effect of foliar spray of zinc oxide on some antioxidant enzymes activity of sunflower under salt stress. J Agric Sci Tech 18:1013–1025

Tripathi DK, Singh S, Singh S et al (2017) An overview on manufactured nanoparticles in plants: uptake, translocation, accumulation and phytotoxicity. Plant Physiol Biochem 110:2–12

Tripathi DK, Singh S et al (2016) Silicon nanoparticles more efficiently alleviate arsenate toxicity than silicon in maize cultivar and hybrid differing in arsenate tolerance. Front Environ Sci 4:23. https://doi.org/10.3389/fenvs.2016.00046

Tripathi DK, Singh VP, Prasad SM et al (2015) Silicon nanoparticles (Si Np) alleviate chromium (VI) phytotoxicity in *Pisum sativum* (L.) seedlings. Plant Physiol Biochem 96:189–198

Upadhyaya H, Shome S, Tewari S et al (2015) Effect of Zn nanoparticles on growth responses of rice. In: Singh B, Kaushik A, Mehta SK, Tripathi SK (eds) Nanotechnology: novel perspectives and prospects. McGraw Hill Education, New Delhi, India, pp 508–512

Vadlapudi V, Kaladhar DSVGK (2014) Review: green synthesis of silver and gold nanoparticles. Middle East J Sci Res 19(6):834–842

Van Aken B (2015) Gene expression changes in plants and microorganisms exposed to nanomaterials. Curr Opin Biotechnol 33:206–219

Venkatachalam P, Jayaraj M, Manikandan R et al (2016) Zinc oxide nanoparticles (ZnONPs) alleviate heavy metal-induced toxicity in *Leucaena leucocephala* seedlings: A physiochemical analysis. Plant Physiol Bioch 110:59–69

Venkateswarlu S, Rao YS, Balaji T et al (2013) Biogenic synthesis of Fe_3O_4 magnetic nanoparticles using plantain peel extract. Mater Lett 100:241–244

Verma SK, Das AK, Patel MK et al (2018) Engineered nanomaterials for plant growth and development: a perspective analysis. Sci Total Environ 630:1413–1435. https://doi.org/10.1016/j.scitotenv.2018.02.313

Wahid A (2007) Physiological implications of metabolites biosynthesis in net assimilation and heat stress tolerance of sugarcane (*Saccharum officinarum*) sprouts. J Plant Res 120:219–228

Wang H, Kou X, Pei Z et al (2011) Physiological effects of magnetite (Fe_3O_4) nanoparticles on perennial ryegrass (*Lolium perenne*) and pumpkin (*Cucurbita mixta*) plants. Nanotoxicology 5:30–42

Wang T, Lin J, Chen Z et al (2014a) Green synthesized iron nanoparticles by green tea and eucalyptus leaves extracts used for removal of nitrate in aqueous solution. J Clean Prod 83:413–419

Wang T, Jin X, Chen Z et al (2014b) Green synthesis of Fe nanoparticles using eucalyptus leaf extracts for treatment of eutrophic wastewater. Sci Total Environ 466–467:210–213

Wang WN, Tarafdar JC, Biswas P (2013) Nanoparticle synthesis and delivery by an aerosol route for watermelon plant foliar uptake. J Nanopart Res 15(1):1417. https://doi.org/10.1007/s11051-013-1417-8

Wang Y, Stass A, Horst WJ (2004) Apoplastic binding of aluminum is involved in silicon-induced amelioration of aluminum toxicity in maize. Plant Physiol 136:3762–3770. https://doi.org/10.1104/pp.104.045005

Wang ZY, Xie XY, Zhao J et al (2012) Xylem- and phloem-based transport of CuO nanoparticles in maize (*Zea mays* L.). Environ Sci Technol 46(8):4434–4441 doi:10.1021/ es204212z

Wasaya A, Yasir TA, Sarwar N et al (2020) 76. Improving growth and yield of mungbean (*Vigna radiata* L.) through foliar application of silver and zinc nanoparticles. Pure Appl Biol PAB 9:790–797. https://doi.org/10.19045/bspab.2020.90085

Watts-Williams SJ, Turney TW, Patti AF et al (2014) Uptake of zinc and phosphorus by plants is affected by zinc fertiliser material and arbuscular mycorrhizas. Plant Soil 376:165–175

White PJ (2012) Ion uptake mechanisms of individual cells and roots: short-distance transport. In: Marschner P (ed) Marschner's mineral nutrition of higher plants. Elsevier, London, pp 7–47

Xiao L, Wang S, Yang D et al (2019) Physiological Effects of MgO and ZnO Nanoparticles on the Citrus maxima. J Wuhan Univ Technol Mat Sci Edit 34:243–253

Xu Y, Qin Y, Palchoudhury S et al (2011) Water-soluble iron oxide nanoparticles with high stability and selective surface functionality. Langmuir 27(14):8990–8997

Yassen A, Abdallah E, Gabalah M, Zaghloul S (2017) Role of silicon dioxide nanofertilizer in mitigating salt stress on growth, yield and chemical composition of cucumber (*Cumis sativus* L.). Inter J Agric Res 12:130–135

Zahedi SM, Abdelrahman M, Hosseini MS et al (2019) Alleviation of the effect of salinity on growth and yield of strawberry by foliar spray of selenium-nanoparticles. Environ Pollut 253:246–258. https://doi.org/10.1016/j.envpol.2019.04.078

Zahedi SM, Karimi M, da Silva JAT (2020) The use of nanotechnology to increase quality and yield of fruit crops. J Sci Food Agric 100:25–31. https://doi.org/10.1002/jsfa.10004

Zeng FR, Zhao FS, Qiu BY et al (2011) Alleviation of chromium toxicity by silicon addition in rice plants. Agric Sci China 10:1188–1196

Zhao H, Dai T, Jiang D, Cao W (2008) Effects of high temperature on key enzymes involved in starch and protein formation in grains of two wheat cultivars. J Agron Crop Sci 194:47–54

Zhao S, Zou L, Mulcahy D (2012a) Brackish water desalination by a hybrid forward osmosis – nanofiltration system using divalent draw solute. Desalination 284:175–181. https://doi.org/10.1016/j.desal.2011.08.053

Zhao L, Peralta-Videa JR, Ren M, Varela-Ramirez A et al (2012b) Transport of Zn in a sandy loam soil treated with ZnO NPs and uptake by corn plants: electron microprobe and confocal microscopy studies. Chem Eng J 184:1–8

Zhu H, Han J, Xiao J et al (2008) Uptake, translocation and accumulation of manufactured iron oxide nanoparticles by pumpkin plants. J Environ Monit 10:713–717

Chapter 10
Biosynthesis and Characterization of Microorganisms-Derived Nanomaterials

Maysaa T. Alloosh, Maysoun M. Saleh, Lina M. Alnaddaf, Abdulsalam K. Almuhammady, Khaled F. M. Salem, and Jameel M. Al-Khayri

Abstract Nanotechnology is considered the newest advances in science that provide different methods to manufacture and develop diverse nanoparticles (NPs). Different metals can be prepared as NPs that can be used in various fields including biological systems. Ongoing research focuses on developing green methods for synthesizing NPs by using microorganisms. This chapter presents the importance of Nanotechnology. In addition to the different methods of synthesis NPs. Then emphasizes the various biological methods responsible for producing (silver, gold, copper, zinc,

M. T. Alloosh (✉)
Department of Biochemistry and Microbiology, Faculty of Pharmacy, Al-Baath University, Homs, Syria
e-mail: malloosh@albaath-univ.edu.sy

M. M. Saleh
Genetic Resources Department, General Commission for Scientific Agricultural Research, Damascus, Syria
e-mail: mzainsamasaleh@gcsar.gov.sy

L. M. Alnaddaf
Biotechnology and Molecular Biology, Faculty of Agricultural, Al-Baath University, Homs, Syria
e-mail: lalnaddaf@albaath-univ.edu.sy

A. K. Almuhammady
Arab Center for Nanotechnology, Cairo, Egypt
e-mail: 42020018.Assim@acu.edu.eg

K. F. M. Salem
Department of Plant Biotechnology, Genetic Engineering and Biotechnology Research Institute (GEBRI), University of Sadat City, Sadat City, Egypt

Department of Biology, College of Science and Humanitarian Studies, Shaqra University, Qwaieah, Saudi Arabia

K. F. M. Salem
e-mail: khaled.salem@gebri.usc.edu.eg; khaledfathi@su.edu.sa

J. M. Al-Khayri
Department of Agricultural Biotechnology, College of Agriculture and Food Sciences, King Faisal University, Al-Ahsa, Saudi Arabia
e-mail: jkhayri@kfu.edu.sa

J. M. Al-Khayri et al. (eds.), *Nanobiotechnology*,
https://doi.org/10.1007/978-3-030-73606-4_10

iron, palladium and selenium) NPs including the diverse metallic NPs using various promising microorganisms (virus, bacteria, actinomyces, algae, yeast and fungi) as a biogenic approach. Moreover, it also highlights the related molecular aspects of NPs that acted as reducing, capping and stabilizing agents together with the various factors influencing green syntheses like pH, temperature, as well as the concentrations of metal salts and substrates.

Keywords Biological molecules · Biosynthesis · Microorganisms · Nanoparticles synthesis · Green nanoparticles synthesis

10.1 Introduction

Nanotechnology is one of the most important developments in all of technology fields. It relies on the synthesis and modulation of nanoparticles (NPs), which require significant changes in metal properties (Rao and Gan 2015). In reality, NPs have been used unknowingly for thousands of years for example, gold NPs used to stain drinking glasses, also have cured other diseases (Singh et al. 2016). Research interest in metal NPs, and their production has increased significantly in recent times due to their groundbreaking applications in various fields. Metal NPs most important property is their large surface area to volume ratio, which increases their interaction with other molecules (Gahlawat et al. 2016). Chemical and physical methods for the synthesis of NPs have been used. These methods have disadvantages like the use of harmful materials, hazardous solvents and energy-intensive use (Azandehi and Moghaddam 2015; Yu et al 2016). NPs are usually developed and stabilized either by a technique of top-down or bottom-up (Murphy 2002). NPs are synthesized in the bottom upstream strategy via self-assembly of atoms into nuclei that further develop into particles of the nanoscale. This technique involves chemical and biological processes, while bulk materials are broken down into small particles in the top-down strategy (Shedbalkar et al. 2014). Because of the increased demand synthesis NPs are being explored from different sources and different methods. Among them, the biological method of synthesizing NPs is highly cost-efficient and simpler in comparison with the physical and chemical methods. The use of microorganisms in biological methods is the commonly used source, which is highly effective for nano synthesis (Hari 2020).

Microbes are used as a promising biological source for metal NPs to be synthesized. However, not all of the organisms show the ability to convert metals to nanoforms (Roy et al. 2019). Microbial metal NPs can occur intracellularly or extracellularly (Jain et al. 2011). Intracellular synthesis of NPs requires additional steps to release synthesized NPs, such as ultrasound treatment or reactions with appropriate detergents (Kalimuthu et al. 2008). While, extracellular biosynthesis is cheap, and simple downstream processing is required. This favors large-scale NPs production to explore its potential applications. Because of that, many studies focused on extracellular methods for metal NPs synthesis (Prasad et al. 2016). This chapter highlights the use of diverse microorganisms for the synthesis of NPs together with the biological molecules collaborating in the synthesis of green NPs and the factors influencing them.

10.2 Biosynthesis of NPs Using Microorganisms

Biosynthesis was considered a practice that was efficient, safe and environment friendly. These techniques employ living organisms such as bacteria, fungi, yeast, algae and plants with their tissues and extracts (Fig. 10.1). Normal extract rich

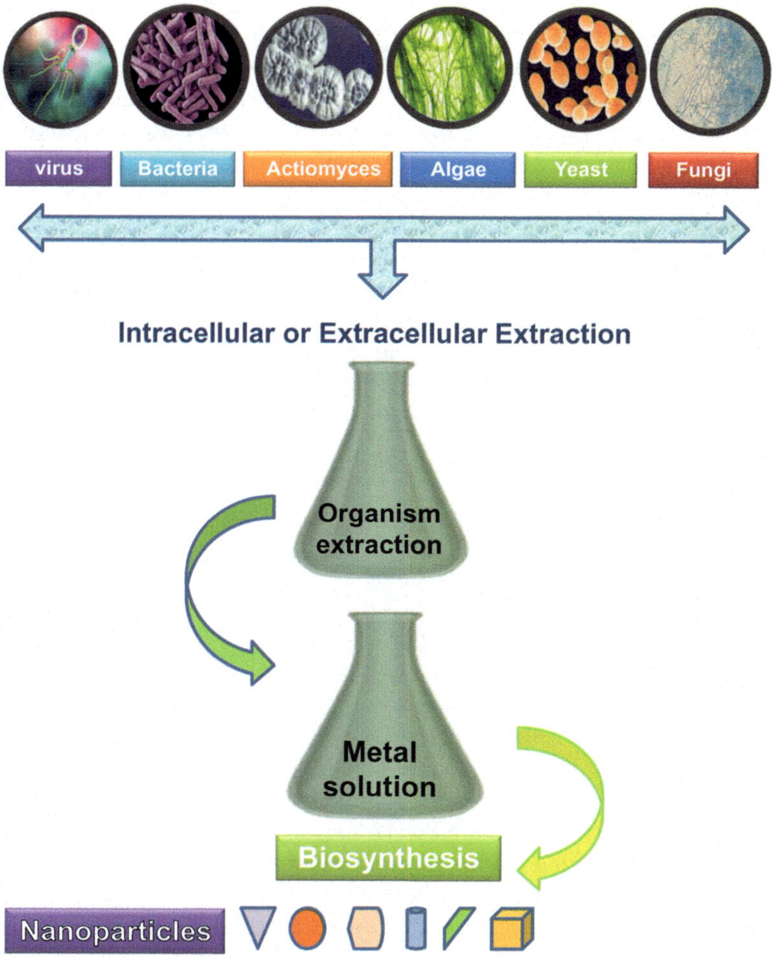

Fig. 10.1 Biosynthesis of NPs using several microorganisms (Constructed by M.T. Alloosh)

in biomolecules such as proteins, flavonoids, phenols and terpenoids can be used as reduction agents of metal ions to NPs (Attia and Elsheery 2020). The web of science database contained more than 159 publications about the green synthesis of NPs, which match the search criteria. The principal publication was in 2003, but till 2009 there was just a limited numbers. The number of papers tripled in 2010 and increasing conduct was seen until 2017. Between 2015 and 2016, the greatest impulse of published work was recorded, hitting a production peak in 2016 at 39 (Ribeiro et al. 2020).

10.2.1 Synthesis of NPs Using Bacteria

Bacteria are an extremely convenient target for green NPs synthesis, due to their diverse variety and ability to adapt to different environmental conditions. There are various bacterial cellular components such as enzymes, proteins, peptides and pigments, which are playing as a factory of NPs. Bacteria used as nanofactories can afford a new platform not only for the removal of metal or metalloid ions, but also the production of materials with distinctive properties (Tsekhmistrenko et al. 2020). Metallic NPs can be made by bacteria both intracellularly and extracellularly. Extracellular creation is more effective and easier for the extraction of NPs. In this case, biosynthetic metal NPs are less affected by oxidation, which makes it possible to use them in many fields (Gahlawat and Choudhury 2019). Some studies showed that not only living bacteria, but also dead forms of those bacteria can be used for NPs biosynthesis (Tsekhmistrenko et al. 2020).

Numerous studies used bacteria as a residual agent for the production of silver NPs (AgNPs). Das et al. (2017) indicated the extracellular synthesis of AgNPs within 24 h using (*Bacillus cereus* Frankland) isolated from heavy metal polluted soil. Also, Ghiuță et al. (2018) reported biosynthesis AgNPs using $AgNO_3$ as a precursor to (*Bacillus amyloliquefaciens* Fukumoto) and (*Bacillus subtillis* Ehrenberg) with spherical shape and 142 nm average diameter. Later, Allam et al. (2019) obtained AgNPs by (*Sphingomonas paucimobilis* Holmes), AgNPs were spherical to oval (4–20 nm), which can use as decontamination of wastewater from harmful dyes. As well as, Divya et al. (2019) used (*Alcaligenes* spp Castellani and Chalmers) as a mediate to synthesize AgNPs (30–50 nm). In this study, AgNPs displayed antimicrobial activity against clinical microbe isolates such as *Bacillus* spp, (*Escherichia coli* Migula), (*Klebsiella pneumonia* Schroeter), (*Pseudomonas aeruginosa* Schroeter), *Staphylococus aureus* Rosenbach) and (*Candida albicans* Berkh). Ameen et al. (2020) described synthesize spherical AgNPs (13 nm) by (*Spirulina platensis* Turpin ex Gomont) extract by heating the mixture of cyanobacterial extract (1%) with 0.5% $AgNO_3$ solution at 40 °C for 7 h. Additionally, the antibacterial activity was considered against seven dissimilar species of clinical bacterial pathogens. In another research work, Ahmed et al. (2020) synthesized AgNPs using *B. cereus* strain SZT1; AgNPs obtained were a spherical shape, its sizes ranging from 18 to 39 nm. Also, Ahsan (2020) employed (*Pseudomonas fluorescens* Flügge) to synthesis AgNPs by mix 90 ml of $AgNO_3$ with 10 ml of extracted broth from *P. fluorescens* at 80 °C, the mixture was kept at 5 pH and stirring for 2 h and then put it over one day. Scanning Electron Microscope (SEM) images showed that AgNPs were spherical and irregular (10–100 nm).

Other study was performed towards the synthesizing gold NPs (AuNPs) using bacteria where Kunoh et al. (2018) used *Pseudomonas stutzeri* Lehmann and Neumann for synthesis AuNPs by reducing gold salt in an aqueous medium. In this study, spherical AuNPs (5 nm diameter) were obtained simply by adding guanine to chloroauric acid ($HAuCl_4$) solution at room temperature. San Diego et al. (2020) utilized bacteria *Lysinibacillus* spp. and *P. stutzeri* for the extracellular synthesis of AuNPs by reduction ($HAuCl_4$) at pH 9, AuNPs were spherical and irregular shapes, it showed no toxicity to *P. aeruginosa*, an increasing inhibition level of pyocyanin production was observed with increasing volumes of NPs used.

Other metallic NPs were synthesized by bacteria where Noman et al. (2020) proved the ability of *Escherichia* sp. to synthesize copper NPs (CuNPs). The particles were spherical shape with size ranging from 22.3 to 39 nm. On the other hand, Yusof et al. (2020) used a strain of (*Lactobacillus plantarum* Orla Jensen) to manufacture zinc oxide NPs (ZnO NPs) with size about 124.2 nm. Sidkey et al. (2020) employed *P. stutzeri*, which isolated from soil and wastewater samples to synthesize magnesium NPs (MgNPs) by both extracellular and intracellular cases. MgNPs intracellular were spherical, its size ranged from (229.3–553.2 nm). Fatemi et al. (2018) successfully used extracellular method to synthesize spherical iron oxide NPs (29.3 nm) using *B. cereus*, which isolated from soil. The experiment was done at the room temperature, where magnetic iron oxide NPs were obtained quickly using $FeCl_3 \cdot 6H_2O$ after 5 min and using $FeCl_2 \cdot 4H_2O$ after 30 min. This study also depicted that these NPs had anticancer effects against some factores of breast cancer cell.

10.2.2 Synthesis of NPs Using Actinomycetes

Actinomycetes are fungi-like bacteria, it is gram-positive, with high G-C content and have the suitability to produce metal NPs. They can produce several kinds of bioactive compounds that have great beneficial values (Omar et al. 2019). Among actinomycetes, species of (*Streptomyces* Waksman and Henrici) (Fig. 10.2) are most generally used in medicinal and enzymatic applications because, out of more than 10,000 identified antibiotics, 55% are formed by them.

Actinomycetes synthesize NPs by both intracellular and extracellular ways, although the extracellular reduction is the most common way. Recently, it was discovered that actinomycetes are capable of manufacturing NPs, which have antimicrobial activities, and that was a milestone in the area of therapeutics. Gahlawat and Choudhury (2019) reviewed several studies, which had been perfprmed between 2012 and 2018 about the effectiveness of actinomycetes (Rhodococcus spp., Streptomyces spp. and Streptacidiphilus spp.) as NPs manufacturing agents, which includes the production of silver, gold and copper NPs. The more recent study of these was by Wypij et al. (2018), where they synthesized spherical and polydispersed AgNPs (5–20 nm) using

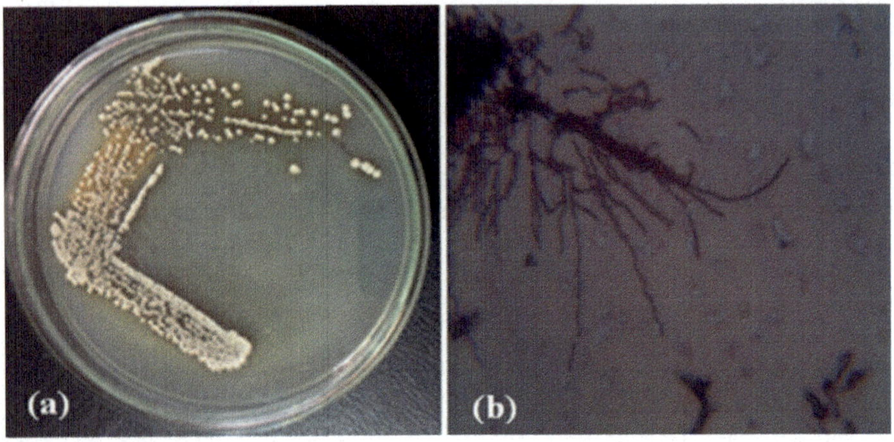

Fig. 10.2 (**a**) *Streptomyces* spp. on a starch casein agar plate showing sporulating white color aerial mycelia; (**b**) under a light microscope with 40× magnification (*Source* Omar et al. 2019)

(*Streptomyces xinghaiensis* Zhao) at room temperature within 2–3 days. Avilala and Golla (2019) used marine actinomycete (*Nocardiopsis alba* Kroppenstedt) to synthesize spherical AgNPs (20–60 nm) in bright conditions within 24 h, AgNPs have had antiviral and antibacterial activities against *E. coli*, *P. aeruginosa*, *K. pneumonia* and (*Streptococcus aureus* Rosenbach). Vairavel et al. (2020) synthesized spherical AuNPs intracellularly by (*Enterococcus* sp. Thiercelin and Jouhaud).

Some studies were done about biosynthesis other NPs elements by actinomycetes like titanium, tellurium, copper and selenium. Ağçeli et al. (2020) used *Streptomyces* spp. to synthesis titanium dioxide NPs (TiO_2 NPs), these particles were spherical (30–70 nm) which have an effective against *Staphylococcus aureus*, *E. coli*, *C. albicans* and (*Aspergillus niger* van Tieghem). On the other hand, El-Sayyad et al. (2020) used (*Streptomyces cyaneus* Krasil'nikov) for biosynthesis tellurium dioxide NPs (TeO_2 NPs) at room temperature, which were spherical (75 nm size), these TeO_2 NPs cause an antimicrobial activity towards (*Aspergillus flavus* Link), *A. niger*, (*Aspergillus fumigatus* Fresenius), *P. aeruginosa*, *S. aureus* and *K. pneumonia*. Hassan et al. (2019) proved the ability of two actinomycete strain (*Streptomyces zaomyceticus* Oc-5Hinuma) and (*Streptomyces pseudogriseolus* Acv-11 Okami and Umezawa) to synthesize CuONPs (size 78 and 80 nm, respectively). Ranjitha and Ravishankar (2018) synthesized selenium NPs SeNPs (100–250 nm) by adding 5 ml of culture (*Streptomyces griseoruber* Yamaguchi and Saburi) to 5 ml of 1 mM Sodium Selenite (Na_2SeO_3) and incubated the mixture at 37 °C for 72 h.

10.2.3 Synthesis of NPs Using Yeast

Yeasts are eukaryotic, monocellular microbes classified in the kingdom of fungi. They use organic compounds to take energy sources (Lachance 2016). The studies on the biosynthesis of AgNPs using yeast strains have gained attention. The benefit of using yeast strains for AgNPs manufacture is that they are easy to the controller in laboratory environments, show quick growth, and are cheap to cultivate (Skalickova et al. 2017). Sowbarnika et al. (2018) employed baker's yeast (*Saccharomyces cerevisiae* Meyen ex E.C. Hansen) to synthesis spherical AgNPs (10–60 nm), which have significant antibacterial action against *E. coli* with an inhibition area of 21 mm. Also, Jalal et al. (2018) used supernatant of (*Candida glabrata* S.A. Mey and Yarrow) isolated for extracellular biosynthesis of spherical AgNPs (2–15 nm).

Researchers also progressed towards the synthesis of different kinds of NPs such as tellurium, palladium and platinum. Sriramulu and Sumathi (2018) used *S. cerevisiae* aqueous extract as a reducing agent of palladium NPs (PdNPs) with an average size of 32 nm. In this study, dry yeast granules (5 g) was dissolved in 100 ml of water and stirred at room temperature for 30 min, then 10 ml of the aqueous yeast extract was added to 90 ml of 1 mM of palladium acetate solution and kept at room temperature for 24 h. Further, SEM images showed hexagonal-shaped of Pd NPs, while atomic force microscopy (AFM) showed highly variable of shape with a rough surface. Faramarzi et al. (2020) synthesized selenium NPs (Se NPs) using *S. cerevisiae* within 4 days in which the size of Se NPs ranging from 75 to 709 nm. Other studies depicted the importance of Se NPs as an antioxidant, anti-inflammatory, antimicrobial and anticancer properties that have gained more attention in the medical field (Wadhwani et al. 2017).

10.2.4 Synthesis of NPs Using Algae

Algae uses for metal NPs synthesizing, their abundance and easy discovery, cost-effective and extensive synthesis highly stable and safe NPs with better biological properties make them a good source for metal NPs synthesis (Azizi et al. 2014). Additionally, the synthesis of NPs with algae occurs in a shorter time than other methods of biosynthesis (Dağlıoğlu and Öztürk 2019). Algae was mediated for biosynthesis of gold (Au), silver (Ag), palladium (Pd), platinum (Pt), iron (Fe), cadmium (Cd), titanium oxide (TiO_2) and zinc oxide (ZnO) bimetallic NPs. Gahlawat and Choudhury (2019) reviewed 23 studies that had been done between 2011 and 2018 concerning algae species in the synthesis of Ag, Au, Pd and ZnO NPs. Likewise, da Silva Ferreira et al. (2017) employed green algae (*Chlorella vulgaris* Beijerinck) for the biosynthesis of spherical silver chloride NPs (AgCl NPs) with the size of 9.8 ± 5.7 nm, where improved effects as antimicrobial agents against *S. aureus* and *K. pneumoniae* were recorded. Also, Arsiya et al. (2017) used for the first time *C. vulgaris* aqueous extract to synthesis PdNPs within 10 min. Transition electron microscope

(TEM) images indicated that PdNPs were spherical and 5–20 nm in size. In other study, Rajeshkumar (2018) employed two brown seaweeds, such as (*Padina tetrastromatica* Hauck) and (*Turbinaria conoid* J. Agardh) algal formulation for ZnO NPs biosynthesis and assessed their antimicrobial ability against fish pathogens.

Dağlıoğlu and Öztürk (2019) used green microalgae (*Desmodesmus* spp. Chodat) as a reducing agent for manufacturing AgNPs intracellularly within 24 h without any aggregates. Further, a TEM image of algae cells showed the presence of spherical AgNPs (15–30 nm) inside them. Mishra et al. (2020) employed green algae *C. vulgaris* extract to synthesis PdNPs, the solutions of palladium chloride ($PdCl_2$) aqueous with algal extract were adjusted at pH 6–7 and stirreded at 60 °C for 2 h, solutions color turned from yellow to dark brown, which indicated the formation of NPs. SEM images showed spherical and triangular PdNPs, with an average size of 70 nm. On the other hand, Yılmaz Öztürk et al. (2020) used extract of red algae (*Gelidium corneum* Huds) as reducing agent for synthesis AgNPs, noticeable color change from light red to dark brown indicated AgNPs formation, also TEM image confirmed that AgNPs were spherical or angular with size ranged 20–40 nm. In this study, AgNPs showed a great antimicrobial activity by very low minimum inhibitory concentrations (MIC) values for both yeast *C. albicans* and bacteria *E. coli* (Fig. 10.3).

In another study, Salaam et al. (2020) successfully synthesized AgNPs using green algae *C. vulgaris*, where the best biosynthesizing conditions were pH 10 and

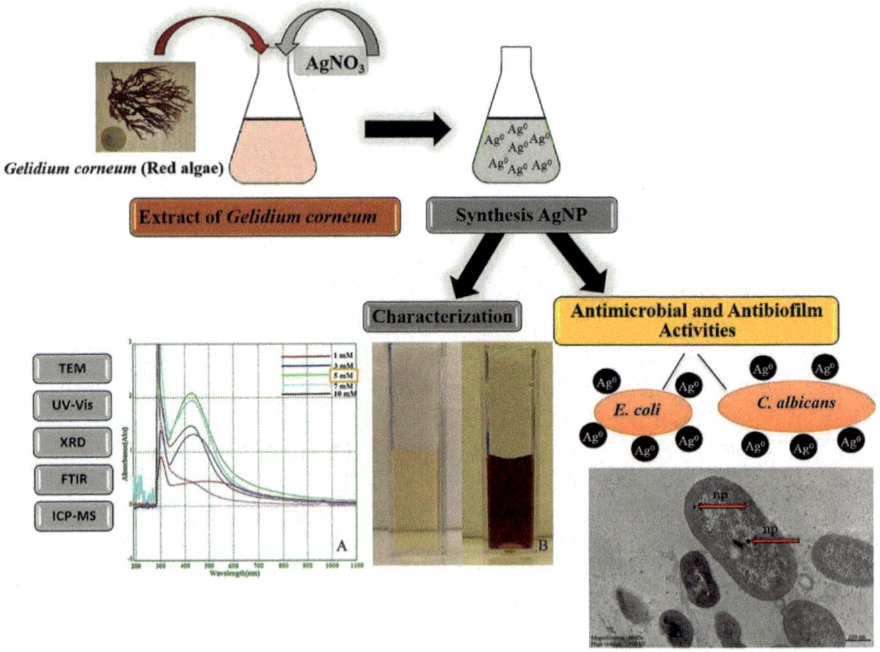

Fig. 10.3 Summarize the biosynthesis of AgNPs using *Gelidiumcorneum*. AgNPs characterization and bacterial activity (*Source* Yılmaz Öztürk et al. 2020)

37 °C. The AgNPs were not only spherical and sized to 10 μm, but also showed antibacterial activity against particular (*Citrobacter* spp. Werkman and Gillen), *S. aureus*, *E. coli* and *P. aeruginosa* strains. Thus these AgNPs can be used safely alternative to antibiotics. Also, El-Naggar et al. (2020) employed *C. vulgaris* for biosynthesized AgNPs in which the algae extract added to 100 mM of $AgNO_3$ and incubated in the dark, so that the pale green color turned brown within 24 h indicating AgNP formation. TEM image showed that AgNPs were spherical with size 3.63 8.68 nm. AgNPs produced in this study had antimicrobial activity against *Bacillus* spp., (*Erwinia* spp Winslow) and *Candida* spp.

10.2.5 Synthesis of NPs Using Fungi

Biosynthesis of NPs by fungi known as mycosynthesis, it is a widespread way due to well-defined dimensions, different chemical structures, sizes and great production of synthesized NPs (Golhani et al. 2020). The mycosynthesis of metal NPs can be occurred by different mechanisms, one of the most common is nitrate reductase by fungal enzymes. Fungi are more versatile in growth and metal tolerance in contrast to bacterial population (Sangappa and Thiagarajan 2012).

Barabadi et al. (2019) reviewed 59 papers about 25 *Penicillium* species that were used as biosynthesis agents of NPs, the only biosynthesis of four kinds of NPs described in all 59 studies including silver, gold, copper, and iron NPs. Meantime, AgNPs with 70.76% of the studies were the first predominant produced by using *Penicillium* species. Notably, a huge 91.22% study reported spherical NPs, while just 8.77% reported other morphologies, including triangular, hexagonal, irregular, cubical, ellipsoidal, and rod shapes. Hamad (2019) studied the synthesized AgNPs by using 50 ml of (5 mM) $AgNO_3$ solution and mixed with 50 ml cell filtrate biomass of (*Penicillium citreonigrum* C. Ramírez and A.T. Martínez) in dark at room temperature until the color change. The color changed from wan yellow to light brown after 24 h of incubation. In other study, Hulikere and Joshi (2019) employed, for the first time, endophytic fungus (*Cladosporium cladosporioides* Fresen), which isolated from brown algae to synthesize AgNPs, the color of the reaction mixture ($AgNO_3$ with aqueous fungi extract) progressively changed from colorless to dark brown, that is confirmed AgNPs formation, SEM images showed that AgNPs were spherical, its sizes ranged from 30 to 60 nm. Interestingly, no aggregation or precipitation happened after two to three weeks AgNPs incubation. According to Noshad et al. (2019) AgNPs were produced by mix 1:1 fungal extracts of (*Trichoderma harzianum* Patouillard) and *A. fumigatus* separately with $AgNO_3$ and heated to 29 °C for 24 h, the change of mixture color from yellowish to dark brown confirmed the formation of AgNPs. In (Fig. 10.4) appeared the formation of AgNPs by the change in the color of the cell-free extract from yellow to dark brown for *Aspergillus* spp. while *Rhizopus* spp. showed a color change from colorless to dark yellow after 24 h (Fig. 10.5).

Salah et al. (2020) employed (*Penicillium chrysogenum* Thom) extract to synthesize AgNPs, SEM images showed that AgNPs were spherical, while AFM showed

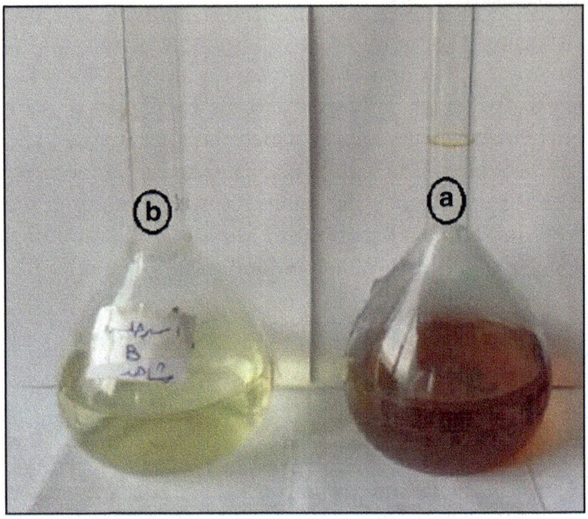

Fig. 10.4 The color change after 24 h in culture filtration of *Aspergillus* spp. **a** Culture contain AgNPs, **b** Control. (Photo by M. T. Alloosh)

Fig. 10.5 The color change after 24 h in culture filtration of *Rhizopus* spp. **a** Culture contain AgNPs, **b** control. (Photo by M. T. Alloosh)

that the particles size were 18.83 nm. This study proved the effect of AgNPs on some *Candida* spp. Also, Noshad et al. (2020) employed mycelial aqueous extract of fungi (*Pythium oligandrum* Dreschler) to synthesis AgNPs from $AgNO_3$ using a magnetic stirrer at 29 °C for 24 h, the X-ray images proved a crystalline structure of AgNPs produced with an average size 12 nm.

Mycosynthesis studied other metal, Mahanty et al. (2019) biosynthesis iron oxide NPs using three fungi (*Trichoderma asperellum* Samuels, Lieckf and Nirenberg), (*Fusarium incarnatum* Desm) and (*Phialemoniopsis ocularis* Gen and Guarro). The color of the reaction mixture (aqueous of three fungi extracts separately with 1: 2 ratio of $FeCl_2$ and $FeCl_3$) changed within 5 min (Fig. 10.6). The images of NPs by

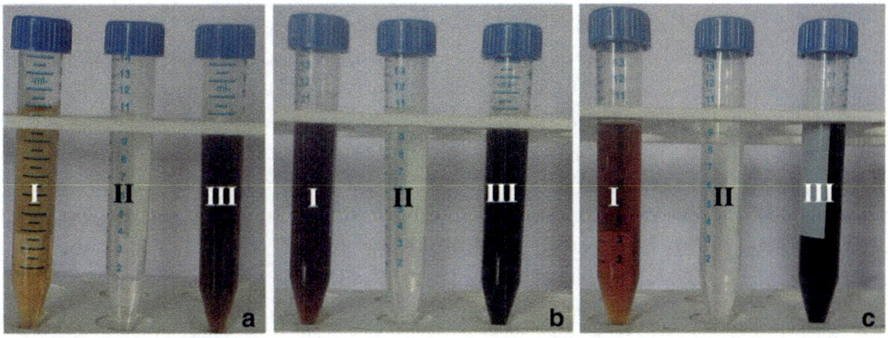

Fig. 10.6 Fungal isolates of *Trichoderma asperellum* (**a**), *Phialemoniopsis ocularis* (**b**) and *Fusarium incarnatum* (**c**) showing color change during reaction with iron precursor salts. **I** = (Positive control), **II** = Iron precursor salt (negative control), **III** = Appearance of black colouration due to addition of fungal extract to 1:2 ratio of $FeCl_2$ and $FeCl_3$ solution after 5 min incubation (*Source* Mahanty et al. 2019)

both SEM and TEM showed that iron nanoparticle was spherical with an average size ranging between 25 ± 3.94 nm for *T. asperellum*, 30.56 ± 8.68 nm for *F. incarnatum* and 13.13 ± 4.32 nm for *P. ocularis*. Salem et al (2020) used (*Penicillium corylophilum* Dierckx) as a reducing agent to synthesize selenium NPs(SeNPs). Spherical shape with average size 29.1–48.9 nm of SeNPs was produced. Also, Abu-Tahon et al. (2020) studied biosynthesis of AuNPs using *A. flavus*, the biosynthesize was by adding 10 mL of $HAuCl_4$ to 90 ml of *A. flavus* culture supernatant adjusted at pH = 7 and heating to 30 °C with shaking for 2 h. AuNPs were producing within 2 min, TEM images indicated that AuNPs were spherical, an average size of 12 nm.

10.2.6 Synthesis of Nanoparticles Using Virus

The dense outer surface capsid coating is an interesting feature of viruses proteins, which provide a highly appropriate platform for metal ion interaction (Kobayashi et al. 2012). Nevertheless, the synthesis of NPs by viruses still faces numerous disadvantages, such as involvement of the host organism for protein expression which restricted research (Gahlawat and Choudhury 2019). Because of its structural and biochemical stability, ease of cultivation, non-toxicity and non-pathogenicity in animals and humans; plant viruses are considered safe for nanotechnology applications. One study indicated that low concentrations of tobacco mosaic virus (TMV Adolf Mayer), which used as additives along with extracts of various plants for example *Nicotiana benthamiana* Domin, *Avena sativa* L. and *Musa pradisiaca* L. the TMV not only led to size reduction but also substantially increased the number of NPs compared to non-virus control (Gahlawat and Choudhury 2019). Also, (Ahiwale et al. 2017) synthesized gold NPs (AuNPs) using bacteriophage. At different physiological parameters, the reaction mixtures showed vivid colors. Au NPs were in the range of 20–100 nm. SEM studies revealed the synthesis of diverse AuNPs such as circles, hexagons, triangles, rhomboids and rectangles. Au NPs antibiofilm activity was measured on a glass sheet, it was noted that *P. aeruginosa* biofilm formation was inhibited at 0.2 mM concentration of these AuNPs induced around 80% inhibitation. Le et al. (2017) employed *potato virus X* for synthesizing elongated filament NPs, these NPs were more penetration capability compared to spherical ones. In the industrial field, Nam et al. (2006) used the viruses to synthesize and assemble cobalt oxide nanowires at ambient temperature. By integrating gold-binding peptides into the filament coat, hybrid gold-cobalt oxide wires were developed to increase battery efficiency. The combination of virus-templated peptide-level synthesis and methods for controlling two-dimensional virus assembly on multilayer polyelectrolytes provides a systematic framework for integrating such nanomaterials (NM) to form small and flexible lithium-ion batteries.

10.3 The Role of Biological Molecules of Microorganisms in Green NPs Synthesis

Biological extracts (plant extracts, microbial extracts and algal extracts) are generally mixed with salt-metal solutions. Biocompounds (including phenolic compounds, alkaloids, enzymes, terpenoids, co-enzymes, sugars and proteins) that exist in extracts diminish metal salts from positive state to zero states of oxidation. The distribution of metallic NPs depends on the biocompounds in the extract. The presence of a strong reducing agent in the extract fosters a rapid reaction rate and favors the formation of smaller NPs (Roy et al. 2019).

The existence of biomolecules acts as a capping sheet, stabilizing agent and a biologically active NPs layer (Ramya et al. 2015). Capping stratum characteristics, clear identification of the capping agents (main peptides such as glutathione, metallothioneins, membrane-associated proteins) and possible purification of NPs are essential for different applications in the future (Prasad et al. 2016; Suresh et al. 2011; Voeikova et al. 2017).

Microorganism synthesis of NPs (in particular, actinomycetes, bacteria, viruses, yeast and fungi) is classified as intracellular and extracellular (Shankar et al. 2016). The intracellular synthesis procedure requires the trapping, capping of diverse NPs and bioreduction (Li et al. 2011). Extracellular synthesis contains secretion of the enzymes, bioreduction and capping of particles (Singh and Singh 2019). Most studied works indicated that extracellular synthesis of NPs is preferable since downstream and purification processes are easier than intracellular methods (Singh 2015). Microbial enzymes changed metal ions to NPs and served to synthesize silver NPs through actinomycetes (Roy et al. 2019). Fouda et al. (2019) succeeded in the biosynthesis of AgNPs by secondary metabolites secreted by endophytic actinomycetes as reducing agents. These enzymes have a major role in electron transport positive ions such as nitrate reductase (Siddiqi et al. 2018; Yin et al. 2016). Ghiuță et al. (2018) utilized (*B. subtillis* Ehrenberg) and (*B. amyloliquefaciens* ex Fukumoto) which can generate the alpha-amylase enzyme to synthesize AgNPs.

Certain functional bacterial protein groups (–OH, $-NH_2$, – COOH and – SH) play an important role in NPs formation and stabilization. These groups offer binding sites for metal ion fixation, further reducing their extracellular concentration and location on the cell wall or in periplasmic space (Tsekhmistrenko et al. 2020). Nangia et al. (2009) indicated that (*Lacto bacillus* Beijerinck) mediated the synthesis of AgNPs from probiotic tablets and yogurt. Two protons generate from NADH to assist in the synthesis of NPs, depending on the lactate dehydrogenase and pyruvate and the combination of thioredoxin and glutathione. The role of peptides and individual amino acids in the microbial synthesis of AgNPs has been proven by Balachandran et al. (2013) and Selvakannan et al. (2013). Peptides including cysteine, arginine, methionine and lysine amino acids can attach to the surface of the nucleus and be used in AgNP production. In alkaline conditions the amino acid tyrosine acts as a reducer. That's because the phenolic tyrosine group may be converted to the quinone group (Dubey et al. 2015). In addition, at the free N-terminus, tyrosine-containing oligopeptides provide stability for the NPs and promote the recovery of metals. These results are coordinated with that tyrosine plays a key role in the reductions (Ali et al. 2019; Daima et al. 2014; Shankar et al. 2016). The tryptophan amino acid can synthesize NPs as a reduction agent due to it has the capability to equip electrons. In this case, tryptophan develops into a tryptophol radical. The presence of carboxyl groups of short yeast peptides of aspartic acid and glutamic acid shares in the synthesis of Ag NPs (Tsekhmistrenko et al. 2020). The components of cytoplasmic act as electron donors for Cu^{2+}. Such as NADH/NADPH, vitamins, and organic acids that exist in (Shewanella loihica PV-4 Gao) used to the biosynthesis of extracellular Cu-NPs. Cytochrome C was the primary reduction factor for electron transfer (Lv et al. 2018). In their study, Ahmed et al. (2020) revealed the presence of coating proteins considered essential for the long-term stabilization of biogenic nano-materials. Roy et al. (2019) indicated that (*Bacillus licheniformis* Weigmann) use silver NPs synthesis at a concentration of 1 mM without cell death. The organism undergoes cell death within minutes when increasing the concentration to 10 mM.

Algae assists to form NPs by the pigments, carbohydrates, fat, proteins, nucleic acid and secondary metabolites that it contains. This synthesis does not fabricate any toxic byproducts. Some agents such as ambient temperature, pH, ion concentration and solution incubation time affect the NPs size (Nadaroğlu et al. 2017). Numerous methodologies for synthesizing metallic NPs have been developed using microalgae from their corresponding aqueous salt solutions. That has an impact on the size and shape of superior quality nanocrystals. Microalgae NPs are synthesized in four ways: a) Direct use of extracted biomolecules from torn microalgae cells, b) Taking advantage of cell-free supernatants made from cultural microalgae media, c) Biosynthesis of NPs of varying nature from whole microalgae cells and d) Uses living microalgae cells. The microalgae biomass is first lyophilized to obtain gold nanoplates and then subjected to reverse-phase high-performance liquid chromatography (RP-HPLC) until it isolates the gold shape-directing protein (GSP). It is absolutely vital for shaping NPs. This protein is then brought into contact with the aqueous solution $HAuCl_4$, thus producing gold NPs of various forms. Low molecular weight proteins (PLW) and high molecular weight proteins (PHW) which exist in microalgae biomass are responsible for the synthesis of silver NPs. Synthesis of algal NPs takes a comparatively shorter time compared to other biosynthesizing methods. Microalgae can be considered a powerful nanofactory capable of producing not only silver ion NPs, as well as other metal ions such as gold, cadmium and platinum (Agarwal et al. 2019). Ahmad et al. (2019) mentioned in their review that the AgNPs can synthesize using microalgae (*Tetraselmis gracilis* F. Stein, *Chaetoceros calcitrans* Paulsen, *Isochrysis galbana* Pascher).

Diatoms are called frustules, which considered unicellular microalgae. it has a very peculiar biomineralized silica cell wall. The diatom frustules have fucoxanthin xanthophyll pigment. Numerous studies have proven fucoxanthin's active role in stabilizing silver NPs. It behaviors as the photo-reducing agent of metal ions (Grasso et al. 2020).

Filamentous fungi have some distinctive advantages over bacteria because of their high metal tolerance, wall binding capacity and intracellular metal absorption capabilities. The fungal mycelial mesh can resist flow, pressure, irritation and other conditions as compared to plant materials and bacteria, in bioreactors or in other chambers. These are fast to grow, fabricable and easy to handle (Velusamy et al. 2016). Therefore, competent fungi can synthesize greater quantities of NPs compared with bacteria (Singh et al. 2018). The fungi have specific enzymes like reductases and secrete much higher protein amounts. Used during NPs biosynthesis. Banu and Rathod (2011) provided information on the extracellular synthesis of silver NPs using soil-isolated fungi. (*Rhizopus Stolonifer* Ehrenberg) efficiently produced silver NPs. *R. stolonifer* has secreted protein to the cape and stabilized the AgNPs at alkaline (pH 7). However, NPs aggregation confirmed to acid (pH 4). likewise, the temperatures impacted on silver nanoparticle production by *R. stolonifer*. The study temperature range is from 25 to 45 °C maximum production was achieved at 40 °C.

Some factors impact the characteristics of microbial biosynthesized NM. These include pH, microbial cell concentration, temperature and precursor concentration. The optimum pH, temperature, and concentration of NaCl were studied with the bacterium (*P. aeruginosa* Schroeter) strain RB to realize a high rate of synthesis of cadmium selenide (CdSe)NPs and high purity. The data show that *P. aeruginosa* strain RB has optimal growth conditions that do not fit the optimum conditions for NPs synthesis (Ayano et al. 2015). The influence on the type of synthesized gold nanostructures and their relative size in yeast (*Yarrowia lipolytica* Harrison) strain NCIM 3589 by changing the concentration of gold precursor salt and the proportion of cell concentration (Pimprikar et al. 2009). The impact of temperature on gold nanostructures releasing from cell walls into the aqueous phase and to obtain shape anisotropy of silver NPs the bacterial growth kinetics of the bacterium (*Morganella psychrotolerans* Emborg) was controlled (Ramanathan et al. 2010). The temperature impacted the NPs size, synthesis speed and stability when synthesis of silver NPs through fungi (Elamawi et al. 2018). When the temperature rose to 40°C the rate of synthesis increased by using (*T. harzianum* Patouillard) in synthesis (Ahluwalia et al. 2014). Whereas higher fungal biomass protein secretion was observed at temperatures ranging from 60 to 80 °C, with radial increases in surface plasmon absorbance and synthesis rate as the temperature increased when using (*Fusarium oxysporum* Gordon) in some fungal species, high temperatures indicate by a NPs synthesis that electrons can be transferred from free amino acids to silver ions. So, increasing temperatures between 80 and 100 °C, cause denaturation of the proteins that compose the nanoparticle capping. This denaturation leads to NPs aggregating, increasing in size and changes the nucleation of Ag^+ ions (Birla et al. 2013). This agrees with Husseiny et al. (2015) which reported that the activity of the enzymes is come down as a result of unsuitable temperatures. This leads to increased nanoparticle size and loss of stability. Du et al. (2015) noticed that a more alkaline pH resulted in the smaller distribution of the size of NPs, shorter synthesis time and the intensity of the dispersion increased. Due to the electrostatic repulsion of anions present in dispersion, these characteristics indicate improved stability (Balakumaran et al. 2015). In addition, Nayak et al. (2011) told that the conformation of nitrate reductase enzymes could be adjusted by the concentration of protons in the reaction medium. Thus changes in morphology and NPs size occur. At higher pH, there is greater competition between protons and metal ions to establish bonds with regions with negative charges, resulting in greater success of synthesis at alkaline pH (Sintubin et al. 2009). The concentration of metal salts has an impact on extracellular silver NPs synthesis using fungi. To get NPs with favorable physicochemical properties. Most of the studies used $AgNO_3$ at a concentration of 1 mM (Saxena et al. 2016; Xue et al. 2016). Reduced metal salt concentration to better dispersion and smaller NPs size (Kaviya et al. 2011; Phanjom and Ahmed 2017). By contrast, when the concentration of the metal precursor increased, this produced a high dispersion color intensity and very huge, irregular forms of NPs due to the competition between the silver ions and the available functional reaction groups. In addition, a higher concentration of $AgNO_3$ may cause greater toxicity (Phanjom and Ahmed 2017; Shahzad et al. 2019). Microorganism perhaps mediated biosynthesis of NPs. Thus it can describe these NPs through growing conditions and the medium of culture. Changes in those conditions result in the synthesis of various metabolites and proteins (Costa Silva et al. 2017). When the culture medium containing a specific substrate which induces the enzymes to produce and release a NPs. The NPs are synthesis with smaller sizes and high concentrations (Husseiny et al. 2015). The amount of biomass utilized can affect the synthesis and characteristics of silver NPs. Some studies have notified that lower biomass concentrations give higher NPs production. While others have found higher synthesis rates using higher concentrations (Elamawi et al. 2018). Birla et al. (2013) showed that there was a relationship between the release of the synthesizing biomolecules and the amount of biomass. Thus the successful synthesis of NPs requires an adequate balance between the amount of metal and the amount of organic material derived from the fungus.

10.4 Conclusion and Prospects

Living organisms used to obtain metal NPs with easy preparation protocols, less toxicity, and a wide variety of applications by size and shape. Further, seek should concentrate on an area that remains largely unexplored is the biosynthesis of metals and their oxide materials/NPs using microorganisms. Accordingly, there remain ample opportunities to explore new green preparedness strategies based on biogenic synthesis. Therefore it must be understanding the contribution of biological molecules in green NPs synthesis. Unfortunately, NPs biosynthesis has some difficulties such as poor product quality control, low production, biological cell contamination and hard segregation of NPs from biological materials. Understanding of the mechanisms associated with NPs formation is necessary to reproduce biomaterials and control morphology, size and dispersity. One of the main challenges in microbial nanobiosynthesis is controlling the dispersity of nanostructured materials, which greatly influences electronic and optical properties as well as plural form isolation and purification. Biosynthesis of microbial NPs could be improved by selecting suitable microbial strains (in terms of growth rate and biocatalytic activity), optimizing crop conditions and using genetic engineering tools could help overcome the disadvantages of slower production rates and polydispersity. The properties of nanoparticles differ substantially from similar macroparticles and from the substances were obtained from. These properties depend on their composition, size, nature, charge, shape, structural, characteristics, both the surface of the nanoproduct itself and the methods of preparation. Such unique features make nanoparticles more practical in biology and medicine. Likewise, for extending laboratory-based work to an industrial scale and to be applied extensively in the field of agriculture, environmental, pharmaceutical, food and cosmetic industries, there is need to know the interaction between nanoparticles and biological molecules such as proteins, carbohydrates and lipids. Future research must, therefore, focus on the various interactions between them. Also, it is promising to study the possibility of using one type of microorganism to synthesize nanoparticles of different elements. Furthermore, the subsequent studies should concern about production of nanoparticles for biocompatible substances, not as a final product but as feedstock. These have certain unique properties and can be used in many areas without side effects.

References

Abu-Tahon MA, Ghareib M, Abdallah WE (2020) Environmentally benign rapid biosynthesis of extracellular gold nanoparticles using *Aspergillus flavus* and their cytotoxic and catalytic activities. Process Biochem 95:1–36. https://doi.org/10.1016/j.procbio.2020.04.015

Agarwal P, Gupta R, Agarwal N (2019) Advances in synthesis and applications of microalgal nanoparticles for wastewater treatment. J Nanotechnol 2019: ID7392713. https://doi.org/10.1155/2019/7392713

Ağçeli GK, Hammachi H, Kodal SP et al (2020) A novel approach to synthesize TiO2 nanoparticles: Biosynthesis by using *streptomyces* sp. HC1. J Inorg Organomet Polym. https://doi.org/10.1007/s10904-020-01486-w

Ahiwale SS, Bankar AV, Tagunde S et al (2017) A bacteriophage mediated gold nanoparticles synthesis and their anti-biofilm activity. Indian J Microbiol 57:188–194. https://doi.org/10.1007/s12088-017-0640-x

Ahluwalia V, Kumar J, Sisodia R et al (2014) Green synthesis of silver nanoparticles by *Trichoderma harzianum* and their bio-efficacy evaluation against *Staphylococcus aureus* and *Klebsiella pneumonia*. Ind Crops Prod 55:202–206. https://doi.org/10.1016/j.indcrop.2014.01.026

Ahmad S, Munir S, Zeb N et al (2019) Green nanotechnology: a review on green synthesis of silver nanoparticles—an ecofriendly approach. Int J Nanomed 14:5087–5107. https://doi.org/10.2147/ijn.s200254

Ahmed T, Shahid M, Noman M et al (2020) Silver nanoparticles synthesized by using *Bacillus cereus* SZT1 ameliorated the damage of bacterial leaf blight pathogen in rice. Pathogens 9:1–17. https://doi.org/10.3390/pathogens9030160

Ahsan T (2020) Biofabrication of silver nanoparticles from *Pseudomonas fluorescens* to control tobacco mosaic virus. Egypt J Biol Pest Control 30:1–4. https://doi.org/10.1186/s41938020-00268-3

Ali J, Ali N, Wang L et al (2019) Revisiting the mechanistic pathways for bacterial mediated synthesis of noble metal nanoparticles. J Microbiol Methods 159:18–25

Allam NG, Ismail GA, El-Gemizy WM, Salem MA (2019) Biosynthesis of silver nanoparticles by cell-free extracts from some bacteria species for dye removal from wastewater. Bio technol Lett 41:379–389. https://doi.org/10.1007/s10529-019-02652-y

Ameen F, Abdullah M, Al-Homaidan A et al (2020) Fabrication of silver nanoparticles employing the cyanobacterium *Spirulina platensis* and its bactericidal effect against opportunistic nosocomial pathogens of the respiratory tract. J MolStruct 1217: https://doi.org/10.1016/j.molstruc.2020.128392

Arsiya F, Sayadi MH, Sobhani S (2017) Green synthesis of palladium nanoparticles using *Chlorella vulgaris*. Mater Lett 186:113–115. https://doi.org/10.1016/j.matlet.2016.09.101

Attia TMS, Elsheery NI (2020) Nanomaterials: Scope, applications, and challenges in agriculture and soil reclamation. In: Hayat S, Pichtel J, Faizan M, Fariduddin Q (eds) Sustainable agriculture reviews 41. Springer International Publishing, Cham, pp 1–39

Avilala J, Golla N (2019) Antibacterial and antiviral properties of silver nanoparticles synthesized by marine actinomycetes. Int J Pharm Sci Res 10:1223–1228. https://doi.org/10.13040/IJPSR.0975-8232

Ayano H, Kuroda M, Soda S et al (2015) Effects of culture conditions of *Pseudomonas aeruginosa* strain RB on the synthesis of Cd Se nanoparticles. J Biosci Bioeng 119:440–445

Azandehi PK, Moghaddam J (2015) Green synthesis, characterization and physiological stability of gold nanoparticles from *Stachys lavandulifolia* Vahl extract. Particuology 19:22–26

Azizi S, Ahmad MB, Namvar F et al (2014) Green biosynthesis and characterization of zinc oxide nanoparticles using brown marine macroalga *Sargassum muticum* aqueous extract. Mater Lett 116:275–277

Balachandran YL, Girija S, Selvakumar R et al (2013) Differently environment stable bio-silver nanoparticles: study on their optical enhancing and antibacterial properties. PLoS ONE 8(10):

Balakumaran MD, Ramachandran R, Kalaicheilvan PT (2015) Exploitation of endophytic fungus, *Guignardia mangiferae* for extracellular synthesis of silver nanoparticles and their in vitro biological activities. Microbiol Res 178:9–17. https://doi.org/10.1016/j.micres.2015.05.009

Banu A, Rathod V (2011) Synthesis and characterization of silver nanoparticles by *Rhizopus stolonier*. Int J Biomed Adv Res 2:148–158

Barabadi H, Tajani B, Moradi M et al (2019) Penicillium family as emerging nanofactory for biosynthesis of green nanomaterials: a journey into the world of microorganisms. J ClustSci 30:843–856. https://doi.org/10.1007/s10876-019-01554-3

Birla SS, Gaikwad SC, Gade AK et al (2013) Rapid synthesis of silver nanoparticles from *Fusarium oxysporum* by optimizing physicocultural conditions. Sci World J 2013: https://doi.org/10.1155/2013/796018

Costa Silva LP, Oliveira JP, Keijok WJ et al (2017) Extracellular biosynthesis of silver nanoparticles using the cell-free filtrate of nematophagus fungus *Duddingtonia flagans*. Int J Nanomed 12:6373–6381. https://doi.org/10.2147/IJN.S137703

da Silva Ferreira V, ConzFerreira ME, Lima LMT et al (2017) Green production of microalgae-based silver chloride nanoparticles with antimicrobial activity against pathogenic bacteria. Enzyme Microb Technol 97:114–121. https://doi.org/10.1016/j.enzmictec.2016.10.018

Dağlıoğlu Y, Öztürk BY (2019) A novel intracellular synthesis of silver nanoparticles using *Desmodesmus* sp. (Scenedesmaceae): different methods of pigment change. RendicontiLinceiSciFis E Nat 30:611–621. https://doi.org/10.1007/s12210-019-00822-8

Daima HK, Selvakannan PR, Kandjani AE et al (2014) Synergistic influence of polyoxometalate surface corona towards enhancing the antibacterial performance of tyrosine-capped Ag nanoparticles. Nanoscale 6(2):758–765

Das B, Dash SK, Mandal D et al (2017) Green synthesized silver nanoparticles destroy multidrug resistant bacteria via reactive oxygen species mediated membrane damage. Arab J Chem 10:862–876

Divya M, Kiran GS, Hassan S et al (2019) Biogenic synthesis and effect of silver nanoparticles (AgNPs) to combat catheter-related urinary tract infections. Biocatal Agric Biotechnol 18:1–8. https://doi.org/10.1016/j.bcab.2019.10103

Du L, Xu Q, Huang M et al (2015) Synthesis of small silver nanoparticles under light radiation by fungus *Penicillium oxalicum* and its application for the catalytic reduction of methylene blue. Mater ChemPhys 160:40–47. https://doi.org/10.1016/j.matchemphys.2015.04.003

Dubey K, Anand BG, Badhwar R et al (2015) Tyrosine-and tryptophan-coated gold nanoparticles inhibit amyloid aggregation of insulin. Amino Acids 47(12):2551–2560

Elamawi RM, Al-Harbi RE, Hendi AA (2018) Biosynthesis and characterization of silver nanoparticles using *Trichoderma longibrachiatum* and their effect on phytopathogenic fungi. Egypt J Biol Pest Control 28:28. https://doi.org/10.1186/s41938-018-0028-1

El-Naggar NE-A, Hussein MH, Shaaban-Dessuuki SA et al (2020) Production, extraction and characterization of *Chlorella vulgaris* soluble polysaccharides and their applications in AgNPs biosynthesis and biostimulation of plant growth. Sci Rep 10:1–19

El-Sayyad GS, Mosallam FM, El-Sayed SS et al (2020) Facile biosynthesis of tellurium dioxide nanoparticles by *Streptomyces cyaneus* melanin pigment and gamma radiation for repressing some *Aspergillus* pathogens and bacterial wound cultures. J ClustSci 31:147–159. https://doi.org/10.1007/s10876-019-01629-1

Faramarzi S, Anzabi Y, Jafarizadeh-Malmiri H (2020) Nanobiotechnology approach in intracellular selenium nanoparticle synthesis using *Saccharomyces cerevisiae*—fabrication and characterization. Arch Microbiol 202:1203–1209. https://doi.org/10.1007/s00203-020-01831-0

Fatemi M, Mollania N, Momeni-Moghaddam M et al (2018) Extracellular biosynthesis of magnetic iron oxide nanoparticles by *Bacillus cereus* strain HMH1: Characterization and in vitro cytotoxicity analysis on MCF-7 and 3T3 cell lines. J Biotechnol 270:1–11. https://doi.org/10.1016/j.jbiotec.2018.01.021

Fouda A, Hassan S, Abdo A et al (2019) Antimicrobial, antioxidant and larvicidal activities of spherical silver nanoparticles synthesized by endophytic *Streptomyces* spp. Biol Trace Elem Res 195:707–724. https://doi.org/10.1007/s12011-019-01883-4

Gahlawat G, Choudhury AR (2019) A review on the biosynthesis of metal and metal salt nanoparticles by microbes. RSC Adv 9:12944–12967. https://doi.org/10.1039/C8RA10483B

Gahlawat G, Shikha S, Chaddha BS et al (2016) Microbial glycolipoprotein-capped silver nanoparticles as emerging antibacterial agents against cholera. Microb Cell Fact 15:25. https://doi.org/10.1186/s12934-016-0422-x

Ghiuţă I, Cristea D, Croitoru C et al (2018) Characterization and antimicrobial activity of silver nanoparticles, biosynthesized using *Bacillus* species. Appl Surf Sci 438:66–73

Golhani DK, Khare A, Burra GK et al (2020) Microbes induced biofabrication of nanoparticles: a review. Inorg Nano-Met Chem. https://doi.org/10.1080/24701556.2020.1731539

Grasso G, Zane D, Dragone R (2020) Microbial nanotechnology: challenges and prospects for green biocatalytic synthesis of nanoscale materials for sensoristic and biomedical applications. Nanomaterials 10:11. https://doi.org/10.3390/nano10010011

Hamad MT (2019) Biosynthesis of silver nanoparticles by fungi and their antibacterial activity. Int J Environ SciTechnol 16:1015–1024. https://doi.org/10.1007/s13762-018-1814-8

Hari S (2020) Biosynthesis of nanoparticles from microorganisms. Res J Pharm Technol 13:2024–2028

Hassan SE-D, Fouda A, Radwan AA et al (2019) Endophytic actinomycetes *Streptomyces* spp mediated biosynthesis of copper oxide nanoparticles as a promising tool for biotechnological applications. J Biol Inorg Chem 24:377–393. https://doi.org/10.1007/s00775-019-01654-5

Hulikere MM, Joshi CG (2019) Characterization, antioxidant and antimicrobial activity of silver nanoparticles synthesized using marine endophytic fungus-*Cladosporium cladosporioides*. Process Biochem 82:199–204. https://doi.org/10.1016/j.procbio.2019.04.011

Husseiny SM, Salah TA, Anter HA (2015) Biosynthesis of size controlled silver nanoparticles by *Fusarium oxysporum*, their antibacterial and antitumoral activities. Beni Suef Univer J Basic ApplSci 4:225–231. https://doi.org/10.1016/j.bjbas.2015.07.004

Jain N, Bhargava A, Majumdar S et al (2011) Extracellular biosynthesis and characterization of silver nanoparticles using *Aspergillus flavus* NJP08: a mechanism perspective. Nanoscale 3(2):635–641

Jalal M, Ansari MA, Alzohairy MA et al (2018) Biosynthesis of silver nanoparticles from oropharyngeal *Candidaglabrata* isolates and their antimicrobial activity against clinical strains of bacteria and fungi. Nanomaterials 8:1–12. https://doi.org/10.3390/nano8080586

Kalimuthu K, Babu RS, Venkataraman D et al (2008) Biosynthesis of silver nanocrystals by *Bacillus licheniformis*. Colloids Surf B Biointerfaces 65(1):150–153

Kaviya S, Santhanalakshmi J, Viswanathan B (2011) Green synthesis of silver nanoparticles using *Polyalthia longifolia* leaf extract along with D-sorbitol: study of antibacterial activity. J Nanotechnol 2011: https://doi.org/10.1155/2011/152970

Kobayashi M, Tomita S, Sawada K et al (2012) Chiral meta-molecules consisting of gold nanoparticles and genetically engineered tobacco mosaic virus. Opt Express 20:24856–24863

Kunoh T, Takeda M, Matsumoto S et al (2018) Green synthesis of gold nanoparticles coupled with nucleic acid oxidation. ACS Sustain Chem Eng 6:364–373

Lachance MA (2016) Paraphyly and (yeast) classification. Int J Syst Evol Microbiol 66:4924–4929

Le DH, Lee KL, Shukla S et al (2017) Potato virus X, a filamentous plant viral nanoparticle for doxorubicin delivery in cancer therapy. Nanoscale 9:2348–2357

Li L, Hu Q, Zeng J et al (2011) Resistance and biosorption mechanism of silver ions by *Bacillus cereus* biomass. J Environ Sci 23(1):108–111

Lv Q, Zhang B, Xing X et al (2018) Biosynthesis of copper nanoparticles using *Shewanella loihica* PV-4 with antibacterial activity: Novel approach and mechanisms investigation. J Hazard Mater 347:141–149

Mahanty S, Bakshi M, Ghosh S et al (2019) Green synthesis of iron oxide nanoparticles mediated by filamentous fungi isolated from Sundarban Mangrove ecosystem, India. BioNanoSci 9:637–651. https://doi.org/10.1007/s12668-019-00644-w

Mishra V, Arya A, Chundawat TS (2020) High catalytic activity of Pd nanoparticles synthesized from green alga *Chlorella vulgaris* in buchwald-hartwig synthesis of N-aryl piperazines. Curr Organo Catalysis 7:23–33

Murphy CJ (2002) Materials science: nanocubes and nanoboxes. Science 298:2139–2141

Nadaroğlu H, Güngör AA, İnce S (2017) Synthesis of nanoparticles by green synthesis method. J Innov Res Rev 1(1):6–9

Nam KT, Kim D-W, Yoo PJ et al (2006) Virus-enabled synthesis and assembly of nanowires for lithium ion battery electrodes. Science 312:885–888

Nangia Y, Wangoo N, Goyal N et al (2009) A novel bacterial isolate *Stenotrophomonas maltophilia* as living factory for synthesis of gold nanoparticles. Microb Cell Fact 8:39. https://doi.org/10.1186/1475-2859-8-39

Nayak RR, Pradhan N, Behera D et al (2011) Green synthesis of silver nanoparticle by *Penicillium purpurogenum* NPMF: the process and optimization. J Nanopart Res 13:3129–3137. https://doi.org/10.1007/s11051-010-0208-8

Noman M, Shahid M, Ahmed T et al (2020) Use of biogenic copper nanoparticles synthesized from a native *Escherichia* sp. as photocatalysts for azo dye degradation and treatment of textile effluents. Environ Pollut 257:1–34. https://doi.org/10.1016/j.envpol.2019.113514

Noshad A, Hetherington C, Iqbal M (2019) Impact of AgNPs on seed germination and seedling growth: A focus study on its antibacterial potential against *Clavibacter michiganensis* subsp. *michiganensis* infection in *Solanum lycopersicum*. J Nanomaterials 2019: ID 6316094. https://doi.org/10.1155/2019/6316094

Noshad A, Iqbal M, Hetherington C, Wahab H (2020) Biogenic AgNPs—A Nano Weapon against Bacterial Canker of Tomato (BCT). Adv Agric 2020:1–10. https://doi.org/10.1155/2020/9630785

Omar R, Afreen S, Talreja N et al (2019) Impact of nanomaterials on the microbial system. In: Prasad R (ed) Microbial nanobionics. Nanotechnology in the Life Sciences, Springer, Cham, pp 141–158

Phanjom P, Ahmed G (2017) Effect of different physicochemical conditions on the synthesis of silver nanoparticles using fungal cell filtrate of *Aspergillus oryzae* (MTCC No. 1846) and their antibacterial effects. Adv Nat Sci Nanosci Nanotechnol 8:1–13. https://doi.org/10.1088/2043-6254/aa92bc

Pimprikar PS, Joshi SS, Kumar AR et al (2009) Influence of biomass and gold salt concentration on nanoparticle synthesis by the tropical marine yeast *Yarrowia lipolytica* NCIM 3589. Colloids Surf B Biointerfaces 74:309–316

Prasad R, Pandey R, Barman I (2016) Engineering tailored nanoparticles with microbes: quo vadis. WIREs Nanomed Nanobiotechnol 8:316–330. https://doi.org/10.1002/wnan.1363

Rajeshkumar S (2018) Synthesis of zinc oxide nanoparticles using algal formulation (*Padina tetrastromatica* and *Turbinaria conoides*) and their antibacterial activity against fish pathogens. Res J Biotechnol 13:15–19

Ramanathan R, O'Mullane AP, Parikh RY et al (2010) Bacterial kinetics- controlled shape- directed biosynthesis of silver nanoplates using *Morganella psychrotolerans*. Langmuir 27:714–719

Ramya S, Shanmugasundaram T, Balagurunathan R (2015) Biomedical potential of actinobacterially synthesized selenium nanoparticles with special reference to anti-biofilm, anti-oxidant, wound healing, cytotoxic and anti-viral activities. J Trace Elem Med Biol 32:30–39

Ranjitha VR, Ravishankar VR (2018) Extracellular synthesis of selenium nanoparticles from an actinomycetes *Streptomyces griseoruber* and evaluation of its cytotoxicity on HT-29 cell line. Pharm Nanotechnol 6:61–68

Rao PV, Gan SH (2015) Recent advances in nanotechnology-based diagnosis and treatments of diabetes. Curr Drug Metab 16:371–375

Ribeiro JJK, da Silva Porto PS, Pereira RD et al (2020) Green synthesis of nanomaterials: most cited papers and research trends. Res Soc Dev 9:1–20. https://doi.org/10.33448/rsd-v9i1.1593

Roy A, Bulut O, Some S et al (2019) Green synthesis of silver nanoparticles: biomolecule-nanoparticle organizations targeting antimicrobial activity. RSC Adv 9:2673–2702

Salaam A, Adebayo-Tayo BC, Ajibade A (2020) Phycosynthesis of silver nanoparticles using *Chlorella vulgaris* metabolites: its antibacterial, anti-biofilm and in-vitro cytotoxicity potential and effect of optimized conditions on biosynthesis: *Chlorella vulgaris* silver nanoparticles. Afr J Biomed Res 23:17–23

Salah NS, Muhsen TA, Risan MH (2020) Antifungal activity of silver nanoparticles using *Penicillium chrysogenum* extract against the formation of biofilm for *candida glabrata*. Indian J Forensic Med Toxicol 14:306–311

Salem SS, Fouda MM, Fouda A et al (2020) Antibacterial, cytotoxicity and larvicidal activity of green synthesized selenium nanoparticles using *Penicillium corylophilum*. J Clust Sci. https://doi.org/10.1007/s10876-020-01794-8

San Diego KD, Alindayu JIA, Baculi RQ (2020) Biosynthesis of gold nanoparticles by bacteria from hyperalkaline spring and evaluation of their inhibitory activity against pyocyanin production. J Microbiol Biotechnol Food Sci 9:781–787

Sangappa M, Thiagarajan P (2012) Mycobiosynthesis and characterization of silver nanoparticles from *Aspergillus niger*: a soil fungal isolate. Int J Life Sci Biotechnol Pharma Res 1:282–289

Saxena J, Sharma PK, Sharma MM et al (2016) Process optimization for green synthesis of silver nanoparticles by *Sclerotinia sclerotiorum* MTCC 8785 and evaluation of its antibacterial properties. SpringerPlus 5:861. https://doi.org/10.1186/s40064-016-2558-x

Selvakannan PR, Ramanathan R, Plowman BJ et al (2013) Probing the effect of charge transfer enhancement in off resonance mode SERS via conjugation of the probe dye between silver nanoparticles and metal substrates. Phys Chem Chem Phys 15(31):12920–12929

Shahzad A, Saeed H, Iqtedar M (2019) Size-controlled production of silver nanoparticles by *Aspergillus fumigatus* BTCB10: likely antibacterial and cytotoxic effects. J Nanomater 2019:5168698. https://doi.org/10.1155/2019/5168698

Shankar PD, Shobana S, Karuppusamy I et al (2016) A review on the biosynthesis of metallic nanoparticles (gold and silver) using bio-components of microalgae: Formation mechanism and applications. Enzyme Microb Tech 95:28–44. https://doi.org/10.1016/j.enzmictec.2016.10.015

Shedbalkar U, Singh R, Wadhwani S et al (2014) Microbial synthesis of gold nanoparticles: Current status and future prospects. Adv Colloid Interface Sci 209:40–48. https://doi.org/10.1016/j.cis.2013.12.011

Siddiqi KS, Husen A, Rao RA (2018) A review on biosynthesis of silver nanoparticles and their biocidal properties. J Nanobiotechnol 16(1):1–14. https://doi.org/10.1186/s12951-018-0334-5

Sidkey NM, Arafa RA, Moustafa YM et al (2020) Biosynthesis of Mg and Mn intracellular nanoparticles via extremo-Metallotolerant *Pseudomonas stutzeri*, B4 Mg/W and *Fusarium nygamai*, F4 Mn/S. J Microbiol Biotechnol Food Sci 9:1181–1187. https://doi.org/10.15414/jmbfs.2017.6.5.1181-1187

Singh J, Dutta T, Kim KH et al (2018) Green synthesis of metals and their oxide nanoparticles: applications for environmental remediation. J Nano biotechnol 16:1–84. https://doi.org/10.1186/s12951-018-0408-4

Singh OV (2015) Bio-nanoparticles: biosynthesis and sustainable biotechnological implications. Wiley, New York

Singh P, Kim Y-J, Zhang D et al (2016) Biological synthesis of nanoparticles from plants and microorganisms. Trends Bio Technol 34:588–599

Singh VK, Singh AK (2019) Role of microbially synthesized nanoparticles in sustainable agriculture and environmental management. In: Kumar A, Singh AK, Choudhary KK (eds) Role of plant growth promoting microorganisms in sustainable agriculture and nanotechnology. Wood head Publishing, pp 55–73

Sintubin L, Windt WD, Dick J et al (2009) Lactic acid bacteria as reducing and capping agent for the fast and effcient production of silver nanoparticles. Appl Microbiol Biotechnol 84:741–749. https://doi.org/10.1007/s00253-009-2032-6

Skalickova S, Baron M, Sochor J (2017) Nanoparticles biosynthesized by yeast: A review of their application. KVASNÝ PR\UUMYSL 63:290–292 https://doi.org/10.18832/kp201727

Sowbarnika R, Anhuradha S, Preetha B (2018) Enhanced antimicrobial effect of yeast mediated silver nanoparticles synthesized from baker's yeast. Int J Nanosci Nanotechnol 14:33–42

Sriramulu M, Sumathi S (2018) Biosynthesis of palladium nanoparticles using *Saccharomyces cerevisiae* extract and its photocatalytic degradation behaviour. Adv Nat Sci Nanosci Nanotechnol 9:1–6. https://doi.org/10.1088/2043-6254/aac506

Suresh AK, Pelletier DA, Wang W et al (2011) Biofabrication of discrete spherical gold nanoparticles using the metal-reducing bacterium *Shewanella oneidensis*. Acta Biomater 7:2148–2152

Tsekhmistrenko S, Bityutskii V, Tsekhmistrenko O et al (2020) Bacterial synthesis of nanoparticles: A green approach. Biosystems Diversity 28(1):9–17. https://doi.org/10.15421/012002

Vairavel M, Devaraj E, Shanmugam R (2020) An eco-friendly synthesis of *Enterococcus* sp.–mediated gold nanoparticle induces cytotoxicity in human colorectal cancer cells. Environ SciPollut Res 27:8166–8175

Velusamy P, Kumar GV, Jeyanthi V et al (2016) Bio-Inspired green nanoparticles: synthesis, mechanism, and antibacterial application. Toxicol Res 32:95–102

Voeikova TA, Zhuravliova OA, Bulushova NV et al (2017) The protein corona of silver-sulfide nanoparticles obtained using gram-negative and-positive bacteria. Mol Genet Microbiol Virol 32:204–211

Wadhwani SA, Gorain M, Banerjee P et al (2017) Green synthesis of selenium nanoparticles using *Acinetobacter* sp. SW30: optimization, characterization and its anticancer activity in breast cancer cells. Int J Nanomedicine 12:6841–6855

Wypij M, Czarnecka J, Świecimska M et al (2018) Synthesis, characterization and evaluation of antimicrobial and cytotoxic activities of biogenic silver nanoparticles synthesized from *Streptomyces xinghaiensis* OF1 strain. World J Microbiol Biotechnol 34:23–34. https://doi.org/10.1007/s11274-017-2406-3

Xue B, He D, Gao S et al (2016) Biosynthesis of silver nanoparticles by the fungus *Arthroderma fulvum* and its antifungal activity against genera of *Candida*, *Aspergillus* and *Fusarium*. Int J Nanomed 11:1899–1906. https://doi.org/10.2147/IJN.S98339

Yılmaz Öztürk B, Yenice Gürsu B, Dağ İ (2020) Antibiofilm and antimicrobial activities of green synthesized silver nanoparticles using marine red algae *Gelidium corneum*. Process Biochem 89:208–219. https://doi.org/10.1016/j.procbio.2019.10.027

Yin Y, Yang X, Hu L et al (2016) Superoxide-mediated extracellular biosynthesis of silver nanoparticles by the fungus *Fusarium oxysporum*. Environ Sci Technol Lett 3(4):160–165

Yu J, Xu D, Guan HN et al (2016) Facile one-step green synthesis of gold nanoparticles using *Citrus maxima* aqueous extracts and its catalytic activity. Mater Lett 166:110–112

Yusof HM, Mohamad R, Zaidan UH (2020) Sustainable microbial cell nanofactory for zinc oxide nanoparticles production by zinc-tolerant probiotic *Lactobacillus plantarum* strain TA4. Microb Cell Factories 19:1–17. https://doi.org/10.1186/s12934-020-1279-6

Chapter 11
Utilization of Nanofertilizers in Crop Tolerance to Abiotic Stress

Khaled F. M. Salem, Maysaa T. Alloosh, Maysoun M. Saleh, Lina M. Alnaddaf, Abdulsalam K. Almuhammady, and Jameel M. Al-Khayri

Abstract Abiotic stresses severely affect plant growth, development, production and quality. These stresses are the main reason for decreased productivity worldwide accompanied by an increasing human population. This necessitates developing novel solutions to achieve sustainability and overcome these problems. Recently, a new era has begun to emerge, the era of nanotechnology. It improved the ability crops to deal with abiotic stress and primary or secondary metabolic function. The present chapter provides insight on the relationship between abiotic stresses and nanotechnology together with nanofertilizers, their characteristics/roles as well as their comparison

---

K. F. M. Salem (✉)
Department of Biology, College of Science and Humanitarian Studies, Shaqra University, Qwaieah, Saudi Arabia
e-mail: khaled.salem@gebri.usc.edu.eg; khaledfathi@su.edu.sa

Department of Plant Biotechnology, Genetic Engineering and Biotechnology Research Institute (GEBRI), University of Sadat City, Sadat City, Egypt

M. T. Alloosh
Fungal of Biochemistry and Microbiology, Faculty of Pharmacy, Al-Baath University, Homs, Syria
e-mail: malloosh@albaath-univ.edu.sy

M. M. Saleh
Genetic Resources Department, General Commission for Scientific Agricultural Research (GCSAR), Damascus, Syria
e-mail: mzainsamasaleh@gcsar.gov.sy

L. M. Alnaddaf
Biotechnology and Molecular Biology, Faculty of Agricultural, Al-Baath University, Homs, Syria
e-mail: lalnaddaf@albaath-univ.edu.sy

A. K. Almuhammady
Arab Center for Nanotechnology, Cairo, Egypt
e-mail: lalnaddaf@albaath-univ.edu.sy

J. M. Al-Khayri
Department of Agricultural Biotechnology, King Faisal University, Al Hofuf, Al-Ahsa, Saudi Arabia
e-mail: jkhayri@kfu.edu.sa

J. M. Al-Khayri et al. (eds.), *Nanobiotechnology*,
https://doi.org/10.1007/978-3-030-73606-4_11

with conventional fertilizers. Moreover, this chapter not only highlights the interaction between NPs and plants on several growth stages, but also the importance of nanotechnology in the purification of irrigation water, the effect of different nanoparticles on crops cultivated under abiotic stress with their possible toxicity impact on plants.

Keywords Abiotic stress · Biosynthesis · Mitigation of abiotic stresses · Nanofertilizer · Nanoparticles · Sustainability development

11.1 Introduction

Plants are constantly subjected to various abiotic stress, from the germination of seeds through to the entire life cycle. These stresses are generally divided into two groups, biotic (biological) and abiotic (non-biological). The most significant abiotic stresses are drought, salinity, temperatures and low soil fertility which limit agricultural productivity, almost everywhere in the world (Hussain et al. 2020; Linh et al. 2020; Nawara et al. 2017; Salem and Matter 2014; Salem et al. 2007; Savita and Singh 2020; Sun et al. 2020; Zhao et al. 2020). Among these problems, drought and nutrient deficiency are major problems, mostly in developing countries, where the income of rural people depends on agriculture (Verma 2016). Farooq et al. (2009) reported that drought caused the yield to decrease between 13 and 94% in many crops. Depending on future scenarios, adaptation and mitigation are necessary to increase the resilience of agricultural systems and ensure crop productivity and quality (Mariani and Ferrante 2017). Food production needs to be increased to meet the demands of an increasing population (Population Institute 2017), so there is a need to use new methods in small concentrations to overcome all these challenges.

Nanotechnology seems to be the best solution (Panwar et al. 2012). As a result of their distinctive properties, NPs emphasize crop growth even in harsh conditions (Giraldo et al. 2014). In general, NPs are materials with dimensions ranging from 1 to 100 nm (Ali et al. 2017; Ball 2002; Khan et al. 2017). The size of NPs that passes and accumulates within plant cells is often between 40 and 50 nm (Corredor et al. 2009; González-Melendi et al. 2008; Sabo-Attwood et al. 2012; Taylor et al. 2014). It happens in the transition region between individual and bulk materials, and hence the NPs have uniquely different properties from their molecular and bulk counterpart (Campbell et al. 1996; Kroto et al. 1985; Taylor and Walton 1993). However, NPs shape and chemical composition are factors that affect absorption (Ma et al. 2010; Pérez-de-Luque 2017; Rico et al. 2011), while in some cases morphology is considered specific (Cunningham et al. 2018; Pérez-de-Luque 2017; Raliya et al. 2016). Nanomaterial versatility and surface coating can change and modify the properties of plant absorption and accumulation (García-Gómez et al. 2018; Judy et al. 2012; Moon et al. 2016; Song et al. 2013; Vidyalakshmi et al. 2017; Zhu et al. 2012). As a result, nanomaterials have special properties such as large specific surface area, high surface energy and quantitative restriction. These improve the solubility properties of biologically active components in the cellular matrix. The use of nanomaterials in agriculture improves agricultural practices efficiency and sustainability by adding fewer chemicals and creating less waste compared to traditional products and approaches.

More interest is devoted to the role of nano-fertilizers in rising nutrient use efficiency and purification of water (Rasouli et al. 2013), in addition to their positive roles in reducing the negative effects on a plant caused by different abiotic stresses, resulting in rising crop yields. Defining the exact benefit concentration of NPs is very important to avoid their toxicity on plants Landa et al. (2012), which differs according to the type of NPs itself, plant stage during the application and nanoparticle concentration.

This chapter presents an overview of the applications of nanoparticles under abiotic stress as nanofertilizers, characteristics of nanofertilizer, responses of different crop growth stages to nanoparticles, purification of irrigation water, the toxicity of nanomaterials to crops and its effects on different growth stages of plants under abiotic stresses.

11.2 Nano Fertilizers

11.2.1 Nanofertilizers Role

Plants essentially require sunlight, water (H_2O), carbon dioxide (CO_2) and numerous chemical elements to grow and develop. The plant may acquire chemical elements from the soil through roots or shoots between those components (Marschner 1995). Those obtained from the soil are considered nutrients in mineral form. Some gaseous form mineral nutrients, ammonia (NH_3) or sulfur dioxide (SO_2) join the leaves through the stomata. Carbon (C), hydrogen (H) and oxygen (O) are CO_2 and H_2O derived and are not regarded as mineral nutrients. Of sixteen critical plant growth elements, those needed at low concentrations are referred to as micronutrients (<0.5 g/kg of plant dry weight), iron (Fe), copper (Cu), zinc (Zn), manganese (Mn), boron (B), molybdenum (Mo), nickel (Ni), sodium (Na), chlorine (Cl) and silicon (Si) and those necessary at high concentrations are referred to as macronutrients (>0.5 g/kg of plant dry weight), nitrogen (N), phosphorus (P), potassium (K), magnesium (Mg), calcium (Ca) and sulfur (S) (Fig. 11.1). After reaching the plant cell, its metabolism involves the translocation of mineral nutrients to different locations. The list of all macro and micronutrients and their roles and deficiency symptoms within the plant system are summarized in (Tables 11.1 and 11.2). It can apply chemical fertilization to provide the plant with the macro and micronutrients it needs. This chemical fertilization may increase pollution of the environment once overuse. These include leaching, runoff, emission and eutrophication of aquatic ecosystems (Adesemoye and Kloepper 2009; Adesemoye et al. 2009; Flessa et al. 2002; Ma et al. 2007; Vessey 2003; Yang et al. 2009).

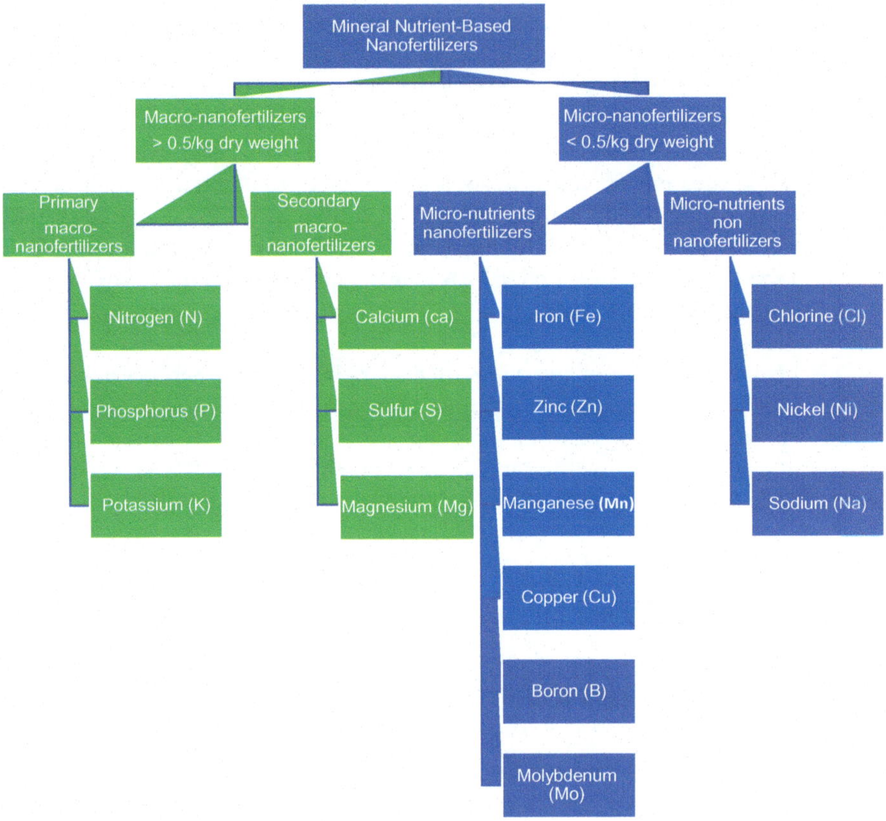

Fig. 11.1 List of all macro and micronutrients (Constructed by K.F.M. Salem)

Nanofertilizers can surpass conventional fertilizers, as nutrients are encapsulated/coated with a thin film of nanomaterials in nanofertilizers or provided as emulsions or as NPs (Derosa et al. 2010). Nanomaterials, particularly metallic and carbon-based NPs, are manufactured for improved crop growth and development and high yield. Also, these are used for assimilation purposes, nutrient translocation and compound storage (Nair et al. 2010). Positive effects of nanotechnology have been documented on many crops, including wheat (*Triticum aestivum* L.) (Storozhenko et al. 2019), soybean (*Glycine max* L.) (Agrawal and Rathore 2014; Linh et al. 2020), spinach (*Spinacia oleracea* L.) (Gao et al. 2006), maize (*Zea mays* L.) (Sun et al. 2020) and peanut (*Arachis hypogaea* L.) (Giraldo et al. 2014), both physiological and morphological. Some impacts of nano-fertilizers can be through: (a) Increase germination level by raising its ability to absorb water, (b) Improve root and shoot length, (c) Establish vegetative biomass, (d) Greater photosynthetic efficiency due to increased light retention and light diffusion in the plant and (e) Assimilation of nitrogen enhanced by nitrate reductase action.

Table 11.1 Role of different macronutrients and their roles and their scarcity indicators

Mineral	Mineral available type	Mineral roles	Scarcity indicators
Nitrogen (N)	Found in soils in organic type (98%), and both of (immo^{-4} bile) NH^{+}, and (highly mobile) NO_3 types as inorganic types. When plants are harvested, nitrogen is often lost (Preetha and Balakrishnan 2017)	An essential element of genetic material and proteins, so plant require it in high quantity	Falling yellowish leaves, stunted growth, low Protein level, poor fruit development
Phosphorous (P)	Found in soils as organic and mineral types. According to Preetha and Balakrishnan (2017), Plants attain Phosphorous as HPO_4 and H_2PO_4 which are found in small quantities in soils	Essential element membranes, nucleic acid, ATP and phospholipids	Uncompleted growth, Necrotic spots and leaves changes to purple
Potassium (K)	Preetha and Balakrishnan (2017) mentioned that potassium consider as an essential mineral in plants which play the main role in biochemical and Physiological processes to keep a plant alive, since it maintains water use efficiency, enzymatic functions, synthesis of protein, photosynthesis and carbohydrates translocation	Enhance cell turgor pressure, increase plant resistance to some diseases, maintain photosynthesis enzymes, play a role in proteins, starch and simple sugar synthesis, carbohydrates translocation, stomatal movements and nitrates reduction	Mottle spots lead to increase of plant susceptibility to diseases
Calcium (Ca)	Found as Ca^{+2} ions	Control membrane function and structure, and play role in cell division, cell wall synthesis and intracellular messenger in the cytosol	If deficient, leads to curled leaves, reduction in root growth, cracked fruits and insufficient fruit storage
Magnesium (Mg)	Found as Mg^{+2} (divalent cation), chlorophyll molecule element	Many functions in plant respiration, photosynthesis and other biochemical and physiological processes, and enzymes activations	Leaves Chlorosis, mostly in the older ones

(continued)

Table 11.1 (continued)

Mineral	Mineral available type	Mineral roles	Scarcity indicators
Sulfur (S)	Found in many types: basic sulfur (S), sulfides (S^-), and sulfate (SO_2^-) types in 4 soil where SO_2^- types are 4 up taken by plants	The element of vitamin A and amino acids	Reduction in the synthesis of protein and content of leaf chlorophyll

Source Solanki et al. (2015)

11.2.2 Characteristics of Nanofertilizer

Some of the properties of nanofertilizers that make it useful and beneficial over conventional fertilizers are:

(a) Higher surface area: Improving nutrient absorption and nutrient efficiency. This is a result of increased surface area and enhancing nanofertilizer reactivity with other compounds.
(b) High solubility: The nanofertilizer is highly soluble in a variety of solvents including water. This nanofertilizer property assists in the solubilization and dispersion of insoluble nutrients in the soil and thus increases nutrient bioavailability (Singh et al. 2017).
(c) Small particle size: The particle size of nanofertilizers is less than 100 nm, which increases the ability of nanofertilizers to penetrate plants from applied surfaces such as soil or leaves and thus increases plant nutrient absorption (Liscano et al. 2000).
(d) Fertilizer encapsulation at NPs: There is an increase in nutrient availability and uptake in crop plants (Tarafdar et al. 2012a, b), zeolite-based nanofertilizers increase nutrient availability in crops during the growth cycle and prevent nutrient loss through denitrification, volatilization and leaching and fix nutrient in the soil.
(e) Fast penetration and controlled fertilizer release: Due to the high penetration rate, nanofertilizers play an important role in increased plant nutrient availability, subsequently, healthy growth of seedlings and minimizes fertilizer toxicity. Nano-ZnO usage provides a higher percentage of germination and root growth peanut seed compared to zinc sulfate in bulk (Prasad et al. 2012).
(f) Nanofertilizers efficiency of nutrient absorption: This improves the proportion of soil nutrient absorption in crop production. Also, nano-fertilizers reduce fertilizer leaching loss (Cui et al. 2010).
(g) Nanofertilizers release period: The activated nutrient release period in bulk fertilizers are effective in the short-term; however, the nutrient release period may be extended with nanofertilizers.

Table 11.2 Role of different micronutrients and their roles and their deficiency symptoms

Mineral	Mineral nutrient and its availability	Physiological role	Mineral deficiency symptoms
Iron (Fe)	Present as Fe^{+2} (ferrous) and Fe^{+3} (ferric) ions	Involved in redox reactions, required for the synthesis of chloroplast protein complexes in chloroplast	Intravenous chlorosis, whitening of leaves
Zinc (Zn)	Present as Zn^{+2} ions	An important element of many enzymes (alkaline phosphatase, alcohol dehydrogenase and carbonic anhydrase), consider as a structural factor of ribosomes, enhance the bio-membranes integrity	Reduction in internodes, and leave area which leads to stunted growth, death of shoot apices when zinc is extremely insufficient
Manganese (Mn)	Ions of Mn^{+2}	The enzymes of photosynthetic reactions and Krebs cycle are activated by manganese	Necrotic spots with chlorosis between veins
Copper (Cu)	Found as Cu^{+2}	Bound with enzymes of redox reactions (plastocyanin)	Dark green leaves, necrotic spots arising from the tip and extending toward the margin
Silicon (Si)	Found in soils as SiO_2	The hydrated amorphous form of silica ($SiO_2 \cdot NH_2O$) plays an important role in the cell wall, intercellular spaces and endoplasmic reticulum	Fungal infection and lodging susceptibility are increased
Chlorine (Cl)	(Cl^-) chlorine ions	Necessary for both photosynthesis and cell division	Leads rarely to wilted leaf and later necrosis and chlorosis
Boron (B)	Found as boric acid (H_3BO_3) and borate (H_2BO_3)	Enhance many processes like cell elongation, synthesis of nucleic acid and membrane functions, regulate cell cycle	Immature leaves with black necrosis, loss of apical dominance is lost gradually leads to raise the number of branches

(continued)

Table 11.2 (continued)

Mineral	Mineral nutrient and its availability	Physiological role	Mineral deficiency symptoms
Sodium (Na)	Found as Na^+ ion	Stimulates growth by affecting cell expansion and water balance of plants replaces potassium (K^+) as a solute, participates in C4 and crassulacean acid meta bolism CAM pathways	Chlorosis, necrosis
Molybdenum (Mo)	Found as Ions of MoO_4^-	Play role in assimilation of nitrate and fixation of nitrogen via being a constituent of nitrogenase and nitrate reductase enzymes	Necrosis, abscission of flowers and chlorosis
Nickel (Ni)	Predominantly found as Ni^{+2}	Urease component	Urea accumulates in leaf and then necrosis

Source Solanki et al. (2015)

11.2.3 Comparison Between Nanofertilizers and Conventional Fertilizers

Nanofertilizers are utilized to increase nutrient efficiency and to reduce pollution (Chinnamuthu and Boopathi 2009). Using nanofertilizers appears to be an important option for satisfying the nutritional needs of field crops. This provides for the gradual release of plant-absorbed nutrients. Nanofertilizer types include nano-coated nutrient encapsulations, or nano-material emulsions (DeRosa et al. 2010). NPs are more efficient in nutrient retention because of their unique surface properties than those of the usual materials used in chemical fertilizer manufacturing (Sasson et al. 2007) (Table 11.3).

11.3 Responses of Crop Growth Stages to Nanoparticles

The interaction between NPs and plants causes several modifications in plant structures (Cox et al. 2017) which differ according to some factors like NPs efficient concentration, reactivity and plant stage (Khodakovskaya et al. 2012).

Table 11.3 Comparison of the use of conventional fertilizers and nanofertilizers

No	Properties	Nano-fertilizers	Conventional fertilizers
1	Solubility	High	Low
2	Adsorption capacity	Low	High
3	Bioavailability	High	Low
4	Nutrient uptake efficiency	High	Low
5	Release of nutrients	Slow	Rapid
6	Loss rate	Low	High

Source Miransari (2011)

11.3.1 Germination

The usage of iron oxide (Fe_2O_3) NPs enhances the root germination and growth of plants (Liu et al. 2016; Sarvendra-Kumar et al. 2015). Also, lettuce and cucumber germination rate were increased by iron oxide NPs usage (Antisari et al. 2015), as well as for each of barley and ryegrass (El-Temsah and Joner 2012). The NPs of Fe_2O_3 reduced the roots conductivity in sunflower (*Helianthus annuus* L.) (Martínez-Fernández et al. 2015). The treatment of maize with TiO_2 NPs resulted in better root growth (Andersen et al. 2016). Boonyanitipong et al. (2011) reported that titanium dioxide (TiO_2) NPs had a better positive effect than zinc oxide (ZnO) NPs on rice roots number and root length. Ruffini Castiglione et al. (2011) mentioned that grains of maize treated with high doses of TiO_2 NPs solutions caused root elongation. Also, Feizi et al. (2012) declared that TiO_2 NPs in specific concentrations supported wheat grains germination. Seedling development of broad bean (*Vicia faba* L.) was enhanced with the applying of TiO_2 NPs (Ruffini Castiglione et al. 2016). Larue et al. (2013) studied the effects of TiO_2 NPs on wheat plant growth, they found that root elongation was motivated. Mattiello et al. (2015) found that the barley germination was not influenced by applying cerium oxide (CeO_2) NPs even at high levels, and the roots elongation of the treated seedling was reduced (Fig. 11.2). López-Moreno et al. (2010b) reported a reduction in root elongation in alfalfa (*Medicago sativa* L.) and the germination of corn plants treated with moderate concentrations of CeO_2 NPs. According to López-Moreno et al. (2010a), the root growth of soybean plantlets increased in treated plants with cubic CeO_2 NPs.

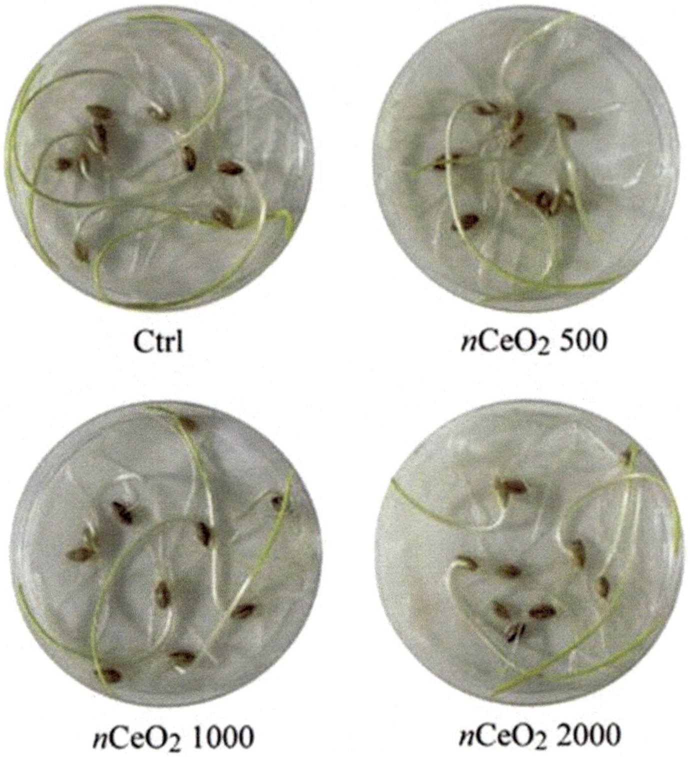

Fig. 11.2 Impact of CeO_2 NPs on barley germination (*Source* Mattiello et al. [2015])

11.3.2 Vegetative Stage

Plant shoot growth can be improved when using iron oxide NPs (Wang et al. 2011). Scientist results indicated a positive impact of applying iron oxide NPs in specific concentrations for vegetative growth in many plants (Bombin et al. 2015). Results of Asli and Neumann (2009) indicated negative effects on leave growth and transpiration after applying TiO_2 NPs to maize. An investigation about the impact of using TiO_2 NPs on mung bean (*Vigna radiata* L.) revealed a significant increase in plant growth, stem length, proteins and chlorophyll content (Raliyaa et al. 2015). Some scientists examined the potential impact of TiO_2 NPs relating to symbiotic systems in medic and pea plants responsible for nitrogen fixation which leads to improving total biomass (Chen et al. 2015; Fan et al. 2014). Chlorophyll content in corn plants was increased by using Fe_2O_3 NPs (Li et al. 2016). The usage of Fe_2O_3 in transgenic wheat led to an increase in the antioxidant enzyme activities (Gui et al. 2015). The chlorophyll content, plant height, total biomass and the accumulated nutrients (copper, zinc, magnesium, iron, phosphorus and potassium) were all increased in barley treated with CeO_2 NPs (Rico et al. 2015a).

11.3.3 Reproduction Stage

Total biomass of many crops like wheat, rice, soybean and peanut, is raised as a result of applying iron oxide NPs (Chittaranjan et al. 2013). Regarding barley, the result revealed increasing the synthesis of both lysine and proline due to the formation of the high level of amino acids in grains grown in soil amended with a specific concentration of CeO_2 NPs, while the higher concentration of CeO_2 NPs harmed grains formation (Rico et al. 2015b), accompanying with the reduction of amylase content (Poščić et al. 2016). Changes in wheat nutritional value concerning the number and size of the endosperm starch granules confirmed the impact of CeO_2 NPs on final wheat quality (Du et al. 2015). Regarding maize (*Zea mays* L.) the final yield (quantity and quality) was negatively affected by adding CeO_2 NPS (Zhao et al. 2015). The results of Majumdar et al. (2015) clarify the possible negative effects of CeO_2 NPs on kidney beans quality.

11.4 Purification of Irrigation Water

Inadequate sources of pure water, with the increasing demand for some sectors to water, affect agriculture sustainability. Applying new irrigation methods to reduce water loss is not always an effective way to overcome water shortage especially in the dry areas, the purification of non-traditional water sources seems to be a promising solution (Ghermandi and Messalem 2009). NPs are used in purifying water and in environmental management (Chekli et al. 2013; Zuverza-Mena et al. 2017).

For sufficient irrigation, NPs can be implemented as nanofilters to eliminate turbidity, unsafe microorganisms (fungi, viruses, bacteria) organic solids and specific ions to avoid their negative impacts on crops like yield reduction (Rasouli et al. 2013), unfavorable crop quality (Bernstein et al. 2011), damaging crop adaptability (Liu et al. 2013) and decreasing cultivation choices of crops (Levy and Tai 2013) with no need to use any disinfectants (Mrayed et al. 2011; Riera et al. 2013). Sharma et al. (2009) referred to the green synthesized silver (Ag) NPs, which can be used to filter water from pollinated materials.

Also, many scientists declared that wastewater used for irrigation could be purified by applying specific NPs (Li et al. 2014; Zhao et al. 2013). The natural pollutants could be perfectly thrown away from water when treated with iron oxide NPs (Li et al. 2015; Zhao et al. 2015). NPs of TiO_2 and Fe are both used to purify contaminated water (Mc Murray et al. 2006; Mueller et al. 2012). Heavy metals accumulate in water cause a risk of contamination (Rastmanesh et al. 2018), which threatens the entire environment (Santhosh et al. 2016) as well as all creatures (Dalzochio et al. 2018).

For water purification, the usage of conventional methods is not sufficient anymore, Qu et al. (2013) and Zhang et al. (2016) declared that nanomaterials can purify water successfully because of their high reactivity and large surface area. The oxidized carbon nanotubes have a good ability to adsorb heavy metals from polluted water (Lau et al. 2015; Liu et al. 2012).

11.5 Nanomaterials Toxicity to Crops

According to many scientists, NPs may revise the plant physiology, morphology, biochemistry and modify gene expression (Shweta et al. 2016; Siddiqui et al. 2015; Singh et al. 2016; Tripathi et al. 2016) (Fig. 11.3).

DeRosa et al. (2010) proved that carbon nanotubes and NPs disrupting the plant surface tissues of tomato (*Solanum lycopersicum* L.) roots and seeds. Also, the toxicity of copper oxide (CuO) NPs in pea (*Pisum sativum* L.) was documented by Ochoa et al. (2017). Tripathi et al. (2017) found in pea (*Pisum sativum* L.), that the investment of nitric oxide could ameliorate the toxicity of (ZnO and Ag) NPs. Growth of

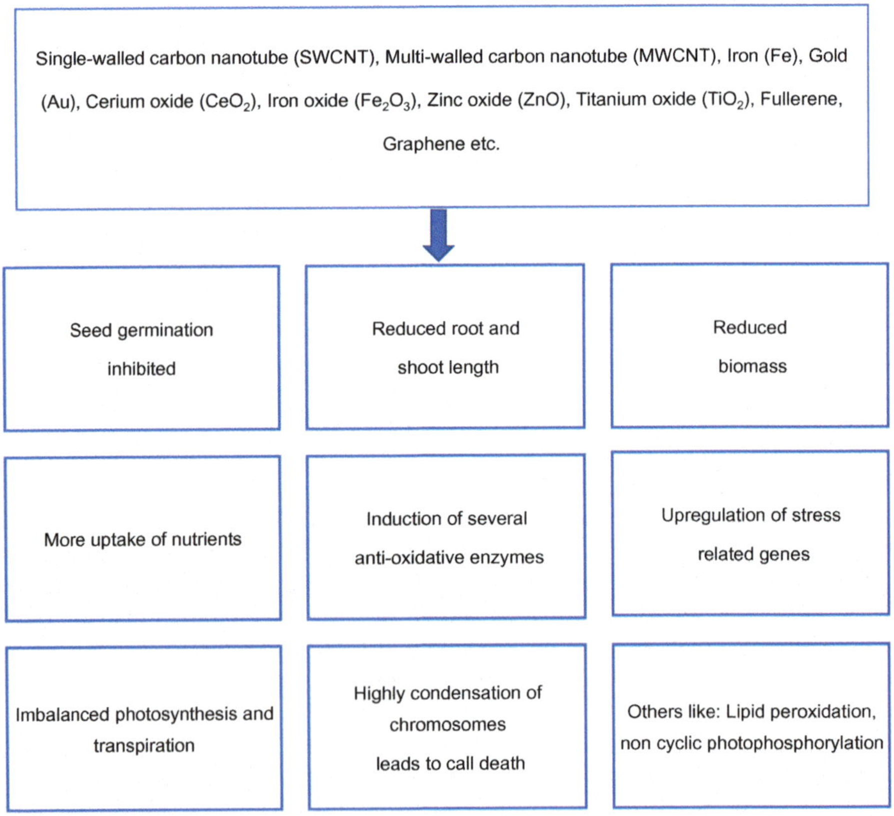

Fig. 11.3 The toxicity of different NPs in the plant (*Source* Adapted from Shweta et al. [2018])

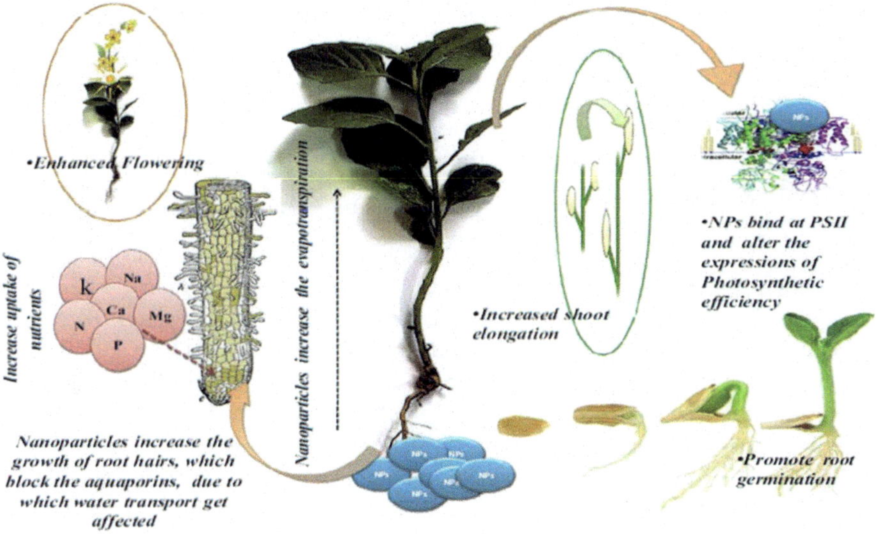

Fig. 11.4 The toxicity of NPs in the entire plant (*Source* Shweta et al. [2018])

maize root hairs is inhibited when treated with single-walled carbon nanotubes (Yan et al. 2013). Also, Aken (2015) mentioned the effect of NPs on the photosynthetic pathway genes. Shweta et al. (2016) discovered the effect of both gold (Au) and Ag NPs on the reaction center of photosynthesis and quantum yield (Fig. 11.4). A significant reduction in (*Arabidopsis thaliana* L.) growth was observed when treated by Ag NPs at specific doses (Kaveh et al. 2013). Results of genome analysis by Landa et al. (2012) revealed that the usage of 100 mg/L of ZnO NPs on (*Arabidopsis thaliana* L.) reduced the biomass and resulted in phytotoxic effects. Carbon NPs raised the plant susceptibility to some abiotic stresses (Wang et al. 2014).

Growth of wheat (*Triticum aestivum* L.) was disrupted, and gene expression was modified in the presence of Ag NPs (Dimkpa et al. 2013). In broad bean (*Vicia faba* L.), the ascorbate peroxidase and glutathione reductase activities were declined in the existence of TiO_2 NPs (Foltête et al. 2011; Rico et al. 2015b). Shen et al. (2010) noticed that treating the cells of the rice plant to single-walled NPs caused the death of cells and DNA damage. Homologous recombination, mutation and DNA damage were observed by López-Moreno et al. (2010a) when he treated soybean with CeO_2 and ZnO NPs. Increasing in chromosomal aberrations and reduction of the mitotic index was noticed by Patlolla et al. (2012) in broad bean (*Vicia faba* L.) after the treatment of Ag NPs.

11.6 Effect of Nanofertilizers on Different Growth Stages of Plants Under Abiotic Stresses

Nanofertilizers (NFs) are widely used as soil and spray-based applications in fruit crop nutrition which provide high-efficiency nutrients and low waste due to their faster and higher translocation to different parts of plants. NFs move through different pathways (apoplastic, symplastic, lipophilic, and hydrophilic) after penetration of the leaf or root tissue, affecting their efficacy, final fate and may also change their properties and thus they communicate, distribute and move inside plant tissues, which may result in different reactions of different plant sections of the same NP. Mineral nutrients play an important role in plant growth and metabolism. There are two major nanonutrient fertilizer types: (a) macronanofertilizers and (b) micronanofertilizers. Recent studies in which positive effects of nanonutrients fertilizer on plants were observed under different abiotic stresses (Table 11.4).

Drought, salinity, heat and toxic metals are considered as main threatened factors for plants and food security (Calanca 2017; Sha Valli Khan et al. 2018) and had effects on the extent and structure of minor metabolism in crops (Javed et al. 2017; Masarovičovà et al. 2019) (Fig. 11.5). Improvement of crop tolerance to specific abiotic stress is an essential objective for breeders (Bechtold and Field 2018; Sha Valli Khan et al. 2018). To alleviate the effects of abiotic stresses on crops, the usage of metal NPs can be a successful solution (Khan et al. 2017).

11.6.1 Drought

Water shortage, water deficit or inadequate available water, are all refer to drought which causes morphological, physiological and biochemical changes ending with decreasing in the final crop yield (Kumar et al. 2018). Ghassemi and Farahvash (2018) found in the wheat flowering stage, that the usage of Zn NPs on leaves, had an affirmative result on plant height, and led to rising the relative water content in wheat, yield and yield traits. Mozafari et al. (2018) reported the possibility of using a combination of salicylic acid and Fe NPs to guarantee better quality and higher quantity in the strawberries in vitro culture and improve strawberry adaptation to drought in the first stages of life. Cotton traits and productivity under drought stress can be emphasized by foliar usage of selenium oxide (SiO_2) and TiO_2 NPs (Shallan et al. 2016). The usage of CeO_2 NPs increased the rates of carbon assimilation and grain yield under drought stress in sorghum (*Sorghum bicolor* L.) (Djanaguiraman et al. 2018b). Results of Haghighi et al. (2013) indicated the improvement of the germination rate of tomato cultivated under water stress when using Si NPs.

Table 11.4 Some recent studies in which positive effects of NPs on plants were observed under different abiotic stresses

Abiotic stress	Crop	NPs used	Reference
Post-harvest	Parsley (*Petroselinum crispum* L.)	Copper (Cu)	Ouzounidou and Gaitis (2011)
Salinity	Tomato (*Lycopersicon esculentum* L.)	Silicon (SiO_2)	Haghighi et al. (2012)
Chilling	Wheat (*Triticum aestivum* L.)	Biogenic silver NPs	Bhati-Kushwaha et al. (2013)
Cold	Chickpea (*Cicer arietinum* L.)	Titanium dioxide (TiO2)	Mohammadi et al. (2013a, b)
Drought	Soybean (*Glycine max* L.)	Zinc oxide (ZnO)	Sedghi et al. (2013)
Drought	Safflower (*Carthamus tinctorious* L.)	Iron (Fe)	Davar et al. (2014)
Heat and cold	Tomato (*Lycopersicon esculentum* L.)	Sodium selenate (Na_2SeO_4)	Haghighi et al. (2014)
Mineral nutrient	Pearl millet (*Pennisetum americanum* L.)	Zinc oxide (ZnO)	Tarafdar et al. (2014)
Drought	Maize (*Zea mays* L.) and Wheat (*Triticum aestivum* L.)	NPs of analcite	Zaimenko et al. (2014)
Waterlogging	Soybean (*Glycine max* L.)	Al_2O_3	Mustafa et al. (2015)
UV radiation	Waterweed (*Elodea nuttallii* (Planch.) H. St. John)	Copper (Cu)	Regier et al. (2015)
Salinity	Maize (*Zea mays* L.)	Nitric oxide-releasing chitosan	Bruna et al. (2016)
Drought	Lentil (*Lens culinaris* L.)	Silver (AgNPs)	Hojjat (2016)
High CO_2	Rice (*Oryza sativa* L.)	Titanium dioxide (TiO_2)	Du et al. (2017)
Salinity	Wheat (*Triticum aestivum* L.)	Zinc oxide (ZnO)	Fathi et al. (2017b)
Salinity	Fenugreek (*Trigonella foenum-graecum*)	Silver (AgNPs)	Hojjat and Kamyab (2017)
Heat	Wheat (*Triticum aestivum* L.)	Silver (AgNPs)	Iqbal et al. (2017)
Mineral nutrient	Rice (*Oryza sativa* L.)	Nano-potassium	Lemraski et al. (2017)
Drought	Wheat (*Triticum aestivum* L.)	Zinc and Copper	Taran et al. (2017)

(continued)

Table 11.4 (continued)

Abiotic stress	Crop	NPs used	Reference
Salinity	Broad bean (*Vicia faba* L.)	Titanium dioxide (TiO_2)	Abdel Latef et al. (2018)
High temperature	Sorghum (*Sorghum bicolor* L.)	Selenium (Se)	Djanaguiraman et al. (2018a)
Heavy metals	Wheat (*Triticum aestivum* L.)	Zinc oxide (ZnO)	Hussain et al. (2018)
Drought	Wheat (*Triticum aestivum* L.)	Zinc and Copper	Storozhenko et al. (2019)
Drought	Maize (*Zea mays* L.)	Zinc oxide (ZnO)	Sun et al. (2020)

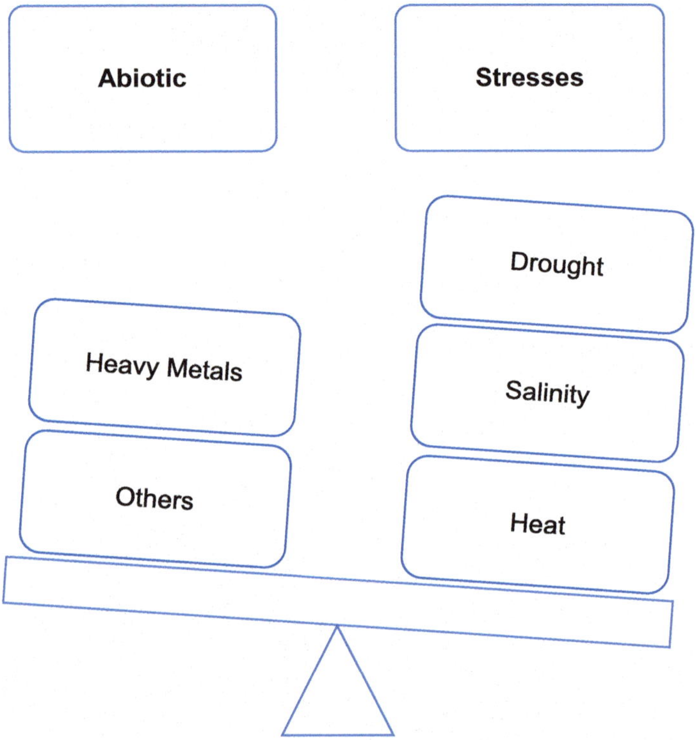

Fig. 11.5 Main types of abiotic stress that threaten plants (Constructed by M.M. Saleh)

11.6.2 Salinity

Globally, soil salinity puts the most important food crops in danger (Majeed et al. 2018) (Fig. 11.6). Salinity inhibits plant ability to absorb water and reduces plant growth (Parihar et al. 2015). Babaei et al. (2017) found that grain yield of treated wheat plants with (Zn, Fe) oxide NPs under salt stress was increased. The negative effects of salinity were reduced when Alharby et al. (2016) applied ZnO NPs in callus culture of five tomato varieties exposed to NaCl stress. Thuesombat et al. (2016) concluded that the treatment of rice (*Oryza sativa* L.) grains with Ag NPs revealed an increase in root growth after the cultivation of plantlets in a salty nutrient solution. In tomato, the usage of Cu NPs can enhance salinity tolerance Hernandez-Hernandez et al. (2018).

Results of Farhangi-Abriz and Torabian (2018) indicated that applying SiO_2 NPs to soybean was useful in enhancing plant growth under salt stress. Alsaeedi et al. (2017) alleviated the negative effects of salinity on bean plantlets when they used SiO_2 NPs. Hussein and Abou-Baker (2018) reported that fertilizing cotton with Zn NPs led to the mitigation of salinity influence.

Latef et al. (2017) found that the usage of ZnO NPs insured the best development of (*Lupinus termis* L.) plants under salt stress. Concerning maize (*Zea mays* L.), the foliar applying of Fe_2O_3 and ZnO improved root growth and played a role in mitigation salinity effects (Fathi et al. 2017a). Almutairi (2016) suggested using of Si NPs as a contributor factor to increase tomato salinity tolerance. Siddiqui et al. (2014) applied SiO_2 NPs on squash plants to develop their resistance to salinity. Concerning sunflower crop, Torabian et al. (2017) found that $FeSO_4$ NPs could increase plant biomass under saline conditions. Usage of TiO_2 NPs led to enhancing root and shoot length in cultivated maize (*Zea mays* L.) under salinity conditions (Mutlu et al. 2018). Rossi et al. (2016) declared that despite the salinity could not reduce, but the applied CeO_2 NPs in (*Brassica napus* L.) plants improved their response to salt stress.

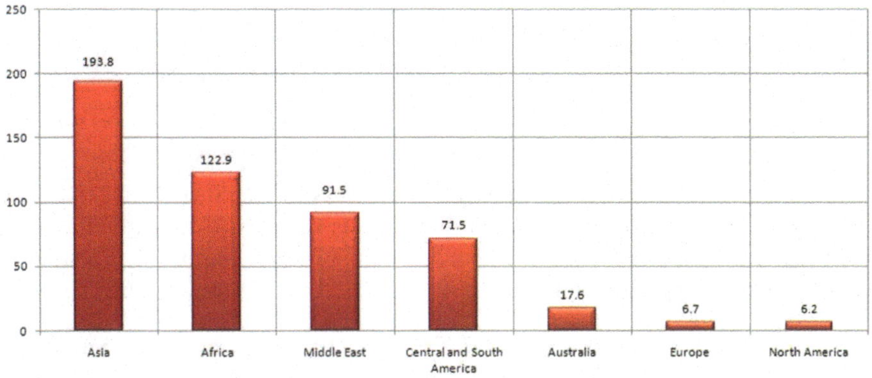

Fig. 11.6 Distribution of affected lands by salinity in the world (This diagram was constructed by M.M. Saleh based on FAO data 2015)

11.6.3 Heavy Metals

Higher concentrations of heavy metals in agricultural soils (like arsenic As, chromium Cr, lead Pb, Cadmium Cd) are destructive to plants and all living system and lead to raising their uptake by crops resulting in economic losses, yield reduction, and significant risks concerning food safety (Kràľovà et al. 2019). Specialists affirmed the effective utilization of NPs to counteract the various effects of heavy metals on crops (Liu et al. 2018; Rizwan et al. 2018). Huang et al. (2018) declared that Fe_3O_4 NPs caused an effective reduction in As absorption of rice under low As concentrations. According to Mohammadi et al. (2018) reduction of the uptake of Cr in sunflower root and shoot can be achieved when using Fe NPs. The reduction of accumulated Ca in rice be attained when using Fe_3O_4 NPs (Sebastian et al. 2017).

The harmful impact of rice Japonica seedlings had mitigated by applying CuO NPs (Liu et al. 2018). A spray of Si NPs to rice variety Xiangzaoxian diminished the accumulated Cd in grains and decreased its accumulation in roots (Chen et al. 2018). Optimistic effects of ZnO NPs foliar usage on the wheat planted under Cd stress were noticed by Hussain et al. (2018). CeO_2 NPs reduced Cd translocation from roots to vegetative parts of soybean crop (Rossi et al. 2018). Levels of absorbed Cd in the root, shoot and grains of wheat cultivated in high concentration Cd soil, were significantly reduced when priming wheat seeds with ZnO NPs (Rizwan et al. 2018).

Cai et al. (2017) applied 1000 mg/L of TiO_2 NPs to rice plants exposed to a high concentration of Pb, their results referred to the reduction of Pb in roots and shoots. Relating to rice, the TiO_2 NPs considerably diminished Cd in plants and enriched each of plant growth, photosynthesis, and decreased Cd absorption and translocation in all plant parts (Ji et al. 2017). Results of Tripathi et al. (2015) clarified the role of Si NPs in protecting pea (*Pisum sativum* L.) seedlings from the harmful impact of Cr via decreasing its accumulations in plant and increasing plant growth. Singh and Lee (2016) confirmed the function of TiO_2 NPs in declining the Cd toxicity and enhancement growth parameters in soybean.

11.6.4 Heat

Some variations in plant traits occur under heat stress (Waraich et al. 2012), besides modification in lipids structure and proteins-lipids relations (Yue and Yun 2018). Plants regulate their physiological response and homeostasis during their adaptation procedures when exposed to high temperatures stress (Nievola et al. 2017).

Valuable results were indicated by Iqbal et al. (2017) about the availability of protection wheat plants under high-temperature stress up by using Ag NPs which improved root and shoot length and other traits. Djanaguiraman et al. (2018a) indicated that the foliar treatment of Se NPs on sorghum cultivated under high-temperature stress 38 days/28 °C night improved the activity of antioxidant enzymes and diminished the oxidants content, which led to protect plants from negative effects of oxidative destruction as a result of high-temperature stress.

López-Moreno et al. (2017) referred that although heat may revise the interaction between plant and applied NPs, root growth in maize increased when applying ZnO NPs which enhances their plant response to heat stress. In tomato leaves, Qi et al. (2013) guarantied the beneficial effect of TiO_2 NPs in protecting the photosynthesis process under mild high temperatures stress.

11.7 Conclusion and Prospects

There is no doubt that nanotechnology is an emerging field. The use of nanotechnology technology and NPs in various agricultural fields are still in the initial stage, especially in bearing different environmental stresses. Also, due to the unique properties of NPs, the poisonous effects of some NPs have been confirmed on different crops. But so far, research focusing on understanding the positive effects of NPs on economic crops grown as an alternative to chemical fertilizers used in agriculture remains unfinished. Although, in recent studies, several preliminary studies have been conducted on this subject that has given many hopes to agricultural scientists of the possibility of using nanotechnology to improve the production of economic crops under normal and stress conditions. However, more important studies are still needed to know the different important applications of NPs and to understand the way NPs work and their effects on the genetic expression of economic crops grown in the presence of abiotic stresses. More stringent work regarding plans to understand their physiological, biochemical and molecular mechanisms in plant economic crops cultivated in the presence of non-vital environmental pressures is needed. Moreover, additional studies are needed to explore the way NPs work. NP's also face possible biological challenges. Consequently, the harmful impact assessment of using this technology should not be fully overlooked to avoid risks to the environment with increasing use in agriculture. There are currently few reports of NPs toxic to plants. Numerous research indicated both the positive and negative effects of NPs on growth, germination, high yields, increased abiotic stress tolerance and reduced chemical fertilizer pollution. Thus further studies at the molecular level need to realize the role of NPs. Furthermore, know the size, shape and concentrations of NPs which are extremely important for understanding the interaction between economic crops and NPs before recommending their field applications.

References

Abdel Latef AAH, Srivastava AK, El-Sadek MSA et al (2018) Titanium dioxide nanoparticles improve growth and enhance tolerance of broad bean plants under saline soil conditions. Land Degrad Dev 29(4):1065–1073

Adesemoye A, Kloepper J (2009) Plant-microbes interactions in enhanced fertilizer-use efficiency. Appl Microbiol Biotechnol 85:1–12

Adesemoye AO, Torbert HA, Kloepper JW (2009) Plant growth promoting rhizobacteria allow reduced application rates of chemical fertilizers. Microbial Ecol 58:921–929

Agrawal S, Rathore P (2014) Nanotechnology pros and cons to agriculture: a review. Int J Curr Microbiol App Sci 3(3):43–55

Aken VB (2015) Gene expression changes in plants and microorganisms exposed to nanomaterials. Curr Opin Biotechnol 33:206–219

Alharby HF, Metwali EMR, Fuller MP et al (2016) Induction, plant regeneration, element content and antioxidant enzyme activity in tomato (*Solanum lycopersicum* MILL.) under salt stress. Arch Biol Sci 68(4):723–735. https://doi.org/10.2298/abs151105017a

Ali S, Khan I, Khan SA et al (2017) Electrocatalytic performance of Ni@ Pt core–shell nanoparticles supported on carbon nanotubes for methanol oxidation reaction. J Electroanal Chem 795:17–25

Almutairi ZM (2016) Effect of nano-silicon application on the expression of salt tolerance genes in germinating tomato (*Solanum lycopersicum* L.) seedlings under salt stress. Plant Omics 9(1):106–114

Alsaeedi AH, El-Ramady H, Alshaal T et al (2017) Engineered silica nanoparticles alleviate the detrimental effects of Na+ stress on germination and growth of common bean (*Phaseolus vulgaris*). Environ Sci Pollut Res 24(27):21917–21928. https://doi.org/10.1007/s11356-017-9847-y

Andersen CP, King G, Plocher M et al (2016) Germination and early plant development of ten plant species exposed to titanium dioxide and cerium oxide nanoparticles. Environ Toxicol Chem 35:2223–2229. https://doi.org/10.1002/etc.3374

Antisari LV, Carbone S, Gatti A et al (2015) Uptake and translocation of metals and nutrients in tomato grown in soil polluted with metal oxide (CeO2, Fe3O4, SnO2, TiO2) or metallic (Ag Co, Ni) engineered nanoparticles. Environ Sci Pollut Res 22(3):1841–1853

Asli S, Neumann M (2009) Colloidal suspensions of clay or titanium dioxide nanoparticles can inhibit leaf growth and transpiration via physical effects on root water transport. Plant Cell Environ 32:577. https://doi.org/10.1111/j.1365-3040.2009.01952.x

Babaei K, Sharifi RS, Pirzad A et al (2017) Effects of bio fertilizer and nano Zn-Fe oxide on physiological traits, antioxidant enzymes activity and yield of wheat (*Triticum aestivum* L.) under salinity stress. J Plant Interact 12(1):381–389. https://doi.org/10.1080/17429145.2017.1371798

Ball P (2002) Natural strategies for the molecular engineer. Nanotechnology 13(5):R15

Bechtold U, Field B (2018) Molecular mechanisms controlling plant growth during abiotic stress. J Exp Bot 69(11):2753–2758. https://doi.org/10.1093/jxb/ery157

Bernstein N, Ioffe M, Luria G et al (2011) Effects of K and N nutrition on function and production of *Ranunculus asiaticus*. Pedosphere 21:288–301. https://doi.org/10.1016/S1002-0160(11)60129-X

Bhati-Kushwaha H, Kaur A, Malik CP (2013) The synthesis and role of biogenic nanoparticles in overcoming chilling stresses. Ind J Plant Sci 2:54–62

Bombin S, LeFebvre M, Sherwood J et al (2015) Developmental and reproductive effects of iron oxide nanoparticles in *Arabidopsis thaliana*. Int J Mol Sci 16(10):4174–24193

Boonyanitipong B, Kositsup B, Kumar P et al (2011) Toxicity of ZnO and TiO_2 nanoparticles on germinating rice seed *Oryza sativa* L. Int J Biosci Biochem Bioinform 1:282–285

Bruna HCO, Gomes CR, Milena T et al (2016) Nitric oxide-releasing chitosan nanoparticles alleviate the effects of salt stress in maize plants. Nitric Oxide 61:10–19

Cai F, Wu XY, Zhang HY et al (2017) Impact of TiO_2 nanoparticles on lead uptake and bioaccumulation in rice (*Oryza sativa* L.). Nanoimpact 5:101–108. https://doi.org/10.1016/j.impact.2017.01.006

Calanca PP (2017) Effects of abiotic stress in crop production. In: Ahmed M, Stockle C (eds) Quantification of climate variability, adaptation and mitigation for agricultural sustainability. Springer, Cham, pp 165–180

Campbell EE, Fowler PW, Mitchell D, Zerbetto F (1996) Increasing cost for Pentagon adjacency for higher fullerenes. Chem Phys Lett 250:544–548

Chekli L, Phuntsho S, Roy M et al (2013) Characterisation of Fe-oxide nanoparticles coated with humic acid and Suwannee River natural organic matter. Sci Total Environ 461–462:19–27

Chen C, Unrine JM, Judy JD et al (2015) Toxicogenomic responses of the model legume *Medicago truncatula* to aged biosolids containing a mixture of nanomaterials (TiO_2, Ag, and ZnO) from a pilot wastewater treatment plant. Environ Sci Technol 49:8759–8768. https://doi.org/10.1021/acs.est.5b01211

Chen R, Zhang CB, Zhao YL et al (2018) Foliar application with nano-silicon reduced cadmium accumulation in grains by inhibiting cadmium translocation in rice plants. Environ Sci Pollut Res 25(3):2361–2368. https://doi.org/10.1007/s11356-017-0681-z

Chinnamuthu CR, Boopathi PM (2009) Nanotechnology and agroecosystem. Madras Agric J 96:17–31

Chittaranjan K, Kole P, Randunu KM et al (2013) Nanobiotechnology can boost crop production and quality: First evidence from increased plant biomass, fruit yield and phytomedicine content in bitter melon (*Momordica charantia*). BMC Biotechnol 13:37. https://doi.org/10.1186/1472-6750-13-37

Corredor E, Testillano PS, Coronado MJ et al (2009) Nanoparticle penetration and transport in living pumpkin plants: *in situ* subcellular identification. BMC Plant Biol 9:45. https://doi.org/10.1186/1471-2229-9-45

Cox A, Venkatachalam P, Sahi S et al (2017) Reprint of: silver and titanium dioxide nanoparticle toxicity in plants: a review of current research. Plant Physiol Biochem 110:33–49. https://doi.org/10.1016/j.plaphy.2016.08.007

Cui H X, Sun C J, Liu Q et al (2010) Applications of nanotechnology in agrochemical formulation, perspectives, challenges and strategies. In: International Conference on Nano Agri, Sao Pedro, Brazil, pp 28–33

Cunningham FJ, Goh NS, Demirer GS et al (2018) Nanoparticle-mediated delivery towards advancing plant genetic engineering. Trends Biotechnol 36(9):882–897

Dalzochio T, Rodrigues GZP, Simões LAR et al (2018) In situ monitoring of the Sinos River, southern Brazil: water quality parameters, biomarkers, and metal bioaccumulation in fish. Environ Sci Pollut Res 25:9485–9500

Davar F, Zareii AR, Amir H (2014) Evaluation the effect of water stress and foliar application of Fe nanoparticles on yield, yield components and oil percentage of safflower (*Carthamus tinctorious* L.). Int J Adv Biol Biomed Res 2:1150–1159

DeRosa MC, Monreal C, Schnitzer M et al (2010) Nanotechnology in fertilizers. Nat Nanotechnol 5:91. https://doi.org/10.1038/nnano.2010.2

Dimkpa CO, Latta DE, McLean JE et al (2013) Fate of CuO and ZnO nano and microparticles in the plant environment. Environ Sci Technol 47:4734–4742

Djanaguiraman M, Belliraj N, Bossmann SH et al (2018a) High-temperature stress alleviation by selenium nanoparticle treatment in grain sorghum. ACS Omega 3(3):2479–2491. https://doi.org/10.1021/acsomega.7b01934

Djanaguiraman M, Nair R, Giraldo JP et al (2018b) Cerium oxide nanoparticles decrease drought-induced oxidative damage in sorghum leading to higher photosynthesis and grain yield. ACS Omega 3(10):14406–14416. https://doi.org/10.1021/acsomega.8b01894

Du W, Gardea-Torresdey JL, Ji R et al (2015) Physiological and biochemical changes imposed by CeO_2 nanoparticles on wheat: a life cycle field study. Environ Sci Technol 49:11884–11893. https://doi.org/10.1021/acs.est.5b03055

Du W, Gardea-Torresdey JL, Xie Y et al (2017) Elevated CO_2 levels modify TiO_2 nanoparticle effects on rice and soil microbial communities. Sci Total Environ 578:408–416

El-Temsah YS, Joner EJ (2012) Impact of Fe and Ag nanoparticles on seed germination and differences in bioavailability during exposure in aqueous suspension and soil. Environ Toxicol 27:42–49

Fan R, Huang YC, Grusak MA et al (2014) Effects of nano-TiO_2 on the agronomically relevant *Rhizobium*-legume symbiosis. Sci Total Environ 466:503–512. https://doi.org/10.1016/j.scitotenv.2013.07.032

FAO (2015) The state of food and agriculture. Social protection and agriculture: Breaking the cycle of rural poverty. UN Food and Agriculture Organization, Rome, Italy

Farhangi-Abriz S, Torabian S (2018) Nano-silicon alters antioxidant activities of soybean seedlings under salt toxicity. Protoplasma 255(3):953–962. https://doi.org/10.1007/s00709-017-1202-0

Farooq M, Wahid A, Kobayashi N et al (2009) Plant drought stress: effects, mechanisms and management. In: Lichtfouse E, Navarrete M, Debaeke P et al (eds) Sustainable agriculture. Springer, Dordrecht

Fathi A, Zahedi M, Torabian S (2017a) Effect of interaction between salinity and nanoparticles (Fe_2O_3 and ZnO) on physiological parameters of *Zea mays* L. J Plant Nutr 40(19):2745–2755. https://doi.org/10.1080/01904167.2017.1381731

Fathi A, Zahedi M, Torabian S, Khoshgoftar A (2017b) Response of wheat genotypes to foliar spray of ZnO and Fe_2O_3 nanoparticles under salt stress. J Plant Nutr 40:1376–1385

Feizi H, Moghaddam PR, Shahtahmassebi N et al (2012) Impact of bulk and nano-sized titanium dioxide on wheat seed germination and seedling growth. Biol Trace Elem Res 146:101. https://doi.org/10.1007/s12011-011-9222-7

Flessa H, Ruser R, Dörsch P et al (2002) Integrated evaluation of greenhouse gas emissions (CO_2, CH_4, N_2O) from two farming systems in southern Germany. Agric Ecosys Environ 91:175–189

Foltête AS, Masfaraud JF, Bigorgne E et al (2011) Environmental impact of sunscreen nanomaterials: ecotoxicity and genotoxicity of altered TiO_2 nanocomposites on *Vicia faba*. Environ Pollut 159(10):2515–2522

Gao F, Hong F, Liu C et al (2006) Mechanism of nano-anatase TiO_2 on promoting photosynthetic carbon reaction of spinach. Biol Trace Element Res 111(1–3):239–253

García-Gómez C, Obrador A, González D et al (2018) Comparative study of the phytotoxicity of ZnO nanoparticles and Zn accumulation in nine crops grown in a calcareous soil and an acidic soil. Sci Total Environ 644:770–780. https://doi.org/10.1016/j.scitotenv.2018.06.356

Ghassemi A, Farahvash F (2018) Effect of nano-zinc foliar application on wheat under drought stress. Fresen Environ Bull 27(7):5022–5026

Ghermandi A, Messalem R (2009) Solar-driven desalination with reverse osmosis: the state of the art. Desalin Water Treat 7(1–3):285–296

Giraldo JP, Landry MP, Faltermeier SM et al (2014) Plant nanobionics approach to augment photosynthesis and biochemical sensing. Nat Mater 13(4):400–408

González-Melendi P, Fernández-Pacheco R, Coronado MJ et al (2008) Nanoparticles as smart treatment-delivery systems in plants: assessment of different techniques of microscopy for their visualisation in plant tissues. Ann Bot 101:187–195. https://doi.org/10.1093/aob/mcm283

Gui X, Zhang Z, Liu S et al (2015) Fate and phytotoxicity of CeO_2 nanoparticles on lettuce cultured in the potting soil environment. PLoS One 10(8):1–10. https://doi.org/10.1371/journal.pone.0134261

Haghighi M, Abolghasemi R, Teixeira da Silva JA (2014) Low and high temperature stress affect the growth characteristics of tomato in hydroponic culture with Se and nano-Se amendment. Sci Hort 178:231–240

Haghighi M, Afifipour Z, Mozafariyan M (2012) The effect of N-Si on tomato seed germination under salinity levels. J Biol Environ Sci 6:87–90

Haghighi M, Da Silva JAT, Mozafarian M et al (2013) Can Si and nano-Si alleviate the effect of drought stress induced by PEG in seed germination and seedling growth of tomato? Minerva Biotechnol 25(1):17–22

Hernandez-Hernandez H, Gonzalez-Morales S, Benavides-Mendoza A et al (2018) Effects of chitosan-PVA and Cu nanoparticles on the growth and antioxidant capacity of tomato under saline stress. Molecules 23(1):178. https://doi.org/10.3390/molecules23010178
Hojjat SS (2016) The effect of silver nanoparticle on lentil seed germination under drought stress. Intl J Farm Alli Sci 5(3):208–212
Hojjat SS, Kamyab M (2017) The effect of silver nanoparticle on fenugreek seed germination under salinity levels. Rus Agric Sci 43(1):61–65
Huang Q, Liu Q, Lin LN et al (2018) Reduction of arsenic toxicity in two rice cultivar seedlings by different nanoparticles. Ecotox Environ Saf 159:261–271. https://doi.org/10.1016/j.ecoenv.2018.05.008
Hussain A, Ali S, Rizwan M et al (2018) Zinc oxide nanoparticles alter the wheat physiological response and reduce the cadmium uptake by plants. Environ Pollut 242:1518–1526
Hussain HA, Men S, Hussain S et al (2020) Maize tolerance against drought and chilling stresses varied with root morphology and antioxidative defense system. Plants 9(6):720
Hussein MM, Abou-Baker NH (2018) The contribution of nano-zinc to alleviate salinity stress on cotton plants. Royal Soc Open Sci 5(8): https://doi.org/10.1098/rsos.171809
Iqbal M, Raja NI, Zia-Ur-Rehman M et al (2017) Effect of silver nanoparticles on growth of wheat under heat stress. Iran J Sci Technol Trans Sci 2017:1–9. https://doi.org/10.1007/s40995-017-0417-4
Javed R, Usman M, Yucesan B et al (2017) Effect of zinc oxide (ZnO) nanoparticles on physiology and steviol glycosides production in micropropagated shoots of *Stevia rebaudiana* Bertoni. Plant Physiol Biochem 110:94–99. https://doi.org/10.1016/j.plaphy.201605.032
Ji Y, Zhou Y, Ma CX et al (2017) Jointed toxicity of TiO_2 NPs and Cd to rice seedlings: NPs alleviated Cd toxicity and Cd promoted NPs uptake. Plant Physiol Biochem 110:82–93. https://doi.org/10.1016/j.plaphy.2016.05.010
Judy JD, Unrine JM, Rao W et al (2012) Bioavailability of gold nanomaterials to plants: importance of particle size and surface coating. Environ Sci Technol 46(15):8467–8474
Kaveh R, Li YS, Ranjbar S et al (2013) Changes in *Arabidopsis thaliana* gene expression in response to silver nanoparticles and silver ions. Environ Sci Technol 47:10637–10644
Khan MN, Mobin M, Abbas ZK et al (2017) Role of nanomaterials in plants under challenging environments. Plant Physiol Biochem 110:194–209. https://doi.org/10.1016/j.plaphy.2016.05.038
Khodakovskaya MV, de Silva K, Biris AS et al (2012) Carbon nanotubes induce growth enhancement of tobacco cells. ACS Nano 6(3):2128–2135
Kràl̆ovà K, Masarovičovà E, Jampílek J (2019) Plant responses to stress induced by toxic metals and their nanoforms. In: Pessarakli M (ed) Handbook of plant and crop stress, 4th edn. CRC Press, Boca Raton
Kroto HW, Heath JR, O'Brien SC et al (1985) C60: Buckminsterfullerene. Nature 318(6042):162–163
Kumar S, Sachdeva S, Bhat KV et al (2018) Plant responses to drought stress: physiological, biochemical and molecular basis. In: Vats S (ed) Biotic and abiotic stress tolerance in plants. Springer, Singapore, pp 1–25. https://doi.org/10.1007/978-981-10-9029-5_1
Landa P, Vankova R, Andrlova J et al (2012) Nanoparticle specific changes in *Arabidopsis thaliana* gene expression after exposure to ZnO_2TiO_2 and fullerene soot. J Hazard Mater 241:55–62
Larue C, Veronesi G, Flank AM et al (2013) Comparative uptake and impact of TiO_2 nanoparticles. J Toxicol Env Heal A 75:722–734. https://doi.org/10.1080/15287394.2012.689800
Latef AAHA, Abu Alhmad MF, Abdelfattah KE (2017) The possible roles of priming with ZnO nanoparticles in mitigation of salinity stress in lupine (*Lupinus termis*) plants. J Plant Growth Regul 36(1):60–70. https://doi.org/10.1007/s00344-016-9618-x
Lau WJ, Gray S, Matsuura T et al (2015) A review on polyamide thin film nanocomposite (TFN) membranes: history, applications, challenges and approaches. Water Res 80:306–324
Lemraski MG, Normohamadi G, Madani H et al (2017) Two Iranian rice cultivars' response to nitrogen and nano-fertilizer. Open J Ecol 7:591–603

Levy D, Tai GCC (2013) Differential response of potatoes (*Solanum tuberosum* L.) to salinity in an arid environment and field performance of the seed tubers grown with fresh water in the following season. Agric Water Manag 116:122–127

Li J, Hu J, Ma C et al (2016) Uptake, translocation and physiological effects of magnetic iron oxide (γ-Fe2O3) nanoparticles in corn (*Zea mays* L.). Chemosphere 159:326–334

Li R, Jin X, Megharaj M et al (2015) Heterogeneous Fenton oxidation of 2,4-dichlorophenol using iron-based nanoparticles and persulfate system. Chem Eng J 264:587–594

Li X, Gui X, Rui Y et al (2014) Bt-transgenic cotton is more sensitive to CeO_2 nanoparticles than its parental non-transgenic cotton. J Hazard Mater 274:173–180

Linh TM, Mai NC, Hoe PT et al (2020) Metal-based nanoparticles enhance drought tolerance in soybean. J Nanomaterials 2020:ID 4056563. https://doi.org/10.1155/2020/4056563

Liscano JF, Wilso CE, Norman-Jr RJ, Slaton NA (2000) Zinc availability to rice from seven granular fertilizers. Arkansas Agric Exp Stat Res B 963:1–31

Liu J, Dhungana B, Cobb GP (2018) Copper oxide nanoparticles and arsenic interact to alter seedling growth of rice (*Oryza sativa japonica*). Chemosphere 206:330–337. https://doi.org/10.1016/j.chemosphere.2018.05.021

Liu L, Liu J, Sun DD (2012) Graphene oxide enwrapped Ag_3PO_4 composite: towards a highly efficient and stable visible-light-induced photocatalyst for water purification. Catal Sci Technol 2:2525–2532

Liu R, Zhang H, Lal R (2016) Effects of stabilized nanoparticles of copper, zinc, manganese, and iron oxides in low concentrations on lettuce (*Lactuca sativa*) seed germination: nanotoxicants or nanonutrients? Water Air Soil Pollut 227(1):1–14

Liu Y, Li X, Xing Z et al (2013) Responses of soil microbial biomass and community composition to biological soil crusts in the revegetated areas of the Tengger Desert. Appl Soil Ecol 65:52–59

López-Moreno ML, De La Rosa G, Cruz-Jimenez G et al (2017) Effect of ZnO nanoparticles on corn seedlings at different temperatures: x-ray absorption spectroscopy and ICP/OES studies. Microchem J 134:54–61. https://doi.org/10.1016/j.microc.2017.05.007

López-Moreno ML, De La Rosa G, Hernandez-Viezcas JA et al (2010a) Evidence of the differential biotransformation and genotoxicity of ZnO and CeO_2 nanoparticles on soybean (*Glycine max*) plants. Environ Sci Technol 44:7315–7320. https://doi.org/10.1021/es903891g

López-Moreno ML, De La Rosa G, Hernandez-Viezcas JA et al (2010b) X-ray absorption spectroscopy (XAS) corroboration of the uptake and storage of CeO_2 nanoparticles and assessment of their differential toxicity in four edible plant species. J Agric Food Chem 58:3689–3693

Ma J, Li XL, Xu H et al (2007) Effects of nitrogen fertilizer and wheat straw application on CH4 and N2O emissions from a paddy rice field. Aust J Soil Res 45:359–367

Ma X, Geisler-Lee J, Deng Y, Kolmakov A (2010) Interactions between engineered nanoparticles (ENPs) and plants: phytotoxicity, uptake and accumulation. Sci Total Environ 408:3053–3061. https://doi.org/10.1016/j.scitotenv.2010.03.031

Majeed A, Muhammad Z, Islam S et al (2018) Salinity imposed stress on principal cereal crops and employing seed priming as a sustainable management approach. Acta Ecol Sin 39(4):280–283. https://doi.org/10.1016/j.chnaes.2018.09.004

Majumdar S, Almeida IC, Arigi EA et al (2015) Environmental effects of nanoceria on seed production of common bean (*Phaseolus vulgaris*): a proteomic analysis. Environ Sci Technol 49:3283–13293. https://doi.org/10.1021/acs.est.5b03452

Mariani L, Ferrante A (2017) Agronomic management for enhancing plant tolerance to abiotic stresses—drought, salinity, hypoxia, and lodging. Horticult 3:52. https://doi.org/10.3390/horticulturae3040052

Marschner H (1995) Mineral nutrition of higher plants, 2nd edn. Academic Press, London, p 889

Martínez-Fernández D, Vitkova M, Bernal MP et al (2015) Effects of nano-maghemite on trace element accumulation and drought response of *Helianthus annuus* L. in a contaminated mine soil. Water Air Soil Pollut 226:101. https://doi.org/10.1007/s11270-015-2365-y

Masarovičovà E, Kràl̆ovà K, Vykoukovà I et al (2019) Responses of medicinal plants to abiotic stresses. In: Pessarakli M (ed) Handbook of plant and crop stress, 4th edn. CRC Press, Boca Raton

Mattiello A, Poščić F, Musetti R et al (2015) Evidence of genotoxicity and phytotoxicity in *Hordeum vulgare* exposed to CeO_2 and TiO_2 nanoparticles. Front Plant Sci 6:1043. https://doi.org/10.3389/fpls.2015.01043

Mc Murray TA, Dunlop PSM, Byrne JA (2006) The photocatalytic degradation of atrazine on nanoparticulate TiO_2 films. J Photochem Photobiol A Chem 182(1):43–51

Miransari M (2011) Soil nutrients. Nova, New York. ISBN 978-1-61324-785-3

Mohammadi H, Hatami M, Feghezadeh K et al (2018) Mitigating effect of nano-zerovalent iron, iron sulfate and EDTA against oxidative stress induced by chromium in *Helianthus annuus* L. Acta Physiol Plant 40(4):69. https://doi.org/10.1007/s11738-018-2647-2

Mohammadi R, Amiri NM, Mantri L (2013a) Effect of TiO_2 nanoparticles on oxidative damage and antioxidant defense systems in chickpea seedlings during cold stress. Russ J Plant Physiol 61:768–775

Mohammadi R, Amiri RM, Abbasi A (2013b) Effect of TiO_2 nanoparticles on chickpea response to cold stress. Biol Trace Elem Res 152:403–410

Moon JW, Phelps TJ, Fitzgerald CL et al (2016) Manufacturing demonstration of microbially mediated zinc sulfide nanoparticles in pilot-plant scale reactors. Appl Microbiol Biotechnol 100:7921–7931. https://doi.org/10.1007/s00253-016-7556-y

Mozafari AA, Havas F, Ghaderi N (2018) Application of iron nanoparticles and salicylic acid in *in vitro* culture of strawberries (*Fragaria* x *ananassa* Duch.) to cope with drought stress. Plant Cell Tissue Organ Cult 132(3):511–523. https://doi.org/10.1007/s11240-017-1347-8

Mrayed SM, Sanciolo P, Zou I et al (2011) An alternative membrane treatment process to produce low-salt and high-nutrient recycled water suitable for irrigation purposes. Desalination 274(19):144–149

Mueller NC, Braun J, Bruns J et al (2012) Application of nanoscale zero valent iron (NZVI) for groundwater remediation in Europe. Environ Sci Pollut Res 19(2):550–558

Mustafa G, Sakata K, Hossain Z, Komatsu S (2015) Proteomic analysis of flooded soybean root exposed to aluminum oxide nanoparticles. J Proteome 128:280–297

Mutlu F, Yurekli F, Mutlu B et al (2018) Assessment of phytotoxic and genotoxic effects of anatase TiO_2 nanoparticles on maize cultivar by using RAPD analysis. Fresen Environ Bull 27(1):436–445

Nair R, Varghese SH, Nair BG et al (2010) Nanoparticulate material delivery to plants. Plant Sci 179(3):154–163

Nawara HM, Mattar MZ, Salem KFM, Eissa OA (2017) Diallel study on some *in vitro* traits of bread wheat (*Triticum aestivum* L.) under salt stress. Inter J Agric Environ Res 3(1):1988–2006

Nievola CC, Carvalho CP, Carvalho V et al (2017) Rapid responses of plants to temperature changes. Temperature 4(4):371–405. https://doi.org/10.1080/23328940.2017.1377812

Ochoa L, Medina-Velo IA, Barrios AC et al (2017) Modulation of CuO nanoparticles toxicity to green pea (*Pisum sativum* Fabaceae) by the phytohormone indole-3-acetic acid. Sci Total Environ 598:513–524

Ouzounidou G, Gaitis F (2011) The use of nano-technology in shelf life extension of green vegetables. J Innov Econ Manag 2:163–171

Panwar J, Jain N, Bhargaya A et al (2012) Positive effect of zinc oxide nanoparticles on tomato plants: a step Towards developing "Nano-fertilizers". Proceeding of 3rd international conference on environmental research and technology. University of Sains, Penang, Malaysia, pp 248–352

Parihar P, Singh S, Singh R et al (2015) Effect of salinity stress on plants and its tolerance strategies: A review. Environ Sci Pollut Res 22:4056–4075. https://doi.org/10.1007/s11356-014-3739-1

Patlolla AK, Berry A, May L et al (2012) Genotoxicity of silver nanoparticles in *Vicia faba*: A pilot study on the environmental monitoring of nanoparticles. Int J Environ Res Public Health 9:1649–1662

Pérez-de-Luque A (2017) Interaction of nanomaterials with plants: what do we need for real applications in Agriculture? Front Environ Sci 5:12. https://doi.org/10.3389/fenvs.2017.00012

Population Institute (2017) FAO says food production must rise by 70%. Available online at: www.populationinstitute.org/resources/populationonline/issue/1/8

Pošćić F, Mattiello A, Fellet G et al (2016) Effects of cerium and titanium oxide nanoparticles in soil on the nutrient composition of barley (*Hordeum vulgare* L.) kernels. Int J Environ Res Public Health 13:577. https://doi.org/10.3390/ijerph13060577

Prasad TNVKV, Sudhakar P, Sreenivasulu Y et al (2012) Effect of nanoscale zinc oxide particles on the germination, growth and yield of peanut. Plant Nutr 35:905–927

Preetha PS, Balakrishnan N (2017) A review of nano fertilizers and their use and functions in soil. Int J Curr Microbiol App Sci 6(12):3117–3133

Qi MF, Liu YF, Li TL (2013) Nano-TiO_2 improve the photosynthesis of tomato leaves under mild heat stress. Biol Trace Elem Res 156(1–3):323–328. https://doi.org/10.1007/s12011-013-9833

Qu X, Brame J, Li Q et al (2013) Nanotechnology for a safe and sustainable water supply: enabling integrated water treatment and reuse. Acc Chem Res 46(3):834–843

Raliya R, Franke C, Chavalmane S et al (2016) Quantitative understanding of nanoparticle uptake in watermelon plants. Front Plant Sci 7:1288. https://doi.org/10.3389/fpls.2016.01288

Raliyaa R, Biswasa P, Tarafdarm JC (2015) TiO_2 nanoparticle biosynthesis and its physiological effect on mung bean (*Vigna radiate* L.). Biotechnol Rep 5:22–26. https://doi.org/10.1016/j.btre.2014.10.009

Rasouli F, Pouya AK, Karimian N (2013) Wheat yield and physico-chemical properties of a sodic soil from semiarid area of Iran as affected by applied gypsum. Geoderma 193:246–255

Rastmanesh F, Safaie S, Zarasv AR et al (2018) Heavy metal enrichment and ecological risk assessment of surface sediments in Khorramabad river, west Iran. Environ Monit Assess 190:273. https://doi.org/10.1007/s10661-018-6650-2

Regier N, Cosio C, von Moos N, Slaveykova VI (2015) Effects of copper-oxide nanoparticles, dissolved copper and ultraviolet radiation on copper bioaccumulation, photosynthesis and oxidative stress in the aquatic macrophyte *Elodea nuttallii*. Chemosphere 128:56–61

Rico CM, Barrios AC, Tan W et al (2015a) Physiological and biochemical response of soil-grown barley (*Hordeum vulgare* L.) to cerium oxide nanoparticles. Environ Sci Pollut Res 22:10551–10558. https://doi.org/10.1007/s11356-015-4243-y

Rico CM, Majumdar S, Duarte-Gardea M et al (2011) Interaction of nanoparticles with edible plants and their possible implications in the food chain. J Agric Food Chem 59(8):3485–3498

Rico CM, Peralta-Videa JR, Gardea-Torresdey JL (2015b) Chemistry, biochemistry of nanoparticles, and their role in antioxidant defense system in plants. In: Siddiqui MH, Al-Whaibi MH, Mohammad F (eds) Nanotechnology and plant sciences: nanoparticles and their impact on plants. Springer, pp 1–17

Riera FA, Suarez A, Muro C et al (2013) Nanofiltration of UHT flash cooler condensates from a dairy factory: characterization and water reuse potential. Desalination 309:52–63

Rizwan M, Ali S, Ali B et al (2018) Zinc and iron oxide nanoparticles improved the plant growth and reduced the oxidative stress and cadmium concentration in wheat. Chemosphere 214:269–277. https://doi.org/10.1016/j.chemosphere.2018.09.120

Rossi L, Sharifan H, Zhang WL et al (2018) Mutual effects and in planta accumulation of co-existing cerium oxide nanoparticles and cadmium in hydroponically grown soybean (*Glycine max* L. Merr.). Environ Sci-Nano 5(1):150–157. https://doi.org/10.1039/c7en00931c

Rossi L, Zhang WL, Lombardini L et al (2016) The impact of cerium oxide nanoparticles on the salt stress responses of *Brassica napus* L. Environ Pollut 219:28–36. https://doi.org/10.1016/j.envpol.2016.09.060

Ruffini Castiglione M, Giorgetti L, Bellani L et al (2016) Root responses to different types of TiO_2 nanoparticles and bulk counterpart in plant model system *Vicia faba* L. Environ Exp Bot 130:11–21. https://doi.org/10.1016/j.envexpbot.2016.05.002

Ruffini Castiglione M, Giorgetti L, Geri C et al (2011) The effects of nano-TiO_2 on seed germination, development and mitosis of root tip cells of *Vicia narbonensis* L. and *Zea mays* L. J Nanopart Res 1:2443–2449. https://doi.org/10.1007/s11051-010-0135-8

Sabo-Attwood T, Unrine JM, Stone JW et al (2012) Uptake, distribution and toxicity of gold nanoparticles in tobacco (*Nicotiana xanthi*) seedlings. Nanotoxicol 6:353–360. https://doi.org/10.3109/17435390.2011.579631

Salem KFM, Matter M (2014) Identification of microsatellite alleles for salt tolerance at seedling stage in wheat (*Triticum aestivum* L.). Life Sci J 11(12s):1064–1073

Salem KFM, Röder MS, Börner A (2007) Identification and mapping quantitative trait loci for stem reserve mobilisation in wheat (*Triticum aestivum* L.). Cereal Res Common 35:1367–1374

Santhosh C, Velmurugan V, Jacob G et al (2016) Role of nanomaterials in water treatment applications: a review. Chem Eng J 306:1116–1137

Sarvendra-Kumar PA, Datta SC, Rosin KG et al (2015) Phytotoxicity of nanoparticles to seed germination of plants. Int J Adv Res 3:854–865

Sasson Y, Levy-Ruso G, Toledano O, Ishaaya I (2007) Nanosuspensions: emerging novel agrochemical formulations. In: Ishaaya I, Horowitz AR, Nauen R (eds) *Insecticides design using advanced technologies*. Springer, Berlin, Heidelberg. https://doi.org/10.1007/978-3-540-46907-0_1

Savita TA, Singh SK (2020) Drought stress tolerance in legume crops. In: Hasanuzzaman M (ed) Agronomic crops. Springer, Singapore, pp 149–155

Sebastian A, Nangia A, Prasad MN (2017) Carbon-bound iron oxide nanoparticles prevent calcium-induced iron deficiency in *Oryza sativa* L. J Agric Food Chem 65(3):557–564. https://doi.org/10.1021/acs.jafc.6b04634

Sedghi M, Hadi M, Toluie SG (2013) Effect of nano zinc oxide on the germination parameters of soybean seeds under drought stress. Ann West Uni Timisoara 16:73–78

Sha Valli Khan PS, Nagamallaiah GV, Dhanunjay RM et al (2018) Abiotic stress tolerance in plants: insights from proteomics. In: Ahmad P, Rosool S (eds) Emerging technologies and management of crop stress tolerance, v.2: a sustainable approach. Academic Press, San Diego, pp 23–68

Shallan MA, Hassan HMM, Namich AAM et al (2016) Biochemical and physiological effects of TiO_2 and SiO_2 nanoparticles on cotton plant under drought stress. Res J Pharm Biol Chem Sci 7(4):1540–1551

Sharma VK, Yngard RA, Lin Y (2009) Silver nanoparticles: green synthesis and their antimicrobial activities. Adv Colloid Interface Sci 145:83–96

Shen CX, Zhang QF, Li J et al (2010) Induction of programmed cell death in *Arabidopsis* and rice by single-wall carbon nanotubes. Am J Bot 97:1602–1609

Shweta S, Tripathi DK, Chauhan DK et al (2018) Availability and risk assessment of nanoparticles in living systems: a virtue or a peril? In: Tripathi DK, Ahmad P, Sharma S et al (eds) Nanomaterials in plants, algae and microorganisms, Academic Press, London, 1:1–31

Shweta S, Tripathi DK, Singh S et al (2016) Impact of nanoparticles on photosynthesis: Challenges and opportunities. Mater Focus 5(5):405–411

Siddiqui MH, Al-Whaibi MH, Faisal M et al (2014) Nano-silicon dioxide mitigates the adverse effects of salt stress on *Cucurbita pepo* L. Environ Toxicol Chem 33(11):2429–2437. https://doi.org/10.1002/etc.2697

Siddiqui MH, Al-Whaibi MH, Firoz M et al (2015) Role of nanoparticles in plants. In: Al-Whaibi M, Mohammad F, Siddiqui MH (eds) Nanotechnology and plant sciences. Springer, Basel, Switzerland, pp 19–35

Singh J, Lee BK (2016) Influence of nano-TiO_2 particles on the bioaccumulation of Cd in soybean plants (*Glycine max*): A possible mechanism for the removal of Cd from the contaminated soil. J Environ Manage 170:88–96. https://doi.org/10.1016/j.jenvman.2016.01.015

Singh MD, Chirag G, Prakash PO et al (2017) Nano fertilizers is a new way to increase nutrients use efficiency in crop production. Int J Agri Sci 9:3831–3833

Singh S, Tripathi DK, Dubey NK et al (2016) Effects of nano-materials on seed germination and seedling growth: striking the slight balance between the concepts and controversies. Mater Focus 5(3):195–201

Solanki P, Bhargava A, Chhipa H et al (2015) Nano-fertilizers and their smart delivery system. In: Rai M, Ribeiro C, Mattoso L, Duran N (eds) Nanotechnologies in food and agriculture. Springer, Cham, pp 81–101

Song U, Jun H, Waldman B et al (2013) Functional analyses of nanoparticle toxicity: a comparative study of the effects of TiO2 and Ag on tomatoes (*Lycopersicon esculentum*). Ecotoxicol Environ Saf 93:60–67. https://doi.org/10.1016/j.ecoenv.2013.03.033

Storozhenko V, Svietlova N, Kovalenko M et al (2019) Induction of wheat seedlings resistance of different ecotypes to the effect of a drought simulated by a colloidal solution of Cu^{2+} and Zn^{2+} nanoparticles. Bull T Shevchenko Nat Uni Kyiv-Biol 76(2):79–84

Sun L, Song F, Guo J et al (2020) Nano-ZnO-induced drought tolerance is associated with melatonin synthesis and metabolism in maize. Int J Mol Sci 21(3):782. https://doi.org/10.3390/ijms21030782

Tarafdar JC, Agarwal A, Raliya R et al (2012a) ZnO nanoparticles induced synthesis of polysaccharides and phosphatase by Aspergillus fungi. Adv Sci Eng Med 4:1–5

Tarafdar JC, Raliya R, Mahawar H, Rathore I (2014) Development of zinc nanofertilizer to enhance crop production in pearl millet (*Pennisetum americanum*). Agric Res 3:257–262

Tarafdar JC, Raliya R, Rathore I (2012b) Microbial synthesis of phosphorous nanoparticles from tri-calcium phosphate using *Aspergillus tubigenesis* TFR-5. J Bionanosci 6:84–89

Taran N, Storozhenko V, Svietlova N et al (2017) Effect of zinc and copper nanoparticles on drought resistance of wheat seedlings. Nanoscale Res Lett 12:60. https://doi.org/10.1186/s11671-017-1839-9

Taylor AF, Rylott EL, Anderson CW, Bruce NC (2014) Investigating the toxicity, uptake, nanoparticle formation and genetic response of plants to gold. PLoS One 9(4): https://doi.org/10.1371/journal.pone.0093793

Taylor R, Walton DR (1993) The chemistry of fullerenes. Nature 363(6431):685

Thuesombat P, Hannongbua S, Ekgasit S et al (2016) Effects of silver nanoparticles on hydrogen peroxide generation and antioxidant enzyme responses in rice. J Nanosci Nanotechnol 16(8):8030–8043. https://doi.org/10.1166/jnn.2016.12754

Torabian S, Zahedi M, Khoshgoftar AH (2017) Effects of foliar spray of nano-particles of $FeSO_4$ on the growth and ion content of sunflower under saline condition. J Plant Nutr 40(5):615–623. https://doi.org/10.1080/01904167.2016.1240187

Tripathi DK, Mishra RK, Singh S et al (2017) Nitric oxide ameliorates zinc oxide nanoparticles phytotoxicity in wheat seedlings: Implication of the ascorbate–glutathione cycle. Front Plant Sci 8:1. https://doi.org/10.3389/fpls.2017.00001

Tripathi DK, Singh S, Singh VP et al (2016) Silicon nanoparticles more efficiently alleviate arsenate toxicity than silicon in maize cultivar and hybrid differing in arsenate tolerance. Front Environ Sci 4:46. https://doi.org/10.3389/fenvs.2016.00046

Tripathi DK, Singh VP, Prasad SM et al (2015) Silicon nanoparticles (SiNP) alleviate chromium (VI) phytotoxicity in *Pisum sativum* (L.) seedlings. Plant Physiol Biochem 96:189–198. https://doi.org/10.1016/j.plaphy.2015.07.026

Verma A (2016) Abiotic stress and crop improvement: current scenario. Adv Plants Agric Res 4:345–346

Vessey KV (2003) Plant growth promoting rhizobacteria as biofertlizers. Plant Soil 255:571–586

Vidyalakshmi N, Thomas R, Aswani R et al (2017) Comparative analysis of the effect of silver nanoparticle and silver nitrate on morphological and anatomical parameters of banana under in vitro conditions. Inorg Nano Metal Chem 47:1530–1536. https://doi.org/10.1080/24701556.2017.1357605

Wang H, Kou X, Pei Z et al (2011) Physiological effects of magnetite (Fe_3O_4) nanoparticles on perennial ryegrass (*Lolium perenne* L.) and pumpkin (*Cucurbita mixta*) plants. Nanotoxicology 5:30–42

Wang Q, Zhao S, Zhao Y et al (2014) Toxicity and translocation of graphene oxide in *Arabidopsis* plants under stress conditions. RSC Adv 4:60891–60901

Waraich EA, Ahmad R, Halim A, Aziz T (2012) Alleviation of temperature stress by nutrient management in crop plants: a review. J Soil Sci Plant Nutr 12(2):221–244. https://doi.org/10.4067/S0718-95162012000200003

Yan S, Zhao L, Li H et al (2013) Single-walled carbon nanotubes selectively influence maize root tissue development accompanied by the change in the related gene expression. J Hazard Mater 246:110–118

Yang J, Kloepper J, Ryu C (2009) Rhizosphere bacteria help plants tolerate abiotic stress. Trend Plant Sci 14:1–4

Yue N, Yun X (2018) An overview of biomembrane functions in plant responses to high-temperature stress. Front Plant Sci 9:915. https://doi.org/10.3389/fpls.2018.00915

Zaimenko NV, Didyk NP, Dzyuba OI et al (2014) Enhancement of drought resistance in wheat and corn by nanoparticles of natural mineral analcite. Ecol Balk 6(1):1–10

Zhang Y, Wu B, Xu H et al (2016) Nanomaterials-enabled water and wastewater treatment. Nano Impact 3:22–39

Zhao L, Hernandez-Viezcas JA, Peralta-Videa JR et al (2013) ZnO nanoparticles fate in soil and zinc bioaccumulation in corn plants *Zea mays* influenced by alginate. Environ Sci 15:260–266

Zhao L, Lu L, Wang A et al (2020) Nano-biotechnology in agriculture: use of nanomaterials to promote plant growth and stress tolerance. J Agric Food Chem 68(7):1935–1947. https://doi.org/10.1021/acs.jafc.9b06615

Zhao L, Sun Y, Hernandez-Viezcas JA et al (2015) Monitoring the environmental effects of CeO_2 and ZnO nanoparticles through the life cycle of corn (*Zea mays*) plants and in situ μ-XRF mapping of nutrients in kernels. Environ Sci Technol 49:2921–2928. https://doi.org/10.1021/es5060226

Zhu ZJ, Wang H, Yan B et al (2012) Effect of surface charge on the uptake and distribution of gold nanoparticles in four plant species. Environ Sci Technol 46:12391–12398. https://doi.org/10.1021/es301977w

Zuverza-Mena N, Martinez-Fernandez D, Du W et al (2017) Exposure of engineered nanomaterials to plants: insights into the physiological and biochemical responses—a review. Plant Physiol Biochem 110:236–264

Chapter 12
Role of Nanomaterials in Regulating Reactive Species as a Signaling Molecule of Abiotic Stress in Plants

Syed Uzma Jalil and Mohammad Israil Ansari

Abstract Nanotechnology is a promising field of science that contributes innovative approach to understand and develop suitable mechanisms to regulate the production of reactive species in plants based on nanoparticles. During abiotic stress condition, reactive oxygen species (ROS) are constantly produced in the cell organelles viz. mitochondria, peroxisomes and cytoplasm that can devastate the metabolism by oxidative damage of macromolecules such as lipids, proteins and nucleic acids.The distinctive physiochemical properties of nanoparticles have numerous applications in agricultural sector. Abiotic stress conditions present severe problems limiting crop productivity. Abiotic stresses can cause nutrient deficiency or toxicity symptoms leading to modification of normal processes of plants. This increases the production of reactive species, which leads to oxidative damage in the cells. Plants regulate various metabolic pathways to alleviate the generation of ROS inside the plant cell for improving the abiotic stress tolerance in plants. Antioxidant enzymes are important for the defense system in respond to reactive species in plants. Nanoparticles treatment provides enhanced performance of tolerance in plants against adverse environmental conditions through enhancing the free radical scavenging potential and antioxidant enzymatic activity. This chapter emphasizes the mechanism of nanoparticles involved in regulating stress tolerance to minimize the implications of abiotic stress in plants.

Keywords Abiotic stress · Nanomaterials · Reactive species · Signaling mechanism

---

S. U. Jalil
Amity Institute of Biotechnology, Amity University Uttar Pradesh, Lucknow Campus, Lucknow 226028, India

M. I. Ansari (✉)
Department of Botany, University of Lucknow, Lucknow University Road, Lucknow 226007, Uttar Pradesh, India
e-mail: ansari_mi@lkouniv.ac.in

J. M. Al-Khayri et al. (eds.), *Nanobiotechnology*,
https://doi.org/10.1007/978-3-030-73606-4_12

12.1 Introduction

Adequate abiotic environment are required by the plants for epigenetic mechanisms to mediate plant physiology and developmental progression. Unfavorable abiotic conditions are that set of abiotic stresses, which restrict plant production. Plants sense and respond to the stress conditions by several manners that maintain their sustenance (He et al. 2018; Jiang et al. 2016). Plants perceive the prior experience of stresses as well as the metabolism engaged with stress tolerance, and again when a similar stress exposed, they can adjusted accordingly (Hilker et al. 2016).The utmost apparent effect of unfavorable environmental conditions primarily appeared at the cell levels and then adverse physiological effects are observable. Water-logging stress destructively impacts on physiological parameters of plants including the photosynthetic complex (Ansari and Lin 2010; Xu and Zhou 2006). In plants, prolong exposure of water-logging stress impacts on morphological and physiological characters of plants (Xu et al. 2016).Therefore, plants can delicately improve the characteristic mechanisms to reduce the utilization of optimum water resources and regulated their growth prior to the exposure of abiotic stresses (Ansari and Silva da 2012; Genc et al. 2019; Iqbal et al. 2020; Osakabe et al. 2013, 2014; Yolcu et al. 2016). Exposure of dark stress slower down the physiological processes and negatively effects on plant development. High light intensity encourages photooxidation that elevates the generation of ROS that influences enzymes and other macromolecules (Pareek et al. 2010).

Plants acclimatize with the sudden changes and difficulty of environmental conditions by their metabolic systems (Simontacchi et al. 2015). Variation in environmental conditions prepared the plants metabolism accessible for homeostasis (Foyer and Noctor 2005). Plants maintain various defense mechanisms for mitigating with adverse environmental effects (Gill and Tuteja 2010) by metabolic reprogramming of cellular systems to facilitate the physicochemical process of the peripheral conditions (Mickelbart et al. 2015).

Nanoparticles are range of 1–100 nm dimensions particles (Kah et al. 2019; Roco 2003), with various physiochemical characteristics including highly reactive, high surface area, acquiescent pore size and varied shapes (Nel et al. 2006). In current era, nanotechnology is achieving momentum to involve the promising situation to alleviate the limitations related with adverse environmental conditions to acquire protected future of agriculture around the globe (Zhao et al. 2020). Plants are frequently exposed to ecological variations and different adverse environment in their life span (Torney et al. 2007). However, plants activates different mechanism to develop resistance against adverse environmental conditions, while responses might be differ significantly even in the similar plant varieties. Thus, classification of tolerant plants or improvement of resistance in plants is constantly a primary concern in agricultural sector as well as crops productivity. Nanobiotechnology includes advanced approaches, which allows pioneering studies in different extents, and nano technological researches pave new way in the area of agricultural sector (Lowry et al. 2019; Siddiqui et al. 2015). In the current consequence, nanoparticles are probable to improves plant's growth (Giraldo et al. 2019; Kah et al. 2019), and are involved in ROS scavenging mechanism that help in protecting plant from adverse environmental environment (Jalil and Ansari 2019; Zhao et al. 2020).

12.2 Production of ROS During Stress Conditions

Various abiotic stresses affected plants' development and decreases crops profitability (Jalil and Ansari 2020a), acidic condition horribly impact on soil supplements, which results to lacking of essential supplements in plants and disturb physiological metabolism (Bromham et al. 2013; Emamverdian et al. 2015). Prolonged exposure to extreme salinity environment leads to harmful effects in the cells and interference of osmotic regulation (Fig. 12.1). Outcome of ionic followed with osmotic stresses results to hamper plant development (Munns and Tester 2008). Tolerance to salt stress required to control ionic and osmotic condition in the plant cell (Jalil and Ansari 2020b). Additionally, plants protect weak plants' tissues from extreme salt condition by emitting ions from roots (Silva et al. 2010). Under chilling environment, various plants activated the system for combating cold temperatures

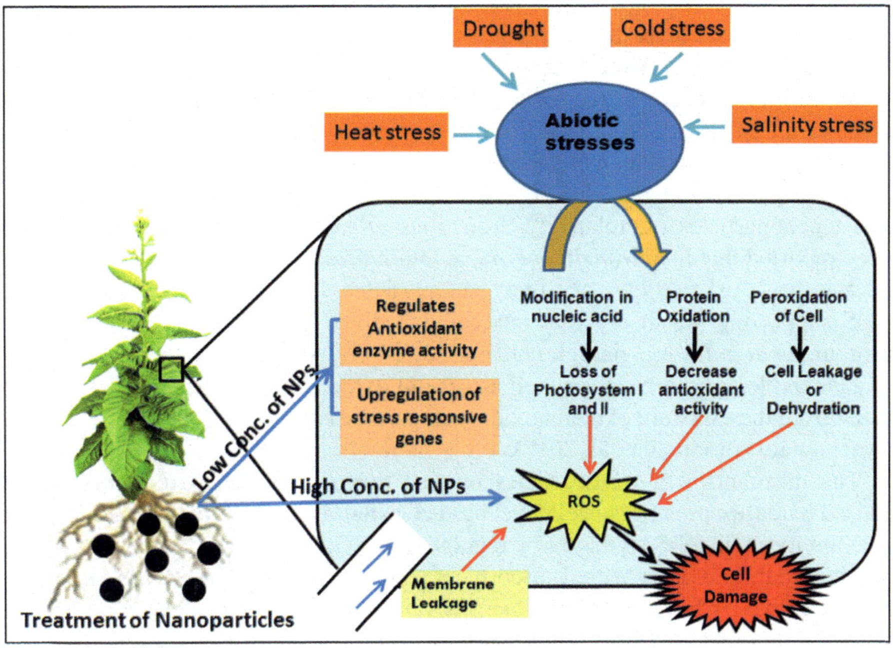

Fig. 12.1 Effect of abiotic stress on plant cell and mechanism of nanoparticles in the regulation of ROS in mitigation of influence of abiotic stress conditions

by increasing their defense ability by cold acclimation (Thomashow 2010). Subsequent to detecting the stresses, plants express an instant and convincing response to start complexed stress-specific signaling via producing phytohormones (Tiwari et al. 2011).

Production of reactive species is the primary reaction of abiotic stresses in plants. The production and annihilation of reactive species are at equilibrium in favorable environment, whereas under adverse environmental condition, it hinders this balance via enhancing the generation of reactive species that are harmful for the plants as they unfavorably affect the macromolecules. The ROS produced in different compartment of plant's cell amongst them chloroplast it the main site for the production of ROS (Davletova et al. 2005), because of elevated oxygen pressure and decreased atomic oxygen in different site of electron transport chain inside photosystem-I. ROS involve in the degradation of the macromolecules, for example, proteins, fats and sugars, which causes cell damage and ultimately diminished the crops production (Foyer and Noctor 2005).

12.3 Process of Nanoparticles Mechanism Under Stress Conditions

Treatment of toxic level of nanoparticles causes oxidative stress in plants (Xia et al. 2015). Treatment of different nanoparticles, viz. Silver, Zinc oxide and Aluminum oxide, also produced reactive species in duckweed (Thwala et al. 2013) and maize (Zhao et al. 2012). Al_2O_3nanoparticles also provoked highly reactive forms of reactive oxygen intermediates i.e. superoxide anion in tobacco cells (Poborilova et al. 2013).Despite of the fact that it is discussed that ROS production caused from intact nanoparticles or, relatively, from ions released from nanoparticles. It has been reported that in *Spirodela polyrhiza*, internalized Ag, whether or not the introduction was Ag^+ particles or Silver nanoparticles, had a similar ability to create ROS supporting the theory that intracellular silver nanoparticles dissociates into profoundly harmful Ag^+ particles (Jiang et al. 2017). Additionally, the accumulation of Zinc oxide, copper oxide as well as cerium oxide and their ions has been recognized from the exposure of metal oxide nanoparticles in carrot (Ebbs et al. 2016) and sweet potato (Bradfield et al. 2017).

The mechanism of nanoparticles by that they produced reactive species and caused oxidative pressure at cell level has been studied. Silver nanoparticles enacted calcium ions and ROS signaling by regulating the Calcium ions channels as well as direct oxidation of apoplastic L-ascorbic acid (Sosan et al. 2016). *Arabidopsis* root hair defective 2 (rhd2) mutant deficient NADPH oxidase-Respiratory burst oxidase homolog protein C (RBOHC) gene showed a fundamentally lower level of reactive species production as comparison to wild type plants after the application of silver nanoparticles (Sosan et al. 2016). This is revealing that the enhanced level of reactive species into the cells is interceded via plasma membrane bound NADPH oxidases

(RBOH) enzyme, which produced reactive species in the apoplast (Mittler 2017). Whereas, the production of reactive species in chloroplast was seen in *Spirodela polyrhiza*, in light of the capability of silver nanoparticles to hinder photosynthetic system (Jiang et al. 2017).

The destructive levels of reactive species are the reason of degradation of macromolecules as well as depletion membrane lipids which leads to cell damage (Van Breusegem and Dat 2006). Exposure of different concentration of Copper oxide nanoparticles (CuONPs) causes growth restriction followed by lipid peroxidation in *Oryza sativa* (Shaw and Husain 2013) and exposure of Titanium oxide nanoparticles (TiO_2NPs) in *Nitzschia closterium* (Xia et al. 2015). Nanoparticles may also degradenucleic acids and proteins of plant cells. Silver and Gold nanoparticles influenced cell division in onion root tip cells (Rajeshwari et al. 2016), and causes chromosomal changes in cells (Kumari et al. 2009). DNA degradation, deterioration of mitochondria, and cell death were reported in eggplants due to oxidative stress provoked by Cobalt oxide (Co_3O_4) (Faisal et al. 2016).

Plants activated enzymatic as well as non enzymatic antioxidant defense responses for scavenging excessive reactive species to alleviate the oxidative stress on plants (Sewelam et al. 2016; Jalil and Ansari 2020c). Furthermore, nanoparticles induced stresses and also initiate antioxidant defense mechanism in plants (Fig. 12.1). It has been studied that exposure of nanoparticles on plants involve in the up-regulation/downregulation of some antioxidant enzymes such as superoxide dismutase (SOD), which detoxifies superoxides into oxygen and water and ascorbate peroxidase (APX) that detoxifies peroxides by ascorbic acid (Asc) were upregulated (Fu et al. 2014). While, dehydro ascorbate reductase (DHAR) and monodehydro ascorbate reductase (MDAR) that regulated the Asc redox reaction in cells were down-regulated (Fu et al. 2014). It has been observed that the exposure of silver nanoparticles on rice shows enhanced activities of SOD, APX as well as glutathione transferase by proteomic analysis (Mirzajani et al. 2014). In Pea seedlings, application of silver nanoparticles increases SOD and APX activities whereas hindering glutathione reductase (GR) and DHAR activities (Tripathi et al. 2017). Catalase, which prevents the cells from oxidative stress, was increased after the exposure of copper oxide nanoparticles on wheat roots (Dimkpa et al. 2012). Dose dependent increase in the reactive species has been observed in Maize plants grown on soils supplemented with different concentration of cerium oxide nanoparticles after 10 days whereas after 20 days did not show any distinction (Zhao et al. 2012). Similarly, cerium oxide nanoparticles treated maize seedlings shows increased antioxidant enzyme activities that protect the plants from lipid per oxidation (Zhao et al. 2012).

The disturbance of generation of reactive species, hinders plants growth, while balancing of reactive species concentration in proportional manners support plant physical condition (Mittler 2017), it has been reported in several studies that it is promising to stimulates the antioxidant machinery by nanoparticles may encourage plant growth condition (Kumar et al. 2013) as long as the toxic level of reactive species is not get to the cells, while, it enters, which can prompt to disturbed organelle functions, membranes harm, and finally causes toxicity in plants.

12.4 Effect of Nanomaterials on Plants During Abiotic Stresses

Plants are able to adjust or adapt up to abiotic stress conditions viz. chilling, salt, heat, and drought stress and so forth. The plant responses at cell and molecular level against these abiotic stresses (Duque et al. 2013; Jalil and Ansari 2020b, c). The fundamental responses of plants for stress tolerance contains the temporary acceleration of Ca^{2+} in cytoplasm that increases the intracellular secondary messengers, reactive species, abscisic acid (ABA) and acceleration in signaling pathways (Baxter et al. 2013). The developmental stage of stresses response regulated the proteins that engaged inprevention from cell death, and regulated the stresses specific-gene expressions (Mahalingam and Fedoroff 2003). Secondary metabolites significantly involves in stress tolerance in plants by maintaining cell structure, protection of photo-system from reactive species as well as signaling mechanism (Oh et al. 2009). During adverse environmental conditions, plant cell play role as physical obstruction against stresses and involvesin plant adaptation for adverse environmental conditions (Degenhardt and Gimmler 2000). Extracellular peroxidases modified cell wall, involve in the accumulation of reactive species and causes oxidative stress while experiencing adverse environmental conditions (Daudi et al. 2012; Rouet et al. 2006).

The effect of nanoparticles treatment over plants metabolism is based on its concentration used that was demonstrated in a few investigations. A nanoparticle also involves in the upregulation of the antioxidant activities as shown in Fig. 12.2 (Laware and Raskar 2014). Furthermore, it has been reported that the application of TiO_2 nanoparticles on onion seedlings increased the superoxide dismutase enzyme activity as increased concentration of nanoparticles whereas, there were significant generation of other enzymes such as amylase, catalase and peroxidase activities were higher at low dose of TiO_2 nanoparticles and declined at higher concentration. Conversely, germination and seedlings development in onion were enhanced in lower

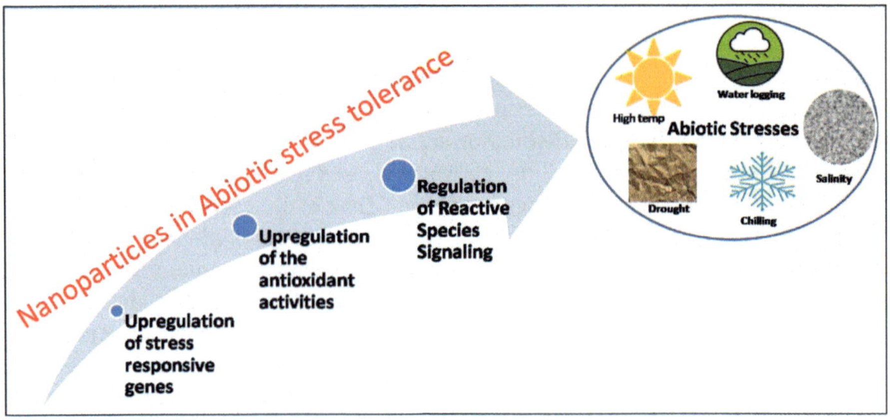

Fig. 12.2 Role of nanoparticle in abiotic stress tolerance in plants

dose of TiO_2 nanoparticles while suppressed at higher concentrations. A few investigations recommended that application of TiO_2 and SiO_2 nanoparticles can possibly improve the germination and development of *Glycine max* (Lu et al. 2002).

12.5 Regulation of Reactive Species Signaling by Nanoparticles Under Abiotic Stresses

As indicated by the current information on nanoparticles-plant communication and translocation in plant development and during abiotic stress conditions (Avellan et al. 2019), it revealed that generation of reactive species is a typical counter of plants to the adverse environmental condition. Plants have defense system for recognition and stimulation of specific stress. ROS involved in stress signaling to activate plants defense system and elevate cell death. (Dat et al. 2000). However, nanoparticles just prompted the production of reactive species (Simon et al. 2013; Qi et al. 2013), yet additionally mimicking antioxidants enzymes activity for scavenging reactive species (Wei and Wang 2013).Treatment of biochemically synthesized Gold Nanoparticles on Tobacco seedlings involved in improving plant development and ROS scavenging responses under different stress conditions (Jalil and Ansari 2018).

The different studies on role of nanoparticles reported that nanoparticles involved in defense mechanism against reactive species and also causes oxidative stress in plant cells. This study may be dealt by investigating about the role of nanoparticles in plant signaling under adverse environmental conditions. While, the mechanism of nanoparticles in reactive species signaling during stress condition is not well known, however, by means of proteomic and genomic studies it will be possible to comprehend the mechanisms of nanoparticles in plant under different stresses. Treatment of Silicon Dioxide nanoparticles enhanced salt stress resistance in Strawberry plants (Avestan et al. 2019). Furthermore, the exposure of silver and silver nitrate nanoparticles modified the proteins that regulated the redox in *Eruca sativa* roots (Vannini et al. 2013).Treatment of silver and silver nanoparticles imbibed with polyvinylpyrrolidone (PVP) controlled stress responsive genes expressions in *Arabidopsis* (Kaveh et al. 2013). Exposure of zinc and iron oxide nanoparticles on wheat enhanced the plant growth and decreased the oxidative stress and cadmium concentration (Rizwan et al. 2019). In another study application of silica nanoparticles inhibited the uptake of arsenic uptake in rice suspension cells by improving pectin synthesis and the mechanical strength of the cell wall (Cui et al. 2020).

The miRNA regulated the biological metabolism in organisms as well as also involved in plant responses against adverse environmental conditions (Macovei et al. 2012). Relationship of nanoparticles with miRNAs uncovers the mechanism of nanoparticles during abiotic stresses. It has been observed that the miRNAs upregulated in response to heavy metal stress on tobacco plants after the application of TiO_2 and Al_2O_3 nanoparticles, however high concentration of these nanoparticles shown negative effect on physiological traits of plants (Frazier et al. 2014). Similarly, the exposure of FeNPs on Arabidopsis involved in upregulation of the *AHA2* (associated with stomatal opening procedure) gene that improved the tolerance of plants against drought stresses (Kim et al. 2015). Moreover, the application of TiO_2, and silver nanoparticle on *Arabidopsis* downregulated the genes involved in roots growth and phosphate starvation during adverse environmental conditions (García-S-anchez et al. 2015). It has been observed in the study that application of nano-NiO on *Hordeum vulgare* actuates the overproduction of reactive species, which elevates the lipid peroxidation and leads to oxidative stress in plants. However, co-treatment of SiO_2nanoparticle on nano-NiO exposed plants prompted antioxidant potential, decreases the lipid peroxidation, regulated the redox reactions that involve in the mitigation of toxicity of nano-NiO on plants. This investigation suggested that SiO_2 nanoparticles act as a protective agent against the stress caused by the exposure of nano-NiO on *Hordeum vulgare* (Soares et al. 2018). In another study, it has been observed that Zinc oxide nanoparticles alleviate drought-induced modification in sorghum by improving the plant development, nutrient acquisition, and grain enrichment (Dimpka et al. 2019).

Signaling networks involves in triggering of defense mechanism, which activates the molecular mechanism against specific stresses. Ca^+ associated with signal transduction as secondary messenger during adverse environmental conditions. Affectability of stress signals prompts the mobilization of Ca^+ to cytosol through Ca^+ channels, which increases the level of Ca^+ in the cytosol to recognize by Ca^+ binding proteins which involved in the regulation of gene expression and plants tolerance against adverse environmental conditions (Khan et al. 2014). Moreover, the nitric oxide (NO) increases the level of Ca^+ in cytosol of the plants under different stresses and consequently, Ca^+ involve in the synthesis of NO (Corpas et al. 2004). The exposure of silver nanoparticle on rice roots shows regulation and signaling of Ca^+ and others metabolic pathways by nanoparticle responsive genes (Mirzajani et al. 2014). Moreover, silver nanoparticles binds with Ca^+ channel or Ca^+/Na^+ pump by Ca^+ receptors, which influence metabolism of cells (Mirzajani et al. 2014). Similarly, the C60 nanoparticles prompt the activity of (Calcium/calmudulin) CAM protein kinases (Miao et al. 2014). Additionally, the exposure of CdSQuantum dots on *Arabidopsis,* over-expressed the Ca^+ binding protein and CAM kinase (Marmiroli et al. 2015). These Ca^+ binding proteins engaged in the regulation of stress responses and over-expression of these genes improved different stress resistance in plants (Boudsocq and Sheen 2013).

Nanoparticles increase the nitrate reductase activities in plants that elevate the nitric oxide level in cells for modulating stress tolerance (Chandra et al. 2015). Therefore, a study reported that NO induces nanoparticles associated toxicity as well as expressed antioxidants enzyme encoded genes followed by reducing the lipid peroxidation and the production of reactive species. Study of correlation between Ca^+ and nanoparticles revealed that nanoparticles mimicked as Ca^+ and bind with Ca^+ binding proteins, which activates cascade of stress responsive genes (Mirzajani et al. 2014). Moreover, exposure of nanoparticles increases the gene expressions of stress related genes, cell elongation and cell division (Almutairi 2016). Furthermore, the multi-wall carbon nanotubes enters into the cell of the plants because of that plants senses carbon nanotubes as stress stimulus as like as biotic stress (Kwak et al. 2019). In this way, it is important to study further on the mechanism of involvement of nanoparticles in signal transduction in plants (Khodakovskaya et al. 2012). Furthermore, it has been reported that the exposure of nanoparticles increases the production of reactive species that causes phytotoxicity, which act as toxic substance or signaling compound in plant cells.

The diverse function of reactive species represented by their fabrication and scavenging activity; disparity in any of these function will results to increase fabrication or reduction in accessibility of ROS that prompts oxidative stress and disturb signaling respectively. Therefore, this regularity is supported through continuous generation or scavenging of ROS. Furthermore, it has been observed that the exposure of high concentration of nanoparticles causes toxicity in plants, while low concentration showed positive or no effect on the plants. It presumes that low concentration of nanoparticles activates antioxidant system which regulates the production of ROS in defined concentration that sufficient for signaling however inadequate to cause harm (Syu et al. 2014; Zhao et al. 2020).

12.6 Conclusion and Prospects

Abiotic stress are the main existing forms of environmental hazard that results to negative effects in plants and causes foremost environmental problems globally. Abiotic stresses are the key source in accumulation of reactive species that prompts oxidative stress and disturb signaling mechanism in plant cells. Higher plants have evolved an intricate antioxidant defense system to scavenge reactive species during abiotic stress conditions. Nanotechnology incorporate in agriculture can pave the way to modernized agricultural practices promoted by the advances made in the crop protection research, particularly in the abiotic stress tolerance. Nanoparticles improving the stress resistance by enhancing the root hydraulic conductivity and water uptake capacity of plants and increasing differential proteins that regulates oxidation–reduction, detoxification of reactive oxygen species, stress signaling and hormonal pathways. Nanoparticles interact with plant cells, which modified the gene expression and metabolic pathways that involve in plant development. Nanomaterials activate antioxidant system that involves in the regulation of the generation of reactive species in defined concentration that adequate for signaling mechanism. In current era, crop protection research has been focusing more so on approaches to use of nanomaterials empowers its utilization for the management of abiotic stresses that affects agricultural crops. Application of nanotechnology in agricultural field leading to the progress of several inexpensive approaches in agriculture. It required advance research on synthesis, characterization, standardization, biodegradability, ecofriendly nature and uptake and translocation of nanoparticles by plants.

References

Almutairi ZM (2016) Effect of nano-silicon application on the expression of salt tolerance genes in germinating tomato (*Solanum lycopersicum* L.) seedlings under salt stress. Plant Omics J 9:106–114

Ansari MI, Lin TP (2010) Molecular analysis of dehydration in plants. Int Res J Plant Sci 1(2):21–25

Ansari MI, da Silva JAT (2012) Molecular Analysis of TLP18.3gene in response to the abiotic stress in *Arabidopsis thaliana*. Plant Stress 6:22–24

Avellan A, Yun J, Zhang Y et al (2019) Nanoparticle size and coating chemistry control foliar uptake pathways, translocation, and leaf-to-thizospheretransport in wheat. ACS Nano 13:5291–5305

Avestan S, Ghasemnezhad M, Esfahani M, Byrt CS (2019) Application of nano-silicon dioxide improves salt stress tolerance in strawberry plants. Agronomy 9:246. https://doi.org/10.3390/agronomy9050246

Baxter A, Mittler R, Suzuki N (2013) Ros as key players in plant stress signalling. J Exp Botany 65:1229–1240

Boudsocq M, Sheen J (2013) CDPKs in immune and stress signalling. Trends Plant Sci 18:30–40

Bradfield SJ, Kumar P, White JC, Ebbs SD (2017) Zinc, copper, or cerium accumulation from metal oxide nanoparticles or ions in sweet potato: yield effects and projected dietary intake from consumption. Plant Physiol Biochem 110:128–137

Bromham L, Saslis-Lagoudakis CH, Bennett TH, Flowers TJ (2013) Soil alkalinity and salt tolerance: Adapting to multiple stresses. Biol Lett 9:20130642

Chandra S, Chakraborty N, Dasgupta A, Sarkar J, Panda K, Acharya K (2015) Chitosan nanoparticles: a positive modulator of innate immune responses in plants. Sci Rep 5:15195

Corpas FJ, Barroso JB, Carreras A, Quiror M, Leon AM, Romero-Puertas MC, Esteban FJ, Valderrama R, Palma JM, Sandalio LM, Gomez M, del Río LA (2004) Cellular and subcellular localization of endogenous nitric oxide in young and senescent pea plants. Plant Physiol 136:2722–2733

Cui J, Li Y, Jin Q, Li F (2020) Silica nanoparticles inhibit arsenic uptake into rice suspension cells via improving pectin synthesis and the mechanical force of the cell wall. Environ Sci Nano 7:162–171

Dat J, Vandenabeele S, Vranova E et al (2000) Dual action of the active oxygen species during plant stress responses. Cell Mol Life Sci 57:779–795

Daudi A, Cheng Z, O'Brien JA et al (2012) The apoplastic oxidative burst peroxidase in Arabidopsis is a major component of pattern-triggered immunity. Plant Cell Online 24:275–287

Davletova S, Rizhsky L, Liang H et al (2005) Cytosolic ascorbate peroxidase 1 is a central component of the reactive oxygen gene network of *Arabidopsis*. Plant Cell 17:268–281

Degenhardt B, Gimmler H (2000) Cell wall adaptations to multiple environmental stresses in maize roots. J Exp Bot 51:595–603

Dimkpa C, Mclean J, Latta D et al (2012) CuO and ZnO nanoparticles: phytotoxicity, metal speciation, and induction of oxidative stress in sand-grown wheat. J Nanopart Res 14:1–15

Dimkpa CO, Singh U, Bindraban PS et al (2019) Zinc oxide nanoparticles alleviate drought-induced alterations in sorghum performance, nutrient acquisition, and grain fortification. Sci Total Environ 688:926–934

Duque AS, de Almeida AM, da Silva AB et al (2013) Abiotic stress responses in plants: unraveling the complexity of genes and networks to survive. In: Vahdati K, Leslie C (eds) Abiotic stress-plant responses and applications in agriculture. Intech, Rijeka, Croatia, pp 49–101

Ebbs SD, Bradfield SJ, Kumar P et al (2016) Accumulation of zinc, copper, or cerium in carrot (*Daucus carota*) exposed to metal oxide nanoparticles and metal ions. Environ Sci Nano 3:114–126

Emamverdian A, Ding Y, Mokhberdoran F, Xie Y (2015) Heavy metal stress and some mechanisms of plant defense response. Sci World J 2015:75–81

Faisal M, Saquib Q, Alatar AA et al (2016) Cobalt oxide nanoparticles aggravate DNA damage and cell death in eggplant via mitochondrial swelling and NO signaling pathway. Biol Res 49:20

Foyer CH, Noctor G (2005) Oxidant and antioxidant signalling in plants: A re-evaluation of the concept of oxidative stress in a physiological context. Plant, Cell Environ 28:1056–1071

Frazier TP, Burklew CE, Zhang B (2014) Titanium dioxide nanoparticles affect the growth and microRNA expression of tobacco (*Nicotiana tabacum*). Funct Integr Genom 14:75–83

Fu PP, Xia Q, Hwang HM, Ray PC, Yu H (2014) Mechanisms of nanotoxicity: generation of reactive oxygen species. J Food Drug Anal 22:64–75

García-S,_anchez S, Bernales I, Cristobal S(2015) Early response to nanoparticles in the Arabidopsis transcriptome compromises plant defence and root-hair development through salicylic acid signaling. BMC Genom 16:341

Genc Y, Taylor J, Lyons G, Li Y, Cheong J, Appelbee M, Oldach K, Sutton T (2019) Bread wheat with high salinity and sodicity tolerance. Fron Plant Sci 10:1280

Gill SS, Tuteja N (2010) Reactive oxygen species and antioxidant machinery in abiotic stress tolerance in crop plants. Plant Physiol Biochem 48:909–930

Giraldo JP, Wu H, Newkirk GM, Kruss S (2019) Nanobiotechnology approaches for engineering smart plant sensors. Nature Nanotechnol 14:541–553

He M, He C-Q, Ding N-Z (2018) Abiotic stresses: general defenses of land plants and chances for engineering multistress tolerance. Front Plant Sci 9:1771

Hilker M, Schwachtje J, Baier M et al (2016) Priming and memory of stress responses in organisms lacking a nervous system. Biol Rev Camb Philos Soc 91:1118–1133

Iqbal MS, Singh AK, Singh SP, Ansari MI (2020) Nanoparticles and plant interaction with respect to stress response. In: Bhushan I, Singh V, Tripathi D (eds) Nanomaterials and environmental biotechnology. Nanotechnology in the Life Sciences, Springer, Cham, pp 1–15

Jalil SU, Ansari MI (2019) Nanoparticles and abiotic stress tolerance in plants: synthesis, action and signalling mechanisms. In: Iqbal M. Khan R, Reddy PS, Khan NA (eds) Plant signaling molecules—role and regulation under stressful environments. Elsevier Publication, pp 549–561

Jalil SU, Zahera M, Khan MJ, Ansari MI (2018) Biochemical Synthesis of Gold Nanoparticles from Leaf Protein of *Nicotian atabacum* L. cv. xanthi and their Physiological, Developmental and ROS Scavenging Responses on Tobacco Plant under Stress Conditions. IET Nanobiotechnol 13:23–29

Jalil SU, Ansari MI (2020a) Stress Implication and crop productivity. In: Hasanuzzaman M (ed) Plant ecophysiology and adaptation under climate change: mechanisms and perspectives I. Springer Nature, Singapore, pp 73–86

Jalil SU, Ansari MI (2020b) Physiological role of Gamma-Aminobutyric Acid in salt stress tolerance (ed), Salt and drought stress tolerance in plants, Signaling and communication in plants. Springer Nature Switzerland AG, pp 337–350

Jalil SU, Ansari MI (2020c) Isoprenoids in plant protection against abiotic stress, (ed) protective chemical agents in the amelioration of plant abiotic stress: Biochemical and Molecular Perspectives, John Wiley & Sons, Inc., pp 424–436

Jiang HS, Yin LY, Ren NN, Zhao ST, Li Z, Zhi Y et al (2017) Silver nanoparticles induced reactive oxygen species via photosynthetic energy transport imbalance in an aquatic plant. Nanotoxicology 11:157–167

Jiang QY, Zhuo F, Long SH, Zhao HD, Yang DJ, Ye ZH et al (2016) Can arbuscular mycorrhizal fungi reduce cd uptake and alleviate cd toxicity of *Lonicera japonica* grown in cd-added soils? Sci Rep 6:21805

Kah M, Tufenkji N, White JC (2019) Nano-enabled strategies to enhance crop nutrition and protection. Nature Nanotechnol 14:532–540

Kaveh R, Li Y-S, Ranjbar S et al (2013) Changes in *Arabidopsis thaliana* gene expression in response to silver nanoparticles and silver ions. Environ Sci Technol 47:10637e10644

Khan MN, Mohammad F, Mobin M, Saqib MA (2014) Tolerance of plants to abiotic stress: a role of nitric oxide and calcium. In: Khan MN, Mobin M, Mohammad F, Corpas FJ (eds) Nitric oxide in plants: metabolism and Role. Springer International Publishing, Switzerland, pp 225–242

Khodakovskaya MV, de Silva K, Biris AS et al (2012) Carbonnanotubes induce growth enhancement of tobacco cells. ACS Nano 6:2128–2135

Kim JH, Oh Y, Yoon H, Hwang I, Chang Y-S (2015) Iron nanoparticle-induced activation of plasma membrane Hþ-ATPase promotes stomatal opening in *Arabidopsis thaliana.* Environ Sci Technol 49:1113–1119

Kumari M, Mukherjee A, Chandrasekaran N (2009) Genotoxicity of silver nanoparticles in *Allium cepa.* Sci Total Environ 407:5243–5246

Kumar V, Guleria P, Kumar V, Yadav SK (2013) Gold nanoparticle exposure induces growth and yield enhancement in *Arabidopsis thaliana.* Sci Total Environ 46:462–468

Kwak S-Y, Lew TTS, Sweeney CJ et al (2019) Chloroplast- selective gene delivery and expression in planta using chitosan-complexed single-walled carbon nanotube carriers. Nature Nanotechnol 14:447–455

Laware SL, Raskar S (2014) Effect of titanium dioxide nanoparticles on hydrolytic and antioxidant enzymes during seed germination in onion. Int J Curr Microbiol App Sci 3:749–760

Lowry GV, Avellan A, Gilbertson LM (2019) Opportunities and challenges for nanotechnology in the agri-tech revolution. Nature Nanotechnol 14:517–522

Lu CM, Zhang CY, Wen JQ, Wu GR, Tao MX (2002) Research of the effect of nanometer materials on germination and growth enhancement of *Glycine max* and its mechanism. Soybean Sci 21:168–172

Macovei A, Gill SS, Tuteja N (2012) MicroRNAs as promising tools for improving stress tolerance in rice. Plant Signal Behav 7:1296–1301

Mahalingam R, Fedoroff N (2003) Stress response, cell death and signalling: The many faces of reactive oxygen species. Physiol Plant 119:56–68

Marmiroli M, Imperiale D, Pagano L et al (2015) The Proteomic response of *Arabidopsis thaliana* to cadmium sulphide quantum dots, and its correlation with the transcriptomic response. Front Plant Sci 6:1104

Miao Y, Xu J, Shen Y et al (2014) Nanoparticle as signalling protein mimic: robust structural and functional modulation of CaMKII upon specific binding to fullerene C60 nanocrystals. ACS Nano 8:6131–6144

Mickelbart MV, Hasegawa PM, Bailey-Serres J (2015) Genetic mechanisms of abiotic stress tolerance that translate to crop yield stability. Nat Rev Genet 16:237–251

Mittler R (2017) ROS are good. Trends Plant Sci 22:11–19

Mirzajani F, Askari H, Hamzelou S et al (2014) Proteomics study of silver nanoparticles toxicity on *Oryza sativa* L. Ecotoxicol Environ Saf 108:335–339

Munns R, Tester M (2008) Mechanisms of salinity tolerance. Annu Rev Plant Biol 59:651–681

Nel A, Xia T, Madler L, Li N (2006) Toxic potential of materials at the nanolevel. Science 311:622–627

Oh M-M, Trick HN, Rajashekar C (2009) Secondary metabolism and antioxidants are involved in environmental adaptation and stress tolerance in lettuce. J Plant Physiol 166:180–191

Osakabe Y, Arinaga N, Umezawa T et al (2013) Osmotic stress responses and plant growth controlled by potassium transporters in *Arabidopsis.* Plant Cell 25:609–624

Osakabe Y, Osakabe K, Shinozaki K. Tran LS (2014) Response of plants to water stress. Front Plant Sci 5:86–91

Pareek A, Sopory SK, Bohnert HK, Govindjee (2010) Abiotic stress adaptation in plants: physiolgical, molecular and genomic foundation. Springer, Dordrecht

Poborilova Z, Opatrilova R, Babula P (2013) Toxicity of aluminium oxide nanoparticles demonstrated using a BY-2 plant cell suspension culture model. Environ Exp Bot 91:1–11

Qi M, Liu Y, LiT (2013) Nano-TiO2 improve the photosynthesis of tomato leaves under mild heat stress. Biol Trace Elem Res 156: 323–328

Rajeshwari A, Suresh S, Chandrasekaran N, Mukherjee A (2016) Toxicity evaluation of gold nanoparticles using an *Allium cepa* bioassay. RSC Adv 6:24000–24009

Rizwan M, Ali S, Ali B et al (2019) Zinc and iron oxide nanoparticles improved the plant growth and reduced the oxidative stress and cadmium concentration in wheat. Chemosphere 214:269–277

Roco MC (2003) Broader societal issue of nanotechnology.J Nanoparticle Re. 5: 181–189

Rouet M, Mathieu Y, Barbier-Brygoo H, Laurière C (2006) Characterization of active oxygen-producing proteins in response to hypo-osmolarity in tobacco and *Arabidopsis* cell suspensions: Identification of a cell wall peroxidase. J Exp Bot 57:1323–1332

Sewelam N, Kazan K, Schenk PM (2016) Global plant stress signaling: reactive oxygen species at the cross-road. Front Plant Sci 7:187

Shaw AK, Hossain Z (2013) Impact of nano-CuO stress on rice (*Oryza sativa* L.) seedlings. Chemosphere 93:906–915

Siddiqui MH, Al-Whaibi MH, Firoz M, Al-Khaishany MY (2015) Role of nanoparticles in plants. In: Siddiqui M, Al-Whaibi M, Mohammad F (eds) Nanotechnology and plant sciences. Springer, Cham, pp 19–35

Silva EN, Ribeiro RV, Ferreira-Silva SL, Viégas RA, Silveira JA (2010) Comparative effects of salinity and water stress on photosynthesis, water relations and growth of *Jatropha curcas*plants. J Arid Environ 74:1130–1137

Simon DF, Domingos RF, Hauser C et al (2013) Transcriptome sequencing (RNA-seq) analysis of the effects of metal nanoparticle exposure on the transcriptome of *Chlamydomonas reinhardtii*. Appl Environ Microbiol 79:4774–4785

Simontacchi M, Galatro A, Ramos-Artuso F, Santa- María GE (2015) Plant survival in a changing environment: The role of nitric oxide in plant responses to abiotic stress. Front Plant Sci 6:977

Soares C, Branco-Neves S, Sousa A et al (2018) SiO_2 nanomaterial as a tool to improve *Hordeum vulgare* L. tolerance to nano-NiO stress. Sci Total Environ 622–623:517–525

Sosan A, Svistunenko D, Straltsova D et al (2016) Engineered silver nanoparticles are sensed at the plasma membrane and dramatically modify the physiology of *Arabidopsis thaliana* plants. Plant J 85:245–257

Syu YY, Hung JH, Chen JC, Chuang HW (2014) Impact of size and shape of silver nanoparticles on *Arabidopsis* plant growth and gene expression. Plant Physiol Biochem 83:57–64

Thomashow MF (2010) Molecular basis of plant cold acclimation: Insights gained from studying the CBF cold response pathway. Plant Physiol 154:571–577

Thwala M, Musee N, Sikhwivhilu L, Wepener V (2013) The oxidative toxicity of Ag and ZnO nanoparticles towards the aquatic plant *Spirodela punctata* and the role of testing media parameters. Environ Sci Process Impacts 15:1830–1843

Tiwari S, Singh P, Tiwari R et al (2011) Salt-tolerant rhizobacteria-mediated induced tolerance in wheat (*Triticum aestivum*) and chemical diversity in rhizosphere enhance plant growth. Biol Fertil Soils 47:907–916

Torney F, Trewyn BG, Lin VSY, Wang K (2007) Mesoporous silica nanoparticles deliver DNA and chemicals into plants. Nat Nanotechnol 2:295–300

Tripathi DK, Singh S, Singh S, Srivastava PK, Singh VP, Singh S et al (2017) Nitric oxide alleviates silver nanoparticles (AgNps)-induced phytotoxicity in *Pisum sativum* seedlings. Plant Physiol Biochem 110:167–177

Van Breusegem F, Dat JF (2006) Reactive oxygen species in plant cell death. Plant Physiol 141:384–390

Vannini C, Domingo G, Onelli E, Prinsi B, Marsoni M, Espen L, Bracale M (2013) Morphological and proteomic responses of *Eruca sativa* exposed to silver nanoparticles or silver nitrate. PLoS ONE 8:68752

Wei H, Wang E (2013) Nanomaterials with enzyme-like characteristics (nanozymes): next-generation artificial enzymes. Chem Soc Rev 42:6060–6093

Xia B, Chen B, Sun X, Qu K, Ma F, Du M (2015) Interaction of TiO2 nanoparticles with the marine microalga *Nitzschiaclosterium*: growth inhibition, oxidative stress and internalization. Sci Total Environ 508:525–533

Xu Z, Jiang Y, Jia B, Zhou G (2016) Elevated-CO2 response of stomata and its dependence on environmental factors. Front Plant Sci 7:657

Xu ZZ, Zhou GS (2006) Combined effects of water stress and high temperature on photosynthesis, nitrogen metabolism and lipid peroxidation of a perennial grass *Leymuschinensis*. Planta 224:1080–1090

Yolcu S, Ozdemir F, Güler A, Bor M (2016) Histone acetylation influences the transcriptional activation of POX in beta vulgaris L. and beta *Maritima* L. under salt stress. Plant Physiol Biochem 100:37–46

Zhao L, Lu L, Wang A et al (2020) Nanobiotechnology in Agriculture: Use of nanomaterials to promote plant growth and stress tolerance. J Agric Food Chem. https://doi.org/10.1021/acs.jafc.9b06615

Zhao L, Peng B, Hernandez-Viezcas JA et al (2012) Stress response and tolerance of *Zea mays* to CeO2 nanoparticles: cross talk among H_2O_2, heat shock protein, and lipid peroxidation. ACS Nano 6:9615–9622

Chapter 13
Role of Nanomaterials in Regulating Oxidative Stress in Plants

Swati Sachdev and Shamshad Ahmad

Abstract Production of reactive oxygen species (ROS) in living cells or tissue is a normal and an important phenomenon. However, in routine life plants encounter various abiotic and biotic stresses such as extreme temperature events, salinity, drought, pathogenic attack that trigger excess production of ROS. The exponential increase in ROS over scavenging antioxidants disturbs cellular homeostasis and induces oxidative stress. Oxidative stress damages biomolecules including proteins, DNA and lipids that causes lipid peroxidation, electrolyte leakage, denaturation of proteins, enzymes inactivation, inhibition of replication of genetic material and finally cell death under severe conditions. Nanomaterials (NMs) are small particles with a size range of 1–100 nm. Due to their small size and high surface area to volume ratio they show unique properties. NMs display ambiguous effects on plants i.e. either trigger or mitigate oxidative stress. NMs regulate ROS production in stressed cell by mimicking ROS quenching antioxidants or eliciting antioxidants production. These properties of NMs can be efficiently utilized to alleviate impacts of abiotic stress. The present chapter discusses how different NMs demonstrate regulating effect on production of ROS in cells, thereby modulate oxidative stress. It also outlines the factors such as particle size, coating materials, concentration that governs NMs activity and plant interaction.

Keywords Abiotic stress · Antioxidants · Electrolyte leakage · Lipid peroxidation · Reactive oxygen species

13.1 Introduction

Nanomaterials (NMs) are ultrafine structure with at least one dimension size ranging from 1 to 100 nm (Zhao et al. 2020). NMs on the basis of origin are categorized as natural, incidental and engineered NMs. Natural NMs (soot, volcanic dust, viruses)

S. Sachdev (✉) · S. Ahmad
Department of Environmental Science, School for Environmental Sciences, Babasaheb Bhimrao Ambedkar University, VidyaVihar, Rae Bareli Road, Lucknow 226025, UP, India

J. M. Al-Khayri et al. (eds.), *Nanobiotechnology*,
https://doi.org/10.1007/978-3-030-73606-4_13

are structures, which are originated through natural processes or activities such as forest fires, volcanic eruptions, ocean spray, weathering of metallic rocks, radioactive decay of radon gas, etc. (Khan et al. 2017; Monica and Cremonini 2009). Incidental NMs are produced unintentionally as a byproduct of intentional activities or operations including combustion of fossil fuel; burning of candles as well as biomass; wearing, tearing and corrosion of materials and industrial operations producing fumes (Khan et al. 2017; Monica and Cremonini 2009). Engineered NMs (carbon nanotubes, quantum dots, nanorods) are nano-scale structures that synthesized deliberately by humans to modify the existing characteristic of materials and produce new ones (Khan et al. 2017). Engineered NMs possess unique and novel physical, chemical, magnetic, electronic and optical properties relative to bulk material and have very high surface area to volume ratio therefore research and use of NMs has gained much focus in last decade (Ghosh et al. 2016; Jefferson 2000; Kumar et al. 2018). NMs found their application in field of biomedical (drug delivery, as biosensor), food industries, cosmetics, textile, electronics as well as in agriculture (Fu et al. 2014; Kumar et al. 2018).

The nano-scale size and large surface to volume ratio of NMs increase their scope for multifarious application, while on other hand these properties makes them highly reactive and enhances penetration power in living cells rendering nanotoxicity (Kumar et al. 2018). The phytotoxicity associated with NMs is of great concerns in terms of plant growth, food productivity and human health and is also a major challenge for the advancement of nanotechnology. Agricultural crops get exposed to NMs directly on application of nano-based fertilizers and pesticides and/or indirectly through leaching from landfills, atmospheric fallout, NMs used for remediation process and contaminated water used for irrigation (Kumar et al. 2018; Rizwan et al. 2017; Tripathi et al. 2017). The NMs present in agricultural soil interact with plant and either act as precursor of reactive oxygen species (ROS) production in cells that induce oxidative stress (Iqbal et al. 2020; Ma et al. 2015; Siddiqui et al. 2015) or provide protection against oxidative stress induced by other factors such as temperature, salinity, etc. (Khan et al. 2017). Onset of oxidative stress in plants affect structural and functional property of biomolecules, which causes membrane and tissue damage, alteration of genetic material and plant metabolism and ultimately leads to cell death (Kumar et al. 2017, 2018; Wani et al. 2016).

NMs have become integral part of our daily life from industries to home, from medical practices to agriculture activities (Nile et al. 2020). Plants being sessile organisms have no choice other than being exposed to NMs that have controversial effect on living cells. At some instance, NMs alleviates oxidative stress (Zhao et al. 2020), while in some cases trigger oxidative stress (Prasad et al. 2017) leading to phytotoxicity (Zhang et al. 2020). Due to the extensive potential of nanotechnology to assist plant to cope abiotic stress, it become extremely important to decipher and understand the mechanism as well as fate of NMs in plants at cellular and molecular level to minimize the potential risk and make maximum use of this new and advance technology (Tian et al. 2020) to render the effect of abiotic stress on plants in a sustainable and efficient modus operandi. The present chapter is dedicated to explore the insight on instrumental role of NMs towards regulation of oxidative stress in plants and the degree of intrinsic or induced (by NMs) defense strategies (antioxidants) deployed by plants for detoxification of oxidative stress.

13.2 Nanomaterials and Oxidative Stress

Oxidative stress is a phenomenon that occurs due imbalance in equilibrium between ROS (free radicals) and antioxidants in cellular compartment where production of ROS exceeds dramatically to that of antioxidants possessing potential to detoxify these free radicals (Pizzino et al. 2017; Sharma et al. 2012) (Fig. 13.1). ROS are oxygen containing molecules that include free radicals and non-radicals such as superoxide radical (O_2^-), hydrogen peroxide (H_2O_2), hydroxyl radical (•OH), singlet oxygen ($^1O_2$) (Dumont and Rivoal 2019; Kumar et al. 2018; Sharma et al. 2012). These molecules have uneven electron numbers that increases their reactivity and causes oxidation of large number of other molecules. ROS are produced as a by-product of oxygen metabolism or cellular activities of biological system such as electron transport chain of mitochondria, chloroplast, plasma membrane and endoplasmic reticulum; photorespiration; enzymatic activity of lipoxygenases and cyclooxygenases during metabolism of fatty acid like arachidonic acid and NADPH oxidase during plant development (Dumont and Rivoal 2019; Marino et al. 2012; Pizzino

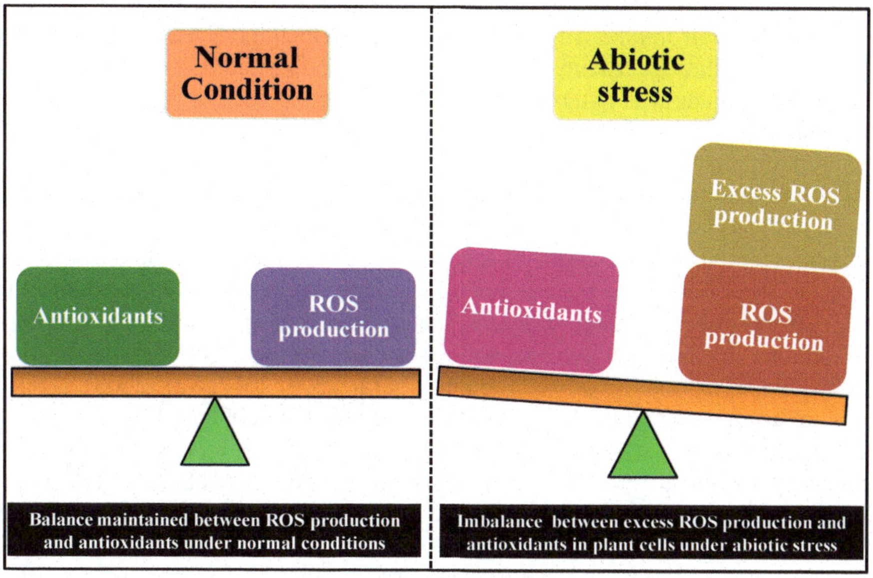

Fig. 13.1 Abiotic stress induced imbalance between ROS and antioxidants trigger oxidative stress in plants (Figure constructed by: Swati Sachdev and Shamshad Ahmad)

et al. 2017; Sharma et al. 2012). External stimuli including UV and ionizing radiation, heavy metals, xenobiotic compounds, heat shock, pathogens and so forth result in generation of ROS in plant cells due to disruption of cellular homeostasis (Sharma et al. 2012).

ROS show both beneficial and detrimental effect on living cells and tissues (Pizzino et al. 2017; Sharma et al. 2012). Production of ROS in cellular compartment in low or moderate level is essential and inevitable part of aerobic organisms (Pizzino et al. 2017). They actively participate in regulation of several processes such as cell signaling, protection against pathogens, apoptosis, protein phosphorylation, activation of various transcriptional factors and many more (Ismail et al. 2014; Pizzino et al. 2017; Rajendran et al. 2014). However, overproduction of these molecules on exposure to external stress stimuli, negatively affects macromolecules (proteins, lipids, carbohydrates and nucleic acids) causing lipid peroxidation, change in membrane fluidity and ion exchange, functional and structural loss of proteins, alteration in gene expression, inhibition of enzymatic activity, apoptosis, which subsequently leads to cell death (Dumont and Rivoal 2019; Pizzino et al. 2017; Sharma et al. 2012). Biological system deploys antioxidant based defense system to scavenge free radicals. The antioxidant system comprises two types of components: (a) Enzymatic including superoxide dismutase (SOD), catalase (CAT), ascorbate peroxidase (APX), guaiacol peroxidase (GPX) and glutathione reductase (GR); (b) Non-enzymatic such as flavonoids, tocopherol, carotenoids, polyamines, phenolic compounds, and glutathione (Dumont and Rivoal 2019; Pizzino et al. 2017).

During oxidative stress situation, ROS generated for instance hydroxyl radical (•OH) act as major initiator of lipid peroxidation, which damages lipoproteins and cell membrane through a series of chain reaction and eventually leads to the formation of malondialdehyde (MDA) and other end products such as conjugated dienes (Sharma et al. 2012). These end products are responsible for cytotoxic and mutagenic effects. Similarly, in presences of free radicals, protein molecules show conformational modification that ultimately impairs their enzymatic activity (Pizzino et al. 2017). ROS cause protein oxidation via four possible pathways: Oxidation of protein backbone; protein-protein cross linkage; oxidation of amino acid residue side chain; and oxidative cleavage of peptide bonds, resulting in protein fragmentation (Sitte 2003). Alteration in protein structure due to oxidation causes protein unfolding with increase hydrophobicity of its surface that in turn increases degradation property as compared to non-oxidized protein (Sharma et al. 2012; Sitte 2003). Nucleic acids are also prone to oxidative stress. Free radicals result in base lesions; break in DNA strands; and cross-linkages. A base lesion of DNA is a phenomenon, which commonly occurs as a part of normal biological functioning (Sharma et al. 2012). Cells have an intrinsic mechanism of base excision repair for defense against DNA lesions, however during oxidative stress conditions, such activity exceeds to a level beyond control and result in mutagenesis (Nishida et al. 2013; Pizzino et al. 2017). Oxidation also causes impairment of CpG island methylation in gene promoter that leads to loss of epigenetic information (Yasui et al. 2014). Analogous to MDA production as end product of lipid peroxidation, oxidation of protein and DNA result in formation of carbonyl derivatives (Sharma et al. 2012; Sitte 2003) and 8-hydroxydeoxyguanosine (8-OHdG) (Fu et al. 2014), respectively.

Metals (Au-gold, Ag-silver), metallic oxides (CuO-copper oxide, Fe_2O_3/Fe_3O_4-iron oxides, CeO_2-cerium oxide, TiO_2-titanium dioxide, MgO-magnesium oxide, SiO_2-silicon oxide, Al_2O_3-aluminium oxide), carbon (carbon nanotubes, fullerenes) and Quantum dots (cadmium selenide and cadmium sulfide) are four categories of NMs that are used in wide array of applications (Fu et al. 2014; Ju-Nam and Lead 2008). Nanotechnology has emerged as effective tool for managing agriculture related activities (Tripathi et al. 2015). Nanoformulations of ZnO, Ag, SiO_2 and many others are used in agriculture as fungicides, pesticides, herbicides, fertilizers, and abiotic stress regulators; nanosensor for identification of diseases such as CdSe/ZnS quantum dots (Santos et al. 2010), soil condition including residue concentration, etc., and nanodevices (carbon nanotubes) for genetic engineering (Demirer et al. 2018; Marslin et al. 2017). Due to advancement of nanotechnology in agriculture plants often get directly exposed to NMs, but they are also encounter with nano-scale materials through indirect means. The NMs waste generated from electronics, household and healthcare products, textile, food packaging and medicines released into environmental components (air, water and soil) and reaches to agricultural ecosystem (Marslin et al. 2017). Exploitation of nano-fertilizers in agriculture is encouraged as large surface area facilitates absorption of fertilizers by plants and reduces their leaching or emission into air or water ecosystem (Khan et al. 2017). Regulation of oxidative stress elicited by abiotic stresses in various plants mediated through NMs is presented in Table 13.1.

Impact of NMs on plants is very complicated and contradictory (Tripathi et al. 2015). NMs show ambiguous effect on plant growth and productivity. Some NMs participate in regulation of oxidative stress in plants, while other induces oxidative stress. Several studies have clearly illustrated effective role of NMs in alleviating oxidative stress in plant induced by abiotic factors such as UV-B radiation, heavy metals, drought, salinity, temperature, etc. (Jalil and Ansari 2019; Khan et al. 2017; Tripathi et al. 2017). Disruption of ROS homeostasis triggers oxidative stress that results in reduced plant growth and productivity, whereas its maintenance promotes plant growth (Mittler 2017). Few studies have reported that NMs mimic as antioxidants such as CAT, POX, SOD (Khan et al. 2017) or elevate production of antioxidants such as phenolics (Comotto et al. 2014; Marslin et al. 2017; Vecerova et al. 2016), which scavenge ROS (Marslin et al. 2017) and regulate oxidative stress, which in turn promote plant growth and development (Burman et al. 2013; Kumar et al. 2013). Treatment of wheat seedlings with silicon (Si) and silicon nanoparticle (SiNp) were observed to alleviate oxidative stress induced by UV-B radiation by stimulating production of antioxidants (Tripathi et al. 2017). Tomato (*Lycopersicon esculentum* Mill.) growing under Cadmium (Cd) stress on treatment with Zinc oxide nanoparticles (ZnO-NPs) displayed significant increase in plant growth and biomass, SPAD chlorophyll and photosynthetic activities with reduce metal toxicity and alleviation in oxidative stress which was evident from reduce H_2O_2, $O_2^{-\bullet}$ and malondialdehyde (MDA) content (Faizan et al. 2020). Similarly, amendment of ZnO-NP at 10–100 mg/L alleviated arsenic (As) induced oxidative stress in rice (*Oryza sativa* L.) by

Table 13.1 Regulation of oxidative stress by nanomaterials elicited by various abiotic stresses

Nanomaterial and effective dose	Plant exposed	Abiotic stress	Activity	References
Nano-anatase TiO_2 (Titanium dioxide)	Spinach (*Spinacia oleracea* L.)	UV-B radiation	Reduced oxidative stress by significantly lowering level of superoxide radical ($O_2^{\cdot -}$), hydrogen peroxide radical (H_2O_2) and malondialdehyde (MDA); enhanced activity of antioxidant enzymes SOD, CAT, APX and GPX and increased evolution of oxygen in chloroplast	Lei et al. (2008)
TiO_2-NP @ 5 ppm	Chickpea (*Cicer arietinum* L.)	Cold stress (4 °C)	Stress induced oxidative damage was alleviated and significant reduction in MDA content was observed	Mohammadi et al. (2013)
ZnO-NP (Zinc oxide) @ 1.5 ppm	Chickpea (*C. arietinum* L.)	–	Lowered ROS generation, lipid peroxidation and MDA concentration and improved biomass	Burman et al. (2013)
Ag-NP (Silver nanoparticle) @ 20–60 mg/cm^3	Pelargonium	Dark storage	Chlorophyll and carotenoid content elevated, production of antioxidant-ascorbate peroxidase and guaiacol peroxidase boosted that significantly reduced oxidative stress and MDA content	Hatami and Ghorbanpour (2013)

(continued)

Table 13.1 (continued)

Nanomaterial and effective dose	Plant exposed	Abiotic stress	Activity	References
TiO_2-NP @ 10 mg/l	Flax seed (*Linum usitatissimum* L.)	Drought	Morphological and physiological parameters improved, photosynthetic pigments (chlorophyll and carotenoid) increased; hydrogen peroxide content and lipid peroxidation reduced that resulted in lower production of MDA content	Aghdam et al. (2016)
TiO_2-NP @ 10 ppm	Dragon head (*Dracocephalum moldavica* L.)	Water deficit stress (50% field capacity irrigation done)	Mitigated oxidative stress and membrane damage by decreasing H_2O_2 and MDA and increasing proline content	Mohammadi et al. (2016)
ZnO-NP @ 60 mg/l	Lupin (*Lupinus termis* Forssk.)	Salinity (0–150 mM NaCl)	Abatement of stress and MDA and Na^+ content in tissues; increased total phenol and antioxidant content (SOD, POX (peroxidase) and APX); improved photosynthetic pigments and osmoregulation	Latef et al. (2017)

(continued)

Table 13.1 (continued)

Nanomaterial and effective dose	Plant exposed	Abiotic stress	Activity	References
γ-Fe_2O_3-NP (Iron oxide nanoparticle-Maghemite)	*Brassica napus* L.	Drought stress	Rate of leaf growth and chlorophyll content was increased from 33 to 50% and 47 to 52%, respectively whereas hydrogen peroxide radical and MDA concentration was downregulated from 151 to 83 μM/g and 36 to 26 mM/g, respectively	Palmqvist et al. (2017)
Poly (acrylic acid) nanoceric (CeO-NP)- PNC	*Arabidopsis thaliana* L.	Light (2000 μmol/m^{2}s for 1.5 h), Heat (35 °C for 2.5 h), and dark chilling (4 °C for 5 days) stress	PNC with low Ce^{3+}/Ce^{4+} ratio scavenged ROS and reduced content of H_2O_2 radical, O_2^- and OH^- by 52%	Wu et al. (2017)
ZnO-NP	*Laucaena leucocephala* (Lam.) de wit	Cadmium and Lead (Heavy metal stress)	Seedling growth improved; photosynthetic pigments and total soluble protein was increased, MDA level in leaves reduced, oxidative stress was lowered via activation of antioxidant machinery that resulted in elevation in SOD, CAT and POX level in leaves	Venkatachalam et al. (2017)

(continued)

Table 13.1 (continued)

Nanomaterial and effective dose	Plant exposed	Abiotic stress	Activity	References
Ag-NP @ 2–5 mM	Wheat (*Triticum aestivum* L.)	Salinity (150 mMNaCl)	Salinity and oxidative stress reduced; MDA and H_2O_2 content decreased in leaves; peroxidase activity, soluble sugar and proline content increased	Mohamed et al. (2017)
TiO_2-NP	Broad bean (*Vicia faba* L.)	Soil salinity	Reduced oxidative stress; improved antioxidant system; lowered H_2O_2 and MDA content; improved proline and other metabolites level resulting in regulation of osmotic stress and significant increase in plant growth	Abdel Latef et al. (2018)
SiO_2-NP @ 3 mg/kg	Barley (*Hordeum vulgare* L.)	Nano-NiO (120 mg/kg)	Elevated oxidative stress generated by NiO by scavenging ROS, reducing lipid peroxidation. Antioxidant machinery was activated and revived photosynthetic activity	Soares et al. (2018)

(continued)

Table 13.1 (continued)

Nanomaterial and effective dose	Plant exposed	Abiotic stress	Activity	References
Si-NP (Silicon nanoparticle) @ 1200 mg/l	Wheat (*T. aestivum* L.)	Cadmium (Heavy metal stress)	Plant growth, photosynthetic pigments and stomatal conductance was improved; oxidative stress reduced due to positive stimulation of antioxidant enzyme activity; Cd content in roots, shoots and grains (12–75% in grains) was reduced while that of Si increased	Hussain et al. (2019)
TiO_2-NP and SiO_2-NP @ 30 mg/l	Rice (*Oryza sativa* L.)	Cadmium (Heavy metal stress)	Oxidative stress was reduced with improved antioxidant status, lower electrolyte leakage and MDA content recorded with enhanced activity of SOD, POD, CAT and APX in shoot. Plant biomass, chlorophyll content and gas exchange attributes of leaves were improved; concentration of Cd in plant tissue was decreased	Rizwan et al. (2019)
TiO_2 25 and 50 mg/l	Saffron (*Crocus sativus* L.)	UV-B radiation	Antioxidant activity increased with total phenolic and flavonoid content in stigma	Rikabad et al. (2019)

(continued)

Table 13.1 (continued)

Nanomaterial and effective dose	Plant exposed	Abiotic stress	Activity	References
Mn-NPs	Jalapeno (*Capsicum annuum* L.)	Salinity	Induced expression of MnSOD in stressed plant and controlled oxidative stress	Ye et al. (2020)
ZnO-NPs	Rice (*O. sativa* L.)	Water stress	Regulated plant growth, reduced ROS content and improved antioxidant level	Upadhyaya et al. (2020)
TiO_2-NPs	Wheat (*T. aestivum* L.)	Drought	TiO_2NP improved antioxidant content in stressed plant while foilar application as sodium nitroprusside in presence of nanoparticle reduced H_2O_2 and MDA content and alleviated oxidative stress	Faraji and Sepehri (2020)
Cu-NPs	Wheat (*T. aestivum* L.)	Chromium (Cr) stress	Prevented translocation of metal by immobilization, increased plant growth, reduced ROS level and oxidative stress	Noman et al. (2020)
Fe-NPs	Wheat (*T. aestivum* L.)	Cadmium (Cd) and Drought stress	Improved photosynthesis, plant growth, yield and Fe content in grains whereas reduced Cd concentration in grains and alleviate oxidative stress	Adrees et al. (2020)

reducing MDA content (17.5–30.8% in shoot) and As accumulation in root and shoot by 8.4–72.3 and 10.2–56.6%, respectively by adsorption of ZnO-NP and increasing antioxidant (SOD and CAT) content (Wu et al. 2020). SiNp had more pronounced protecting effect against UV-B as compared to silicon, which was probably due to nitric oxide mediated triggering of antioxidant defense system. Similarly, treatment of squash seeds with nano-silicon dioxide reduced oxidative damage in presence of salt stress by lowering MDA and H_2O_2 concentration and electrolyte leakage; and enhancing antioxidant (SOD, CAT, POX, APX and GR) system (Siddiqui et al. 2014). Engineered NMs due to their small size and large surface area have good affinity for metals and proactively penetrate in contaminated zones, thus highly effective in alleviation of metal induce toxicity in plants (Khan et al. 2017). Tripathi and co-workers (2015) documented role of SiNp in alleviating oxidative stress and phytotoxicity in pea seedling in presence of Cr (VI). The possible mechanism involved by SiNp was up-regulation of antioxidant based defense system, reduced accumulation of Cr (VI) and enhanced uptake of nutrients, thereby reducing generation of ROS.

The process of photosynthesis is highly susceptible to abiotic stresses especially temperature and radiation or light stress (Shen et al. 2010). Continuous exposure of plants to UV-B, chilling and high temperature stresses induce generation and accumulation of ROS in plant cells that damages PS (photosystem) II, disturbs electron transport chain, reduces thylakoid membrane stability (Eva 1999), decrease chlorophyll content and rate of photosynthesis (Prasad et al. 2011), disintegrate lipid membrane and denature macromolecules (Karuppanapandian et al. 2011; Moller et al. 2007). NMs have been reported to protect photosynthetic machinery by rendering oxidative and osmotic stress (Haghighi and Pessarakli 2013; Khan et al. 2017; Siddiqui et al. 2014) increasing chlorophyll biosynthesis, Rubisco activity, transfer and transportation of light energy and absorbance of harmful radiation without scattering useful visible radiations (Gao et al. 2006; Hong et al. 2005; Sicard et al. 2011). The threefold increase in photosynthetic activity and improved maximum electron transport rate with reduced ROS generation in chloroplast by single-walled carbon nanotubes was reported by Giraldo et al. (2014). Analogously, TiO_2 nanoparticles enhanced activity of antioxidant SOD, APX and CAT in plants experiencing cold stress (Mohammadi et al. 2014).

NMs have been reported to alleviate oxidative stress in plants, but on contrary many cases have shown NMs triggered oxidative stress responsible for phytotoxicity (Begum and Fugetsu 2012; Khan et al. 2017) (Fig. 13.2). Studies have demonstrated that presence of NMs in plant growth medium reduce rate of seed germination, photosynthetic activity, chlorophyll content, plant biomass and qualitative production (Barhoumi et al. 2015; Da Costa and Sharma 2016; Khan et al. 2017; Peralta-Videa et al. 2014). Yanik and Vardar (2018) demonstrated Al_2O_3 induced oxidative stress in wheat after 96 h of exposure. The effect was dose dependent that elevated hydrogen peroxide and proline content, lipid peroxidation and superoxide dismutase activity with decreased catalase activity. Reduced photosynthetic activity by TiO_2 was reported in *Ulmus elongata* L. K. Fu & C. S. Ding seedlings (Gao et al. 2013). Some workers have suggested that application of high concentration of NMs corresponds to nanotoxicity, while low concentration alleviate various

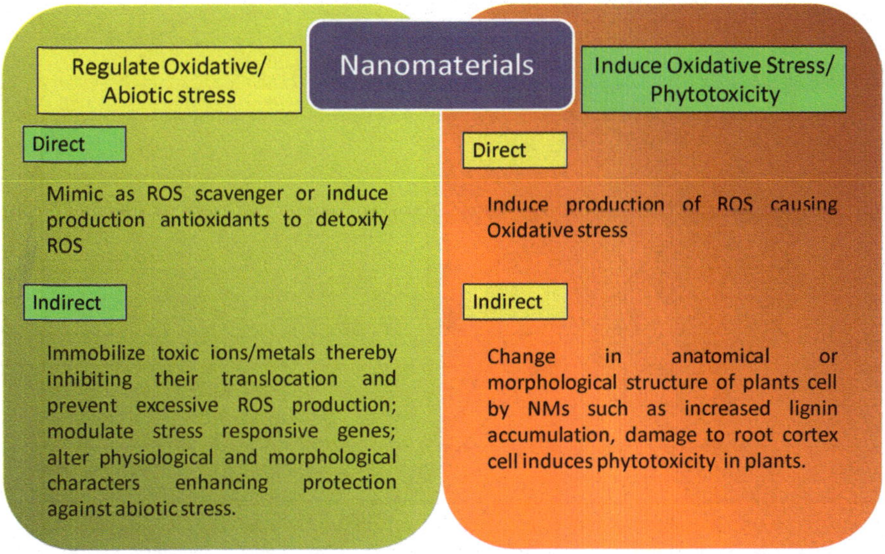

Fig. 13.2 Ambiguous activity of nanomaterials demonstrating both beneficial and detrimental effects on plant either by alleviating oxidative and abiotic stress or by inducing oxidative stress and phytotoxicity (Figure constructed by Swati Sachdev and Shamshad Ahmad)

abiotic stresses and also promote plant growth and development (Khan et al. 2017; Soliman et al. 2015). Chen et al. (2014) studied effect of Cadmium telluride quantum dots (CdTe-QD) alone and in combination with UV-B radiation on wheat seedlings. The findings showed, CdTe-QD was accumulated in roots and resulted in apoptosis. Further exposure of seedling to CdTe-QD in combination with UV-B resulted in synergistic inhibitory effects.

13.3 Mechanism Implicated by NMs to Alleviate Oxidative Burst in Plants

Plants have intrinsic defense system that fights against oxidative stress, known as antioxidant system (Denaxa et al. 2020). Antioxidants help plants to quench excessive level of ROS generated and withstand under stressful situations (Denaxa et al. 2020). Under extreme circumstances, plants are unable to cope up with existing conditions due to increase concentration of ROS than antioxidants, which create imbalance and hinder plant growth, development and productivity ultimately leading to plant death in many cases (Denaxa et al. 2020; Taha et al. 2020). NMs have demonstrated their significant role in mitigation of stress in plants, possessing abilities to alleviate deteriorating effects and enhancing plant health (Adrees et al. 2020; Noman et al. 2020). NMs as discussed earlier have ambiguous identity in plant system. On one

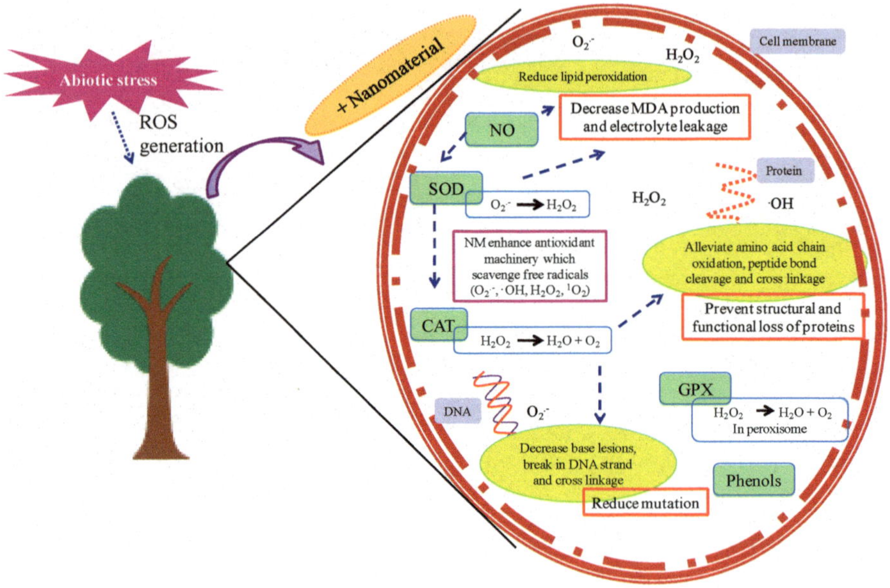

Fig. 13.3 Mechanism of alleviation of oxidative stress in plant by nanomaterials. Amendment of Nanomaterials induces Nitric oxide (NO) production that initiate signal and trigger generation of enzymatic and non-enzymatic antioxidant superoxide dismutase (SOD), guaiacol peroxidase (GPX), catalase (CAT) and phenols that scavenge ROS (O_2^{-}•, H_2O_2, OH^-) generated in different cell organelles due to abiotic stress. SOD dismutate superoxide radical (O_2^{-}•) into hydrogen peroxide (H_2O_2), which on action of CAT and GPX converted into water molecule (H_2O) and molecular oxygen (O_2), thereby reducing ROS concentration and oxidative damage to proteins, lipids and nucleic acid (Figure constructed by Swati Sachdev and Shamshad Ahmad)

hand, NMs mitigate oxidative stress induced via other stresses by mimicking as antioxidant or enhancing production of antioxidant (Zhao et al. 2020). On the other hand, they act as oxidative stress inducers (Prasad et al. 2017). Both protection and predation are unique traits of NMs. Due to this dual characteristic, it becomes extremely important to understand the mechanism deployed by NMs on interaction with plants. The exact mechanism of NMs induced cell signaling is not yet fully deciphered as limited number of studies has been carried out in order to deduce the underlying mechanistic action of NMs in alleviation of oxidative stress induced by abiotic stress.

Treatment of wheat seedlings exposed to UV-B radiation with SiNP resulted in reduced oxidative damage and improved photosynthetic activity (Tripathi et al. 2017). SiNP alleviated oxidative damage via production of NO (nitric oxide) in stressed seedlings. NO act as a signaling molecule that trigger antioxidant defense system in plant. Tripathi et al. (2017) reported SiNP induced NO generation which in turn triggered production of enzymatic antioxidant SOD and APX and non-enzymatic antioxidants flavonoid and phenolic in leaves. The antioxidants were generated that quenched ROS level, thereby reducing lipid peroxidation and electrolyte leakage.

Similar results were obtained in a study conducted by Rizwan et al. (2019) who reported alleviated oxidative stress in rice under Cd stress on foliar application of Si- and TiO_2-NPs that reduced lipid peroxidation and electrolyte leakage by enhancing antioxidant enzyme (SOD, POX, CAT and APX) activities. Analogously, Singh et al. (2020) reported alleviation of Cd toxicity and oxidative stress in *T. aestivum* L. exposed to Cd stress on treatment of SiNP in combination with NO via triggered antioxidant defense system. Figure 13.3 depicts the mechanism of alleviation of oxidative stress in plant by NMs.

13.4 Attributes Governing Activities of Nanomaterials

The potential of NMs to alleviate oxidative burst in plants induced by abiotic stress depend on physico-chemical properties, which in turn govern by factors such as particle size, shape, surface chemistry, coating material, dissolution potential and degree of aggregation/agglomeration (Aslani et al. 2014; Cui et al. 2017; Fu et al. 2014; Nair and Chung 2014; Perez-Labrada et al. 2020). Plant material, concentration of NMs in growth media/soil and external environmental stimuli (abiotic stress, soil microbial communities and other soil characteristics) also contributes to activity of NMs (Aslani et al. 2014; Fu et al. 2014; Reddy et al. 2016). The factors that affect activity of NMs on interaction with plants are represented in Fig. 13.4. The effect of TiO_2NP on two different genotypes ILC533 and Sel 11439 of chickpea seedlings was studied by Mohammadi et al. (2014), where it was reported that NPs caused decline in production of H_2O_2 in Sel 11439 genotype under cold storage stress condition, while no effect on genotype ILC533 was documented in comparison to control. Considering this, it can be deduce that alleviation of oxidative stress by NMs is genotype specific characteristic.

Effect of different size of SiNP (19, 48 and 202 nm) on toxicity induced by Cd (VI) in rice was studied by Cui et al. (2017). They reported ameliorating effect of SiNP on number of cells living with increase in particle size in presence of Cd stress. The proportion of live cells as well as decrease in movement of Cd^{2+} in cells on treatment with 19, 48 and 202 nm size of SiNP was recorded as 95.4, 78.6 and 66.2% and 15.7, 11.1 and 4.6 times, respectively. Reduced influx of $Cd^{2+}$was influenced by the presence of SiNP, which improved Cd transporter activity. The surface of nanoceria (nanoparticle of cerium oxide) consists of $Ce^{3+}$and Ce^{4+} oxidation states and oxygen vacancy in lattices (Djanaguiraman et al. 2018). The ratio of Ce^{3+}/Ce^{4+} sites on nanoceria determines its antioxidant-enzyme mimicking potential (Djanaguiraman et al. 2018). The large surface area increase its redox potential that facilitates redox reaction and scavenging of ROS generated during stress conditions (Djanaguiraman et al. 2018). Similarly the oxygen vacancy in lattice structure of nanoceria catalyzes scavenging of ROS in chloroplast, thereby improving photosynthetic activity (Djanaguiraman et al. 2018). Parallelly, Pulido-Reyes et al. (2015) reported that nanoceria with low Ce^{3+}/Ce^{4+} ratio exhibit catalase and superoxide

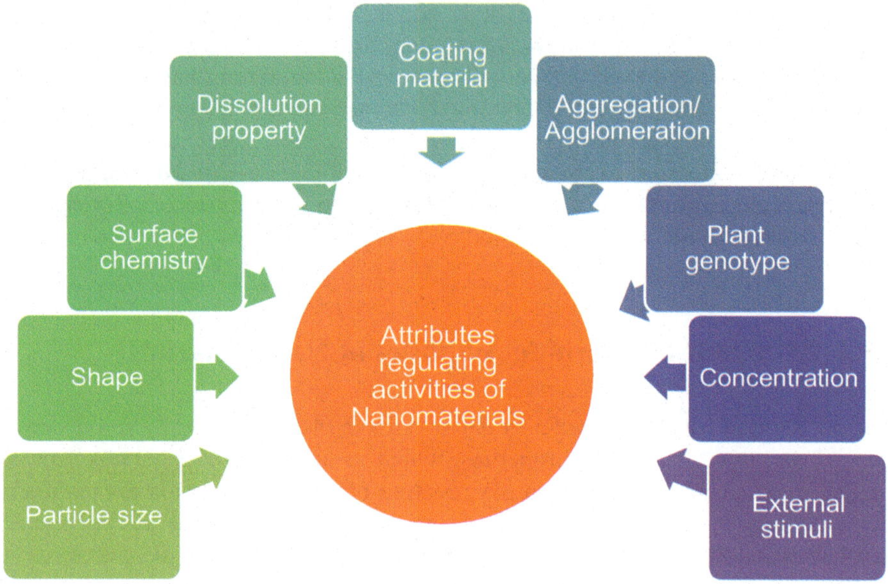

Fig. 13.4 Physical, chemical and biological factors that govern ambiguous activity of nanomaterials (stress inducer or stress regulator) on interaction with plants (Figure constructed by Swati Sachdev and Shamshad Ahmad)

dismutase antioxidant enzyme mimetic activities and participate in quenching of superoxide and H_2O_2 radicals.

The cell damaging effect of several NMs at high concentration via production of ROS has been reported in several studies (Cox et al. 2016; Laware and Raskar 2014; Mohammadi et al. 2013). Such studies have demonstrated that interaction of NMs beyond certain threshold concentration induce toxic impact on plant growth and activity. On the other hand, the study conducted by Mohammadi et al. (2013) outlined that TiO_2NPs at concentration ~ 5 mg/l alleviated oxidative damages induced by cold storage in chickpea seedlings. However, increase in doses of TiO_2, i.e., 10 mg/l was found to induce oxidative stress in seedling, suggesting concentration dependent effect of NMs in plants. Nair and Chung (2014) observed similar concentration dependent oxidative burst in root of *Vigna radiata* (L.) R. Wilczek (mung bean). The exposure of mung bean to AgNPs induced production of H_2O_2 and lipid peroxidation in roots at concentration 20 and 50 mg/l, however no significant change in two parameters was observed at 5 and 10 mg/l of AgNPs as compared to control. Further, increase in AgNP concentration was documented to be positively correlated with production of superoxide radical and change in mitochondrial membrane potential in plant roots over control. The gene expression studies revealed no significant change in gene expression level of CuZn-SOD and CAT in roots at 5 mg/l AgNPs, although exposure of AgNPs at 10 and 20 mg/l upregulated expression of CuZn-SOD and CAT gene. Further increase in concentration i.e., at 50 mg/l CuZn-SOD expression was upregulated and CAT gene expression downregulated. Correspondingly, negative correlation between TiO_2 concentration and activity of antioxidant (CAT and POX) enzymes in onion seedlings was reported by Laware and Raskar (2014). Exposure of onion seeds with graded concentration of TiO_2 (0, 10, 20, 30, 40 and 50 μl/ml), initially at lower concentration, improved germination rate, seedling growth, activity of hydrolytic enzymes (amylase and protease) and antioxidants (SOD, CAT and POX), but with increase in concentration (30 μl/ml or above depending on parameter) showed declined trend in aforementioned activities except SOD, which found highest at maximum concentration of TiO_2.

13.5 Conclusions and Prospects

Nanotechnology is an emerging field of science that growing its application in every nook and corner. Use of NMs in agriculture is the most trending and beneficial approach for smart crop cultivation. Engineered NMs of essential metals and other materials not only fulfill the demand of nutrients in plants, but also aid in protection against several biotic and abiotic stresses. Though, NMs are efficient player in area of agriculture, still their implementation in actual sites possess dilemma due to ambiguity in their activities. Being an efficient oxidative stress regulator, sometimes NMs themselves act as precursor of oxidative stress in plants. Additional, information on long term exposure or residence time of most of the NMs in different ecosystems is still lacking. Therefore, before entering into new advanced phase of agrotechnology based on NMs, it is essentially important to clearly and precisely understand the mechanism or events that occur on interaction of NMs with plants under different existing natural conditions, their residence time in various ecological compartments and effects on long term exposure. Beside this, characteristics of engineered NMs governing their activity on living cells are the overall concern of agronanotechnology. Thus, decoding different mechanisms of NMs at cellular/molecular level and factors governing their activities enable us to avoid production and use of NMs with undesirable effects on plants, ecosystem and human beings. Though, it is difficult to assess the actual toxicological implications of NMs by inspecting mechanisms under influence of individual or couple of physical, chemical and/or biological variables nevertheless, scrutinizing actual field simulations considering all possible variables can help to deduce quite accurate impact of NMs. Thus it necessitates future research to recognize and design such models immediately to advance knowledge on NMs induced phytotoxicity and plant growth promoting attributes. Further, there is also a need to formulate stringent regulatory guidelines for proper and safe use as well as disposal of NMs to reduce their ecotoxicological imprints.

References

Abdel Latef AAH, Srivastava AK, Elsadek MSA (2018) Titanium dioxide nanoparticles improve growth and enhance tolerance of broad bean plants under saline soil conditions. Land Degrad Dev 29(4):1065–1073

Adrees M, Khan ZS, Ali S et al (2020) Simultaneous mitigation of cadmium and drought stress in wheat by soil application of iron nanoparticles. Chemosphere 238:124681

Aghdam MTB, Mohammadi H, Ghorbanpour M (2016) Effects of nanoparticulate anatase titanium dioxide on physiological and biochemical performance of *Linum usitatissimum* (Linaceae) under well-watered and drought stress conditions. Braz J Bot 39(1):139–146

Aslani F, Bagheri S, MuhdJulkapli N et al (2014) Effects of engineered nanomaterials on plants growth: an overview. The Scientific World J. https://doi.org/10.1155/2014/641759

Barhoumi L, Oukarroum A, Taher LB et al (2015) Effects of superparamagnetic iron oxide nanoparticles on photosynthesis and growth of the aquatic plant *Lemnagibba*. Arch Environ Contam Toxicol 68(3):510–520

Begum P, Fugetsu B (2012) Phytotoxicity of multi-walled carbon nanotubes on red spinach (Amaranthus tricolor L) and the role of ascorbic acid as an antioxidant. J Hazard Mater 243:212–222

Burman U, Saini M, Kumar P (2013) Effect of zinc oxide nanoparticles on growth and antioxidant system of chickpea seedlings. Toxicol Environ Chem 95(4):605–612

Chen H, Gong Y, Han R (2014) Cadmium telluride quantum dots (CdTe-QDs) and enhanced ultraviolet-B (UV-B) radiation trigger antioxidant enzyme metabolism and programmed cell death in wheat seedlings. PloS one 9:10

Comotto M, Casazza AA, Aliakbarian B et al (2014) Influence of TiO_2 nanoparticles on growth and phenolic compounds production in photosynthetic microorganisms. Scientific World J. https://doi.org/10.1155/2014/961437

Cox A, Venkatachalam P, Sahi S et al (2016) Silver and titanium dioxide nanoparticle toxicity in plants: a review of current research. Plant Physiol Biochem 107:147–163

Cui J, Liu T, Li F et al (2017) Silica nanoparticles alleviate cadmium toxicity in rice cells: Mechanisms and size effects. Environ Pollut 228:363–369

Da Costa MVJ, Sharma PK (2016) Effect of copper oxide nanoparticles on growth, morphology, photosynthesis, and antioxidant response in *Oryza sativa*. Photosynthetica 54:110–119

Demirer GS, Zhang H, Matos J et al (2018) High aspect ratio nanomaterials enable delivery of functional genetic material without DNA integration in mature plants. Nature Nanotechnol 14(5):456–464

Denaxa NK, Damvakaris T, Roussos PA (2020) Antioxidant defense system in young olive plants against drought stress and mitigation of adverse effects through external application of alleviating products. Sci Hortic 259:108812

Djanaguiraman M, Nair R, Giraldo JP et al (2018) Cerium oxide nanoparticles decrease drought-induced oxidative damage in sorghum leading to higher photosynthesis and grain yield. ACS omega 3(10):14406–14416

Dumont S, Rivoal J (2019) Consequences of oxidative stress on plant glycolytic and respiratory metabolism. Front. Plant Sci 10:166. https://doi.org/10.3389/fpls.2019.00166

Eva H (1999) Utilizing new adamantly spin traps in studying UV-B-induced oxidative damage of photosystem II. J Photo chem Photobiol B Biol 48:174–179

Faizan M, Faraz A, Mir AR, Hayat S (2020) Role of zinc oxide nanoparticles in countering negative effects generated by cadmium in *Lycopersicon esculentum*. J Plant Growth Regul 1–15. https://doi.org/10.1007/s00344-019-10059-2

Faraji J, Sepehri A (2020) Exogenous nitric oxide improves the protective effects of TiO_2 nanoparticles on growth, antioxidant system, and photosynthetic performance of wheat seedlings under drought stress. J Soil Sci Plant Nutr 20:703–714

Fu PP, Xia Q, Hwang HM et al (2014) Mechanisms of nanotoxicity: generation of reactive oxygen species. J Food Drug Anal 22:64–75

Gao FQ, Hong FS, Liu C et al (2006) Mechanism of nanoanatase TiO_2 on promoting photosynthetic carbon reaction of spinach: inducing complex of Rubisco-Rubiscoactivase. Biol Trace Elem Res 11:239–254

Gao J, Xu G, Qian H et al (2013) Effects of nano-TiO_2 on photosynthetic characteristics of Ulmus elongata seedlings. Environ Pollut 176:63–70

Ghosh M, Jana A, Sinha S et al (2016) The effect of ZnO nanoparticles in plants: cytotoxicity, genotoxicity, deregulation of antioxidant defenses, and cell-cycle arrest. Mutat Res 807:25–32

Giraldo JP, Landry MP, Faltermeier SM et al (2014) Plant nanobionics approach to augment photosynthesis and biochemical sensing. Nat Mater 13(4):400–408

Haghighi M, Pessarakli M (2013) Influence of silicon and nano-silicon on salinity tolerance of cherry tomatoes (*Solanum lycopersicum* L.) at early growth stage. Scientia Horti 161:111–117

Hatami M, Ghorbanpour M (2013) Effect of nano silver on physiological performance of pelargonium plants exposed to dark storage. J Hortic Res 21(1):15–20

Hong F, Liu C, Zheng L et al (2005) Formation of complexes of rubisco-rubiscoactivase from La^{3+}, Ce^{3+} treatment spinach. Sci China Series B: Chem 48(1):67

Hussain A, Rizwan M, Ali Q et al (2019) Seed priming with silicon nanoparticles improved the biomass and yield while reduced the oxidative stress and cadmium concentration in wheat grains. Environ Sci Pollut R26(8):7579–7588

Iqbal MS, Singh AK, Singh SP, Ansari MI (2020) Nanoparticles and plant interaction with respect to stress response. In: Bhushan et al (eds) Nanomaterials and Environmental Biotechnology, Springer, Cham, pp 1–15

Ismail A, Takeda S, Nick PR (2014) Life and death under salt stress: same players, different timing? J Exp Bot 65:2963–2979

Jalil SU, Ansari MI (2019) Nanoparticles and abiotic stress tolerance in plants: synthesis, action and signalling mechanisms, In: Khan MIR, Reddy PS, Ferrante A, Khan NA (eds) Plant signaling molecules-role and regulation under stressful environments. Elsevier Publication, United Kingdom, pp 549–561. https://doi.org/10.1016/B978-0-12-816451-8.00034-4

Jefferson DA (2000) The surface activity of ultrafine particles. Philos Trans R Soc Lond A 358:2683–2692

Ju-Nam Y, Lead JR (2008) Manufactured nanoparticles: an overview of their chemistry, interactions and potential environmental implications. Sci Total Environ 400(1–3):396–414

Karuppanapandian T, Wang HW, Prabakaran N et al (2011) 2,4-dichlorophenoxyacetic acid-induced leaf senescence in mung bean (*Vigna radiata* L. Wilczek) and senescence inhibition by co-treatment with silver nanoparticles. Plant Physiol Biochem 49:168–217

Khan MN, Mobin M, Abbas ZK et al (2017) Role of nanomaterials in plants under challenging environments. Plant Physiol Biochem 110:194–209

Kumar V, Guleria P, Kumar V et al (2013) Gold nanoparticle exposure induces growth and yield enhancement in *Arabidopsis thaliana*. Sci Total Environ 461:462–468

Kumar V, Sharma M, Khare T et al (2018) Impact of nanoparticles on oxidative stress and responsive antioxidative defense in plants. In: Tripathi DK, Ahmad P, Sharma S et al (eds) Nanomaterials in Plants, Algae, and Microorganisms, vol 1. Academic Press, pp 393–406

Kumar V, Shriram V, Hoque, TS et al (2017) Glycinebetaine-mediated abiotic oxidative-stress tolerance in plants: physiological and biochemical mechanisms. In: Sarwat M et al (eds) Stress Signaling in Plants: Genomics and Proteomics Perspective, vol 2. Springer International, Switzerland, pp 111–133. http://dx.doi.org/10.1007/978-3-319-42183-4-5

Latef AAHA, Alhmad MFA, Abdelfattah KE (2017) The possible roles of priming with ZnO nanoparticles in mitigation of salinity stress in lupine (*Lupinus termis*) plants. J Plant Growth Regul 36(1):60–70

Laware SL, Raskar S (2014) Effect of titanium dioxide nanoparticles on hydrolytic and antioxidant enzymes during seed germination in onion. Int J Curr Microbiol App Sci 3(7):749–760

Lei Z, Mingyu S, Xiao W et al (2008) Antioxidant stress is promoted by nano-anatase in spinach chloroplasts under UV-B radiation. Biol Trace Elem Res 121(1):69–79

Ma C, White JC, Dhankher OP et al (2015) Metal-based nanotoxicity and detoxification pathways in higher plants. Environ Sci Technol 49:7109–7122

Marino D, Dunand C, Puppo A et al (2012) A burst of plant NADPH oxidases. Trends Plant Sci 17:9–15

Marslin G, Sheeba CJ, Franklin G (2017) Nanoparticles alter secondary metabolism in plants via ROS burst. Front Plant Sci 8:832

Mittler R (2017) ROS are good. Trends Plant Sci 22(1):11–19

Mohamed AKS, Qayyum MF, Abdel-Hadi AM et al (2017) Interactive effect of salinity and silver nanoparticles on photosynthetic and biochemical parameters of wheat. Arch Agron Soil Sci 63(12):1736–1747

Mohammadi H, Esmailpour M, Gheranpaye A (2016) Effects of TiO_2 nanoparticles and water-deficit stress on morpho-physiological characteristics of dragonhead (*Dracocephalum moldavica* L.) plants. Acta Agric Slov 107(2):385–396

Mohammadi R, Maali-Amiri R, Abbasi A (2013) Effect of TiO_2 nanoparticles on chickpea response to cold stress. Biol Trace Elem Res 152(3):403–410

Mohammadi R, Maali-Amiri R, Mantri N (2014) Effect of TiO_2 nanoparticles on oxidative damage and antioxidant defense systems in chickpea seedlings during cold stress. Russ J Plant Physiol 61:768–775

Moller IM, Jensen PE, Hansson A (2007) Oxidative modifications to cellular components in plants. Annu Rev Plant Biol 58:459–481

Monica RC, Cremonini R (2009) Nanoparticles and higher plants. Caryologia 62(2):161–165

Nair PMG, Chung IM (2014) Physiological and molecular level studies on the toxicity of silver nanoparticles in germinating seedlings of mung bean (*Vigna radiata* L.). Acta Physiol Plant 37(1):1719

Nile SH, Baskar V, Selvaraj D et al (2020) Nanotechnologies in food science: applications, recent trends, and future perspectives. Nanomicro Lett 12(1):45

Nishida N, Arizumi T, Takita M et al (2013) Reactive oxygen species induce epigenetic instability through the formation of 8-hydroxydeoxyguanosine in human hepatocarcinogenesis. Dig Dis 31(5–6):459–466

Noman M, Shahid M, Ahmed T et al (2020) Green copper nanoparticles from a native *Klebsiella pneumoniae* strain alleviated oxidative stress impairment of wheat plants by reducing the chromium bioavailability and increasing the growth. Ecotoxicol Environ Saf 192:110303

Palmqvist NM, Seisenbaeva GA, Svedlindh P et al (2017) Maghemite nanoparticles acts as nanozymes, improving growth and abiotic stress tolerance in *Brassica napus*. Nanoscale Res Lett 12(1):631

Peralta-Videa JR, Hernandez-Viezcas JA, Zhao L et al (2014) Cerium dioxide and zinc oxide nanoparticles alter the nutritional value of soil cultivated soybean plants. Plant Physiol Biochem 80:128–135

Perez-Labrada F, Hernández-Hernández H, López-Pérez MC, et al (2020) Nanoparticles in plants: Morphophysiological, biochemical, and molecular responses. In: Tripathi DK, Singh VP, Chauhan DK et al (eds) Plant life under changing environment. Elsevier publication, AcademicPress. 289–322. https://doi.org/10.1016/C2018-1-02300-8

Pizzino G, Irrera N, Cucinotta M et al (2017) Oxidative stress: harms and benefits for human health. Oxid Med Cell Longev. https://doi.org/10.1155/2017/8416763

Prasad PVV, Pisipati SR, Momcilovic I et al (2011) Independent and combined effects of high temperature and drought stress during grain filling on plant yield and chloroplast EF-Tu expression in spring wheat. J Agron Crop Sci 197(6):430–441

Prasad R, Bhattacharyya A, Nguyen QD (2017) Nanotechnology in sustainable agriculture: recent developments, challenges, and perspectives. Front Microbiol 8:1014

Pulido-Reyes G, Rodea-Palomares I, Das S et al (2015) Untangling the biological effects of cerium oxide nanoparticles: the role of surface valence states. Scientific Rep 5(1):1–14

Rajendran P, Nandakumar N, Rengarajan T et al (2014) Antioxidants and human diseases. Clin Chim Acta 436:332–347

Reddy PVL, Hernandez-Viezcas JA, Peralta-Videa JR et al (2016) Lessons learned: are engineered nanomaterials toxic to terrestrial plants? Sci Total Environ 568:470–479. https://doi.org/10.1016/j.scitotenv.2016.06.042

Rikabad MM, Pourakbar L, Moghaddam SS et al (2019) Agrobiological, chemical and antioxidant properties of saffron (*Crocus sativus* L.) exposed to TiO2 nanoparticles and ultraviolet-B stress. Ind Crop Prod 137:137–143

Rizwan M, Ali S, Qayyum MF (2017) Effect of metal and metal oxide nanoparticles on growth and physiology of globally important food crops: a critical review. J Hazard Mater 322:2–16. https://doi.org/10.1016/j.jhazmat.2016.05.061

Rizwan M, Ali S, Rehman MZ et al (2019) Effect of foliar applications of silicon and titanium dioxide nanoparticles on growth, oxidative stress, and cadmium accumulation by rice (Oryza sativa). Acta Physiol Plant 41(3):35

Santos AR, Miguel AS, Tomaz L et al (2010) The impact of CdSe/ZnS quantum dots in cells of *Medicago sativa* in suspension culture. J Nanobiotechnol 8:24. https://doi.org/10.1186/1477-3155-8-24

Sharma P, Jha AB, Dubey RS et al (2012) Reactive oxygen species, oxidative damage, and antioxidative defense mechanism in plants under stressful conditions. J Bot 217–237

Shen C, Zhang Q, Li L et al (2010) Induction of programmed cell death in Arabidopsis an rice by single-wall carbon nanotubes. Am J Bot 97:1602–1609

Sicard C, Perullini M, Spedalieri C et al (2011) CeO_2 nanoparticles for the protection of photosynthetic organisms immobilized in silica gels. Chem Mater 23(6):1374–1378

Siddiqui MH, Al-Whaibi MH, Faisal M et al (2014) Nano-silicon dioxide mitigates the adverse effects of salt stress on *Cucurbita pepo* L. Environ Toxicol Chem 33(11):2429–2437

Siddiqui MH, Al-Whaibi MH, Mohammad F (eds) (2015) Nanotechnology and plant sciences: nanoparticles and their impact on plants. Springer International Publishing, Switzerland, pp 19–35

Singh S, Prasad SM, Sharma S et al (2020) Silicon and nitric oxide-mediated mechanisms of cadmium toxicity alleviation in wheat seedlings. Physiol Plant. https://doi.org/10.1111/ppl.13065

Sitte N (2003) Oxidative damage to proteins. In: Aging at the molecular level, pp 27–45

Soares C, Branco-Neves S, de Sousa A et al (2018) SiO_2 nanomaterial as a tool to improve *Hordeum vulgare* L. tolerance to nano-NiO stress. Sci Total Environ 622:517–525

Soliman AS, El-feky SA, Darwish E (2015) Alleviation of salt stress on *Moringa peregrina* using foliar application of nanofertilizers. J Hortic For 7:3647

Taha RS, Alharby HF, Bamagoos AA et al (2020) Elevating tolerance of drought stress in *Ocimum basilicum* using pollen grains extract; a natural biostimulant by regulation of plant performance and antioxidant defense system. S Afr J Bot 128:42–53

Tripathi DK, Singh S, Singh VP et al (2017) Silicon nanoparticles more effectively alleviated UV-B stress than silicon in wheat (*Triticum aestivum*) seedlings. Plant Physiol Biochem 110:70–81

Tripathi DK, Singh VP, Prasad SM et al (2015) Silicon nanoparticles (SiNp) alleviate chromium (VI) phytotoxicity in *Pisum sativum* (L.) seedlings. Plant Physiol Biochem 96:189–198

Vecerova K, Vecera Z, Docekal B et al (2016) Changes of primary and secondary metabolites in barley plants exposed to CdO nanoparticles. Environ Pollut 218:207–218

Venkatachalam P, Jayaraj M, Manikandan R et al (2017) Zinc oxide nanoparticles (ZnONPs) alleviate heavy metal-induced toxicity in *Leucaena leucocephala* seedlings: a physiochemical analysis. Plant Physiol Biochem 110:59–69

Wani SH, Sah SK, Hussain MA et al (2016) Transgenic approach for abiotic stress tolerance in crop plants. In: Al-Khayri JM, Jain SM, Johnson DV (eds) Advances in plant breeding strategies, Vol. 2: agronomic, abiotic and biotic stress traits. Springer International Publishing, Switzerland, pp 345–396

Wu H, Tito N, Giraldo JP (2017) Anionic cerium oxide nanoparticles protect plant photosynthesis from abiotic stress by scavenging reactive oxygen species. ACS Nano 11(11):11283–11297

Yanık F, Vardar F (2018) Oxidative stress response to aluminum oxide (Al_2O_3) nanoparticles in Triticum aestivum. Biologia 73(2):129–135

Yasui M, Kanemaru Y, Kamoshita N et al (2014) Tracing the fates of site-specifically introduced DNA adducts in the human genome. DNA Repair 15:11–20

Tian L, Zhang H, Zhao X et al (2020) Cads nanoparticles in soil induce metabolic reprogramming in broad bean (*Vicia faba L.*) roots and leaves. Environmental Science: Nano 7(1):93–104

Upadhyaya H, Shome S, Tewari S et al (2020) Responses to ZnO nanoparticles during water stress in *Oryza sativa* L. J Stress Physiol Biochem 16(2):67–76

Wu F, Fang Q, Yan S et al (2020) Effects of zinc oxide nanoparticles on arsenic stress in rice *(Oryza sativa* L.): germination, early growth, and arsenic uptake. Environ Sci Pollut Res Int. https://doi.org/10.1007/s11356-020-08965-0

Ye Y, Cota-Ruiz K, Hernandez-Viezcas JA et al (2020) Manganese nanoparticles control salinity-modulated molecular responses in *Capsicum annuum* L. through priming: a sustainable approach for agriculture. ACS Sustain Chem Eng 8(3):1427–1436

Zhang P, Guo Z, Luo W et al (2020) Graphene oxide-induced pH alteration, iron overload, and subsequent oxidative damage in rice (Oryza sativa L.): a new mechanism of nanomaterial phytotoxicity. Environ Sci Technol 54(6):3181–3190

Zhao L, Lu L, Wang A et al (2020) Nano-biotechnology in agriculture: use of nanomaterials to promote plant growth and stress tolerance. J Agric Food Chem 68(7):1935–1947

Chapter 14
Plant Stress Enzymes Nanobiotechnology

Paras Porwal, Sashi Sonkar, and Akhilesh Kumar Singh

Abstract Abiotic and biotic stresses are significantly affects the plant growth, thereby limiting agricultural productivity of crops. Agricultural plants/crops should be able to cope-up with both biotic and abiotic stresses by their innate biological mechanisms, failing which affect their growth, development and productivity. As per FAO, there is a need to foster the crop productivity factor greater than 70% by 2050 to feed additional 2.3 billion people worldwide. Moreover, sustainable agriculture acts as a main pillar for the development of the mankind and national economy as well as fulfills the food demand in developing countries. Realizing these critical facts, it becomes necessary for the scientific arena to generate harmless stress-mitigating mechanisms in plants, so that the plants/crop productivity is improved. In today's world, nanobiotechnology receiving an increasing attention towards the mitigation of biotic and/or abiotic stresses of agricultural plants/crops including the challenges in the yield barriers with the development of eco-friendly technologies. Although, there exists a huge gap in our understanding of the eco-toxicity, tolerable limit, and uptake capability of various nanoparticles in plants. This chapter encapsulates the promises as well as progress in plant nanobiotechnology especially with respect to promoting plant growth factors and ways to overcome abiotic stresses.

Keywords Abiotic and biotic stresses · Antioxidant · Nanobiotechnology · Nanoparticles · Plants/crop productivity · Reactive oxygen species · Salinity

---

P. Porwal
Amity Institute of Biotechnology, Amity University Uttar Pradesh, Lucknow Campus, Lucknow, Uttar Pradesh 226028, India
e-mail: pporwal@lko.amity.edu

S. Sonkar
Department of Botany, Bankim Sardar College, Tangrakhali 743329, South 24 Parganas, West Bengal, India

A. K. Singh (✉)
Department of Biotechnology, School of Life Sciences, Mahatma Gandhi Central University, Motihari 845401, Bihar, India
e-mail: akhileshsingh@mgcub.ac.in

J. M. Al-Khayri et al. (eds.), *Nanobiotechnology*,
https://doi.org/10.1007/978-3-030-73606-4_14

14.1 Introduction

Sustainable agriculture acts as a backbone for the development of the national economy as well as fulfills the aspiration of food demand in developing countries. To satisfy the demand for food supply for the upcoming future with the changing environmental conditions as well as rapidly increasing population, there is urgent need to increase crop yield and stability of plants in adverse conditions by exploiting the advance approaches like nanobiotechnology (Eckardt et al. 2009; Zhang 2007). Agricultural plants/production of crops together with their protection are reliant on the various parameters such as type as well as the quantity of applied fertilizers and pesticides. The growth and development of agricultural plants/crops entirely depend on ease of availability of optimal environmental as well as nutritional factors and any deviation from it leads to plant stress. Stress is a condition in which plants are not able to fully express their genetic potential for growth, development, and reproduction, thereby limits productivity owing to damage to biomass. Being sessile, plants cannot escape from adverse climatic conditions and, thus have to meet both the stresses, i.e., biotic stresses, for instance, interactions among organisms like microbial pathogens and so on and abiotic stresses that involve interactions among organisms with their physical environment. Abiotic (physical) stresses include temperature alteration (high or low), nutrient starvation, water deficit (drought), anoxia (during the flood), salinity and alkalinity of the soil, light intensity, submergence, mineral, and metal toxicity/deficiency (Cramer et al. 2011; Hirel et al. 2007; Wang et al. 2003). These stresses are unpredictable in nature in terms of their intensity, duration, and occurrence, so sustaining the development and survival of plants in an unfavorable environment turns out to be a difficult task. So, plants need to respond distinctly to protect themselves from physical stresses like cold, drought, heat, etc., that ultimately lead to the development of some adaptative mechanism in plants (Mittler 2002). Plants have the ability to sense abiotic stress and respond accordingly as per their past exposure so that in further repetitive stress can be adjusted (Ahmad et al. 2015; Hilker et al. 2015; Jiang et al. 2016). However, on the other hand, transgenic plants/crops are still not popular among the grower or farmers owing to their high level of safety concern. Therefore, in the current scenario, plant nanobiotechnology offers promising technological approaches for achieving food safety and security by increasing the efficiency of plants/crops, protecting them from different types of biotic as well as abiotic stresses via. modulating the mechanisms of different pathways, apart from those achieved through genetic and chemical production (Giraldo et al. 2019; Iqbal et al. 2020; Kah et al. 2019). Nanobiotechnology involves the cutting edge application-oriented research in the area of Nanoscience together with biotechnology. Nanomaterials (NMs) can be defined as materials depicting diameter in the range of 1–100 nm (Porwal and Sharma 2016; Pandey et al. 2018; Porwal et al. 2020; Rani et al. 2020; Singh and Porwal 2020; Singh, Pal, et al. 2018; Singh, Yadav, et al. 2018; Singh et al. 2020). The effect of various kinds of nanomaterials on plants under normal and/or abiotic stressed environment is presented in Table 14.1.

Table 14.1 Impact of nanomaterials on plants under normal and/or abiotic stressed condition

Nanoparticle type	Abiotic stress	Plant name	Impact	Reference
Ag	Dark stress	Horseshoe pelargonium (*Pelargonium zonale* (L.) L'Hér. ex Aiton)	Elevated antioxidative enzymes activities, petal longevity, leaf carotenoids and chlorophyll content. Decreased the peroxidation of lipid and petal abscission	Ghorbanpour and Hatami (2014)
Ag	Flooding	Soybean (*Glycine max* (L.) Merr.)	Promotes seedling growth and abundance of stress-related proteins. Decreases the cytotoxic by-products of glycolysis	Mustafa et al. (2015b)
Ag	Flooding	Saffron (*Crocus sativus* L.)	Promotes root growth. Blocks signaling pathway of ethylene	Rezvani et al. (2012)
Al_2O_3	Flooding	Soybean (*Glycine max* (L.) Merr.)	Controls energy metabolism and cell death	Mustafa et al. (2015a)
Al_2O_3	Nanotoxicity	Onion (*Allium cepa* L.)	Increases the activities of CAT and SOD	Rajeshwari et al. (2015), Riahi-Madvar et al. (2012)
CeO_2	Nanotoxicity	Maize (*Zea mays* L.)	Up-regulation of heat shock protein such as HSP70 and improved generation of H_2O_2	
CeO_2	Nanotoxicity	Soybean (*Glycine max* (L.) Merr.)	Stimulates plant growth. Rubisco carboxylase activity and photosynthesis rate increases	Zhao et al. (2012)
CuO	Nanotoxicity	Chickpea (*Cicer arietinum* L.)	Increase the activity of POD	Nair and Chung (2015)
CuO	Nanotoxicity	Wheat (*Triticum aestivum* L.)	Increase the activity of CAT and POD	Dimkpa et al. (2012)
Fe_2O_3	Nanotoxicity	Watermelon *Citrullus lanatus* (Thunb.) Matsum & Nakai	Increase the activities of CAT, POD, and SOD. Changes in the root activity, ferric reductase activity as well as chlorophyll, root apoplastic iron, and MDA contents were observed	Li et al. (2013)

(continued)

Table 14.1 (continued)

Nanoparticle type	Abiotic stress	Plant name	Impact	Reference
Fe_3O_4	Nanotoxicity	Wheat (*Triticum aestivum* L.)	Increases the activities of CAT, APX, GPOX, and SOD	Iannone et al. (2016)
SiO_2	Cold	Tall wheatgrass (*Agropyron elongatum* L.)	Overcome seed dormancy. Improved seed germination and seedling weight	Azimi et al. (2014)
SiO_2	Drought	Hawthorn (*Crataegus* sp.)	Increase photosynthetic rate, plant biomass, and stomatal conductance while insignificant effect on carotenoid and chlorophyll content	Ashkavand et al. (2015)
SiO_2	Salinity	Basil (*Ocimum basilicum* L.)	Increased chlorophyll and proline content. Improves dry and fresh weight	Kalteh et al. (2014)
SiO_2	Salinity	Broad bean (*Vicia faba* L.)	Increased the activity of antioxidant enzymes. Stimulates seed germination, water content and total yield	Qados and Moftah (2015), Qados (2015)
SiO_2	Salinity	Tomato (*Lycopersicon esculentum* Mill.)	Nano-SiO_2 at low concentration improved seed germination, dry weight, and root length whereas at higher concentration suppressed seed germination	Haghighi et al. (2012)
SiO_2	Salinity	Tomato (*Solanum lycopersicum* L.)	Downregulation of six genes RBOH1, APX2, MAPK2, ERF5, MAPK3, and DDF2 and upregulation of four salt stress genes AREB, TAS14, NCED3, and CRK1 thereby suppressing the effect of salinity stress on seed germination rate, root length, and fresh weight	Almutairi (2016)
SiO_2	Salinity	Tomato (*Solanum lycopersicum* L.)	Eliminate the effect of stress on photosynthetic rate, leaf water, and chlorophyll content	Haghighi and Pessarakli (2013)

(continued)

Table 14.1 (continued)

Nanoparticle type	Abiotic stress	Plant name	Impact	Reference
TiO_2	Drought	Basil (*Ocimum basilicum* L.)	Ameliorate negative effects of stress on the plant	Kiapour et al. (2015)
TiO_2	Drought	Flax (*Linum usitatissimum* L.)	Improve growth, carotenoids, and chlorophyll contents. Reduces H_2O_2 and MDA contents	Aghdam et al. (2016)
TiO_2	Drought	Wheat (*Triticum aestivum* L.)	Increase in gluten and starch content. Improves the overall growth and yield of the plant	Jaberzadeh et al. (2013)
TiO_2	Cold	Chickpea (*Cicer arietinum* L.)	Enhanced the activity of antioxidant enzymes, phosphoenolpyruvate carboxylase, and expression of Rubisco and chlorophyll-binding protein genes. Decreased in H_2O_2 content and electrolyte leakage	Hasanpour et al. (2015), Mohammadi et al. (2013, 2014)
TiO_2	Heat	Tomato (*Lycopersicon esculentum* Mill.)	Induced stomatal opening and cooling of leaves	Qi et al. (2013)
TiO_2	Nanotoxicity	Broad bean (*Vicia faba* L.)	Decreased the activity of GR and APX	Foltete et al. (2011)
TiO_2	Nanotoxicity	Duckweed (*Lemna minor* L.)	Increased the activity of SOD, CAT, and POD	Song et al. (2012)
TiO_2	Nanotoxicity	Hydrilla (*Hydrilla verticillata* (L.f.) Royle)	The activity of enzymes such as CAT and GR are increased	Okupnik and Pflugmacher (2016)
TiO_2	Nanotoxicity	Peppermint (*Mentha piperita* L.)	Increase the amount of chlorophyll (a and b) and carotenoid	Samadi et al. (2014)
TiO_2	Nanotoxicity	Spinach (*Spinacia oleracea* L.)	Increased the activity of SOD, CAT, APX, and GPOX↑	Lei et al. (2008)
ZnO and Fe_3O_4	Salinity	Ben tree *Moringa peregrine* (Forssk.) Fiori	Increased enzymatic and non-enzymatic antioxidants. Promotes the chlorophyll, carotenoids, proline, N, P, K, Ca^{2+}, Mg^{2+} carbohydrates, and crude protein content. Decreased Na^+ and Cl^- content	Soliman et al. (2015)

(continued)

Table 14.1 (continued)

Nanoparticle type	Abiotic stress	Plant name	Impact	Reference
ZnO	Nanotoxicity	Green pea (*Pisum sativum* L.)	Increased the elongation of root	Mukherjee et al. (2014)
ZnO	Nanotoxicity	Mouse-ear cress (*Arabidopsis thaliana* (L.) Heynh.)	Increase in lateral root formation.	Nair and Chung (2017)
ZnO	Nanotoxicity	Wheat (*Triticum aestivum* L.)	Reduced the activity of CAT	Dimkpa et al. (2012)
ZnO	Salinity	White lupin (*Lupinus termis* L.)	Increased the activity of ascorbic acid, phenols, organic solutes, and SOD, CAT, POD, and APX whereas decreased the content of MDA	Latef et al. (2017)

APX: Ascorbate peroxidase; CAT: Catalase; GPOX: Guaiacol peroxidase; GR: Glutathione reductase; MDA: Malondialdehyde; POD: Peroxidase; SOD: Superoxide dismutase

14.2 ROS Scrounging Antioxidants of Plants

ROS (reactive oxygen species) are short-lived, unstable, and reactive (Halliwell 2006), which includes singlet oxygen ($^1O_2$), hydroxyl radical (OH•), superoxide radical ($O_2^{\bullet -}$) as well as hydrogen peroxide (H_2O_2), etc. These are generated in different cellular compartments such as chloroplast, mitochondria, peroxisomes, plasma membrane (Apel and Hirt 2004) as a regular (unavoidable) by-product of aerobic metabolism such as photosynthesis and respiration in plants (Miller et al. 2010; You and Chan 2015) which is regulated by both enzymatic and non-enzymatic antioxidant defense system of the plant. The low or moderate level of ROS is responsible for plant growth (reproductive and senescence) and development including leaf shape, root hair elongation, trichome development (Gapper and Dolan 2006), stomatal closure, programmed cell death (Petrov et al. 2015), gravitropism (Wassim et al. 2013) as well as act as the second messenger in mediating different series of reactions in plant cells, and promotes the tolerance from biotic and abiotic stress conditions (Nath et al. 2017). However, excessive production of ROS due to both biotic and abiotic stresses (Bhattacharjee 2012; Khare et al. 2014; Kumar and Khare 2014) was not removed then results in damage to cell membranes (lipid peroxidation), proteins, nucleic acid (DNA as well as RNA), and several other cellular components of the plants, thereby affecting plant growth including development and ultimately yield (Demidchik 2015; Mittler 2002). Various abiotic stresses induced ROS generation and the role of nanomaterials enhancing stress tolerance in the plant is depicted in Fig. 14.1.

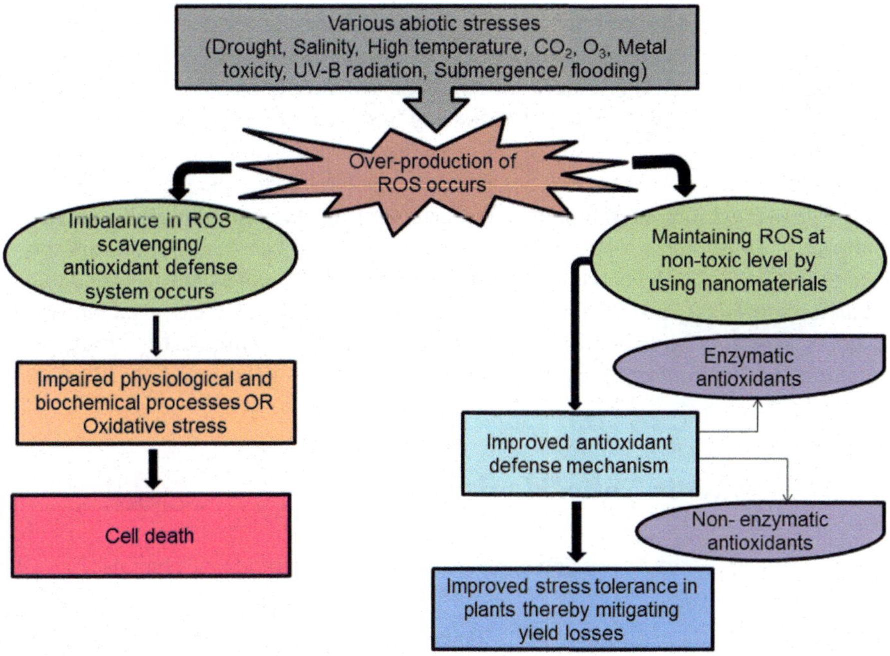

Fig. 14.1 An overview of the abiotic stress-induced ROS generation in agricultural plants/crops and the role of nanomaterials in improving stress tolerance (*Source* Modified from Meena et al. 2017; Xie et al. 2019)

14.3 Stimulation of Antioxidant Mechanism in Response to Nanoparticle Exposure

Plants make use of enzymatic and non-enzymatic antioxidative systems and/or pathways to mitigate oxidative stress. The key enzymes involved in the ROS-scrounging include catalase (CAT), ascorbate peroxidase (APX), superoxide dismutase (SOD), peroxidase (POD), glutathione reductase (GR), glutathione peroxidase (GPX), glutathione S-transferase (GST), alternative oxidases (AOX), peroxiredoxin (PRX), monodehydroascorbate reductase (MDHAR), dehydroascorbate reductase (DHAR), and many more (Catalá and Díaz 2016; Jaleel et al. 2009; Maxwell et al. 1999; Mittler et al. 2004). Non-enzymatic antioxidants comprised of low molecular weight metabolites such as flavonoids, polyphenols, glutathione (GSH), ascorbic acid (AsA), β-carotene, α-tocopherol, proline, glycine betaine, and many more (Gill and Tuteja 2010; Pandey et al. 2017). During stressed conditions plants protect themselves from ROS toxicity (leads to oxidative damage) by changing gene expressions as well as adapting ROS-scrounging antioxidant metabolic pathways such as ascorbate, aldarate, and shikimate phenylpropanoid biosynthesis routes (Zhang et al. 2018), using ROS as signaling molecules (Dietz 2015; Foyer and Noctor 2013; Ismail et al. 2014; Mignolet-Spruyt et al. 2016). Ascorbate-glutathione cycle (AsA-GSH) is a major ROS-scrounging pathway in plants (chloroplast, mitochondria, apoplast, and peroxisomes), which involves successive oxidation and reduction of ascorbate, glutathione, and NADPH catalyzed by APX, MDHAR, DHAR, and GR, thereby helps in combating oxidative damages triggered by abiotic stresses (Mittler 2002). Association of ROS in signaling reveals that there must be some regulation of network to maintain ROS at non-toxic level, needs a precise balance between ROS production (during cellular metabolism), ROS generating enzyme and ROS-scrounging pathways. Thus, stress tolerance of the plants/crop can be improved remarkably by manipulating the ROS levels. Numerous, research studies have demonstrated the role of nanomaterials (CeO_2, C60 as well as Fe_2O_3) in scrounging the over-accumulation of ROS, generated during abiotic stress in plants, thereby improving abiotic stress tolerance in the plant and finally mitigating yield losses (Zhao et al. 2020).

14.4 Enzymatic Antioxidants

The agricultural plants/crops depict different types of antioxidants systems (Fig. 14.2) which are as follows:

(a) Superoxide dismutases (SOD): SOD enzymes are present naturally in different living organisms like agricultural plants/crops and so on. They speed-up the dismutation of $O_2^{\bullet-}$ to H_2O_2, so act as the first line of defense against ROS

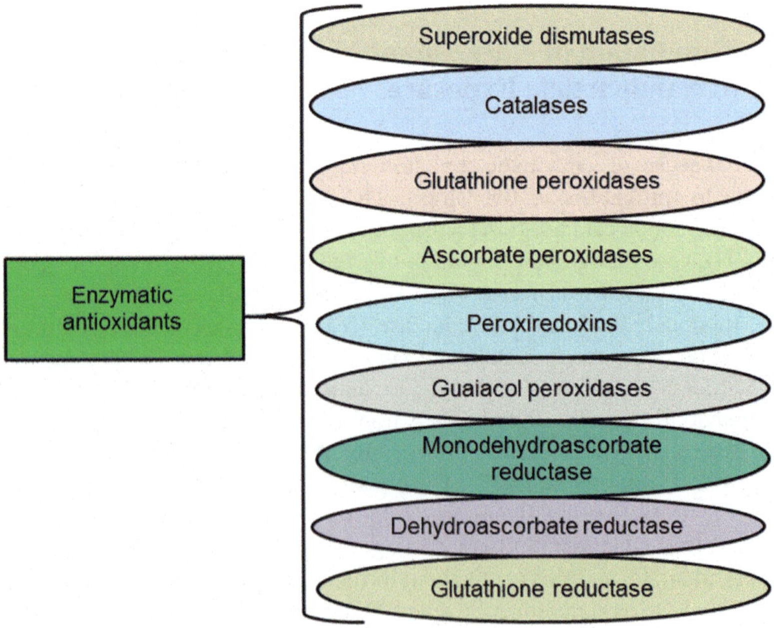

Fig. 14.2 Various types of antioxidant systems in plants

(Moustaka et al. 2015). Generally, due to the attachment of SODs to a metal ion such as Cu, Zn, Fe, Mn, and Ni, are distinguished based on their subcellular location and metal cofactor. In agricultural plants/crops, SODs encoding genes can be controlled and managed by development, tissue-specific, and abiotic stresses/signals (Scandalios 2005).

(b) Catalases: These enzymes are mostly confined to peroxisomes, known for the exclusion of H_2O_2 by reducing it into $2H_2O$. The specific gene that encodes for CATs responds separately to each abiotic stress known to produce ROS (Scandalios 2005).

(c) *Glutathione peroxidases* (GPX): These proteins are mostly confined to mitochondria, cytosol, and chloroplast. GPX is nonheme thiol peroxidases responsible for speed-up the reduction of organic H_2O_2 to H_2O (Margis et al. 2008).

(d) *Ascorbate peroxidases* (APX): These enzymes utilize ascorbate as an electron donor and are responsible for catalyzing the conversion of hydrogen peroxide into water. Different isomers of APX are present in the subcellular compartment of the plants like mitochondria, chloroplast, peroxisomes, and cytosol. The APX gene in plants is modulated by several environmental stresses (Caverzan et al. 2014) whereas the balance between APX, SOD, and CAT determines the intracellular level of $O_2^{\bullet-}$ and H_2O_2. Any alteration in the balance of these three enzymes seems to induce defense-mechanism pathways (Scandalios 2005).

(e) *Peroxiredoxins*: These antioxidant enzymes (thiol specific) are responsible for ROS detoxification in the chloroplast (Foyer and Shigeoka 2010), cell defense of plants by protecting them from oxidative damage, speed-up the reduction of peroxynitrite and various organic H_2O_2 to their corresponding alcohols (Wood et al. 2003).

(f) *Guaiacol peroxidases*: These are heme-containing enzymes known to detoxify H_2O_2 and belong to class III or secreted plant peroxidases. Guaiacol peroxidases can also carry out hydroxylic reaction (second cyclic reaction), different from the peroxidative reaction. These class III peroxidases support many activities in plants such as auxin metabolism, germination to senescence, cell wall elongation, and protection from pathogens (Passardi et al. 2004).

(g) *Monodehydroascorbate reductase* (MDAR): Different isomers of MDAR are found in the different subcellular compartments of plants such as mitochondria, peroxisomes, and cytosol. MDAR (flavin adenine dinucleotide enzyme) maintains the ascorbate pool in plants by catalyzing the regeneration of monodehydroascorbate radical utilizing NAD(P)H as an electron donor (Asada 1999; Leterrier et al. 2005).

(h) *Dehydroascorbate reductase* (DHAR): It helps to maintain ascorbate (AsA) in its reduced form and speed up dehydroascorbate reduction into ascorbate by utilizing glutathione as reducing substrate (Gratão et al. 2005).

(i) *Glutathione reductase* (GR): These enzymes are NAD(P)H dependent, speed-up the reduction of oxidized glutathione (GSSG) into reduced glutathione (GSH), and high GSH/GSSG ratio is required to protect the plant from oxidative damage (Foyer and Noctor 2005). GR plays a significant role in the ascorbate-glutathione cycle and maintains an appropriate level of reduced glutathione.

14.5 Impact of Nanoparticles on Plant Growth

Nanoparticles such as platinum (Pt), gold (Au), fullerene C60, Fe_3O_4, CeO_2, Mn_3O_4 and many more are reported to improve in the functional activities of antioxidant enzymes like SOD, CAT, and POD, that results in more improved adaptation of plants to different abiotic stresses (Chen et al. 2018; Upadhyaya et al. 2018). The fabricated nanosheets of MoS_2 resemble SODs, CATs, and PODs like activities. The nanoparticles of CeO_2 at low concentration (5 μM) efficiently decrease ROS level and protect chloroplast (Boghossian et al. 2013), whereas CeO_2 nanoparticles, when coated with polyacrylic acid, shows SOD and CAT activities, and successfully retained the photosynthetic capability of *Arabidopsis* plants under saline conditions (Wu et al. 2018). Foliar-sprayed CeO_2 nanoparticles in sorghum under drought conditions mitigate the effect of oxidative damage (Djanaguiraman et al. 2018). γ-Fe_2O_3 nanoparticles in *Brassica napus* under drought conditions protect plants from oxidative stress by efficiently reducing H_2O_2 and malondialdehyde (Palmqvist et al. 2017). In the investigation conducted by Yao et al. (2018), suggested that Mn_3O_4 may be used to enhance plant stress resistance (as Mn is micronutrient for plants) due to their stronger ROS-scrounging ability over Ce nanoparticles. When Fe_2O_3 nanoparticles are applied on watermelon in different concentrations, the activities of SOD, CAT, POD, and seedling germination were found to significantly increase and, therefore help to mitigate abiotic stress (Li et al. 2013). Nanoparticles have shown a concentration-dependent impact on the growth and development of plants (Mishra et al. 2017). For instance, onion seedlings, when exposed to TiO_2 nanoparticles, the SOD activity was increases with the increase in the concentration of TiO_2 nanoparticles, whereas onion seed germination as well as seedling growth was enhanced at low concentration and suppressed at higher concentration of TiO_2 nanoparticles (Dimkpa et al. 2017). Shallan et al. (2016) in their study, discovered that foliar spray of SiO_2 (3200 mg L^{-1}) or TiO_2 (50 mg L^{-1}) nanoparticles were found to enhance the drought tolerance of cotton plants. Siddiqui et al. (2014) reported that the application of SiO_2 nanoparticles (1.5–7.5 g L^{-1}) on squash (*Cucurbita pepo* L.) under saline condition upregulated the gene expression of SOD, CAT, POD, APX, and GR as well as increase the chlorophyll concentration, photosynthesis and biomass content of the plant. Under saline conditions, SOD and GPX gene expression are downregulated in tomato (*Solanum lycopersicum*), while on application of ZnO nanoparticles (15 and 30 mg L^{-1}) showed positive growth response (Alharby et al. 2016). On similar lines, foliar spray of ZnO in finger millet (*Eleusine coracana* (L.) Gaertn) improved salinity stress tolerance (Sathiyanarayanan 2018). Dimkpa et al. (2019), reported positive effect on drought tolerance when ZnO nanoparticles (18 nm, 5 mg kg^{-1}) are applied to soil-grown sorghum. However, several reports confirmed the negative impact of nanoparticles/engineered nanoparticles (Rico et al. 2015; Singh et al. 2016) on seed quality of plants like wheat (Rico et al. 2014) and common bean (Majumdar et al. 2015).

14.6 Effect of Nanoparticles on Plant Growth Under Salinity

Excessive accumulation of NaCl in the soil increases the salinity of soil and it affects the growth, development, and productivity of the plants in two ways: osmotic stress and ionic toxicity. Generally, osmotic pressure in the plant cell is more than the osmotic pressure in soil solution. Under high osmotic pressure, plant cell take-up water as well as other requisite minerals from soil solution into the root cells, but during saline conditions, this situation gets reversed and plant ability to take-up water and requisite minerals such as K^+ and Ca^{2+} also disturbed, meanwhile Na^+ and Cl^- ions enter into cytosol that leads to low K^+/Na^+ ratio which is responsible for increased ROS production, electrolytes leakage, toxicity to cell membranes, and also affects metabolic activities in the cytosol (Khan et al. 2012; Kumar 2013; Kumar and Khare 2014). Overall, salinity has a negative effect on various biological and physiological processes of the plant. Some major negative effects of salinity stress on the plant include nutritional imbalance, increased ionic toxicity, ROS overproduction, reduced osmotic potential, the decline in photosystem II efficiency, and stomatal conductance (Negrão et al. 2017). Recently, nanoparticles have been reported to enhance the antioxidative defense mechanism of plants. This potential approach is being exploited to mitigate the salinity stress of the plants (Sabaghnia and Janmohammad 2015). Derosa et al. (2010) reported that SiO_2 nanoparticles enunciate a layer inside the cell wall that facilitates them to conquer salinity stress and uphold yield. Silicon nanoparticles increase the rate of photosynthesis, proline accretion, seed germination, leaf water content, and antioxidant enzymes activities (Qados 2015). On the application of SiO_2 nanoparticles, improvement in salinity stress was observed in *Ocimum basilicum* (Kalteh et al. 2014), *Lens culinaris* (Sabaghnia and Janmohammadi 2014) and *Vicia faba* (Qados 2015). Similarly, SiO_2 nanoparticles were reported to enhance seed germination and antioxidant system in squash and tomato (Siddiqui et al. 2014). Further, mitigation in salinity stress was observed by the application of the foliar spray of Fe_3O_4 as well as ZnO (60 mg L^{-1}) as nano-fertilizers on *Moringa peregrina* (Soliman et al. 2015). The efficiency of a chloroplast, as well as biomass, were increased in treating *Brassica napus* L. with CeO_2 nanoparticles under both fresh and saline water irrigation (Rossi et al. 2016).

14.7 Impact of Nanoparticles on Plant Growth Under Drought Stress

Water is a prerequisite necessity for plant growth and survival, essentially needed for transporting nutrients, thus its crises result in drought stress. Drought stress affects the growth of plants, thereby ultimately influencing the agricultural plants/crops yield globally. During water crises situation, plants limit their various activities such as stomatal closure to prevent additional water loss, reduce CO_2 fixation (photosynthesis), and $NADP^+$ regeneration through the Calvin cycle (Gunjan et al. 2014). Drought stress tolerance of plant varies from species to species and depend to a larger extent on time and intensity they spend under stressful surroundings. Research studies confirm that during drought conditions plants overproduce ROS (H_2O_2, $O_2^{\bullet}$, $^1O_2$, and $OH^{\bullet}$) which causes lipid peroxidation, denaturation of protein, DNA mutation, and eventually cell death (Molassiotis et al. 2016). However, plants protect themselves from negative effects of ROS by its several antioxidant enzymes like SOD, CAT, APX, and GR, while the degree of cellular oxidative damage depends on the capacity of their antioxidant defense system (enzymatic or non-enzymatic). Drought stress can be modulated by the application of different nanoparticles such as silica, silver, copper, ZnO, CeO_2, and many more. On the application of silica nanoparticles improvement in drought tolerance was observed in two sorghum (*Sorghum bicolor* L. Moench) cultivars (Hattori et al. 2005), *Crataegus* sp., and hawthorns (Ashkavand et al. 2015). Similar results were observed in wheat on the application of 1.0 mM sodium silicate (Pei et al. 2010). Sedghi et al. (2013) reported an increased rate of germination on the application of ZnO nanoparticles in soybean under drought-stressed conditions. Foliar application of some micronutrients like iron and titanium nanoparticles were reported to improve drought stress in safflower cultivars and wheat, correspondingly (Davar et al. 2014). Further, Zn and Cu nanoparticles reported improving drought stress by enhancing SOD and CAT enzymes in wheat that results in limiting lipid peroxidation and increasing relative water content by enhancing photosynthesis (Taran et al. 2017). CeO_2 nanoparticles when applied at 100 mg kg^{-1} reported enhancing photosynthesis and Rubisco carboxylase activity (Cao et al. 2017), while composite of CuO, ZnO, and B_2O_3 improve drought stress in *Glycine max* (Dimkpa et al. 2017). Encapsulated abscisic acid (ABA) was delivered successfully to *Arabidopsis thaliana* plant through glutathione-responsive mesoporous silica nanoparticles and their controlled release in plants increased the expression of ABA inducible marker gene (AtGALK2), ultimately improved drought resistance (Sun et al. 2018).

14.8 Impact of Nanoparticles on Plant Growth Under Metallic Stress

Excessive accumulation of metals in plants causes phytotoxicity, alters plant growth, and causes oxidative damage. Metal toxicity interferes with plant growth by suppressing activities of different plant enzymes, interrupting uptake of essential elements which leads to deficiency symptoms. Metals in growth medium are responsible for the overproduction of ROS, which leads to oxidative damage to biomolecules, cell structure, and cell membrane denaturation (Sharma et al. 2012). Biophysical barriers form the first line of defense against metallic stress. If metal passes through this barrier and enters cells, then plants resist metal uptake by its accumulated biomolecules such as organic acids, metal-chelates, and polyphosphates by activating cellular defense system which is responsible for ROS scrounging. However, timely and target-oriented stimulation of these antioxidant defense systems is essential to remove the effects of metallic stress. Nanoparticles (such as nano-selenium, nano-oxides of iron, manganese, and cerium) enters the contamination zone easily due to their smaller size and large surface area, possess a strong affinity towards metal/metalloids adsorption. Nanoparticles in plants retard metal-induced oxidative stress by regulating their energy metabolism, antioxidants, ROS production, and thereby mitigating abiotic stresses. Nanoparticles immobilize metal/metalloids in soil and improve the growth and development of plants during phytoremediation (Martínez-Fernández et al. 2017). Nano-TiO_2 has been reported to limit cadmium (Cd) toxicity and enhance photosynthesis and plant growth rate (Singh and Lee 2016), nano-scale hydroxyapatite mitigates Cd toxicity in *Brassica juncea* (Li and Huang 2014), and ZnO nanoparticles attenuate uptake of Cd in plants (Venkatachalam et al. 2017). Tripathi et al. (2015) demonstrated that silicon nanoparticles hampers Cr accumulation in growth medium and prevents pea seedlings against Cr (VI) phytotoxicity by enhancing the antioxidant defense system. However, research studies reveal that nanoparticles may yield good or bad effects on plants at any level. Toxicological studies of nanomaterials done so far provide a great understanding of nanoparticle interaction with the plants and their potential risk hazards associated with the abiotic stress management and crop productivity improvement (Mustafa and Komatsu 2016; Venkatachalam et al. 2017).

14.9 Impact of Nanoparticles on Plant Growth Under Ultraviolet Radiation Stress

Sunlight together with the UV-B radiation (280-315 nm) is unavoidable abiotic stress for photosynthetic organisms due to the continuous depletion of the ozone layer in the stratosphere. On exposure to such non-ionizing radiation, structural changes occur in cellular components such as DNA, protein, chloroplast, and also induces accumulation of ROS, and free radical scrounging enzymes like SOD (Hideg et al. 2013). Moreover, plants also accumulate phenolic compounds which absorb detrimental UV-radiations (Shen et al. 2010). Nanoparticles are known to intensify the harmful effects of UV-B radiation on plants such as the application of CuO nanoparticles alone on *Elodea nuttallii* (waterweed species) shows no detrimental effects but in combination with the UV-B radiation, induces considerable negative effects on biochemical and physiological traits (Regier et al. 2015).

14.10 Effect of Nanoparticles on Plant Growth Under Flooding Stress

During flooding state, plants suffer from hypoxia conditions because the rate of diffusion of O_2 is slower in water than in air. Flooding stress/hypoxia condition inhibits respiration, seed germination, root, vegetative and reproductive growth, hypocotyl pigmentation, and up-regulation of genes for ethylene synthesis (Komatsu et al. 2012). ATP formation is suppressed under hypoxic conditions, thus to sustain cellular energy level, flooded plants are required to shift their carbohydrate metabolism towards fermentation (Banti et al. 2013), and up-regulation of genes for alcohol dehydrogenase and pyruvate decarboxylase (Mustafa et al. 2015a). Nanoparticles mitigate flooding stress and improve plant growth by inhibiting ethylene biosynthesis (Syu et al. 2014). For instance, the silver nanoparticle treated plant shows less O_2 distress under flooding stress. Besides, employing a gel-free proteomic technique by Mustafa et al. (2015b), reported that Al_2O_3 nanoparticles treated soybean plant under flooding stress has shown better growth performance as compared to plant treated with Ag and ZnO by regulating metabolic pathways and cell death.

14.11 Conclusion and Prospects

Globally, in the arena of agriculture, nanobiotechnology has been used to improve the productivity of crops with quality enhancement by improving cultivation methods. Plants being sessile encounter a variety of abiotic stresses such as salinity, drought, extreme low/high temperature, metal toxicity, UV-B radiation, flooding, and many more in their whole life-span. They accommodate themselves at the biochemical, physiological, and molecular levels by regulating their genes and enzymes responsible for the antioxidant defense system as well as maintaining homeostasis. Plenty of nanoparticles have been exploited for up-regulating various genes and enzymes to mitigate different abiotic stresses but still in its early stage. So far, very little work has been done on the phytotoxicity of nanoparticles on plants, and there exists a huge gap in our understanding of the eco-toxicity, tolerable limit, and uptake capability of various nanoparticles in plants. Therefore, to prevent negative effects of nanoparticles on the environment and living commodity (flora and fauna), and to harness best peculiar attributes of nanoparticles for improving plant growth, development and productivity in stressed conditions, further research is urgently needed to have a clear-cut understanding of the nanoparticle interaction with the plants and environments. Moreover, there is a need to develop a regulatory framework established on the various research evidence which will limit mankind's exposure to undesirable bioengineered nanoparticles to a harmless level, although the application of nanoparticles had increase the productivity of crops. The remarkable applications of nanomaterials presents an optimistic prospect of nanobiotechnology with well understanding of their ecotoxicity and by including all the aspects like reutilizing, feasibility, manufacturing, and framework of policy to handle them securely and utilize them in an eco-friendly manner.

References

Aghdam MTB, Mohammadi H, Ghorbanpour M (2016) Effects of nanoparticulate anatase titanium dioxide on physiological and biochemical performance of *Linum usitatissimum* (*Linaceae*) under well watered and drought stress conditions. Rev Bras Bot 39:139–146

Ahmad P, Hashem A, Abd-Allah EF et al (2015) Role of *Trichoderma harzianum* in mitigating NaCl stress in Indian mustard (*Brassica juncea* L) through antioxidative defense system. Front Plant Sci 6:868. https://doi.org/10.3389/fpls.2015.00868

Alharby HF, Metwali EM, Fuller MP, Aldhebiani AY (2016) The alteration of mRNA expression of SOD and GPX genes, and proteins in tomato (*Lycopersicon esculentum* Mill) under stress of NaCl and/or ZnO nanoparticles. Saudi J Biol Sci 23(6):773–781

Almutairi ZM (2016) Effect of nano-silicon application on the expression of salt tolerance genes in germinating tomato (*Solanum lycopersicum* L.) seedlings under salt stress. Plant Omics J 9:106–114

Apel K, Hirt H (2004) Reactive oxygen species: metabolism, oxidative stress, and signal transduction. Annu Rev Plant Biol 55:373–399

Asada K (1999) The water-water cycle in chloroplasts: scavening of active oxygens and dissipation of excess photons. Annu Rev Plant Physiol Plant Mol Biol 50(NaN):601–639

Ashkavand P, Tabari M, Zarafshar M et al (2015) Effect of SiO_2 nanoparticles on drought resistance in hawthorn seedlings. For Res Pap 76(4):350–359. https://doi.org/10.1515/frp-2015-0034

Azimi R, Borzelabad MJ, Feizi H, Azimi A (2014) Interaction of SiO_2 nanoparticles with seed prechilling on germination and early seedling growth of tall wheatgrass (*Agropyron elongatum* L.). Pol J Chem Tech 16:25–29

Banti V, Giuntoli B, Gonzali S et al (2013) Low oxygen response mechanisms in green organisms. Int J Mol Sci 14:4734–4761

Bhattacharjee S (2012) The language of reactive oxygen species signaling in plants. J Bot 2012:1–22. https://doi.org/10.1155/2012/985298

Boghossian AA, Sen F, Gibbons BM et al (2013) Application of nanoparticle antioxidants to enable hyperstable chloroplasts for solar energy harvesting. Adv Energy Mater 3(7):881–893. https://doi.org/10.1002/aenm.201201014

Cao Z, Stowers C, Rossi L et al (2017) Physiological effects of cerium oxide nanoparticles on the photosynthesis and water use efficiency of soybean (*Glycine max* (L.) Merr.). Environ Sci Nano 4(5):1086–1094. https://doi.org/10.1039/c7en00015d

Catalá A, Díaz M (2016) Editorial: impact of lipid peroxidation on the physiology and pathophysiology of cell membranes. Front Physiol 7:423. https://doi.org/10.3389/fphys.2016.00423

Caverzan A, Bonifacio A, Carvalho FEL et al (2014) The knockdown of chloroplastic ascorbate peroxidases reveals its regulatory role in the photosynthesis and protection under photo–oxidative stress in rice. Plant Sci 214:74–87. https://doi.org/10.1016/j.plantsci.2013.10.001

Chen T, Zou H, Wu X et al (2018) Nanozymatic antioxidant system based on mos_2 nanosheets. ACS Appl Mater Interfaces 10(15):12453–12462. https://doi.org/10.1021/acsami.8b01245

Cramer GR, Urano K, Delrot S et al (2011) Effects of abiotic stress on plants: a systems biology perspective. BMC Plant Biol 11:163. https://doi.org/10.1186/1471-2229-11-163

Davar F, Zareii AR, Amir H (2014) Evaluation the effect of water stress and foliar application of Fe nanoparticles on yield, yield components and oil percentage of safflower (*Carthamus tinctorious* L.). Int J Adv Biol Biomed Res 2(4):1150–1159

Demidchik V (2015) Mechanisms of oxidative stress in plants: from classical chemistry to cell biology. Environ Exp Bot 109:212–228. https://doi.org/10.1016/j.envexpbot.2014.06.021

DeRosa MC, Monreal C, Schnitzer M et al (2010) Nanotechnology in fertilizers. Nat Nanotechnol 5:91–94. https://doi.org/10.1038/nnano.2010.2

Dietz KJ (2015) Efficient high light acclimation involves rapid processes at multiple mechanistic levels. J Exp Bot 66:2401–2414

Dimkpa CO, Bindraban PS, Fugice J et al (2017) Composite micronutrient nanoparticles and salts decrease drought stress in soybean. Agron Sustain Dev 37:5. https://doi.org/10.1007/s13593-016-0412-8

Dimkpa CO, McLean JE, Latta DE et al (2012) CuO and ZnO nanoparticles: phytotoxicity, metal speciation, and induction of oxidative stress in sand-grown wheat. J Nanopart Res 14(9):1125. https://doi.org/10.1007/s11051-012-1125-9

Dimkpa CO, Singh U, Bindraban PS et al (2019) Zinc oxide nanoparticles alleviate drought-induced alterations in sorghum performance, nutrient acquisition, and grain fortification. Sci Total Environ 688:926–934. https://doi.org/10.1016/j.scitotenv.2019.06.392

Djanaguiraman M, Nair, R, Giraldo, JP, Venkata Vara Prasad, P. (2018) Cerium oxide nanoparticles decrease drought-induced oxidative damage in sorghum leading to higher photosynthesis and grain yield. ACS Omega 3(10):14406–14416. https://doi.org/10.1021/acsomega.8b01894

Eckardt NA, Cominelli E, Galbiati M, Tonelli C (2009) The future of science: food and water for life. Plant Cell 21:368–372. https://doi.org/10.1105/tpc.109.066209

Foltete AS, Masfaraud JF, Bigorgne E et al (2011) Environmental impact of sunscreen nanomaterials: ecotoxicity and gentoxicity of altered TiO_2 nanocomposites on *Vicia faba*. Environ Pollut 159(10):2515–2522. https://doi.org/10.1016/j.envpol.2011.06.020

Foyer CH, Noctor G (2005) Redox homeostasis and antioxidant signaling: a metabolic inter-face between stress perception and physiological responses. Plant Cell 17(7):1866–1875. https://doi.org/10.1105/tpc.105.033589

Foyer CH, Noctor G (2013) Redox signaling in plants. Antioxid Redox Signal 18(16):2087–2090. https://doi.org/10.1089/ars.2013.5278

Foyer CH, Shigeoka S (2010) Understanding oxidative stress and antioxidant functions to enhance photosynthesis. Plant Physiol 155(1):93–100. https://doi.org/10.1104/pp.110.166181

Gapper C, Dolan L (2006) Control of plant development by reactive oxygen species. Plant Physiol 141:341–345

Ghorbanpour M, Hatami M (2014) Spray treatment with silver nanoparticles plus thidiazuron increases antioxidant enzyme activities and reduces petal and leaf abscission in four cultivars of geranium (*Pelargonium zonale*) during storage in the dark. J Hort Sci Biotech 89:712–718

Gill SS, Tuteja N (2010) Reactive oxygen species and antioxidant machinery in abiotic stress tolerance in crop plants. Plant Physiol Biochem 48(12):909–930. https://doi.org/10.1016/j.plaphy.2010.08.016

Giraldo JP, Wu H, Newkirk GM, Kruss S (2019) Nanobiotechnology approaches for engineering smart plant sensors. Nat Nanotechnol 14(6):541–553. https://doi.org/10.1038/s41565-019-0470-6

Gratão PL, Polle A, Lea PJ, Azevedo RA (2005) Making the life of heavy metal-stressed plants a little easier. Function Plant Biol 32(6):481–494

Gunjan B, Zaidi MGH, Sandeep A (2014) Impact of gold nanoparticles on physiological and biochemical characteristics of *Brassica juncea*. J Plant Biochem Physiol 2(3):133–138. https://doi.org/10.4172/2329-9029.1000133

Haghighi M, Afifipour Z, Mozafarian M (2012) The effect of N-Si on tomato seed germination under salinity levels. J Biol Environ Sci 6(16):87–90

Haghighi M, Pessarakli M (2013) Influence of silicon and nano-silicon on salinity tolerance of cherry tomatoes (*Solanum lycopersicum* L.) at early growth stage. Sci Hortic 161:111–117. https://doi.org/10.1016/j.scienta.2013.06.034

Halliwell B (2006) Reactive species and antioxidants Redox biology is a fundamental theme of aerobic life. Plant Physiol 141:312–322

Hasanpour H, Maali-Amiri R, Zeinali H (2015) Effect of TiO_2 nanoparticles on metabolic limitations to photosynthesis under cold in *chickpea*. Russ J Plant Physiol 62:779–787

Hattori T, Inanaga S, Araki H et al (2005) Application of silicon enhanced drought tolerance in *Sorghum bicolour*. Physiol Plant 123:459–466

Hideg E, Jansen MA, Strid A (2013) UV-B exposure, ROS, and stress: inseparable companions or loosely linked associates? Trends Plant Sci 18(2):107–115. https://doi.org/10.1016/j.tplants.2012.09.003

Hilker M, Schwachtje J, Baier M et al (2015) Priming and memory of stress responses in organisms lacking a nervous system. Biol Rev 91:1118–1133. https://doi.org/10.1111/brv.12215

Hirel B, Le-Gouis J, Ney B, Gallais A (2007) The challenge of improving nitrogen use efficiency in crop plants: towards a more central role for genetic variability and quantitative genetics within integrated approaches. J Exp Bot 58:2369–2387. https://doi.org/10.1093/jxb/erm097

Iannone MF, Groppa MD, Desousa ME et al (2016) Impact of magnetite iron oxide nanoparticles on wheat (*Triticum aestivum* L.) development: evaluation of oxidative damage. Environ Exp Bot 131:77–88

Iqbal MS, Singh AK, Singh SP, Ansari MI (2020) Nanoparticles and Plant Interaction with Respect to Stress Response. In: Bhushan I, Singh V, Tripathi D (ed) Nanomaterials and environmental biotechnology. Nanotechnology in the life sciences. Springer, Cham, p 1–15. https://doi.org/10.1007/978-3-030-34544-0_1

Ismail A, Takeda S, Nick PR (2014) Life and death under salt stress: same players, different timing. J Exp Bot 65(12):2963–2979. https://doi.org/10.1093/jxb/eru159

Jaberzadeh A, Moaveni P, Reza H et al (2013) Influence of bulk and nanoparticles titanium foliar application on some agronomic traits, seed gluten and starch contents of wheat subjected to water deficit stress. Not Bot Horti Agrobot Cluj Napoca 41:201–207. https://doi.org/10.15835/nbha4119093

Jaleel CA, Riadh K, Gopi R et al (2009) Antioxidant defense responses: Physiological plasticity in higher plants under abiotic constraints. Acta Physiol Plant 31(3):427–436. https://doi.org/10.1007/s11738-009-0275-6

Jiang QY, Zhuo F, Long SH et al (2016) Can arbuscular mycorrhizal fungi reduce Cd uptake and alleviate Cd toxicity of *Lonicera japonica* grown in Cd–added soils? Sci Rep 6(1):21805. https://doi.org/10.1038/srep21805

Kah M, Tufenkji N, White JC (2019) Nano-enabled strategies to enhance crop nutrition and protection. Nat Nanotechnol 14(6):532–540

Kalteh M, Alipour ZT, Ashraf S et al (2014) Effect of silica nanoparticles on basil (*Ocimum basilicum*) under salinity stress. J Chem Health Risks 4:49–55

Khan MN, Siddiqui MH, Mohammad F, Naeem M (2012) Interactive role of nitric oxide and calcium chloride in the tolerance of plants to salt stress. Nitric Oxide 27:210–218. https://doi.org/10.1016/j.niox.2012.07.005

Khare T, Kumar V, Kishor PBK (2014) Na^+ and Cl^- ions show additive effects under NaCl stress on induction of oxidative stress and the responsive antioxidative defense in rice. Protoplasma 252:1149–1165. https://doi.org/10.1007/s00709-014-0749-2

Kiapour H, Moaveni P, Habibi D, Sani B (2015) Evaluation of the application of gibbrellic acid and titanium dioxide nanoparticles under drought stress on some traits of basil (*Ocimum basilicum* L.). Int J Agr and Agrl Res 6(4):138–150

Komatsu S, Hiraga S, Yanagawa Y (2012) Proteomics techniques for the development of flood tolerant crops. J Proteome Res 11:68–78. https://doi.org/10.1021/pr2008863

Kumar M (2013) Crop plants and abiotic stresses. J Biomol Res Ther 3:125. https://doi.org/10.4172/2167-7956.1000e125

Kumar V, Khare T (2014) Individual and additive effects of Na^+ and Cl^- ions on rice under salinity stress. Arch Agron Soil Sci 61(3):381–395. https://doi.org/10.1080/03650340.2014.936400

Latef AAHA, Alhmad MFA, Abdelfattah KE (2017) The possible roles of priming with ZnO nanoparticles in mitigation of salinity stress in lupine (*Lupinus termis*) plants. J Plant Growth Regul 36(1):60–70

Lei Z, Mingyu S, Xiao W et al (2008) Antioxidant stress is promoted by nano-anatase in spinach chloroplasts under UV-B radiation. Biol Trace Elem Res 121:69–79. https://doi.org/10.1007/s12011-007-8028-0

Leterrier M, Corpas FJ, Barroso JB et al (2005) Peroxisomal monodehydroascorbate reductase Genomic clone characterization and functional analysis under environmental stress conditions. Plant Physiol 138(4):2111–2123. https://doi.org/10.1104/pp.105.066225

Li J, Chang PR, Huang J et al (2013) Physiological effects of magnetic iron oxide nanoparticles towards watermelon. J Nanosci Nanotechnol 13(8):5561–5567. https://doi.org/10.1166/jnn.2013.7533

Li Z, Huang J (2014) Effects of nanoparticle hydroxyapatite on growth and antioxidant system in pakchoi (*Brassica chinensis* L.) from cadmium-contaminated soil. J Nanomater 2014:1–7. https://doi.org/10.1155/2014/470962

Majumdar S, Almeida IC, Arigi EA et al (2015) Environmental effects of nanoceria on seed production of common bean (*Phaseolus vulgaris*): a proteomic analysis. Environ Sci Technol 49:13283–13293

Margis R, Dunand C, Teixeira FK, Margis-Pinheiro M (2008) Glutathione peroxidase family-an evolutionary overview. FEBS J 275(15):3959–3970. https://doi.org/10.1111/j.1742-4658.2008.06542.x

Martínez-Fernández D, Vitkova M, Vaňková Z, Komárek M (2017) Engineered nanomaterials for phytoremediation of metal/metalloid-contaminated soils: implications for plant physiology. In: Ansari A, Gill S, Gill R et al (eds) Phytoremediation. Springer, Cham, p 369–403. https://doi.org/10.1007/978-3-319-52381-1_14

Maxwell DP, Wang Y, Mcintosh L (1999) The alternative oxidase lowers mitochondrial reactive oxygen production in plant cells. Proc Natl Acad Sci USA 96(14):8271–8276. https://doi.org/10.1073/pnas.96.14.8271

Meena KK, Sorty AM, Bitla UM et al (2017) Abiotic Stress Responses and Microbe-Mediated Mitigation in Plants: The Omics Strategies. Front Plant Sci 8:172. https://doi.org/10.3389/fpls.2017.00172

Mignolet-Spruyt L, Xu E, Idanheimo N et al (2016) Spreading the news: subcellular and organellar reactive oxygen species production and signalling. J Exp Bot 67:3831–3844

Miller G, Suzuki N, Ciftci-Yilmaz S, Mittler R (2010) Reactive oxygen species homeostasis and signalling during drought and salinity stresses. Plant, Cell Environ 33:453–467. https://doi.org/10.1111/j.1365-3040.2009.02041.x

Mishra S, Keswani C, Abhilash PC et al (2017) Integrated approach of agri-nanotechnology: challenges and future trends. Front Plant Sci 8:471. https://doi.org/10.3389/fpls.2017.00471

Mittler R (2002) Oxidative stress, antioxidants and stress tolerance. Trends Plant Sci 7(9):405–410. https://doi.org/10.1016/s1360-1385(02)02312-9

Mittler R, Vanderauwera S, Gollery M, Breusegem VF (2004) Reactive oxygen gene network of plants. Trends Plant Sci 9(10):490–498. https://doi.org/10.1016/j.tplants.2004.08.009

Mohammadi R, Maali-Amiri R, Abbasi A (2013) Effect of TiO_2 nanoparticles on chickpea response to cold stress. Biol Trace Elem Res 152(3):403–410. https://doi.org/10.1007/s12011-013-9631-x

Mohammadi R, Maali-Amiri R, Mantri N (2014) Effect of TiO_2 nanoparticles on oxidative damage and antioxidant defense systems in chickpea seedlings during cold stress. Russ J Plant Physiol 61(6):768–775. https://doi.org/10.1134/s1021443714050124

Molassiotis A, Job D, Ziogas V, Tanou G (2016) Citrus plants: a model system for unlocking the secrets of NO and ROS-inspired priming against salinity and drought. Front Plant Sci 7:229. https://doi.org/10.3389/fpls.2016.00229

Moustaka J, Tanou G, Adamakis ID et al (2015) Leaf age-dependent photoprotective and antioxidative response mechanisms to paraquat-induced oxidative stress in *Arabidopsis thaliana*. Int J Mol Sci 16(12):13989–14006. https://doi.org/10.3390/ijms160613989

Mukherjee A, Peralta-Videa JR, Bandopadhyay S et al (2014) Physiological effects of nanoparticulate ZnO in green peas (*Pisum sativum* L.) cultivated in soil. Metallomics 6(1):132–138. https://doi.org/10.1039/c3mt00064h

Mustafa G (1864) Komatsu S (2016) Toxicity of heavy metals and metal-containing nanoparticles on plants. Biochim Biophys Acta Bioenerg 8:932–944. https://doi.org/10.1016/j.bbapap.2016.02.020

Mustafa G, Sakata K, Hossain Z, Komatsu S (2015a) Proteomic analysis of flooded soybean root exposed to aluminum oxide nanoparticles. J Proteom 128:280–297. https://doi.org/10.1016/j.jprot.2015.08.010

Mustafa G, Sakata K, Hossain Z, Komatsu S (2015b) Proteomic study on the effects of silver nanoparticles on soybean under flooding stress. J Proteom 122:100–118. https://doi.org/10.1016/j.jprot.2015.03.030

Nair PMG, Chung IM (2015) Changes in the growth, redox status and expression of oxidative stress related genes in chickpea (*Cicer arietinum* L.) in response to copper oxide nanoparticle exposure. J Plant Growth Regul 34:350–361. https://doi.org/10.1007/s00344-014-9468-3

Nair PMG, Chung IM (2017) Regulation of morphological, molecular and nutrient status in *Arabidopsis thaliana* seedlings in response to ZnO nanoparticles and Zn ion exposure. Sci Total Environ 575:187–198. https://doi.org/10.1016/j.scitotenv.2016.10.017

Nath M, Bhatt D, Prasad R, Tuteja N (2017) Reactive oxygen species (ROS) metabolism and signaling in plant-mycorrhizal association under biotic and abiotic stress conditions. In: Varma A, Prasad R, Tuteja N. (eds) Mycorrhiza-eco-physiology, secondary metabolites, nanomaterials. Springer, Cham, Switzerland, p 223–232. https://doi.org/10.1007/978-3-319-57849-1_12

Negrão S, Schmöckel SM, Tester M (2017) Evaluating physiological responses of plants to salinity stress. Ann Bot 119(1):1–11

Okupnik A, Pflugmacher S (2016) Oxidative stress response of the aquatic macrophyte *Hydrilla verticillata* exposed to TiO_2 nanoparticles. Environ Toxicol Chem 35:2859–2866. https://doi.org/10.1002/etc.3469

Palmqvist NGM, Seisenbaeva GA, Svedlindh P, Kessler VG (2017) Maghemite nanoparticles acts as nanozymes, improving growth and abiotic stress tolerance in *Brassica napus*. Nanoscale Res Lett 12(1):631

Pandey B, Singh AK, Singh SP (2018) Nanoparticles mediated gene knock out through miRNA replacement: recent progress and challenges. In: Mishra RK, Thomas S, Mohapatra S et al (eds) Applications of targeted nano-drugs and delivery systems. Elsevier, p 469–497

Pandey S, Fartyal D, Agarwal A (2017) Abiotic stress tolerance in plants: myriad roles of ascorbate peroxidase. Front Plant Sci 8:581. https://doi.org/10.3389/fpls.2017.00581

Passardi F, Longet D, Penel C, Dunand C (2004) The class III peroxidase multigenic family in rice and its evolution in land plants. Phytochem 65(13):1879–1893

Pei S, Zhao J, Du J et al (2010) Direct reduction of graphene oxide films into highly conductive and flexible graphene films by hydrohalic acids. Carbon 48(15):4466–4474

Petrov V, Hille J, Mueller-Roeber B, Gechev TS (2015) ROS-mediated abiotic stress-induced programmed cell death in plants. Front Plant Sci 18(6):69. https://doi.org/10.3389/fpls.2015.00069

Porwal P, Porwal S, Singh SP et al (2020) Improving futuristic nanomaterial researches in forestry sector: An overview. In: Husen A, Jawaid M (eds.), In micro and nano technologies, Nanomaterials

for agriculture and forestry applications, Elsevier, pp 505–518. https://doi.org/10.1016/B978-0-12-817852-2.00021-4
Porwal P, Sharma A (2016) Improving water quality through nanotechnology. Int J Appl Res Technol 1(2):119–133
Qados AMSA (2015) Mechanism of nanosilicon-mediated alleviation of salinity stress in faba bean (*Vicia faba* L.) plants. Am J Exp Agric 7:78–95
Qados AMSA, Moftah AE (2015) Influence of silicon and nano-silicon on germination, growth and yield of faba bean (*Vicia faba* L.) under salt stress conditions. Am J Exp Agric 5:509–524
Qi M, Liu Y, Li T (2013) Nano-TiO_2 improve the photosynthesis of tomato leaves under mild heat stress. Biol Trace Elem Res 156:323–328
Rajeshwari A, Kavitha S, Alex SA et al (2015) Cytotoxicity of aluminum oxide nanoparticles on *Allium cepa* root tip-effects of oxidative stress generation and biouptake. Environ Sci Pollut Res 22(14):11057–11066. https://doi.org/10.1007/s11356-015-4355-4
Rani H, Singh SP, Yadav TP et al (2020) In-vitro catalytic, antimicrobial and antioxidant activities of bioengineered copper quantum dots using *Mangifera indica* (L.) leaf extract. Mater Chem Phys 239:122052. https://doi.org/10.1016/j.matchemphys.2019.122052
Regier N, Cosio C, Moos VN, Slaveykova VI (2015) Effects of copper-oxide nanoparticles, dissolved copper and ultraviolet radiation on copper bio–accumulation, photosynthesis and oxidative stress in the aquatic macrophyte *Elodea nuttallii*. Chemosphere 128:56–61
Rezvani N, Sorooshzadeh A, Farhadi N (2012) Effect of nano-silver on growth of saffron in flooding stress. Int J Biol Biomol Agric Food Biotechnol Eng 6:11–16
Riahi-Madvar A, Rezaee F, Jalili V (2012) Effects of alumina nanoparticles on morphological properties and antioxidant system of *Triticum aestivum*. Iran J Plant Physiol 3:595–603
Rico CM, Lee SC, Rubenecia R et al (2014) Cerium oxide nanoparticles impact yield and modify nutritional parameters in wheat (*Triticum aestivum* L.). J Agric Food Chem 62:9669–9675
Rico CM, Peralta-Videa JR, Garadea-Torresdey JL (2015) Chemistry, biochemistry of nanoparticles, and their role in antioxidant defense system in plants. In: Siddiqui HM et al (eds) Nanotechnology and plant sciences. Springer International Publishing, Switzerland, pp 1–17
Rossi L, Zhang W, Lombardini L, Ma X (2016) The impact of cerium oxide nanoparticles on the salt stress responses of *Brassica napus* L. Environ Pollut 219:28–36
Sabaghnia N, Janmohammadi M (2014) Effect of nanosilicon particles application on salinity tolerance in early growth of some lentil genotypes. Ann. UMCS Biol 69:39–55
Sabaghnia N, Janmohammad M (2015) Effect of nano-silicon particles application on salinity tolerance in early growth of some lentil genotypes. Ann UMCS Biol 69(2):39–55
Samadi N, Yahyaabadi S, Rezayatmand Z (2014) Effect of TiO_2 and TiO_2 nanoparticle on germination, root and shoot length and photosynthetic pigments of *Mentha piperita*. Int J Plant Soil Sci 3:408–418
Sathiyanarayanan A (2018) Foliar Spray of Zinc Oxide Nanoparticles Improves salt tolerance in finger millet crops under glasshouse condition. SCIOL Biothnol 1(1):20–29
Scandalios JG (2005) Oxidative stress: molecular perception and transduction of signals triggering antioxidant gene defenses. Brazilian J Medic Biologic Res 38(7):995–1014
Sedghi M, Hadi M, Toluie SG (2013) Effect of nano zinc oxide on the germination parameters of soybean seeds under drought stress. Ann WUT-ser Biol XVI(2):73–78
Shallan M, Hassan H, Namich AM, Ibrahim A (2016) Biochemical and physiological effects of tio_2 and sio_2 nanoparticles on cotton plant under drought stress. Res J Pharm Bio and Chem Sci 7(4):1540–1551
Sharma P, Jha AB, Dubey RS, Pessarakli M (2012) Reactive oxygen species, oxidative damage, and antioxidant defense mechanisms in plants under stressful conditions. J Bot 1–26. https://doi.org/10.1155/2012/217037
Shen X, Zhou Y, Duan L et al (2010) Silicon effects on photosynthesis and antioxidant parameters of soybean seedlings under drought and ultraviolet-B radiation. J Plant Physiol 167:1248–1252
Siddiqui MH, Whaibi AL, Faisal M, Alsahli AA (2014) Nano-silicon dioxide mitigates the adverse effects of salt stress on *Cucurbita pepo* L. Environ Toxicol Chem 33:2429–2437

Singh AK, Pal P, Gupta V et al (2018) Green synthesis, characterization and antimicrobial activity of zinc oxide quantum dots using Eclipta alba. Mater Chem Phys 203:40–48
Singh AK, Porwal P (2020) Nanotechnology as Potential and innovative platform toward wastewater treatment: an overview. In: Bhushan I, Singh V, Tripathi D (eds) Nanomaterials and environmental biotechnology. Nanotechnology in the life sciences. Springer Nature Switzerland, p 201–220. https://doi.org/10.1007/978-3-030-34544-0_12
Singh AK, Yadav TP, Pandey B et al (2018b) Engineering nanomaterials for smart drug release: recent advances and challenges. In: Mishra RK, Thomas S, Mohapatra S et al (eds) Applications of targeted nano-drugs and delivery systems. Elsevier, p 411–449. https://doi.org/10.1016/B978-0-12-814029-1.00015-6
Singh J, Lee BK (2016) Influence of nano-TiO_2 particles on the bioaccumulation of Cd in soybean plants (*Glycine max*): a possible mechanism for the removal of Cd from the contaminated soil. J Environ Manag 170:88–96
Singh S, Tripathi DK, Dubey NK, Chauhan DK (2016) Effects of nano-materials on seed germination and seedling growth: striking the slight balance between the concepts and controversies. Mater Focus 5(3):195–201
Singh SP, Ansari MI, Pandey B et al (2020) Recent trends and advancement toward phyto-mediated fabrication of noble metallic nanomaterials: focus on silver, gold, platinum and palladium. In: Singh VK, Tripathi DK (eds) Bhushan I. Nanomaterials and environmental biotechnology, Springer, pp 87–105
Soliman ASH, El-feky SA, Darwish E (2015) Alleviation of salt stress on *Moringa peregrina* using foliar application of nanofertilizers. J Hortic For 7:36–47
Song G, Gao Y, Wu H et al (2012) Physiological effect of anatase TiO_2 nanoparticle on *Lemna minor*. Environ Toxicol Chem 31:2147–2152
Sun D, Hussain HI, Yi Z et al (2018) Delivery of abscisic acid to plants using glutathione responsive mesoporous silica nanoparticles. J Nanosci Nanotechnol 18(3):1615–1625
Syu YY, Hung JH, Chen JC, Chuang HW (2014) Impact of size and shape of silver nanoparticles on *Arabidopsis* plant growth and gene exp-ression. Plant Physiol Biochem 83:57–64
Taran N, Storozhenko V, Svietlova N et al (2017) Effect of zinc and copper nanoparticles on drought resistance of wheat seedlings. Nanoscale Res Lett 12(1):60
Tripathi DK, Singh VP, Prasad SM et al (2015) Silicon nanoparticles (SiNp) alleviate chromium (VI) phytotoxicity in *Pisum sativum* (L.) seedlings. Plant Physiol Biochem 96:189–198
Upadhyaya H, Dutta BK, Panda SK (2018) Impact of zinc on dehydration and rehydration responses in tea. Biol Plant 62(2):395–399. https://doi.org/10.1007/s10535-017-0758-z
Venkatachalam P, Jayaraj M, Manikandan R et al (2017) Zinc oxide nanoparticles (ZnONPs) alleviate heavy metal-induced toxicity in *Leucaena leucocephala* seedlings: a physiochemical analysis. Plant Physiol Biochem 110:59–69
Wang W, Vinocur B, Altman A (2003) Plant responses to drought, salinity and extreme temperatures: towards genetic engineering for stress tolerance. Planta 218:1–14. https://doi.org/10.1007/s00425-003-1105-5
Wassim A, Ichrak BR, Sa¨ıda A (2013) Putative role of proteins involved in detoxification of reactive oxygen species in the early response to gravitropic stimulation of poplar stems. Plant Signal Behav 8(1):1–6
Wood ZA, Schroder E, Harris JR, Poole LB (2003) Structure, mechanism and regulation of peroxiredoxins. Trends Biochem Sci 28(1):23–40
Wu H, Shabala L, Shabala S, Giraldo JP (2018) Hydroxyl radical scavenging by cerium oxide nanoparticles improves *Arabidopsis* salinity tolerance by enhancing leaf mesophyll potassium retention. Environ Sci Nano 5:1567–1583
Xie X, He Z, Chen N et al (2019) The Roles of Environmental Factors in Regulation of Oxidative Stress in Plant. Biomed Res Int 2019:1–11. https://doi.org/10.1155/2019/9732325
Yao J, Cheng Y, Zhou M et al (2018) ROS scavenging Mn_3O_4 nanozymes for *in vivo* anti-inflammation. Chem Sci 9(11):2927–2933

You J, Chan Z (2015) ROS regulation during abiotic stress responses in crop plants. Front Plant Sci 6:1092. https://doi.org/10.3389/fpls.2015.01092

Zhang H, Du W, Peralta-Videa JR (2018) Metabolomics reveals how cucumber (*Cucumis sativus*) reprograms metabolites to cope with silver ions and silver nanoparticle-induced oxidative stress. Environ Sci Technol 52(14):8016–8026

Zhang Q (2007) Strategies for developing Green Super Rice. Proc Natl Acad Sci USA 104:16402–16409. https://doi.org/10.1073/pnas.0708013104

Zhao L, Lu L, Wang A et al (2020) Nanobiotechnology in agriculture: use of nanomaterials to promote plant growth and stress tolerance. J Agric Food Chem. https://doi.org/10.1021/acs.jafc.9b06615

Zhao L, Peng B, Hernandez-Viezcas JA et al (2012) Stress response and tolerance of *Zea mays* to CeO_2 nanoparticles: cross talk among H_2O_2, heat shock protein and lipid peroxidation. ACS Nano 6:9615–9622

Chapter 15
Plant Stress Hormones Nanobiotechnology

Sashi Sonkar, Laxuman Sharma, Rishi Kumar Singh, Brijesh Pandey, Saurabh Singh Rathore, Akhilesh Kumar Singh, Paras Porwal, and Sujeet Pratap Singh

Abstract In the epoch of global warming with climate change, various unprecedented challenges were encountered by the agricultural systems globally. The same was reflected in the response by the agricultural scientists and practitioners to combat the challenges. Many findings at the level of lab and improvising practices at the level of farm have influenced the outcome. Inventions and discoveries in nanotechnology is also such factor which has influenced every arena of agricultural science. Consequently, nanotechnology is considered a handy tool for improving crop productivity as well as promising sustainability to alleviate food insecurity. Despite being at nascent phase of its development, role of nanotechnology towards improvement in

S. Sonkar
Department of Botany, Bankim Sardar College, Tangrakhali, South 24 Parganas 743329, West Bengal, India

L. Sharma
Department of Horticulture, Sikkim University, Gangtok 737102, India
e-mail: lsharma@cus.ac.in

R. K. Singh
School of Biotechnology, Mangalmay Institute of Management and Technology, Greater Noida 201310, Uttar Pradesh, India

B. Pandey · S. S. Rathore · A. K. Singh (✉)
Department of Biotechnology, School of Life Sciences, Mahatma Gandhi Central University, Motihari 845401,
Bihar, India
e-mail: akhileshsingh@mgcub.ac.in

B. Pandey
e-mail: brijeshpandey@mgcub.ac.in

S. S. Rathore
e-mail: ssrathore@mgcub.ac.in

P. Porwal · S. P. Singh
Amity Institute of Biotechnology, Amity University Uttar Pradesh,
Lucknow Campus, Lucknow 226028, Uttar Pradesh, India
e-mail: pporwal@lko.amity.edu

S. P. Singh
e-mail: spsingh3@amity.edu

J. M. Al-Khayri et al. (eds.), *Nanobiotechnology*,
https://doi.org/10.1007/978-3-030-73606-4_15

crop production has been significant. Nanomaterials support the improvement of agricultural crop productivity due to the alleviation of adverse influences of biotic as well as abiotic stresses by the activation of plant defense response, quality enhancement, and growth regulation via the alteration of the phytohormones contents. Although, there are still vast gaps in our understanding of the eco-toxicity, permissible limit, and uptake capacity of various nanomaterials in plant systems. Therefore, this chapter aims at the acquisition of information on basic features of plant hormones and their interaction with nanomaterials under various stresses together with exploitation of nanomaterials in farming and, more in general, in-plant biological investigation.

Keywords Abiotic stress · Biotic stress · Crop productivity · Nanomaterials · Nanotechnology · Phytohormones

15.1 Introduction

In an ecosystem, there exist close interactions as well as associations amongst living organisms and abiotic factors. Plants thrive best at optimum of different abiotic factors and deviations from the optimum adversely affect the plant growth, development as well as productivity that are generally referred to as stress (Verma et al. 2013). Levitt (1972) has defined stress as "any ecological factor able to induce a potentially injurious strain in living systems". Abiotic stresses arise due to ultraviolet as well as ionizing radiations, salinity, very low or high temperature, drought, flood, elevated carbon dioxide, low or high soil pH, and contamination of soil with high content of heavy metals, which in turn would lead to the huge loss of economically important plants (Arun-Chinnappa et al. 2017; Compant et al. 2010; Gull et al. 2019; Reis et al. 2012). The loss in the crop yield due to abiotic factors can be more than 50%. Globally, about 91% of agriculture land is affected by one or other forms of abiotic stresses, which may worsen further due to changing climate (Minhas et al. 2017). Furthermore, most of the crops of commercial importance and those indispensable for food security are grown in tropical as well as subtropical areas. These crops do not have the ability of withstand low temperature. An exposure of such plants to low temperature, which is generally known as chilling temperature leads to dysfunction of different physiological processes (Lukatkin et al. 2012). Low temperature (chilling or freezing) causes cellular dehydration due to the formation of ice in the plant tissues and the leakage of the intracellular solute from the plasma membrane (Chinnusamy et al. 2007). Dehydration conditions in the plant are also attributed to the limited water uptake by the plant during cold stress (Chinnusamy and Zhu 2009). On the other hand, heat stress is expected to be more severe due to a constant increase in temperature. A hypothesis of alteration in physiological processes, plant growth, development, and yield still holds true. Under heat stress, the plant undergoes oxidative stress owing to enhance the generation of reactive oxygen species (ROS) in the plant tissues (Hasanuzzaman et al. 2013). Besides, there has been a continuous increase in the contamination of agriculture land with heavy metals, due to several anthropogenic activities, excess use of chemicals like pesticides, mining as well as industrialization (Tiwari and Lata 2018). Heavy metals affect the crop yield, basically due to the adverse effect on physiological as well as molecular activities (Amari et al. 2017; Hassan et al. 2017). Heavy metals interact with biomolecules like protein and DNA, thereby elevating the free ROS (Emamverdian et al. 2015). The redox-active metals like Cr, Cu, Mn and Fe cause severe oxidative injury and results in defragmentation of proteins, breakage of the DNA strands, cell homeostasis as well as ultimately the cell death (Schutzendubel and Polle 2002). Nevertheless, oxidative stress is inflicted indirectly by non-redox metals like Cd, Ni and Hg through various mechanisms like preventing antioxidative enzymes or by eliciting enzymes such as NADPH oxidases (Bielen et al. 2013), binding of sulfhydryl groups of protein molecules or diminution of glutathione (Valko et al. 2005). Likewise, in the condition of low soil as well as atmospheric humidity and at high air temperature, the evapotranspiration flux and water intake by the plant from the soil being severely affected, which leads to drought stress (Lipiec et al. 2013). Usually, drought stress occurs as a result of water deficiency as well as high temperatures. Some of the major effects imposed by drought stress in plants are stomatal closure and decreased cell growth including enlargement (Farooq et al. 2009; Iqbal et al. 2020a). Nutrient uptake processes, photosynthetic activities, chlorophyll synthesis, carbohydrate metabolism, and respiration are disturbed (Farooq et al. 2009; Jaleel et al. 2008; Limbu et al. 2018) and causes huge loss to the plant productivity and at the extreme case, drying or death of the plants occur. Also, flooding of the soil is a universal problem and has been a major threat to food security. Two-third of the damages and the loss to crops at the global scale are due to floods in the period between 2006 and 2016 (Food and Agriculture Organization of the United Nations [FAO] 2017). Flooding creates complex stress (Fukao et al. 2019), due to submergence as well as partial submergence (Zhou et al. 2020), thereby imposing the hypoxia (deficiency of oxygen) or anoxia (absence of oxygen) and obstructs growth, development as well as the survival of plants (Ashraf 2012). Physiologically, in waterlogged conditions, the stomatal conductance, carbon dioxide assimilation rate together with root hydraulic conductivity are hampered. Salinity is another abiotic stress having a severe impact on agriculture production. World-wide around 800 million hectares is affected by salinity or sodicity (FAO 2009). The salt stress accounts for impairment of crop growth together with development (Isayenkov and Maathuis 2019). Salinity is known to elicit water stress, cytotoxicity as well as nutritional imbalance and imparts oxidative stress caused by the formation of ROS (Isayenkov 2012; Tsugane et al. 1999).

Plants fundamentally develop certain adaptive measures once exposed to the stresses. Several tolerance mechanisms employed by the plants that manifest at subcellular level have been studied thoroughly. Despite own puissant mechanisms, plant sustains severe loss due to abiotic stress. The stress can be externally managed by inducing the tolerance by various methods viz. exogenous application of plant hormones, adopting appropriate agronomic techniques, use of arbuscular mycorrhizal fungal strains, exploitation of plant growth-promoting rhizobacteria (PGPR), use of genetically improved plants, use of elicitor and tolerance inductor (Hernánde et al. 2018). Furthermore, it is well-studied fact that plant hormones like abscisic acid, salicylic acid, jasmonates as well as ethylene mediate the plant defense responses during stress conditions. Abscisic acid is typically known to be responsible for defending the abiotic stress, though there are a plethora of literature citing the interaction of abscisic acid with other hormones like gibberellin, ethylene as well as auxin. Auxin like indole acetic acid regulates the growth/development in stress environments (Kazan 2013), particularly under salinity (Fahad et al. 2015b), heavy metal (Egamberdieva 2009; Hu et al. 2013) and aluminum toxicity (Wang et al. 2016). Brassinosteroids are studied for having a role during stresses like high temperature, chilling, salinity, drought (Wani et al. 2016) and so the jasmonates (Pauwels et al. 2009; Wani et al. 2016). External application of methyl jasmonates was found to minimize the stress caused due to salinity (Yoon et al. 2009) and heavy metals like copper and cadmium (Maksymiec et al. 2007; Yan et al. 2013). Salicylic acid in combination with abscisic acid can regulate the responses during drought stress (Rivas-San and Plasencia 2011). Apart from these, various agronomic practices are recommended for inducing tolerance to plants during stresses (Hernández et al. 2018; Mariani and Ferrante 2017). The tolerance against abiotic stresses like drought and high temperature can be stimulated with proper soil management practices before sowing of the crops and during the growth period. Amendment of soil applying organic manure and minerals, growing cover crops before sowing, proper crop rotation, low tillage, and application of nanofertilizer is generally practiced. Seed treatments with arbuscular mycorrhizal fungi as well as rhizobacteria are known to induce tolerance against drought, high temperature, salinity, and mineral deficiency. There have been immense efforts of plant breeders across the globe for developing genetically improved plants through traditional breeding methods or by developing the transgenic for inducing the tolerance for abiotic stresses. Marker-assisted selection using biotechnological tools makes the breeding program more practical. Besides agronomic practices, root-associated microbes (Khan et al. 2013), endophytic microorganisms (Berg et al. 2013; Lubna et al. 2018) are known to have a beneficial effect on plant stress management. Exogenous application of plant hormones of microbial origin (Egamberdieva et al. 2017), use of biostimulants from microalgae (Sharma et al. 2014) or application of stress-specific protectant like osmoprotectants can induce the tolerance to plants.

Interestingly, apart from the aforementioned facts, nanobiotechnology has now emerged as one of the potential strategies for abiotic stress management. Nanotechnology deals with the science of manipulation/regulation of substances in the size-range of 1–100 nm with unique characteristics, which enable them for various potential applications (Pandey et al. 2018; Iqbal et al. 2020b; Porwal et al. 2020; Rani et al. 2020; Singh and Porwal 2020; Singh et al. 2018a, b; Singh et al. 2020). Some studies were performed to elucidate the function of nanoparticles towards mitigating the stress. The plant tolerance towards drought stress was found to be increased with the exploitation of nanoparticles at different concentrations like nano-TiO_2 in crops like flax and basil (Aghdam et al. 2016; Kiapour et al. 2015), nano-ZnO in *Brassica napus* L. (Mahmoodzadeh et al. 2013), and many others. The particles like nano-SiO_2, nano-ZnO were used in different crops for mitigating salinity (Almutairi 2016a; Siddiqui and Al-Whaibi 2014; Soliman et al. 2015; Torabian et al. 2016) and the flooding stress was managed by nano-Ag as well as nano-Al_2O_3 (Mustafa et al. 2015; Rezvani et al. 2012). However, it is worth mention that phytohormones are the major players in combating various abiotic stresses. If any strategy is to be thought of about increased tolerance to stress, factors interacting, and triggering the synthesis of stress hormones are to be considered with immense priority. Even though the application of nanoparticles for stress management gives new rays of hope, it is of utmost importance to revisit the cross-talk regarding the possible effect of nanoparticles in plants including phytotoxicity. In light of the plant stress management, it is quite imperative to understand the interaction of nanoparticle with stress hormones and possible mechanisms of action on the biosynthetic pathway of stress hormones. Several principles, theories, and practices have been put forth so far for understanding the basics of abiotic stress. The literatures on management are also seen in abundance. Intense knowledge on the use of various strategies for abiotic stress management is infused and instilled amongst the scientific fraternity. The use of nanoparticles, though comparatively new, seems to be an area of deep study. It is also understood that the stress to plant system and likely loss is cause and effect of several factors and it is really difficult to address it as one-step solution. Nevertheless, the availability of literature, compiled facts, and insightful review of the findings is limited particularly on the use of nanoparticles in stress management and their effects on stress hormones. This chapter attempts to compile the knowledge scaffolding on diverse nanoparticles with their interactive role in boosting various plant hormones under abiotic stresses that might lead to obtaining higher crop productivity in near futures.

15.2 Plant Growth Hormones and Their Physiological Significance

Plant hormones or phytohormones are a group of small organic molecules that occur naturally and affect various physiological processes in the plant at low concentrations (Davies 2010). The main physiological processes affected were development, differentiation, and growth of the plant including other processes like stomatal movement (Su et al. 2017). To date, nine categories of plant hormones viz. ethylene, abscisic acid, gibberellins, cytokinins, auxin, strigolactones, jasmonates, salicylic acid, and brassinosteroids have been recognized so far (Fig. 15.1). The first five are regarded as the "classical" plant hormones, whereas the last four are recently added to the growing family of plant hormones (Su et al. 2017). Auxin considered being the first plant hormone to be discovered. It plays an important role in flowering, promotes femaleness in dioecious plants, growth of floral parts, fruit ripening, assimilate partitioning, fruit setting and growth, leaf and fruit abscission, leaf senescence, apical dominance, tropistic responses, root initiation, vascular tissue differentiation, cell division, and cell enlargement (Davies 2010; Wani et al. 2016). Gibberellins are carboxylic acid which regulates various physiological processes in plants which

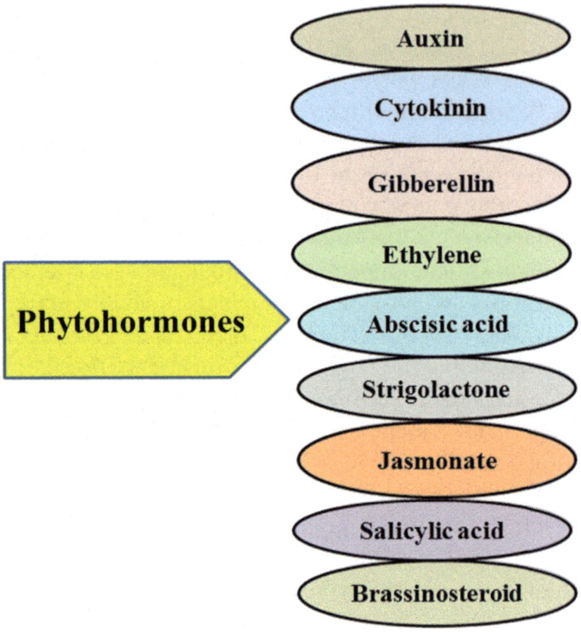

Fig. 15.1 Types of phytohormones

include elicitation of maleness in dioecious flowers, fruit setting, and growth, enzyme (α-amylase) production during germination, induction of seed germination, bolting in long-day plants, and stem growth (Davies 2010; Wani et al. 2016). Cytokinins are adenine derivatives that are characterized by the capability to prompt cell division with auxin in plant tissue culture. Its physiological role includes chloroplast development, enhance stomatal opening, delay leaf senescence, morphogenesis and cell division (Davies 2010). Cytokinins also alleviate seed dormancy (Fahad et al. 2015c) and act as abscisic acid antagonists (Pospíšilová 2003). Ethylene is a gaseous plant hormone that produces in response to stress by most tissue and regulates various activities such as fruit ripening, flower, leaf senescence, flower opening and release from dormancy (Davies 2010; Wani et al. 2016). Abscisic acid also called abiotic stress hormone (Wani et al. 2016), affects induction and maintenance of dormancy in buds and seeds, counteracts the consequence of gibberellin on the α-amylase synthesis, induces storage protein synthesis in seeds, promote transport of photosynthate to the developing seeds and its uptake by growing embryos, and stomatal closure (Davies 2010). On the other hand, salicylic acid is a phenolic compound that controls the expression of pathogenesis-related proteins (El-Esawi 2017; Miura and Tada 2014). Salicylic acid performs a significant function in the growth/development of plants as well as in fruit ripening (Rivas-San and Plasencia 2011; Wani et al. 2016). Jasmonates are the cyclo-pentanone multifunctional plant hormones obtained from the metabolism of the fatty acids of cell membrane comprising mainly jasmonic acid and methyl jasmonate. It carries out vital functions concerning various plant physiological processes like indirect as well as direct defense responses, secondary metabolism, senescence, fruiting, flowering, and reproductive processes (Fahad et al. 2015c; Wani et al. 2016). Brassinosteroids are a polyhydroxy steroidal phytohormone that promotes growth and development of plants such as fruits and flower development, floral initiation, root, and stem growth (Bajguz and Hayat 2009; Wani et al. 2016). Strigolactones consist of carotenoid-derived molecules formed in small quantities mainly in roots as well as other parts of the plant (Koltai and Beveridge 2013). They promote root and shoot development (Kapulnik and Koltai 2014), nodulation during the legume and rhizobium symbiosis (Foo and Davies 2011; Soto et al. 2010), and parasitic plants seed germination (Harrison 2012; Wani et al. 2016). Strigolactones also regulates seedling development and seed germination (Stanga et al. 2013) as well as leaf senescence (Akiyama et al. 2005).

15.3 Phytohormones as Regulators of the Stress Responses

The phytohormones like abscisic acid, ethylene, salicylic acid as well as jasmonates are well known for regulating the protection response of plants towards abiotic stress as well as pathogenic organisms (Bari and Jones 2009; Nakashima and Yamaguchi-Shinozaki 2013; Verma et al. 2016). Abscisic acid is accountable for phyto defense towards abiotic stresses like heat, cold, salinity as well as drought stresses including wounding, which are responsible to increase its concentration in plant cells (Lata and Prasad 2011; Zhang et al. 2006). Abscisic acid promotes the ability of the plant to signal its shoots about the water stress conditions, which results in anti-transpiration activity by reducing leaf expansion and enhancing stomatal closure (Wilkinson et al. 2012). Abscisic acid also encourages the modification of plant architecture under nitrogen limitation (Zhang et al. 2007) as well as drought stress (Giuliani et al. 2005). It also controls the production of dehydrins, late embryogenesis abundant proteins, and other protective proteins as well as the expression of various stress-responsive genes (Sreenivasulu et al. 2012; Verslues et al. 2006). It also promotes the production of antioxidant enzymes and osmoprotectants as well as cellular maintenance of turgor pressure, which enhances the tolerance capacity of plants against desiccation (Chaves et al. 2003). However, ethylene, jasmonates, and salicylic acid are mainly responsible for plant defense against biotic stress such as pathogen infection, which triggers to increase their levels in plant cells (Bari and Jones 2009; Verma et al. 2016). The abiotic stresses like salinity, lower temperature, and others, increase the content of ethylene endogenously which promotes the tolerance level in plants (Shi et al. 2012). Ethylene also alleviates heat stress in the plant by activating plant defense response (Larkindale et al. 2005). Ethylene also with other plant hormones viz. jasmonate and salicylic acid promote plant defense against pathogens as well as pests (Kazan 2015; Wani et al. 2016). Jasmonates promote defense responses of the plant under various abiotic stresses (like ultra-violet radiation and drought) as well as pathogenic attacks (Demkura et al. 2010; Du et al. 2013; Pauwels et al. 2009; Seo et al. 2011; Wani et al. 2016). Jasmonates also eliminate the stress caused by heavy metals by triggering the antioxidant systems (Wani et al. 2016; Yan et al. 2013).

Salicylic acid performs a substantial function in abiotic (heat, chilling, salinity, and drought) and biotic stress responses (Wani et al. 2016). The higher concentration of salicylic acid in plants either makes it prone to abiotic stress or causes cell death. However, its adequate amount promotes the antioxidant capacity of the plant (Jumali et al. 2011). Salicylic acid is associated with genes accountable towards the formation of secondary metabolites (cinnamyl alcohol dehydrogenase and sinapyl alcohol dehydrogenase) and cytochrome P450 as well as gene encoding antioxidants, heat shock proteins, and chaperones (Jumali et al. 2011; Wani et al. 2016). Auxin takes a crucial part in the alleviation of heavy metal (Egamberdieva 2009) and salinity stresses (Fahad et al. 2015b) in plants. Moreover, it is also considered as a vital component of protection/defense responses through the regulation of various genes and intermediation of crosstalk among biotic besides abiotic stress responses (Fahad et al. 2015a).

Recently, considerable evidence have emerged towards the cross-talk of phytohormones such as abscisic acid, jasmonates and salicylic acid for controlling plant protection/defense response (Bari and Jones 2009; Navarro et al. 2008; Nishiyama et al. 2013; Verma et al. 2016). Cytokinins also take part in abiotic stress responses (O'Brien and Benkova 2013) like drought (Kang et al. 2012) as well as salinity stresses (Nishiyama et al. 2011). During water stress conditions, the concentration of abscisic acid increases, whereas cytokinin content decreases, which increased the ratio of abscisic acid as well as cytokinin. The decreased concentration of cytokinin promotes apical dominance, which in combination with the abscisic acid regulates stomatal aperture, helps in adaptation towards drought stress (O'Brien and Benkova 2013; Wani et al. 2016). Gibberellins carry out an important function in abiotic stress alleviation (Colebrook et al. 2014) like osmotic stress (El-Esawi 2017). Gibberellins interrelate with all plant hormones in various developmental and stimulus-response processes (Munteanu et al. 2014). Its interaction with ethylene shows both positive and negative mutual regulation dependent on the signaling and tissue perspective (Munteanu et al. 2014; Wani et al. 2016). Brassinosteroids promotes defense responses of the plant towards different abiotic stresses like organic pollutants (Ahammed et al. 2013), metals or metalloids (Bajguz 2010), flooding (Liang and Liang 2009), drought (Mahesh et al. 2013), light (Kurepin et al. 2012), soil salinity (Abbas et al. 2013), chilling (Wang et al. 2014), and high temperature (Janeczko et al. 2011). Strigolactones are also found to take part in abiotic and biotic stress responses (El-Esawi 2017; Wani et al. 2016). Strigolactones regulates plant responses and acclimation to numerous abiotic stresses, mainly chilling, salinity, drought, and nutrient deficiency (Yang et al. 2019).

15.4 Phytohormone Signaling Under Stress

Phytohormones control various functions in plants both at molecular as well as cellular levels. There are diverse signaling routes and relationships identified with phytohormones, amongst which the function of phytohormone signaling in stress

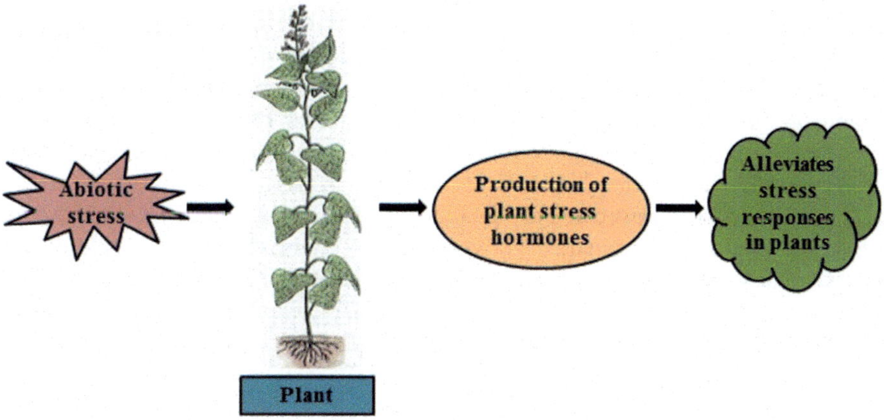

Fig. 15.2 Plant hormones can alleviates stress response in plants

conditions can be of utmost significance (Hirayama and Shinozaki 2010; Miransari 2012). Under stressful environmental conditions, the reaction of plants is directed by phytohormones demonstrating that the occurrence of phytohormones can expand plant resilience to stress (Fig. 15.2). The phytohormones produced in plants may bring about actuation of various genes and consequently promotes various exercises, which include plant response to stress, plant water behavior, cell cycling and triggering of various signaling pathways (Rahman 2013; Tuteja 2007; Wang et al. 2007).

The impacts of auxin in stress can be executed by the initiation of transcription factors associated with genes, for example, small auxin-up RNA, GH3, and Aux/IAA genes. The signaling route of auxin is generally activated and controlled by transcription factors involving the Aux/IAA repressors and auxin response factor (Han et al. 2009; Jain and Khurana 2009). On the other hand, the vital functions of ethylene in plants consist of tissue senescence, abscission, and germination of seeds. Ethylene usually interacts with ethylene receptors generally present on the cell membrane which consists of two-component histidine protein kinases (Miransari 2014; Miransari and Smith 2014; Mount and Chang 2002). The signaling route of ethylene is amongst the well-identified routes and has the significant ETHYLENE INSENSITIVE3 transcription factor. The synthesis of ethylene increases under stress, which decreases the growth/development of the plant. Moreover, the utilization of plant growth-promoting rhizobacteria reduces the synthesis of ethylene due to the biogenesis of aminocyclopropane carboxylate deaminase (Glick et al. 2007; Jalili et al. 2009; Miransari 2014). In the case of cytokinin, the *ipt* gene is responsible for the biogenesis of cytokinin, which produces isopentenyl adenosine-5'-monophosphate and isopentyl transferase (McGraw 1987; Miransari 2014). One of the major roles of cytokinin is the fortification of the photosynthetic process under stress in association with receptor protein molecules and the initiation of the interrelated signaling route. Consequently, the genes are expressed, and rubisco, photosynthesis-related proteins, carbon, electrons, and microRNAs are formed. The response of plants can be altered genetically using *ipt* gene under drought stress as the leaf senescence process is postponed (Miransari 2014; Rivero et al. 2007, 2009). Similarly, the functions of abscisic acid under stress have been also reported. Abiotic stress viz. drought and salinity cause the synthesis of abscisic acid in the plant. The opening and closing of the stomatal pore are controlled by abscisic acid under various conditions as well as stresses (Jia and Davies 2007; Miransari 2014). Among phytohormones, abscisic acid is considered as the most vital signaling molecule in the plant due to its numerous functions. Plants can eliminate the effects of stress owing to the activation of various genes by abscisic acid, which regulates the defense mechanism, for instance, the activation of nced genes in the plant (Miransari 2014; Wan and Li 2006). The antagonistic impacts on sRNA prompt the synthesis of abscisic acid, demonstrating that there occurs a connection between abscisic acid and sRNA signaling pathways in the plant (Miransari 2014; Zhang et al. 2008). Likewise, the synthesis of gibberellins in plants is accelerated by the cyclases, dioxygenases, and monooxygenases enzymes. The increasing impact of gibberellins on the growth of the plant is due to the breakdown of DELLA protein (Griffiths et al. 2006; Miransari 2014). DELLA protein molecules are accountable for the modification of plant response to stress by influencing the combining response of phytohormones to stress (Miransari 2012, 2014). On the other hand, one of the vital function of salicylic acid on the growth of the plant is to regulates the systemic resistance by the processes, which includes the expression of the various gene including the priming and phenylalanine ammonia lyase genes, triggering of phytoalexin related pathways, phenolic products, and callose deposition and influencing the signaling pathway for auxin (Chen et al. 2009). Brassinosteroids synthesis in plants utilized oxygen, which signifies its ability to alter the impacts of hypoxia on the development as well as the growth of the plant. It also overcomes the adverse effects in plants arises due to various stresses (Miransari 2012). Strigolactones influences the germination of parasitic weed *Striga*, shoot branching, and fungi mycorrhizal association with its host plant. The vital feature influencing the synthesis of this phytohormone in the plant is phosphate deficiency (Akiyama et al. 2005; Lopez-Raez et al. 2008; Miransari 2011, 2014). Jasmonates in combination with other phytohormones affect the growth of plants in stress, activates nitrogen protein kinase, the influx of calcium, and regulates the synthesis of ROS (Hu et al. 2009; Miransari 2014).

15.5 Impact of Nanoparticles on the Content of Phytohormones

The concentration and mode of action of phytohormones are viewed as a significant criterion of toxicity in plants (Yang et al. 2017). The consequence of nanoparticles in plants mainly based on their concentration (Vankova et al. 2017). Hence, small concentrations of nanoparticles have positive or even insignificant effects on the growth of the plant, e.g., in the case of peanut, the application of ZnO nanoparticle at the concentration of 1000 ppm (Prasad et al. 2012; Vankova et al. 2017). According to Le Van et al. (2015), the presence of CeO_2 nanoparticles does not have a noteworthy influence on gibberellic acid, abscisic acid as well as indole-3-acetic acid content in the leaves of conventional cotton as well as Bt-transgenic cotton in comparison to the control one. Moreover, the treatment of conventional cotton with 0.5 g L^{-1} of CeO_2 nanoparticles results in the decrement of transzeatin-riboside content in the leaves by 25% over the control condition. Gui et al (2015) reported that the abscisic acid and indole-3-acetic acid concentration were increases in the roots of non-transgenic and transgenic rice plants on the implication of Fe_2O_3 nanoparticles indicating that it had a substantial impact on the synthesis of phytohormones. The decrease in the concentration of plant hormones was reported with the carbon nanotubes exposure of rice seedling (Hao et al. 2016). An iron oxide nanoparticle stimulates the growth of *Arachis hypogaea* L. when applied as fertilizer by controlling the plant hormones such as indole propionic acid, dihydrozeatin, zeatinriboside and gibberellin in root and shoot (Rui et al. 2016). According to Vankova et al. (2017), the exposure of *Arabidopsis thaliana* (L.) Heynh. to ZnO nanoparticles repressed the synthesis of auxins and cytokinins in shoot apices, whereas cis-zeatin (cytokinin) was increased in the root. However, upregulation of abscisic acid was observed in leaves and apices, as well as stimulation of salicylic acid, were found in roots and leaves. Moreover, the concentration of jasmonic acid was repressed due to the occurrence of ZnO nanoparticles. Nano-stressed in pepper (*Capsicum annuum* L.) occurs due to the exposure to silver nanoparticles result in the increment of phytohormones such as cytokinins (Vinković et al. 2017).

15.6 Role of Nanomaterials and Phytohormones to Cope with Plants Stresses

The treatment of plant with nanomaterials may alter the gene transcription in plants involved in signal transduction and biogenesis of phytohormones like ethylene signaling components, the gene responsible of synthesis of abscisic acid and auxin response or repressor genes (Kaveh et al. 2013; Syu et al. 2014) to alleviate biotic as well as abiotic stresses. Reports on the effects of nanomaterials and plant hormones to alleviate plant stress are presented in Table 15.1. Rezvani et al. (2012) investigates the effects of silver nanoparticles in *Crocus sativus* and demonstrated that silver nanoparticles promote the growth of root by blocking the ethylene signaling pathway. García-Sánchez et al. (2015) conducted a study on the effects of multi-walled carbon nanotubes, Ag nanoparticles, or TiO_2 nanoparticles on *Arabidopsis thaliana* (L.) Heynh. during abiotic (wounding, drought or saline) or biotic (microbial pathogens) stresses. The exogenous application of salicylic acid prevented few phenotypic and nano-specific transcriptional effects, comprising the bacterial establishment of distal leaves as well as reduction of root hair formation. This study shows the result of the interaction of nanoparticles on gene expression with plant reactions

Table 15.1 Effect of nanomaterials and plant hormones to alleviates plants stress

Plant	Nanomaterial used	Stress	Stress response	References
Saffron (*Crocus sativus* L.)	Nano silver	Flooding	Blocking of ethylene signaling, promotion of root growth	Rezvani et al. (2012)
Mouse-ear cress (*Arabidopsis thaliana* L.)	Multi-walled carbon nanotubes, Ag nanoparticles or TiO_2 nanoparticles	Abiotic (wounding, drought or saline) or biotic (microbial pathogens) stresses	The exogenous application of salicylic acid prevented few phenotypic and nano-specific transcriptional effects, comprising the bacterial establishment of distal leaves as well as reduction of root hair formation	García-Sánchez et al. (2015)
Tomato (*Solanum lycopersicum* L.)	Nano-silicon	Salt	Enhanced seed germination and plant growth. Salt stress genes such as ethylene response factor 5 (ERF5) are downregulated whereas abscisic acid and environmental stress-inducible protein (TAS14), and abscisic acid-responsive element-binding protein (AREB) were upregulated	Almutairi (2016a)
Tomato (*Solanum lycopersicum* L.)	Silver nanoparticles	Salt	Enhanced seed germination and plant growth. Salt stress genes such as TAS14 and AREB are upregulated	Almutairi (2016b)

(continued)

Table 15.1 (continued)

Plant	Nanomaterial used	Stress	Stress response	References
Tomato (*Solanum lycopersicum* L.)	Magnesium oxide	Soil-borne pathogen (*Ralstonia solanacearum*)	Promotes systemic resistance against bacterial wilt disease. Systemic resistance-related GluA, ethylene-inducible Osm, jasmonic acid-inducible LoxA, and salicylic acid-inducible PR1 were up-regulated in both the hypocotyls and roots after application of the plant roots with magnesium oxide nanoparticles	Imada et al. (2016)
Tobacco (*Nicotiana Tabacum* L.)	Rooting phytohormones (indole-3 acetic acid or indole-3-butyric acid) capped silver nanoparticles as nanobullets	Soil-borne pathogen	Hormone capped Ag-nanoparticles inhibit pathogens (*Colletotrichum gloeosporioides*, *Rhizoctonia solani*, and *Curvularia lunata*) to cause root diseases and promote root growths	Thangavelu et al. (2016)
Wheat (*Triticum aestivum* L.)	Silver nanoparticles	Salt	Promotes germination and growth. Reduces abscisic acid and stimulates cytokinins and auxins contents	Abou-Zeid and Ismail (2018)

(continued)

Table 15.1 (continued)

Plant	Nanomaterial used	Stress	Stress response	References
Tobacco (*Nicotiana benthamiana* L.)	Metal-based (Fe_2O_3 or TiO_2) and carbon-based (C_{60} or multi-walled carbon nanotubes) nanomaterials	Pathogen stress (Turnip mosaic virus)	Activates defense mechanisms and stimulates the growth of the plant. Increased the level of cytokinins and brassinolide as well as decreases the concentration of abscisic acid	Hao et al. (2018)
Strawberry (*Fragaria* × *ananassa* Duch.)	Iron nanoparticles	Drought	Salicylic acid in combination with nanoparticles alleviates the harmful effects of drought stress and promotes growth	Mozafari et al. (2018)
Wheat (var. Pishgam) (*Triticum aestivum* L.)	TiO_2 nanoparticles	Drought	TiO_2 nanoparticles in combination with sodium nitroprusside promote seed germination and growth by regulating auxin, cytokinin, and gibberellin	Faraji and Sepehri (2019)

to chief sources of environmental stress and shows the path to overcome the effect of these possibly detrimental molecules via hormonal priming. Likewise, Almutairi et al. (2016a) studied the effects of various concentrations of nano-silicon on the germination of tomato (*Solanum lycopersicum* L.) during salt stress. Generally, the germination and the growth of seedling were inhibited by salinity stress. Conversely, these effects were reversed by the application of nano-silicon, which results in the increment of germination percentage and germination rate of seeds and the fresh weight and root length of seedlings under salinity stress. Moreover, the salinity stress genes such as dwarf and delayed flowering 2 (DDF2), mitogen-activated protein kinase 3 (MAPK3) ethylene response factor 5 (ERF5), mitogen-activated protein kinase 2 (MAPK2), cytosolic ascorbate peroxidase 2 (APX2), and respiratory burst oxidase (RBOH1) were downregulated whereas cysteine-rich receptor-like protein kinase 42-like (CRK1), 9-cis-epoxycarotenoid dioxygenase (NCED3), abscisic acid and environmental stress-inducible protein (TAS14), and abscisic acid-responsive element-binding protein (AREB) were upregulated with the application of nano-silicon during salinity stress. In another study, Almutairi et al. (2016b) examined the effects of various concentrations of silver nanoparticles on the germination of tomato (*Solanum lycopersicum* L.) during salt stress. Tomato seeds were treated with various concentrations of silver nanoparticles (0.05, 0.5, 1.5, 2 and 2.5 mg L^{-1}) and incubated under NaCl (salt) stress (150 and 100 mM). The germination of seed and growth were repressed by salinity stress, and these consequences were eliminated by the treatment of silver nanoparticles. The germination of seed and growth of seedling was promoted due to the application of nanoparticles under salinity stress. The genes responsible for salinity stress such as zinc finger homeodomain transcription factor family (ZFHD1), TAS14, and DDF2 were down-regulated, whereas CRK1, delta-1-pyrroline-5-carboxylate synthetase (P5CS), MAPK2, and AREB were up-regulated by silver nanoparticles in response to salt stress. Therefore, expression patterns for genes related to the application of nanoparticles suggested a possible contribution and involvement of nanoparticles (nano-silicon and silver nanoparticles) in the plant defense reaction to alleviate the salt stress. Imada et al. (2016) investigated the effects of magnesium oxide nanoparticles on resistance ability in tomato plants against the soil-borne pathogen (*Ralstonia solanacearum* (Smith) Yabuuchi et al.). The soaking of the roots in the suspension of nanoparticles before the inoculation of pathogen reduces the incidence of bacterial wilt significantly. However, the treatment of seedling roots with nanoparticles after the inoculation with pathogen cause very slight prevention of disease. Treatment of root with nanoparticles promotes ROS; however, the addition of polyphenols or tomato plant extracts to the nanoparticle suspension results in the generation of ROS was more quickly, signifying the promotion of ROS in roots of tomato plant probably due to a response of polyphenols and nanoparticles presence in roots. Systemic resistance-related GluA, ethylene-inducible Osm, jasmonic acid-inducible LoxA, and salicylic acid-inducible PR1 were up-regulated in both the hypocotyls and roots after application of the plant roots with magnesium oxide nanoparticles. Consequently, it promotes systemic resistance against bacterial wilt disease. Thangavelu et al. (2016) studied the effects of rooting phytohormones (indole-3 acetic acid or indole-3-butyric acid) capped silver nanoparticles as nanobullets on tobacco (*Nicotiana tabacum* L.) plant against the soil-borne pathogen. Hormone capped Ag-nanoparticles inhibit pathogens, such as *Colletotrichum gloeosporioides* Penz., *Rhizoctonia solani* Kühn, and *Curvularia lunata* (Wakker) Boedijn., to cause root diseases and promote root growths about 3-fold in comparison to control condition as well as it increases the rooting abilities against root growth-inhibiting phytopathogens. Abou-Zeid and Ismail (2018) examined the production of silver nanoparticles utilizing the extract of *Capparis spinose* L. as a stabilizing and reducing agent and explores its effects as a priming agent (1 mg L^{-1}) in the elimination of salinity stress (25 and 100 mM NaCl) during the germination of wheat. Silver nanoparticles promote seed sprouting and plant growth. Moreover, it influences the balance of phytohormones in the wheat plant by decreasing abscisic acid as well as promoting 6-benzyl-amino-purine, 1-naphthalene acetic acid, and indole-3-butyric acid contents. Salinity stress (100 mM NaCl) stimulates the synthesis of abscisic acid content and reduces cytokinins and auxins contents as well as chlorophyll stability index, pigment contents, growth index, and germination percentage. These factors particularly photosynthetic efficiency, growth parameters, and plant hormone equilibrium were significantly improved by the application of silver nanoparticles. Thus, signifying the role of silver nanoparticles priming in the development of plant tolerance against salt stress. Hao et al. (2018) showed the exposure of tobacco (*Nicotiana benthamiana* L.) to metal-based (Fe_2O_3 or TiO_2) and carbon-based (C_{60} or multi-walled carbon nanotubes) nanomaterials provide protection against pathogen stress, such as Turnip mosaic virus, by activating defense mechanisms and stimulate plant growth due to the increased production of cytokinins and brassinolide as well as decreases the concentration of abscisic acid. Escalations in the concentration of plant hormones by 40% also imply that nanomaterials hold an immense signification in promoting the growth of the plant and triggering defense mechanisms. Mozafari et al. (2018) studied the role of salicylic acid and iron nanoparticles on strawberry (*Fragaria* × *ananassa* Duch.) plants under drought stress. Their results pointed out that treatment of salicylic acid in combination with iron nanoparticles can be a favorable technique for providing a higher quantity of strawberries and could be used for alleviates the harmful effects of drought stress and promote growth. Also, Faraji and Sepehri (2019) investigated the effects of sodium nitroprusside (0 and 100 μM), as NO donor and TiO_2 nanoparticles (0, 500, 1000 and 2000 mg L^{-1}) on seed germination and seedling growth of wheat (var. Pishgam) plant to overcome drought stress induced by polyethylene glycol (0, −0.4 and −0.8 MPa). The outcome of their study indicates the increase in germination percentage (23.72%), germination energy (50%), germination rate (33.74%), vigor index (91.04%), shoot fresh weight (91.91%), root fresh weight (73%), shoot length (93.28%), and root length (85.38%) but reduced mean germination time (up to 28.36%) in wheat seeds due to the exposure of sodium nitroprusside and TiO_2 nanoparticles alone or in combination under severe drought stress by regulating phytohormones such as auxin, cytokinin, and gibberellin.

15.7 Conclusions and Prospects

To the best of our knowledge, limited studies are depicting the consequences of using nanomaterials with plant hormones to alleviates biotic and abiotic stress in the plant. Phytohormones control various functions in plants both at molecular as well as cellular levels. There are diverse signaling pathways and relationships identified with plant hormones, amongst which the role of phytohormone signaling under stress can be of the utmost importance. Under stressful environmental conditions, the reaction of plants is directed by phytohormones demonstrating that the occurrence of phytohormones can expand plant resilience to stress. The phytohormones produced in plants may bring about actuation of various genes and consequently promote various exercises which include plant response to stress, plant water behavior, cell cycling and triggering of various signaling pathways. The treatment of plants with nanomaterials may alter the gene transcription in plants involved in signal transduction and biogenesis of phytohormones. These include ethylene signaling components, the gene responsible for the synthesis of abscisic acid, and auxin response or repressor genes to alleviate biotic and abiotic stresses. Therefore, there is an urgent need for more investigations with different crops to evaluate the combined effect of nanomaterials and phytohormones by designing the system in different agro-climatic zones under various stresses over a longer period to obtain higher productivity without deteriorating the environmental and health status. It is not only essential to the farmers but also holds immense significance to the policymakers. The importance of hormonal signaling under abiotic stress is of great significance and must be explained so that the creation of stress tolerant plants may be possible at the large-scale. Nevertheless, for upcoming prospects, researchers require to make the outcomes of their investigation more relevant: (a) The investigation being offered by scientists is operative as well as helps in the development of systematic knowledge; (b) The accuracy of suggested and new techniques must be confirmed frequently so that the associated signaling pathways are accurately revealed; (c) the utilization of supplementary precise and sophisticated devices can be an advantageous apparatus to alter plants more promising under stress conditions; (d) the interactions, as well as cross-talk among several signaling pathways, can prominently specify the responses of the plant under the biotic and abiotic stress as well as the consequent usage of required and effective techniques; and (e) proteins and expressed genes, as well as cellular behavior, were evaluated and investigated accurately for the utilization of improving approaches more pertinent.

References

Abbas S, Latif HH, Elsherbiny EA (2013) Effect of 24-epibrassinolide on the physiological and genetic changes on two varieties of pepper under salt stress conditions. Pak J Bot 45:1273–1284

Abou-Zeid HM, Ismail GSM (2018) The role of priming with biosynthesized silver nanoparticles in the response of *Triticum aestivum* L. to salt stress. Egypt J Bot 58(1):73–85

Aghdam MTB, Mohammadi H, Ghorbanpour M (2016) Effects of nanoparticulate anatase titanium dioxide on physiological and biochemical performance of *Linum usitatissimum* (Linaceae) under well watered and drought stress conditions. Braz J Bot 39:139–146. https://doi.org/10.1007/s40415-015-0227-x

Ahammed GJ, Choudhary SP, Chen S et al (2013) Role of brassinosteroids in alleviation of phenanthrene-cadmium co-contamination-induced photosynthetic inhibition and oxidative stress in tomato. J Exp Bot 64:199–213

Akiyama K, Matsuzaki K, Hayashi H (2005) Plant sesquiterpenes induce hyphal branching in arbuscular mycorrhizal fungi. Nature 435:824–827

Almutairi ZM (2016a) Effect of nanosilicon application on the expression of salt tolerance genes in germinating tomato (*Solanum lycopersicum* L.) seedlings under salt stress. Plant Omics J 9:106–114

Almutairi ZM (2016b) Influence of silver nano-particles on the salt resistance of tomato (*Solanum lycopersicum*) during germination. Int J Agric Biol 18:449–457

Amari T, Ghnaya T, Abdelly C (2017) Nickel, cadmium and lead phytotoxicity and potential of halophytic plants in heavy metal extraction. S Afr J Bot 111:99–110. https://doi.org/10.1016/j.sajb.2017.03.011

Arun-Chinnappa KS, Ranawake L, Seneweera S (2017) Impacts and management of temperature and water stress in crop plants. In: Minhas PS (ed) Abiotic stress management for resilient agriculture. Springer, Australia, pp 221–233

Ashraf MA (2012) Waterlogging stress in plants: a review. African J Agric Res 7(13):1976–1981

Bajguz A (2010) An enhancing effect of exogenous brassinolide on the growth and antioxidant activity in *Chlorella vulgaris* cultures under heavy metals stress. Environ Exp Bot 68:175–179

Bajguz A, Hayat S (2009) Effects of brassinosteroids on the plant responses to environmental stresses. Plant Physiol Biochem 47:1–8

Bari R, Jones JD (2009) Role of plant hormones in plant defence responses. Plant Mol Biol 69(4):473–488

Berg G, Alavi M, Schmidt CS et al (2013) Biocontrol and osmoprotection for plants under saline conditions. In: de Bruijn FJ (ed) Molecular microbial ecology of the rhizosphere. Wiley-Blackwell, USA. https://doi.org/10.1002/9781118297674.ch55

Bielen A, Remans T, Vangronsveld J et al (2013) The influence of metal stress on the availability and redox state of ascorbate and possible interference with its cellular functions. Int J Mol Sci 14(3):6382–6413

Chaves MM, Maroco JP, Pereira JS (2003) Understanding plant responses to drought—from genes to the whole plant. Funct Plant Biol 30:239–264

Chen R, Hu Z, Zhang H (2009) Identification of microRNAs in wild soybean (*Glycine soja*). J Integr Plant Biol 51:1071–1079

Chinnusamy V, Zhu J, Zhu JK (2007) Cold stress regulation of gene expression in plants. Trends Plant Sci 12:444–451. https://doi.org/10.1016/j.tplants.2007.07.002

Chinnusamy V, Zhu JK (2009) Epigenetic regulation of stress responses in plants. Curr Opin Plant Biol 12:1–7. https://doi.org/10.1016/j.pbi.2008.12.006

Colebrook EH, Thomas SG, Phillips AL et al (2014) The role of gibberellin signaling in plant responses to abiotic stress. J Exp Biol 217:67–75

Compant S, Van Der Heijden MG, Sessitsch A (2010) Climate change effects on beneficial plant–microorganism interactions. FEMS Microbiol Ecol 73(2):197–214

Davies PJ (2010) The plant hormones: their nature, occurrence, and functions. In: Davies PJ (ed) Plant hormones. Springer, Dordrecht, pp 1–15. https://doi.org/10.1007/978-1-4020-2686-7_1

Demkura PV, Abdala G, Baldwin IT et al (2010) Jasmonate-dependent and -independent pathways mediate specific effects of solar ultraviolet B radiation on leaf phenolics and antiherbivore defense. Plant Physiol 152(1084):1095

Du H, Liu HB, Xiong LZ (2013) Endogenous auxin and jasmonic acid levels are differentially modulated by abiotic stresses in rice. Front Plant Sci 4:397. https://doi.org/10.3389/fpls.2013.00397

Egamberdieva D (2009) Alleviation of salt stress by plant growth regulators and IAA producing bacteria in wheat. Acta Physiol Plant 31:861–864

Egamberdieva D, Wirth SJ, Alqarawi AA et al (2017) Phytohormones and beneficial microbes: essential components for plants to balance stress and fitness. Front Microbiol 8:2104. https://doi.org/10.3389/fmicb.2017.02104

El-Esawi MA (2017) Introductory chapter: hormonal regulation in plant development and stress tolerance, phytohormones. Signaling mechanisms and crosstalk in plant development and stress responses. IntechOpen, 1–9. https://doi.org/10.5772/intechopen.69806

Emamverdian A, Ding Y, Mokhberdoran F et al (2015) Heavy metal stress and some mechanisms of plant defense response. Sci World J 15:1–18

Fahad S, Hussain S, Bano A et al (2015a) Potential role of phytohormones and plant growth-promoting rhizobacteria in abiotic stresses: consequences for changing environment. Environ Sci Pol Res 22:4907–4921

Fahad S, Hussain S, Matloob A et al (2015b) Phytohormones and plant responses to salinity stress: a review. Plant Growth Regul 75:391–404

Fahad S, Nie L, Chen Y et al (2015c) Crop plant hormones and environmental stress. Sustain Agric Rev 15:371–400

FAO. (2009). FAO land and plant nutrition management service. http://www.fao.org/ag/agl/agll/spush

Faraji J, Sepehri A (2019) Ameliorative effects of TiO_2 nanoparticles and sodium nitroprusside on seed germination and seedling growth of wheat under PEG-stimulated drought stress. J Seed Sci 41(3):309–317

Farooq M, Wahid A, Kobayashi N, Fujita D, Basra SMA (2009) Plant drought stress: effects, mechanisms and management. Agron Sustain Dev 29:185–212. https://doi.org/10.1051/agro:2008021

Foo E, Davies NW (2011) Strigolactones promote nodulation in pea. Planta 234:1073–1081

Food and Agriculture Organization of the United Nations (FAO) (2017) The impact of disasters and crises on agriculture and food security. FAO, Rome, p 168

Fukao T, Barrera-Figueroa BE, Juntawong P et al (2019) Submergence and waterlogging stress in plants: a review highlighting research opportunities and understudied aspects. Front Plant Sci 10:340. https://doi.org/10.3389/fpls.2019.00340

García-Sánchez S, Bernales I, Cristobal S (2015) Early response to nanoparticles in the *Arabidopsis* transcriptome compromises plant defence and root-hair development through salicylic acid signaling. BMC Genom 16:341. https://doi.org/10.1186/s12864-015-1530-4

Giuliani S, Sanguineti MC, Tuberosa R et al (2005) Root-ABA1 a major constitutive QTL affects maize root architecture and leaf ABA concentration at different water regimes. J Exp Bot 56:3061–3070

Glick BR, Todorovic B, Czarny J et al (2007) Promotion of plant growth by bacterial ACC deaminase. Crit Rev Plant Sci 26:227–242

Griffiths J, Murase K, Rieu I et al (2006) Genetic characterization and functional analysis of the GID1 gibberellin receptors in *Arabidopsis*. Plant Cell 18:3399–3414

Gui X, Deng Y, Rui Y et al (2015) Response difference of transgenic and conventional rice (*Oryza sativa*) to nanoparticles (γFe_2O_3). Environ Sci Pollut Res 22:17716–17723

Gull A, Lone AA, Wani N (2019) Biotic and abiotic stresses in plants. IntechOpen, 1–6. http://dx.doi.org/10.5772/intechopen.85832

Han H, Zhang S, Sun X (2009) A review on the molecular mechanism of plants rooting modulated by auxin. Afr J Biotechnol 8:348–353

Hao Y, Yu F, Lv R et al (2016) Carbon nanotubes filled with different ferromagnetic alloys affect the growth and development of rice seedlings by changing the C: N ratio and plant hormones concentrations. PLoS ONE 11:

Hao Y, Yuan W, Ma C et al (2018) Engineered nanomaterials suppress Turnip mosaic virus infection in tobacco (*Nicotiana benthamiana*). Environ Sci Nano 5(7):1685–1693. https://doi.org/10.1039/C8EN00014J

Harrison MA (2012) Cross-talk between phytohormone signaling pathways under both optimal and stressful environmental conditions. In: Khan NA, Nazar R, Iqbal N et al (eds) Phytohormones and abiotic stress tolerance in plants. Springer-Verlag, Berlin, pp 49–76

Hasanuzzaman M, Nahar K, Alam Md et al (2013) Physiological, biochemical, and molecular mechanisms of heat stress tolerance in plants. Int J Mol Sci 14:9643–9684. https://doi.org/10.3390/ijms14059643

Hassan TU, Bano A, Naz I (2017) Alleviation of heavy metals toxicity by the application of plant growth promoting rhizobacteria and effects on wheat grown in saline sodic field. Int J Phytoremed 19:522–529. https://doi.org/10.1080/15226514.2016.1267696

Hernández HH, Fabián PL, Ema LGE et al (2018) Tolerance-induction techniques and agronomical practices to mitigate stress in extensive crops and vegetables. In: Andjelkovic V (ed), Plant, abiotic stress and responses to climate change. InTech, Rijeka, Croatia, pp 145–181. https://doi.org/10.5772/intechopen.71771

Hirayama T, Shinozaki K (2010) Research on plant abiotic stress responses in the postgenome era: past, present and future. Plant J 61:1041–1052

Hu X, Li W, Chen Q et al (2009) Early signal transduction linking the synthesis of jasmonic acid in plant. Plant Signal Behav 4:696–697

Hu YF, Zhou G, Na XF et al (2013) Cadmium interferes with maintenance of auxin homeostasis in *Arabidopsis seedlings*. J Plant Physiol 170(11): 965–975. https://doi.org/10.1016/j.jplph.2013.02.008

Imada K, Sakai S, Kajihara H et al (2016) Magnesium oxide nanoparticles induce systemic resistance in tomato against bacterial wilt disease. Plant Pathol 65:551–560

Iqbal MS, Singh AK, Ansari MI (2020a) Effect of drought stress on crop production. In: Rakshit A, Singh H, Singh A, Singh U, Fraceto L (eds) New frontiers in stress management for durable agriculture. Springer, Singapore, pp 35–47

Iqbal MS, Singh AK, Singh SP, Ansari MI (2020b) Nanoparticles and plant interaction with respect to tress response. In: Bhushan I, Singh VK, Tripathi DK (eds) Nanomaterials and environmental biotechnology. Springer, Switzerland, pp 1–15

Isayenkov SV (2012) Physiological and molecular aspects of salt stress in plants. Cytol Genet 46:302–318. https://doi.org/10.3103/S0095452712050040

Isayenkov SV, Maathuis FJM (2019) Plant salinity stress: many unanswered questions remain. Front Plant Sci 10:80. https://doi.org/10.3389/fpls.2019.00080

Jain M, Khurana J (2009) Transcript profiling reveals diverse roles of auxin-responsive genes during reproductive development and abiotic stress in rice. FEBS J 276:3148–3162

Jaleel CA, Gopi R, Sankar B et al (2008) Differential responses in water use efficiency in two varieties of *Catharanthus roseus* under drought stress. Comp Rend Biol 331:42–47. https://doi.org/10.1016/j.crvi.2007.11.003

Jalili F, Khavazi K, Pazira E et al (2009) Isolation and characterization of ACC deaminase producing fluorescent pseudomonads, to alleviate salinity stress on canola (*Brassica napus* L.) growth. J Plant Physiol 166:667–674

Janeczko A, Oklestkova J, Pociecha E et al (2011) Physiological effects and transport of 24-epibrassinolide in heat-stressed barley. Acta Physiol Plant 33:1249–1259

Jia W, Davies WJ (2007) Modification of leaf apoplastic pH in relation to stomatal sensitivity to root-sourced abscisic acid signals. Plant Physiol 143:68–77

Jumali SS, Said IM, Ismail I et al (2011) Genes induced by high concentration of salicylic acid in *Mitragyna speciose*. Aust J Crop Sci 5:296–303

Kang NY, Cho C, Kim NY et al (2012) Cytokinin receptor-dependent and receptor-independent pathways in the dehydration response of *Arabidopsis thaliana*. J Plant Physiol 169:1382–1391

Kapulnik Y, Koltai H (2014) Strigolactone involvement in root development, response to abiotic stress, and interactions with the biotic soil environment. Plant Physiol 166:560–569

Kaveh R, Li YS, Ranjbar S et al (2013) Changes in *Arabidopsis thaliana* gene expression in response to silver nanoparticles and silver ions. Environ Sci Technol 47:10637–10644

Kazan K (2013) Auxin and integration of environmental signals into plant root development. Ann Bot 112:1655–1665

Kazan K (2015) Diverse roles of jasmonates and ethylene in abiotic stress tolerance. Trends Plant Sci 20:219–229

Khan AL, Hussain J, Al-Harrasi A et al (2013) Endophytic fungi: a source of gibberellins and crop resistance to abiotic stress. Crit Rev Biotechnol 35:62–74. https://doi.org/10.3109/07388551.2013.800018

Kiapour H, Moaveni P, Habibi D, Sani B (2015) Evaluation of the application of gibbrellic acid and titanium dioxide nanoparticles under drought stress on some traits of basil (*Ocimum basilicum* L.). Int J Agro Agric Res 6:138–150

Koltai H, Beveridge CA (2013) Strigolactones and the coordinated development of shoot and root. In: Baluska F (ed) Long-distance systemic signaling and communication in plants. Springer, Berlin, pp 189–204

Kurepin LV, Joo SH, Kim SK et al (2012) Interaction of brassinosteroids with light quality and plant hormones in regulating shoot growth of young sunflower and *Arabidopsis* seedlings. J Plant Growth Regul 31:156–164

Larkindale J, Hall DJ, Knight MR et al (2005) Heat stress phenotypes of Arabidopsis mutants implicate multiple signaling pathways in the acquisition of thermo-tolerance. Plant Physiol 138:882–897

Lata C, Prasad M (2011) Role of DREBs in regulation of abiotic stress responses in plants. J Exp Bot 62(14):4731–4748

Le Van N, Ma C, Rui Y et al (2015) Phytotoxic mechanism of nanoparticles: destruction of chloroplasts and vascular bundles and alteration of nutrient absorption. Sci Rep 5:116–118

Levitt J (1972) Responses of plants to environmental stresses. Academic, New York and London, p 697

Liang JQ, Liang Y (2009) Effects of plant growth substances on water-logging resistance of oilseed rape seedling. J Southwest China Norm Univ (Nat Sci Ed) 34:58–62 (in Chinese with English abstract)

Limbu S, Sharma L, Rao AP (2018) Growth and photosynthetic gas exchange characteristics in *Solanum aethiopicum* under water stress in organic production system. J Pharmacol Phytochem 7(2):1180–1182

Lipiec J, Doussan C, Nosalewicz A, Kondracka K (2013) Effect of drought and heat stresses on plant growth and yield: a review. Int Agrophys 27:463–477. https://doi.org/10.2478/intag-2013-0017

Lopez-Raez JA, Charnikhova T, Gomez-Roldon V et al (2008) Tomato strigolactones are derived from carotenoids and their biosynthesis is promoted by phosphate starvation. New Phytol 178:863–874

Lubna N, Asaf S, Hamayun M et al (2018) Salt tolerance of *Glycine max* L. induced by endophytic fungus *Aspergillus flavus* CSH1, via regulating its endogenous hormones and antioxidative system. Plant Physiol Biochem 128:13–23. https://doi.org/10.1016/j.plaphy.2018.05.007

Lukatkin AS, Brazaityte A, Bobinas C et al (2012) Chilling injury in chilling-sensitive plants: a review. Zem Agric 99(2):111–124

Mahesh B, Parshavaneni B, Ramakrishna B et al (2013) Effect of brassinosteroids on germination and seedling growth of radish (*Raphanus sativus* L.) under PEG-6000 induced water stress. Am J Plant Sci 4:2305–2313

Mahmoodzadeh H, Nabavi M, Kashefi H (2013) Effect of nanoscale titanium dioxide particles on the germination and growth of canola (*Brassica napus*). J Orn Hort Plant 3:25–32

Maksymiec W, Wojcik M, Krupa Z (2007) Variation in oxidative stress and photochemical activity in *Arabidopsis thaliana* leaves subjected to cadmium and excess copper in the presence or absence of jasmonate and ascorbate. Chemos 66:421–427

Mariani L, Ferrante A (2017) Agronomic management for enhancing plant tolerance to abiotic stresses-drought, salinity, hypoxia, and lodging. Horticulturae 3:52. https://doi.org/10.3390/horticulturae3040052

McGraw BA (1987) Cytokinin biosynthesis and metabolism. In: Davies PJ (ed) Plant hormones and their role in plant growth and development. Martinus Nijhoff, Dordrecht, pp 76–93
Minhas PS, Rane J, Pasala RK (2017) Abiotic stresses in agriculture: an overview. In: Minhas P, Rane J, Pasala R (eds) Abiotic stress management for resilient agriculture. Springer, Singapore, pp 3–8
Miransari M (2011) Interactions between arbuscular mycorrhizal fungi and soil bacteria. Appl Microbiol Biotechnol 89:917–930
Miransari M (2012) Role of phytohormone signaling during stress. In: Ahmad P, Prasad MNV (eds) Environmental adaptations and stress tolerance of plants in the era of climate change. Springer, New York, pp 381–393. https://doi.org/10.1007/978-1-4614-0815-4_17
Miransari M (2014) Plant signaling under environmental stress. In: Ahmad P (ed) Oxidative damage to plants: antioxidant networks and signaling. Elsevier, Philadelphia, pp 541–550
Miransari M, Smith DL (2014) Plant hormones and seed germination. Environ Exp Bot 99:110–121
Miura K, Tada Y (2014) Regulation of water, salinity, and cold stress responses by salicylic acid. Front Plant Sci 5:4. https://doi.org/10.3389/fpls.2014.00004
Mount SM, Chang C (2002) Evidence for a plastid origin of plant ethylene receptor genes. Plant Physiol 130:10–14
Mozafari AA, Havas F, Ghaderi N (2018) Application of iron nanoparticles and salicylic acid in in vitro culture of strawberries (*Fragaria × ananassa* Duch.) to cope with drought stress. Plant Cell Tissue Org Cult 132:511–523
Munteanu V, Gordeev V, Martea R et al (2014) Effect of gibberellin cross talk with other phytohormones on cellular growth and mitosis to endoreduplication transition. Int J Adv Res Biol Sci 1(6):136–153
Mustafa G, Sakata K, Komatsu S (2015) Proteomic analysis of flooded soybean root exposed to aluminum oxide nanoparticles. J Proteom 128:280–297
Nakashima K, Yamaguchi-Shinozaki K (2013) ABA signaling in stress-response and seed development. Plant Cell Rep 32(7):959–970
Navarro L, Bari R, Achard P et al (2008) DELLAs control plant immune responses by modulating the balance of jasmonic acid and salicylic acid signaling. Curr Biol 18(9):650–655
Nishiyama R, Watanabe Y, Fujita Y et al (2011) Analysis of cytokinin mutants and regulation of cytokinin metabolic genes reveals important regulatory roles of cytokinins in drought, salt and abscisic acid responses, and abscisic acid biosynthesis. Plant Cell 23:2169–2183
Nishiyama R, Watanabe Y, Leyva-Gonzalez MA et al (2013) *Arabidopsis* AHP2, AHP3, and AHP5 histidine phosphotransfer proteins function as redundant negative regulators of drought stress response. Proc Natl Acad Sci U S A 110(12):4840–4845
O'Brien JA, Benkova E (2013) Cytokinin cross-talking during biotic and abiotic stress responses. Front Plant Sci 4:451. https://doi.org/10.3389/fpls.2013.00451
Pandey B, Singh AK, Singh SP (2018) Nanoparticles mediated gene knock out through miRNA replacement: recent progress and challenges. In: Mishra RK, Thomas S, Mohapatra S, Dasgupta N, Ranjan S (eds) Applications of targeted nano-drugs and delivery systems. Elsevier, pp 469–497
Pauwels L, Inze D, Goossens A (2009) Jasmonate-inducible gene: what does it mean? Trends Plant Sci 14:87–91. https://doi.org/10.1016/j.tplants.2008.11.005
Porwal P, Porwal S, Singh SP et al (2020) Improving futuristic nanomaterial researches in forestry sector: an overview. In: Husen A, Jawaid M (eds) Nanomaterials for agriculture and forestry applications. Elsevier, Netherlands, pp 505–518
Pospíšilová J (2003) Participation of phytohormones in the stomatal regulation of gas exchange during water stress. Biol Plant 46:491–506
Prasad TNVKV, Sudhakar P, Sreenivasulu Y et al (2012) Effect of nanoscale zinc oxide particles on the germination, growth and yield of peanut. J Plant Nutr 35:905–927
Rahman R (2013) Auxin: a regulator of cold stress response. Physiol Plant 147:28–35
Rani H, Singh SP, Yadav TP, Khan MS, Ansari MI, Singh AK (2020) In-vitro catalytic, antimicrobial and antioxidant activities of bioengineered copper quantum dots using *Mangifera indica* (L.) leaf extract. Mater Chem Phys 239:122052

Reis DR, Lima AM, de Souza CRB (2012) Recent molecular advances on downstream plant responses to abiotic stress. Int J Mol Sci 13(7):8628–8647

Rezvani N, Sorooshzadeh A, Farhadi N (2012) Effect of nano-silver on growth of saffron in flooding stress. Int J Biol Biomol Agri Food Biotechnol Eng 6:11–16

Rivas-San VM, Plasencia J (2011) Salicylic acid beyond defense: its role in plant growth and development. J Exp Bot 62:3321–3338

Rivero RM, Kojima M, Gepstein A et al (2007) Delayed leaf senescence induces extreme drought tolerance in a flowering plant. Proc Natl Acad Sci USA 104:19631–19636

Rivero RM, Shulaev V, Blumwald E (2009) Cytokinin dependent photorespiration and the protection of photosynthesis during water deficit. Plant Physiol 150:1530–1540

Rui M, Ma C, Hao Y et al (2016) Iron oxide nanoparticles as a potential iron fertilizer for peanut (*Arachis hypogaea*). Front Plant Sci 7:815. https://doi.org/10.3389/fpls.2016.00815

Schutzendubel A, Polle A (2002) Plant responses to abiotic stresses: heavy metal-induced oxidative stress and protection by mycorrhization. The J Expt Bot 53:1351–1365

Seo JS, Joo J, Kim MJ et al (2011) OsbHLH148, a basic helix-loop-helix protein, interacts with OsJAZ proteins in a jasmonate signaling pathway leading to drought tolerance in rice. Plant J 65:907–921

Sharma HSS, Fleming C, Selby C et al (2014) Plant biostimulants: a review on the processing of macroalgae and use of extracts for crop management to reduce abiotic and biotic stresses. J Appl Phycol 26:65–490. https://doi.org/10.1007/s10811-013-0101-9

Shi Y, Tian S, Hou L et al (2012) Ethylene signaling negatively regulates freezing tolerance by repressing expression of CBF and Type-A ARR genes in Arabidopsis. Plant Cell 24:2578–2595

Siddiqui MH, Al-Whaibi MH (2014) Role of nano-SiO_2 in germination of tomato (*Lycopersicum esculentum* seeds Mill.). Saudi J Biol Sci 21:13–17

Singh AK, Pal P, Gupta V et al (2018a) Green synthesis, characterization and antimicrobial activity of zinc oxide quantum dots using Eclipta alba. Mater Chem Phys 203:40–48

Singh AK, Porwal P (2020) Nano-technology as potential and innovative platform toward waste water treatment: an overview. In: Bhushan I, Singh VK, Tripathi DK (eds) Nanomaterials and environmental biotechnology. Springer, Switzerland, pp 201–220

Singh AK, Yadav TP, Pandey B et al (2018b) Engineering nanomaterials for smart drug release: recent advances and challenges. In: Mishra RK, Thomas S, Mohapatra S, Dasgupta N, Ranjan S (eds) Applications of targeted nano-drugs and delivery systems. Elsevier, Netherlands, pp 411–449

Singh SP, Ansari MI, Pandey B et al (2020) Recent trends and advancement towards phyto-mediated fabrication of noble metallic nanomaterials: focus on silver, gold, platinum and palladium. In: Bhushan I, Singh VK, Tripathi DK (eds) Nanomaterials and environmental biotechnology. Springer, Switzerland, pp 87–105

Soliman AS, El-Feky SA, Darwish E (2015) Alleviation of salt stress on Moringa peregrine using foliar application of nanofertilizers. J Hort Forest 7(2):36–47

Soto MJ, Fernandez-Aparicio MN, Castellanos-Morales V et al (2010) First indications for the involvement of strigolactones on nodule formation in alfalfa (*Medicago sativa*). Soil Biol Biochem 42:383–385

Sreenivasulu N, Harshavardhan VT, Govind G et al (2012) Contrapuntal role of ABA: does it mediate stress tolerance or plant growth retardation under long-term drought stress? Gene 506:265–273

Stanga JP, Smith SM, Briggs WR et al (2013) SUPPRESSOR OF MORE AXILLARY GROWTH2 1 controls seed germination and seedling development in Arabidopsis. Plant Physiol 163:318–330

Su Y, Xia S, Wang R et al (2017) Phytohormonal quantification based on biological principles. In: Li J, Li C, Smith SM (eds) Hormone metabolism and signaling in plants. Academic Press, London, UK, pp 431–470. https://doi.org/10.1016/b978-0-12-811562-6.00013-x

Syu YY, Hung JH, Chen JC et al (2014) Impacts of size and shape of silver nanoparticles on *Arabidopsis* plant growth and gene expression. Plant Physiol Biochem 83:57–64

Thangavelu RM, Gunasekaran D, Jesse MI et al (2016) Nanobiotechnology approach using plant rooting hormone synthesized silver nanoparticle as "nanobullets" for the dynamic applications in

horticulture—an in vitro and ex vitro study. Arab J Chem 11(1):48–61. https://doi.org/10.1016/j.arabjc.2016.09.022
Tiwari S, Lata C (2018) Heavy metal stress, signaling, and tolerance due to plant-associated microbes: an overview. Front Plant Sci 9:1–12. https://doi.org/10.3389/fpls.2018.00452
Torabian S, Zahedi M, Khoshgoftarmanesh A (2016) Effect of foliar spray of zinc oxide on some antioxidant enzymes activity of sunflower under salt stress. J Agril Sc Tech 18:1013–1025
Tsugane K, Kobayashi K, Niwa Y et al (1999) A recessive Arabidopsis mutant that grows photoautotrophically under salt stress shows enhanced active oxygen detoxification. Plant Cell 11:1195–1206. https://doi.org/10.2307/3870742
Tuteja N (2007) Abscisic acid and abiotic stress signalling. Plant Signal Behav 2:135–138
Valko M, Morris H, Cronin MTD (2005) Metals, toxicity and oxidative stress. Curr Med Chem 12(10):1161–1208
Vankova R, Landa P, Podlipna R et al (2017) ZnO nanoparticle effects on hormonal pools in *Arabidopsis thaliana*. Sci Total Environ 593–594:535–542
Verma S, Nizam S, Verma PK (2013) Biotic and abiotic stress signalling in plants. Stress Signal Plants: Genom Proteom Perspect 1:25–49
Verma V, Ravindran P, Kumar PP (2016) Plant hormone-mediated regulation of stress responses. BMC Plant Biol 16:86
Verslues PE, Agarwal M, Katiyar-Agarwal S et al (2006) Methods and concepts in quantifying resistance to drought, salt and freezing, abiotic stresses that affect plant water status. Plant J 45:523–539
Vinković T, Novák O, Strnad M et al (2017) Cytokinin response in pepper plants (*Capsicum annuum* L.) exposed to silver nanoparticles. Environ Res 156:10–18
Wan X, Li L (2006) Regulation of ABA level and water stress tolerance of *Arabidopsis* by ectopic expression of a peanut 9-cisepoxycarotenoid dioxygenase gene. Biochem Biophys Res Commun 347:1030–1038
Wang D, Pajerowska-Mukhtar K, Culler AH et al (2007) Salicylic acid inhibits pathogen growth in plants through repression of the auxin signaling pathway. Curr Biol 17:1784–1790
Wang S, Ren X, Huang B et al (2016) Aluminium induced reduction of plant growth in alfalfa (*Medicago sativa*) is mediated by interrupting auxin transport and accumulation in roots. Sci Rep 6:30079. https://doi.org/10.1038/srep30079
Wang XH, Shu C, Li HY et al (2014) Effects of 0.01% brassinolide solution application on yield of rice and its resistance to autumn low-temperature damage. Acta Agric Jiangxi 26:36–38 (in Chinese with English abstract)
Wani SH, Kumar V, Shriram V, Sah SK (2016) Phytohormones and their metabolic engineering for abiotic stress tolerance in crop plants. Crop J 4:162–176
Wilkinson S, Kudoyarova GR, Veselov DS et al (2012) Plant hormone interactions: innovative targets for crop breeding and management. J Exp Bot 63:3499–3509
Yan Z, Chen J, Li X (2013) Methyl jasmonate as modulator of Cd toxicity in *Capsicum frutescens* var. *fasciculatum* seedlings. Ecotoxicol Environ Saf 98:203–209
Yang J, Cao W, Rui Y (2017) Interactions between nanoparticles and plants: phytotoxicity and defense mechanisms. J Plant Interact 12(1):158–169. https://doi.org/10.1080/17429145.2017.1310944
Yang T, Lian Y, Wang C (2019) Comparing and contrasting the multiple roles of butenolide plant growth regulators: strigolactones and karrikins in plant development and adaptation to abiotic stresses. Int J Mol Sci 20:6270. https://doi.org/10.3390/ijms20246270
Yoon J, Hamayun M, Lee SK et al (2009) Methyl jasmonate alleviated salinity stress in soybean. J Crop Sc Biotech 12:63–68
Zhang J, Jia W, Yang J et al (2006) Role of ABA in integrating plant responses to drought and salt stresses. Field Crops Res 97(1):111–119
Zhang JF, Yuan LJ, Shao Y et al (2008) The disturbance of small RNA pathways enhanced abscisic acid response and multiple stress responses in *Arabidopsis*. Plant, Cell Environ 31:562–574

Zhang S, Hu J, Zhang Y et al (2007) Seed priming with brassinolide improves lucerne (*Medicago sativa* L.) seed germination and seedling growth in relation to physiological changes under salinity stress. Aust J Agric Res 58:811–815

Zhou W, Chen F, Meng Y et al (2020) Plant waterlogging/flooding stress responses: from seed germination to maturation. Plant Physiol Biochem 148:228–236

Chapter 16
Application of Nanobiotechnology in Overcoming Salinity Stress

Shalini Tiwari, Charu Lata, and Vivek Prasad

Abstract Soil salinity is the challenging environmental threat impose on plants that hinder its development and productivity. Salinity obstruct the absorption of water by plant roots that leads to accumulation of hazardous salt ions and hence impairment in plant mechanisms by altering numerous biochemical and molecular parameters. Nowadays, nanotechnology gained the interest of researchers and agronomist to combat the environmental stress conditions. In agriculture, nanotechnology helps to improve growth, productivity and nutritional value of both monocotyledonous and dicotyledonous crop plants by the use of various nano scale products including, nanofertilizers, nanoherbicides, nanofungicides and nanopesticides. Use of nano scale products also have the capability to increase the vegetative and reproductive traits of various fruits and vegetables. Thus, the present chapter discusses the nanotechnology approaches used to mitigate the effect of salinity in crop plants. Hence, the role of biosensor and use of nanoparticles in cultivated land having high soil content have been described here to understand the impact of nanotechnology in overcoming salt stress for sustainable production of cereal and fruit crop plants. Lastly, the use of green nanoparticles has also been described against the adverse effects and limitations of chemically and physically synthesize nanoparticle.

Keywords Biosensors · Crop · Soil salinity · Nanoparticles · Nanotechnology

S. Tiwari (✉)
Council of Scientific and Industrial Research-National Botanical Research Institute (CSIR-NBRI), Microbial Technologies Division, Rana Pratap Marg, Lucknow 226001, India

C. Lata
CSIR-National Institute of Science Communication and Information Resources, New Delhi 110067, India
e-mail: charulata@niscair.res.in

V. Prasad
Department of Botany, University of Lucknow, Lucknow 226007, India

J. M. Al-Khayri et al. (eds.), *Nanobiotechnology*,
https://doi.org/10.1007/978-3-030-73606-4_16

16.1 Introduction

Plants are constantly confronted by a wide range of environmental challenges due to its sessile nature that leads to reduction in their growth and yield. These stresses may be biotic or abiotic in nature. Among abiotic stress, salinity is the major constraint impose on plants that limits the crop growth globally and cause severe damage in plants throughout their life cycle (Chauhan et al. 2019; Hirayama and Shinozaki 2010). To combat the effect of saline condition, plant execute several strategies, such as (a) adaptation mechanisms, that shows plant endurance to adverse conditions, or (b) avoidance mechanisms that is specific growth habits by plants to confront stress (Chauhan et al. 2019; Tiwari et al. 2016). Plants shows alteration in a large number of genes expression against salinity and, subsequently the after reaction cause stress resistance and tolerance in plants. Saline condition affect plants in different ways including, limits water absorption, ion accumulation and toxicity, nutritional imbalance, oxidative damage, alteration in metabolism, membrane instability, reduced cell division and cell expansion (Munns and Tester 2008). Altogether, these factors limit the crop productivity by imposing adverse impact on their germination, development and crop yield (Munns 2002; Zhu 2007). Apart from major contribution by increased sea level mineral weathering, use of inorganic fertilizers, and irrigation water also induce soil salinity (Amacher et al. 2000). High soil salinity levels, damage more than 77 million hectares of cultivated land of 1.5 billion hectares of total world's cultivated land (Kamran et al. 2020). Saline soils have mostly chloride (Cl^-), and sulphate (SO_4^{2-}) salts in them. While in sodic soil, bicarbonate (HCO_3^-) and, carbonate (CO_3^{2-}) of sodium is mostly abundant. The excessive quantity of Na cations in soils makes them firm and non-porous. Nowadays, numerous technologies have been used to combat the adverse effect of soil salinity that would facilitate sustained agricultural productivity. Due to serious threats of salinity and fewer acceptances for breeding approach and transgenic crops, an alternate technology is, the use of nanotechnology for stress amelioration that holds quite significance nowadays.

Nanotechnology has been suggested as a significant method and approach for nano-scale implementation of novel products in agricultural sector to boost yield of cereal crops at sustainable and productive manner to contribute to agriculture security (Fincheira et al. 2020). Besides cereals, nanotechnology has also been used for vegetables and fruit crop to enhance vegetative growth and nutrition values (Zahedi et al. 2020). Recently, a review by Fincheira et al. (2020) reported the use of nanoparticles on various monocotyledonous and dicotyledonous plants. Till date various investigations have been performed to combat the adverse effect of salinity in different plants by using the different nano-scales products (Lu et al. 2020; Ye et al. 2020). Thus, in current chapter we summarize the effect of salinity on plants and describes the various nanotechnologicals tools and products including nano-biosensors, and nanoparticles to mitigate the adverse effect of salinity to improve growth and productivity of crop plants.

16.2 Effect of Salinity on Plants

Plant tolerance to soil salinity comprises complex metabolic, and molecular interconnection depending on the extent of the stress that eventually leads to decline in crop production (Fig. 16.1). The salt stress effects plants by inducement of ion toxicity and imbalance, osmotic damage, and nutrient deficiencies that result in membrane destruction, decreased cell expansion and division, changes in metabolic processes (de Azevedo Neto and da Silva 2015). Thus, salt tolerance by plants involves all these highly complex phenomena that leads to variations in numerous physiological and biochemical traits, which ultimately result in morphological and developmental modifications.

Salt stress is firstly perceived by the plant roots that leads to reduction in numerous plant traits including seedling biomass, root and shoot length (Acosta-Motos et al. 2017). Mainly in the sensitive genotypes, initial symptoms of salinity detected in the old leaves that start to dehydrate and roll inward (Gholizadeh and Navabpour 2011). This phenomenon includes alteration in morphology, anatomy, photosynthesis, phytohormone content, distribution of toxic ion and biochemical modulation.

Initially, the root system of plants perceives the salinity that ultimately leads to impairment of plant growth, primarily by inducing osmotic imbalance caused by ion toxicity due to nutrient imbalance (Acosta-motos et al. 2017). Thus, major salinity impact is due to induction of osmotic stress and salt ion toxicity, leading to nutrient deficiencies and imbalances. All these responses of plants towards excessive salt uptake leads to deterioration of plant health (Gupta and Huang 2014).

Various mechanisms govern gene expression regulation during the central dogma of living cells, i.e., from the initiation step of RNA-transcription and up to the final

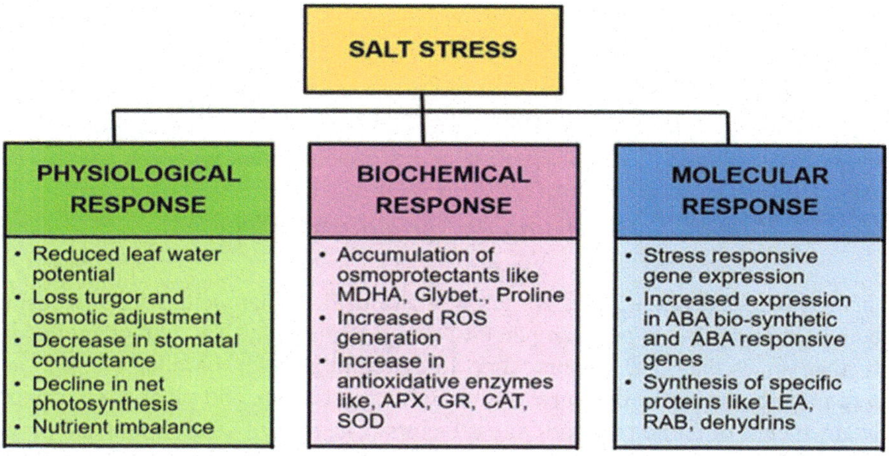

Fig. 16.1 Physiological, biochemical and molecular alterations in plants under salt stress (Figure constructed by Shalini Tiwari)

step of the proteins post-translational modification. Under the salinity stress, plants use numerous mechanisms for gene expression regulation to alter (i.e., upregulate or downregulate) the production of specific gene products. These mechanisms for gene regulation comprise molecules such as heat-shock proteins, antifreeze proteins, late embryogenesis abundant (LEA) proteins, transporters, water channel proteins, osmolyte biosynthesis, detoxification enzymes, and proteases (Bhardwaj et al. 2013; de Souza Filho et al. 2003). Salinity triggers salt overly sensitive (SOS) signaling, thus activates SOS1 Na^+ antiporter and SOS2 kinase (Yang and Guo 2018). Salt stress is also known to trigger ABA phytohormone pathways and osmotic stress signaling in plants. Osmoprotectants such as osmolytes and LEA proteins in plants cell, stabilize cell membrane and protein structure for stress tolerance. Among LEA proteins gene family, synthesis of Group 2 LEA proteins is mainly due to salinity (Allagulova et al. 2003; Bhardwaj et al. 2013). LEA proteins comprise of regulatory proteins that further regulate the signal cascading and hence genes expression of cell under stress. Several other transcription factors (TFs), enzymes, kinases, phosphatases, and signaling factors are induced by stress suggesting its complex transcriptional regulatory mechanisms under salinity stress. However, nowadays, nanobiotechnology has been used widely to combat the challenges caused by soil salinity and helps in improving plant growth and productivity in crops.

16.3 Application of Nanobiotechnology in Agriculture

In present scenario, nanobiotechnology gaining momentum in modern agriculture. In agricultural sector, this technology has diverse applications including nanofertilizers, nanoherbicides, nanofungicides, nanopesticides, nanoparticles, nanosensors and exhibit promising role in crop improvement, enhanced crop productivity, maintaining soil health, and precision farming (Amjad et al. 2018, Fig. 16.2). Hence, in this chapter, from these applications of nanobiotechnology, we here discussed the detailed role of nanotechnology in the context to salinity stress, i.e., as nanosensor and the use of nanoparticles for combating salt stress.

16.3.1 Role of Nanosensors in Agricultural Field

Nano-biosensors/Nanosensors are those sensor devices that were assembled using nanoscale components and can perceive the chemical, physical, and/or biological fluctuations as the other sensors work. In agricultural field, these nanosensors are used to detect water level, nutrient availability, pest attack and plant stress caused by any means of biotic or abiotic stress factors (Liu and Lal 2015; Singh and Singh 2019). Thus, by this beneficial use of nanosensors, farmers and agronomist would be capable to enhance the crop production by reducing nutrient input or water and also maintaining their delivery at temporal and spatial scales, i.e., only when and where

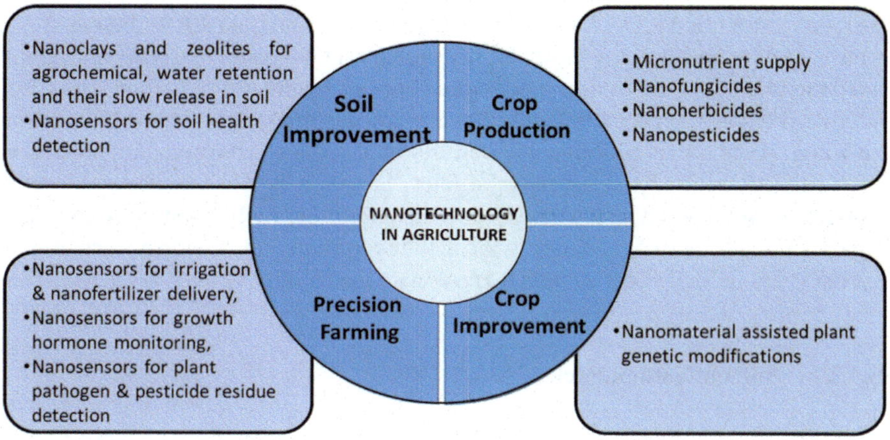

Fig. 16.2 Various applications of nanotechnology for sustainable agriculture (Figure constructed by Shalini Tiwari)

it is necessary. Monreal et al. (2015) showed that desired root signals interact with incorporated biosensor that leads to change their permeability allowing urea delivery to plants only in need. A recent review article by Kaushal and Wani (2017) mentioned several reports indicating the role of nanosensors that protect plants against various plant pathogens leading to enhancement in crop yield. Kaushal and Wani (2017) also reported the rejuvenation of salt affected soil and detection of soil contaminants by the use of microscopic or submicroscopic nanosensors. However, no such studies reported till date on the plants grown under salt stressed condition or on any cultivating land having high salt concentration.

16.3.2 Use of Nanoparticles for Mitigating Soil Salinity

Nanoparticles are the foundation of nanotechnology discipline that play crucial role in numerous field including electronics, industry, medicine, environment and agriculture (Ahmad and Akhtar 2019; Akhtar et al. 2013; Zhang et al. 2015). In agricultural sector, nanoparticles are used for crop improvement by precision farming, controlled nutrient and water delivery, detection and protection of crop against pest and pathogens (Panwar et al. 2012). Because of distinct physical and chemical characteristics, nanoparticles easily penetrate plant cells, can alter their metabolism and also can transport chemicals and DNA into the plant cells (Giraldo et al. 2014; Torney et al. 2007). Apart from these, nanotechnology also provides potential approach to the researchers to incorporate nanoparticles into the plants to extend its functions and also create new ones. Beside traditional DNA delivery approach i.e. transgene integration method, nowadays genome editing technologies are the globally accepted approach for the crop improvement program as they overcome the limitations that arises due to traditional DNA-delivery method (Tiwari and Lata 2019). Interestingly, by the help of nanotechnology in 2019, researchers at University of California manipulated the plant genome of non-model plant species by the use of carbon nanotubes and CRISPR-cas9 gene editing without integrating transgene resulting comparatively high protein expression level (Demirer et al. 2019). Abovementioned studies showed the diversified role of nanoparticles, however, below we discussed about some commonly known nanoparticles used in agriculture for alleviation of oxidative damage in plants caused by salinity. We also incorporated the nanoparticles used in past ten years in Table 16.1 in plants to mitigate the salinity stress.

16.3.2.1 Silicon Nanoparticles

Silica (Si) is one of the common and abundantly found element in the earth's crust. For plants, this element is beneficial and plays crucial role in plant stress tolerance including salt stress (Ahmad and Akhtar 2019). Numerous studies reported the plant growth promotion, i.e., enhanced chlorophyll concentration and, thus photosynthesis, declined evapotranspiration, and improved product quality upon exogenous application of silica in various plants under both stressed and unstressed condition (Ahmad and Akhtar 2019; Rodrigues et al. 2009). During stress, silicon form cellulose silicon layer by combining with calcium and pectin present in cell wall to increase its stability (Ahmad and Akhtar 2019). Under salt stress, silica enhanced the antioxidant enzyme activities, reduce membrane permeability, increase root absorption of nutrients, limit transport of Na^+ towards leaves and store them in roots for reduction of the salinity toxicity and, thus helps to improve plant growth (Gong et al. 2005; Liang et al. 2003). Gao et al. 2006, observed the increment in the plant biomass upon silicon application under salinity stress. Hence, researchers shifted towards nano-form of beneficial elements for better utilization and crop improvement. Specifically, by the use of nanosilica (nanoSi) for salinity tolerance in plants, several studies have been performed till date. In a comparative study between silica fertilizer and silica nanoparticles applied on basil plants grown under saline condition, Kalteh et al. (2014) found that tolerance capability of basil plants increased in silica-nanoparticles treated plants than silica fertilizer. Increased biomass, chlorophyll and proline content was found under salinity in basil plants applied with nanoparticles. Similarly, Qados (2015) investigated the effect of Si and nanoSi on faba bean plants treated with salt and concluded that nanoSi treatment ameliorate the salt stress by positively modulating membrane stability, and inducing antioxidant enzyme activity, and sugar accumulation. Avestan et al. (2019) treated the strawberry plants with combinations of salt and nano-silicon and stated that application of nano-silicon alleviates the non-favorable conditions of salt by imposing the anatomical and biochemical changes in plants. Alsaeedi et al. (2017) also examined the consequences of nanoSi on germination and the growth of common bean seedlings. This study showed that bean seeds could not able to grow under salt stress while nanoSi treatment helps them to grow well. Treatment of nanoSi also enhanced the germination percentage and germination speed of bean seedlings as well that leads to enhanced biomass of the seedlings.

Table 16.1 List of various nanoparticles to study the effect of salinity on plant growth of different plant species

Nanoparticle name	Nanoparticle concentration	Plant common name	Plant scientific nomenclature	Nanoparticle application method	Response	References
Silica/Silicon dioxide(SiO_2)	1% suspension	Basil	*Ocimum basilicum* L.	Irrigation and spraying on seedlings	Increased fresh and dry weight, increased chlorophyll and proline content	Kalteh et al. (2014)
Silica/Silicon dioxide (SiO_2)	0.15%, 0.3%, 0.45%, 0.6%, 0.75% suspension	Squash	*Cucurbita pepo* L.	Suspension applied in petri dishes	Improved plant germination, net photosynthetic rate, stomatal conductance, transpiration rate, water use efficiency, total chlorophyll, proline, and carbonic anhydrase activity	Siddiqui et al. (2014)
Silicon	1, 2, and 3 mM suspension	Faba bean	*Vicia faba* L.	Suspension applied to pot	Increased ascorbate peroxidase (APX), catalase (CAT) and peroxidase activity (POD), decreased superoxide dismutase (SOD)	Qados (2015)
Silica (SiO_2)	1 and 2 mM suspension	Tomato	*Lycopersicon esculentum* L.	Suspension applied to petri dishes	Enhancement in germination rate, root length and dry weight	Haghighi et al. (2012)
Silica (SiO_2)	15% suspension	Rice	*Oryza sativa* L.	Suspension applied to pot	No significant difference was observed	Abdel-Haliem et al. (2017)

(continued)

Table 16.1 (continued)

Nanoparticle name	Nanoparticle concentration	Plant common name	Plant scientific nomenclature	Nanoparticle application method	Response	References
Silica (SiO_2)	100, 200, and 300 mg/L	Common bean	*Phaseolus vulgaris* L.	Seeds soaked in suspension (Seed priming)	Enhanced germination percentage, vigor index, germination rate for seeds, shoot and root lengths, shoot and root dry masses	Alsaeedi et al. (2017)
Cerium oxide (CeO_2)	200 and 1000 mg/kg dry sand and clay mixture	Canola	*Brassica napus* L.	Nanoparticles applied to soil	Higher plant biomass, and higher efficiency of the photosynthetic apparatus	Rossi et al. (2016)
Cerium oxide (CeO_2)	500 mg/kg dry sand and clay mixture	Canola	*Brassica napus* L.	Nanoparticles applied to soil	Shortened the root apoplastic barriers hence allowed more Na^+ transport to shoots and less accumulation of Na^+	Rossi et al. (2017)
Cerium oxide (CeO_2)	–	Mouse-ear cress	*Arabidopsis thaliana* L.	Applied in powder form and infiltrated in leaves also	Improve chlorophyll content, biomass, photosynthesis, and leaf mesophyll K^+ retention	Wu et al. (2018)
Zinc oxide (ZnO)	15 and 30 mg/L	Tomato	*Lycopersicon esculentum* L.	Irrigation on seedlings	Higher expression levels of mRNA in the salt tolerant cultivars	Alharby et al. (2016)

(continued)

Table 16.1 (continued)

Nanoparticle name	Nanoparticle concentration	Plant common name	Plant scientific nomenclature	Nanoparticle application method	Response	References
Zinc oxide (ZnO)	20, 40 and 60 mg/L	Lupine	*Lupinus termis* Forssk.	Seeds soaked in suspension (Seed priming)	Enhanced levels of photosynthetic pigments, organic solutes, total phenols, ascorbic acid and Zn. Enhanced activities of SOD, CAT, POD, and APX. Decrement in the MDA and Na^+ content	Latef et al. (2017)
Zinc oxide (ZnO)	2 g/L	Sunflower	*Helianthus annuus* L.	Foliar application	Positive effect on biomass production. Highest proline content and SOD activity	Torabian et al. (2018)
Zinc oxide (ZnO)	1000 and 3000 ppm	Fenugreek	*Trigonella foenum-graecum* L.	Foliar application	Increased proline levels and trigonelline content	Noohpisheh et al. (2020)
Zinc oxide (ZnO)	20 and 40 mg/L	Tomato	*Lycopersicon esculentum* L.	Seed priming	Alteration in cytosine methylation that induced a positive anti-genotoxic effect	Haliloglu et al. (2020)

(continued)

Table 16.1 (continued)

Nanoparticle name	Nanoparticle concentration	Plant common name	Plant scientific nomenclature	Nanoparticle application method	Response	References
Zinc oxide (ZnO) and Iron oxide (Fe_3O_4)	30, 60 and 90 mg/L	Drumstick tree (Sahajna)	*Moringa peregrina* Fiori.	Spraying on seedlings	Reduction in Na^+ and Cl^- and an increase in N, P, K^+, Mg^{2+}, Mn^{2+}, Fe, Zn, total chlorophyll, carotenoids, proline, carbohydrates, crude protein levels, antioxidant non-enzymes (vitamin A and C), and enzymes (POD and SOD)	Soliman et al. (2015)
Zinc oxide (ZnO) and Iron oxide (Fe_3O_4)	2 g/L	maize	*Zea mays* L.	Foliar application	Enhancement in the net CO2, assimilation rate, sub-stomatal CO2 concentration, and root growth	Fathi et al. (2017)
Iron (Fe)	0.08 and 0.8 ppm	Grape	–	Irrigated with suspension	Increased total protein content and reduced proline, enzymatic antioxidant activity and hydrogen peroxide. Lower sodium content and increase the potassium content	Mozafari et al. (2018a)

(continued)

Table 16.1 (continued)

Nanoparticle name	Nanoparticle concentration	Plant common name	Plant scientific nomenclature	Nanoparticle application method	Response	References
Iron (Fe)	0.08, and 0.8 ppm	Strawberry	*Fragaria × ananassa* Duch.	Plant cuttings soaked in suspension	Increased pigment content, RWC, membrane stability index (MSI), iron and potassium contents. Decreased sodium content	Mozafari et al. (2018b)
Iron (Fe)	3 mM	Ajowan	*Trachyspermum ammi* L.	Foliar treatment	Enhanced K + uptake, K +/Na + ratio, Fe content, endogenous level of SA, activities of antioxidant enzymes (SOD, CAT, POD, PPO). Improved MSI, RWC, leaf pigments, root and shoot growth	Abdoli et al. (2020)
Iron oxide (Fe_2O_3)	30, 60 and 90 ppm	Moldavian balm	*Dracocephalum moldavica* L.	Foliar application	Affected amino acids content, MSI PPO, PAL, and SOD activities. Enhanced gene expression level of tyrosine aminotransferase (TAT), rosmarinic acid synthase (RAS) and RA content	Moradbeygi et al. (2020a)
Iron (Fe_2O_3)	60 ppm	Moldavian balm	*Dracocephalum moldavica* L.	Foliar application	Increased leaf area, leaf length, fresh and dry weight of the shoot and root	Moradbeygi et al. (2020b)

(continued)

Table 16.1 (continued)

Nanoparticle name	Nanoparticle concentration	Plant common name	Plant scientific nomenclature	Nanoparticle application method	Response	References
Iron (Fe_2O_3)	10, 50, 100 and 500 mg/L	Sorghum	*Sorghum bicolor* L.	Seed priming	Enhanced photosystem (PSII) efficiency, photosynthetic rate, chlorophyll index, RWC with decreased lipid peroxidation	Maswada et al. (2018)
Silver (Ag)	2, 5, and 10 mM	Wheat	*Triticum aestivum* L.	Seeds soaked in suspension (Seed priming)	Increased shoot fresh and dry weight, increased soluble sugars and proline contents, decreased CAT activity and increased POD activity.	Mohamed et al. (2017)
Silver (Ag)	10, 20, 30 and 40 μg/mL	Fenugreek	*Trigonella foenum-graecum* L.	Irrigated with suspension	Improved seed germination	Hojjat and Kamyab (2017)
Silver (Ag)	20, 40, 60, 80 and 100 mg/Kg	Fennel	*Foeniculum vulgare* Mill.	Seeds soaked in suspension (Seed priming)	Higher germination percentage and germination speed	Ekhtiyari et al. (2011)
Silver (Ag)	20, 40, 60, 80 and 100 mg/Kg	Basil	*Ocimum basilicum* L.	Seeds soaked in suspension (Seed priming)	Enhanced seed germination	Darvishzadeh et al. (2015)
Silver (Ag)	5, 10 and 15 ppm	Grass Pea	*Lathyrus Sativus* L.	Irrigated with suspension	Improved germination percentage, shoot and root length, seedling fresh weight and seedling dry weight and seedling dry contents	Hojjat (2019)

(continued)

Table 16.1 (continued)

Nanoparticle name	Nanoparticle concentration	Plant common name	Plant scientific nomenclature	Nanoparticle application method	Response	References
Silver (Ag)	0.05, 0.5, 1.5, 2 and 2.5 mg/L	Tomato	*Lycopersicon esculentum* L.	Seeds soaked in suspension (Seed priming)	Improved germination percentage, germination rate, root length and seedling fresh and dry weight. Alteration in gene expression	Almutairi (2016)
Silver (Ag)	10, 20, 40 ppm	Tomato	*Lycopersicon esculentum* L.	Irrigated with suspension	Reduced fruit number per plant, fruit diameter, average fruit weight, number of branches per plant and plant height	Younes and Nassef (2015)
Nitric oxide (NO) chitosan (CS)	50 and 100 μM	Maize	*Zea mays* L.	Suspension applied to pot	Induced alterations in photosystem II maximum quantum yield, relative electron transport rate, chlorophyll content and chlorophyll a/b ratio	Oliveira et al. (2016)
Manganese (Mn)	0.1, 0.5 and 1 mg/L		*Capsicum annuum* L.	Seed priming	Improved seed germination, root growth. Alteration in molecular responses	Ye et al. (2020)
Manganese oxide (Mn_3O_4)	1 and 5 mg	Cucumber	*Cucumis sativus* L.	Directly applied to soil	Increased leaf photosynthetic pigment content, net photosynthesis, and biomass. Alteration in metabolomes	Lu et al. (2020)

16.3.2.2 Cerium Oxide Nanoparticles

In recent years, nanoceria (nanoCe) has been used widely due to its multifarious role in medical, agriculture, industry and cosmetic field (Ayub et al. 2019). Earlier investigations confirmed that nanoCe impose both positive and negative impact on human and plant health. In context to plants, effect of cerium oxide nanoparticles depends on degree and extent of exposure, plant species and its developmental stage (Ma et al. 2016; Wang et al. 2012). A study justified, cerium oxide nanoparticles upon certain exposure conditions augments the growth and yield in tomato while, at other concentrations no visible changes were observed (Wang et al. 2012). Rossi et al. (2016) explained the effect of nanoCe treatment in alteration in physio-biochemical responses in canola (*Brassica napus* L.) during salinity stress. Plants treated with nanoCe had enhanced the photosynthetic apparatus proficiency, thus plant biomass, and experienced less stress in both standard and saline irrigated environments. Furthermore, Rossi et al. (2017) examined the impact of cerium nanoparticles under salinity and showed that nanoCe modified the plant root anatomy by amending salt stress tolerance of canola plant. NanoCe are also known as potent catalytic scavengers of reactive oxygen species (ROS), i.e., hydroxyl ions, hydrogen peroxide (H_2O_2), and superoxide anion in plants under abiotic stress. Wu et al. (2018) investigated the scavenging property of cerium oxide nanoparticles in *Arabidopsis* under salt stress and showed that hydroxyl radical scavenging alleviates the effect of salinity.

16.3.2.3 Zinc Nanoparticles

Zinc is the most crucial micronutrient required by plants for their vital metabolic functioning that lead to their better growth and development (Auld 2001; Saxena et al. 2016). Recently, zinc nanoparticles (nanoZn) has been used in agricultural field as nanofertilizer, nanopesticides, antimicrobial agent and drugs (Ahmad and Akhtar 2019; Prasad et al. 2012). Previous findings reported on legumes viz. chickpea, peanut, soybean and on cereals, i.e., wheat showed beneficial effect of zinc nanoparticles on germination of their seeds (Burman et al. 2013; Prasad et al. 2012; Ramesh et al. 2014; Sedghi et al. 2013). Several investigations also confirmed that nanoZn positively modulates the plant mechanism to mitigate the adverse effect caused by abiotic stresses. A study by Alharby et al. (2016) on tomato under salt stress revealed that application of nanoZn enhanced the mRNA expression level of antioxidant enzyme related genes, i.e., superoxide dismutase (SOD) and glutathione peroxidase (GPX) genes indicating the positive response of zinc nanoparticles on plants to combat the adverse effect of salinity. Latef et al. (2017) demonstrated the effect of nanoZn particles in lupine (*Lupinus termis* Forssk.) plants during saline condition and denoted that nanoZn stimulate its growth under salt by strengthening the levels of chlorophyll, total phenols, and organic solutes content. Application of ZnO nanoparticles on *Trigonella foenum-graecum* altered various antioxidant enzymes that lead to increment in the trigonelline content of the plant under the salt stress (Noohpisheh et al. 2020). Torabian et al. (2018) comparatively studied the effects of foliar application of ZnO particles and nanoparticles on sunflower plants, and showed that better absorption and high mobility due to nano size of nanoparticles lead to increased biomass, proline content and SOD activity that helps plant to alleviate salt stress. In tomato seedlings, combination of nanoZn with rhizobacteria altered the cytosine methylation that induced a positive anti-genotoxic effect in response to salinity (Haliloglu et al. 2020).

16.3.2.4 Iron Nanoparticles

Iron is also one of the most crucial micronutrient of the plants that is required for their better growth under both unstressed and stressed condition. Earlier studies reported that iron supplement to plants also can compensate for salt stress (Uauy et al. 2006). Therefore, in recent years, agronomists used to apply nano form of iron, i.e., iron nanoparticles (nanoFe) in the field to enhance crop productivity and tolerance capacity against environmental challenges. Taiz et al. (2015) stated that use of nanoFe-fertilizer in plants increase the permeability of plasma membrane, decrease the absorption as well as accumulation of sodium by root and, thus enhance the plant resistance towards salt stress. Improved seedling growth via enhanced photosystem (PSII) efficiency, photosynthetic rate, chlorophyll index, RWC with decreased lipid peroxidation was observed in sorghum treated with iron oxide nanopartcles under salinity stress (Maswada et al. 2018). Effect of nanoFe on rosmarinic acid (RA) production in old Moldavian balm (*Dracocephalum moldavica* L.) plants was investigated, and it was observed that NP application significantly affected amino acids content [tyrosine (Tyr), phenylalanine (Phe), proline (Pro)], and MSI PPO, PAL, and SOD activities. Enhanced gene expression level of tyrosine aminotransferase (TAT), rosmarinic acid synthase (RAS) and ultimately RA content was also recorded (Moradbeygi et al. 2020a). Similar researchers group in another study, reported increased leaf area, leaf length, biomass and antioxidant activities of *D. moldavica* upon nanoFe application during salt stress (Moradbeygi et al. 2020b). Moreover, combination of nanoFe along with other nanoparticles used to enable more favourable conditions to plants for additional crop yield. An experiment performed by Soliman et al. (2015) *Moringa peregrina* grown under salinity were treated with combination of ZnO and Fe_3O_4 nanoparticles showed improvement in plant growth parameters under both controlled and salt stressed conditions. Combination of these nanoparticles reduce the accumulation of sodium ion and increase the level of total photosynthetic pigments, proline, sugars, and antioxidant nonenzymes that leads to enhancement in salt tolerance by plants. Likewise, combination of nanoFe-particles and potassium silicate reduced the proline content, malondialdehyde (MDA) content, antioxidant enzyme activity, and increase the membrane stability index (MSI) in grape plants to compensate the detrimental effect caused by salinity stress (Mozafari et al. 2018a). To evaluate the effects of this combination in maize, Fathi et al. (2017) conducted a pot experiment using two maize cultivars under the saline condition and observed the enhancement in the net CO_2, assimilation rate, sub-stomatal CO_2 concentration and root growth. In another study, combination of salicylic acid with iron nanoparticles was applied in salt stressed strawberry plant. An improvement in all growth-related parameters and increment in pigment content, RWC, MSI and declination in sodium content was observed (Mozafari et al. 2018b). Similar combination was also applied by Abdoli et al. (2020) in ajowan (*Trachyspermum ammi* L.) plants for mitigating salt toxicity by various growth parameters, element content and anti-oxidant enzyme activities.

16.3.2.5 Silver Nanoparticles

Similar to the most of the nanoparticles, silver nanoparticles (nanoAg) are also widely use in the field of food packaging, medical drug delivery, pesticides, and electronics. Application of silver nanoparticles is also well studied in agricultural sector to augment resistance against stress and for better plant growth and productivity. Application of nanoAg on wheat plant grown under saline condition were observed to alleviate salt stress by positively modulating the antioxidant enzymes (Mohamed et al. 2017). The use of nanoAg also reported for better germination percentage, plant growth and resistance under salt stress in basil (*Ocimum basilicum* L.), fenugreek and *Foeniculum vulgare* seeds (Darvishzadeh et al. 2015; Ekhtiyari et al. 2011; Hojjat and Kamyab 2017). Almutairi (2016) also reported improved seed germination percentage, germination rate, root length and seedling fresh and dry weight. By the help of semi-quantitative RT-PCR he also examined the salt stress responsive genes, and showed that four genes, abscisic acid response element-binding protein (AREB), mitogen-activated protein kinase 2 (MAPK2), delta-1-pyrroline-5-carboxylate synthetase (P5CS) and cysteine-rich receptor-like protein kinase 42-like (CRK1), were upregulated in response to nanoAg under salinity, and three genes, abscisic acid and environmental stress-inducible protein (TAS14), dwarf and delayed flowering 2 (DDF2) and zinc finger homeodomain transcription factor (ZFHD1), were downregulated by nanoAg treatment. A recent study by Hojjat (2019) demonstrated that exogenous application of nanoAg on *Lathyrus sativus* L. (Grass pea) grown under salinity, improved the seedlings germination, plant height and root growth, biomass, and osmotic adjustment to mitigate the effect of salinity on pea seedlings. Apart from this, Younes and Nassef (2015) studied the effect of nanoAg on tomato plants under salt stress and reported the reduction in fruit diameter and average fruit weight, number of fruit per plant, branches number per plant and plant height. Hence reported no significant effect of nanoAg treatment on tomato plants under salt stress.

16.4 Biogenic Nanoparticle to Mitigate Salt Stress in Plants

Synthetic nano-fertilizers are the easily available nutrients in nanoscale range. Its application in cultivated field increases the solubility and diffusion of insoluble nutrients present in soil, increases the bio-availability, and decrease nutrient immobilization to deliver nutrients in soil and in plants for longer duration (Naderi and Danesh-Shahraki 2013; Rameshaiah et al. 2015). Apart from these beneficial role of synthetic nano-fertilizers, its production is cost intensive process and involves challenges related to environment as well as health due to use of toxic chemicals. Therefore, the need of hour is to synthesize the nanoparticles naturally and eco-friendly, i.e., biogenic or green synthesis of nanoparticles. Nowadays, researchers use numerous biological entities viz. bacteria, actinomycetes, algae, fungi, seaweeds, plants for the synthesis of nanoparticles (Pandey et al. 2019; Tripathi et al. 2017). Plant-mediated green synthesis to synthesize nano-particles are considered more efficient, non-hazardous and economical (Khandel et al. 2018). Plant extracts or secondary metabolites such as alkaloids, flavonoids, saponins, steroids, tannins act as precursor for nanoparticles synthesis. Studies reported that the plants are widely used to synthesize various biogenic nano-particles such as copper, cobalt, gold, silver, platinum, palladium, zinc oxide (Kuppusamy et al. 2016). Abdel-Haliem et al. (2017) separated silica nanoparticles from rice straw and studied its effect on rice under different condition of salt stress. Result revealed alteration in physio-biochemical and molecular parameters in rice to alleviate the salt stress. Plant growth promoting rhizobacteria (PGPR) or other rhizoccupants, globally used as biofertilizers for growth promotion and yield enhancement under both unstressed and stressed conditions including salinity (Tiwari et al. 2016, 2017). Extraction of nanoparticles from rhizoccupants is also studied earlier. Several studies reported the separation of gold and silver nanoparticle from *Trichoderma*, a rhizosphere occupant have been used for bioremediation, antimicrobial efficacy, and against biotic stress (Kumari et al. 2016, 2017; Mishra et al. 2014). Wang et al. (2019) reported the immobilization of mercury (Hg) element by biogenic selenium (Se) nanoparticles in varying concentrations of soil salinity. A study by Khalkhal et al. (2020) reported the effect of physical and chemical properties, i.e., salt contaminant of soil on immobilization by biogenic nanoparticle, however very less direct approach has been made till date to study the role of biogenic nanoparticle to mitigate salinity stress in plants.

16.5 Conclusions and Prospects

Nanotechnological approaches have shown various potential prospects for improvement of sustainable agriculture by enhancing crop production and protection against biotic and abiotic stresses. In present scenario, use of nanoparticles in agricultural sector gaining momentum due to its regulative, responsive, spatial and temporal delivery of nutrients to plants and its use as of nanoherbicides, nanofungicides, nanopesticides. Thus, this chapter comprehensively evaluate the role of nanobiotechnology applications including nanobiosensors and nanoparticles in agriculture sector under the effect of soil salinity. In the case of soil salinity, nanosensors helps in the detection of soil contaminants and hence rejuvenating the salt affected soil for better crop performance. Additionally, application of various nanoparticles mainly silica, zinc oxide, iron oxide, cerium oxide, and silver have been widely reported for imparting beneficial effect on plant growth and enhancing the stress tolerance capability during soil salinity. Long-term application of synthetic nanoparticles leads to toxicity and hence induce environmental and human health risks. Therefore, nowadays synthesis of green or biogenic nanoparticles i.e. synthesis of nanoparticles from biological entities came into occurrence. The biogenic nanoparticles combat the hazardous effect caused by synthetic nanoparticles. Their long-term exposure is non-hazardous as they are biodegradable in nature. Overall, this field achieving great interest by researchers to alleviate environmental stress issues and increase crop yield. However, more investigations are requisite to investigate mode and mechanisms of action of nanoparticles for better understanding the regulation of metabolism and gene expression in plants. A recent method for DNA-delivery in plants without transgene integration using carbon nanotubes takes the research one step closure towards crop improvement. These various ways of nanoscale approaches may be used in the future to make farming systems smart (Fig. 16.3).

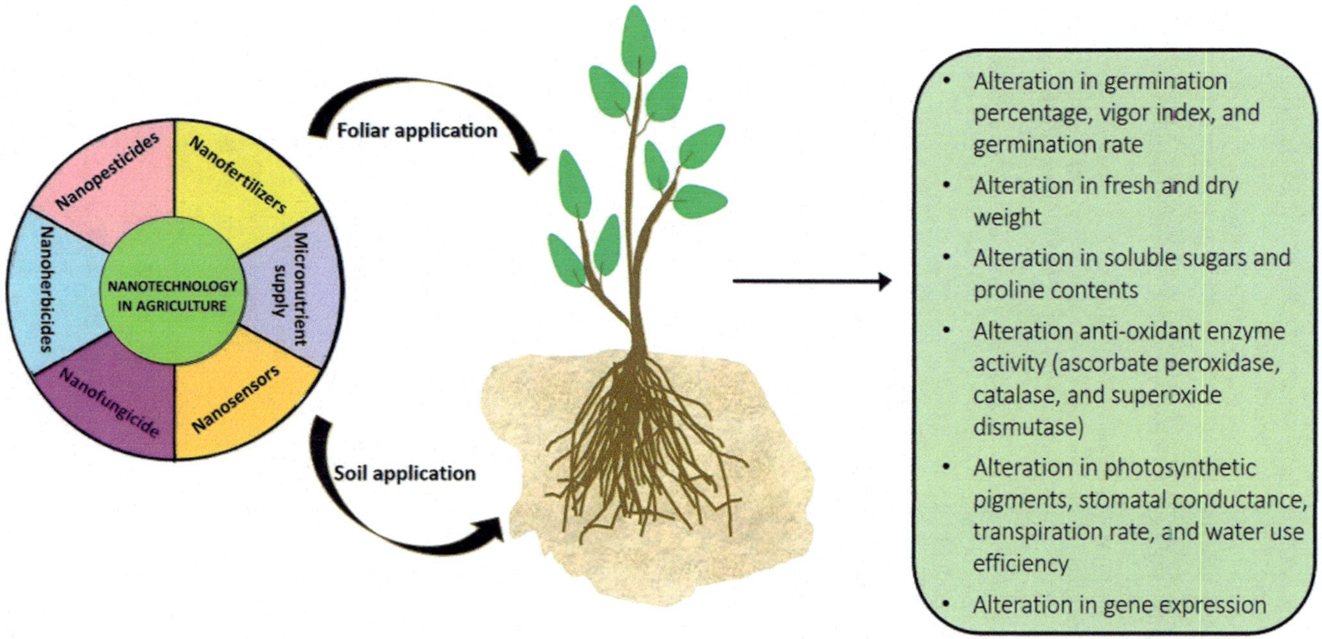

Fig. 16.3 A model of the physiological, biochemical, and molecular basis of salinity stress tolerance operating in plants upon different nanoparticles treatment is created based on the basis of well-known concepts reported in the previously published articles (Figure constructed by Shalini Tiwari)

References

Abdel-Haliem ME, Hegazy HS, Hassan NS et al (2017) Effect of silica ions and nano silica on rice plants under salinity stress. Ecol Eng 99:282–289
Abdoli S, Ghassemi-Golezani K, Alizadeh-Salteh S (2020) Responses of ajowan (*Trachyspermum ammi* L.) to exogenous salicylic acid and iron oxide nanoparticles under salt stress. Environ Sci Pollut Res 27(29):36939–36953
Acosta-Motos JR, Ortuño MF, Bernal-Vicente A et al (2017) Plant responses to salt stress: adaptive mechanisms. Agronomy 7(1):18
Ahmad I, Akhtar MS (2019) Use of nanoparticles in alleviating salt stress. In: Akhtar M (ed) Salt stress, microbes, and plant Interactions: causes and solution. Springer, Singapore, pp 199–215
Akhtar MS, Panwar J, Yun YS (2013) Biogenic synthesis of metallic nanoparticles by plant extracts. ACS Sustain Chem Eng 1:591–602
Alharby HF, Metwali EMR, Fuller MP et al (2016) The alteration of mRNA expression of SOD and GPX genes, and proteins in tomato (*Lycopersicon esculentum* Mill) under stress of NaCl and/or ZnO nanoparticles. Saudi J Biol Sci 23:773–781
Allagulova CR, Gimalov FR, Shakirova FM et al (2003) The plant dehydrins: structure and putative functions. Biochemistry (Moscow) 68(9):945–951
Almutairi ZM (2016) Influence of silver nano-particles on the salt resistance of tomato (*Solanum lycopersicum*) during germination. Int J Agric Biol 18(2):449–457
Alsaeedi AH, El-Ramady H, Alshaal T et al (2017) Engineered silica nanoparticles alleviate the detrimental effects of Na^+ stress on germination and growth of common bean (*Phaseolus vulgaris*). Environ Sci Pollut Res 24:21917–21928
Amacher JK, Koenig R, Kitchen B (2000) Salinity and plant tolerance. Electronic Publishing. Utah State University Extension, Logan. AG-SO 3:1–8
Amjad S, Sharma AK, Serajuddin M (2018) Toxicity assessment of cypermethrin nanoparticles in *Channa punctatus*: Behavioural response, micronuclei induction and enzyme alteration. Regul Toxicol Pharmacol 100:127–133
Auld DS (2001) Zinc coordination sphere in biochemical zinc sites. Biometals 14:271–313
Avestan S, Ghasemnezhad M, Esfahani M et al (2019) Application of nano-silicon dioxide improves salt stress tolerance in strawberry plants. Agronomy 9(5):246
Ayub MA, Sohail MI, Umair M et al (2019) Cerium oxide nanoparticles: advances in synthesis, prospects and application in agro-ecosystem. In: Verma SK, Das AK (eds) Engineered nanomaterials and phytonanotechnology: challenges for plant sustainability. Comprehensive analytical chemistry, vol 87. Elsevier, Amsterdam, Netherlands, pp 209–250
Bhardwaj R, Sharma I, Kanwar M et al (2013) LEA proteins in salt stress tolerance. In: Ahmad P, Azooz MM, Prasad MNV (eds) Salt stress in plants. Springer, New York, pp 79–112
Burman U, Saini M, Kumar P (2013) Effect of zinc oxide nanoparticles on growth and antioxidant system of chickpea seedlings. Toxicol Environ Chem 95(4):605–612
Chauhan PS, Lata C, Tiwari S et al (2019) Transcriptional alterations reveal *Bacillus amyloliquefaciens*—rice cooperation under salt stress. Sci Rep 9:11912
Darvishzadeh F, Nejatzadeh F, Iranbakhsh AR (2015) Effects of silver nanoparticles on salinity tolerance in basil plant (*Ocimum basilicum* L.) during germination in vitro. New Cell Mol Biotechnol J 2(10):63–70
de Azevedo Neto A, da Silva E (2015) Physiology and biochemistry of salt stress tolerance in plants. In: Chakraborty B, Chakraborty U (eds) Abiotic stresses in crop plants. CABI, Wallingford, UK, pp 81–101
de Souza Filho GA, Ferreira BS, Dias JM et al (2003) Accumulation of SALT protein in rice plants as a response to environmental stresses. Plant Sci 164(4):623–628
Demirer GS, Zhang H, Goh NS et al (2019) Carbon nanotube-mediated DNA delivery without transgene integration in intact plants. Nat Protoc 14(10):2954–2971
Ekhtiyari R, Mohebbi H, Mansouri M (2011) The study of the effects of nano silver technology on salinity tolerance of (*Foeniculum vulgare* Mill.). Plant Ecosyst 7:55–62

Fathi A, Zahedi M, Torabian S (2017) Effect of interaction between salinity and nanoparticles (Fe_2O_3 and ZnO) on physiological parameters of *Zea mays* L. J Plant Nutr 40(19):2745–2755
Fincheira P, Tortella G, Duran N et al (2020) Current applications of nanotechnology to develop plant growth inducer agents as an innovation strategy. Critical Rev Biotechnol 40(1):15–30
Gao X, Zou CH, Wang L et al (2006) Silicon decreases transpiration rate and conductance from stomata of maize plants. J Plant Nutr 29:1637–1647
Gholizadeh F, Navabpour S (2011) Effect of salinity on morphological and physiological characteristics in correlation to selection of salt tolerance in rice (*Oryza sativa* L.). Int J Agri Res 6:780–788
Giraldo JP, Landry MP, Faltermeier SM et al (2014) Plant nanobionics approach to augment photosynthesis and biochemical sensing. Nat Mater 13:400–408
Gong H, Zhu X, Chen K et al (2005) Silicon alleviates oxidative damage of wheat plants in pots under drought. Plant Sci 169:313–321
Gupta B, Huang B (2014) Mechanism of salinity tolerance in plants: physiological, biochemical, and molecular characterization. Int J Gen, 2014
Haghighi M, Afifipour Z, Mozafarian M (2012) The effect of N-Si on tomato seed germination under salinity levels. Int J Environ Sci 6:87–90
Haliloglu K, Hosseinpour A, Cinisli KT et al (2020) Investigation of the protective roles of zinc oxide nanoparticles and plant growth promoting bacteria on DNA damage and methylation in tomato (*Solanum lycopersicum* L.) under salinity stress. Hortic environ biotechnol
Hirayama T, Shinozaki K (2010) Research on plant abiotic stress responses in the post-genome era: past, present and future. Plant J 61(6):1041–1052
Hojjat SS (2019) Effect of interaction between Ag nanoparticles and salinity on germination stages of *Lathyrus Sativus* L. Open Acc J Envi Soi Sci 2(2):193–198
Hojjat SS, Kamyab M (2017) The effect of silver nanoparticle on Fenugreek seed germination under salinity levels. Russ Agric Sci 43(1):61–65
Kalteh M, Alipour ZT, Ashraf S, Aliabadi MM, Nosratabadi AF (2014) Effect of silica nanoparticles on basil (*Ocimum basilicum*) under salinity stress. J Chem Health Risks 4:49–55
Kamran M, Parveen A, Ahmar S et al (2020) An overview of hazardous impacts of soil salinity in crops, tolerance mechanisms, and amelioration through selenium supplementation. Int J Mol Sci 21(1):148
Kaushal M, Wani SP (2017) Nanosensors: frontiers in precision agriculture. In: Prasad R, Kumar M, Kumar V (eds) Nanotechnology. Springer, Singapore, pp 279–291
Khalkhal K, Lajayer BA, Ghorbanpour M (2020) An overview on the effect of soil physicochemical properties on the immobilization of biogenic nanoparticles. In: Ghorbanpour M, Bhargava P, Varma A, Choudhary D (eds) Biogenic nano-particles and their use in agro-ecosystems. Springer, Singapore, pp 133–160
Khandel P, Yadaw RK, Soni DK et al (2018) Biogenesis of metal nanoparticles and their pharmacological applications: present status and application prospects. J Nanostructure Chem 8(3):217–254
Kumari M, Mishra A, Pandey S et al (2016) Physico-chemical condition optimization during biosynthesis lead to development of improved and catalytically efficient gold nano particles. Sci Rep 6(1):1–14
Kumari M, Pandey S, Bhattacharya A et al (2017) Protective role of biosynthesized silver nanoparticles against early blight disease in *Solanum lycopersicum*. Plant Physiol Biochem 121:216–225
Kuppusamy P, Yusoff MM, Maniam GP et al (2016) Biosynthesis of metallic nanoparticles using plant derivatives and their new avenues in pharmacological applications—an updated report. Saudi Pharma J 24(4):473–484
Latef AAHA, Alhmad MFA, Abdelfattah KE (2017) The possible roles of priming with ZnO nanoparticles in mitigation of salinity stress in lupine (*Lupinus termis*). Plant Growth Regul 36:60–70

Liang Y, Chen Q, Liu Q et al (2003) Exogenous silicon (Si) increases antioxidant enzyme activity and reduces lipid peroxidation in roots of salt-stressed barley (*Hordeum vulgare* L.). J Plant Physiol 160:1157–1164

Liu R, Lal R (2015) Potentials of engineered nanoparticles as fertilizers for increasing agronomic productions. Sci Total Environ 514:131–139

Lu L, Huang M, Huang Y et al (2020) Mn_3O_4 nanozymes boost endogenous antioxidant metabolites in cucumber (*Cucumis sativus*) plant and enhance resistance to salinity stress. Environ Sci Nano 7:1692–1703

Ma X, Wang Q, Rossi L et al (2016) Multigenerational exposure to cerium oxide nanoparticles: physiological and biochemical analysis reveals transmissible changes in rapid cycling *Brassica rapa*. Nano Impact 1:46–54

Maswada HF, Djanaguiraman M, Prasad PVV (2018) Seed treatment with nano-iron (III) oxide enhances germination, seeding growth and salinity tolerance of sorghum. J Agro Crop Sci 204(6):577–587

Mishra A, Kumari M, Pandey S et al (2014) Biocatalytic and antimicrobial activities of gold nanoparticles synthesized by *Trichoderma* sp. Bioresource Technol 166:235–242

Mohamed AKS, Qayyum MF, Abdel-Hadi AM et al (2017) Interactive effect of salinity and silver nanoparticles on photosynthetic and biochemical parameters of wheat. Archiv Agronom Soil Sci 63(12):1736–1747

Monreal CM, DeRosa M, Mallubhotla SC et al (2015) The application of nanotechnology for micronutrients in soil-plant systems, VFRC Report 2015/3. Virtual Fertilizer Research Center, Washington, DC, p 44

Moradbeygi H, Jamei R, Heidari R, Darvishzadeh R (2020a) Fe_2O_3 nanoparticles induced biochemical responses and expression of genes involved in rosmarinic acid biosynthesis pathway in Moldavian balm under salinity stress. Physiol Plantarum 169:555–570

Moradbeygi H, Jamei R, Heidari R, Darvishzadeh R (2020b) Investigating the enzymatic and non-enzymatic antioxidant defense by applying iron oxide nanoparticles in *Dracocephalum moldavica* L. plant under salinity stress. Sci Hortic 272:109537

Mozafari A, Ghadakchiasl A, Ghaderi N (2018a) Grape response to salinity stress and role of iron nanoparticle and potassium silicate to mitigate salt induced damage under in vitro conditions. Physiol Mol Biol Plants 24:25–35

Mozafari AA, Dedejani S, Ghaderi N (2018b) Positive responses of strawberry (*Fragaria* × *ananassa* Duch.) explants to salicylic and iron nanoparticle application under salinity conditions. Plant Cell Tiss Org Cult 134(2):267–275

Munns R (2002) Comparative physiology of salt and water stress. Plant Cell Environ 25(2):239–250

Munns R, Tester M (2008) Mechanisms of salinity tolerance. Annual Rev Plant Biol 59:651–681

Naderi MR, Danesh-Shahraki A (2013) Nanofertilizers and their roles in sustainable agriculture. Int J Agri Crop Sci 5(19):2229

Noohpisheh Z, Amiri H, Mohammadi A, Farhadi S (2020) Effect of the foliar application of zinc oxide nanoparticles on some biochemical and physiological parameters of *Trigonella foenum-graecum* under salinity stress. Plant Biosyst—Int J Dealing Aspects Plant Biol. https://doi.org/10.1080/11263504.2020.1739160

Oliveira HC, Gomes BCR, Pelegrino MT et al (2016) Nitric oxide-releasing chitosan nanoparticles alleviate the effects of salt stress in maize plants. Nitric Oxide 61:10–19

Pandey S, Mishra A, Giri VP et al (2019) A green nano-synthesis to explore the plant microbe interactions. In: Singh JS, Singh DP (eds) New and future developments in microbial biotechnology and bioengineering. Elsevier, Amsterdam, pp 85–105

Panwar J, Jain N, Bhargaya A et al (2012) Positive effect of zinc oxide nanoparticles on tomato plants: a step towards developing "Nano-fertilizers". Proceeding of 3rd international conference on environmental research and technology. University of Sains, Penang, Malaysia, pp 248–352

Prasad TNVKV, Sudhakar P, Sreenivasulu Y et al (2012) Effect of nanoscale zinc oxide particles on the germination, growth and yield of peanut. J Plant Nutr 35:905–927

Qados AMSA (2015) Mechanism of nanosilicon-mediated alleviation of salinity stress in faba bean (*Vicia faba* L.) plants. Am J Exp Agric 7:78–95
Ramesh M, Palanisamy K, Babu K et al (2014) Effects of bulk and nano-titanium dioxide and zinc oxide on physio-morphological changes in *Triticum aestivum* Linn. J Glob Biosci 3:415–422
Rameshaiah GN, Pallavi J, Shabnam S (2015) Nano fertilizers and nano sensors–an attempt for developing smart agriculture. Int J Eng Res Gen Sci 3(1):314–320
Rodrigues F, Duarte H, Domiciano G et al (2009) Foliar application of potassium silicate reduces the intensity of soybean rust. Aust Plant Pathol 38:366–372
Rossi L, Zhang W, Lombardini L et al (2016) The impact of cerium oxide nanoparticles on the salt stress responses of *Brassica napus* L. Environ Pollut 219:28–36
Rossi L, Zhang W, Ma X (2017) Cerium oxide nanoparticles alter the salt stress tolerance of *Brassica napus* L. by modifying the formation of root apoplastic barriers. Environ Pollut 229:132–138
Saxena R, Tomar RS, Kumar M (2016) Exploring nanobiotechnology to mitigate abiotic stress in crop plants. J Pharmaceutical Sci Res 8(9):974–980
Sedghi M, Hadi M, Toluie SG (2013) Effect of nano zinc oxide on the germination of soybean seeds under drought stress. Ann West Univ Timişoara Ser Biol XV I(2):73–78
Siddiqui MH, Al-Whaibi MH, Faisal M et al (2014) Nanosilicon dioxide mitigates the adverse effects of salt stress on *Cucurbita pepo* L. Environ Toxicol Chem 33(11):2429–2437
Singh VK, Singh AK (2019) Role of microbially synthesized nanoparticles in sustainable agriculture and environmental management. In: Kumar A, Singh AK, Choudhary KK (eds) Role of plant growth promoting microorganisms in sustainable agriculture and nanotechnology. Woodhead Publishing, Cambridge, UK, pp 55–73
Soliman AS, El-feky SA, Darwish E (2015) Alleviation of salt stress on *Moringa peregrina* using foliar application of nanofertilizers. J Hortic For 7:36–47
Taiz L, Zeiger E, Møller IM, Murphy A (2015) Plant physiology and development. Sinauer Associates, Sunderland, CT, p 6
Tiwari S, Lata C (2019) Genome engineering in rice: applications, advancements and future perspectives. In: Singh SP, Upadhyay SK, Pandey A, Kumar S (eds) Molecular approaches in plant biology and environmental challenges. Springer, Singapore, pp 323–337
Tiwari S, Lata C, Chauhan PS et al (2016) *Pseudomonas putida* attunes morphophysiological, biochemical and molecular responses in *Cicer arietinum* L. during drought stress and recovery. Plant Physiol Biochem 99:108–117
Tiwari S, Prasad V, Chauhan PS et al (2017) *Bacillus amyloliquefaciens* confers tolerance to various abiotic stresses and modulates plant response to phytohormones through osmoprotection and gene expression regulation in rice. Front Plant Sci 8:1510
Torabian S, Zahedi M, Khoshgoftarmanesh A (2018) Effect of foliar spray of zinc oxide on some antioxidant enzymes activity of sunflower under salt stress. J Agr Sci Tech 18:1013–1025
Torney F, Trewyn BG, Lin VSY et al (2007) Mesoporous silica nanoparticles deliver DNA and chemicals into plants. Nat Nanotechnol 2:295–300
Tripathi DK, Ahmad P, Sharma S, Chauhan DK, Dubey NK (eds) (2017) Nanomaterials in plants, algae, and microorganisms: concepts and controversies, vol 1. Academic Press, London
Uauy C, Distelfeld A, Fahima T et al (2006) A NAC gene regulating senescence improves grain protein, zinc, and iron content in wheat. Science 314:1298–1301
Wang Q, Ma X, Zhang W et al (2012) The impact of cerium oxide nanoparticles on tomato (*Solanum lycopersicum* L.) and its implications for food safety. Metallomics 4:1105–1112
Wang X, Pan X, Gadd GM (2019) Immobilization of elemental mercury by biogenic Se nanoparticles in soils of varying salinity. Sci Total Environ 668:303–309
Wu H, Shabala L, Shabala S et al (2018) Hydroxyl radical scavenging by cerium oxide nanoparticles improves *Arabidopsis* salinity tolerance by enhancing leaf mesophyll potassium retention. Environ Sci Nano 5(7):1567–1583
Yang Y, Guo Y (2018) Elucidating the molecular mechanisms mediating plant salt-stress responses. New Phytol 217(2):523–539

Ye Y, Cota-Ruiz K, Hernández-Viezcas JA et al (2020) Manganese nanoparticles control salinity-modulated molecular responses in *Capsicum annuum* L. through priming: a sustainable approach for agriculture. ACS Sustain Chem Eng 8(3):1427–1436
Younes NA, Nassef DM (2015) Effect of silver nanoparticles on salt tolerancy of tomato transplants (*Solanum lycopersicom* L. Mill.). Assiut J Agric Sci 46:76–85
Zahedi SM, Karimi M, Teixeira da Silva JA (2020) The use of nanotechnology to increase quality and yield of fruit crops. J Sci Food Agri 100(1):25–31
Zhang M, Gao B, Chen J et al (2015) Effects of graphene on seed germination and seedling growth. J Nanopart Res 17:73–80
Zhu JK (2007) Plant salt stress. Wiley, Hoboken, NJ

Chapter 17
Applications of Nanobiotechnology in Overcoming Drought Stress in Crops

Saima Amjad, Shalini Tiwari, and Mohammad Serajuddin

Abstract Drought is a major environmental constraint that hinders the yield and production of crops worldwide. It arises from complex gene networking in plants leading to alterations in metabolism for better growth and development. The adverse effect of environmental stress can be mitigated through nanobiotechnology applications in agriculture, which opens up new opportunities for crop management and development. The application of nanobiotechnology in agriculture is the most appropriate way for the better development and yield of food crops under plant stress conditions. Nanoparticles exhibit novel properties and thereby enabling advanced agriculture research in crop management as well as dealing with various environmental stresses. Nanoparticle-based plant modification has the potential to improve crop plants through genetic modification to augment traditional technology. Nanoparticles also enhance the water use efficiency in the plant by improving water retention capacity in food crops. Nanoparticles such as titanium, silicon, iron, zinc, copper and chitosan have no effect on cellular physiological and biochemical functions, but assist in controlling plant development under water deficit conditions. Therefore, this chapter discusses the role of nanobiotechnology in the agricultural sector with special emphasis on nanoparticles-mediated drought stress tolerance in various crops to illustrate its contemporary and future needs for crop improvement programs particular under the global climate change.

Keywords Abiotic stress · Agriculture · Drought stress · Environment · Nanoparticles

S. Amjad · M. Serajuddin (✉)
Department of Zoology, University of Lucknow, Lucknow, UP 226007, India
e-mail: serajuddin_m@lkouniv.ac.in

S. Tiwari
CSIR-National Botanical Research Institute, Rana Pratap Marg, Lucknow 226001, India

J. M. Al-Khayri et al. (eds.), *Nanobiotechnology*,
https://doi.org/10.1007/978-3-030-73606-4_17

17.1 Introduction

Plant stress is a condition in which plant grows in unsuitable environment that enhance the demand made upon it. The impact of stress leads to a decline in growth, reduction in crop yields, and/or plant death if the plant exceeds the stress tolerance limit. Plant stress primarily categorized into abiotic and biotic stress (Verma et al. 2013). The abiotic stress enforced on plants by nonliving physical or chemical variables such as radiations, floods, salinity, extreme temperature, heavy metals, drought, etc., whereas biotic stress occurs as a consequence of the damage caused to the crop plants by the biological factor such as diseases, insects, microorganisms (Gull et al. 2019; Iqbal et al. 2020a; Verma et al. 2013). Additionally, plant stress can be divided into various types on the basis of several factors as summarize in Fig. 17.1.

The study conducted by Rosegrant and Cline (2003) revealed that water reservoirs are diminishing for the existing and forthcoming human population for societal requirements and putting pressure towards the sustainable use of water. The rigorousness of water shortage is uncertain as it also dependent on several environmental conditions such as evaporative demands the moisture-retaining capacity of the soil, and lack of rainfall (Wery et al. 1994). The water shortage not only leads to drought but also critical hazard for world food security (Somerville and Briscoe 2001).

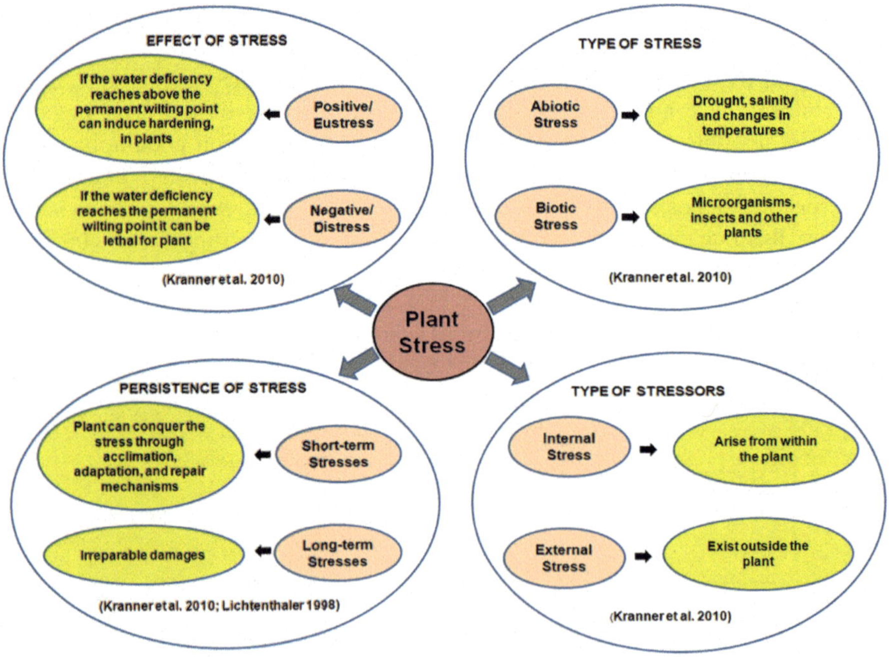

Fig. 17.1 Plant stress classification in accordance of responsible factors (Figure constructed by Saima Amjad on the basis of classification given by Kranner et al. 2010; Lichtenthaler 1998)

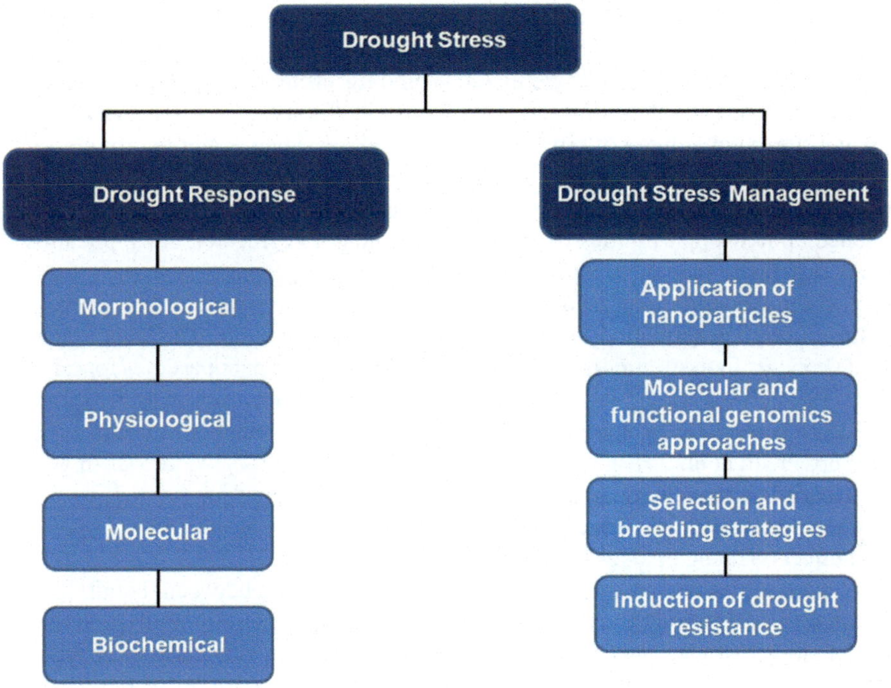

Fig. 17.2 Plant stress responses to drought stress and the strategies use for stress management (Figure constructed by Saima Amjad)

Drought is regarded as the most significant and common stress in plants that affect its development and yield specifically in the arid and semiarid regions due to weather conditions (Madhava Rao 2006; Misra et al. 2016; Tiwari et al. 2017a). It is multivariable stress, which stimulates changes in the physio-biochemical traits and molecular responses of plants (Salehi-Lisar and Bakhshayeshan-Agdam 2016; Tiwari et al. 2016) as shown in Fig. 17.2.

Sessile nature of plants evolved complex functioning to acclimatize in the adverse environment to survive in short-term and long-term drought stress conditions (Tiwari et al. 2017b). When the drought stress is imposed on crop plants, the first response subjected to the plant is the growth arrest that reduces the growth of plant shoots and their metabolic demands. These defensive mechanisms of plants provide an understanding of resistance and tolerance in plants against stress (Harb et al. 2010). Drought tolerant plant development can be an advantageous approach to overcome the problem of decreasing worldwide food production (Das and Das 2019). The mode of action of drought is a complex process and involves alteration in several metabolic pathways such as stomatal conductance, anthocyanin aggregation, osmoprotectants intervention, carotenoid degradation and ROS-scavenging enzymes, through which plants react to particular sets of the condition at any given time (Varshney et al. 2018).

The other factors which influence drought at the molecular level are transcriptional factors, specifically abscisic acid (ABA) responsive element-binding factor (ABRE), dehydration-responsive element-binding protein (DREB), and no apical meristem (NAM) (Nakashima and Suenaga 2017; Tiwari et al. 2016). Additionally, the genetic improvement of crop plants at the cellular level is an opportunity in the agricultural region to overcome drought stress by diverting the genes to the target site. Apart from these technologies, nanobiotechnology applications in agriculture opened up new opportunities for crop improvement. Specifically, the application of nanoparticles to manage with biotic and abiotic stress proposes new ways for crop protection. Nanoparticle based plant modification has the potential to improve crop plants through genetic modification and replace old technology. The latest scientific achievements in tissue engineering are used of engineered nanoparticles for targeted delivery of clustered regularly interspaced short palindromic repeats (CRISPR)/Cas (CRISPR-associated protein) mRNA, and siRNA for the genetic modification (GM) of crops (Kim et al. 2017; Miller et al. 2017; Ran et al. 2017). Moreover in genetic engineering silicon dioxide nanoparticles have been designed as a vehicle to deliver DNA sequences to the desired plant (tobacco and corn plants) without any adverse side effects (Cheng et al. 2016; Galbraith 2007).

Nanoparticles are an ultra-fine unit in which size lies between 1 and 100 nm and possess different physico-chemical properties, high reactivity, and active surface morphology (Monica and Cremonini 2009). Numerous previous studies also reported the positive effects of nanoparticles on crop yield during drought stress. Furthermore, the promising benefits of nanotechnology in agricultural field can be understood by analyzing the mechanism of diffusion and the movement of nanoparticles inside the plants (Hojjat and Ganjali 2016).

17.2 Global Climate Change and Influence on Water Availability

Climate is a deciding factor for the availability of water resources in many regions in the world. Global warming is driving changes in climate in the last few years; it has brought a hike in ocean water temperature and global average air, extensively melting of snow and ice (Pathak et al. 2014). Global climate change has also inseparably connected with rainfall patterns (Bates et al. 2008; Ghosh and Misra 2010) and consequently, led at-risk state of the water resources universally (Iglesias and Garrote 2015). The climate-vulnerable sectors such as agriculture have been highly affected by water scarcity (Mendelsohn 2001) and agriculture is becoming increasingly difficult day by day due to water crisis. Currently, agriculture uses 70% of the total freshwater basically for irrigation and in the twentieth century, a frequent number of irrigated lands were introduced (Chartzoulakis and Bertaki 2015). The climate change challenges have to lead plants to adapt under water-scarce conditions (Iglesias and Garrote 2015). Water accessibility is a key factor for better food crop productivity (Siebert and Döll 2010). The insufficient water supply has already limited the crop productivity in arid regions, which is prognosticated to generate severe unfavorable conditions for crop yield due to future climate changes (Vörösmarty et al. 2000). Nanotechnology application in the form of nanoparticles can provide an adequately high quality of water for the agriculture sector by replacing the traditional water resources with nanoparticle treated municipal wastewater, or industrial wastewater in water-limited regions (Alvarez et al. 2018). According to the experimental study of Wang (2018), nano-enabled solar distillation and crystallization processes can be used for the treatment of high-salinity wastewater and make it suitable for agriculture. Nanoparticles also enhance the water use efficiency in the plant by improving water retention capacity in food crops (Villagarcia et al. 2012; Zhou et al. 2015) and its application can be utilized to create 'smart plants' that can convey their water requirements to cultivator (Lowry et al. 2019).

17.3 Potential Applicable Tools of Nanobiotechnology in Agriculture

Nanotechnology application has a multidimensional approach and applicable in all fields of science. Nanobiotechnology is a term that denotes the merging of nanotechnology and biology (Gazit 2007), which specifically designed and developed devices so as to study the biological functions, for instance, different types of nanoparticles can work as a vehicle and probe for the delivery of biomolecules at a cellular level. The applications of agricultural nanobiotechnology are progressively stepping out from theoretical and experimental laboratories to the agriculture fields for crop safety and increased production (Ahmed et al. 2013). Such nanotechnology help to deliver the pesticides in a specific, precise and targeted manner, genes as well as nutrient molecules delivery (nanofertilizers) to particular sites at the atomic level. Additionally, for stress management, such as abiotic and biotic stress,gene technologies such as nano array used in plants for gene expressions (Ahmed et al. 2013; Kuzma 2007; Maysinger 2007; Scott 2007). Nanoparticles can work as a "magic bullet" and it enhances the seed germination and growth, and plant physiological activities such as nitrogen metabolism and photosynthesis, chlorophyll contents, carbohydrate contents, protein and yield of the crop (Siddiqui et al. 2015). Nanoparticles increase water stress tolerance through increasing the hydraulic conductance of root and uptake of water in plant pathways (Das and Das 2019). Nanoparticles large surface area increases their functionality and biological activity (Dubchak et al. 2010) and its application influences different plant development and growth stages in both positive and negative ways. Nanoparticles consisting of novel properties that assists in advancing agricultural research in crop management as well as dealing with stresses (Moraru et al. 2003).

17.4 Sustainable Approach of Nanoparticles to Mitigate Drought Stress

The nanotechnology revolution will change the whole food industry on every stage from production, handling, packaging, transportation, and consumption of agricultural materials (Amjad et al. 2018). Plant physiology, morphology and biochemical aspects has severely influenced by the effect of drought stress. Therefore, it is necessary to identify the causes of interferences linked with drought stress for better crop management (Iqbal et al. 2020b). According to Iqbal et al. (2020a) nanoparticles interaction with plants can be affected by many factors such as plant species, nanoparticles type, and their physical and chemical characteristics. Moreover, plant nanoparticle interactions leading to stress in plants at each stage from uptake to translocation within the plant which result desired or undesired outcome. Some recent experimental studies discussed that, nanoparticles in plants affect several physio-biochemical processes controlling plant yield and at the same time also dealt with plant environmental stress responses (Arora et al. 2012; Aghdam et al. 2016; Burke et al. 2015; Jalil and Ansari 2019; Ngo et al. 2014; Regier et al. 2015). Several studies listed in Table 17.1 reported that nanoparticles promise a significant effort to mitigate drought stress (Lee et al. 2010).

17.4.1 Titanium Dioxide Nanoparticles

Titanium nanoparticles have significant biological impacts on plant development at size and concentration dependent manner. The positive effect of titanium dioxide nanoparticles (TiO_2NPs) have been reported, where 0.02% of TiO_2NPs under water stress condition showed improved agronomic traits of the wheat plant such as enhanced crop yield, gluten and starch content (Jaberzadeh et al. 2013). The study carried out by Kiapour et al. (2015),depicted that different concentrations of TiO_2NPs increases the crop plant resistance against drought stress by the application of nanoparticles and gibberellic acid hormone on basil plants. TiO_2NPs also improve the photosynthetic mechanism and increases the plant's capability to absorb sunlight, which affects the biosynthesis of photosynthetic pigments and transformation of daylight energy to the active electron followed by a chemical activity and, as a consequence enhances the photosynthetic efficiency in maize crop plant under water deficit condition (Akbari et al. 2014; Morteza et al. 2013).

Table 17.1 Application of nanoparticles and their mechanism of action for overcoming the drought stress in plants

S. no	Nanoparticles	Scientific name	Common name	Mechanism of action	References
1.	Titanium dioxide nanoparticles	*Ocimumbasilicum* L.	Basil	Increases the crop plant resistance against drought stress, improve photosynthetic mechanism	Kiapour et al. (2015)
		Triticumaestivum L.	Wheat	Enhancing yield, growth, starch and gluten content	Jaberzadeh et al. (2013)
2.	Anatase titanium dioxide nanoparticles	*Linumusitatissimum* L.	Flax	Increasing amount of chlorophyll and carotenoids, enhancing flax development and yield, declining MDA and H_2O_2 content	Aghdam et al. (2016)
3.	Iron nanoparticles	*Carthamustinctorius* L.	Safflower	Crop yield improved	Zareii et al. (2014)
		Arabidopsis thaliana L.	Mouseear cress	Enhances plasma membrane proton pump ATPase activity, enhancing plant biomass and Chlorophyll content, maintaining normal water deficit condition, enhancing CO_2 assimilation in plant	Kim et al. (2015)
4.	Silicon nanoparticles	*Crataegusmonogyna* Jacq.	Hawthorn	Improved photosynthetic rate which increased stomatal conductance and plant biomass	Ashkavand et al. (2015)
5.	Zinc oxide nanoparticles	*Glycine max* L.	Soybean	Enhanced germination rate, Length and fresh weight of radical, reduces in seed residual fresh and dry weight	Sedghi et al. (2013)

(continued)

Table 17.1 (continued)

S. no	Nanoparticles	Scientific name	Common name	Mechanism of action	References
6.	Composition of titanium dioxide and silicon dioxide nanoparticles	*Gossypiumbarbadense* L.	Cotton	Improved total content of phenols, free amino acids, soluble proteins, antioxidant capacity, proline content, all antioxidant enzyme activities	Magdy et al. (2016)
7.	Silver nanoparticles	*Lens culinaris* Medic.	Lentil seed	Increased germination rate, shoot and root length	Hojjat and Ganjali (2016)
8.	Chitosan nanoparticles	*Hordeum vulgare* L.	Barley	Improved relative water content (RWC), grain weight, protein, proline content, and enzyme activities CAT and the SOD	Behboudi et al. (2018)
		Triticum aestivum L.	Wheat	Improved chlorophyll content, enhanced photosynthesis rate and leaf photochemical efficiency	Behboudi et al. (2019)
9.	Copper nanoparticles	*Zea mays* L.	Maize	Maintains leaf water status, and chlorophyll and carotenoid content, and enhances plant growth and biomass	Van Ha et al. 2020
10.	Fullerenol Nanoparticles	*Beta vulgaris* L.	Sugar Beets	Bind and reserve intracellular water in cell compartments of leaf and root tissue, reduced Osmolyte proline, increased antioxidant enzyme activities (APx, CAT, and GPx), GSH and MDA content	Borišev et al. (2016)

17.4.2 Anatase Titanium Dioxide Nanoparticles

Effect of anatase titanium dioxide nanoparticles ($AnTiO_2NP$) on the flax plant was studied under water scarce conditions. Different doses and sizes of $AnTiO_2NP$ (10–25 nm) showed a positive impact on flax plants such as on growth, development,seed oil, yield, hydrogen peroxide (H_2O_2), photosynthetic pigment, and malondialdehyde contents, and protein contents. Moreover, it also revealed that a low concentration of exogenous application $AnTiO_2NP$ improved physiological and morphological traits of plant viz. enhanced carotenoids and chlorophyll contents in plant leaves as compared to other concentrations under both normal and water deficit environments (Aghdam et al. 2016).

17.4.3 Iron Nanoparticles

Iron is a vital micronutrient for different enzymatic activities and has a significant role in various plant mechanisms and its deficiency leads to chlorosis in the plant (Bameri et al. 2013). The foliar application of iron oxide nanoparticles decreases the harmful impact of drought stress and also stimulates the yield, growth, and development parameters in *Carthamustinctorius* (Zareii et al. 2014). Similarly, Martínez-Fernández et al. (2015) experimental studies showed maghemite nanoparticles (member of iron oxide) that noticeably reduces drought stress traits in Sunflower. The hypothesis of the effect of zerovalent iron nanoparticles enhances the stomatal aperture of *Arabidopsis thaliana* plants and tolerance to drought which would trigger the proton pump ATPase (H+ -ATPase) of the plasma membrane in leaves. The hypothesis of Kim et al. (2015) had shown the effect of zerovalent iron nanoparticles (nZVI) on the increased activity of stomatal aperture of *Arabidopsis thaliana* plants which would triggered the activation of proton pump ATPase (H+ -ATPase) in plasma membrane of the leaves to deal with drought stress. The rise in stomatal activity caused an enhancement in leaf area, and stomata aperture and decreases theapoplastic pH. The increased rate of stomatal opening by the exposure of zerovalent iron nanoparticles enhanced the chlorophyll amount and plant biomass, maintaining drought vulnerability and improved CO_2 uptake in *Arabidopsis thaliana* plant (Kim et al. 2015).

17.4.4 Silicon Nanoparticles

Earth's crust largely composed of Silicon (Si) element (Ma 2004) and only some studies have reported the biological activity of silica element (Ma and Yamaji 2006). Ashkavand et al. (2015) studied the plant tolerance towards drought stress via the treatment of silicon nanoparticles (SiNPs) on hawthorns plant (*Crataegus* sp.) under different levels of water deficit condition from temperate to extreme state. The findings of the study demonstrate SiNPs play a vital role in sustaining significant biochemical and physiological functions of hawthorn seedlings under water deficit stress conditions. SiNPs also increased the plant biomass, malondialdehyde content, and xylem water potential and the positive effect on the rate of photosynthetic and stomatal conductance was evident especially under water shortage condition. Pre-treated seedlings of SiNPs had no significant impact on total chlorophyll and carotenoid content.

17.4.5 Zinc Oxide Nanoparticles

Micronutrients are necessary for plant growth and development and also increase the resistance in plants against environmental stresses (Baybordi 2005). Seed germination percentage was significantly declined in *Glycine max* L. and germination characteristics in safflower plant by the effect of abiotic stresses (Abedi Baba–Arabi 2008; Dornbas et al. 1989; Tatic et al. 2004). According to Cakmak (2008), iron and zinc have the potential to prepare plants resistant to drought stress.

Zinc oxide nanoparticles (ZnONPs) improved plant growth parameters in *Glycine max* L. seed under water scarce environments. The experimental studies showed a significant impact on plants by the combined treatment of PEG and ZnONPs at different concentrations (Sedghi et al. 2013). The exposure of ZnONPs on plants enhances the germination rate, root fresh and dry weight, root length, and also reduces seed residual fresh and dry weight as compared to untreated plants of *Glycine max* L.

17.4.6 Titanium Dioxide and Silicon Dioxide Nanoparticles

Titanium dioxide nanoparticles (TiO_2NPs) are the most commonly applicable nanoparticle in the world. It has wide range of biological applications not only in animals, but also in plants. It is found to be advantageous at a very low concentration, but lethal at higher concentrations (Magdy et al. 2016). Similarly, SiO_2has a positive effect on the plant system, where it protects plants in high temperatures and drought environment (Asadzade et al. 2015; Zhu et al. 2004).

Magdy et al. (2016) investigated the effect of titanium dioxide (TiO_2NPs) and silicon dioxide nanoparticles (SiO_2NPs) on the *Gossypium* plant under water deficit condition. Before exposure to drought stress, *Gossypium* was exposed with four different concentrations of TiO_2NPs (25, 50, 100, and 200 ppm) or SiO_2NPs (400, 800, 1600, and 3200 ppm). The exposure of TiO_2NPs and SiO_2NPs to *Gossypium* resulted in the improvement of yield characteristics, enhancement of pigment contents, proline content, total phenols, total soluble sugars, free amino acids, total reducing power, entire soluble proteins and all antioxidant enzyme activities.

17.4.7 Sliver Nanoparticles

Silver is the most commonly used nanoparticle for improving the quality of product and some experimental studies showed the positive impact of silver nanoparticles (AgNPs) on plant metabolic activities. Sharon et al. (2010) found that in agricultural soils and hydroponics systems, silver removes unnecessary microorganisms, stimulate plant growth and its foliar application protects the plant from various diseases. Silver has a positive impact on hydraulic conductivity of plant stem (Van Ieperen 2007), and the application of AgNPs on the Zucchini plant decreased its biomass and transpiration but prolonged their growth (Stampoulis et al. 2009). Silver nanoparticles (AgNPs) had an important impact on seed growth and development of *Lens esculenta* (lentil seed). The application of AgNPs has the potential to alter the quality and yield of lentil seed in a field environment. AgNPs enhanced the germination percentage, length, and fresh weight of radicle in lentil under drought stress (Hojjat and Ganjali 2016).

17.4.8 Chitosan Nanoparticles

Chitosan is a naturally found polysaccharide constituent synthesized from chitin shells of crawfish, shrimp, and crab (Orgaz et al. 2011). Some previous studies showed that chitosan application has been used as a growth stimulator to improve yield and germination in various crop species such as *Zea mays* (Lizarraga-Paulin et al. 2011) and *Viciafaba* (El-Sawy et al. 2010). The chitosan nanoparticle at two concentrations of 60 and 90 ppm, is equally applied in irrigation regimes of barley plant, which improved the relative water content status, grain weight, proline amount, grain protein, as well as enzymatic activities catalase (CAT), and superoxide dismutase (SOD) and reduces the harmful effect of drought stress condition. However, chitosan nanoparticles may also generate several metabolites that are helpful to retain water for better development and production of barley (Behboudi et al. 2018).

The foliar or soil application of chitosan nanoparticles (90 ppm) in wheat crop diminishes the negative effect of drought stress condition and generated several metabolites, which improved chlorophyll content, photosynthesis rate, leaf photochemical efficiency, however it reduces the transpiration in the plant (Behboudi et al. 2019).

17.4.9 Copper Nanoparticles

Copper micronutrient is required in the plant for normal development because it has a significant role in several physiological (Yruela 2005) and structural processes. It acts as an active component in regulatory proteins and participates in mitochondrial respiration, oxidative stress responses, hormone signaling, photosynthetic electron transport and cell wall metabolic processes (Raven et al. 1999; Solymosi and Bertrand 2012; Da Costa and Sharma 2016).

The study of Van Ha et al. (2020) demonstrated the treatment of copper nanoparticles (CuNPs) in maize crops under drought stress which improved maize growth and development. The CuNPs maintains chlorophyll and carotenoid content under water deficit environment and also maintain leaf water condition which enhanced the crop development and biomass. Additionally, CuNPs enhance anthocyanin content, enzyme functionalities such as SOD, and ascorbate peroxidase activities. The outcome reduced the excess of ROS (reactive oxygen species) production and consequently enhanced the adaptation of maize under drought stress condition.

17.4.10 Fullerenol Nanoparticles

Fullerenol $C_{60}(OH)_{24}$ nanoparticles are derivatives of polyhydroxylated fullerene and are highly soluble in water which makes it attractive tool of nanobiotechnology for various biological applications (Verma et al. 2019). According to the study of Borišev et al. (2016) foliar application of fullerenol nanoparticles (FNPs) on *Beta vulgaris* L. help to bind and reserve intracellular water in drought stress condition. FNPs bind with water in cell compartments through leaf and root tissue penetration. Additionally, FNPs also alleviated the oxidative effects of drought stress in plants by enhancing the antioxidant enzyme activities (APx, CAT, and GPx), GSH and MDA content. Osmolyte proline significantly enhanced in control plant (leaves and roots) under water deficit condition as compared to FNP treated plants (Borišev et al. 2016). Moreover, according to Ahmad et al. (2020), direct application of FNPs on different agricultural practices could be more effectual where water supply is limited.

17.5 Conclusion and Prospects

Plant growth under stress environment has become an increasingly important issue for crop protection and production. The diverse studies on nanoparticles depict their important roles in plant growth and development under stress conditions.The application of manufactured nanoparticles such as titanium, iron, silicon, zinc oxide, copper, chitosan and silver not only improved crop production but also enhances the drought tolerance. A lot of experimental studies are required in future to understand the interaction of nanoparticles with plants, so as to explore the mechanism, gene expression and regulation, of nanoparticles under scarce water conditions. However, some previous studies reported the negative impact of nanoparticles on the environment. Therefore, on the basis of consolidated experimental studies, it is necessary to use nanoparticles strategically to reduce environmental contamination and researchers should also focus on plant nanoparticle interaction for better crop yield. The approach of nanobiotechnology would be socially and ethically acceptable worldwide for crop improvement. Hence, in the near future by using nanobiotechnology approach in agriculture would improve crop variety which would ultimately help to fulfill the gap of food requirement across the globe under drought stress conditions.

References

Abedi Baba–Arabi S (2008) Effects of Zn and K foliar application on physiological traits and yield of spring safflower under drought stress. MSc Thesis, Yasouj University, Yasuj, Iran (In Persian), p 92

Aghdam MTB, Mohammadi H, Ghorbanpour M (2016) Effects of nanoparticulate anatase titanium dioxide on physiological and biochemical performance of *Linum usitatissimum* (Linaceae) under well watered and drought stress. Braz J Bot 39:139–146

Ahmad J, Qamar S, Kausar N, Qureshi MI (2020) Nanoparticles: the magic bullets in mitigating drought stress in plants. In: Hakeem K, Pirzadah T (eds) Nanobiotechnology in agriculture. Nanotechnology in the life sciences. Springer, Cham, pp 145–161

Ahmed F, Arshi N, Kumar S et al (2013) Nanobiotechnology: scope and potential for crop improvement. In: Tuteja N, Gill S (eds) Crop improvement under adverse conditions. Springer, New York, pp 245–269

Akbari GA, Morteza E, Moaveni P et al (2014) Pigments apparatus and anthocyanins reactions of borage to irrigation, methylalchol and titanium dioxide. Int J Biosci 4(7):192–208

Alvarez PJJ, Chan CK, Elimelech M et al (2018) Emerging opportunities for nanotechnology to enhance water security. Nat Nanotechnol 13:634–641

Amjad S, Sharma AK, Serajuddin M (2018) Toxicity assessment of cypermethrin nanoparticles in *Channa punctatus*: behavioural response, micronuclei induction and enzyme alteration. Regul Toxicol Pharmacl 100:127–133

Arora S, Sharma P, Kumar S et al (2012) Gold-nanoparticle induced enhancement in growth and seed yield of *Brassica juncea*. Plant Growth Regul 66:303–310

Asadzade N, Moosavi SG, Seghatoleslami MJ (2015) Effect of low irrigation and Zn and SiO_2 nano-fertilizers and conventional fertilizers on morphophysiological traits and seed yield of sunflower. Biol Forum 7(1):357–364

Ashkavand P, Tabari M, Zarafshar M et al (2015) Effect of SiO_2 nanoparticles on drought resistance in hawthorn seedlings. For Res Pap 76(4):350–359

Bameri M, Abdolshahi R, Mohammadi-Nejad G et al (2013) Effect of different microelement treatment on wheat *(Triticum aestivum)* growth and yield. Int Res J Appl Basic Sci 3(1):219–223

Bates BC, Kundzewicz ZW, Wu S, (2008) Climate change and water, Technical paper of the inter-governmental panel on climate change secretariat. Available via http://hdl.handle.net/123456789/552

Baybordi A (2005) Effect of zinc, iron, manganese and copper on wheat quality under salt stress conditions. J water Soil 140:150–170

Behboudi F, Tahmasebi Sarvestani Z, Kassaee MZ et al (2018) Evaluation of chitosan nanoparticles effects on yield and yield components of barley (*Hordeum vulgare* L.) under late season drought stress. J Water Environ Nanotechnol 3(1):22–39

Behboudi F, Tahmasebi Sarvestani Z, Kassaee MZ et al (2019) Evaluation of chitosan nanoparticles effects with two application methods on wheat under drought stress. J Plant Nutr 42(13):1439–1451

Borišev M, Borišev I, Župunski et al (2016) Drought impact is alleviated in sugar beets (*Beta vulgaris L.*) by foliar application of fullerenol nanoparticles. PLoS One 11(11):e0166248

Burke DJ et al (2015) Iron oxide and titanium dioxide nanoparticle effects on plant performance and root associated microbes. Int J Mol Sci 16:23630–23650

Cakmak I (2008) Enrichment of cereal grains with zinc: agronomic or genetic biofortification. Plant Soil 30(2):1–17
Chartzoulakis K, Bertaki M (2015) Sustainable water management in agriculture under climate change. Agric Agric Sci Procedia 4:88–98
Cheng HN, Klasson KT, Asakura T, Wu Q (2016) Nanotechnology in agriculture. In: Chen HN, Doemeny L, Geraci CL, Schmidt DG (eds) Nanotechnology: delivering on the Promise, vol 2. American Chemical Society, USA, pp 233–242
Da Costa MVJ, Sharma PK (2016) Effect of copper oxide nanoparticles on growth, morphology, photosynthesis, and antioxidant response in *Oryza sativa*. Photosynthetica 54(1):110–119
Das A, Das B (2019) Nanotechnology a potential tool to mitigate abiotic stress in crop plants. In: de Oliveira AB (ed) Abiotic and biotic stress in plants. IntechOpen. https://doi.org/10.5772/intechopen.83562
Dornbos DL, Mullen RE, Shibles RE (1989) Drought stress effects during seed fill on soybean seed germination and vigor. Crop Sci 29(2):476–480
Dubchak S, Ogar A, Mietelski JW et al (2010) Influence of silver and titanium nanoparticles on arbuscular mycorrhiza colonization and accumulation of radiocaesium in *Helianthus annuus*. Span J Agric Res 8:103–108
Gazit E (2007) Plenty of room for biology at the bottom: an introduction to bionanotechnology. Imperial College Press. http://www.worldscibooks.com/lifesci/p465.html
El-Sawy NM, El-Rehim HAA, Elbarbary AM et al (2010) Radiation-induced degradation of chitosan for possible use as a growth promoter in agricultural purposes. Carbohydr Polym 79(3):555–562
Galbraith DW (2007) Silica breaks through in plants. Nat Nanotechnol 2:272–273
Ghosh S, Misra C (2010) Assessing hydrological impacts of climate change: modeling techniques and challenges. Open Hydrol J 4:115–121
Gull A, Lone AA, Wani NUI (2019) Biotic and abiotic stresses in plants. In: de Oliveira AB (ed) Abiotic and biotic stress in plants. IntechOpen, pp 1–19
Harb A, Krishnan A, Ambavaram MM et al (2010) Molecular and physiological analysis of drought stress in *Arabidopsis* reveals early responses leading to acclimation in plant growth. Plant Physiol 154(3):1254–1271
Hojjat SS, Ganjali A (2016) The effect of silver nanoparticle on lentil seed germination under drought stress. Int J Farm Allied Sci 5(3):208–212
Iglesias A, Garrote L (2015) Adaptation strategies for agricultural water management under climate change in Europe. Agric Water Manag 155:113–124
Iqbal MS, Singh AK, Ansari MI (2020a) Effect of drought stress on crop production. In: Rakshit A, Singh H, Singh A, Singh U, Fraceto L (eds) New frontiers in stress management for durable agriculture. Springer, Singapore, pp 35–47
Iqbal MS, Singh AK, Singh SP, Ansari MI (2020b) Nanoparticles and plant interaction with respect to stress response. In: Bhushan I, Singh V, Tripathi D (eds) Nanomaterials and environmental biotechnology. Springer, Cham, pp 1–15
Jaberzadeh A, Moaveni P, Moghadam HRT et al (2013) Influence of bulk and nanoparticles titanium foliar application on some agronomic traits, seed gluten and starch contents of wheat subjected to water deficit stress. Not Bot Horti Agrobot Cluj Napoca 41:201–207
Jalil SU, Ansari MI (2019) Nanoparticles and abiotic stress tolerance in plants: synthesis, action and signaling mechanisms. In: Khan MIR, Reddy PS, Ferrante A, Khan NA (eds) Plant signaling molecules, role and regulation under stressful environment, 1st edn. Woodhead Publishing, Elsevier, pp 549–561
Kiapour H, Moaveni P, Habibi D (2015) Evaluation of the application of gibbrellic acid and titanium dioxide nanoparticles under drought stress on some traits of basil (*Ocimum basilicum* L.). Int J Agric Res 6:138–150
Kim DH, Gopal J, Sivanesan I (2017) Nanomaterials in plant tissue culture: the disclosed and undisclosed. RSC Adv 7:36492–36505

Kim JH, Oh Y, Yoon H et al (2015) Iron nanoparticle-induced activation of plasma membrane H+ -ATPase promotes stomatal opening in *Arabidopsis thaliana*. Environ Sci Technol 49(2):1113–1119

Kranner I, Minibayeva FV, Beckett RP et al (2010) What is stress? Concepts, definitions and applications in seed science. New Phytol 88:655–673

Kuzma J (2007) Moving forward responsibly: oversight for the nanotechnology-biology interface. J Nanopart Res 9(1):165–182

Lee CW, Mahendra S, Zodrow K et al (2010) Developmental phytotoxicity of metal oxide nanoparticles to *Arabidopsis thaliana*. Environ Toxicol Chem 29(3):669–675

Lichtenthaler HK (1998) The stress concept in plants: an introduction. Ann N Y Acad Sci 851(1):187–198

Lizarraga-Paulin EG, Torres-Pacheco I, Moreno-Martinez E et al (2011) Chitosan application in maize (*Zea mays*) to counteract the effects of abiotic stress at seedling level. Afr J Biotechnol 10(34):6439–6446

Lowry GV, Avellan A, Gilbertson LM (2019) Opportunities and challenges for nanotechnology in the agri-tech revolution. Nat Nanotechnol 14(6):517–522

Ma JF (2004) Role of silicon in enhancing the resistance of plants to biotic and abiotic stresses. Soil Sci Plant Nutr 50(1):11–18

Ma JF, Yamaji N (2006) Silicon uptake and accumulation in higher plants. Trends Plant Sci 11(8):392–397

Madhava Rao KV (2006) Introduction. In: Madhava Rao KV, Raghavendra AS, Janardhan Reddy K (eds) Physiology and molecular biology of stress tolerance in plants. Springer, Dordrecht, Netherlands, pp 1–14

Magdy AS, Hazem MMH, Alia AMN et al (2016) Biochemical and physiological effects of TiO_2 and SiO_2 nanoparticles on cotton plant under drought stress. Res J Pharm Biol Chem Sci 7(4):1540–1551

Martínez-Fernández D, Vítková M, Bernal MP, Komárek M (2015) Effects of nano-maghemite on trace element accumulation and drought response of *Helianthus annuus* L. in a contaminated mine soil. Water Air Soil Pollut 226(4):1–9

Maysinger D (2007) Nanoparticles and cells: good companions and doomed partnerships. Org Biomol Chem 5(15):2335–2342

Mendelsohn RO (2001) Global warming and the American economy: a regional assessment of climate change impacts. New horizons in environmental economics series. Edward Elgar Publishing Limited, UK

Miller JB, Zhang S, Kos P et al (2017) Non-viral CRISPR/Cas gene editing in vitro and in vivo enabled by synthetic nanoparticle co-delivery of Cas9 mRNA and sgRNA. Angew Chem Int Ed 56(4):1059–1063

Misra V, Solomon S, Ansari MI (2016) Impact of drought on post-harvest quality of sugarcane crop. Adv Life Sci 20(5):9496–9505

Monica RC, Cremonini R (2009) Nanoparticles and higher plants. Caryologia 62(2):161–165

Moraru CI, Panchapakesan CP, Huang Q et al (2003) Nanotechnology: a new frontier in food science understanding the special properties of materials of nanometer size will allow food scientists to design new, healthier, tastier, and safer foods. Nanotechnol 57(12):24–29

Morteza E, Moaveni P, Farahani HA et al (2013) Study of photosynthetic pigments changes of maize (*Zea mays L.*) under nano TiO_2 spraying at various growth stages. Springerplus 2(1):1–5

Nakashima K, Suenaga K (2017) Toward the genetic improvement of drought tolerance in crops. Jpn Agric Res Q 51(1):1–10

Ngo BQ, Dao TT, Nguyen CH et al (2014) Effects of nanocrystalline powders (Fe, Co and Cu) on the germination, growth, crop yield and product quality of soybean (Vietnamese species DT-51). Adv Nat Sci-Nanosci 5(1):015016

Orgaz B, Lobete MM, Puga CH et al (2011) Effectiveness of chitosan against mature biofilms formed by food related bacteria. Int J Mol Sci 12(1):817–828

Pathak H, Pramanik P, Khanna M, Kumar A (2014) Climate change and water availability in Indian agriculture: impacts and adaptation. Indian J Agr Sci 84:671–679

Ran Y, Liang Z, Gao C (2017) Current and future editing reagent delivery systems for plant genome editing. Sci China Life Sci 60(5):490–505

Raven JA, Evans MC, Korb RE (1999) The role of trace metals in photosynthetic electron transport in O_2 evolving organisms. Photosynth Res 60(2):111–150

Regier N, Cosio C, von Moos N et al (2015) Effects of copper-oxide nanoparticles, dissolved copper and ultraviolet radiation on copper bioaccumulation, photosynthesis and oxidative stress in the aquatic macrophyte *Elodea nuttallii*. Chemosphere 128:56–61

Rosegrant MW, Cline SA (2003) Global food security: challenges and policies. Science 302:1917–1919

Salehi-Lisar SY, Bakhshayeshan-Agdam H (2016) Drought stress in plants: causes, consequences, and tolerance. In: Hossain M, Wani S, Bhattacharjee S, Burritt D, Tran LS (eds) Drought stress tolerance in plants, vol 1. Springer, Cham, pp 1–16

Scott NR (2007) Nanoscience in veterinary medicine. Vet Res Commun 31(1):139–144

Sedghi M, Hadi M, Toluie SG (2013) Effect of nano zinc oxide on the germination parameters of soybean seeds under drought stress. Ann West Univ Timis Ser Biol 16(2):73–78

Sharon M, Choudhary AK, Kumar R (2010) Nanotechnology in agricultural diseases and food safety. J Phytol 2(4):83–92

Siddiqui MH, Al-Whaibi MH, Firoz M, Al-Khaishany MY (2015) Role of nanoparticles in plants. In: Siddiqui M, Al-Whaibi M, Mohammad F (eds) Nanotechnology and plant sciences. Springer, Cham, pp 19–35

Siebert S, Döll P (2010) Quantifying blue and green virtual water contents in global crop production as well as potential production losses without irrigation. J Hydrol 384(3–4):198–217

Solymosi K, Bertrand M (2012) Soil metals, chloroplasts, and secure crop production: a review. Agron Sustain Dev 32:245–272

Somerville C, Briscoe J (2001) Genetic engineering and water. Science 292(5525):2217

Stampoulis D, Sinha SK, White JC (2009) Assay-dependent phytotoxicity of nanoparticles to plants. Environ Sci Technol 43(24):9473–9479

Tatic M, Mladen J, Balesevi – Tubic S et al (2004) Effect of drought caused stress on the quality and yield of soybean seed. Paper presented at the 27th ISTA congress on the seed symposium, p 14

Tiwari S, Lata C, Chauhan PS et al (2016) *Pseudomonas putida* attunes morpho-physiological, biochemical and molecular responses in *Cicer arietinum* L. during drought stress and recovery. Plant Physiol Biochem 99:108–117

Tiwari S, Lata C, Chauhan PS et al (2017a) A functional genomic perspective on drought signalling and its crosstalk with phytohormone-mediated signalling pathways in plants. Curr Genom 18(6):469–482

Tiwari S, Prasad V, Chauhan PS et al (2017b) *Bacillus amyloliquefaciens* confers tolerance to various abiotic stresses and modulates plant response to phytohormones through osmoprotection and gene expression regulation in rice. Front Plant Sci 8:1510

Van Ha C, Van Nguyen D, Nguyen HM et al (2020) Copper nanoparticle application enhances plant growth and grain yield in maize under drought stress conditions. bioRxiv 02.24.963132

Van Ieperen W (2007) Ion-mediated changes of xylem hydraulic resistance in plant: fact or fiction? Trends Plant Sci 12(4):137–142

Varshney RK, Roberto Tuberosa, Francois Tardieu (2018) Progress in understanding drought tolerance: from alleles to cropping systems. J Exp Bot 69(13):3175–3179

Verma S, Nizam S, Verma PK (2013) Biotic and abiotic stress signaling in plants. In: Sarwat M, Ahmad A, Abdin MZ (eds) Stress signaling in plants: genomics and proteomics perspective, vol 1. Springer Science, New York, pp 25–49

Verma SK, Das AK, Gantait S et al (2019) Applications of carbon nanomaterials in the plant system: a perspective view on the pros and cons. Sci Total Environ 667:485–499

Villagarcia H, Dervishi E, De Silva K et al (2012) Surface chemistry of carbon nanotubes impacts the growth and expression of water channel protein in tomato plants. Small 8:2328–2334

Vörösmarty CJ, Green P, Salisbury J, Lammers RB (2000) Global water resources: vulnerability from climate change and population growth. Science 289:284–288

Wang P (2018) Emerging investigator series: the rise of nano-enabled photothermal materials for water evaporation and clean water production by sunlight. Environ Sci Nano 5:1078–1089

Wery J, Silim SN, Knights EJ et al (1994) Screening techniques and sources and tolerance to extremes of moisture and air temperature in cool season food legumes. In: Muehlbauer FJ, Kaiser WJ (eds) Expanding the production and use of cool season food legumes. Current plant science and biotechnology in agriculture, vol 19. Springer, Dordrecht, pp 439–456

Yruela I (2005) Copper in plants. Braz J Plant Physiol 17:145–156

Zareii FD, Roozbahani A, Hosnamidi A (2014) Evaluation the effect of water stress and foliar application of Fe nanoparticles on yield, yield components and oil percentage of safflower (*Carthamus tinctorious* L.). Int J Adv Biol Biomed Res 2:1150–1159

Zhou L et al (2015) Fabrication of a high-performance fertilizer to control the loss of water and nutrient using micro/nano networks. ACS Sustain Chem Eng 3:645–653

Zhu J, Wei G, Li J et al (2004) Silicon alleviates salt stress and increases antioxidant enzymes activity in leaves of salt-stressed cucumber (*Cucumis sativus* L.). Plant Sci 167:527–533

Chapter 18
Applications of Nanobiotechnology in Overcoming Temperature Stress

Sashi Sonkar, Brijesh Pandey, Saurabh Singh Rathore, Laxuman Sharma, and Akhilesh Kumar Singh

Abstract Nanoparticles application in plant science requires technological and theoretical amalgamation governed by features of nanoparticles as well as their molecular and physiological responses in each plant. Agriculture is the most important sector for human wellbeing as it provides resources for food and feed industries. Factors contributing to abiotic stress are regarded as one of the strong constraints for the sustainable production of crops as it negatively influence the vegetative as well as reproductive physiology of plants, thereby reducing the yield. Heat and cold stresses in plants cause diverse as well as often detrimental variations in growth, development, biological processes as well as yields. Nanotechnology has improved crop production, though, it is still in the infancy of its development. In recent years, the exploitation of nanomaterials has been used to diminish the damage of plant systems due to temperature stress by stimulating the protective/tolerance physiology in plants. This journey has long way to go regarding our understanding of the eco-toxicity, permissible limit, and uptake capacity of various nanomaterials in plant systems. This chapter explores the range of uses of various nanoparticles for alleviating the temperature stress in plant science.

Keywords Abiotic stress · Agriculture · Nanomaterials · Nanotechnology · Stress · Temperature stress

---

S. Sonkar
Department of Botany, Bankim Sardar College, Tangrakhali South 24 Parganas, 743329, West Bengal, India

B. Pandey · S. S. Rathore · A. K. Singh (✉)
Department of Biotechnology, School of Life Sciences, Mahatma Gandhi Central University, Motihari 845401, Bihar, India
e-mail: akhileshsingh@mgcub.ac.in

L. Sharma
Department of Horticulture, Sikkim University, Gangtok 737102, India

J. M. Al-Khayri et al. (eds.), *Nanobiotechnology*,
https://doi.org/10.1007/978-3-030-73606-4_18

18.1 Introduction

Abiotic stresses (temperature, drought, heavy metal, and salinity) are regarded as one of the strong constraints for the sustainable production of crops as it diminishes the development of plant along with its yield (Kerchev et al. 2020; Iqbal et al. 2020). To surmount all these issues, the development of plants that can withstand abiotic stress can be useful to solve the food insecurity issues. Nevertheless, conventional plant breeding methods (e.g., polyploidy, hybridization, selection, mutation and so on) for increasing crop productivity has probably touched the climax and further improvements by such practices seem difficult. Even with such circumstances, the development of agriculture is essential for the elimination of hunger and poverty in developing countries (Prasad et al. 2017). Under the scenario as mentioned above, the application of nanotechnology is considered as the most promising for mitigating the constraints associated with crop productivity worldwide due to abiotic stress. The word 'nanotechnology' was introduced in 1974 by Norio Taniguchi, a professor at Tokyo University of Science (Shang et al. 2019). Nanotechnology is a developing arena of the twenty-first Century with its influence on the global economy, industry as well as the lives of people by introducing nanosensors, nano drugs, nanocarriers, quantum dots, nanotubes, nanorods, nanoparticles, and so on. Nanotechnology involves the science of understanding as well as regulation of materials possess distinctive physical characteristics which give rise to potential uses as detailed previously (Mukhopadhyay 2014; Pandey et al. 2018; Singh et al. 2018a, b; Rani et al. 2020; Singh and Porwal 2020; Singh et al. 2020; Porwal et al. 2020). The use of nanotechnology could open up innovative techniques in the arena of agriculture and plant biotechnology (Perez-de-Luque and Diego 2009). The primary agricultural uses of nanotechnology (Fig. 18.1) include soil enhancement (nanomaterials), precision farming (nanosensors), crop protection (nanopesticides), crop improvement (nanobiotechnology), crop growth (nanofertilizers), and stress tolerance (nanoparticles) (Shang et al. 2019). This chapter provides an overview on the various abiotic stresses mediated responses in plant systems with special focus on temperature including their nanotechnology based mitigation apart from the acclimatization mechanism of plant systems. It also stressed upon the uptake, translocation and accumulation of nanomaterials/nanoparticles in plants together with their uses for alleviating the temperature stress like heat and cold.

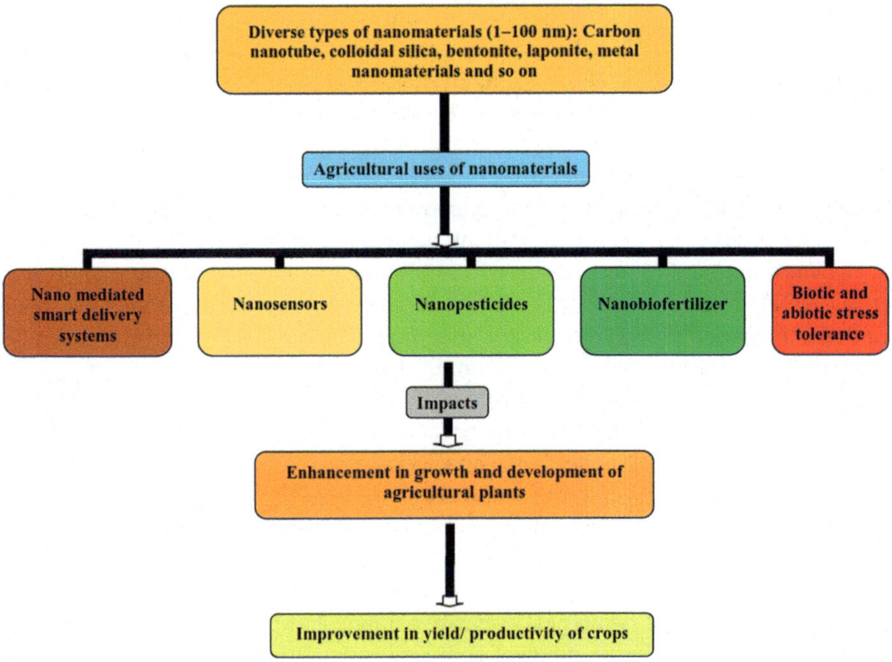

Fig. 18.1 Agricultural applications of nanotechnology (Figure constructed by AK Singh)

18.2 Nanoparticles in Plant Science

Synthetic nanoparticles can be of two types: metal- and carbon-based. Metal-based nanoparticles are divided into quantum dots, metal oxides, and metals; whereas carbon-based nanoparticles are grouped into carbon nanotubes as well as fullerenes (Peralta-Videa et al. 2011). Nanoparticles depict distinctive optical characteristics, greater surface-to-volume ratio as well as size-dependent features, which confer to them excellent biological as well as chemical properties together with physical properties (Iqbal et al. 2017). Owing to physico-chemical characteristics, it can be exploited in various arenas like chemical engineering, electronics as well as life science (Jeevanandam et al. 2018). It is worth mentioning here that the nanotechnology is used in plant system, because of the necessity to developed miniaturized effective approaches so as to overcome biotic as well as abiotic stresses including contribution towards plant protection and growth, including seed germination (Wang et al. 2016).

Metal oxide such as TiO_2, ZnO, CaO as well as MgO have been commonly introduced in plant science due to their superior light absorption, catalytic as well as electrical properties (Jahan et al. 2018). Metals like silver as well as gold have usually been used in the plant as nanoparticles for diverse applications. Their chemical synthesis utilizes harmful chemicals, and is quite costly (Rastogi et al. 2019; Viswanath and Kim 2015). Furthermore, eco-friendly methods based on the usage of plant extract in addition to ionizing radiation chemistry in aqueous solutions have also been developed (Abedini et al. 2013). Now a days, the utilization of polymeric nanoparticle is enhancing because of their cost-effective production, bio-compatibility as well as the ability to respond to exterior stimuli (Baskar et al. 2018). The shell/core nanoparticles can be produced by various combinations of materials like organic-organic, organic-inorganic materials and so on. The selection of the shell of nanoparticles was based on its utilizations as well as applications (Ghosh Chaudhuri and Paria 2012). The mesoporous silica nanoparticles with nanostructured shells have been produced from a mesoporous structure having extremely functionalizable surface area (Torney et al. 2007). Nanogels are the nano-sized, non-ionic as well as ionic hydrogels composed of natural or synthetic polymer chain, which are physically/chemically cross-linked (Molina et al. 2015; Neamtu et al. 2017; Singh et al. 2018a). They have outstanding physicochemical properties, stimuli-responsiveness (temperature and pH), high encapsulation ability of biomolecules (bio-conjugation), and colloidal stability. Nanogels have higher load capacity, degree of porosity, and water content of 70–90% of whole structure). Nanogels such as poly (N-isopropylacrylamide), poly (vinylpyrrolidone), poly (ethyleneimine), poly (ethylene oxide), poly (vinyl alcohol), alginate, chitosan and so on are usually exploited. Moreover, nanogels with hybrid structures can be produced from non-polymeric or polymeric materials. Hybrid nanogels are of two types: polymer-nanogel composites as well as nanomaterial-nanogel. The polymer-nanogel composites comprise core-shell particles, co-polymer, and interpenetrated networks, whereas nanomaterial-nanogel produced by the combination of nano-sized materials viz. carbonaceous or magnetic nanoparticles (Molina et al. 2015).

Nanomaterials of Fe, Mn, Zn, Cu, Mo, and so on may be supplemented into the soil as nano-fertilizers or may be exploited as improved delivery systems for increasing the performance as well as absorption of traditional fertilizers (Liu and Lal 2015). Although nanofertilizers are very favorable to agriculture, the exploitation of nanomaterials as fertilizer is infrequent (DeRosa et al. 2010; Sanzari et al. 2019).

18.3 Uptake, Translocation and Accumulation of Nanomaterials in Plants

The factors responsible for the uptake of nanoparticle in the plant cell include growth conditions for the plant, plant types, and age (Snehal and Lohani 2018). Moreover, the physicochemical features of the nanoparticles (such as chemical composition, size, dimension, and its stability in solution) are also responsible for the uptake, translocation, and accumulation of nanoparticles in the plant system (Rico et al. 2011; Sanzari et al. 2019; Snehal and Lohani 2018). Generally, nanoparticle intrudes into the plant root employing the lateral root connections and reaches the xylem through the cortex as well as the pericycle (Dietz and Herth 2011). The interaction between the plant system and nanoparticles is mainly based on chemical processes that result in the lipid peroxidation, oxidative damage, ion transport activity as well as generation of reactive oxygen species (ROS). Upon introduction in the plant system, nanoparticles undergo reaction with carboxyl as well as sulfhydryl groups, which alter the protein activities (Kurepa et al. 2009; Watanabe et al. 2008).

The uptake of the nutrients as well as minerals in the plants is mainly controlled by the transporter or pumps present in the cytoplasmic membrane of the roots. In several cases, the nanoparticles first bind to the carrier proteins and then pass through the ions channels, aquaporin, or engulfed by endocytosis (Snehal and Lohani 2018). Metals like silicon in its silicic acid form are mainly absorbed by the plants through diffusion (apoplastic transport); nevertheless, for symplastic transport-specific aquaporin (NIP2) is essential. Xylem is accountable towards the upward movement of silicic acid to the aerial tissue system, including shoot and the leaves (Deshmukh et al. 2013; Gregoire et al. 2012; Snehal and Lohani 2018).

The uptake or absorption of nanoparticles or its aggregate by the plant cells depends upon its size, which should be less than pore diameter (5–20 nm) so that it could easily reach to the cell membrane after passing through the cell wall (Fleischer et al. 1999; Kumar et al. 2016). The nanoparticles were taken to the plant systems once establishing complexes within membrane transporters/root exudates. The uptake of nanomaterials by the plant systems either through the stomata or base of the trichome in leaves has also been reported (Snehal and Lohani 2018). After the entry of nanoparticle into the cell membrane, further transportation take place either by apoplastic or symplastic pathways. The mobilization of nanoparticles, after absorption within the plant cell, can also be facilitated by plasmodesmata from one cell to another (Rico et al. 2011; Sanzari et al. 2019).

18.4 Abiotic Stress Response of Plant

Plant systems are continuously come across to different hostile ecological conditions like stimuli causing abiotic stress (UV radiation, waterlogging, heavy metals, heat, chilling, drought, and salinity), which is mainly responsible for decreasing the crop productivity. On an average, it led to the reduction of about 50% yield losses mostly due to high temperature (20%), low temperature (7%), salinity (10%), drought (9%) and other abiotic stresses (4%) (Kajla et al. 2015). In response to the environmental stress, ROS are produced as well as accumulates, which ultimately result in oxidative burst in plant systems (Jalil and Ansari 2019). In plants, ROS include free radical of lipid peroxidation, singlet oxygen, super anion, etc. (Mittler 2002), which increases cell toxicity (Yadav et al. 2014), damage membrane lipids and macromolecules (Foyer and Noctor 2000), and reduces plant growth (Khan et al. 2016). Therefore, to alleviate the effect of ROS, the plant systems have developed their defense mechanism called the antioxidant system by producing antioxidant enzymes like glutathione reductase, glutathione peroxidase, and peroxidase as well as non-enzymatic antioxidants viz. tocopherol and so on (Mittler 2002), which scavenge ROS (Khan and Khan 2017). Moreover, the plant produces amino acids, polyols,

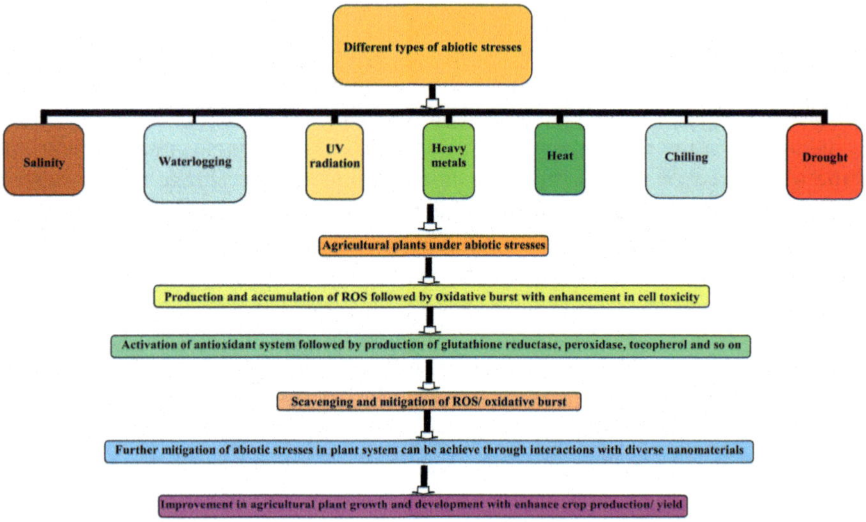

Fig. 18.2 Schematical illustration of abiotic stress response in plants (Figure constructed by AK Singh)

and trehalose, which help to maintain the osmotic level in the plant cell (Jalil and Ansari 2019). Plants accumulate organic acids, metal chelates, and polyphosphates due to heavy metal stress, which lead to requisitioning and limiting of toxic metals in the cell membrane. There are several studies, which show the importance of nanoparticles towards overcoming the harmful effect of abiotic stresses in plants and promoting its development as well as growth (Khan et al. 2017), protecting the photosynthetic system of plants (Siddiqui et al. 2014), scavenging ROS (Wei and Wang 2013), and reducing the heavy metals toxicity (Worms et al. 2012). Furthermore, the nanoparticles also modify gene expression for energy pathways in responses to abiotic as well as biotic stress, electron transport, cell organization, cell biosynthesis, and metabolism (Aken 2015; Jalil and Ansari 2019). The abiotic stress response in plant system is schematically illustrated in Fig. 18.2.

18.4.1 Temperature Stress

Temperature is the essential ecological factors responsible for the development/growth of plant systems (Ashraf et al. 2010; Monjardino et al. 2005). The plant systems are subjected to the temperature stress when the surrounding temperature is above (high temperature stress) or below (low-temperature stress) the optimum values (tolerance window) for growth. Both are detrimental to the development as well as growth of plants as they are adapted to grow in a narrow range of temperatures (Singh and Grover 2008). Moreover, the anthropogenic activities have played a vital role in escalating the global warming and climate change effects by increasing the level of atmospheric CO_2 including other greenhouse gases, which adversely affects agriculture through its direct as well as indirect consequences on crop production (Grover et al. 2013). Further, plant nanotechnology employs nanomaterials, which upon interaction with the plant parts such as cells and tissues help to improve the functions. This field has the prospective to enhance tolerance level to abiotic stresses like temperature stress of plants by implanting nanomaterials within cell organelles and photosynthetic tissues. Though substantial improvement has been made on the way to considerate the nanoparticle and plants interactions, several opportunities and challenges remain to utilize nanobiotechnology as a device to investigate and engineer plant function (Wu et al. 2017).

18.4.1.1 Heat Stress

The growth of the plants within the optimum range of temperature produces high yields. However, temperature variation due to seasonal change might expose the plant to an extended period of supra-optimal temperature, which is often higher than the optimum temperature. Such temperature causes heat stress in plants, which is detrimental to accumulation as well as growth as it inhibits photosynthesis, chlorophyll synthesis, enzyme activity, and protein synthesis, as well as increases the transpiration rate (Gibson and Paulsen 1999). Heat stress in plants causes different, and often detrimental variations in growth, development, biological processes, and yield (Hasanuzzaman et al. 2013) (Fig. 18.3).

Heat stress in plants stimulates the generation of ROS as well as creates oxidative stress resulting in ion leakage and lipid disintegration of the cell membrane, followed by the breakdown of protein molecules (Karuppanapandian et al. 2011; Moller et al. 2007; Savicka and Skute 2010) besides the reduction of chlorophyll molecules as well as photosynthetic rate (Prasad et al. 2011). Nanotechnology plays an important function towards the mitigation of heat stress. Some investigations have revealed the role of nanomaterials towards decreasing the impacts of heat stress in the plant system (Table 18.1).

Selenium, a trace element, not vital for the plant, is demonstrated as a protectant under varying abiotic stresses conditions, including heat stress (Hasanuzzaman et al. 2013). Haghighi et al. (2014) examined the impacts of nano-selenium as well as selenium upon *Lycopersicum esculentum* Mill. cv. 'Halil' (tomato) during heat stress. The concentration of selenium used was 0, 2.5, 5 and 8 μM, whereas, for nano-selenium, it was 1, 4, 8 and 12 μM. The plants were incubated at optimum (25/17 ± 2 °C day/night), then to high (40 °C for 1 day), and low (10 °C for 1 day) temperature stress followed by incubation for 10 days at optimum temperature. Plant growth was found to be promoted after the incubation at both high and low-temperature stresses with 2.5 μM of selenium. The diameter of the shoot and dry weight was increased after high-temperature stress, whereas shoot diameter, fresh, and dry weight increased after low-temperature stress. Moreover, root volume, fresh weight, and dry weight increase after both high and low-temperature stresses. Nevertheless, no effects on

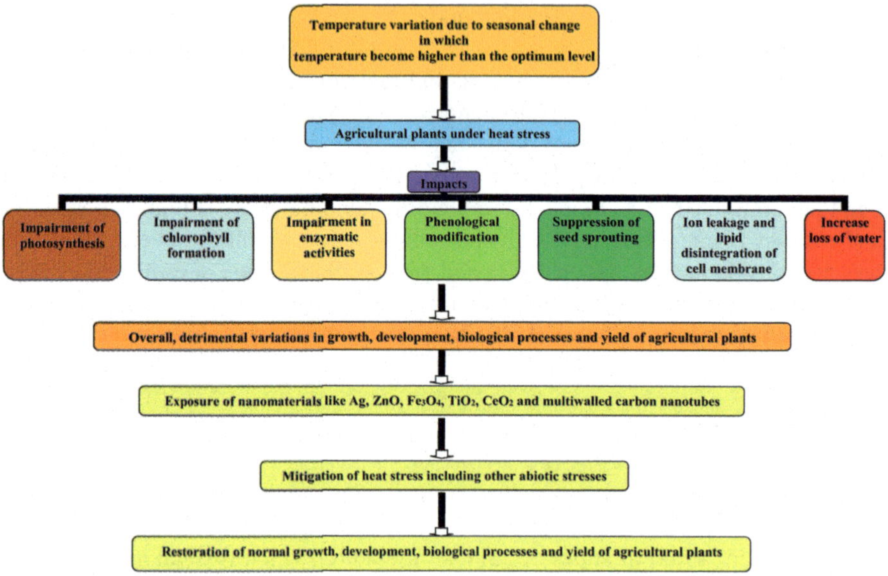

Fig. 18.3 Key impacts as well as the mitigation of heat stress in plant systems (Figure constructed by AK Singh)

plant growth were recorded with the application of nano-selenium. After incubation at low-temperature stress, the chlorophyll was increased in leaves by 19.2 and 27.5% with 2.5 and 1 μM application of selenium and nano-selenium, correspondingly. The volume of roots was also found to have increased by 33.3 and 60% after the incubation at high and low-temperature stress, correspondingly, with the application of selenium at 2.5 μM. Nano-selenium and selenium increase the water content after incubation at high as well as lower temperature stress. Therefore, nano-selenium including selenium may increase the growth features of the selected tomato plant after a small exposure to high as well as lower temperature stress.

Djanaguiraman et al. (2018) investigated the impacts of selenium nanoparticles (10–40 nm) to mitigate the influences of heat stress (38/28 °C day/night). The selenium nanoparticles were applied in sorghum via a foliar spray of 10 mg L^{-1} in the course of booting phase. It stimulates the antioxidant defense system through enhancing the antioxidant enzymatic activities. Selenium nanoparticles are responsible for enabling the increment of unsaturated phospholipids as well as decreasing the oxidants concentrations. Under heat stress, selenium nanoparticles enhanced the pollen germination percentage, responsible for increased seed quantity considerably. The high amount of Se nanomaterials induces oxidative stress in plant systems, while a lower level of Se nanomaterials serves like antioxidant (Hartikainen et al. 2000; Jalil and Ansari 2019).

The TiO_2 nanomaterials were found to depict different impacts on the biochemical, physiological as well as morphological characteristics of several plant species

Table 18.1 Impact of different nanomaterials towards mitigation of heat stress in plants

Plant species	Nanomaterial used	Stress response	References
Tomato (*Lycopersicon esculentum* L.)	Multi-walled carbon nanotubes	Upregulated the expression of different stress linked genes together with HSP90	Khodakovskaya et al. (2011)
Maize (*Zea mays* L.)	CeO_2	Enhanced the production of H_2O_2 and upregulation of HSP70	Zhao et al. (2012)
Tomato (*L. esculentum* L.)	TiO_2	Enhanced photosynthesis by regulating energy dissipation triggered cooling of leaves by stimulating stomatal opening	Qi et al. (2013)
Tomato (*L. esculentum* L.)	Selenium	Increase chlorophyll content as well as increase hydration ability	Haghighi et al. (2014)
Wheat (*Triticum aestivum* L.)	Silver	Promotes morphological growth in plant which includes root length and shoots length. Likewise, substantial enhancement in leaf area and leaf number was also observed	Iqbal et al. (2017)
Sorghum (*Sorghum Bicolor* (L.) Moench)	Selenium	Promotes antioxidant defense, improved the pollen germination percentage as well as improved seed quantity	Djanaguiraman et al. (2018)
Wheat (*T. aestivum* L.)	ZnO Fe_3O_4	Yield quantity increased and promotes antioxidant defense system	Hassan et al. (2018)

(Gohari et al. 2020). The exploitation of TiO_2 nanoparticles enhanced chlorophyll formation, photosynthetic rate, antioxidant enzymes as well as rubisco activities that consequently enhanced crop production (Lei et al. 2008). Qi et al. (2013) studied the impacts of TiO_2 nanoparticles on photosynthetic rate of a tomato crop in mild heat stress. The study revealed an enhancement towards the transpiration and photosynthetic rate as well as water conductance after the application of TiO_2 nanoparticles (0.05 g L^{-1}). Moreover, its use also decreases the relative electron transport and minimum chlorophyll fluorescence substantially. The TiO_2 nanoparticles reduce the non-regulated PS II energy dissipation; however, it enhances the regulated PS II energy dissipation (Qi et al. 2013).

Iqbal et al. (2017) studied the consequence of silver nanomaterials for growth in the wheat crop during heat stress. Silver nanomaterials have been synthesized by means of *Moringa oleifera* Lam. extract. The silver nanoparticles were applied at the trifoliate stage to the wheat plant at various concentrations (25, 50, 75, and 100 mg L^{-1}). The heat stress (35–40 °C for 3 h day^{-1}) was given for 3 days resulting in a reduction of leaf number, dry leaf weight, fresh leaf weight, leaf area, root number, shoot length, root length, dry weight, and fresh weight by 2, 0.01, 0.02, 12.1, 1.8, 6.2, 2.5, 0.16 as well as 1.2%, correspondingly. While the application of silver nanoparticles results in an increase of growth at all concentrations used; nevertheless, significant results were recorded at the level of 50 and 75 mg L^{-1} silver nanoparticles in heat stress. Conversely, the use of silver nanoparticles defends wheat plant systems towards heat stress and increases plant dry weight (0.36 and 0.60%), fresh plant weight (1.3 and 2%), root number (6.6 and 7.5%), shoot length (22.2 and 26.1%), and plant root length (5 and 5.4%) and in 50 and 75 mg L^{-1} of silver nanoparticles, correspondingly compared to control. Similarly, a notable increase in leaf dry weight (0.06 and 0.18%), fresh leaf weight (0.09 and 0.15%), leaf number (4 and 4.8%), and leaf area (18.3 and 33.8%) has been observed in 50 and 75 mg L^{-1} of silver nanoparticles over control.

Hassan et al. (2018) stated that the cultivation of two wheat cultivars Gimmeza7 (heat sensitive) and Sids1 (heat tolerant) in El Wadi El Gadeed governorate, Egypt (a hot climate area) with zinc oxide (80 nm) with several concentrations (0, 1.40, 2.80, 4.20, 5.61, and 56.06 mg L^{-1}) and iron oxide (50 nm) nanoparticles with several concentrations (0, 1.31, 2.62, 3.94, 5.25, and 52.50 mg L^{-1}) were conducted. The best survival of wheat plant systems in heat stress environments was detected at 1.40 and 52.50 mg L^{-1} of iron oxide and zinc oxide nanoparticles, respectively, by Gimmeza7 cultivar in terms of yield quantity of grain. The increasing effect of such nanoparticles during heat stress was related to increasing antioxidant enzyme activities.

A plant produces various heat shock proteins (10–200 KD), characterized as molecular chaperones, during heat stress (Al-Whaibi 2011). Heat shock proteins are involved in heat stress resistance as well as support other proteins in sustaining their constancy in stress environments (Wahid 2007). It is found that multiwall carbon nanotubes capable to upregulate the gene expression responsible for the production of heat shock proteins, e.g., HSP90 in tomato (Khodakovskaya et al. 2011). Moreover, the application of CeO_2 nanoparticles in maize causes upregulation of HSP70 and excessive production of hydrogen peroxide (Zhao et al. 2012).

18.4.1.2 Cold and Freezing Stress

Like heat stress, damages in the plant cells can also be caused by very low-temperatures (Hasanuzzaman et al. 2013) (Fig. 18.4). The temperature rang 0–15 °C is cool enough to cause damage without developing ice crystals in plant cells (Hasanuzzaman et al. 2013), whereas, freezing stress is caused at the temperature below 0 °C. It causes permanent injury to plant systems mainly by mechanical forces generated due to the development of the high concentration of intracellular salts, cellular dehydration, and extracellular ice crystals (Ashraf et al. 2010). The

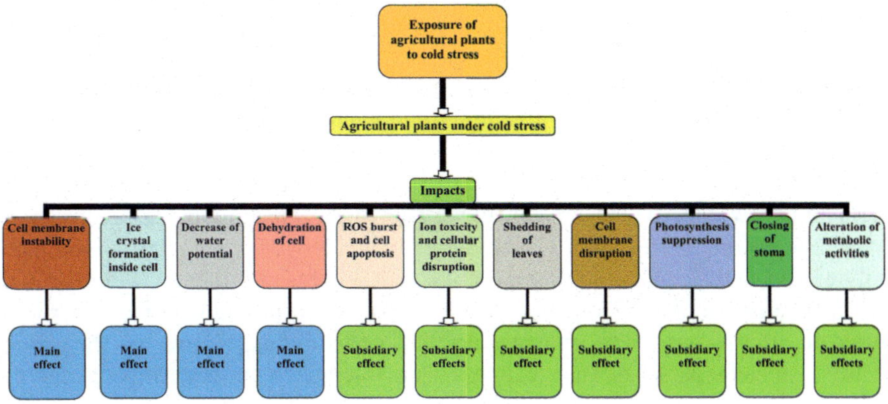

Fig. 18.4 Major impact of cold stress in plants (Figure constructed by AK Singh)

generation of ROS increases in plant cells because of freezing stress, which elevated lipid peroxidation results in membrane damage and developed injury symptoms from freezing (McKersie et al. 1993) and chilling stress (Senaratna et al. 1988). At lower temperatures, the fluidity of the cell membrane was generally affected, which changes its structure (Marschner 1995). The leakages of the ion, as well as alteration of membrane permeability, are the significant consequences of chilling stress, which adversely disturb the growth as well as germination of the plant (Suzuki et al. 2008; Welti et al. 2002). At low-temperature stress, absorbed light energy surpasses the ability of chloroplasts to exploit it for CO_2 fixation, and the surplus energy is instead utilized towards triggering of O_2 into ROS (Ashraf et al. 2010). Moreover, the susceptibility to freezing stress varies from species to species; the plants with better tolerance capacity slow less injury of the cell membrane over sensitive plants (Jalil and Ansari 2019).

Photosynthesis is the essential process of a plant, which is found to be vulnerable under low-temperature stress. It has been reported that CO_2 fixation, transpiration rate, chlorophyll content, rubisco activity and so on are the chief targets impaired in plants under chilling stress (Ashraf et al. 2010; Liu et al. 2012; Yordanova and Popova 2007). Further, it was demonstrated that the upregulation of *Me*APX2 as well as *MeCu*/*Zn*SOD genes occurs during chilling stress and enhanced the activities of glutathione reductase, dehydroascorbate reductase as well as monodehydroascorbate reductase, thereby scavenging ROS that could lead to the repression of oxidative stress via H_2O_2 generation, chlorophyll degradation, and lipid peroxidation to confirmed stress tolerance (Xu et al. 2014). Nanoparticles have capability for mitigating the harmful impacts of chilling environment through decreasing membrane damage as well as leakage of elec-trolyte (Mohammadi et al. 2013). Nanoparticles increase the protein genes expression responsible for rubisco- and chlorophyll-binding (Hasanpour et al. 2015), ascorbate peroxidase, catalase as well as superoxide dismutase activities (Mohammadi et al. 2014) preserve the steadiness of chlorophyll as well as carotenoid molecules together with increasing tolerance towards low-temperature stress (Hasanpour et al. 2015). The use of nanomaterials improves the growth, physiological as well as biochemical characteristics of plant systems subjected to cold stress (Haghighi et al. 2014; Kohan-Baghkheirati and Geisler-Lee 2015). Various studies have reported the use of nanomaterials towards mitigation of injuries triggered through low-temperature stress (Table 18.2).

To surmount the problems associated with the low-temperature stress in plants, Mohammadi et al. (2013) investigate the impact of TiO_2 nanoparticles on alterations in membrane impairment index such as malondialdehyde accumulation and electrolyte leakage throughout low-temperature (4 °C) stress in a tolerant (Sel 11439) as well as a sensitive (ILC 533) genotypes of chickpea. Accumulation of nanoparticles inside the chloroplast and vacuole revealed that, under thermal treatments, the permeability to the nanoparticles was much less in the tolerant one over sensitive genotype.

Table 18.2 Effect of nanomaterials towards alleviation of cold stress in plants

Plant species	Nanomaterial	Stress response	References
Rice (*Oryza sativa* L.)	SiO_2	Promotes the root length, root volume, as well as dry weight of shoot/root	Adhikari et al. (2013)
Chickpea (*Cicer arietinum* L.)	TiO_2	Diminished H_2O_2 concentration/electrolyte leakage and improved antioxidative enzymatic activities including increased TiO_2 gathering in sensitive genotype over tolerant one	Mohammadi et al. (2013, 2014)
Wheatgrass (*Agropyron elongatum* L.)	ZnO	Alleviate seed dormancy, promotes seed germination as well as increased seedling weight	Azimi et al. (2014)
Tomato (*Lycopersicon esculentum* Mill.)	Selenium	Promotes leaf-relative water contents, chlorophyll, and plant growth during the heat and cold stress	Haghighi et al. (2014)
Chickpea (*Cicer arietinum* L.)	TiO_2	Increased expression level of chlorophyll- and rubisco-binding protein genes as well as decreased H_2O_2 content	Hasanpour et al. (2015)
Arabidopsis (*Arabidopsis thaliana* (L.) Heynh.)	Ag	Stimulated as well as enhanced antioxidant genes such as *MeCu/Zn*SOD as well as *Me*APX2, 35% of identical genes were controlled by both cold stress together with Ag nanoparticles	Kohan-Baghkheirati and Geisler-Lee (2015)
Arabidopsis (*Arabidopsis thaliana* (L.) Heynh.)	CeO_2	Improve the photosynthesis by facilitating higher rubisco carboxylation rates	Wu et al. (2017)

During low-temperature stress, the TiO_2 content was significantly increased in sensitive genotype in comparison to the optimum temperature. Moreover, the treatment of TiO_2 nanoparticles prevented membrane damage and oxidative damage in chickpea under low-temperature stress. Further, Mohammadi et al. (2014) established the previous findings with suggestion that TiO_2 nanoparticles help to enhance the resistance of chickpea plant systems under cold stress through stimulating the defense systems as well as decreasing the injuries.

Similarly, Hasanpour et al. (2015) studied the role of TiO_2 nanomaterials on molecular as well as metabolic traits took part in photosynthesis of two genotypes of chickpea such as cold susceptible (ILC533), and cold tolerant (Sel96Th11439) genotypes during low-temperature stress (4 °C). It states that H_2O_2 content was significant increase in cold susceptible as compared to tolerant ones under low-temperature stress. TiO_2 nanoparticles decrease the H_2O_2 content considerably subsequently tolerant genotype exhibited lesser H_2O_2 content with the higher metabolic potential for photosynthesis in comparison to susceptible genotype. TiO_2 nanoparticles substantially promote the rubisco activity, in contrast, to control while its activity reduced considerably under low temperature as compared to optimum temperature. Besides decreasing H_2O_2, particularly in plants treated with TiO_2 nanoparticles, greater photosynthetic activity observed at the transcription levels of *Cachlorophyll a/b*, *CaSRubisco*, as well as *CaLRubisco-* binding protein genes in a synchronized fashion confirm the adaptation of plants to recovery or survival. Under such a condition, the activities of phosphoenolpyruvate carboxylase increased mainly in tolerant plants in comparison to susceptible genotype along with plants supplemented with TiO_2 nanoparticles over control plants, signifying possibly an enhancement in energy effectiveness via diverse mechanisms such as malate. Accordingly, the alleviation of cold stress after the treatment with TiO_2 nanoparticles in chickpea plants occurs by changed metabolism for plant growth and improves the burden of damage that arises due to low-temperature stress.

Mo and SiO_2 nanoparticles are some of the frequently used engineered oxide nanoparticles. The effects of Mo (<100 nm) and SiO_2 (10–20 nm) nanoparticles on germination of rice seed were investigated by Adhikari et al. (2013). In the presence of nanoparticles rice seeds were germinated well. The optimal growth was detected under 101.4 mg L^{-1} of Mo nanoparticles as well as 106 mg L^{-1} of SiO_2 nanoparticles for rice plants. SiO_2 nanoparticles had revealed no toxic effect on the growth of rice, whereas Mo nanoparticles detained elongation and growth of root after 50 mg L^{-1}. In several cases, root necrosis has occurred because of toxicity due to the accumulation of Mo nanoparticles at a higher amount into the root system. Rice seedlings absorbed both Mo and SiO_2 nanoparticles. The enhancement in the root, shoot dry weight, root volume as well as length of the rice crop observed with SiO_2 nanoparticle. Therefore, this investigation revealed that straightforward exposure to particular kinds of nanomaterials produced both negative and positive consequences towards growth of plant.

Azimi et al. (2014) evaluated the effects of SiO_2 nanoparticles with different concentrations (0, 5, 20, 40, 60 and 80 mg L^{-1}) on seed germination as well as seedling growth of tall wheatgrass by applying three different treatments (control, seed pre-chilling before SiO_2 nanomaterials treatment, and application of seed with SiO_2 nanoparticles before pre-chilling). The use of SiO_2 nanomaterials was found to considerably enhance the germination of seed from 58% in control to 86.3 and 85.7% in 40 and 60 mg L^{-1}, correspondingly. Moreover, the treatment of SiO_2 nanomaterials enhanced dry weight of seedling, shoot, and root of tall wheatgrass. Nevertheless, seedling weight increased nearly 49% over control by increasing the nanoparticles concentration in the range of 0–40 mg L^{-1}; however, it decreases under supplementation of 60 and 80 mg L^{-1}. Therefore, the combination of SiO_2 nanoparticles and pre-chilling of seed in tall wheatgrass mostly overcomes the seed dormancy.

Kohan-Baghkheirati and Geisler-Lee (2015) reported the activation and enrichment of antioxidant genes (such as *MeCu/Zn*SOD as well as *Me*APX2) and 35% of identical genes were controlled by both cold stress as well as Ag nanoparticles in *Arabidopsis*. Wu et al. (2017) demonstrated that CeO_2 nanoparticles (nanoceria) promotes photosynthesis as well as ROS scavenging in *Arabidopsis thaliana* (L.) Heynh. under excess light (2000 μmol m^{-2} s^{-1} for 1.5 h), heat (35 °C for 2.5 h), and dark chilling (4 °C for 5 days). Poly (acrylic acid) nanoceria are transported via non-endocytic pathways into chloroplasts. PNC with a low Ce^{3+}/Ce^{4+} ratio (35%) decreases leaf ROS levels by 52%, including hydroxyl radicals, superoxide anion, and hydrogen peroxide. Plants entrenched with poly (acrylic acid) nanoceria were exposed to abiotic stress, display a rise up to 61% in rubisco carboxylation, 67% in carbon assimilation, and 19% in quantum yield of PS II in comparison to plants without nanoparticles. However, poly (acrylic acid) nanoceria with high Ce^{3+}/Ce^{4+} ratio (60.8%) escalates ROS levels of leaf and does not defend the photosynthesis from oxidative damage during abiotic stress.

18.5 Conclusions and Prospects

In the arena of plant science, nanotechnology has improved crop production though it is still in the nascent phase of its development. Nanomaterials reduce the damage of plants due to temperature stress through triggering the defense plant system. The nanoparticles can easily penetrate the plant cell due to its small size and regulate ion channels that promote seed germination as well as growth. The large surface area of nanoparticle supports high absorption and the transport of molecules. Conversely, nanomaterials are also caused the generation of ROS that result in phyto-toxicity. Thus the perspective of nanotechnology with reduced agricultural risks motivates another green revolution. Though, there are still vast gaps in our understanding of the eco-toxicity, permissible limit, and uptake capacity of various nanomaterials. Therefore, to gain a clear-cut understanding of the nanoparticle interaction with the plants and environments, there is an urgent need to thoroughly conduct experimental work/investigations on more plant species in different agro-climatic zones with different nanomaterials. Such extensive investigations will be helpful to understand their responses of alleviating abiotic stress like temperature stress. This also assist to develop a regulatory framework established on the various research evidence, which will limit mankind's exposure to undesirable bioengineered nanoparticles to a harmless level. With the remarkable applications of nanomaterials, we can make an optimistic prospect of nanobiotechnology with not only the better understanding of their ecotoxicity, but also all the aspects like reutilizing, feasibility, manufacturing, and framework of policy to handle them securely and utilize them in an eco-friendly manner.

References

Abedini A, Daud AR, Hamid MAA et al (2013) A review on radiation-induced nucleation and growth of colloidal metallic nanoparticles. Nanoscale Res Lett 8:1–10. https://doi.org/10.1186/1556-276X-8-474

Adhikari T, Kundu S, Rao AS (2013) Impact of SiO2 and Mo nano particles on seed germination of rice (*Oryza Sativa* L.). Int J Res Agric Food Sci 4(8):809–816

Aken BV (2015) Gene expression changes in plants and microorganisms exposed to nanomaterials. Curr Opin Biotechnol 33:206–219. https://doi.org/10.1016/j.copbio.2015.03.005

Al-Whaibi MH (2011) Plant heat-shock proteins: a mini review. J King Saud Univ Sci 23(2):139–150. https://doi.org/10.1016/j.jksus.2010.06.022

Ashraf M, Afzal M, Ahmad R et al (2010) Silicon management for mitigating abiotic stress effects in plants. Plant Stress 4(2):104–114

Azimi R, Borzelabad MJ, Feizi H et al (2014) Interaction of SiO2 nanoparticles with seed prechilling on germination and early seedling growth of tall wheatgrass (*Agropyron elongatum* L.). Pol J Chem Tech 16:25–29. https://doi.org/10.2478/pjct-2014-0045

Baskar V, Meeran S, Shabeer TK et al (2018) Historic review on modern herbal nanogel formulation and delivery methods. Int J Pharm Pharm Sci 10:1–10. https://doi.org/10.22159/ijpps.2018v10i10.23071

DeRosa MC, Monreal C, Schnitzer M et al (2010) Nanotechnology in fertilizers. Nature Nanotechnol 5:91. https://doi.org/10.1038/nnano.2010.2

Deshmukh RK, Vivancos J, Guérin V et al (2013) Identification and functional characterization of silicon transporters in soybean using comparative genomics of major intrinsic proteins in Arabidopsis and rice. Plant Mol Biol 83:303–315. https://doi.org/10.1007/s11103-013-0087-3

Dietz KJ, Herth S (2011) Plant nanotoxicology. Trends Plant Sci 16(11):582–589. https://doi.org/10.1016/j.tplants.2011.08.003

Djanaguiraman M, Belliraj N, Bossmann SH et al (2018) High-temperature stress alleviation by selenium nanoparticle treatment in grain sorghum. ACS Omega 3(3):2479–2491. https://doi.org/10.1021/acsomega.7b01934

Fleischer A, Neill MA, Ehwald R (1999) The pore size of non-graminaceous plant cell walls is rapidly decreased by borate ester cross-linking of the pectic polysaccharide rhamnogalacturonan II. Plant Physiol 121:829–838. https://doi.org/10.1104/pp.121.3.829

Foyer CH, Noctor G (2000) Oxygen processing in photosynthesis: regulation and signalling. New Phytol 146:359–388. https://doi.org/10.1046/j.1469-8137.2000.00667.x

Ghosh Chaudhuri R, Paria S (2012) Core/shell nanoparticles: classes, properties, synthesis mechanisms, characterization, and applications. Chem Rev 112:2373–2433. https://doi.org/10.1021/cr100449n

Gibson LR, Paulsen GM (1999) Yield components of wheat grown under high temperature sress during reproductive growth. Crop Sci 39:1841–1846. https://doi.org/10.2135/cropsci1999.3961841x

Gohari G, Mohammadi A, Akbari A et al (2020) Titanium dioxide nanoparticles (TiO2 NPs) promote growth and ameliorate salinity stress effects on essential oil profile and biochemical attributes of

Dracocephalum moldavica. Scientific Reports 10:912. https://doi.org/10.1038/s41598-020-57794-1

Gregoire C, Remus-Borel W, Vivancos J et al (2012) Discovery of a multigene family of aquaporin silicon transporters in the primitive plant *Equisetum arvense*. Plant J 72:320–330. https://doi.org/10.1111/j.1365-313X.2012.05082.x

Grover A, Mittal D, Negi M et al (2013) Generating high temperature tolerant transgenic plants: achievements and challenges. Plant Sci 205–206:38–47. https://doi.org/10.1016/j.plantsci.2013.01.005

Haghighi M, Abolghasemi R, Teixeira da Silva JA (2014) Low and high temperature stress affect the growth characteristics of tomato in hydroponic culture with Se and nano-Se amendment. Sci Hortic 178:231–240. https://doi.org/10.1016/j.scienta.2014.09.006

Hartikainen H, Xue T, Piironen V (2000) Selenium as an anti-oxidant and pro-oxidant in ryegrass. Plant Soil 225:193–200. https://doi.org/10.1023/A:1026512921026

Hasanpour H, Maali-Amiri R, Zeinali H (2015) Effect of TiO2 nanoparticles on metabolic limitations to photosynthesis under cold in chickpea. Russ J Plant Physiol 62:779–787. https://doi.org/10.1134/S1021443715060096

Hasanuzzaman M, Nahar K, Alam MM et al (2013) Physiological, biochemical, and molecular mechanisms of heat stress tolerance in plants. Int J Mol Sci 14:9643–9684. https://doi.org/10.3390/ijms14059643

Hassan NS, Din TASE, Hendawey MH et al (2018) Magnetite and zinc oxide nanoparticles alleviated heat stress in wheat plants. Curr Nanomate 3:32–43. https://doi.org/10.2174/2405461503666180619160923

Iqbal M, Raja NI, Mashwani Z et al (2017) Effect of silver nanoparticles on growth of wheat under heat stress. Iran J Sci Technol Trans Sci 43:387–395. https://doi.org/10.1007/s40995-017-0417-4

Iqbal MS, Singh AK, Ansari MI (2020) Effect of drought stress on crop production. In: Rakshit A, Singh H, Singh A, Singh U, Fraceto L (eds) New frontiers in stress management for durable agriculture. Springer, Singapore, pp 35–47

Jahan S, Alias YB, Bakar AF et al (2018) Toxicity evaluation of ZnO and TiO2 nanomaterials in hydroponic red bean (*Vigna angularis*) plant: physiology, biochemistry and kinetic transport. J Environ Sci 72:140–152. https://doi.org/10.1016/j.jes.2017.12.022

Jalil SU, Ansari MI (2019) N. Nanoparticles and abiotic stress tolerance in plants: synthesis, action, and signaling mechanisms. In: Khan MIR, Reddy PS, Ferrante A et al (eds) Plant signaling molecules. Woodhead Publishing, pp 549–561

Jeevanandam J, Barhoum A, Chan YS et al (2018) Review on nanoparticles and nanostructured materials: history, sources, toxicity and regulations. Beilstein J Nanotechnol 9:1050–1074. https://doi.org/10.3762/bjnano.9.98

Kajla M, Yadav VK, Khokhar J et al (2015) Increase in wheat production through management of abiotic stresses: a review. J Appl Nat Sci 7:1070–1080

Karuppanapandian T, Wang HW, Prabakaran N et al (2011) 2,4-dichlorophenoxyacetic acid-induced leaf senescence in mung bean (*Vigna radiate* L. Wilczek) and senescence inhibition by co-treatment with silver nanoparticles. Plant Physiol Biochem 49:168–217. https://doi.org/10.1016/j.plaphy.2010.11.007

Kerchev P, van der Meer T, Sujeeth N et al (2020) Molecular priming as an approach to induce tolerance against abiotic and oxidative stresses in crop plants. Biotechnol Adv 40:107503. https://doi.org/10.1016/j.biotechadv.2019.107503

Khan MIR, Khan N (2017) Reactive oxygen species and antioxidant system in plants: role and regulation under abiotic stress. Springer Nature, New York

Khan MIR, Khan NA, Masood A et al (2016) Hydrogen peroxide alleviates nickel-inhibited photosynthetic responses through increase in use-efficiency of nitrogen and sulfur, and glutathione production in mustard. Front. Plant Sci. 7:44. https://doi.org/10.3389/fpls.2016.00044

Khan MN, Mobin M, Abbas ZK et al (2017) Role of nanomaterials in plants under challenging environments. Plant Physiol Biochem 110:194–209. https://doi.org/10.1016/j.plaphy.2016.05.038

Khodakovskaya MV, de Silva K, Nedosekin DA et al (2011) Complex genetic, photo thermal, and photo acoustic analysis of nanoparticle-plant interactions. Proc Natl Acad Sci USA 108(3):1028–1033. https://doi.org/10.1073/pnas.1008856108

Kohan-Baghkheirati E, Geisler-Lee J (2015) Gene expression, protein function and pathways of *Arabidopsis thaliana* responding to silver nanoparticles in comparison to silver ions, cold, salt, drought, and heat. Nano 5:436–467. https://doi.org/10.3390/nano5020436

Kumar V, Tiwari P, Krishnia L et al (2016) Green route synthesis of silicon/silicon oxide from bamboo. Adv Mater Lett 7(4):271–276. https://doi.org/10.5185/amlett.2016.6151

Kurepa J, Paunesku T, Vogt S et al (2009) Uptake and distribution of ultrasmall anatase TiO2 alizarin red S nanoconjugates in *Arabidopsis thaliana*. Nano Lett 10(7):2296–2302. https://doi.org/10.1021/nl903518f

Lei Z, Mingyu S, Xiao W (2008) Antioxidant stress is promoted by nano-anatase in spinach chloroplasts under UV-B radiation. Biol Trace Elem Res 121(1):69–79. https://doi.org/10.1007/s12011-007-8028-0

Liu R, Lal R (2015) Potentials of engineered nanoparticles as fertilizers for increasing agronomic productions. Sci Total Environ 514:131–139. https://doi.org/10.1016/j.scitotenv.2015.01.104

Liu YF, Qi MF, Li TL (2012) Photosynthesis, photoinhibition, and antioxidant system in tomato leaves stressed by low night temperature and their subsequent recovery. Plant Sci 196:8–17. https://doi.org/10.1016/j.plantsci.2012.07.005

Marschner H (1995) Mineral nutrition of higher plants, 2nd edn. Academic Press, London, UK, pp 258–316

McKersie BD, Chen Y, de Beus M et al (1993) Superoxide dismutase enhances tolerance of freezing stress in Transgenic Alfalfa (*Medicago sativa* L.). Plant Physiol 103(4):1155–1163.https://doi.org/10.1104/pp.103.4.1155

Mittler R (2002) Oxidative stress, antioxidants and stress tolerance. Trends Plant Sci 7:405–410. https://doi.org/10.1016/s1360-1385(02)02312-9

Mohammadi R, Amiri RM, Abbasi A (2013) Effect of TiO2 nanoparticles on chickpea response to cold stress. Biol Trace Elem Res 152:403–410. https://doi.org/10.1007/s12011-013-9631-x

Mohammadi R, Amiri NM, Mantri L (2014) Effect of TiO2 nanoparticles on oxidative damage and antioxidant defense systems in chickpea seedlings during cold stress. Russ J Plant Physiol 61:768–775. https://doi.org/10.1134/S1021443714050124

Molina M, Asadian-Birjand M, Balach J et al (2015) Stimuli-responsive nanogel composites and their application in nanomedicine. Chem Soc Rev 44:6161–6186. https://doi.org/10.1039/C5CS00199D

Moller IM, Jensen PE, Hansson A (2007) Oxidative modifications to cellular components in plants. Ann Rev Plant Biol 58:459–481. https://doi.org/10.1146/annurev.arplant.58.032806.103946

Monjardino P, Smith AG, Jones RJ (2005) Heat stress effects on protein accumulation of maize endosperm. Crop Sci 45:1203–1210. https://doi.org/10.2135/cropsci2003.0122

Mukhopadhyay SS (2014) Nanotechnology in agriculture: prospects and constraints. Nanotechnol Sci Appl 7:63–71. https://doi.org/10.2147/NSA.S39409

Neamtu I, Rusu AG, Diaconu A et al (2017) Basic concepts and recent advances in nanogels as carriers for medical applications. Drug Deliv 24:539–557. https://doi.org/10.1080/10717544.2016.1276232

Pandey B, Singh AK, Singh SP (2018) Nanoparticles mediated gene knock out through miRNA replacement: recent progress and challenges. In: Mishra RK, Thomas S, Mohapatra S, Dasgupta N, Ranjan S (eds) Applications of targeted nano-drugs and delivery systems. Elsevier, Amsterdam, Netherlands, pp 469–497

Peralta-Videa JR, Zhao L, Lopez-Moreno ML et al (2011) Nanomaterials and the environment: a review for the biennium 2008–2010. J Hazard Mater 186:1–15. https://doi.org/10.1016/j.jhazmat.2010.11.020

Perez-de-Luque A, Diego R (2009) Nanotechnology for parasitic plant control. Pest Manag Sci 65:540–545. https://doi.org/10.1002/ps.1732

Porwal P, Porwal S, Singh SP, Husen A, Singh AK (2020) Improving futuristic nanomaterial researches in forestry sector: an overview. In: Husen A, Jawaid M (eds) Nanomaterials for agriculture and forestry applications. Elsevier, Amsterdam, Netherlands, pp 505–518

Prasad PVV, Pisipati SR, Momcilovic I et al (2011) Independent and combined effects of high temperature and drought stress during grain filling on plant yield and chloroplast EF-Tu expression in spring wheat. J Agron Crop Sci 197(6):430–441. https://doi.org/10.1111/j.1439-037X.2011.00477.x

Prasad R, Bhattacharyya A, Nguyen QD (2017) Nanotechnology in sustainable agriculture: recent developments, challenges, and perspectives. Front Microbiol 8:1014. https://doi.org/10.3389/fmicb.2017.01014

Qi M, Liu Y, Li T (2013) Nano-TiO2 improves the photosynthesis of tomato leaves under mild heat stress. Biol Trace Elem Res 156:323–328. https://doi.org/10.1007/s12011-013-9833-2

Rani H, Singh SP, Yadav TP et al (2020) In-vitro catalytic, antimicrobial and antioxidant activities of bioengineered copper quantum dots using *Mangifera indica* (L.) leaf extract. Mater Chem Phys 239:122052

Rastogi A, Tripathi DK, Yadav S et al (2019) Application of silicon nanoparticles in agriculture. 3 Biotech 9:90. https://doi.org/10.1007/s13205-019-1626-7

Rico CM, Majumdar S, Duarte-Gardea M et al (2011) Interaction of nanoparticles with edible plants and their possible implications in the food chain. J Agric Food Chem 59:3485–3498. https://doi.org/10.1021/jf104517j

Sanzari I, Leone A, Ambrosone A (2019) Nanotechnology in plant science: to make a long story short. Front Bioeng Biotechnol 7:120. https://doi.org/10.3389/fbioe.2019.00120

Savicka M, Skute N (2010) Effects of high temperature on malondialdehyde content, superoxide production and growth changes in wheat seedlings (*Triticum aestivum* L.). Ekologija 56:26–33. https://doi.org/10.2478/v10055-010-0004-x

Senaratna T, Mackay CE, McKersie BD et al (1988) Uniconazole-Induced Chilling Tolerance in Tomato and its Relationship to Antioxidant Content. Plant Physiol 133(1):56–61. https://doi.org/10.1016/S0176-1617(88)80084-1

Shang Y, Hasan MK, Ahammed GJ et al (2019) Applications of Nanotechnology in plant growth and crop protection: a review. Molecules 24(14):2558. https://doi.org/10.3390/molecules24142558

Siddiqui MH, Al-Whaibi MH, Faisal M et al (2014) Nano-silicon dioxide mitigates the adverse effects of salt stress on *Cucurbita pepo* L. Environ Toxicol Chem 33(11):2429–2437. https://doi.org/10.1002/etc.2697

Singh A, Grover A (2008) Genetic engineering for heat tolerance in plants. Physiol Mol Biol Plants 14(1–2):155–166. https://doi.org/10.1007/s12298-008-0014-2

Singh AK, Porwal P (2020) Nano-technology as potential and innovative platform toward waste water treatment: an overview. In: Bhushan I, Singh VK, Tripathi DK (eds) Nanomaterials and environmental biotechnology. Springer, Switzerland, pp 201–220

Singh AK, Pal P, Gupta V et al (2018a) Green synthesis, characterization and antimicrobial activity of zinc oxide quantum dots using *Eclipta alba*. Mater Chem Phys 203:40–48

Singh AK, Yadav TP, Pandey B et al (2018b) Engineering nanomaterials for smart drug release: recent advances and challenges. In: Mishra RK, Thomas S, Mohapatra S, Dasgupta N, Ranjan S (eds) Applications of targeted nano-drugs and delivery systems. Elsevier, Amsterdam, Netherlands, pp 411–449

Singh SP, Ansari MI, Pandey B et al (2020) Recent Trends and Advancement towards phyto-mediated fabrication of noble metallic nanomaterials: focus on silver, gold, platinum and palladium. In: Bhushan I, Singh VK, Tripathi DK (eds) Nanomaterials and environmental biotechnology. Springer, Switzerland, pp 87–105

Snehal S, Lohani P (2018) Silica nanoparticles: Its green synthesis and importance in agriculture. J Pharmacogn Phytochem 7(5):3383–3393

Suzuki K, Nagasuga K, Okada M (2008) The chilling injury induced by high root temperature in the leaves of rice seedlings. Plant Cell Physiol 49(3):433–442. https://doi.org/10.1093/pcp/pcn020

Torney F, Trewyn BG, Lin VSY et al (2007) Mesoporous silica nanoparticles deliver DNA and chemicals into plants. Nat Nanotechnol 2:295–300. https://doi.org/10.1038/nnano.2007.108

Viswanath B, Kim S (2015) Influence of nanotoxicity on human health and environment: the alternative strategies. Rev Environ Contam Toxicol 240:77. https://doi.org/10.1007/398

Wahid A (2007) Physiological implications of metabolites biosynthesis in net assimilation and heat stress tolerance of sugarcane (*Saccharum officinarum*) sprouts. J Plant Res 120:219–228. https://doi.org/10.1007/s10265-006-0040-5

Wang P, Lombi E, Zhao FJ et al (2016) Nanotechnology: a new opportunity in plant sciences. Trends Plant Sci 21:699–712. https://doi.org/10.1016/j.tplants.2016.04.005

Watanabe T, Misawa S, Hiradate S et al (2008) Root mucilage enhances aluminum accumulation in *Melastoma malabathricum*, an aluminum accumulator. Plant Signal Behav 3(8):603–605. https://doi.org/10.4161/psb.3.8.6356

Wei H, Wang E (2013) Nanomaterials with enzyme-like characteristics (nanozymes): next-generation artificial enzymes. Chem Soc Rev 42:6060–6093. https://doi.org/10.1039/c3cs35486e

Welti R, Li W, Li M et al (2002) Profiling membrane lipids in plant stress responses. Role of phospholipase D alpha in freezing-induced lipid changes in Arabidopsis. J Biol Chem 277(35):31994–32002. https://doi.org/10.1074/jbc.m205375200

Worms IAM, Boltzman J, Garcia M et al (2012) Cell-wall-dependent effect of carboxyl-CdSe/ZnS quantum dots on lead and copper availability to green microalgae. Environ Pollut 167:27–33. https://doi.org/10.1016/j.envpol.2012.03.030

Wu H, Tito N, Giraldo JP (2017) Anionic cerium oxide nanoparticles protect plant photosynthesis from abiotic stress by scavenging reactive oxygen species. ACS Nano 11(11):11283–11297. https://doi.org/10.1021/acsnano.7b05723

Xu J, Yang J, Duan X et al (2014) Increased expression of native cytosolic Cu/Zn superoxide dismutase and ascorbate peroxidase improves tolerance to oxidative and chilling stresses in cassava (Manihot esculenta Crantz). BMC Plant Biol 14:208. https://doi.org/10.1186/s12870-014-0208-4

Yadav T, Mungray AA, Mungray AK (2014) Fabricated Nanoparticles: current Status and Potential Phytotoxic Threats. In: Whitacre D (ed) Reviews of Environmental Contamination and Toxicology volume. Reviews of Environmental Contamination and Toxicology (Continuation of Residue Reviews). Springer, Cham, pp 83–110. https://doi.org/10.1007/978-3-319-04411-8_4

Yordanova R, Popova L (2007) Effect of exogenous treatment with salicylic acid on photosynthetic activity and antioxidant capacity of chilled wheat plants. Gen Appl Plant Physiol 33(3–4):155–170

Zhao L, Peng B, Hernandez-Viezcas JA et al (2012) Stress response and tolerance of *Zea mays* to CeO2 nanoparticles: cross talk among H2O2, heat shock protein and lipid peroxidation. ACS Nano 6(11):9615–9622. https://doi.org/10.1021/nn302975u

Chapter 19
Applications of Nanobiotechnology to Mitigate Mineral Nutrients Deficiency Stress in Crop Plants

Saima Amjad and Mohammad Serajuddin

Abstract Environmental challenges adversely affect plant growth and their productivity worldwide. Therefore, there is a need for the adoption of new technologies. The rapid potential development in nanotechnology influences the agriculture and food industry, which sustainably increases crop productivity by using new techniques, i.e., use of nanofertilizers and nanopesticides for better nutrient efficiency and pest management. Nanofertilizers are an excellent fertilizer that improves nutrient use efficiency of plants by slow and specific release of nutrient minerals and replacing the overuse of conventional fertilizers. Previous studies on nanoform of mineral particles such as calcium carbonate, cerium oxide, molybdenum, zinc, titanium dioxide, manganese hollow core shell and magnesium nanoparticles showed that they enhanced the nutrient uptake in plants and it also reduces metal accumulation in crop plants. It improved new set of devices to develop a genetically based tool using nanocapsules, nanofibers and nanoparticles. Hence, based on the research data available so far, the present chapter provides an overview on the various nanoform of mineral nutrients, which are beneficial for crop productivity. Moreover, this chapter also focuses on challenges and function of nanoparticles so as to understand the mode of action of nanoparticles on plants to overcome the plant nutrient stress.

Keywords Agriculture · Nanobiotechnology · Nanoparticles · Nutrient deficiency · Nutrient stress

19.1 Introduction

Plant nutrition deficiency is an important limiting aspect for plant growth and productivity after drought and salinity (Rajemahadik et al. 2018). The plant requires nutrients for their growth and development and absorbs most mineral nutrients present in the soil through their roots. However, to compensate the deficiency of nutrients, fertilizers

S. Amjad · M. Serajuddin (✉)
Department of Zoology, University of Lucknow, Lucknow, UP 226007, India
e-mail: serajuddin_m@lkouniv.ac.in

J. M. Al-Khayri et al. (eds.), *Nanobiotechnology*,
https://doi.org/10.1007/978-3-030-73606-4_19

are applied to the soil for healthy development of plants (Baloch et al. 2008; Bernal et al. 2007). Plants require seventeen elements for their growth and development out of which 14 are essential nutrients required. Among these essential nutrients, six are macronutrients that required in large amounts viz. calcium (Ca), potassium (K), magnesium (Mg), nitrogen (N), phosphorus (P), and sulfur (S) (Maathuis 2009). The micronutrients viz. chloride (Cl), copper (Cu), manganese (Mn), iron (Fe), zinc (Zn), cobalt (Co), molybdenum (Mo), and nickel (Ni) are required in a small amount (Marschner 2012; Zeng 2014). The nutrition deficiency found in commercially available economic crops influence human health, especially the people that belong to the rural areas, but the sustainable approach of nanotechnology diminishing these challenges.

During past few decades, biotechnological approach used for bioremediation or phytoremediation to restore agro-chemically damaged soils (Ghormade et al. 2011) to raise the use of nutrient efficiency in crops and to inhibit mineral losses. Several research studies revealed that plants grow by adapting a specific mechanism to uptake their required level and acclimate variation of nutrient availability (Ohkama-Ohtsu and Wasaki 2010; Schachtman and Shin 2007). Several strategies have attempted to provide protection and nutrition to the plants. The plant growth was hampered by two causes, the lack of micronutrients or either the pore size of roots is so small that it is unable to uptake and translocate the nutrients inside the plant. Therefore, it is essential to get better nutrient uptake competency strategies to enhance the quality and production of the crop (Elemike et al. 2019). The required necessities of nutrients in plants are provided by the fertilizers, with the certainty that minerals could be absorbed from the soil. The widespread contributions of biotechnology from conventional breeding to improve crop nutrient efficiency have been made through the molecular technique approach, but these achievements are limited (Ashraf et al. 2011). The transporters and enzymes efficiently involve in nutrient absorption and are necessary for acquiring elevated nutrient uptake, and it directly affect the status of crop yield. In an instance, the accumulation of nitrogen in shoot and grains of wheat plants increased by the over-expression of the glutamine synthetase gene (GS1) (Hu et al. 2018), while in maize plants the number of kernels enhanced by the over-expression of GS13 (Martin et al. 2006). The transgenic rice plants required a high amount of nitrogen for growth, OsAMT1 efficiently working as ammonium transporter function in elevating the nutrient uptake efficiency under optimal as well as suboptimal nitrogen conditions (Ranathunge et al. 2014). However, these approaches are inadequate and it is necessary to resolve the limited approach of biotechnology by involving the application of nanotechnology towards agriculture (Ghormade et al. 2011). The main purpose of the application of nanomaterial in agriculture is to overcome nutrient stress, pest control, reduce the effect of hazardous chemicals and crop protection as shown in Fig. 19.1 (Thakur et al. 2018).

The ongoing application of nanobiotechnology provides an opportunity to overcome the mineral nutrient stress in plants. The engineered nanoparticles hold an immense promise in the growing use of nanofertilizers, which utilize competence along with growing agriculture production.

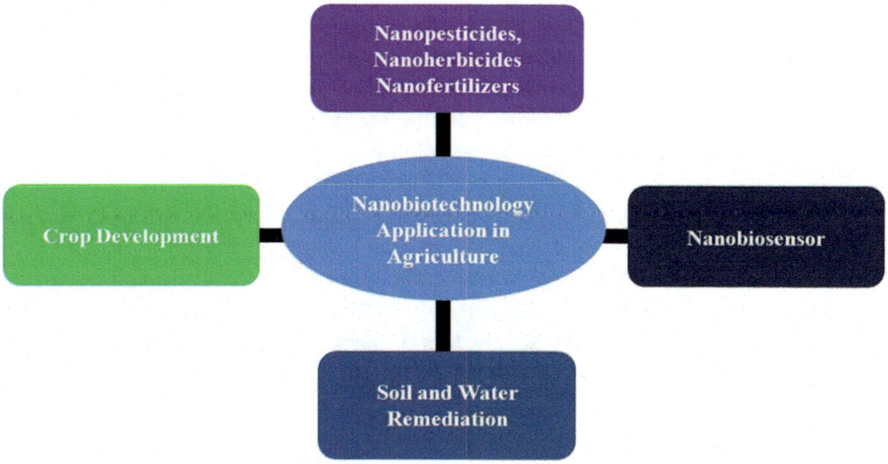

Fig. 19.1 Schematic representation of application of nanobiotechnology in agriculture for better crop productivity (Fig constructed by Saima Amjad)

19.2 Sustainable Approach of Nanobiotechnology in Agriculture

Nanotechnology is the upcoming innovatory technology after biotechnology revolution. It has an extensive application in various disciplines such as electronics, optics, pharmaceuticals, agriculture, and the environment. In the last few decades, it is growing with time and become a significant application in agriculture (Chhipa 2017). Nanotechnology can be used to design, develop and synthesize minerals in nanoform and this application gives considerable assurance for agricultural crop productivity (Baruah and Dutta 2009; Kuzma 2007; Navrotsky 2000). It is the procedure to design, develop and synthesize nanomaterials to represents an area holding considerable assurance for agricultural field in the form of nanofertilizers (Baruah and Dutta 2009; Kuzma 2007; Navrotsky 2000). Nanotechnologies using next generation of pesticides and fertilizers in agriculture by using functionalize nanomaterial and nanoparticle. It also minimizes the loss of pesticides and fertilizers by controlled and targeted delivery through nanocarriers (Ghormade et al. 2011; Joseph and Morrison 2006; Khot et al. 2012; Nair et al. 2010; Robinson and Morrison 2009; Scott and Chen 2013). The plant protection material (nanopesticides) and nutrients (nanofertilizers) have advanced properties because of their unique physical, chemical and rapidly dispersible properties of nanoparticles, which also enhanced the uptake of nutrition in the plant (Ghormade et al. 2011). There are several wide varieties of materials used to synthesize nanoparticles are magnetic materials, metal oxides, semiconductors, quantum dots, lipids, ceramics, polymers (synthetic or natural), emulsions and dendrimers (Puoci et al. 2008). However, the limitation of conventional fertilizers is that it does not contain all the required nutrients for plant augmentation and development. Therefore, it becomes an interesting venture for nanofertilizers that can address nutrient deficiency and environmental issues related with fertilizers (Dimkpa and Bindraban 2017). Sustainable and productive crop management can also be achieved by using an important tool of foliar fertilization in nanotechnology. The present knowledge of the factors, which influence the efficacy of foliar applications, may be determined by the art of nanotechnology (Eichert et al. 2008; Mortvedt 1992; Millán et al. 2008; Pandey et al. 2010; Roco 2011).

Leaching and loss of harmful substances have also reduced by using the encapsulated nanominerals, it plays a vital role in environmental protection (Zheng et al. 2005). The other accomplishment of nanobiotechnology is to develop nano-carriers for smart delivery systems and nanosensors, which are broadly used to assess environmental pollution in soil and water system of the agricultural field. Several sensors already developed on the concept of nano-detection to detect the traces of heavy metals such as electrochemical sensors, optical sensors, biosensors and so on (Handforda et al. 2015; Ion et al. 2010; Parisi et al. 2014). Nanobiotechnology is also being explored to improve crop productivity in plant reproduction through hereditary transfection (Prasad et al. 2014; Torney et al. 2007). It offers a new set of devices to develop a genetically based tool using nanocapsules, nanofibers and nanoparticles (Gutiérrez et al. 2011 and Nair et al. 2010). A Chitosan nanoparticle has versatile properties and emerged as a valuable carrier for genetic transfer of material; it can be modified by PEGylated to control the transfer of genetic material (Kashyap et al. 2015).

19.3 Strategies to Improve Plant Nutrient Uptake by Nanobiotechnology

The nanobiotechnology plays a noteworthy role in the plant production through the controlled release of mineral nutrients (Gruère 2012; Mukhopadhyay 2014). Nanoparticles or nanominerals size lies between 1–100 nm and different from their corresponding parental materials, which produce both useful and harmful biological effects in a living cell (Nel et al. 2006). Fertilizer nanoparticles are also known as magic bullets which enhance crop productivity to deal with global food problems (Bhatt et al. 2020). Several research studies reported the toxicological interaction of engineered nanoparticles on plants even though these studies implicated the exposure of nanoparticles in a certain specific situation with a short period of time at high doses. On the contrary, very few studies paying attention on the favorable impact of nanoparticles on plant development and productivity. Furthermore, a mineral nanoparticle integrated into conventional fertilizers and pesticides to increase the production, also reduces the chance of diseases and improves nutritional augmentation (Elmer and White 2018; Prasad et al. 2017). The potential benefits of nanotechnology can understand by application of some mineral nanoparticles on plants and analyze the mechanism of transport and nutrient uptake. The mechanism of some nanoparticles in nutrient stress has given in Table 19.1 to understand the function of mineral nanoparticles in plants.

Table 19.1 Nanomineral nutrients and their application on different type of crops and their effect on plant growth and physiology

Nutrients	Crop		Function	References
	Botanical name	Common name		
Ca	*Arachis hypogaea* L.	Peanut	Enhanced the nutrient content in shoot and root of plant	Liu et al. (2004)
	Ziziphus mauritiana Lam.	Indian plum	Increased in Ca^{2+} uptake	Hua et al. (2015)
$CaCO_3$	*Arachis hypogaea* L.	Peanut	Enhanced Ca^{2+} uptake and growth	Liu et al. (2005)
nano U-NPK	*Triticum durum* Desf	Durum wheat	Increased efficiency of fertilization, yields of grains, decline lower 40 wt% of N amount	Ramírez-Rodrígu et al. (2020)
CaP	*Zea mays* L.	Maize	Promote plant growth efficiency, enhanced root, and improved vitality properties	Rane et al. (2015)
CeO_2	*Cucumis sativus* L.	Cucumber	Uptake of Mg^{2+} ion, Improved starch and globulin content	Zhao et al. (2014)
Cu	*Lactuca sativa* L.	Lettuce	Amplified the shoot and root length	Shah and Belozerova (2009)
CuO	*Zea mays* L.	Maize	Improved plant growth	Adhikari et al. (2016)
Cu	*Triticum aestivum* L.	Millat-2011 (wheat)	Enhanced growth and yield of chlorophyll content, leaf area, fresh and dry weight, and root dry weight of plant	Hafeez et al. (2015)
	Cajanus cajan L.	Pigeon pea	Enhances plant growth and seedlings of the plant	Shende et al. (2017)
Hydroxyapatite	*Glycine max* (L.) Merr	Soybean	Improved the growth rate and seed yield	Liu and Lal (2014)

(continued)

Table 19.1 (continued)

Nutrients	Crop		Function	References
	Botanical name	Common name		
SPIONs	*Glycine max* (L.) Merr	Soybean	Increased chlorophyll content	Ghafariyan et al. (2013)
	Pisum sativum L.	Pea	Enhanced seed weight and chlorophyll content	Delfani et al. (2014)
Fe_2O_3	*Arachis hypogaea* L.	Peanut	Enhanced root length, biomass, and SPAD	Rui et al. (2016)
FeO	*Medicago falcata* L.	Yellow medick	Increased chlorophyll α fluorescence, plant root length and miR159c expression	Kokina et al. (2020)
Mn	*Vigna radiata* (L.) R. Wilczek	Mung bean	Enhanced the growth of shoot length and chlorophyll content. It also increased the rate of photosynthesis	Pradhan et al. (2013)
	Oryza sativa L.	Rice	Improved Zn uptake in plant	Yuvaraj and Subramanian (2015)
Mg	*Vigna unguiculata* (L.) Walp.	Cowpea	Mg content increased in stem and plasma and it also accelerated the enzyme activity	Delfani et al. (2014)
	Triticum aestivum L.	Common wheat	Increased growth and yield of plant and minerals uptake	Rathore and Tarafdar (2015)
MgO	*Arachis hypogaea* L.	Peanut	Enhanced seed germination, growth rate mechanism and biomass production	Jhansi et al. (2017)
Mo	*Cicer arietinum* L.	Chickpea	Plant mass and nodules number increased	Taran et al. (2014)
			Enhanced root area, diameter, length, perimeter, and tips number, improved microbial activities, increased biomass and grain yield	Thomas et al. (2017)

(continued)

Table 19.1 (continued)

Nutrients	Crop		Function	References
	Botanical name	Common name		
Si	*Ocimum basilicum* L.	Basil	Chlorophyll content increased and reduced proline content	Kalteh et al. (2018)
	Pisum sativum L.	Pea	Reduced the accumulation of Cr in roots and shoots, enhanced antioxidant activity	Tripathi et al. (2015)
	Triticum aestivum L.	Wheat	Improved development, photosynthesis of plant and reduced the oxidative stress, inhibited metal accumulation	Ali et al. (2019)
TiO_2	*Spinacia oleracea* L.	Spinach	Increased plant dry weight	Zheng et al. (2005)
	Spinacia oleracea L.	Spinach	N_2 fixation improvement	
	Vigna radiata (L.) R. Wilczek	Mung bean	Improvement in plant growth and nutrient content	
	Vigna unguiculata (L.) Walp.	Cowpea	Cowpea yield increased up to 26–51%	Owolade and Ogunleti (2008)
	Coriandrum sativum L.	Coriander	Improved nutritional quality, enhanced root and shoot fresh biomass	Hu et al. (2020)
Zn	*Lolium* L.	Ryegrass	Root extension	Lin and Xing (2008)
ZnO	*Vigna radiata* (L.) R. Wilczek and *Cicer arietinum* L.	Mung bean and chickpea	Plant growth increased in mung bean and in chickpea plant	Mahajan et al. (2011)
	Cucumis sativus L.	Cucumber	Increased root dry weight and fruit gluten	

(continued)

Table 19.1 (continued)

Nutrients	Crop		Function	References
	Botanical name	Common name		
	Brassica napus L.	Rapeseed	Root extension	Lin and Xing (2007)
	Arachis hypogaea L.	Peanut	34% enhancement in pod yield for per plant	Prasad et al. (2012)
	Cicer arietinum L.	Chickpea	Increased shoot dry weight and rate of antioxidant activity	Burman et al. (2013)
	Zea Maize L.	Maize	Improved plant length and dry weight	Adhikari et al. (2015)
	Cyamopsis tetragonoloba (L.) Taub.	Lond bean	Enhancement in plant growth and nutrient content	Raliya and Tarafdar (2013)
	Lactuca sativa L.	Lettuce	Stimulated catalase enzyme activity, enhanced seed germination, and biomass	Rawashdeh et al. (2020)

19.3.1 Calcium Nanoparticles

The calcium in the nanoforms was more effective compared to the chelated form; it improves plant growth and production (Liu et al. 2005). Effect of calcium carbonate nanoparticles have studied on *Arachis hypogeae*, L. it showed that the nano form of calcium enhanced branch number and increased 15% weight (fresh and dry) of plants (Liu et al. 2004; Tantawy et al. 2014). Liu et al. (2005) have reported a study that showed improvement in the physiological process for instance chlorophyll content of tomato plant increased by the effect of nano calcium. Biomimetic calcium phosphate nanoparticles with the composition of potassium (K) and nitrogen (N, as nitrate and urea) was used as a multinutrient nanofertilizear (nano U-NPK) for the cultivation of *Triticum durum* Desf. The result showed the application of the slow-release nano U-NPK were a promising strategy towards increasing the competency of the fertilization, and yields of grains were obtained, and the additional advantage of using a much lower N amount (a decline of 40 wt%) (Ramírez-Rodrígu et al. 2020). *Zea mays* L. crop has been treated with calcium phosphate nanoparticles (CaPNPs) along with *Piriformospora indica* and *Glomus mosseaec* which promote plant growth efficiency, root enhancement, and improved vitality properties (Rane et al. 2015).

19.3.2 Cerium Oxide Nanoparticles

Zhao et al. (2014) conducted a study on cucumber (*Cucumis sativus* L.) plants grown in soil treated with Cerium Oxide (CeO_2) nanoparticles. The results indicated that CeO_2NPs influenced the fruit flavor decreased the antioxidant capacity and increased starch and globulin content. It considerably enhanced the uptake of Mg^{2+} ion, which is a vital constituent of the chlorophyll molecule. It also decreased the uptake of molybdenum (Mo) concentration and altered the non-reducing sugars, phenolic content and changed the protein fractionation.

19.3.3 Copper Nanoparticles

Copper (Cu) metal is an important micronutrient required for plants enzymatic activity, which functions as a regulatory co-factor or catalyst for a large number of enzymes or acts as a functional structural. Adhikari et al. (2016) investigated the effect of copper oxide (CuO) nanoparticles (< 50 nm) on the development, bioaccumulation and enzymatic action of maize (*Zea mays* L.) plant. The experimental studies showed the easy assimilation of CuO nanoparticles through plant cells and increase the growth of maize by regulating the different enzyme activities. The glucose-6-phosphate dehydrogenase enzymatic activity was extremely influenced by copper oxide (CuO) nanoparticles and affected the pentose phosphate pathway in maize plants. Hafeez et al. (2015) have studied the effect of the concentration-dependent copper nanoparticles (CuNPs) on Millat-2011 (*Triticum aestivum* L.) crop which significantly enhanced growth and yield of the plant, chlorophyll content, leaf area, fresh and dry weight, and root dry weight. Similarly, the exposure of biogenic CuNPs (20 ppm) on pigeon pea (*Cajanus cajan* L.) plant was evaluated which enhances growth such as height, root length, fresh, and dry weights and seedlings of the plant (Shende et al. 2017).

19.3.4 Hydroxyapatite Nanoparticles

Urea based nitrogen fertilizers viz. Urea-coated zeolite chips and hydroxyapatite has been used in nanoparticle form as a source of nitrogen (N) to study the controlled release of N for a long duration of time (Kottegoda et al. 2011; Millán et al. 2008). Similarly, hydroxyapatite ($Ca_5(PO_4)_3OH$) nanoparticles have a significant impact on seed yield of *Glycine max* (L.) Merr., 20% and 33% increment in seed yield as compared with conventional phosphorus treated plant (Liu and Lal 2014).

19.3.5 Iron Oxide Nanoparticles

Delfani et al. (2014) analyzed the iron nanoparticles (FeNPs) effect on blacked eyed pea plants which not only improved the pods number per plant but also enhanced the weight of seeds and improved the chlorophyll biosynthesis. FeNPs enhanced the seed protein content by 2% compared to Fe. In another study comparative effect of iron oxide nanoparticles (Fe_2O_3NPs) and a chelated-Fe fertilizer (ethylenediaminetetraaceticacid-Fe; EDTA-Fe) were studied on the development and growth of *Arachis hypogaea* L. plant. The obtained results showed Fe_2O_3NPs enhanced root length, biomass, and Soil Plant Analysis Development (SPAD) chlorophyll meter values of peanut (*Arachis hypogaea* L.) plants. It also regulated phytohormone contents and antioxidant enzyme activity which promote plant growth (Rui et al. 2016). Also, in *Glycine max* (L.) Merr. plant chlorophyll content was enhanced by superparamagnetic iron oxide nanoparticles (SPIONs) without any phytotoxicity. Additionally, the physicochemical characteristics of SPIONs had an essential role in sub-apical leaves of soybean for an increment of chlorophyll content (Ghafariyan et al. 2013). Kokina et al. (2020) studied the exposure of different sizes of iron oxide nanoparticles (FeONPs) on yellow medick (*Medicago falcata* L.) plants. The observed results indicated a significant increase in chlorophyll α fluorescence, plant root length, induced genotoxicity, and reduced genome stability compared to the control plant. Moreover, enhanced miR159c expression indicated enhanced plant resistance against fungal pathogens.

19.3.6 Manganese Nanoparticles

Manganese (Mn) micronutrient is primarily essential for the photosynthesis process in plants. Pradhan et al. (2013) observed the effect of manganese nanoparticles (MnNPs) on leguminous plant *Vigna radiate* (L.) R. Wilczek at a specific concentration as compared to conventionally available manganese salt, $MnSO_4$ under laboratory conditions. The higher concentration of MnNP had not induced phytotoxicity to the plant and the size of MnNP possibly helped plants to uptake these nanoparticles more easily and translocated itself in the leaves using xylem. The MnNPs treated chloroplasts increased the function of CP43 protein in the reaction center of photosystem II (PS II) and had shown higher photophosphorylation, where oxygen was generated by water molecule splitting and also enhanced the electron transport chain.

Manganese hollow core shell nanoparticles have used for the controlled and targeted release of zinc (Zn) to the plant (*Oryza sativa* L.) soil. The result indicated that the nano-sized manganese hollow core shell enhances and improve Zn uptake by rice plant and reduce the loss of nutrients (Yuvaraj and Subramanian 2015).

19.3.7 Magnesium Nanoparticles

Magnesium nanoparticles (MgNPs) used as an alternate of Mg in the *Vigna unguiculata* (L.) Walp. plant, which enhanced the seed weight and yield (Delfani et al. 2014). Further, a study was conducted by Rathore and Tarafdar (2015), on the controlled delivery of magnesium nanoparticles (MgNPs) to wheat plants. The foliar application of MgNPs on plant increased the light absorption on the leaf surface and it also improved different enzyme activities which resulted in the enhanced mobilization of nutrients (Fe, Cu, Zn, P, and Mg) uptake. It also significantly improved the wheat plant root length and biomass. Magnesium oxide nanoparticles (MgONPs) (15 nm) enhanced seed germination and growth rate mechanism in *Arachis hypogaea* L. at 0.5 mg/L concentrations by penetrating seeds coat and internally support water retention in seeds which increased biomass production for the plant (Jhansi et al. 2017).

19.3.8 Molybdenum Nanoparticles

The colloidal solution of molybdenum nanoparticles (MoNPs) enhances plant resistance and also increased crops productivity due to the active uptake of nanoparticle into the plant cells. The experimental studies showed when the *Cicer arietinum* L. seeds were treated with MoNPs (colloidal solution) it enhanced the formation of nodule per plant by four fold as compared to control (Taran et al. 2014). Thomas et al. have also studied (2017), the effect of MoNPs (2–7 nm) on *Cicer arietinum* L. plant at 4 ppm concentration. The results showed a significant improvement in the root area, diameter, length, perimeter, and tips number. It also enhanced microbial activities and useful enzymes along with increased biomass and grain yield.

19.3.9 Silicon Nanoparticles

Silicon nanoparticles (SiNPs) application has used for Basil plant to reduce the pollution caused by salinity (*Ocimum basilicum* L.). SiNPs significantly increased the chlorophyll and reduced proline content in basil (Kalteh et al. 2018). Tripathi et al. 2015 studied the shielding effect of SiNP for *Pisum sativum* L. against chromium Cr(VI) phytotoxicity. The results showed that SiNPs reduced the accumulation of Cr in roots and shoots and enhances the intake of mineral nutrients which possess antioxidant activity and also reduces the ROS level. According to the study of Ali et al. (2019) SiNPs have the ability to restrain the accumulation of cadmium (Cd) concentrations in *Triticum aestivum* L. grain and other parts of the plant. Moreover, SiNPs improved the development, photosynthesis and also reduced the oxidative stress of wheat grain and in future, it can also be used as a fertilizer for controlling metal accumulation in crop plants.

19.3.10 Titanium Dioxide Nanoparticles

The comparative study of nano-TiO_2 and TiO_2 was investigated on the enlargement and germination of naturally-aged spinach seed (*Spinacia oleracea* L.) (Zheng et al. 2005). The experimental studies indicated that at a certain concentration nano-TiO_2 treated spinach seed enhances the germination of the aged seeds. The 30 days nano-TiO2 effect on spinach plants showed a 73% increase in dry weight, improves the chlorophyll formation up to 45% and enhances the photosynthetic rate three times as compared to the control during the germination stage. This effect of the development rate of spinach seeds was inversely proportional to the size of TiO_2 signifying, smaller the nanoparticle different the germination growth. Furthermore, the researcher defines the possibility of entry of nano-TiO_2 into cells those have increased oxidation–reduction reactions via the superoxide ion radical during germination in the dark and resulted in the quenching of free radicals during the germination of seeds,which also increased the photosynthetic rate. Effect of TiO_2NP on coriander plant (*Coriandrum sativum* L.) at concentration dependent manner improved the nutritional quality, enhanced root and shoot fresh biomass (Hu et al. 2020).

19.3.11 Zinc Nanoparticles

Zinc oxide nanoparticles (ZnONPs) have been used in industry for the last many years. Zn is the most important vital micronutrient that is mandatory to enhance the crop productivity. Prasad et al. (2012) investigated the effect of Zn micronutrient into peanut seeds through ZnONPs (25 nm mean particle size). It improves the uptake of Zn through leaf and kernel and a high content of Zn was found in the seed and increase chlorophyll amount in leaf. These nanoparticles proved helpful in enhancing stem and root growth (Prasad et al. 2012). Lin and Xing (2007), reported ZnONPs significantly adhered to the root surface and enter inside the plant cell through penetration and found in the apoplast and protoplast of the root endodermis and stele. Rawashdeh et al. (2020) studied effect of two different concentrations of ZnONPs (25 ppm or 50 ppm) on Lettuce seeds (*Lactuca sativa* L.). The obtained results showed enhanced seed germination, and biomass of seeds due to stimulated catalase enzyme activity.

19.4 Conclusion and Prospects

The sustainable agriculture, nutrient safety, and food accessibility are included the key sustainable development objectives to cope the crop nutrient deficiency. Hence, it is necessary to exploit the benefits of nanotechnology in reaching the feat by enhancing plant nutrient accessibility and reducing their losses on agricultural field. Nanobiotechnology application plays a notable role in agriculture to manage nutrient deficiency stress for crop management which is necessary for sustainable agriculture. It could be significantly enhanced nutritional health and sanitation of crops simultaneously, improved food security and sustainability in the coming times. Nanobiotechnology has a multidirectional approach therefore, it is necessary to exploit the benefits of its application in reaching the feat by enhancing plant nutrient accessibility and reducing their losses on the agricultural field. Previous studies indicated that the mineral nanoparticles are beneficial for plant growth efficiency, root enhancement, chlorophyll content, uptake of mineral, regulation of phytohormone contents and antioxidant enzyme activity, enhanced gene expression, and seed development. Additionally, it is also beneficial for inhibition of accumulation of heavy metals concentration inside the plants. Although the obtained results are limited to the experimental level, therefore it is necessary to introduce the application of nanofertilizer in the nursery stage to proceed towards a large-scale agriculture field. More scientific research studies required to analyze the environmental risks of nanoparticles to encourage the safe development of nanofertilizer. Hence, nanotechnology is a promising sector to provide commercialized nanofertilizers for better crop productivity and soon that will be available in the market.

References

Adhikari T, Kundu S, Biswas AK et al (2015) Characterization of zinc oxide nano particles and their effect on growth of maize (Zea mays L.) plant. J of Pla Nut 38(10):1505–1515

Adhikari T, Sarkar D, Mashayekhi H, Xing B (2016) Growth and enzymatic activity of maize (Zea mays L.) plant: solution culture test for copper dioxide nano particles. J of P Nut 39(1):99–115

Ali S, Rizwan M, Hussain A et al (2019) Silicon nanoparticles enhanced the growth and reduced the cadmium accumulation in grains of wheat (Triticum aestivum L.). Pla Phys and Biochem 140:1–8

Ashraf M, Akram NA, Al-Qurainy F, Foolad MR (2011) Drought tolerance: roles of organic osmolytes, growth regulators and mineral nutrients. Adv in Agr 111:249–296

Baloch OB, Chacar OI, Tareen MN (2008) Effect of foliar application of macro and micronutrients on production of green chilies (Capcicum annuum L.) J Agric Tech 4(2):177–184

Baruah S, Dutta J (2009) Nanotechnology applications in pollution sensing and degradation in agriculture: a review. EnvChem Let. 7(3):191–204

Bernal M, Cases R, Picorel R, Yruela I (2007) Foliar and root Cu supply affect differently Fe and Zn-uptake and photosynthetic activity in soybean plants. Environ Exp Bot 60:145–150

Bhatt D, Bhatt MD, Nath M et al (2020) Application of nanoparticles in overcoming different environmental stresses. Protective chemical agents in the amelioration of plant abiotic stress: Biochemical and molecular perspectives, 635–654

Burman U, Saini M, Kumar P (2013) Effect of zinc oxide nanoparticles on growth and antioxidant system of chickpea seedlings. Tox & Env Chem 95(4):605–612
Chhipa H (2017) Nanofertilizers and nanopesticides for agriculture. Env Chem Let 15(1):15–22
Delfani M, Firouzabadi MB, Farrokhi N, Makarian H (2014) Some physiological responses of black-eyed pea to iron and magnesium nanofertilizers. Commun Soil Sci Plant Anal 45:11
Dimkpa CO, Bindraban PS (2017) Nanofertilizers: new products for the industry?. J Agric Food Chem 66(26):6462–6473
Eichert T, Kurtz A, Steiner U, Goldbach HE (2008) Size exclusion limits and lateral heterogeneity of the stomatal foliar uptake pathway for aqueous solutes and water-suspended nanoparticles. Phy Pla 134(1):151–160
Elemike EE, Uzoh IM, Onwudiwe DC, Babalola OO (2019) The role of nanotechnology in the fortification of plant nutrients and improvement of crop production. App Sci 9(3):499
Elmer W, White JC (2018) The future of nanotechnology in plant pathology. Ann Rev Phyto 56:111–133
Ghafariyan MH, Malakouti MJ, Dadpour MR et al (2013) Effects of magnetite nanoparticles on soybean chlorophyll. Envir Sci Tec 47(18):10645–10652
Ghormade V, Deshpande MV, Paknikar KM (2011) Perspectives for nano-biotechnology enabled protection and nutrition of plants. Biotec Adva 29(6):792–803
Gruère GP (2012) Implications of nanotechnology growth in food and agriculture in OECD countries. F Pol 37(2):191–198
Gutiérrez FJ, Mussons ML, Gatón P, Rojo R (2011) Nanotechnology and food industry. In: Valdez B, Schorr M, Zlatev R (eds) Scientific, health and social aspects of the food industry. InTech, pp 95–128
Hafeez A, Razzaq A, Mahmood T, Jhanzab HM (2015) Potential of copper nanoparticles to increase growth and yield of wheat. J Nanosci Adv Technol 1(1):6–11
Handford CE, Dean M, Henchion M et al (2014) Implications of nanotechnology for the agri-food industry: opportunities, benefits and risks. Tre in F Sci Tech 40(2):226–241
Hu J, Wu X, Wu F et al (2020) Potential application of titanium dioxide nanoparticles to improve the nutritional quality of coriander (Coriandrum sativum L.). J of hazar mater 389:121837
Hu M, Zhao X, Liu Q et al (2018) Transgenic expression of plastidic glutamine synthetase increases nitrogen uptake and yield in wheat. Pla Biotec J 16(11):1858–1867
Hua KH, Wang HC, Chung RS, Hsu JC (2015) Calcium carbonate nanoparticles can enhance plant nutrition and insect pest tolerance. J of Pest Sci 40(4):208–213
Ion AC, Ion I, Culetu A (2010) Carbon-based nanomaterials: Environmental applications. Uni Pol Buch 38:129–132
Jhansi K, Jayarambabu N, Reddy KP et al (2017) Biosynthesis of MgO nanoparticles using mushroom extract: effect on peanut (Arachis hypogaea L.) seed germination. 3 Biotec 7(4):263
Joseph T, Morrison M (2006) Nanotechnology in agriculture and food. Nano Rep 2:2–3
Kalteh M, Alipour ZT, Ashraf S et al (2018) Effect of silica nanoparticles on basil (Ocimum basilicum) under salinity stress. J Chem Health Risks 4(3):49–55
Kashyap PL, Xiang X, Heiden P (2015) Chitosan nanoparticle based delivery systems for sustainable agriculture. Int J Bio Mac 77:36–51
Khot LR, Sankaran S, Maja JM et al (2012) Applications of nanomaterials in agricultural production and crop protection: a review. C Pro 35:64–70
Kokina I, Plaksenkova I, Jermaļonoka M, Petrova A (2020) Impact of iron oxide nanoparticles on yellow medick (Medicago falcata L.) plants. J of Pla Inter 15(1):1–7
Kottegoda N, Munaweera I, Madusanka N, Karunaratne V (2011) A green slow-release fertilizer composition based on urea-modified hydroxyapatite nanoparticles encapsulated wood. Cuur Sci 101(1):73–78
Kuzma J (2007) Moving forward responsibly: Oversight for the nanotechnology-biology interface. J Nano Res 9(1):165–182
Lin D, Xing B (2007) Phytotoxicity of nanoparticles: inhibition of seed germination and root growth. Env Pol 150(2):243–250

Lin D, Xing B (2008) Root uptake and phytotoxicity of ZnO nanoparticles. Env Sci Tech 42(15):5580–5585

Liu R, Lal R (2014) Synthetic apatite nanoparticles as a phosphorus fertilizer for soybean (Glycine max). Sci Rep 4:5686

Liu X, Zhang F, Zhang S et al (2004) Responses of peanut to nano-calcium carbonate. Plant Nutr Fertil Sci 11:385–389

Liu X, Zhang F, Zhang S et al (2005) Responses of peanut to nano calcium.carbonate. P Nut Fer Sci 11(3):385–389

Maathuis FJM (2009) Physiological functions of mineral macronutrients. Curr Opin Plant Bio 12:250–258

Mahajan P, Dhoke SK, Khanna AS (2011) Effect of nano-ZnO particle suspension on growth of mung (*Vigna radiata*) and gram (*Cicer arietinum*) seedlings using plant agar method. J Nano 2011:1–7

Marschner H (2012) Mineral nutrition of higher plants, 3rd edn. Academic, London, UK

Martin A, Leeb J, Kicheyc T et al (2006) Two cytosolic glutamine synthetase isoforms of maize are specifically involved in the control of grain production. Pla C 18:3252–3274

Millán G, Agosto F, Vázquez M (2008) Use of clinoptilolite as a carrier for nitrogen fertilizers in soils of the Pampean regions of Argentina. Int J Agr Nat Res 35(3):293–302

Mortvedt JJ (1992) Crop response to level of water-soluble zinc in granular zinc fertilizers. Fert Res 33(3):249–255

Mukhopadhyay SS (2014) Nanotechnology in agriculture: prospects and constraints. Nano Sci App 7:63

Nair R, Varghese SH, Nair BG et al (2010) Nanoparticulate material delivery to plants. Pla Sci 179(3):154–163

Navrotsky A (2000) Nanomaterials in the environment, agriculture, and technology (NEAT). J of Nano Res 2(3):321–323

Nel A, Xia T, Madler L, Li N (2006) Toxic potential of materials at the nanolevel. Sci 311: 622–627

Ohkama-Ohtsu N, Wasaki J (2010) Recent progress in plant nutrition research: cross-talk between nutrients, plant physiology and soil microorganisms. Pl Cel Phy 51(8):1255–1264

Owolade O, Ogunleti D (2008) Effects of titanium dioxide on the diseases, development and yield of edible cowpea. J of P Prot Res 48(3):329–336

Pandey AC, Sanjay SS, Yadav SR (2010) Application of ZnO nanoparticles in influencing the growth rate of Cicer arietinum. J of Exp Nano 5(6):488–497

Parisi C, Vigani M, Rodríguez-Cerezo E (2015) Agricultural nanotechnologies: what are the current possibilities? Nano Tod 10(2):124–127

Pradhan S, Patra P, Das S et al (2013) Photochemical modulation of biosafe manganese nanoparticles on Vignaradiata: a detailed molecular, biochemical, and biophysical study. Env Sci Tech 47(22):13122–13131

Prasad R, Bhattacharyya A, Nguyen QD (2017) Nanotechnology in sustainable agriculture: recent developments, challenges, and perspectives. Fron in Mic 8:1014

Prasad R, Kumar V, Prasad KS (2014) Nanotechnology in sustainable agriculture: present concerns and future aspects. AfrJou of Bio 13(6):705–713

Prasad TNVKV, Sudhakar P, Sreenivasulu Y et al (2012) Effect of nanoscale zinc oxide particles on the germination, growth and yield of peanut. J P Nut 35(6):905–927

Puoci F, Lemma F, Spizzirri UG et al (2008) Polymer in agriculture: A review. Am J Agri Bio Sci 3:299–314

Rajemahadik VA (2018) Nanotechnology: Innovative approach in crop nutrition management. Int J Agr Sci ISSN: 0975–3710

Raliya R, Tarafdar JC (2013) ZnO nanoparticle biosynthesis and its effect on phosphorous-mobilizing enzyme secretion and gum contents in Cluster bean (Cyamopsistetragonoloba L.). Agr Res 2(1): 48–57

Ramírez-Rodríguez GB, Dal Sasso G, Carmona FJ et al (2020) Engineering biomimetic calcium phosphate nanoparticles: A green synthesis of slow-release multinutrient (NPK) Nanofertilizers. ACS App Bio Mat 3(3):1344–1353

Ranathunge K, El-kereamy A, Gidda S et al (2014) OsAMT1;1 transgenic rice plants with enhanced NHþ4 permeability show superior growth and higher yield under optimal and suboptimal NHþ4 conditions. J Exp Bot 65:965–979

Rane M, Bawskar M, Rathod D et al (2015) Influence of calcium phosphate nanoparticles, Piriformospora indica and Glomus mosseae on growth of Zea mays. Advances in natural sciences: Nanosci Nanotec 6(4):

Rathore I, Tarafdar JC (2015) Perspectives of biosynthesized magnesium nanoparticles in foliar application of wheat plant. J Bionano 9(3):209–214

Rawashdeh RY, Harb AM, AlHasan AM (2020) Biological interaction levels of zinc oxide nanoparticles; lettuce seeds as case study. Heli 6(5):

Robinson DKR, Morrison MJ (2009) Nanotechnology developments for the agrifood sector - report of the observatory, NANO. May 2009. www.observatorynano.eu

Roco MC (2011) The long view of nanotechnology development: the National Nanotechnology Initiative at 10 years. J Nanopart Res 13:427–445

Rui M, Ma C, Hao Y et al (2016) Iron oxide nanoparticles as a potential iron fertilizer for peanut (Arachis hypogaea). Fron in Plan Sci 7:815

Schachtman DP, Shin R (2007) Nutrient sensing and signaling: NPKS. Ann Rev Plant Biol 58:47–69

Scott N, Chen H (2013) Nanoscale science and engineering for agriculture and food systems. Ind Biotechnol 9:17–18

Shah V, Belozerova I (2009) Influence of metal nanoparticles on the soil microbial community and germination of lettuce seeds. Wat A S Poll 197(1–4):143–148

Shende S, Rathod D, Gade A, Rai M (2017) Biogenic copper nanoparticles promote the growth of pigeon pea (Cajanus cajan L.). IET Nanobiotec 11(7):773–781

Tantawy AS, Salama YAM, Abdel-Mawgoud MR, Ghoname AA (2014) Comparison of chelated calcium with nano calcium on alleviation of salinity negative effects on tomato plants. Mid E Jou of Ag Res 3(4):912–916

Taran NY, Gonchar OM, Lopatko KG et al (2014) The effect of colloidal solution of molybdenum nanoparticles on the microbial composition in rhizosphere of Cicer arietinum L. Nanos Res Let 9(1):289

Thakur S, Thakur S, Kumar R (2018) Bio-nanotechnology and its role in agriculture and food industry. J Mol Genet Med 12:324 https://doi.org/10.4172/1747-0862.1000324

Thomas E, Rathore I, Tarafdar JC (2017) Bioinspired Production of Molybdenum Nanoparticles and Its Effect on Chickpea (Cicer arietinum L). J of Bionanosc 11(2):153–159

Torney F, Trewyn BG, Lin VSY, Wang K (2007) Mesoporous silica nanoparticles deliver DNA and chemicals into plants. Nat Nano 2(5):295

Tripathi DK, Singh VP, Prasad SM et al (2015) Silicon nanoparticles (SiNp) alleviate chromium (VI) phytotoxicity in Pisum sativum (L.) seedlings. Pla Phys and Biochem. 96:189–198

Yuvaraj M, Subramanian KS (2015) Controlled-release fertilizer of zinc encapsulated by a manganese hollow core shell. Soil Sci Pl Nut 61(2):319–326

Zeng H, Wang G, Hu X et al (2014) Role of microRNAs in plant responses to nutrient stress. Pla and Soil 374(1–2):1005–1021

Zhao L, Peralta-Videa JR, Rico CM et al (2014) CeO2 and ZnO nanoparticles change the nutritional qualities of cucumber (Cucumis sativus). J Agr Food Chem 62(13):2752–2759

Zheng L, Hong F, Lu S, Liu C (2005) Effect of nano-TiO2 on strength of naturally aged seeds and growth of spinach. Bio Trace Elem Res 104(1):83–91

Chapter 20
Nano-Oxide Materials Combat Heavy Metals Toxicity by Modulating Oxidative Stress Pathways

Fawaz Alasmari, Omer I. Fantoukh, Qasim Alhadidi, Faleh Alqahtani, Abdulkareem A. Alanezi, and Asad Syed

Abstract Heavy metals have been found to exert toxicological effects on living organisms, including plants and animals. Prior studies explored the role of metallic compounds for the induction of pathways that are associated with cardiovascular, neurological and immunological diseases. At acute and chronic doses, these heavy metals showed the ability to promote the oxidative stress and inflammatory pathways processing, which may lead to adverse health consequences such as genotoxicity and cytotoxicity in plants. Dysregulation of antioxidant enzymes activities and inhibition of growth have been documented in plants exposed to heavy metals. This indicates that the heavy metals should be considered as risky compounds once they are in contact with plants, animals and humans. Recently, it has been found that nanoparticles of certain compounds such as cerium oxide could attenuate oxidative stress

Fawaz Alasmari and Omer I. Fantoukh these authors contributed equally to this chapter.

F. Alasmari (✉) · F. Alqahtani
Department of Pharmacology and Toxicology, College of Pharmacy, King Saud University, Riyadh, Saudi Arabia
e-mail: ffalasmari@ksu.edu.sa

F. Alqahtani
e-mail: afaleh@ksu.edu.sa

O. I. Fantoukh
Department of Pharmacognosy, College of Pharmacy, King Saud University, 2457, Riyadh 11451, Saudi Arabia
e-mail: ofantoukh@ksu.edu.sa

Q. Alhadidi
Department of Pharmacy, Salah Al-Din Health Directorate, Salah Al-Din, Iraq

A. A. Alanezi
Department of Pharmaceutics, College of Pharmacy, University of Hafr Al Batin, Hafr Al Batin, Saudi Arabia
e-mail: aalanezi@mail.usciences.edu

A. Syed
Department of Botany and Microbiology, College of Science, King Saud University, Riyadh, Saudi Arabia

J. M. Al-Khayri et al. (eds.), *Nanobiotechnology*,
https://doi.org/10.1007/978-3-030-73606-4_20

induced by heavy metals in the plants. This chapter sheds light on the role of nanomaterials in regulating oxidative stress associated with heavy metals exposure in plants.

Keywords Metallic compounds · Nanomaterials · Oxidative stress · Plants · Secondary metabolites

20.1 Introduction

Myriad toxicological effects could be induced due to exposure to particular metals (Andjelkovic et al. 2019; Nayfeh et al. 2018; Pokorska-Niewiada et al. 2018). These toxicological responses have been observed in plants, animals and humans (An et al. 2004; Hung and Chung 2004; Wen et al. 2017). Studies highlighted the potentially harmful effects of numerous heavy metals on the living organisms, including abundant plant species (An et al. 2004; Cargnelutti et al. 2006). However, some nanomaterials of metals provide beneficial effects on plants (Rizwan et al. 2019). For example, an improvement in the growth of plant exposed to cadmium was reported after exposure to zinc and iron oxide nanomaterials, which indicates that these metals may produce positive effects once they get in contact with a living organism (Rizwan et al. 2019). In this chapter, we discussed the biological effects of metals on plants.

Oxidative stress is one of the substantial pathways that is involved in initiating and maintaining the toxicological reactions induced by several metallic agents (Cargnelutti et al. 2006; Hasanuzzaman et al. 2018). Several studies explored the role of oxidative stress on the induction of toxicity of metals in living organisms (Cargnelutti et al. 2006; Hasanuzzaman et al. 2018; Panda et al. 2016). The metals were found to dysregulate the oxidative stress pathways, and this effect was associated with negative consequences on plants, animals and humans (Cargnelutti et al. 2006; Khan and Parvez 2015; Pujalté et al. 2011). For instance, glutathione system has been impaired in wheat due to the effect of particular metals such as lead (Hasanuzzaman et al. 2018). Another prior study showed that mercury exposure induced dysregulation in the length of cucumber root and shoot in part by alteration of the activity/content of catalase, chlorophyll and lipid peroxides (Cargnelutti et al. 2006). In human renal cells, exposure to nanoparticles of metallic compounds such as zinc oxide and cadmium sulfide could dose-dependently increase the death and toxicity of the cells (Pujalté et al. 2011). This study found that the physicochemical properties of these nanoparticles are the reason for increased reactive oxygen species productions. This effect was associated with a marked decreased ratio of glutathione to oxidized glutathione and total glutathione levels. These effects might be induced by the activation of nuclear factor kappa B (NF-KB) (Pujalté et al. 2011). Therefore, oxidative stress biomarkers regulate signaling for the effects of metallic compounds on living organisms, including numerous plant species. Oxidative stress has been observed in *Panax ginseng* exposed to a heavy metal (Huo et al. 2020). Moreover, induction of oxidative stress and increased secondary metabolites accumulation has been found in plant cells following exposure with methyl jasmonate (Ho et al. 2020). In this chapter, we highlighted the role of oxidative stress pathways on the toxicological effects of heavy metals in plant species.

Acute and chronic exposure to metals may induce differential effects on clinical and preclinical subjects as well as living organisms (An et al. 2004; Hung and Chung 2004; Pujalté et al. 2011; Wen et al. 2017; Zhai et al. 2017). To the best of our knowledge, the exact mechanisms for the effects of short-term and long-term exposures of metals to plant species have not been fully discussed. Besides, the differential effects may be also observed using different formulations of specific metals (Rizwan et al. 2019). For instance, a nanoparticle of iron and zinc oxide exposure induced beneficial effects on plant growth (Rizwan et al. 2019), which indicates that the pharmaceutical dosage form of these metals plays a potential role in their biological effects on living organisms, including plant species. This chapter highlights the role of acute and chronic exposure as well as dosage forms of metals on the oxidative stress associated with biological effects on living organisms (Fig. 20.1).

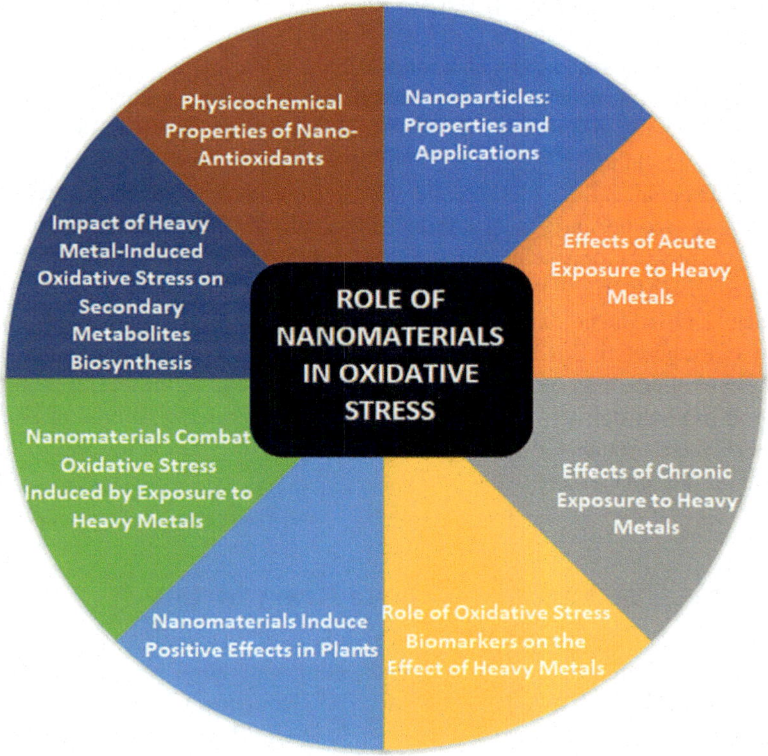

Fig. 20.1 A schematic diagram summarizes the objectives of the chapter (Figure constructed by authors)

20.2 Nanoparticles: Properties and Applications

The field of nanotechnology is expanding rapidly, and it is of high importance in various biomedical applications. The application of nanotechnology in medicine is known as nanomedicine and aims to produce materials on a nanometer scale to be employed in the diagnosis and treatment of some pathological conditions (De Jong and Borm 2008). Anyhow, nanoparticles are the most commonly used nanotechnology platforms in nanomedicine. Because of their small particle size (1–100 nm), nanoparticles are characterized by a large surface area, which enables them to distribute efficiently in the bloodstream and also makes them an efficient drug carrier (Borm et al. 2006; Stapleton and Nurkiewicz 2014). Such a reduction in particle size improved the physicochemical properties of the nanoparticles. Accordingly, the pharmacokinetic parameters of nanoparticles are enhanced by the large surface area, and this property makes them suitable candidates for various biomedical applications (Borm et al. 2006).

According to their chemical compositions, nanoparticles can be classified into three groups: organic, inorganic (metals, metal oxide, ceramic, and quantum dots), and carbon-based nanoparticles (De Matteis and Rinaldi 2018). Among these types, metal oxide nanoparticles are characterized by remarkable antioxidant activities which make them appropriate for many applications including heavy metals-induced oxidative stress (Mauricio et al. 2018). In plants, heavy metals-induced oxidative stress injury is mainly triggered by lipid peroxidation, protein and DNA damage (Manikandan et al. 2015). Metal oxide nanoparticles are used as an adsorbent in the medium to overcome heavy metals-induce phytotoxicity (Venkatachalam et al. 2017). Mechanistically, nanoparticles can work both as free radical scavengers and antioxidants carriers (Mauricio et al. 2018). For example, cerium nanoparticles were able to reduce H_2O_2-induced oxidative stress by 25–50% through free radical scavenging activity (Ciofani et al. 2013). It is important to note that cerium oxide nanoparticles has a role in attenuating oxidative stress (Singh et al. 2020). Furthermore, nanoparticles are used to deliver a targeted drug to a specific site (Ahmad et al. 2010). Therefore, this evidence indicates that nanoparticles could be exploited as promising antioxidant therapeutic agents.

20.3 Effects of Acute Exposure to Heavy Metals

Studies explored the acute effects of metals using specific plant species, animals or humans (An et al. 2004; Andjelkovic et al. 2019; Hung and Chung 2004; Luan et al. 2008; Nayfeh et al. 2018; Pokorska-Niewiada et al. 2018; Wen et al. 2017). These studies found that these metals could lead to severe toxicity to plants or poisoning to animal models and humans. However, the exact mechanism for the toxicity of heavy metals after acute exposure has not been fully determined.

Using plants species, metals exposure was associated with marked injuries to several species (An et al. 2004; Luan et al. 2008; Pokorska-Niewiada et al. 2018). For instance, the germination process in seeds of plant species was inhibited by acute exposure to metals such as lead and mercury (Pokorska-Niewiada et al. 2018). It is essential to consider that these metals are accumulated in plant tissues (Tangahu et al. 2011). Arsenic, for example, was accumulated in *Pteris vittata* L., while the accumulation of mercury was observed in *Brassica juncea* (L.) Czern. (Tangahu et al. 2011). Moreover, soil-exposed metals such as arsenic, cadmium or lead reduced the root and shoot growth in soybean seeds using acute toxicity test (Luan et al. 2008). This study found that a combination of these three metals resulted in phytotoxicity effects on the soybean seeds. Moreover, another study used five days acute exposure toxicity experiment has supported these findings and reported a marked inhibition for the growth of *Cucumis sativus* L. following exposure to a combination of copper, cadmium and lead (An et al. 2004). These studies suggest that acute exposure to metals is associated with a high rate of accumulation and distribution in plant species, which might result in toxicological reactions. These effects could lead to adverse responses to humans/animals who consume these plants in their food.

Additionally, impairments in the histological, biochemical and hematological biomarkers, as well as other parameters of vital organs, have been also observed in preclinical models after exposure to acute concentrations of metals (Andjelkovic et al. 2019; Larsen et al. 2016; Wen et al. 2017). In a prior report, organist has been indicated that acute inhalation of several metal oxides such as zinc oxide, titanium dioxide, aluminum oxide or cerium oxide were able to reduce the tidal volume and induced nasal irritation in mice (Larsen et al. 2016). Additionally, impairments in the hematological biomarkers were documented in rats exposed to acute doses of cadmium and lead (Andjelkovic et al. 2019). Dysregulation in kidney function and marked damages to several organs such as spleen, liver and kidney, have been reported in Sprague-Dawley rats exposed to an acute dose of silver nanoparticles (Wen et al. 2017). Therefore, acute exposure to metals may induce damage to the organ that first contacts.

In addition to plants and animals, acute exposure to metallic compounds might generate adverse health reactions to humans (Hung and Chung 2004; Nayfeh et al. 2018). Importantly, the environmental pollution of metals is increased globally in recent years (Jaishankar et al. 2014). It is reported that a 42-year old man had thrombocytopenia, fever and hypokalemia due to acute exposure to barium and cadmium (Hung and Chung 2004). Another report highlighted the acute effects of mercury in humans which showed that an increase in urine and serum mercury concentrations as well as deposition of mercury in colon were associated with fatigue, irritability and other adverse reactions in 52-year old man (Nayfeh et al. 2018). These findings raised a question regarding the effects of consuming food containing high levels of metals on the body. It is important to consider that high levels of blood mercury and lead, as well as total urine arsenic, were observed in people who consumed a gluten-free diet compared to people who were on a regular diet (Raehsler et al. 2018). Moreover, a high risk of adverse effects occurrence was highlighted in a study reviewed the levels of metals in vegetables in Armenia (Pipoyan et al. 2018). The findings confirm that monitoring the levels of metals in the fruits or vegetables is necessary to avoid any adverse effects resulting from exposure to them.

20.4 Effects of Chronic Exposure to Heavy Metals

The long-term exposure effects of the metals on living organisms have been determined in previous reports. For example, a study found that subchronic oral exposure to lead was involved in the pathology of gut microbiota in mice (Zhai et al. 2017). A marked decrease in the length of the shoot and root in cucumber was demonstrated following exposure to different concentrations of mercury (Cargnelutti et al. 2006). These findings indicated the toxicity effects after exposure to heavy metals and the significance of regular monitoring for the levels of these metals in plants and other living organisms to avoid any long-term negative consequences.

Plants are exposed to soil and water for a long time for their growth and development. Therefore, the purity and quality of plants-derived foods might be influenced by multiple factors, including air pollution. Cadmium is one of the metals that exhibits plant toxicity due to its ability to interchange from soil to plants (Clemens 2006; Fangmin et al. 2006; Jha and Bohra 2016; Khan, Ding, et al. 2018; Satarug et al. 2009; Wojcik and Tukiendorf 2004). Chronic exposure to cadmium increased the formation of phytochelatins in shoots and roots parts of *Arabidopsis thaliana*, which led to toxic effects in part by dysregulating glutathione system (Wojcik and Tukiendorf 2004). In the leaf of *Camellia sinensis* (L.) Kuntze, it was suggested that cadmium exposure had induced plant toxicity through modulating oxidative stress (Mohanpuria et al. 2007). This suggestion was raised due to the upregulatory effects of cadmium on glutathione syntheses genes.

20.5 Role of Oxidative Stress Biomarkers on the Effect of Heavy Metals

Heavy metals exposure can lead to oxidative stress in plants (Çekiç et al. 2017; Haghighi Pak et al. 2017; Mosa et al. 2018; Ruttkay-Nedecky et al. 2017; Yan and Chen 2019). These metals might induce these adverse reactions through modulating various pathways (Bonaventura et al. 2015; Rokadia and Agarwal 2013; Yang et al. 2014). Such pathways include inflammatory and immunological proteins cascade (Bonaventura et al. 2015; Houston 2007; Iannitti et al. 2010; Kataranovski et al. 2009). Thus, targeting cellular or molecular proteins in this cascade can provide a promising strategy, which can be applied therapeutically in future studies. In this section, we summarize the effects of heavy metals on the oxidative stress biomarkers focusing on plants.

It has been found that copper exposure was associated with adverse reactions in plants (De Vos et al. 1992; Drążkiewicz et al. 2004; Mosa et al. 2018; Thounaojam et al. 2012; Zhang et al. 2008). For instance, copper nanoparticles were able to induce oxidative stress in *C. sativus* (Mosa et al. 2018). This study reported dysregulation in superoxide dismutase enzyme gene expression in groups treated with copper nanoparticles. The levels of hydrogen superoxide and lipid peroxidation were increased in the roots and shoots of *C. sativus* after exposure to copper nanoparticles (Mosa et al. 2018). Copper nanoparticles showed the ability to increase the electrolyte leakage suggesting that *C. sativus* membrane integrity may be sensitive to nanoparticles of heavy metals. Besides, the biomass of *C. sativus* was reduced after exposure to copper nanoparticles. Moreover, the chlorophyll contents were also reduced in *C. sativus* treated with copper nanoparticles. These findings confirm the toxicity effects of copper nanoparticles on plants, which may affect their bioactive constituents. Therefore, it is suggested that nanoparticle properties such as surface area, particle size, and presence of metals as well as other physicochemical properties are critical in the development of oxidative stress (Manke et al. 2013). In *Silene cucubalus* Wibel, copper was found to induce depletion of glutathione, leading to oxidative stress (De Vos et al. 1992). Another study reported that copper exposure led to oxidative stress in *A. thaliana* (Drążkiewicz et al. 2004).

In addition to copper, exposure to silver nanoparticles induced adverse reactions to plants (Çekiç et al. 2017; Haghighi Pak et al. 2017; Yan and Chen 2019). A significant decrease of *Dracocephalum moldavica* L. growth and chlorophyll content has been determined following treatment to high concentrations of silver nanoparticles (Haghighi Pak et al. 2017). These effects were associated with dose-dependently overproduction of hydrogen peroxide, indicating that both copper and silver exposure potentially lead to oxidative stress to the plants. Haghighi Pak et al. reported also an increase in the activity of peroxidase and catalase enzymes in addition to antioxidant (Haghighi Pak et al. 2017). It is suggested that the elevations of antioxidant enzymes activity and gene expression is negative feedback to combat the oxidative stress induced by heavy metals exposure. The findings of silver effects on *D. moldavica* were further supported by a prior study which found that silver could also induce oxidative stress in tomato plants, *Solanum lycopersicum* L. (Çekiç et al. 2017). This and Haghighi Pak et al. studies reported similar data about the ability of silver nanoparticles to reduce chlorophyll contents in plants. Interestingly, a decrease in the activity of ascorbate peroxidase and an increase in the activity of sodium dismutase have been observed after the application of different concentrations of silver nanoparticles to *S. lycopersicum* (Çekiç et al. 2017). In addition, DNA damage and increased lipid peroxidation were associated with these negative reactions.

These reports confirm that nanoparticles of heavy metals exhibit the ability to form oxidative stress responses to living organisms. Monitoring the levels of the nanoparticles of heavy metals in plants might provide beneficial results economically. This will minimize the undesirable health effects that resulted from consuming vegetables and fruits in humans and animals. Future research should determine the differential effects of various formulations of heavy metals on plants focusing on oxidative stress pathways.

20.6 Nanomaterials Induce Positive Effects in Plants

Nanomaterials technology has been developed to overcome the challenging of oxidative stress determined in plants (Elsakhawy et al. 2018; Giraldo et al. 2019; Singh and Husen 2019). It was found that nanoparticles of anionic cerium oxide could protect *A. thaliana* from abiotic stress induced by photosynthesis at least in part by attenuating oxygen reactive species (Wu et al. 2017). These protective effects might be developed because these nanoparticles possess high energy and surface/volume ratios characteristics (Singh and Husen 2019). The high surface density of nanomaterials could interact with the plants' cell surface wall charges (Juárez-Maldonado et al. 2019). These properties may lead to an improvement in plant growth since nanomaterials enable plants to be exposed to the environmental needs (Juárez-Maldonado et al. 2019). These suggestions were further supported by findings showing that nanomaterials induce positive effects on plants such as a decrease of leaching rate and improvement in growth rate (Zhu et al. 2019).

Several studies reported beneficial effects for nanomaterials on the plants, and these effects could be applied to enhance plant life-cycle (Iannone et al. 2016). For instance, exposure to nanomaterials of magnetite iron oxide was found to increase the activity of antioxidant enzymes, which might attenuate oxidative stress in the root and aerial parts of *Triticum aestivum* L. (Iannone et al. 2016). The photosynthesis in sorghum plants was increased following exposure to nanoparticles of cerium oxide (Djanaguiraman et al. 2018). This study reported that these nanoparticles could protect the plant from damage and oxidative stress induced by drought. The promising data for the effects of cerium oxide nanoparticles on oxidative stress parameters included a marked reduction in lipid peroxidation, hydrogen peroxide, and superoxide levels (Djanaguiraman et al. 2018). This indicates that nanotechnology could be a valuable strategy to avoid or reduce the abiotic stress (such as temperature or radiation) or oxidative stress (such as free radicals or superoxide) induced by certain substances, including exposure to heavy metals in plants (Fig. 20.2).

Additionally, a prior study on *C. sinensis* reported that chitosan nanoparticles increased the expression and activity of defense genes and enzymes, respectively (Chandra et al. 2015). This effect was associated with an improvement in the immune system. This study found that chitosan nanoparticles increased the activity and gene expression of antioxidant enzymes, catalase and superoxide dismutase. This effect was also associated with an increased nitric oxide level, a molecule that is involved in stimulating defense pathways (Chandra et al. 2015). However, exposure to nanomaterials can lead to toxicological effects in plants based on various factors (Chichiriccò and Poma 2015; Lin and Xing 2007). In this chapter, we emphasized exclusively on the beneficial effects of using nanobiotechnology approaches to combat negative responses induced by heavy metals in plants (Fig. 20.2).

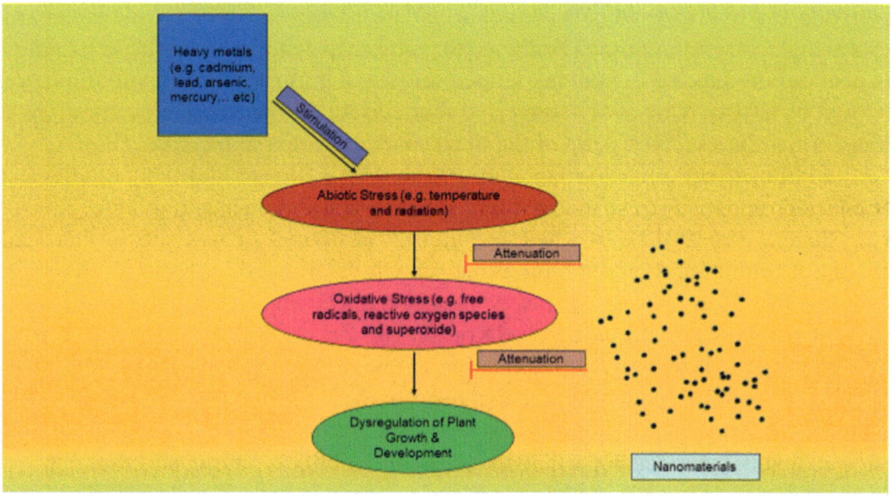

Fig. 20.2 A schematic diagram shows the role of nano-oxide materials in attenuating the effects of abiotic and oxidative stress in plant growth and development (Figure constructed by authors)

20.7 Nanomaterials Combat Oxidative Stress Induced by Exposure to Heavy Metals

Previous studies reported promising findings for the effects of nanomaterials on the oxidative stress induced by heavy metals (Rizwan et al. 2019). Rizwan et al. reported that iron and zinc oxide nanoparticles could improve the plant growth in wheat under the stress of cadmium (Rizwan et al. 2019). This study found that these nanoparticles could modulate oxidative stress pathways, including superoxide dismutase, electrolyte leakage and peroxidase in wheat treated with cadmium. These nanoparticles also could elevate the content chlorophyll in the plant (Rizwan et al. 2019). This hypothesis was supported by a prior study that showed that nanoparticles of cerium oxide could reduce the reactive oxygen species in the leaf of *A. thaliana* through modulating the oxidative stress pathways (Wu et al. 2017). Moreover, nanoparticles of cerium could increase photosynthesis of *Sorghum bicolor* (L.) Moench through attenuating oxidative stress induced by drought (Djanaguiraman et al. 2018). This report found that the nanoparticles could decrease the leaf contents of superoxide radicals and hydrogen peroxide. Another study by Wu., et al., reported that cerium oxide nanoparticles elevated the content of chlorophyll in *A. thaliana* leaf treated with sodium chloride for three days (Wu et al. 2018). The report also showed the ability of cerium oxide nanoparticle to decrease the levels of reactive oxygen species in these leaves exposed to salinity stress, which further indicate the effectiveness of cerium oxide nanoparticles in modulation the oxidative stress parameters. These effects on oxidative stress were determined by measuring the levels of hydrogen peroxide and hydroxy radicals (Wu et al. 2018). However, cerium oxide nanoparticles at high concentrations exposure led to membrane damage and lipid peroxidation in agar cultured media containing lettuce seeds and inhibited the growth of its roots (Cui et al. 2014). This study suggested that cerium (IV) at high concentrations is transformed into specific parts of the plants and converted into cerium (III).

In addition to cerium oxide nanoparticles, magnetic (iron oxide) nanoparticles also attenuated oxidative stress induced by exposure to heavy metals in plants (Konate, He, Zhang, et al. 2017). In wheat seeding (*T. aestivum*), it was shown that lead, copper, cadmium and zinc inhibited the growth determined by reduced shoot and root growth. Moreover, these heavy metals reduced the activities of peroxidase and superoxide dismutase enzymes (Konate, He, Zhang, et al. 2017). These effects were abolished following the application of magnetic nanoparticles (Konate, He, Zhang, et al. 2017), which indicate that nano-oxide like compounds possess activity against oxidative stress in plants. This study was further confirmed by a prior study showed that magnetic nanoparticle induced similar effects on root and shoot growth in cucumber seedlings (*C. sativus*) exposed to heavy metals (Konate, He, Rui, et al. 2017). This study also found that magnetic nanoparticles attenuated heavy metals-decreased peroxidase and superoxide dismutase activities in *C. sativus* Therefore, nanobiotechnology is a recent strategy that has beneficial applications in reducing the plant abiotic stress associated with heavy metals exposure mainly through modulating oxidative stress biomarkers.

Additionally, previous reports pointed out the potential therapeutic effects against oxidative stress in human and animal cells/tissues (Hassanin et al. 2013; Niu et al. 2011; Wardani et al. 2018). Cerium nanoparticles could attenuate oxidative stress induced by exposure to cigarette smoke in rat cardiomyocytes (Niu et al. 2011). This study found that cerium oxide reduces cigarette smoke-activated NF*K*B. Moreover, these nanoparticles were reported to restore cigarette smoke-dysregulated inflammatory cytokines concentrations and gene expression as well as cell viability and apoptosis (Niu et al. 2011).

Another study found a potential efficacy of chitosan against oxidative stress-induced by chronic exposure of cadmium chloride in male rats (Wardani et al. 2018). This study found that a 28-day oral intake of cadmium chloride could induce gastric ulcers and decreased the activities of superoxide dismutase and glutathione peroxidase. These effects were restored with treatments of chitosan nanoparticles (Wardani et al. 2018). These thyroid damage and toxicity through inducing oxidative stress was attenuated after treatments with selenium nanoparticles in rats (Hassanin et al. 2013). Specifically, chromium exposure reduced thyroid hormone levels in part by reducing the level glutathione and increasing the enzyme activities of superoxide dismutase and catalase (Hassanin et al. 2013). This study found that selenium nanoparticles were able to restore impaired thyroid hormones and oxidative stress parameters.

The findings of heavy metals-induced oxidative stress in rats were also observed in other animal models such as mice (Khan, Qureshi, et al. 2018; Zhang et al. 2016). Although iron oxide nanoparticles slightly decreased superoxide dismutase and glutathione peroxidase enzymes activity in the liver of mice, a marked reduction in these enzymes activities has been observed with the treatment of cadmium chloride in livers of mice (Zhang et al. 2016). Interestingly, this study reported that co-exposure to these agents restored the activities of the normal enzymes. It is potential that oxide nanoparticles had alleviated the toxicity effects-induced by cadmium chloride.

Lipid peroxidation was found to be higher in the liver of mice treated with cadmium chloride while it was unchanged in a group received a combination of iron oxide and chromium chloride (Zhang et al. 2016). Exposure to nanoparticles of cerium oxide could attenuate silver-induced oxidative stress in *Labeo rohita* F. Hamilton (Khan, Qureshi, et al. 2018). This report found that nano cerium oxide restored the lipid peroxidation and the level of glutathione as well as the activity of catalase, superoxide dismutase and glutathione-S-transferase. Therefore, the ability of certain nano-like compounds to attenuate diseases associated with oxidative stress might provide potential therapeutic directions for pharmaceutical laboratories to synthesize efficient nano-related compounds.

20.8 Impact of Heavy Metal-Induced Oxidative Stress on Secondary Metabolites Biosynthesis

It is evident that abiotic stressors, including heavy metals, perturb the equilibrium between the generation of reactive oxygen species and the detoxification mechanism in the plants. This imbalance induces the oxidation of crucial biomolecules such as DNA, proteins, and lipids under a condition called oxidative stress, inactivating them and leading ultimately to cell death. The plants are exposed to heavy metals from environmental sources such as contaminated air, water, soil, and food. Therefore, an antioxidative defense system is triggered, which comprises multiple enzymatic and nonenzymatic molecules to battle the adverse effects of oxidative stress (Asgari Lajayer et al. 2017; Bartwal et al. 2013; Berni et al. 2019; Jomova and Valko 2011; Naik and Al-Khayri 2016; Ramakrishna and Ravishankar 2011; Sharma et al. 2012; Zhai et al. 2015).

Among the non-enzymatic molecules, an array of compounds called secondary metabolites are commonly released by the plants to combat the oxidative stress. In contrast to primary metabolites, they have unique functions in interaction with the environment for adaptation and defense but not involved directly in the growth and development of plants. Such secondary metabolites generally include but not limited to polyphenols, terpenoids, and alkaloids. Selected secondary metabolites extensively studied in the literature as antioxidants are illustrated in Fig. 20.3. The structural features of these components play crucial roles in antioxidant and scavenging activities. For instance, the sequence of conjugated double bonds in the carotenoids (tetraterpenoids) such as lycopene is related to the efficiency of quenching while the chelating effect of flavonoids such as quercetin is introduced by three sites of vicinal hydroxy groups which form a complex via coordinate bonds with metal ions

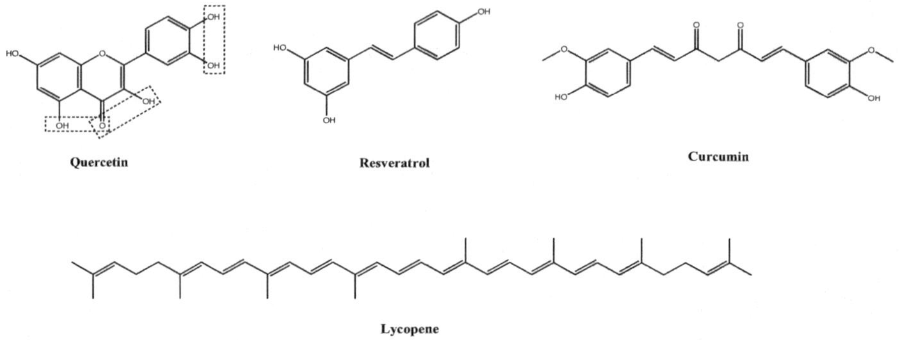

Fig. 20.3 Selected phytochemicals extensively studied in the literature as antioxidants (Figure constructed by authors using ChemDraw)

reducing their catalytic activity to form reactive oxygen species (Berni et al. 2019; Jomova and Valko 2011; Sharma et al. 2012).

Cultivation of medicinal plants in environments surrounded by abiotic stressors such as heavy metals influences biosynthesis of secondary metabolites via induction or inhibition of responsible genes in the biosynthetic pathways or through impaired substrate utilization. This process results in a wide variation in the identity and quantity of these compounds in the marketed botanical products, which disturbs the therapeutic potential, induce toxicity for human organs and raise a substantial quality issue (Asgari Lajayer et al. 2017).

The effects of heavy metals on the production of bioactive compounds in medicinal plants are previously documented. For instance, betalains were produced in higher yields (60% increment) in *Beta vulgaris* L. when treated with Cu^{+2} (1–5 μM) (Trejo-Tapia et al. 2004). Also, cadmium and cupper induced accumulation of sesquiterpenoids such as lubimin and 3-hydroxy lubimin in *Datura stramonium* L. (Furze et al. 1991). Exposure of *Vitis vinifera* L. to silver, cadmium, and cobalt resulted in a remarkable production of a well-known stilbenoid, resveratrol-3-*O*-glucopyranoside (Cai et al. 2013). In addition, exposure to heavy metals such as silver enhanced the biosynthesis of a tropane alkaloid, atropine and a diterpenoid, tanshinone in *Datura metel* L. and *Salvia castanea* Diels, respectively (Li et al. 2016; Zahra Shakeran et al. 2015). The plant production for these secondary metabolites as antioxidants is expected to eliminate the toxicity induced by the heavy metals. Recently, a multitude of medicinal plants and herbal supplements have been exploited as antioxidant agents to prevent or treat toxicities in human body that are induced by diverse toxicants such as heavy metals.

This section sheds light on the significance of monitoring the heavy metals in the environment as a significant factor for assessment of the quality of medicinal plants as well as herbal supplements in the market. On the other hand, the recent developments in the elicitation of plant tissue culture have opened a new avenue to activate cryptic (silent) secondary metabolites or produce certain phytochemicals in higher yields. As a result, heavy metals could be recruited as elicitors to manipulate the genes responsible for production, thus mining for novel bioactive natural compounds. This strategy might be applied for other organisms such as marine and microbial species to be exploited as a cost-effective alternative platform for the total synthetic approach.

20.9 Physicochemical Properties of Nano-antioxidants

It is essential to consider that nanomaterials are protective compounds for plants. They also exhibit an ability to facilitate the action of insecticides, herbicides or fungicides as they work as carriers (Worrall et al. 2018). This chapter discussed the positive characteristics for these nanomaterials to work as carriers, which include their ability to decrease the toxicity, increase the half-life, and enhance the uptake of insecticides, herbicides or fungicides. These actions are critical to combat multiple diseases in plants. We suggest here that the physicochemical characteristics for nanomaterial of certain compounds might aid in attenuating oxidative stress associated with heavy metals exposure.

It is important to note that the delivery strategy of antioxidants has been suggested to attenuate many pathological conditions induced by oxidative stress (Eftekhari et al. 2018). Optimal pharmacokinetic properties, including absorption and distribution rates of antioxidant nanoparticles are essential to combat oxidative stress as compared to conventional antioxidants (Eftekhari et al. 2018). However, the nanoparticles of heavy metals or other environmental toxins can induce oxidative stress due to their physicochemical characteristics such as particle size and surface area (Manke et al. 2013).

20.10 Conclusion and Prospects

The central hypothesis that will be obtained from this chapter focuses on the beneficial effects of nanobiotechnology applications to avoid any damage or injuries that occurred in plants after exposure to heavy metals. Monitoring the levels of heavy metals on the environment is essential to maintain high quality standards of marketed herbal supplements and medicinal plants. The use of nano-oxide particles could mitigate abiotic stress in plants, which provide a clear understanding of the effectiveness of using nan-technology in the future to overcome oxidative stress in plants that is induced by various factors including the presence of heavy metals. Future work should determine the role of nanobiotechnology applications in minimizing the undesirable consequences economically and medically. It is important that the pharmaceutical industry develops nano-antioxidant particles with optimal efficacy and pharmacokinetic profiles that can be used to attenuate oxidative stress in plants. Moreover, nanotechnology will provide beneficial outcomes on the production of phytochemicals in higher yields. This technology might be also applied on marine and microbial species as an alternative strategy for total synthetic approach.

Acknowledgements The authors would like to thank the international scientific partnership program (ISPP-146), King Saud University, Riyadh, Saudi Arabia, for their support.

References

Ahmad MZ, Akhter S, Jain GK et al (2010) Metallic nanoparticles: technology overview and drug delivery applications in oncology. Expert Opin Drug Deliv 7(8):927–942

An Y-J, Kim Y-M, Kwon T-I, Jeong S-W (2004) Combined effect of copper, cadmium, and lead upon *Cucumis sativus* growth and bioaccumulation. Sci Total Environ 326(1–3):85–93

Andjelkovic M, Buha Djordjevic A, Antonijevic E et al (2019) Toxic effect of acute cadmium and lead exposure in rat blood, liver, and kidney. Int J Environ Res Public Health 16(2):274

Asgari Lajayer B, Ghorbanpour M, Nikabadi S (2017) Heavy metals in contaminated environment: destiny of secondary metabolite biosynthesis, oxidative status and phytoextraction in medicinal plants. Ecotoxicol Environ Saf 145:377–390

Bartwal A, Mall R, Lohani P et al (2013) Role of secondary metabolites and brassinosteroids in plant defense against environmental stresses. J Plant Growth Regul 32(1):216–232

Berni R, Luyckx M, Xu X et al (2019) Reactive oxygen species and heavy metal stress in plants: impact on the cell wall and secondary metabolism. Environ Exp Bot 161:98–106

Bonaventura P, Benedetti G, Albarède F, Miossec P (2015) Zinc and its role in immunity and inflammation. Autoimmun Rev 14(4):277–285

Borm PJ, Robbins D, Haubold S et al (2006) The potential risks of nanomaterials: a review carried out for ecetoc. Part Fibre Toxicol 3:11

Cai Z, Kastell A, Speiser C, Smetanska I (2013) Enhanced resveratrol production in *Vitis vinifera* cell suspension cultures by heavy metals without loss of cell viability. Appl Biochem Biotechnol 171(2):330–340

Cargnelutti D, Tabaldi LA, Spanevello RM et al (2006) Mercury toxicity induces oxidative stress in growing cucumber seedlings. Chemosphere 65(6):999–1006

Çekiç FÖ, Ekinci S, İnal MS, Özakça D (2017) Silver nanoparticles induced genotoxicity and oxidative stress in tomato plants. Turk J Biol 41(5):700–707

Chandra S, Chakraborty N, Dasgupta A et al (2015) Chitosan nanoparticles: a positive modulator of innate immune responses in plants. Sci Rep 5:15195

Chichiriccò G, Poma A (2015) Penetration and toxicity of nanomaterials in higher plants. Nanomaterials 5(2):851–873

Ciofani G, Genchi GG, Liakos I et al (2013) Effects of cerium oxide nanoparticles on pc12 neuronal-like cells: proliferation, differentiation, and dopamine secretion. Pharm Res 30(8):2133–2145

Clemens S (2006) Toxic metal accumulation, responses to exposure and mechanisms of tolerance in plants. Biochimie 88(11):1707–1719

Cui D, Zhang P, Ma Y et al (2014) Effect of cerium oxide nanoparticles on asparagus lettuce cultured in an agar medium. Environ Sci Nano 1(5):459–465

De Jong WH, Borm PJ (2008) Drug delivery and nanoparticles: applications and hazards. Int J Nanomedicine 3(2):133–149

De Matteis V, Rinaldi R (2018) Toxicity assessment in the nanoparticle era. Adv Exp Med Biol 1048:1–19

De Vos CR, Vonk MJ, Vooijs R, Schat H (1992) Glutathione depletion due to copper-induced phytochelatin synthesis causes oxidative stress in *Silene cucubalus*. Plant Physiol 98(3):853–858

Djanaguiraman M, Nair R, Giraldo JP, Prasad PVV (2018) Cerium oxide nanoparticles decrease drought-induced oxidative damage in sorghum leading to higher photosynthesis and grain yield. ACS omega 3(10):14406–14416

Drążkiewicz M, Skórzyńska-Polit E, Krupa Z (2004) Copper-induced oxidative stress and antioxidant defence in *Arabidopsis thaliana*. Biometals 17(4):379–387

Eftekhari A, Dizaj SM, Chodari L et al (2018) The promising future of nano-antioxidant therapy against environmental pollutants induced-toxicities. Biomed Pharmacother 103:1018–1027
Elsakhawy T, Omara AE-D, Alshaal T, El-Ramady H (2018) Nanomaterials and plant abiotic stress in agroecosystems. Environ Biodivers Soil Secur 2(2018):73–94
Fangmin C, Ningchun Z, Haiming X et al (2006) Cadmium and lead contamination in japonica rice grains and its variation among the different locations in Southeast China. Sci Total Environ 359(1–3):156–166
Furze JM, Rhodes MJ, Parr AJ et al (1991) Abiotic factors elicit sesquiterpenoid phytoalexin production but not alkaloid production in transformed root cultures of *Datura stramonium*. Plant Cell Rep 10(3):111–114
Giraldo JP, Wu H, Newkirk GM, Kruss S (2019) Nanobiotechnology approaches for engineering smart plant sensors. Nat Nanotechnol 14(6):541–553
Haghighi Pak Z, Karimi N, Abbaspour H (2017) Effects of silver nanoparticle exposure on growth, physiological and biochemical parameters of *Dracocephalum moldavica* l. Plant Physiol 7(4):2173–2183
Hasanuzzaman M, Nahar K, Rahman A et al (2018) Exogenous glutathione attenuates lead-induced oxidative stress in wheat by improving antioxidant defense and physiological mechanisms. J Plant Interact 13(1):203–212
Hassanin KM, El-Kawi SHA, Hashem KS (2013) The prospective protective effect of selenium nanoparticles against chromium-induced oxidative and cellular damage in rat thyroid. Int J Nanomedicine 8:1713
Ho T-T, Murthy HN, Park S-Y (2020) Methyl jasmonate induced oxidative stress and accumulation of secondary metabolites in plant cell and organ cultures. Int J Mol Sci 21(3):716
Houston MC (2007) The role of mercury and cadmium heavy metals in vascular disease, hypertension, coronary heart disease, and myocardial infarction. Altern Ther Health Med 13(2):S128–S133
Hung Y-M, Chung H-M (2004) Acute self-poisoning by ingestion of cadmium and barium. Nephrol Dial Transplant 19(5):1308–1309
Huo Y, Kang JP, Ahn JC et al (2020) Siderophore-producing rhizobacteria reduce heavy metal-induced oxidative stress in *Panax ginseng* meyer. J Ginseng Res
Iannitti T, Capone S, Gatti A et al (2010) Intracellular heavy metal nanoparticle storage: progressive accumulation within lymph nodes with transformation from chronic inflammation to malignancy. Int J Nanomedicine 5:955
Iannone MF, Groppa MD, de Sousa ME et al (2016) Impact of magnetite iron oxide nanoparticles on wheat (*Triticum aestivum* L.) development: evaluation of oxidative damage. Environ Exp Bot 131:77–88
Jaishankar M, Tseten T, Anbalagan N et al (2014) Toxicity, mechanism and health effects of some heavy metals. Interdiscip Toxicol 7(2):60–72
Jha UC, Bohra A (2016) Genomics enabled breeding approaches for improving cadmium stress tolerance in plants. Euphytica 208(1):1–31
Jomova K, Valko M (2011) Advances in metal-induced oxidative stress and human disease. Toxicology 283(2–3):65–87
Juárez-Maldonado A, Ortega-Ortíz H, Morales-Díaz A et al (2019) Nanoparticles and nanomaterials as plant biostimulants. Int J Mol Sci 20(1):162
Kataranovski M, Janković S, Kataranovski D et al (2009) Gender differences in acute cadmium-induced systemic inflammation in rats. Biomed Environ Sci 22(1):1–7
Khan MA, Ding X, Khan S et al (2018) The influence of various organic amendments on the bioavailability and plant uptake of cadmium present in mine-degraded soil. Sci Total Environ 636:810–817
Khan MHA, Parvez S (2015) Hesperidin ameliorates heavy metal induced toxicity mediated by oxidative stress in brain of Wistar rats. J Trace Elem Med Biol 31:53–60
Khan MS, Qureshi NA, Jabeen F (2018) Ameliorative role of nano-ceria against amine coated ag-np induced toxicity in labeo rohita. Appl Nanosci 8(3):323–337

Konate A, He X, Rui Y-K, Zhang Z-Y (2017) Magnetite (Fe_3O_4) nanoparticles alleviate growth inhibition and oxidative stress caused by heavy metals in young seedlings of cucumber (*Cucumis sativus* L). Paper presented at the ITM Web of Conferences

Konate A, He X, Zhang Z et al (2017) Magnetic (Fe_3O_4) nanoparticles reduce heavy metals uptake and mitigate their toxicity in wheat seedling. Sustainability 9(5):790

Larsen ST, Jackson P, Poulsen SS et al (2016) Airway irritation, inflammation, and toxicity in mice following inhalation of metal oxide nanoparticles. Nanotoxicology 10(9):1254–1262

Li B, Wang B, Li H et al (2016) Establishment of *Salvia castanea* diels f. Tomentosa stib. Hairy root cultures and the promotion of tanshinone accumulation and gene expression with Ag^+, methyl jasmonate, and yeast extract elicitation. Protoplasma 253(1):87–100

Lin D, Xing B (2007) Phytotoxicity of nanoparticles: inhibition of seed germination and root growth. Environ Pollut 150(2):243–250

Luan Z, Cao H, Yan B (2008) Individual and combined phytotoxic effects of cadmium, lead and arsenic on soybean in phaeozem. Plant Soil Environ 54(9):403–411

Manikandan R, Sahi SV, Venkatachalam P (2015) Impact assessment of mercury accumulation and biochemical and molecular response of mentha arvensis: a potential hyperaccumulator plant. Sci World J 2015:715217

Manke A, Wang L, Rojanasakul Y (2013) Mechanisms of nanoparticle-induced oxidative stress and toxicity. Biomed Res Int 2013

Mauricio MD, Guerra-Ojeda S, Marchio P et al (2018) Nanoparticles in medicine: a focus on vascular oxidative stress. Oxid Med Cell Longev 2018:6231482

Mohanpuria P, Rana NK, Yadav SK (2007) Cadmium induced oxidative stress influence on glutathione metabolic genes of *Camellia sinensis* (L.) o. Kuntze. Environ Toxicol 22(4):368–374

Mosa KA, El-Naggar M, Ramamoorthy K et al (2018) Copper nanoparticles induced genotoxicty, oxidative stress, and changes in superoxide dismutase (sod) gene expression in cucumber (*Cucumis sativus*) plants. Front Plant Sci 9

Naik PM, Al-Khayri JM (2016) Abiotic and biotic elicitors–role in secondary metabolites production through *in vitro* culture of medicinal plants. In: Shanker A (ed) Abiotic and biotic stress in plants—recent advances and future perspectives. Intech Open, London, pp 247–277

Nayfeh A, Kassim T, Addasi N et al (2018) A challenging case of acute mercury toxicity. Case Rep Med 2018

Niu J, Wang K, Kolattukudy PE (2011) Cerium oxide nanoparticles inhibits oxidative stress and nuclear factor-κb activation in h9c2 cardiomyocytes exposed to cigarette smoke extract. J Pharmacol Exp Ther 338(1):53–61

Panda SK, Choudhury S, Patra HK (2016) Heavy metal induced oxidative stress in plants: physiological and molecular perspectives. In: Tuteja N, Gill SS (ed) Abiotic stress response in plants. Wiley Online Library, pp 219–231

Pipoyan D, Beglaryan M, Costantini L et al (2018) Risk assessment of population exposure to toxic trace elements via consumption of vegetables and fruits grown in some mining areas of armenia. Hum Ecol Risk Assess 24(2):317–330

Pokorska-Niewiada K, Rajkowska-Myśliwiec M, Protasowicki M (2018) Acute lethal toxicity of heavy metals to the seeds of plants of high importance to humans. Bull Environ Contam Toxicol 101(2):222–228

Pujalté I, Passagne I, Brouillaud B et al (2011) Cytotoxicity and oxidative stress induced by different metallic nanoparticles on human kidney cells. Part Fibre Toxicol 8(1):10

Raehsler SL, Marietta EV, Murray JA (2018) Accumulation of heavy metals in people on a gluten-free diet. Clin Gastroenterol Hepatol 16(2):244–251

Ramakrishna A, Ravishankar GA (2011) Influence of abiotic stress signals on secondary metabolites in plants. Plant Signal Behav 6(11):1720–1731

Rizwan M, Ali S, Ali B et al (2019) Zinc and iron oxide nanoparticles improved the plant growth and reduced the oxidative stress and cadmium concentration in wheat. Chemosphere 214:269–277

Rokadia HK, Agarwal S (2013) Serum heavy metals and obstructive lung disease: results from the national health and nutrition examination survey. Chest 143(2):388–397

Ruttkay-Nedecky B, Krystofova O, Nejdl L, Adam V (2017) Nanoparticles based on essential metals and their phytotoxicity. J Nanobiotechnology 15(1):33

Satarug S, Garrett SH, Sens MA, Sens DA (2009) Cadmium, environmental exposure, and health outcomes. Environ Health Perspect 118(2):182–190

Sharma P, Jha AB, Dubey RS, Pessarakli M (2012) Reactive oxygen species, oxidative damage, and antioxidative defense mechanism in plants under stressful conditions. J Bot 2012

Singh KR, Nayak V, Sarkar T, Singh RP (2020) Cerium oxide nanoparticles: properties, biosynthesis and biomedical application. RSC Adv 10(45):27194–27214

Singh S, Husen A (2019) Role of nanomaterials in the mitigation of abiotic stress in plants. In: Husen A, Iqbal M (eds) Nanomaterials and plant potential. Springer, Cham, pp 441–471

Stapleton PA, Nurkiewicz TR (2014) Vascular distribution of nanomaterials. Wiley Interdiscip Rev Nanomed Nanobiotechnol 6(4):338–348

Tangahu BV, Abdullah S, Rozaimah S et al (2011) A review on heavy metals (As, Pb, and Hg) uptake by plants through phytoremediation. Int J Chem Eng 2011

Thounaojam TC, Panda P, Mazumdar P et al (2012) Excess copper induced oxidative stress and response of antioxidants in rice. Plant Physiol Biochem 53:33–39

Trejo-Tapia G, Jimenez-Aparicio A, Rodriguez-Monroy M et al (2004) Influence of cobalt and other microelements on the production of betalains and the growth of suspension cultures of beta vulgaris. Plant Cell, Tissue Organ Cult 67(1):19–23

Venkatachalam P, Jayaraj M, Manikandan R et al (2017) Zinc oxide nanoparticles (znonps) alleviate heavy metal-induced toxicity in *Leucaena leucocephala* seedlings: a physiochemical analysis. Plant Physiol Biochem 110:59–69

Wardani G, Eraiko K, Sudjarwo SA (2018) Protective activity of chitosan nanoparticle against cadmium chloride induced gastric toxicity in rat. J Young Pharm 10(3):303

Wen H, Dan M, Yang Y et al (2017) Acute toxicity and genotoxicity of silver nanoparticle in rats. PLoS ONE 12(9):e0185554

Wojcik M, Tukiendorf A (2004) Phytochelatin synthesis and cadmium localization in wild type of *Arabidopsis thaliana*. Plant Growth Regul 44(1):71–80

Worrall E, Hamid A, Mody K et al (2018) Nanotechnology for plant disease management. Agronomy 8(12):285

Wu H, Shabala L, Shabala S, Giraldo JP (2018) Hydroxyl radical scavenging by cerium oxide nanoparticles improves arabidopsis salinity tolerance by enhancing leaf mesophyll potassium retention. Environ Sci Nano 5(7):1567–1583

Wu H, Tito N, Giraldo JP (2017) Anionic cerium oxide nanoparticles protect plant photosynthesis from abiotic stress by scavenging reactive oxygen species. ACS Nano 11(11):11283–11297

Yan A, Chen Z (2019) Impacts of silver nanoparticles on plants: a focus on the phytotoxicity and underlying mechanism. Int J Mol Sci 20(5):1003

Yang S-N, Hsieh C-C, Kuo H-F et al (2014) The effects of environmental toxins on allergic inflammation. Allergy Asthma Immunol Res 6(6):478–484

Zahra Shakeran MK, Asghari Gholamreza, Ghandian Mustafa (2015) Improvement of atropine production by different biotic and abiotic elicitors in hairy root cultures of datura metel. Turk J Biol 39:111–118

Zhai Q, Li T, Yu L et al (2017) Effects of subchronic oral toxic metal exposure on the intestinal microbiota of mice. Sci Bull 62(12):831–840

Zhai Q, Narbad A, Chen W (2015) Dietary strategies for the treatment of cadmium and lead toxicity. Nutrients 7(1):552–571

Zhang H, Hu LY, Hu KD et al (2008) Hydrogen sulfide promotes wheat seed germination and alleviates oxidative damage against copper stress. J Integr Plant Biol 50(12):1518–1529

Zhang Y, Xu X, Zhu S et al (2016) Combined toxicity of fe 3 o 4 nanoparticles and cadmium chloride in mice. Toxicol Res 5(5):1309–1317

Zhu Y, Xu F, Liu Q et al (2019) Nanomaterials and plants: positive effects, toxicity and the remediation of metal and metalloid pollution in soil. Sci Total Environ 662:414–421

Chapter 21
Nanonutrients: Plant Nutritive and Possible Antioxidant Regulators

Ayoob O. Alfalahi and Fadwa W. Abdulqahar

Abstract There is a growing use of nanotechnology in agriculture, especially in the densely populated countries looking for unconventional sources for feeding their peoples. One of the main concerns considering nutrients application is very low of applied nutrient succeeded in reaching the targeted site, thus the delivered quantity will be much below the required concentration adequate for specific biological activity. Notably, only 20% of the applied nutrients through soil can be uptaken by the plant, whereas the residue either creating stable complexes with soil components or being washed away with water. In both cases, plants will be capable to get advantage only from the minimum limit of the applied nutrients. The nanoparticle-based nutrients have several key advantages over traditional nutrients. Primarily, nanonutrient does not release as fast as the traditional nutrient, hence it will not significantly affect the soil pH due to gradual release. This, in turn, will guarantee a slow and steady release of a specific nutrient that permits plants to continuously take up the nutrient as they grow. Throughout their development, plants face a vigorously shifting in environment conditions falling within either biotic or abiotic factors. Regarding this, nanofertilizers have proven efficiency in reducing the adverse side effects of unfavorable environmental conditions by activating antioxidant enzymes and decreased oxidative processes outputs, primarily reactive oxygen species (ROS) and/or reactive nitrogen species (RNS). Although, plants needed micronutrients in small quantities, they still playing a vital role in several metabolic pathways. Even as plants are cultivated in a variety of stressful conditions, nanoparticles (NPs) can be an effective tool for endorsing a protective antioxidant system. Considerable investigations/studies have to be done before decisively determining the biosafety of nanomaterials, as long as their toxic effects have already been demonstrated on many occasions.

A. O. Alfalahi (✉)
Department of Plant Protection, College of Agriculture, University of Anbar, P.O. Box 55431, Ramadi, Iraq
e-mail: ag.ayoob.obaid@uoanbar.edu.iq

F. W. Abdulqahar
Department of Food Sciences, University of Anbar, P.O. Box 55431, Ramadi, Iraq
e-mail: ag.fadwa.waleed@uoanbar.edu.iq

J. M. Al-Khayri et al. (eds.), *Nanobiotechnology*,
https://doi.org/10.1007/978-3-030-73606-4_21

Keywords Nanoparticles · Nanofertilizers · Nanonutrient · Nanofortifiers · Secondary metabolites

21.1 Introduction

Strategy plans must be taken into consideration with the constant increase in the human population. During the last few decades, conventional fertilizers participate significantly in boosting plant productivity to ensure requirements of the global food basket. However, the shift in the lifestyle of societies and human activities on agricultural lands causes them to lose their fertility, thus, it is more likely to search for nanoscale alternatives in an attempt to restore soil nutritive capabilities and secures an acceptable level of production to bridge the food gap (Henchion et al. 2017; Savci 2012). Although the adoption of different fertilizers is laying behind the tremendous augmentation of crop productivity, particularly during the green revolution, chemical fertilizers have their own drawbacks (Lin et al. 2019). Thereby, there is an urgent need for innovative strategies marked with low waste and cost of the supplemented agrochemicals. Nanomaterials (NMs) are not novel as some peoples think, nanoscale particles are naturally occurred via geological or biological processes in the ecosystem, however precipitation and bioreduction are on the top of the list (Kamle et al. 2020). Definitely, the naturally emanated NMs with a relative difference in terms of physical, chemical and optical characteristics. In this context, volcanoes and hydrothermal activities are the most common examples of natural emanated nanoscopic particles (Jacob 2018). Remarkably, the natural biological system that we part of is generating NPs infrequently (Gupta and Xie 2018). In response to this, human beings are exposed daily to different types of nanoparticles with or without their awareness, at the same time they have a limited ability to control either the generation or distribution of natural nanoparticles (Jeevanandam et al. 2018).

The field of engineered nanotechnology has made numerous innovative progress over the last two decades, especially in agricultural and industrial sectors. Nanoparticles (NPs) involved a wide range of particulate substances with at least one dimension less than 100 nm (Khan et al. 2019). Due to their various biological, pharmaceutical, chemical, food and industrial applications, the engineered nanomaterials (ENMs) will be released in a considerable amount to the local environment, consequently their deposition should be considered (Zoufan et al. 2020). Depending on their distinctive characteristics, nanomaterials can interact directly or indirectly with most components of the biological system, nevertheless the arising threat lies behind unsatisfactory knowledge of either the nature or outputs of these interactions (Bundschuh et al. 2018). Alternatively, the production of plant-derived NPs shows substantial benefits over other bio-systems as plants are readily obtainable and the biogenic synthesis process offering value-effective technique (Sharma et al. 2019). Nanofertilizers can be practiced as macro or micro-nutrients per se or as carriers (Kah et al. 2018), and even as coated nutrients (DeRosa 2010). Furthermore, molecular coatings with various biomolecules have a great significance for their use as smart delivery systems to ensure the slow-release of nutrients at the root zone (Usman et al. 2020). Falling within nanotechnology, nanocarriers may be a common concept in pharmaceutical and animal systems, while it is less shared in the botanical system despite its importance.

The level at which nanomaterial affects plant performance varies according to general and particular features of these materials (origin, synthesis method, size, charge, surface, concentration and plant species). In this context, some nanomaterials retain a purely nutritional effect reflected on improved growth parameters, meanwhile, it may constitute a catalyst influence by generating Reactive oxygen species (ROS) and/or secondary metabolites (Aslani et al. 2014). Despite the great ability of nanomaterials to trigger disruptive and toxic effects through ROS generation, these substances are holding significant promises in enhancing the nutritional value of agricultural products from a fortification viewpoint (Armstead and Li 2016).

This chapter is focusing on the agricultural applications of nanomaterials in what became known as "Agronanotechnology", in which the most common approaches for synthesizing NPs especially the green biosynthesis will be addressed. As a promising technology, the key advantages of using nanomaterials in developing an effective delivery system via nanocarriers, modulating secondary metabolites and oxidative response as well as the possible eco-toxicological effects of their application will be outlined.

21.2 Biosynthesis of Nanonutrients

Nanomaterials are widely used for agricultural applications due to their small sizes and efficient delivery system for nutritive elements. Likewise, NPs exhibit entirely unique or improved properties, meanwhile retaining some distinctive features such as structure, shape, optical properties and nano-size that falling between 1 and 100 nm (Shang et al. 2019). Moreover, the NPs have a high surface-to-volume ratio that qualifies them to incorporate with numerous moieties and in term of size, NPs has become a bridge link between traditional bulk and molecular systems (Henriksen-Lacey et al. 2017; Sharma et al. 2015). Traditionally, several physical and chemical approaches have been employed for the preparation of NPs (Fig. 22.1). Commonly, all fall in either top-down or bottom-up approaches (Khan et al. 2019). However, most of these methods have some drawbacks, e.g., toxicity, labor, high cost and requirement. Hence, in recent years, researchers focused on developing new simple, cheap and safe protocols that guarantee easy preparation and manipulation (Singh et al. 2018). Generally, NPs can be classified into organic and inorganic, and despite their different physical, optical, chemical, electrical, thermal properties, both NPs categories share the same nano sizes. Inorganic NPs integrate metallic, magnetic and semiconductor NPs. In contrast, organic NPs are mainly integrate carbon NPs, e.g., carbon nanotubes, fullerenes and quantum dots (Khan et al. 2019).

The development of more reliable and eco-friendly approaches to synthesize nanonutrients is a significant step in the field of nanotechnology in general, and

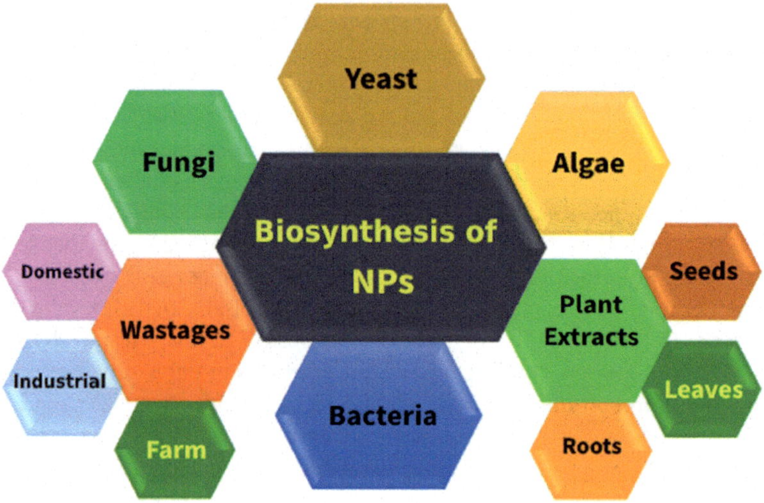

Fig. 22.1 Potential approaches serving biosynthesis of nanoparticles (Figure constructed by Ayoob O. Alfalahi)

agronanotechnology in particular (Prasad et al. 2017). Inorganic metal ions can be transformed into metal nanoparticles through proteins-mediated reductive process by and some other metabolites exist in a wide range of biological systems including eukaryotes such as plants and algae (Makarov et al. 2014), yeasts (Khandel and Shahi 2018), fungi (Chan and Mat Don 2013) and even human cells (Anshup et al. 2005; Khan et al. 2019) and prokaryotes; like bacteria (Das et al. 2017; Shivaji et al. 2011), (Fig. 22.1). However, using microbes to synthesize nanoparticles can be a comparatively challenging technique since maintaining cell cultures will require complicated operations and consecutive purification steps (Kajbafna et al. 2012). Thereby, the using of plant in the production of nanoparticle offer important benefits more than other biological systems where plants are easier to deal with and more available. Furthermore, the procedure of plant-based biogenic synthesis is low-priced and less elaborate as compared to other suggested approaches which are time consuming, elaborate, and require aseptic conditions, such as fungi (Chaudhuri and Malodia 2017).

Interestingly, enzymes and other active components contained by the plant cellular system including alkaloids, flavonoids, glycosides, terpenoids, terpenes, saponins, steroids, tannins and volatiles play an important role as natural capping and reducing agents (Raghunandan et al. 2009). Recently, potassium nano-fertilizer was prepared from banana peels, and the resulted peels-extract was physically and chemically characterized. Although the nanofertilizer was ranged in size between 19 and 55 nm, the majority of the nanoparticles (36%) were in the size of 40 nm. However, only 6% of the synthesized nanoparticles were in a larger size of 55 nm. The prepared nanofertilizer composed of potassium and iron in chelated form, urea, proteins and different amino acids. The results indicated that the increased dosage of banana peels extract improved the germination percentage from 14 to 97%; and from 25 to 93.14% after seven days in tomato and fenugreek crops, respectively (Hussein et al. 2019).

The microwave-assisted hydrothermal technique was successfully adopted by Shebl et al. (2019) to synthesize manganese zinc ferrite nanoparticles using 13 green chemistry techniques. FE-SEM and HR-TEM tests demonstrated the cubic shape of the resulted nanoparticles with 10–12 nm sizes. The efficiency of prepared nanofertilizers was proved in nourishing squash plant (*Cucurbita pepo* L.) and the 23 minerals content. The results showed that lower concentrations had a more positive effect on growth and yield traits, compared to the higher concentrations. Leaf extract of Calotropis (*Calotropis gigantea* L.) was used in green synthesis of zinc oxide nanoparticles in combination with zinc acetate salt mediated by NaOH. An amount of 200 mM zinc acetate found ideal to produce zinc oxide nanoparticles with less than 20 nm size. The prepared crystalline nanoparticles were characterized throughout FTIR (Fourier transform infrared spectroscopy), XRD (X-ray diffraction) and UV–Vis spectroscopy. Biogenic zinc oxide enhances the parameters of seedlings growth and normal development at the nursery stage (Chaudhuri and Malodia 2017).

21.3 Nanofertilizers as a Crop Nutrients

A growing plant needs about eighteen essential elements to grow and develop normally. However, only three of these elements, light, air and water can be obtained naturally from the surrounding environment. Accordingly, the plant depends completely on the soil to ensure the rest of fifteen elements. Nutrients deficiency is a common problem hindering the development of many essential crops (Manwaring et al. 2016). Typically, the use of traditional fertilizers is accompanied by many obstacles, in the forefront of which is the massive additions will lead to; low bioavailability of other microelements, higher accumulation rate of soil and groundwater pollutants, as well as irreversibly impact the soil chemical and/or physical ecology, finally leading to low crops productivity (Meena et al. 2017). To overcome this, nanotechnology has the potential to transfer the agricultural and food industry to a new level, by evolving new insecticide, herbicide and more absorbable nutrients (Duhan et al. 2017).

Nanotechnology started to appeal more attention to develop nanofertilizers with minimum loss to the surrounding environment, slowly and controlled release and improving nutrient use efficiency, thus it became the successful alternative option of improving new forms of fertilizers serving for sustainable agriculture (Zulfiqar et al. 2019). Most of nanonutrients share the same positive effect at relatively low concentrations, meanwhile adverse effects of growth inhibition and deterioration of physiological and morphological indicators are established along with the higher concentrations of NPs (Mahakham et al. 2017; Raliya et al. 2018). For instance, TiO_2 reflected on improved photosynthesis and metabolic activities at very low concentrations (20 mg/L) (Yang et al. 2006). However, higher dosages of TiO_2 adversely affected transcriptomic patterns and root hair development of *Arabidopsis* (García-Sánchez et al. 2015). Nanofertilizers can be categorized into three kinds, nanoscale fertilizers, additive fertilizers, and coated fertilizers (Pandorf et al. 2020). The first category including NPs that contain nutrients. While the second category of nanosize additive fertilizers involved conventional fertilizers combined with nanosize additives. However, loading or coating the conventional fertilizers with NPs represents the nanosize coated fertilizers (Shang et al. 2019).

Commonly, there are two approaches to produce NMs; physical that described as a top-down approach, and chemical that described as a bottom-up approach (Slepička et al. 2019). The desired nutrient can be encapsulated either within nanoporous materials or nanoemulsions. The rapid advancement in the nanotechnology field has shaped alternative classification for nanofertilizers consistent with their actions, control-release or loss fertilizers and nanocomposite fertilizers where nanodevices ensure gradual release of collected micro- and macro-nutrients (Shang et al. 2019). The encapsulated microorganism will be a successful practice to improve the availability of major nutrients around the root area like nitrogen, phosphorus and potassium, thus positively affecting growth and yield attributes (Bargaz et al. 2018). As for the nanoporous, it is an effective option to improve nutrient use efficiency by rationing nutrient supply according to the actual need, furthermore porous nanomaterials can notably increase the solubility of nutritional minerals. For example, ammonium charged zeolites found to be efficient for long-lasting release and diminish leaching losses (Preetha and Balakrishnan 2017). Nevertheless, the synthesized nanofertilizer may be a nano potassium, phosphorus, zinc, silver, silica, iron or titanium dioxide, ZnCdSe/ZnS core shell QDs, Mn/ZnSe QDs, gold nanorods, nanozeolite etc. (Elemike et al. 2019). Particle properties, pH, and kinetics hamper the synthesized fertilizer efficiency. Hence, it is vital to appreciate the nanonutrients mechanism in the plant–soil system (Ruttkay-Nedecky et al. 2017).

Several studies approved the positive significant effect of zinc oxide nanofertilizer ZnO-NPs in improving agronomic, physiologic and yield indices of wheat (Munir et al. 2018) and common bean (Salama et al. 2019). Remarkably, ZnO-NPs found to be more effective in improving germination and growing indicators than ZnSO4, and the latter was more toxic compared to ZnO-NPs, especially in higher dosages (Du et al. 2019). Meanwhile, Khodakovskaya et al. (2013) found that carbon nanoparticles improve growth and yield parameters of tomato. In a pot experiment conducted in growth chamber conditions, Cieschi et al. (2019) applied F, S and M hybrid nanomaterials to synthesize iron-humic nanofertilizers applied in 35, 75 and 150 mmol pot^{-1} on calcareous soils. Treated soybean plants showed a significant increase in iron uptake, reflected on higher shoot fresh weight. The availability of the applied humic nanofertilizers lasted for a long period and was verified in the harvested soybean pods. The applications of nanomaterials are emerging and diversifying rapidly, serving in presenting solutions for the growing challenges (Table 22.1). For example, the development of nanosensors has a promising future in improving plant tolerance to biotic and abiotic stresses known as precision agriculture (Afsharinejad et al. 2016; Kwak et al. 2017). Nanofertilizers can contribute to supporting plant nutritional status in one of two forms. The first is the use of the nanostructured element combined

Table 22.1 The inductive effect of different types and doses of plant nanofertilizers

Nanofertilizer	Applied dose	Targeted plant species	Inductive effect	References
Iron oxide nanoparticles (γ–Fe_2O_3 NPs)	20–100 mg L^{-1}	Pummelo (*Citrus maxima* Burman)	Reduced nutrient loss, inconsequential effect on plant growth	Hu et al. (2017)
Silver nanoparticles (Ag NPs)	10–20 mg L^{-1}	Rice (*Oryza sativa* L.)	Enhanced enzymatic activity, soluble sugar content and seed germination	Mahakham et al. (2017)
Nanozeolite	50 mg L^{-1}	Maize (*Zea mays* L.)	Increased twofold Chlorophyll content, vegetative growth, protein and yield	Khati et al. (2018)
Iron-humic nanofertilizer	35, 75, and 150 mmol pot^{-1}	Soybean (*Glycine max* L.)	Boosted shoot fresh weight	Cieschi et al. (2019)
Zinc oxide nanoparticles (ZnO NPs)	1000 mg L^{-1}	Chili pepper (*Capsicum annuum* L.)	Positively affected Chlorophyll content, vegetative growth, and fruit yield	García-López et al. (2019)
Zinc oxide nanoparticles (ZnO NPs)	50 mg L^{-1}	Common bean (*Phaseolus vulgaris* L.)	Heightened vegetative growth, fresh pod yield, pods physical quality and nutritional value.	Marzouk et al. (2019)

(continued)

Table 22.1 (continued)

Nanofertilizer	Applied dose	Targeted plant species	Inductive effect	References
Nanophosphorus (P NPs)	0.0781 g L^{-1} –0.1563 g L^{-1}	Rice (*Oryza sativa* L.)	Induced higher biomass accumulation and photosynthetic rate	Miranda-Villagómez et al. (2019)
2-D graphite carbon nanoparticles (CNPs)	3000 mg CNP per kg Conventional fertilizer	Lettuce (*Lactuca sativa* L.)	Decreased nitrate leaching, no growth inhibition	Pandorf et al. (2020)
Iron phosphate nanoparticles ($FePO_4$ NPs)	100 μM	Maize (*Zea mays* L.), Cucumber (*Cucumis sativus* L.)	Improved Chlorophyll content and fresh biomass	Sega et al. (2019)
Zinc, iron, and manganese oxide nanoparticles	20 mg L^{-1}	Pumpkin (*Cucurbita pepo* L.)	Enhanced vegetative growth, fruits, yield, and Chlorophyll content	Shebl et al. (2019)

with a carrier that may or may not be in nano scale, like, clay, chitosan, or zeolite (Mohammad et al. 2016). The second is to directly use nutritional elements per se in nanoform supplying through the irrigation system, incorporate with soil or foliar feeding (Fedorenko et al. 2015).

21.4 Nanocarriers Delivering Plant Nutrients

Nanomaterials may be found more frequently in the agricultural and food sectors. Therefore, it can be mentioned that nanotechnology is a rapidly expanding field that provides opportunities for developing nanoscale materials with unique properties as well as creating a wide range of applications (Zoufan et al. 2020). Solutions that secure the nutrients needed for normal plant growth have attracted great attention in the view of industrial and academic prospects in an attempt to guarantee protected and sustainable release of the required nutrients, while minimizing the production cost (Shang et al. 2019). NPs have many unique characteristics that distinguish them from their larger counterparts micro- and macro-scale nutrients, making them more suitable for designing bio-based delivery systems (Jeevanandam et al. 2018). More recently, the application of nanomaterials for the purpose of delivering nutrients and active compounds that promote plant growth has become more popular progressively. The use of nanonutrients in the proper place, time, quantity and composition will determine the efficacy of supplied nutrients (Duhan et al. 2017). In this context, many researchers investigating more effective ways to manufacture and use nanotechnologies to design mechanisms through which an efficient delivery system for agrochemicals can be secured in a streamlined manner (Gunasekaran et al. 2014; Mura et al. 2013). Therefore, to design an efficient nano-transport system for the required nutrient, significant familiarity with the bioavailability of relevant active compounds and their metabolism should be addressed.

Nanoscale delivery systems provide improved nutritional exactness by overwhelming biological obstacles and enhancing nutrient targeting active sites (Vega-Vásquez et al. 2020). Regarding this, it has been confirmed that the plant-derived NPs penetrate the leaf and transmit active components in a two directions, up towards plant leaves and down towards root system (Banerjee et al. 2019). In addition to other features, NPs have an improved penetration property that outperforms their traditional counterparts, thus nearly 33% of the sprayed nanoparticles penetrated into plant leaves, against less than one percent of the ordinary nutrients applied in the identical way (Ruttkay-Nedecky et al. 2017). Treating tomato plants with lipo-nanoparticles loaded with micronutrients (Mg and Fe) were able to bypass acute nutrient deficiency that was not treatable with conventional agricultural nutrients. These applications support the expanded use of nanotechnology to deliver nutrients and increasing crops productivity (Karny et al. 2019).

In the agriculture field, numerous nanostructures have been adopted to deliver macronutrients like N, P, K, and base minerals that stimulate plant growth as well as mono and multilayer carbon nanotubes and non-metallic minerals that positively enhancing plant growth and can be broadly applied in the agriculture sector (Yatim et al. 2018). Clay minerals are natural soil components with modified charge and surface properties capabilities enable them to be a decent option for delivering nutrients and ensuring gradual release at the targeted zones (Jampílek and Kráľová 2017). Alternatively, silica nanoparticles have emerged as potential delivery vehicles for plant nutrients due to its structural resilience in creating nanoparticles of different sizes and shapes, as well as its unique capability to form pores for packing a wide range of biomolecules (Shi et al. 2010). Moreover, silica is a vital micronutrient that significantly supports plant growth and modulates stress response (Campbell et al. 2011; Jang et al. 2013). For example, the absorption and distribution of mesoporous silica nanoparticles (MSNs) have been examined during seeds germination in each of *Arabidopsis*, lupine and wheat grew in a hydroponic system. The nanoparticles were detected in the leaves and roots of plants, however, they did not affect seed germination and had no toxic effects. Nanoparticles are localized within cells and cell walls of the developed root as well as in vascular transport elements, as well as other associated cells. Accordingly, it has been suggested that MSNs can be hired for delivering nanoparticles into plant biosystems (Hussain et al. 2013).

21.5 Nanofortifiers

Agricultural products are vital components in the food basket, especially for rural society and, hence decreasing crop productivity will pose a serious threat to the nutritional security of these societies and may result in starvation (Vijaya Bhaskar et al. 2017). Always, there is a need for concerting efforts to develop plants towards magnifying production and to diminish the adverse effects on plant production that can lead to malnutrition and starvation. Nevertheless, key micronutrient deficiencies has become a persistent issue in resource-poor communities. Meanwhile, many major crops are limited suppliers for essential nutrients necessary for normal human growth and development (Garg et al. 2018).

Fertilizers are enriched with minerals necessary for normal plant growth and development. Thereby, macro and/or micro-nutrients deficiency will be manifested in abnormal organs development, as well as edible parts with low essential nutrients (Etienne et al. 2018). Notably, nutrients shortage is not always related to soil deficiency of such nutrients, rather some roots are with small pores that limit their ability to absorb and transport the needed nutrients (Elemike et al. 2019). Although chemical fertilizers is an old common practice and it has enormously improved the agricultural outputs in terms of quality and quantity, they negatively contribute to soil, fertility, structure, nutrient balance, in addition to its side effects on the local ecosystem that representing a significant threat for the long term (Lin et al. 2019). On the other hand, conventional fertilizers have active particles with higher than 100 nm in size, make them vulnerable to leach (Giroto et al. 2017). Furthermore, nutrients can be depleted from the soil due to continuous farming, therefore there is an urgent need for frequent recovery of agricultural lands using various synthesized chemical and green or bio-fertilizers (Khan et al. 2019). Principally, not all peoples are able to change their lifestyle and diversify their food to ensure as many as it possible of necessary nutrients. Alternatively, biofortification is an emerging feasible solution focusing on enhancement of the nutritional value of plant-derived food via crop breeding and cultivation practices (Jha and Warkentin 2020), in addition to modern technologies.

Due to unique properties, NPs have promising applications in the near future of agricultural systems. The application of nutrients in nanoscale will minimize the wasted costly active substances and allow sustainable release at the targeted area. Thereby, effective uptake of the required nutrients can be achieved (Khan et al. 2019; Sekhon 2014). In this contest, achieving healthy nutrition requires the development of novel varieties forfeited with essential minerals (e.g. iron, zinc, manganese, copper, selenium, and iodine), amino acids (tryptophan and lysine) and vitamins. Affordability and availability are the two keys that confer the nanofortification an advantage over other interventions serving in battle against malnutrition particularly in low-income countries (Jha and Warkentin 2020). Fortified crops are extensively cultivated and consumed by the people globally. Staple food crops like cereals (e.g. wheat, rice, maize and sorghum), pulses (soybean, common bean), vegetables and fruits have been fortified for various nutritional aspects using different agronomical and/or biotechnological approaches (Garg et al. 2018).

The successful application of nanofortifiers is mainly depended on the plant type, physical and chemical properties of the prepared NPs. Therefore, nanotechnology will serve efficiently in fortifying plant with the desired nutrients (Patra and Baek 2014; Patra et al. 2018). Nanofortification will use the nanoporous present on the plant part surfaces, therefore, this technology will be a unique platform serving in modulating sustainable nutrient delivery systems (Elemike et al. 2019). The efficacy of applied nanonutrients, and zeolites can be enhanced through the encapsulated NPs. Ultimately, this in turn will quickly restore soil fertility and minimizing environmental pollution (Mout et al. 2017).

21.6 Nanonutrients Mediating Oxidative Response

During their life cycle, plants may expose to a wide range of inappropriate environmental conditions, typically termed stresses. Under this, stresses are divided into two categories; abiotic stress including salinity, drought, pollutants, toxic metals, extreme temperature, radiations and pesticides; and biotic stress comprising high density, pathogens and insects for instance (Waqas et al. 2019). Throughout their adaptive response, plants develop multiple physiological and molecular techniques mainly excessive ROS that in turn will affect the plant cellular processes and shape the total response (Huang et al. 2019). The balance between generated ROS and scavenging them by antioxidant defense system will determine the negative effect of stress condition, and, thereby the sustained productivity of plants (Xie et al. 2019). Stress-induced free radicals (e.g. hydrogen peroxide, peroxy radical, superoxide radical, perhydroxy radical, hydroxyl radical and singlet oxygen) are capable of damaging the plant cellular components involving proteins, lipids, and nucleic acids, thus triggering programmed cell death (PCD) which ultimately results in plant death (Elsahookie et al. 2009). Consequently, the improvement of plant tolerance to harsh environmental conditions begins with maintaining the antioxidants level in order to enhance the machinery defense and minimize the oxidative damage to the lower limit (Khan et al. 2019).

Several enzymes are involving in the machinery defense system combating oxidative stress in a wide range of plant types including superoxide dismutase (SOD), catalase, peroxidase and the ascorbate glutathione enzymes (Sarker and Oba 2018). Biological systems witness a rapid growing of nanotechnology applications. Under this, there is a strong belief that nanoparticles can improve the plants tolerance to oxidative stress by enhancing the ability of their antioxidant system (Zoufan et al. 2020). Via biochemical investigations, a strong believe have emerged that NPs playing a crucial role in regulating key biological processes in plants such as photosynthesis, antioxidant enzymes, oxidative stress and gene expression (Tan et al. 2018). Like other substances, nanoparticles show different norms of action according to the origin, preparation method, size and applied concentration. Regarding this, nanoparticles found to be highly concentration-dependent materials, in which low concentration resulted in low oxidative stress, and finally reduce the antioxidant activity (Sharma et al. 2019).

The adoption of nanocolloidal solutions as micronutrients is an effective approach to improve plant tolerance to unfavorable environmental conditions and ensures high quantity and quality yields of food crops. The recent reports showed that nanomolybdenum was efficiently reduced the oxidation level by activating antioxidant enzymes including superoxide dismutase in about 15%, thereby enhancing plants' adaptation to stress conditions (Taran et al. 2016). The activity of ROS scavenging enzymatic system including superoxide dismutase, catalase and ascorbate peroxidase was investigated in *Brassica juncea* nourished by two types of nanoparticles micronutrients, titanium dioxide (TiO_2) and copper oxide (CuO) (Sunita and Shekhawat 2016). The increased level of TiO_2 NPs had a positive effect on plant growth, whereas the opposite effect was noticed for CuO NPs. Interestingly, the less bioaccumulated NPs improved the defense mechanism against stress conditions via antioxidative enzymes.

Similar findings were stated by Homaee and Ehsanpour (2016) as they compared two sources of silver (Ag NPs and Ag ions) in terms of oxidative response development in potato plant (*Solanum tuberosum* L.) under in vitro conditions. Although both Ag forms, NPs and ions had elevated the activity of the antioxidant enzymes compared to the control, the higher concentration of NPs and ions significantly diminished the oxidative enzymatic activity. Recently, Zoufan et al. (2020) reported a substantial induction in the oxidative stress in response to the subjected concentrations of Zn oxide nanoparticles applied on *Chenopodium murale* using a hydroponic system. The different treatments of ZnO NPs magnified the activity of superoxide dismutase (SOD), catalase (CAT) and guaiacol peroxidase (GPX) along with a significant reduction in growth indices.

21.7 Nanonutrients Modulating Plant Secondary Metabolites

The possible effects of NPs have been investigated consistently across plant species on different morphological and physiological attributes. Unfortunately, the modulatory effect of NPs is still poorly understood, where NPs can improve the secondary metabolite processes, hence active natural compounds (Ebadollahi et al. 2019). The induction of ROS found to be strongly addicted to the applied NPs through the plant kingdom (Marslin et al. 2017). The importance of ROS cannot be summarized in reflecting the cell fatigue, as it has an important role in several developmental processes. Additionally, many literatures have provided strong evidences of ROS-related signal molecules that mediating plant secondary metabolisms (Singh et al. 2016). In fact, some of these literatures referred to ROS themselves as signaling molecules and can be inductive to secondary metabolism pathways (Jacobo-Velazquez et al. 2015; Simon et al. 2010).

Although several reports established the important role of NPs in physiological, growth and developmental plant aspects (Gohari et al. 2020), the influence of NPs on plant secondary metabolites is not fully discovered, however, numerous studies assured the modulation of NPs towards plant secondary metabolism (Ghorbanpour and Hadian 2015). Plant secondary metabolites are commonly regulated by transcriptional process guided by secondary signaling messengers, and the later has a prevailing link with ROS (Meraj et al. 2020). In this perspective, it have been suggested that NPs may regulate the production of secondary metabolites since ROS burst is a common indication of NPs application (Egea et al. 2017) (Table 22.2). More recently, an alternative scenario has been proposed to explain the relationship between nanoparticles and the overproduction of secondary metabolites, in which it is believed that the latter plays a protection role against oxidative response developed after NPs exposure (Ebadollahi et al. 2019). Regardless of the mechanism by which nanomaterials can regulate the cellular production and accumulation of secondary metabolites, a number of investigations have indicated that the plant shows a pattern of response to nanomaterials largely simulating the response to biotic and abiotic stresses (Khodakovskaya et al. 2011b; Kohan-Baghkheirati and Geisler-Lee 2015). The catalytic effect of nanomaterials in increasing the production of secondary metabolites may include a number of cellular signal transduction pathways, primarily via MAPK cascade (mitogen-activated protein kinase), cytosolic Ca^{2+} and ROS burst (Sosan et al. 2016). Zhang et al. (2013) reported enhanced production of secondary metabolites (artemisinin) in hairy roots of *Artemisia annua* in response to silver nanoparticles (AgNPs), along with an elevated level of oxidative stress and antioxidant enzymatic activity. The biosynthesized silver nanoparticles (AgNPs) had the same positive effect on the synthesis of phytochemical diosgenin in fenugreek seedlings. The inducibility of Ag NPs leads to a profound increase in the produced secondary metabolites that open up new techniques by which natural and medicinal plant products can be magnified (Jasim et al. 2017). Garcia-Sanchez et al. (2015) noticed that AgNPs, TiO_2NPs and carbon nanotubes (CNTs) lead to a

Table 22.2 The modulation of plant secondary metabolites in response to different types and doses of nanoparticles (NPs)

Nanoparticles	Applied dose	Plant species	Modulated secondary metabolites	Reference
Aluminum oxide nanoparticles (Al_2O_3 NPs)	100 μg ml^{-1}	Tobacco (*Nicotiana tabacum* L.)	Accumulated phenolics compounds	Poborilova et al. (2013)
Ag-SiO_2 core-shell nanoparticles	900 mg L^{-1}	Sweet wormwood (*Artemisia annua* L.)	Increased artemisinin about four folds	Zhang et al. (2013)
Titanium dioxide nanoparticles (TiO_2 NPs)	100 mg L^{-1}	Spirulina (*Arthrospira platensis*)	Magnifying the secreted phenolic compounds	Comotto et al. (2014)
TiO_2 NPs, Ag NPs and multi-walled carbon nanotubes (MWCNTs)	0.2–25 μg mL^{-1}	*Arabidopsis* (*Arabidopsis thaliana* L.)	Upregulated anthocyanin and flavonoid gene expression	Garcia-Sanchez et al. (2015)
Multi-walled carbon nanotubes (MWCNTs)	100 μg ml^{-1}	*Satureja khuzestanica*	Maximize flavonoid and phenolics content	Ghorbanpour and Hadian (2015)
Cerium (IV) oxide and Indium(III) oxide nanoparticles (CeO_2 and In_2O_3 NPs)	1000 mg L^{-1}	*Arabidopsis* (*Arabidopsis thaliana* L.)	Provoked synthesis of phenylalanine ammonia lyase (PAL)	Ma et al. (2016)
Cadmium oxide nanoparticles (CdO NPs)	2.03 ± 0.45 105 particles cm^3	*Barley* (*Hordeum vulgar* L.)	Increased ferulic acid and isovitexin	Večeřová et al. (2016)

(continued)

Table 22.2 (continued)

Nanoparticles	Applied dose	Plant species	Modulated secondary metabolites	Reference
Silver nanoparticles (Ag NPs)	200 μL of 1 μg mL^{-1} (w/v)	Fenugreek (*Trigonella foenum-graecum* L.)	Enhanced diosgenin synthesis	Jasim et al. (2017)
Titanium dioxide and Cerium (IV) oxide nanoparticles TiO_2 and CeO_2 NPs	500 mg L^{-1}	*Arabidopsis* (*Arabidopsis thaliana* L.)	Upregulated photosynthesis and ethylene	Tumburu et al. (2017)
Titanium dioxide TiO_2-Perlite nanocomposites (NCs)	200 mg L^{-1}	St. John's wort (*Hypericum perforatum* L.)	Elicited the production of volatile compounds, hypericin and pseudohypericin	Ebadollahi et al. (2019)
Titanium dioxide nanoparticles TiO_2 NPs	100 mg L^{-1}	Moldavian dragonhead (*Dracocephalum moldavica* L.)	Improved geranial, geraniol and z-citral content	Gohari et al. (2020)

marked impact on genes coding anthocyanin and flavonoid in *A. thaliana*. Also, the increased concentration of carbon nanotubes in culture medium improved phenolics and flavonoids content, thereby growth parameters of *Satureja khuzestanica* (Ghorbanpour and Hadian 2015).

21.8 Biosafety of Nanomaterials

Nanomaterials are gaining increased attention for boosting plants' nutrient and agricultural productivity, but then again the safety of these materials should be considered because only a thin line separating shortage and toxicity of nanomaterials (Shafiq et al. 2020). There is no doubt that nano techniques have witnessed a great expansion during the past two decades, and it has imposed itself as one of the fastest growing applications in the pharmaceutical, physical, chemical and agricultural fields (Usman et al. 2020). Accordingly, this huge growth of use has greatly contributed to the increased leakage and accumulation of NMs in the ecosystem (Shang et al. 2019). Soil, water and air are the three major gears of earth ecosystem where plants are growing. Statistics showed that nanomaterials have different accumulation rates in each component, nevertheless, soil has shown the highest rate of accumulated nanomaterials compared to water and air (Yang et al. 2017). Consequently, due to their limited choices in selecting the growing environment, plants are more vulnerable than other organisms to the less apposite component in the ecosystem.

The most important character of nanostructures (NPs) is the low of at least one dimension (1–100 nm), which be responsible for their distinctive properties and biological activity (Ndolomingo et al. 2020). On the other hand, this tiny size offers NPs their destructive ability to the cell components and limits their uses. For that reason, it looks more rational not to exaggerate these materials (Jeevanandam et al. 2018). Subsequently, and for real assessment, it became necessary to study both kinetics and biotoxicity of nanomaterials in the short and long term of use (Ripp and Henry 2011). The time factor is crucial in determining the toxicity level of a specific nanomaterial, as many NPs revealed different toxicity behavior over time. The net effect of nanomaterials is interestingly driven by many variables, such as type of nanomaterial, origin (organic or inorganic), preparation procedure (biosynthesis, physical or chemical), form (ionic or non-ionic), magnetism properties, nano-size and the targeted plant species, however, the applied dose seems the most critical factor in determining NPs toxicity (Jeevanandam et al. 2018). In light of the large number of variables that each nanomaterial holds, it seems difficult to accurately predict its fate in the added environment (Bundschuh et al. 2018) (Fig. 22.2).

Soil represents the largest repository of nanoparticles, that's why biotic (bacteria, mycorrhiza, fungi) and abiotic factors (heat, pH, moisture) have an essential role in the accumulation and release rates of such materials in the soil, which finally determine NPs toxicity (Simonin et al. 2015). Given the concentration, size, solubility, shape, surface and aggregation state that each nanoparticle retains, it seems difficult to accurately predict the biological and chemical and/or physical behavior of those materials in the surrounding environment (Morales-Díaz et al. 2017).

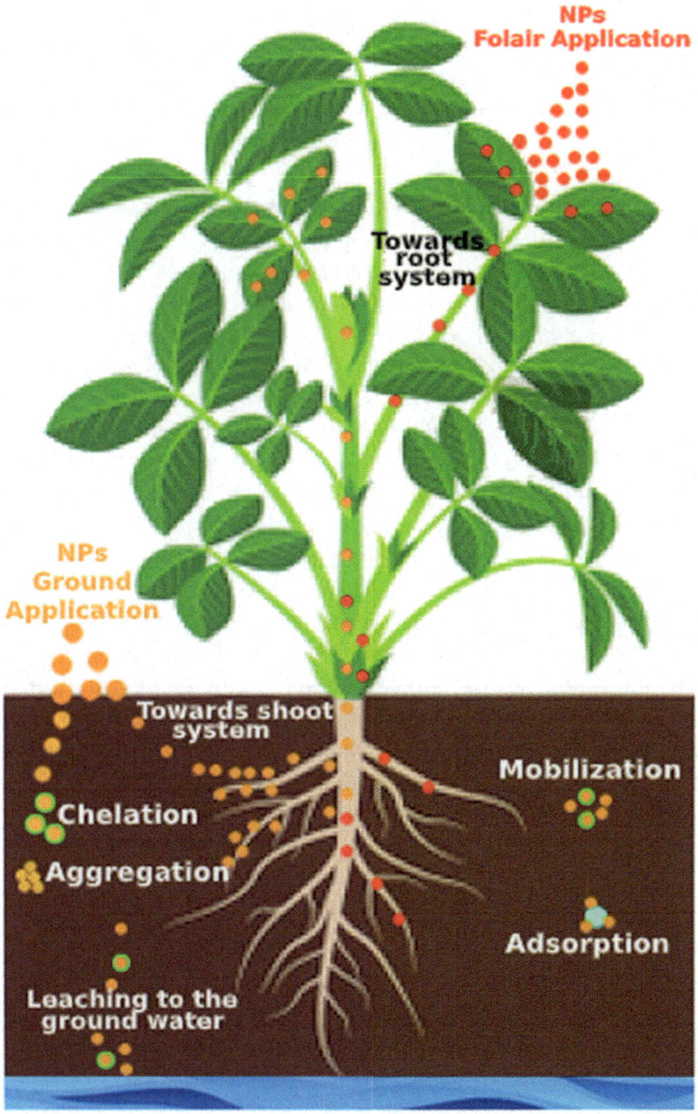

Fig. 22.2 Suggested scenario of the foliar- and soil-applied nanoparticles fate in the ecosystem (Figure constructed by Ayoob O. Alfalahi)

The toxicity of nanomaterials remarkably depends on their surface-to-mass ratio, that gives NPs the higher affinity to adsorb pollutants of the niche, particularly heavy metals like cadmium, cobalt, lead, nickel (Yaqoob et al. 2020).

Several studies have investigated the biotoxicity of nanoparticles using different types, forms and concentrations of NPs (Kalpana and Rajeswari 2018; Zhu et al. 2008). Vicario-Pares et al. (2014) confirmed that the three used metal oxides (TiO_2, CuO and ZnO NPs) were more toxic to zebrafish embryo in the ionic form compared to the NPs. In a part of their biological effects, nano-sized materials have a distinct ability to alter the enzymatic soil content, thereby disrupting their efficacy, and lead to toxic effects (Elemike et al. 2019). It is interesting to note that the nano-scale materials possess features fundamentally different from their larger counterparts (micro- and macro-scale) of the same material (Urban et al. 2016). The sensitivity that revealed by the different organisms is another key factor that may indirectly controls the toxicity effect caused by nanoparticles (Table 22.3). Regarding this, the effect of some nanoparticles may be restricted to the exposed plant, while in other cases the destructive effect can encompass the plant-mycorrhiza and/or rhizobia symbiotic relationships (Tian et al. 2019). In addition, the encapsulation of nanomaterials can significantly alter their properties and solubility, and its effect extends to influence the ecosystem components and toxic level. Yin et al. (2012) stated that the suspension of Ag NPs had a positive effect on the seed germination of several plant species, however the coated Ag NPs showed a higher toxic level and less favorable effect upon seed germination. Physiological and growth parameters of *Eichhornia crassipes* were investigated in response to different concentrations of two Ag nanoforms, biological and synthesized. Although the higher applied concentration of Ag exhibited a higher accumulation rate in different plant parts after 5 days, the synthesized form was more able to inhibit the plant growth (Rani et al. 2016).

21.9 Conclusions and Prospects

The versatility of nanomaterials has become a reality we live in today, yet their use appears to be growing steadily, mainly in areas that do not require a high level of caution. In the agricultural field, nanomaterials have relatively greater flexibility to be used for designing novel fertilizers with enhanced features that enable them to effectively provoke secondary metabolites and plant growth. Although nanomaterials can be disruptive and nanotoxic, in addition to its association with the ROS generation, however, from a fortification viewpoint, it may also be creative in designing agents impacting through a combination of chemical and physical approaches of action. Due to vagueness of nanomaterials biosafety and their complicated environmental interactions, there is a need for extensive investigations before releasing them for prevalent use. It must be said that almost all bulk materials have a corresponding nanoscale, thus the nanomaterials were and still are an integral part of earth biological system. However, nanomaterials are strong candidates for dominating different agricultural sectors, basically for rationalizing the use of expensive agricultural inputs and chemicals whose use in high concentrations pose a real threat to the ecosystem.

Table 22.3 The toxic effect of different nanoparticles against several plant species

Applied nanoparticles	Plant species	Toxic effect	Reference
Zinc oxide nanoparticles (ZnO NPs)	Perennial ryegrass (*Lolium perenne* L.)	Reduced the plant biomass	Lin and Xing (2008)
Carbon nanotubes	Tobacco (*Nicotiana tabacum* L.)	Upregulation of genes responsible for water transport and plant growth	Khodakovskaya et al. (2011a)
Silver nanoparticles (Ag NPs)	*Microstegium vimeneum*	Growth inhibition	Colman et al. (2013)
Silver nanoparticles (Ag NPs)	Wheat (*Triticum aestivum* L.)	Increased oxidative stress	Dimkpa et al. (2013)
Zinc oxide nanoparticles (ZnO NPs)	Onion (*Allium cepa* L.)	Increased chromosomal abnormalities	Raskar and Laware (2014)
Gold nanoparticles (Au NPs)	Arabidopsis (*Arabidopsis thaliana* L.)	Upregulation of genes responsible for oxidative response, glutathione, water transport and plant hormones	Shukla et al. (2014)
Silicon dioxide nanoparticle (SiO_2 NPs)	Cotton (*Gossypium hirsutum* L.)	Reduced plant biomass, SOD activity and IAA concentration	Le et al. (2014)
Iron oxide nanoparticles (Fe_3O_4 NPs)	Duckweed (*Lemna gibba* L.)	Reduced Chlorophyll content, Increased reactive oxygen species (ROS) and growth inhibition	Barhoumi et al. (2015)
TiO_2 nanoparticles, Ag nanoparticles, Multi-walled carbon nanotubes	Arabidopsis (*Arabidopsis thaliana* L.)	Negatively affected transcriptomic patterns and root hair development	García-Sánchez et al. (2015)
Silver nanoparticles (Ag NPs)	Stevia (*Stevia rebaudiana* Bert.)	Inhibits normal development and reduced Chlorophyll content	Castro-González et al. (2019)
Silver nanoparticles (Ag NPs)	Chlamydomonas (*Chlamydomonas reinhardtii* P.A. Dangeard)	Reduced Chlorophyll content and electron transport activity	Dewez and Oukarroum (2012)

(continued)

Table 22.3 (continued)

Applied nanoparticles	Plant species	Toxic effect	Reference
Thin-walled carbon nanotubes (CNTs)	Rice (*Oryza sativa* L.)	Decreasing the concentrations of endogenous plant hormones and inhibited plant growth	Hao et al. (2016)
Iron oxide nanoparticles	Sunflower (*Helianthus annuus* L.)	Reduced the nutrients uptake and root hydraulic conductivity	Martínez-Fernández et al. (2016)
Copper oxide nanoparticles (CuO NPs)	Arabidopsis (*Arabidopsis thaliana* L.)	Increased ROS accumulation, adversely affected chlorophyll contents, stomatal aperture and reduced biomass	Azhar et al. (2020)
Titanium Dioxide Nanoparticles (TiO_2 NPs)	Moldavian dragonhead (*Dracocephalum moldavica* L.)	Increased antioxidant enzyme activity and improved all agronomic traits	Gohari et al. (2020)
Iron oxide nanoparticles (Fe_3O_4)	Yellow alfalfa (*Medicago sativa ssp. falcate* L.)	Increased chlorophyll a fluorescence, miRNA expression, genotoxicity and reduced genome stability	Kokina et al. (2020)
Carbon nanotubes, Carbon nanofibers, Silicon nanotubes	Heterosigma (*Heterosigma akashiwo* Y. Hada)	Inhibited growth	Pikula et al. (2020)
Iron Oxide Nanoparticles (Fe_3O_4)	Rocket (*Eruca sativa* L.)	Induced genotoxicity	Plaksenkova et al. (2019)
Cerium oxide nanoparticles (CeO_2)	Peas (*Pisum sativum* L.)	Reduced Chlorophyll content and plant growth	Skiba and Wolf (2019)
Zinc oxide nanoparticles (ZnO NPs)	Arabidopsis (*Arabidopsis thaliana* L.)	Reduced Chlorophyll content, Growth inhibition	Wang et al. (2016)
Silver nanoparticles (Ag NPs)	Lettuce (*Lactuca sativa* L.)	Blocking nutrient transport, Induce the enzymatic antioxidants activity, Biomass reduction	Wu et al. (2020)

(continued)

Table 22.3 (continued)

Applied nanoparticles	Plant species	Toxic effect	Reference
Zinc oxide nanoparticles (ZnO NPs)	Black mustard (*Brassica nigra* L.)	Adversely affects seed germination and seedling growth, increasing the antioxidative activities and non-enzymatic antioxidants	Zafar et al. (2020)
Zinc oxide nanoparticles (ZnO NPs)	*Nettle-leaved goosefoot* (Chenopodium murale L.)	Reduced Chlorophyll content and Soluble proteins, Increased oxidative stress, SOD and CAT activity, Inhibited growth	Zoufan et al. (2020)

References

Afsharinejad A, Davy A, Jennings B et al (2016) Performance analysis of plant monitoring nanosensor networks at THz frequencies. IEEE Internet Things J 3:59–69. https://doi.org/10.1109/jiot.2015.2463685

Anshup A, Venkataraman JS, Subramaniam C et al (2005) Growth of gold nanoparticles in human cells. Langmuir 21(25):11562–11567. https://doi.org/10.1021/la0519249

Armstead AL, Li B (2016) Nanotoxicity: emerging concerns regarding nanomaterial safety and occupational hard metal (WC-Co) nanoparticle exposure. Int J Nanomedicine 11:6421–6433. https://doi.org/10.2147/ijn.s121238

Aslani F, Bagheri S, Muhd Julkapli N et al (2014) Effects of engineered nanomaterials on plants growth: an overview. Sci World J 641759. https://doi.org/10.1155/2014/641759

Azhar W, Khan AR, Muhammad N et al (2020) Ethylene mediates CuO NP-induced ultrastructural changes and oxidative stress in *Arabidopsis thaliana* leaves. Environ Sci: Nano 7:938–953. https://doi.org/10.1039/c9en01302d

Banerjee K, Pramanik P, Maity A et al (2019) Methods of using nanomaterials to plant systems and their delivery to plants (mode of entry, uptake, translocation, accumulation, biotransformation and barriers). Adv Phytonanotechnol 123–152. https://doi.org/10.1016/b978-0-12-815322-2.00005-5

Bargaz A, Lyamlouli K, Chtouki M et al (2018) Soil microbial resources for improving fertilizers efficiency in an integrated plant nutrient management system. Front Micro 9:1606 https://doi.org/10.3389/fmicb.2018.01606

Barhoumi L, Oukarroum A, Ben Taher L et al (2015) Effects of superparamagnetic iron oxide nanoparticles on photosynthesis and growth of the aquatic plant *lemna gibba*. Arch Environ Contam Toxicology 68:510–520. https://doi.org/10.1007/s00244-014-0092-9

Bundschuh M, Filser J, Lüderwald S et al (2018) Nanoparticles in the environment: where do we come from, where do we go to? Environ Sci Eu 30(1):6. https://doi.org/10.1186/s12302-018-0132-6

Campbell JL, Arora J, Cowell SF et al (2011) Quasi-cubic magnetite/silica core–shell nanoparticles as enhanced MRI contrast agents for cancer imaging. PLoS One 6(7):e21857. https://doi.org/10.1371/journal.pone.0021857

Castro-González CG, Sánchez-Segura L, Gómez-Merino FC et al (2019) Exposure of stevia (*Stevia rebaudiana* B.) to silver nanoparticles in vitro: transport and accumulation. Sci Rep 9:10372 https://doi.org/10.1038/s41598-019-46828-y

Chan YS, Mat Don M (2013) Biosynthesis and structural characterization of Ag nanoparticles from white rot fungi. Mater Sci Eng C 33(1):282–288. https://doi.org/10.1016/j.msec.2012.08.041

Chaudhuri SK, Malodia L (2017) Biosynthesis of zinc oxide nanoparticles using leaf extract of *Calotropis gigantea*: characterization and its evaluation on tree seedling growth in nursery stage. Appl Nanosci 7:501–512. https://doi.org/10.1007/s13204-017-0586-7

Cieschi MT, Polyakov AY, Lebedev VA et al (2019) Eco-friendly iron-humic nanofertilizers synthesis for the prevention of iron chlorosis in soybean (*Glycine max*) grown in calcareous soil. Front Plant Sci 10:413. https://doi.org/10.3389/fpls.2019.00413

Colman BP, Arnaout CL, Anciaux S et al (2013) Low concentrations of silver nanoparticles in biosolids cause adverse ecosystem responses under realistic field scenario. PLoS One 8(2):e57189. https://doi.org/10.1371/journal.pone.0057189

Comotto M, Casazza AA, Aliakbarian B et al (2014) Influence of TiO_2 nanoparticles on growth and phenolic compounds production in photosynthetic microorganisms. Sci World J 2014:9. https://doi.org/10.1155/2014/961437

Das RK, Pachapur VL, Lonappan L (2017) Biological synthesis of metallic nanoparticles: plants, animals and microbial aspects. Nanotechnol Environ Eng 2:18. https://doi.org/10.1007/s41204-017-0029-4

DeRosa MC (2010) Nanotechnology in fertilizers. Nat Nanotechnol 5:91. https://doi.org/10.1038/nnano.2010.2

Dewez D, Oukarroum A (2012) Silver nanoparticles toxicity effect on photosystem II photochemistry of the green alga *Chlamydomonas reinhardtii* treated in light and dark conditions. Toxicological Environ Chem 12:1536–1546. https://doi.org/10.1080/02772248.2012.712124

Dimkpa CO, McLean JE, Martineau N (2013) Silver nanoparticles disrupt wheat (*Triticum aestivum* L.) growth in a sand matrix. Environ Sci Technol 47:1082. https://doi.org/10.1021/es302973y

Du W, Yang J, Peng Q et al (2019) Comparison study of zinc nanoparticles and zinc sulphate on wheat growth: From toxicity and zinc biofortification. Chemosphere 227:109–116. https://doi.org/10.1016/j.chemosphere.2019.03.168

Duhan JS, Kumar R, Kumar N et al (2017) Nanotechnology: the new perspective in precision agriculture. Biotechnol Rep 15:11–23. https://doi.org/10.1016/j.btre.2017.03.002

Ebadollahi R, Jafarirad S, Kosari-Nasab M et al (2019) Effect of explant source, perlite nanoparticles and TiO_2/perlite nanocomposites on phytochemical composition of metabolites in callus cultures of *hypericum perforatum*. Sci Rep 9:1–15. https://doi.org/10.1038/s41598-019-49504-3

Egea J, Fabregat I, Frapart YM et al (2017) European contribution to the study of ROS: a summary of the findings and prospects for the future from the COST action BM1203 (EU-ROS). Redox Bio 13:94–162. https://doi.org/10.1016/j.redox.2017.05.007

Elemike EE, Uzoh IM, Onwudiwe DC et al (2019) The role of nanotechnology in the fortification of plant nutrients and improvement of crop production. Appl Sci 9:499. https://doi.org/10.3390/app9030499

Elsahookie MM, Abdulqahar FW, Alfalahi AO (2009) Cell suicide or programmed cell death. The Iraqi J Agric Sci 40(3):1–26

Etienne P, Diquelou S, Prudent M et al (2018) Macro and micronutrient storage in plants and their remobilization when facing scarcity: the case of drought. Agriculture 8:14. https://doi.org/10.3390/agriculture8010014

Fedorenko VF, Buklagin DS, Golubev IG et al (2015) Review of Russian nanoagents for crops treatment. Nanotechnol Russ 10:318. https://doi.org/10.1134/s199507801502010x

García-López JI, Niño-Medina G, Olivares-Sáenz E et al (2019) Foliar application of zinc oxide nanoparticles and zinc sulfate boosts the content of bioactive compounds in habanero peppers. Plants 8(8):254. https://doi.org/10.3390/plants8080254

Garcia-Sanchez S, Bernales I, Cristobal S (2015) Early response to nanoparticles in the *Arabidopsis* transcriptome compromises plant defence and root-hair development through salicylic acid signalling. BMC Genom 16:341. https://doi.org/10.1186/s12864-015-1530-4

Garg M, Sharma N, Sharma S et al (2018) Biofortified crops generated by breeding, agronomy, and transgenic approaches are improving lives of millions of people around the world. Front Nutr 5:12. https://doi.org/10.3389/fnut.2018.00012

Ghorbanpour M, Hadian J (2015) Multi-walled carbon nanotubes stimulate callus induction, secondary metabolites biosynthesis and antioxidant capacity in medicinal plant Satureja khuzestanica grown in vitro. Carbon 94:749–759. https://doi.org/10.1016/j.carbon.2015.07.056

Giroto A, Guimarães G, Foschini M (2017) Role of slow-release nanocomposite fertilizers on nitrogen and phosphate availability in soil. Sci Rep 7:46032. https://doi.org/10.1038/srep46032

Gohari G, Mohammadi A, Akbari A et al (2020) Titanium dioxide nanoparticles (TiO_2 NPs) promote growth and ameliorate salinity stress effects on essential oil profile and biochemical attributes of *Dracocephalum moldavica*. Sci Rep 10:912. https://doi.org/10.1038/s41598-020-57794-1

Gunasekaran T, Haile T, Nigusse T et al (2014) Nanotechnology: an effective tool for enhancing bioavailability and bioactivity of phytomedicine. Asian Pac J Trop Biomed 4(1):S1–S7. https://doi.org/10.12980/apjtb.4.2014c980

Gupta R, Xie H (2018) Nanoparticles in daily life: applications, toxicity and regulations. J Environ Pathol Toxicol Oncol 37(3):209–230. https://doi.org/10.1615/jenvironpatholtoxicoloncol.2018026009

Hao Y, Yu F, Lv R et al (2016) Carbon nanotubes filled with different ferromagnetic alloys affect the growth and development of rice seedlings by changing the C:N ratio and plant hormones concentrations. PLoS One 10:11(6):e0157264. https://doi.org/10.1371/journal.pone.0157264

Henchion M, Hayes M, Mullen AM et al (2017) Future protein supply and demand: strategies and factors influencing a sustainable equilibrium. Foods 6(7):53. https://doi.org/10.3390/foods6070053

Henriksen-Lacey M, Carregal-Romero S, Liz-Marzan LM (2017) Current challenges toward In vitro cellular validation of inorganic nanoparticles. Bioconjugate Chem 28:212–221. https://doi.org/10.1021/acs.bioconjchem.6b00514

Homaee MB, Ehsanpour AA (2016) Silver nanoparticles and silver ions: oxidative stress responses and toxicity in potato (*Solanum tuberosum* L.) grown in vitro. Hortic Environ Biotechnol 57(6):544–553. https://doi.org/10.1007/s13580-016-0083-z

Hu J, Guo H, Li J et al (2017) Interaction of γ-Fe2O3 nanoparticles with Citrus maxima leaves and the corresponding physiological effects via foliar application. Nanobiotechnol 15:51. https://doi.org/10.1186/s12951-017-0286-1

Huang H, Ullah F, Zhou D-X et al (2019) Mechanisms of ROS regulation of plant development and stress responses. Front Plant Sci 10:800. https://doi.org/10.3389/fpls.2019.00800

Hussain HI, Zhifeng Y, Rookes JE et al (2013) Mesoporous silica nanoparticles as a biomolecule delivery vehicle in plants. J Nanopart Res 15:1676. https://doi.org/10.1007/s11051-013-1676-4

Hussein HS, Shaarawy HH, Hussien NH et al (2019) Preparation of nano-fertilizer blend from banana peels. Bull Nat Res Cen 43:26. https://doi.org/10.1186/s42269-019-0058-1

Jacob C (2018) Inspired by nature: antioxidants and nanotechnology. Antioxidants 18(7):101. https://doi.org/10.3390/antiox7080101

Jacobo-Velazquez DA, Gonzalez-Aguero M, Cisneros-Zevallos L (2015) Cross-talk between signaling pathways: the link between plant secondary metabolite production and wounding stress response. Sci Rep 5:8608. https://doi.org/10.1038/srep08608

Jampílek J, Kráľová K (2017) Nanomaterials for delivery of nutrients and growth-promoting compounds to plants. In: Prasad R, Kumar M, Kumar V (eds) Nanotechnology. Springer, Singapore. https://doi.org/10.1007/978-981-10-4573-8_9

Jang HR, Oh H-J, Kim J-H et al (2013) Synthesis of mesoporous spherical silica via spray pyrolysis: pore size control and evaluation of performance in paclitaxel prepurification. Microporous Mesoporous Mater 165:219–227. https://doi.org/10.1016/j.micromeso.2012.08.010

Jasim B, Thomas R, Mathew J et al (2017) Plant growth and diosgenin enhancement effect of silver nanoparticles in Fenugreek (*Trigonella foenum-graecum* L.). Saudi Pharm J 25:443–447. https://doi.org/10.1016/j.jsps.2016.09.012

Jeevanandam J, Barhoum A, Chan YS et al (2018) Review on nanoparticles and nanostructured materials: history, sources, toxicity and regulations. Beilstein J Nanotech 9:1050–1074. https://doi.org/10.3762/bjnano.9.98

Jha AB, Warkentin TD (2020) Biofortification of pulse crops: status and future perspectives. Plants 9:73. https://doi.org/10.3390/plants9010073

Kah M, Kookana RS, Gogos A et al (2018) A critical evaluation of nanopesticides and nanofertilizers against their conventional analogues. Nat Nanotechnol 13:677–684. https://doi.org/10.1038/s41565-018-0131-1

Kajbafna A, Ghorham H, Parnikar A et al (2012) Effects of morphology on photocatalytic performance of ZnO nanostructures synthesized by rapid microwave irradiation methods. Superlattices Microstruct 4:512–522. https://doi.org/10.1016/j.spmi.2012.01.015

Kalpana VN, Rajeswari VD (2018) A review on green synthesis, biomedical applications, and toxicity studies of ZnO NPs. Bioinorganic Chem Appl 3569758-12. https://doi.org/10.1155/2018/3569758

Kamle MM, Devi DK, Soni SR et al (2020) Nanotechnological interventions for plant health improvement and sustainable agriculture. Biotechnology 10:168. https://doi.org/10.1007/s13205-020-2152-3

Karny A, Zinger A, Kajal A et al (2019) Therapeutic nanoparticles penetrate leaves and deliver nutrients to agricultural crops. Sci Rep 8:1. https://doi.org/10.1038/s41598-018-25197-y

Khan A, Numan M, Khan AL et al (2020) Melatonin: awakening the defense mechanisms during plant oxidative stress. Plants 9(4):407. https://doi.org/10.3390/plants9040407

Khan I, Saeed K, Khan I (2019) Nanoparticles: properties, applications and toxicities. Arabian J Chem 12(7):908–931. https://doi.org/10.1016/j.arabjc.2017.05.011

Khandel P, Shahi SK (2018) Mycogenic nanoparticles and their bio-prospective applications: current status and future challenges. J Nanostruc Chem 8:369–391. https://doi.org/10.1007/s40097-018-0285-2

Khati P, Pankaj P, Nisha B et al (2018) Effect of nanozeolite and plant growth promoting rhizobacteria on maize. Biotechnol 8:141. https://doi.org/10.1007/s13205-018-1142-1

Khodakovskaya MV, de Silva K, Biris AS et al (2011a) Carbon nanotubes induce growth enhancement of tobacco cells. ACS Nano 6(3):2128. https://doi.org/10.1021/nn204643g

Khodakovskaya MV, de Silva K, Nedosekin DA et al (2011b) Complex genetic, photothermal, and photoacoustic analysis of nanoparticle-plant interactions. Proc Nat Acad Sci 108:1028–1033. https://doi.org/10.1073/pnas.1008856108

Khodakovskaya MV, Kim BS, Kim JN et al (2013) Carbon nanotubes as plant growth regulators: effects on tomato growth, reproductive system, and soil microbial community. Small 9(1):115–123. https://doi.org/10.1002/smll.201201225

Kohan-Baghkheirati E, Geisler-Lee J (2015) Gene expression, protein function and pathways of *Arabidopsis* thaliana responding to silver nanoparticles in comparison to silver ions, cold, salt, drought, and heat. Nanomaterials 5:436–467. https://doi.org/10.3390/nano5020436

Kokina I, Plaksenkova I, Jermaļonoka M et al (2020) Impact of iron oxide nanoparticles on yellow medick (*Medicago falcata* L.) plants. J Plant Inter 15(1):1–7. https://doi.org/10.1080/17429145.2019.1708489

Kwak S-Y, Wong MH, Lew TTS et al (2017) Nanosensor technology applied to living plant systems. Annu Rev Anal Chem 10:113–140. https://doi.org/10.1146/annurev-anchem-061516-045310

Le V, Rui Y, Gui X et al (2014) Uptake, transport, distribution and Bio-effects of SiO_2 nanoparticles in Bt-transgenic cotton. J Nanobiotechnol 12:50. https://doi.org/10.1186/s12951-014-0050-8

Lin D, Xing B (2008) Root uptake and phytotoxicity of ZnO nanoparticles. Environ Sci Technol 42:5580. https://doi.org/10.1021/es800422x

Lin W, Lin M, Zhou H et al (2019) The effects of chemical and organic fertilizer usage on rhizosphere soil in tea orchards. PLOS One 14(5):e0217018. https://doi.org/10.1371/journal.pone.0217018

Ma C, Liu H, Guo H et al (2016) Defense mechanisms and nutrient displacement in *Arabidopsis* thaliana upon exposure to CeO_2 and In_2O_3 nanoparticles. Environ Sci Nano 3:1369–1379. https://doi.org/10.1039/c6en00189k

Mahakham W, Sarmah AK, Maensiri S et al (2017) Nanopriming technology for enhancing germination and starch metabolism of aged rice seeds using phytosynthesized silver nanoparticles. Sci Rep 7:8263. https://doi.org/10.1038/s41598-017-08669-5

Makarov VV, Love AJ, Sinitsyna OV et al (2014) "Green" nanotechnologies: synthesis of metal nanoparticles using plants. Acta Naturae 6(1):35–44

Manwaring HR, Bligh HF, Yadav R (2016) The challenges and opportunities associated with biofortification of pearl millet (*Pennisetum glaucum*) with elevated levels of grain iron and zinc. Front Plant Sci 7:1944. https://doi.org/10.3389/fpls.2016.01944

Marslin G, Sheeba CJ, Gregory F (2017) Nanoparticles alter secondary metabolism in plants via ROS burst. Front Plant Sci 8:832. https://doi.org/10.3389/fpls.2017.00832

Martínez-Fernández D, Barroso D, Komárek M (2016) Root water transport of *Helianthus annuus* L. under iron oxide nanoparticle exposure. Environ Sci Pollut Res 23:1732–1741. https://doi.org/10.1007/s11356-015-5423-5

Marzouk NM, Abd-Alrahman HA, EL-Tanahy AM et al (2019) Impact of foliar spraying of nano micronutrient fertilizers on the growth, yield, physical quality, and nutritional value of two snap bean cultivars in sandy soils. Bull Nat Res Cen 43:84. https://doi.org/10.1186/s42269-019-0127-5

Meena S, Gautam C, Patidar P et al (2017) Nano-fertilizers is a new way to increase nutrients use efficiency in crop production. Int J Agric Sci 9(1):3831–3833. https://doi.org/10.1088/1755-1315/388/1/012046

Meraj TA, Fu J, Raza MA et al (2020) Transcriptional factors regulate plant stress responses through mediating secondary metabolism. Genes 11:346. https://doi.org/10.3390/genes11040346

Miranda-Villagómez E, Trejo-Téllez LI, Gómez-Merino FC et al (2019) Nanophosphorus fertilizer stimulates growth and photosynthetic activity and improves P status in rice. J Nanomater 1:11. https://doi.org/10.1155/2019/5368027

Mohammad G, Hossein S, Iraj A et al (2016) Synthesis of highly intercalated urea-clay nanocomposite via domestic montmorillonite as eco-friendly slow-release fertilizer. Agron Soil Sci 63(1):84–95. https://doi.org/10.1080/03650340.2016.1177175

Morales-Díaz AB, Ortega-Ortíz H, Juárez-Maldonado A et al (2017) Application of nanoelements in plant nutrition and its impact in ecosystems. Adv Natural Sci: Nanosci Nanotechnol 8(1): https://doi.org/10.1088/2043-6254/8/1/013001

Mout R, Ray M, Tonga GY et al (2017) Direct cytosolic delivery of CRISPR/Cas9-ribonucleoprotein for efficient gene editing. ACS Nano 11:2452–2458. https://doi.org/10.1021/acsnano.6b07600

Munir T, Rizwan M, Kashif M et al (2018) Effect of zinc oxide nanoparticles on the growth and Zn uptake in wheat (*Triticum aestivum* L.) by seed priming method. Digest J Nanomater Biostr (DJNB) 13:1

Mura S, Seddaiu G, Bacchini F et al (2013) Advances of nanotechnology in agro-environmental studies. Italian J Agro 8(3):e18. https://doi.org/10.4081/ija.2013.e18

Ndolomingo MJ, Bingwa N, Meijboom R (2020) Review of supported metal nanoparticles: synthesis methodologies, advantages and application as catalysts. J Mater Sci 55:6195–6241. https://doi.org/10.1007/s10853-020-04415-x

Pandorf M, Pourzahedi L, Gilbertson L et al (2020) Graphite nanoparticle addition to fertilizers reduces nitrate leaching in growth of lettuce (*Lactuca sativa*). Environ Sci: Nano 7:127–138. https://doi.org/10.1039/c9en00890j

Patra JK, Baek K-H (2014) Green nanobiotechnology: Factors affecting synthesis and characterization techniques. J Nanomater 417305. https://doi.org/10.1155/2014/417305

Patra JK, Das G, Fraceto LF (2018) Nano based drug delivery systems: recent developments and future prospects. J Nanobiotechnol 16(1):71. https://doi.org/10.1186/s12951-018-0392-8

Pikula K, Chaika V, Zakharenko A et al (2020) Comparison of the level and mechanisms of toxicity of carbon nanotubes, carbon nanofibers, and silicon nanotubes in bioassay with four marine microalgae. Nanomaterials 10(3):485. https://doi.org/10.3390/nano10030485

Plaksenkova I, Jermaļonoka M, Bankovska L (2019) Effects of Fe3O4 nanoparticle stress on the growth and development of rocket *Eruca sativa*. J Nanomater 2678247. https://doi.org/10.1155/2019/2678247
Poborilova Z, Opatrilova R, Babula P (2013) Toxicity of aluminium oxide nanoparticles demonstrated using a BY-2 plant cell suspension culture model. Environ Exp Bot 91:1–11
Prasad R, Bhattacharyya A, Nguyen QD (2017) Nanotechnology in sustainable agriculture: recent developments, challenges, and perspectives. Front Microbiol 20(8):1014. https://doi.org/10.3389/fmicb.2017.01014
Preetha S, Balakrishnan N (2017) A review of nano fertilizers and their use and functions in soil. Int J Curr Microbiol Appl Sci 6(12):3117–3133. https://doi.org/10.20546/ijcmas.2017.612.364
Raghunandan D, Basavaraja S, Mahesh B (2009) Biosynthesis of stable polyshaped gold nanoparticles from microwave-exposed aqueous extracellular anti-malignant Guava (*Psidium guajava*) leaf extract. Nanobiotechnol 5:34–41. https://doi.org/10.1007/s12030-009-9030-8
Raliya R, Saharan V, Dimkpa C et al (2018) Nanofertilizer for precision and sustainable agriculture: current state and future perspectives. J Agric Food Chem 66(26):6487–6503. https://doi.org/10.1021/acs.jafc.7b02178
Rani PU, Yasur J, Loke KS et al (2016) Effect of synthetic and biosynthesized silver nanoparticles on growth, physiology and oxidative stress of water hyacinth: *Eichhornia crassipes* (Mart) Solms. Acta Physiol Plant 38:1–9. https://doi.org/10.1007/s11738-016-2074-1
Raskar SV, Laware SL (2014) Effect of zinc oxide nanoparticles on cytology and seed germination in onion. Int J Curr Microb Appl Sci 3(2):467–473
Ripp S, Henry TB (2011) Biotechnology and nanotechnology risk assessment: minding and managing the potential threats around us. ACS Symposium Series. American Chemical Society, Washington, DC. https://doi.org/10.1021/bk-2011-1079
Ruttkay-Nedecky B, Krystofova O, Nejdl L et al 2017. Nanoparticles based on essential metals and their phytotoxicity. J Nanobiotechnol 15(1):33. https://doi.org/10.1186/s12951-017-0268-3
Salama DM, Osman SA, Abd El-Aziz ME (2019) Effect of zinc oxide nanoparticles on the growth, genomic DNA, production and the quality of common dry bean (*Phaseolus vulgaris*). Biocat and Agric Biotech 18:101083. https://doi.org/10.1016/j.bcab.2019.101083
Sarker U, Oba S (2018) Catalase, superoxide dismutase and ascorbate-glutathione cycle enzymes confer drought tolerance of *Amaranthus tricolor*. Sci Rep 8(1):16496. https://doi.org/10.1038/s41598-018-34944-0
Savci S (2012) Investigation of effect of chemical fertilizers on environment. APCBEE Procedia 1:287–292. https://doi.org/10.1016/j.apcbee.2012.03.047
Sega D, Ciuffreda G, Mariotto G et al (2019) FePO4 nanoparticles produced by an industrially scalable continuous-flow method are an available form of P and Fe for cucumber and maize plants. Sci Rep 9:11252. https://doi.org/10.1038/s41598-019-47492-y
Sekhon BS (2014) Nanotechnology in agri-food production: an overview. Nanotechnol Sci Appl 7:31–53. https://doi.org/10.2147/nsa.s39406
Shafiq M, Anjum S, Hano C et al (2020) An overview of the applications of nanomaterials and nanodevices in the food industry. Foods 9(2):148. https://doi.org/10.3390/foods9020148
Shang Y, Hasan MK, Ahammed GJ et al (2019) Applications of nanotechnology in plant growth and crop protection: a review. Molecules 24:2558. https://doi.org/10.3390/molecules24142558
Sharma D, Kanchi S, Bisetty K (2015) Biogenic synthesis of nanoparticles: a review. Arabian J Chem 12:3576–3600. https://doi.org/10.1016/j.arabjc.2015.11.002
Sharma S, Singh VK, Kumar A et al (2019) Effect of nanoparticles on oxidative damage and antioxidant defense system in plants. In: Roychoudhury A, Tripathi D (eds) Molecular plant abiotic stress: biology and biotechnology. Wiley, p 480. https://doi.org/10.1002/9781119463665.ch17
Shebl A, Hassan AA, Salama DM et al (2019) Green synthesis of nanofertilizers and their application as a foliar for *Cucurbita pepo* L. J Nanomater 3476347. https://doi.org/10.1155/2019/3476347

Shi Y-T, Cheng H-Y, Geng Y (2010) The size-controllable synthesis of nanometer-sized mesoporous silica in extremely dilute surfactant solution. Mater Chem Phys 120(1):193–198. https://doi.org/10.1016/j.matchemphys.2009.https://doi.org/10.045

Shivaji S, Madhu S, Singh S (2011) Extracellular synthesis of antibacterial silver nanoparticles using psychrophilic bacteria. Process Biochem 46(9):1800–1807. https://doi.org/10.1016/j.procbio.2011.06.008

Shukla D, Krishnamurthy S, Sahi SV (2014) Genome wide transcriptome analysis reveals ABA mediated response in *Arabidopsis* during gold (AuCl-4) treatment. Front Plant Sci 5:652. https://doi.org/10.3389/fpls.2014.00652

Simon C, Langlois-Meurinne M, Bellvert F et al (2010) The differential spatial distribution of secondary metabolites in *Arabidopsis* leaves reacting hypersensitively to *Pseudomonas syringae* pv. tomato is dependent on the oxidative burst. J Exp Bot 61:3355–3370. https://doi.org/10.1093/jxb/erq157

Simonin M, Guyonnet JP, Martins JMF et al (2015) Influence of soil properties on the toxicity of TiO2 nanoparticles on carbon mineralization and bacterial abundance. J Hazardous Mater 283:529–535. https://doi.org/10.1016/j.jhazmat.2014.https://doi.org/10.004

Singh J, Dutta T, Kim K-H et al (2018) 'Green' synthesis of metals and their oxide nanoparticles: applications for environmental remediation. J Nanobiotechnol 16:84. https://doi.org/10.1186/s12951-018-0408-4

Singh R, Singh S, Parihar P et al (2016) Reactive oxygen species (ROS): beneficial companions of plants' developmental processes. Front Plant Sci 7:1299. https://doi.org/10.3389/fpls.2016.01299

Skiba E, Wolf WM (2019) Cerium oxide nanoparticles affect heavy metals uptake by pea in a divergent way than their ionic and bulk counterparts. Water Air Soil Pollut 230:248. https://doi.org/10.1007/s11270-019-4296-5

Slepička P, Kasálková NS, Siegel J et al (2019) Methods of gold and silver nanoparticles preparation. Materials 13(1):1. https://doi.org/10.3390/ma13010001

Sosan A, Straltsova D, Tsiurkina K et al (2016) Engineered silver nanoparticles are sensed at the plasma membrane and dramatically modify the physiology of *Arabidopsis* thaliana plants. The Plant J 85:245–257. https://doi.org/10.1111/tpj.13105

Sunita R, Shekhawat GS (2016) Phytotoxicity and oxidative stress perspective of two selected nanoparticles in Brassica juncea. Biotechnol 6(2):244. https://doi.org/10.1007/s13205-016-0550-3

Tan BL, Norhaizan ME, Liew WP et al (2018) Antioxidant and oxidative stress: a mutual interplay in age-related diseases. Front Pharmacol 9:1162. https://doi.org/10.3389/fphar.2018.01162

Taran N, Batsmanova L, Kosyk O et al (2016) Colloidal nanomolybdenum influence upon the antioxidative reaction of chickpea plants (*Cicer arietinum* L.). Nanoscale Res Let 11:476. https://doi.org/10.1186/s11671-016-1690-4

Tian H, Kah M Kariman K (2019) Are nanoparticles a threat to mycorrhizal and rhizobial symbioses. Front Microbio 10:1660. https://doi.org/10.3389/fmicb.2019.01660

Tumburu L, Andersen CP, Rygiewicz PT et al (2017) Molecular and physiological responses to titanium dioxide and cerium oxide nanoparticles in *Arabidopsis*. Environ Toxicology Chem 36(1):71–82. https://doi.org/10.1002/etc.3500

Urban DA, Rodriguez-Lorenzo L, Balog S (2016) Plasmonic nanoparticles and their characterization in physiological fluids. Colloids Surf 137:39–49. https://doi.org/10.1016/j.colsurfb.2015.05.053

Usman M, Farooq M, Wakeel A et al (2020) Nanotechnology in agriculture: current status, challenges and future opportunities. Sci Total Environ 15(721):137778. https://doi.org/10.1016/j.scitotenv.2020.137778

Večeřová K, Večeřa Z, Dočekal B et al (2016) Changes of primary and secondary metabolites in barley plants exposed to CdO nanoparticles. Environ Pollut 218:207–218. https://doi.org/10.1016/j.envpol.2016.05.013

Vega-Vásquez P, Mosier NS, Irudayaraj J (2020) Nanoscale drug delivery systems: from medicine to agriculture. Front Bioeng Biotech 8. https://doi.org/10.3389/fbioe.2020.00079

Vicario-Pares U, Castanaga L, Lacave JM et al (2014) Comparative toxicity of metal oxide nanoparticles (CuO, ZnO and TiO_2) to developing zebrafish embryos. J Nanoparticle Res 16(8):2550. https://doi.org/10.1007/s11051-014-2550-8

Vijaya Bhaskar AV, Nithya DJ, Raju S et al (2017) Establishing integrated agriculture-nutrition programmes to diversify household food and diets in rural India. Food Sec 9:981–999. https://doi.org/10.1007/s12571-017-0721-z

Wang X, Yang X, Chen S et al (2016) Zinc oxide nanoparticles affect biomass accumulation and photosynthesis in *Arabidopsis*. Front Plant Sci 6:1243. https://doi.org/10.3389/fpls.2015.01243

Waqas MA, Kaya C, Riaz A et al (2019) Potential mechanisms of abiotic stress tolerance in crop plants induced by Thiourea. Front P Sci 10:1336. https://doi.org/10.3389/fpls.2019.01336

Wu J, Wang G, Vijver MG et al (2020) Foliar versus root exposure of AgNPs to lettuce: phytotoxicity, antioxidant responses and internal translocation. Environ Pollut 114117. https://doi.org/10.1016/j.envpol.2020.114117

Xie X, He Z, Chen N et al (2019) The roles of environmental factors in regulation of oxidative stress in plant. BioMed Res Int 9732325:11. https://doi.org/10.1155/2019/9732325

Yang F, Hong F, You W et al (2006) Influences of nano-anatase TiO_2 on the nitrogen metabolism of growing spinach. Bio Trace Element Res 110(2):179–190. https://doi.org/10.1385/bter:110:2:179

Yang J, Caob W, Ruia Y (2017) Interactions between nanoparticles and plants: phytotoxicity and defense mechanisms. J Plant Interact 2(1):158–169. https://doi.org/10.1080/17429145.2017.1310944

Yaqoob AA, Parveen T, Umar K et al (2020) Role of nanomaterials in the treatment of wastewater: a review. Water 12(2):495. https://doi.org/10.3390/w12020495

Yatim NM, Shaaban A, Dimin MF et al (2018) Effect of functionalised and non-functionalised carbon nanotubes-urea fertilizer on the growth of paddy. Trop Life Sci Res 29(1):17–35. https://doi.org/10.21315/tlsr2018.29.1.2

Yin L, Colman BP, McGill BM et al (2012) Effects of silver nanoparticle exposure on germination and early growth of eleven wetland plants. PLoS One 7:e47674. https://doi.org/10.1371/journal.pone.0047674

Zafar H, Ali A, Ali JS et al (2020) Effect of ZnO nanoparticles on *Brassica nigra* seedlings and stem explants: growth dynamics and antioxidative response. Front Plant Sci 20(7):535. https://doi.org/10.3389/fpls.2016.00535

Zhang B, Zheng LP, Li WY (2013) Stimulation of artemisinin production in *Artemisia annua* hairy roots by Ag-SiO_2 core-shell nanoparticles. Curr Nanosci 9:363–370. https://doi.org/10.2174/1573413711309030012

Zhu X, Zhu L, Duan Z et al (2008) Comparative toxicity of several metal oxide nanoparticle aqueous suspensions to zebra fish (*Danio rerio*) early developmental stage. J Environ Sci Health 43(3):278–284. https://doi.org/10.1080/10934520701792779

Zoufan P, Baroonian M, Zargar B (2020) ZnO nanoparticles-induced oxidative stress in *Chenopodium murale* L., Zn uptake, and accumulation under hydroponic culture. Environ Sci Poll Res (10):11066–11078. https://doi.org/10.1007/s11356-020-07735-2

Zulfiqar F, Navarrob M, Ashrafd M et al (2019) Nanofertilizer use for sustainable agriculture: advantages and limitations. Plant Sci 289:110270. https://doi.org/10.1016/j.plantsci.2019.110270

Chapter 22
Impact of Nanomaterials Stress on Plants

Zahra Iqbal, Mohammad Israil Ansari, Ausaf Ahmad, Ziyaul Haque, and Mohammed Shariq Iqbal

Abstract Response of plants to any stress condition is a complex phenomenon. Biomolecular, biomedical and biochemical approaches have been deployed to gain insights into this complex phenomenon. In the past few years, nanotechnology has also assisted in achieving acumens into response of plants to stress particularly, the stress induced by nanoparticles. Various anthropogenic activities contaminate the environment with xenobiotics like nanomaterials. Nanoparticles induce cellular stress in ecosystems, affecting plants the most. Interaction between the nanoparticle and cell influence the cellular processes of a plant. Several nanomaterials with different reactive affinities influence the cellular responses in relation to stress. The effect of nanomaterials on plants is also determined by various conditions of agro-ecosystems. Moreover, varying concentrations of nanomaterials are toxic to plants by producing reactive oxygen species (ROS) and reactive nitrogen species (RNS). Various research had been accomplished to demonstrate the impact of nanomaterials on medicinal, industrial and agricultural sectors. Further, research at molecular and sub-cellular level can determine the impact of nanomaterials in regulating (inducing/inhibiting) plant responses. This chapter involves impact of nanomaterial stress on plants and consequent response by plants in relation to nanomaterial stress.

Keywords Agriculture · Biotechnology · Cellular processes · Nanoparticles · Nanotechnology · Plant stress

Z. Iqbal
Department of Biochemistry, Faculty of Science, Chulalongkorn University, Bangkok 10330, Thailand

M. I. Ansari
Department of Botany, University of Lucknow, Lucknow 226007, India

A. Ahmad · Z. Haque · M. S. Iqbal (✉)
Amity Institute of Biotechnology, Amity University Uttar Pradesh, Lucknow Campus, Lucknow 226028, India
e-mail: msiqbal@lko.amity.edu

J. M. Al-Khayri et al. (eds.), *Nanobiotechnology*,
https://doi.org/10.1007/978-3-030-73606-4_22

22.1 Introduction

Nanotechnology involves a scientific approach studying extremely small particles that can be applied across other related fields including material science, physics, biology, chemistry and engineering. With the advent of nanotechnology, physical and chemical properties of substances are manipulated. This manipulation occurs at molecular level to develop desired products useful in diverse range of fields (medicine, chemistry, physics, biology, agriculture and so on) (Siddiqi and Husen 2016, 2017). Nanoparticles are described as colloidal particles of size 1–100 nm with a proximate interfacial layer. The interfacial layer comprises of ions, organic and inorganic molecules and is a pivotal part of nanoscale matter. Generally, the nanoparticles possess unique properties specifically, high surface to volume ratio, optical properties and size related properties (Al-Halafi 2014; Dutschk et al. 2014). Nanomaterials are categorized with various features such as nanofires, nanosheets and nanowires. They are also classified as carbonaceous (Baughman et al. 2002), semiconductor, metal oxides (Lang et al. 2011; Rizzello and Pompa 2014), lipids (Diao and Yao 2009; Yang and Ma 2010), zero-valent metals (Diao and Yao 2009), quantum dots, nanopolymer (Ljubimova and Holler 2012), and dendrimers (Astruc 2012). Nanoparticles are either directly or synthetically generated in nanosize, followed by grinding, milling, homogenization and sonication, reactive precipitation and solvent displacements (Mehta et al. 2012; Podsiadlo et al. 2007; Vasquez et al. 2008).

Since, agriculture is the main source of livelihood for at least 60% population in developing countries (Brock et al. 2011). Hence, nanotechnology is generally applied to transform food and agricultural industry by offering novel tools and molecular techniques for managing plant diseases and, thus enhancing crop performance. Nanotechnology can also be applied to plants for enhancing their ability to absorb water/nutrients and survive in adverse environmental conditions (Fincheira et al. 2020; Ocsoy et al. 2018). This could lead to increased crop yields, better nutrient value of staple crops, increased disease resistance of plants and better adaptation to environmental cues (Tarafdar et al. 2013). This particular field also provide robust sensors and delivery systems that may benefit agriculture sector incalculably. Some years down the line, nanostructured catalysts may be implemented in agriculture industry to substantially lower the agricultural inputs and enhance the efficiency of input deployment (Liu and Lal 2015). In addition, nanotechnology can trigger the usage of renewable energy sources, filters and catalysts to uncontaminate the prevailing pollutants in soil or decrease pollution (Adeleye et al. 2016; Mohamed 2017). The agricultural applications encircling nanotechnology comprises of water nano-remediation, plant nano-protection, plant nano-nutrition and so on. However, precise monitoring should be done so that nanoparticles itself do not impose stress on plants. The use of nanotechnology in agriculture industry must be stringently monitored to address agri-sustainability, amelioration of plant stress, crop protection against diseases, remediating environmental pollution, optimizing nutrient uptake and water management. Nanoparticles due to their higher contact surface area are toxic in comparison to their bulk equivalents (Shang et al. 2019). Due to this characteristic, they sometimes possess threat to open agricultural ecosystems (Lang et al. 2011; Rizzello and Pompa 2014). For instance, solid matrices have freely attached nanoparticles that can leach or can detach upon contact with water, air, or mechanical stress (Astruc 2012). The present chapter focusses on impact of nanomaterial stress on plants with respect to its effect on overall plant growth and development. The chapter also summarizes the interaction mechanisms of nanomaterials with plants surrounding; uptake, translocation, transmission and phytotoxicity.

22.2 Application of Nanotechnology to Plant Systems

Nanotechnology research and its subsequent implementation in agriculture, enabled the generation of Genetically Modified Crop's (GMC), strict farming systems and biocides (Bandyopadhyay et al. 2013; Griffitt et al. 2009; Hansen et al. 2009; Lombi et al. 2012; Musee 2011; Nadiminti et al. 2013; Prasad et al. 2017). As is true with other technologies applied to agricultural, cost effective nanoparticles and precise field delivery technologies of nanoparticles are essential (Delgado-Ramos 2014; Dutschk et al. 2014; Neto 2014; Safari and Zarnegar 2014). Nanoparticles due to its unique properties can either improve seed germination or enhance overall plant productivity (Fig. 22.1). However, the impact of nanoparticles on essential staple crops still remains elusive (Lombi et al. 2012; Nadiminti et al. 2013; Nevius et al. 2012).

The size of nanoparticles, its chemical composition and the plant species largely determines the uptake and build-up of nanoparticles (Bagheri et al. 2012; Ghormade et al. 2011; Khiew et al. 2011). Moreover, the effect of nanoparticles on plant system depends upon the developmental stage of the plant, method of delivery, time of exposure, shape, concentration, solubility, aggregation and surface structure of the nanoparticle (Khiew et al. 2011; Nevius et al. 2012). In addition, the delivery techniques must be scrutinized for efficient application of nanoparticles on potential targets. Some plants are more efficient in up-taking and accruing nanoparticles. Plant cell interacts with nanoparticles, which results in altered gene expression and consequentially into modified biological pathways. This affects the overall plant growth and development (Feizi et al. 2013a; Ghormade et al. 2011; Khiew et al. 2011). For the very first time, the effects of nanoparticles were efficiently demonstrated on *Salvia officinalis* L. (sage) (Feizi et al. 2013a). Although nanoparticles are reported to enhance germination of seeds in some cases, but they can have contradictory consequences on overall plant performance (Lombi et al. 2012; Nadiminti et al. 2013). Further, research is needed to assess the possible risk associated with the use of nanoparticles in agriculture (Bandyopadhyay et al. 2013; Griffitt et al. 2009; Hansen et al. 2009; Prasad et al. 2017; Sanzari et al. 2019).

Several articles had indicated the plausible *in vitro* toxicity of nanoparticles on bacterial, aquatic and human cells. Even non-harmful chemicals become toxic at nanoscale level (Huang et al. 2017). Nanoparticles usually persists for days to weeks if

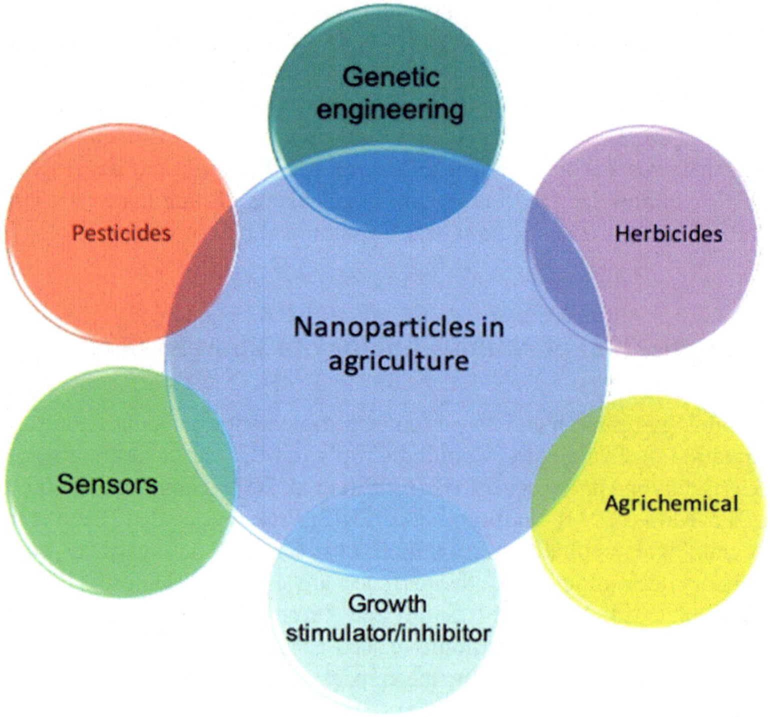

Fig. 22.1 Applications of nanotechnology in agriculture (Figure constructed by Zahra Iqbal)

suspended in air (Elliott 2011; Liden 2011; Potera 2010). Nanoparticles are absorbed and collected by all the parts of the respiratory system of the plants. Since, nanoparticles are extremely small in size, they follow airstream with more ease that the larger particles (Baughman et al. 2002; Lang et al. 2011; Safiuddin et al. 2014; Zhou et al. 2014). Therefore, the capability of the nanoparticle entering the respiratory system of the plant through individual cells and nuclei is a major concern. Yet another concern is bioaccumulation of these nanoparticles in the plant system (Diao and Yao 2009; Rizzello and Pompa 2014; Yang and Ma 2010; Ye et al. 2012a). It becomes difficult to evaluate the health and environmental concerns encircling nanoparticles due to its nanoscale properties, which are sparsely understood (Elliott 2011; Hansen 2010; Liden 2011). This makes it necessary that nanoparticles are designed such that they have effective delivery, stability, solubility, uptake, accumulation, translocation, transformation and degradation (Feizi et al. 2013a; Ghormade et al. 2011). Thus, prior to the application of nanoparticles in agriculture it is important to investigate its toxic effects, to be considered safe for plants.

22.3 Impact of Different Nanoparticles on Plant Growth and Development

Under particular environmental conditions plant may absorb several essential or non-essential compounds at a concentration, which might be toxic to many crops (Lai et al. 2006; Nair et al. 2010; Zhang et al. 2006). Elements of unknown biological significance accumulate in plants causing lethality (Liden 2011; Zhang et al. 2006). Nanoparticle stress on plant growth can happen due to their direct application or accidental release. The harmful effect of these nanoparticles worsens, when they move to higher order in the food chain (plants to humans). In this chapter various categories of nanoparticles and their stress impact on plant will be discussed (Fig. 22.2).

22.3.1 Carbon-Based Nanoparticles

Carbon-based nanoparticles are simply defined as nanoparticles with carbon as the main component. Carbon based nanoparticles have tremendous application in biomedical science due to their electrical, optical, mechanical, chemical and thermal properties (Patel et al. 2019). Due to their implication in biomedical science and other

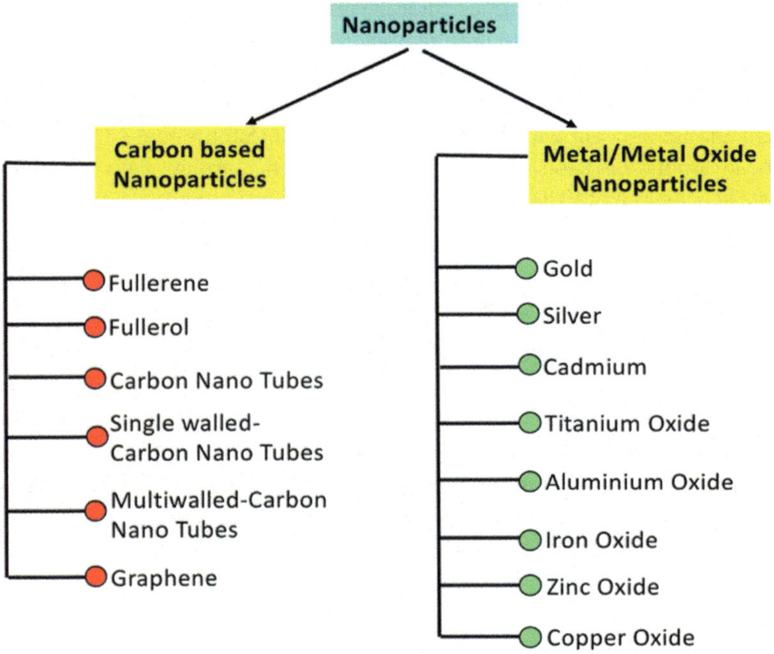

Fig. 22.2 Nanoparticles implicated in agriculture (Figure constructed by Zahra Iqbal)

related fields, there has been an enormous production of carbon-based nanoparticles. Their release into the environment is either intentional (discharge), or unintentional (spillage) posing adverse environmental effects (Baughman et al. 2002). Fullerene C_{70}, fullerol, and carbon nano-tubes (CNTs) are prevalently used nanomaterials. High hydrophobicity of carbon-based nanomaterials results in their aggregation in living systems and interaction with organic substances (De La Torre-Roche et al. 2013; Santos et al. 2013). CNTs can accumulate in few edible crops, with such specific uptake mechanism. Critically generated nanoparticles penetrate through targeted mechanisms, which permits slow and governed discharge, especially for weeds. For example, lipophilic nano-silica induces death by dissection after getting absorbed into the cuticular lipids of insects.

22.3.1.1 Fullerene

Fullerenes are defined as allotropes of carbon. The carbon atoms in fullerenes are linked with single or double bonds to form a partially or fully closed mesh. Fullerenes consists of fused rings of 5–7 atoms which may be ellipsoid, sphere or tobe of different sizes and shapes (Schwerdtfeger et al. 2015). Black aggregates of fullerene have been reported to accumulate in the seeds and roots of rice (De La Torre-Roche et al. 2013; Liu et al. 2013; Santos et al. 2013). Osmotic pressure, cell wall pores, intercellular plasmodesmata, capillary forces and symplastic routes are the major pathways through which fullerenes enter the root system (Liu et al. 2010). Fullerenes with diameter smaller than the pore diameter of cell walls or membranes can enter the roots of plants (Santos et al. 2013). In addition, transport of fullerenes from roots to shoots also happens with mature plants. Thus, generally fullerenes accumulate in the stem and leaves, while the roots are free of them (Santos et al. 2013). Fullerenes translocate from roots to shoot along with the nutrients through the xylem tissue (De La Torre-Roche et al. 2013).

22.3.1.2 Fullerol

Fullerol are basically polyhydroxy fullerenes (functionalized fullerenes). They are water soluble and have been reported to be toxic in some case while beneficial in some other cases (Gao et al. 2011). They permeate through the pores of cell wall because of their small size and hydrophobic properties (Kole et al. 2013). Thus, fullerol aggregate amid the cell membrane and the cell wall (Gao et al. 2011). They also, because of their apoplectic mode of transport tend to accumulate between adjacent epidermal cell walls (Kole et al. 2013).

22.3.1.3 Carbon Nano Tubes (CNTs)

CNTs are cylindrical nanomaterials made up of either a single or multiple layer of carbon atoms (Collins et al. 2000; Hazarika and Maji 2014; Li et al. 1996; Liu et al. 2014; Mani et al. 2014). CNTs have characteristics different from carbon or graphite and act as fibres (Collins et al. 2000). CNTs possess exceptional tensile strength. They are the sturdiest and smallest known fiber (Hazarika and Maji 2014; Liu et al. 2014; Mani et al. 2014). The mechanism of uptake, transport and effect of CNTs in plants has been studied extensively (Chai et al. 2013; Khodakovskaya et al. 2013; Long et al. 2012). CNTs can easily translocate to fruits, roots and leaves accounting to significant variations in gene expression (Khodakovskaya et al. 2013). CNTs are proven to induce cell death in a dose dependent manner and, thus can have phytotoxic effects on plant cells (De La Torre-Roche et al. 2013). Cell death induction by CNTs are proven by electrolyte leakage assays and the local swellings of plant organs. The interplay of CNTs with the proteins and polysaccharides of the cell stimulates hypersensitive response similar to that of pathogen attack. This eventually results in cell death (Collins et al. 2000; Hazarika and Maji 2014; Li et al. 1996; Mani et al. 2014). CNTs are reported to enhance root growth in *Cucumis sativus* (cucumber) and *Allium cepa* (onion) (Canas et al. 2008; Ke and Lamm 2011). CNTs are also shown to inhibit root growth in tomato. However, CNTs can penetrate seed coat of tomato to induce seed germination and seedling growth (Ghodake et al. 2010; Ke and Lamm 2011; Khodakovskaya et al. 2009).

22.3.1.4 Multi Walled Carbon Nano Tubes (MWCNTs)

MWCNTs are basically CNTs which are 1 mm long and 20 nm is diameter (Kong et al. 2004; Li et al. 2005; Lin et al. 2002; Muller et al. 2005). MWCNTs are imbibed by roots or seeds through pores in plant cell or during water uptake. MWCNTs are reported to induce growth of *Solanum* (tomato) plantlets by boosting seed germination and improving seed water uptake (Chekin et al. 2012; Wang et al. 2012). MWCNTs are generally spotted on the root surface rather than on root hair cell walls or root cap (Smirnova et al. 2012). MWCNTs increased the seed germination rates by 90% in treated samples when compared to control which was just 71%. They also enhanced plant biomass (Tan and Fugetsu 2007). However, in few cases such as rice, MWCNTs cannot pass through the cell walls, forming a black clump (Chekin et al. 2012). This clump wraps around the cells; and as the concentration of MWCNTs increases the size and number of the black clumps also increases proportionally (Wang et al. 2012). This might be a hypersensitive response to prevent the entry of MWCNTs into the plant system (Smirnova et al. 2012; Tan and Fugetsu 2007; Wang et al. 2012). On contrary lines, MWCNTs accumulated less in vascular tissues. MWCNTs within the examined range do not possess any adverse effect on germination rates and root growth in Zucchini species (Stampoulis et al. 2009).

22.3.1.5 Single Walled Carbon Nano Tubes (SWCNTs)

SWCNTs, similar to MWCNTs are also CNTs which are 0.1 μm long and 1–2 nm is diameter (Hata et al. 2004; Ren et al. 2007; Warheit et al. 2004). The widespread usage, dispensability, and water column stability of carbon nanoparticles are due to its surface modifications and the associated benefits (Cui et al. 2012; Flores et al. 2013; Hao et al. 2012; Parise et al. 2014; Yan et al. 2013). However, there is no report of uptake of SWCNTs by roots of cucumber seedlings post 84 h of exposure. SWCNTs derived nanosheets can clung to the primary and secondary roots (Lou et al. 2011; Yuan et al. 2011). Further research is needed to establish the transport of SWCNTs from root to shoot (Cui et al. 2012; Hao et al. 2012; Lou et al. 2011; Shen et al. 2010; Yuan et al. 2011). SWCNTs are too big to pass through the cell wall. Nonetheless, in *Arabidopsis thaliana* endocytosis-like structure of the plasma membranes are observed, indicating the presence of carbon nanoparticles (Lin et al. 2009a; Shen et al. 2010). This was corroborated with another study involving *Nicotiana tabacum* (tobacco) cv. Bright Yellow where SWCNTs with 500 nm or lesser length has penetrated the cell wall and cell membrane (Khodakovskaya et al. 2009; Yi et al. 2005).

22.3.1.6 Graphene

Graphene is a single layer of graphite and is a 2D crystalline allotope of carbon. Excessive dosages of graphene (1000 mg L^{-1}) are reported to be inhibitory to root hair growth in cabbage, tomato and spinach (Al-Ghamdi et al. 2014; Anjum et al. 2013; Begum et al. 2011; Lee and Kim 2014; Ye et al. 2012b). Graphene induces oxidative stress in plants by accumulating H_2O_2, electrolyte leakage, cell death and necrotic damage (Al-Ghamdi et al. 2014; Fugetsu and Parvin 2011; Mogharabi et al. 2014). Graphene treatments increases root surface area of cabbage but on contrary, can lead to inflammation in *Origanum vulgare* (oregano) (Al-Ghamdi et al. 2014; Begum et al. 2011). Graphene in a dose dependent manner can also cause phytotoxicity in plant cells (Akhavan and Ghaderi 2010; Sasidharan et al. 2011). Graphene could systematically translocate to leaves, roots and fruits and can substantially alter the gene expression profiles (Anjum et al. 2014; Begum and Fugetsu 2013; Kim 2013; Lee and Kim 2014). Accumulation of nanostructured materials can substantially reduce plant growth and biomass, which might eventually lead to phytotoxicity (Ocsoy et al. 2013). Graphene toxicity is also reported in terrestrial plants exposing them to potential risks (Begum and Fugetsu 2013; Kim 2013).

22.3.2 Metal/Metal Oxide-Based Nanoparticles

The production of metal/metal oxide-based nanoparticles is estimated to skyrocket from 2000 tons in 2004 to 58,000 tons by 2020 (Franke et al. 2006; Kolmakov and Moskovits 2004; Niederberger 2007; Stoimenov et al. 2002). Such nanoparticles exhibit size related properties, viz., magnetism, fluorescence and photocatalytic degradation. With such incredible properties metal/metal oxide-based nanoparticles are utilized in sensor production, agrochemical industry and soil remediation (Franke et al. 2006; Kolmakov and Moskovits 2004). Metal/metal oxide-based nanoparticles effects on plants mainly depends upon chemical colloidal and organic properties of the living system (Moisala et al. 2003; Niederberger et al. 2006). The response of plants to metal/metal oxide-based nanoparticles is also dependent upon the type of metal, species and growth stage. The frequently deployed metal/metal oxide-based nanoparticles are TiO_2, CeO_2, Fe_3O_4, and ZnO nanoparticles. Metal/metal oxide-based nanoparticles in a concentration dependent pattern effect plant growth and development (Stoimenov et al. 2002). Excessive absorption of metals can lead to obvious deleterious effects, including stunted growth and irregular cellular divisions in plants (Franke et al. 2006; Kolmakov and Moskovits 2004; Niederberger 2007).

22.3.2.1 Gold (Au)

Au is the most commonly used nanoparticle. Au in ionic or soluble form in toxic to many organisms (Boisselier and Astruc 2009; Goodman et al. 2004; Karamushka and Gadd 1999; Murphy et al. 2008). Au acts as an antibacterial agent in soaps, shampoos and biocide coating (Goodman et al. 2004; Murphy et al. 2008). The production, usage and eventually the discharge of Au nanoparticles can cause severe environmental issues over a prolonged period of time (Hauck et al. 2008; Johnston et al. 2010; Khlebtsov and Dykman 2011; Murphy et al. 2008). For instance, a substantial increase in Au uptake has been reported in *Brassica juncea* (brown mustard) and *Medicago sativa* (Alpha-alpha) with increase in exposure time (Perreault et al. 2012; Saison et al. 2010; Zhai et al. 2014). Both these species are hyperaccumulators of Au, with the nucleus as the prime site (Arora et al. 2012; Gardea-Torresdey et al. 2000; Green and Renault 2008). Au nanoparticles via plasmodesmeta can also enter the cells. TEM (Transmission Electron Microscopy) analysis of rice roots revealed the presence of varying size Au nanoparticles (Perreault et al. 2012). Sometimes Au nanoparticles bigger than the pore size of cell wall try to enter the cell leading to cellular damage (Saison et al. 2010). In *Allium cepa*, Au nanoparticles have been reported to disrupt root tips. This disruption causes cell disintegration, and cell stickiness leading to hampered cell division process (Feretti et al. 2007).

22.3.2.2 Silver (Ag)

Ag nanoparticles are used in imaging and chemical sensing. They are synthesized by a number of methods including electrochemical, photochemical, laser ablations amongst others (Rizzello and Pompa 2014). Ag nanoparticles if deployed in vivo at large scale, needs to be strictly monitored for biocompatibility and environmental effects (Galvez and Wood 1997; Hogstrand et al. 1996; Wagner et al. 1975). Reports have shown that Ag particles persists in sludge and surface water at low concentrations (0.1 and 2.9 mg L^{-1}) (Ferguson and Hogstrand 1998; Lee et al. 2005; Wood et al. 1996). Ag nanoparticles implement great risk to terrestrial plants (Bianchini and Wood 2003; Lee et al. 2005; McGeer et al. 2000; Wood et al. 1996). In rice (*Oryza sativa*), Mung bean (*Vigna radiata*), and Chinese cabbage (*Brassica campestris*) Ag nanoparticles impose negative effect. It effect seed germinations, root, and shoot growth at concentrations of 4500, 6000, and 3000 μg mL^{-1} respectively (Justin and Armstrong 1991; Nguyen et al. 2003; Mao et al. 2004). Moreover, 40 nm Ag nanoparticles are toxic to *Chlamydomonas reinhardtii* algae and *Cucurbita pepo* (pumpkin) (Cheng et al. 2011; Ratte 1999; Slade and Pegg 1993; Yin et al. 2011, 2012). There is approximately 4.4 to tenfolds decrease in biomass and transpirational rates in comparison to wild type (Cheng et al. 2011; Howe and Merchant 1992; Ouda 2014; Piccapietra et al. 2012; Roessler and Lien 1984). The impact of Ag nanoparticles for their uptake, transport and dispersal was also studied in *Medicago sativa* and *Brassica juncea* species (Beebe and Turgeon 1992; Cairns et al. 1975; Inokuchi et al. 1997; Musante and White 2012; Ouda 2014; Roh et al. 2009; Shoults-Wilson et al. 2011; Unrine et al. 2012; Wang 1986; Kumari et al. 2009; Saxena et al. 2010). Cellular uptake and cytotoxicity of Ag nanoparticles have been explored extensively in plant systems. For example, Zucchini were grown in hydroponics supplemented with Ag nanoparticles and were studied for subsequent effects. There was no adverse impact on seed germination and root growth in the presence of Ag nanoparticles, however, plant biomass and transpiration rates reduced in the treatment in comparison to control. Ag nanoparticles also disrupted cell division resulting into cell disintegration (disturbed metaphase, stickiness, chromatin bridge, amongst other effects) (Cheng et al. 2011; Yin et al. 2012). Ag nanoparticles also affect aquatic plants (Cairns et al. 1975; Inokuchi et al. 1997; Wang 1986). Exposing aquatic plant i.e. *Lemna minor* L. clone to 5 mgL^{-1} concentration of varied size (20–100 nm) Ag nanoparticles resulted in suppression of plant growth (Cairns et al. 1975; Unrine et al. 2012; Wang 1986). Ag nanoparticles of size 29 nm reduced germination rates in lettuce and cucumber. On contrary, there was no such effect of reduced germination rates in by Ag nanoparticles in barley and ryegrass. Ag nanoparticles greater in size than 100 nm decrease transpiration rates and biomass in *Cucurbita pepo* (Ratte 1999). Ag nanoparticle toxicity on germination rate, uptake efficiency and translocation had also been documented well in soil (Musante and White 2012; Roh et al. 2009; Shoults-Wilson et al. 2011). For *Polyboroides radiatus* (Madagascan harrier-hawk) and *Sorghum bicolor* (great millet), the bioavailability and toxicity of Ag nanoparticles was monitored on soil as well as agar medium (Beebe and Turgeon 1992; Kumari et al. 2009; Liu et al. 1994; Saxena et al. 2010). There was concentration dependent inhibition of growth in *Poyboroides radiatus* and *Sorghum bicolor*

(Dimkpa et al. 2013; Lee et al. 2012). Moreso, the phytotoxicity, cellular toxicity, solubility and bioaccumulation of ag nanoparticles in plants strictly depends upon the media of exposure (Dimkpa et al. 2013).

22.3.2.3 Cadmium (Cd)

Toxicity testing was done for Cd nanoparticles in tomato, cucumber, lettuce and carrot. The seedling growth was inversely related to the concentrations of Cd nanoparticles (Clarke and Brennan 1989; Jiang et al. 2003; Kashem and Kawai 2007). To deep dive into the details of root architecture effected by Cd nanoparticles microscopic analysis was carried out. The analysis revealed decline in root diameter along with disintegration of root epidermal cells under Cd nanoparticle stress (Groppa et al. 2008; Kairong 1996). Cd nanoparticles were also found to be accumulated on root surface hindering proper cell growth (Cho and Seo 2005; John and Van Laerhoven 1976).

22.3.2.4 Titanium Oxide (TiO_2)

TiO_2 nanoparticles are frequently used in daily products. Studies on TiO_2 effect (uptake, transport, toxicity etc.) on plants especially, staple food crops is very limited (Castiglione et al. 2011; Dehkourdi and Mosavi 2013; Feizi et al. 2012; 2013b; Elghniji et al. 2012; Hasan et al. 2012; Khataee et al. 2014; Kurepa et al. 2010; Larue et al. 2012; Mahmoodzadeh et al. 2013; Mahmoodzadeh and Aghili 2014; Paret et al. 2013; Qi et al. 2013; Qiu et al. 2013; Song et al. 2012, 2013; Wang et al. 2007). TiO_2 nanoparticles have a small size up to < 5 nm. Exploiting their small size, they form covalent bonds with natural organic matter. They then translocate and distribute in a tissue and cell specific manner (Castiglione et al. 2011; Feizi et al. 2012, 2013b; Qiu et al. 2013; Song et al. 2013). TiO_2 nanoparticles are reported to be toxic to algae *Desmodesmus subspicatus* (Hund-Rinke and Simon 2006). Additionally, TiO_2 nanoparticles form ROS (reactive oxygen species) upon coming in contact with certain organisms and UV rays (Castiglione et al. 2011; Dehkourdi and Mosavi 2013; Elghniji et al. 2012; Feizi et al. 2012; Hasan et al. 2012; Kurepa et al. 2010; Khataee et al. 2014; Larue et al. 2012; Mahmoodzadeh et al. 2013; Mahmoodzadeh and Aghili 2014; Paret et al. 2013; Qiu et al. 2013; Song et al. 2012, 2013). Upon exposure to TiO_2, *Arabidopsis thaliana* roots discharge mucilage and generate a capsule of pectin hydrogel around the root (Kurepa et al. 2010; Wang et al. 2011). In *Glycine max* (Soybean), TiO_2 is reported to enhance nitrate reductase activity, water use efficiency and boost the antioxidant structure. TiO_2 exposed seeds generated plants with dry weight increased up to 73%, photosynthetic rates increased up to 3 times and chlorophyll a content increased by 45% (Hund-Rinke and Simon 2006; Molina-Barahona et al. 2005; Qi et al. 2013). However, with spinach seeds the rate of germination was directly related to the size of TiO_2 nanoparticle (Wu et al. 2012;

Zheng et al. 2005). TiO_2 enhance nitrogen metabolism and hence promotes absorption of nitrate resulting in enhanced plant growth (Gao et al. 2008; Wu et al. 2012; Yang et al. 2006). In *Oryza sativa*, TiO_2 adversely effects germination rates and count of roots formed (Foltête et al. 2011). TiO_2 nanoparticles exhibit low toxic effects to plants and follow a dose dependent pattern. This might be because of agglomeration and eventual sedimentation of particles.

22.3.2.5 Aluminum Oxide (Al_2O_3)

Phytotoxicity of coated (phenanthrene-coated) and uncoated Al_2O_3 has been extensively established through various studies (Blamey et al. 1983; Kollmeier et al. 2000; Yamamoto et al. 2001). Concentrations as low as 2 mg L^{-1} of uncoated Al_2O_3 inhibited root elongation in soybean, carrot, corn, cucumber and cabbage (Blamey et al. 1983; Kollmeier et al. 2000; Yamamoto et al. 2001). The toxic effect is essentially due to solubility of Al_2O_3 nanoparticles. Smaller Al_2O_3 particles were found to be toxic to seedling growth (Kollmeier et al. 2000; Ryan et al. 1992; Tian et al. 2007; Yamamoto et al. 2001). Hence, it becomes evident that the surface properties of Al_2O_3 nanoparticles play a crucial role in imparting toxicity to these nanoparticles (Tian et al. 2007). High concentration of Al_2O_3 reduces root growth in carrot, corn, cabbage, cucumber and soybean (Cumming et al. 1992; Kinraide et al. 1992; Sartain and Kamprath 1978). However, in *Lolium perenne* (reygrass)and *Phaseolus vulgaris* (kidney bean) Al_2O_3 nanoparticles did not impose any stress (Lazof and Holland 1999).

22.3.2.6 Iron Oxide (Fe_3O_4)

Fe_3O_4 is generally used as a magnetic nanomaterial. It can sometimes impose negative effect on plant growth and development (Besson-Bard et al. 2009; Zhu et al. 2008). Chlorophyll *a* content is inversely linked to Fe_3O_4 concentrations. At low concentrations of Fe_3O_4, the chlorophyll *a* content was higher and at higher Fe_3O_4 concentrations the chlorophyll *a* content was lower (Besson-Bard et al. 2009; Zhu et al. 2008). Higher concentration of Fe_3O_4 nanoparticles also resulted in brown spots during the growth of plantlet (Hartley and Lepp 2008; Kraemer 2004). Fe_3O_4 nanoparticles also induces oxidative stress, resulting into decreased photosynthetic and metabolic rates (Becana et al. 1998; Kumar et al. 2018). To reduce the adverse effect of Fe_3O_4 nanoparticles, they are sometimes coated with carbon. This increases the adsorption surface of Fe_3O_4 nanoparticles and imparts them with biocompatible properties (Katsoyiannis and Zouboulis 2002; Lombi et al. 1999; Mahmoudi et al. 2009). For instance, in pumpkin, certain concentrations of carbon coated Fe_3O_4 nanoparticles did not adversely affect plant cells as well as did not contaminate the environment upon discharge (Katsoyiannis and Zouboulis 2002). Similarly, tetramethylammonium hydroxide coated Fe_3O_4 nanoparticles at low concentrations enhanced chlorophyll contents (Greipsson and Crowder 1992).

22.3.2.7 Zinc/Zinc Oxide (Zn/ZnO)

They are frequently used metallic nanoparticles. Zn constitutes an essential micronutrient for plants, animals and humans (Bai et al. 2010; Huang et al. 2002; Ma et al. 2009; Sharma et al. 2009; Wong et al. 2010). Zn is also a pivotal cofactor for a number of enzymes (Wang et al. 2004). It is commonly used in beauty products, biosensors, electrodes etc. (Ma et al. 2009). Due to its widespread use, Zn/ZnO can either accidently or deliberately spill out is environment, effecting terrestrial life (Adamson et al. 2000; bin Hussein et al. 2002; Sharma et al. 2009; Wang et al. 2004). Zn in inappropriate concentrations also affect plant system. Zn/ZnO affects the germination of plants studied in buckwheat (*Fagopyrum esculentum*) and might cause damage to chromosome at cellular level (Lee et al. 2013). Zn/ZnO nanoparticles inhibited root growth and is toxic in ryegrass, rape and radish (Adamson et al. 2000; Bai et al. 2010; bin Hussein et al. 2002; Huang et al. 2002; Lee et al. 2013; Ma et al. 2009; Reddy et al. 2007; Sharma et al. 2009; Wang et al. 2004; Wong et al. 2010). Zn/ZnO can cause stress either due to chemical toxicity (chemical composition) or because of its physical properties (size, surface and shape) (Wang et al. 2004).

22.3.2.8 Copper/Copper Oxide (Cu/Cu_2O)

A study suggests that Cu/Cu_2O nanoparticles are adsorbed by onion roots. This adsorption blocks water channels and results into radical penetration in the roots of onion (Fiskesjó 1993; Geremias et al. 2010). Eventually this hinders cell division and cellular metabolism. In similar lines, Cu nanoparticles concentration is inversely related to the seedling growth in Mung bean (*Phaseolus radiates*) and wheat (*Triticum aestivum*) species (Lee et al. 2008; Munzuroglu and Geckil 2002). Root morphology in wheat results in greater accumulation of Cu nanoparticles. Mung bean is more sensitive to different concentrations of Cu in comparison to wheat (Keltjens and Van Beusichem 1998). Additionally, in zucchini plants Cu nanoparticles decreased the root length at emergence stage (Wang and Zhou 2005). However, in lettuce an increase in root to shoot ratio was observed in the presence of Cu nanoparticles (Mocquot et al. 1996). Thus, Cu nanoparticles affect shoot length/biomass, root length/biomass, germination rate, biomass and growth in a variety of staple crops (Lombardi and Sebastiani 2005).

22.4 Nanoparticle Interaction with Plants

The interaction of nanoparticles with plant system is a matter of great concern since it directly or indirectly impacts food chain. Once applied nanoparticles inevitably enters the plant cell, whereby modifying cellular machinery. Deliberate application of nanoparticles to plants is done to achieve beneficial effects (Pérez-de-Luque 2017). However, accidental leakage can result in free circulation of nanoparticles (not bound to the specific substrate). In such a case, nanoparticles can pose serious toxicity and can decline the overall performance of plant. Hence, it becomes indispensable to understand the interaction of nanoparticles with plant system. Interaction of nanoparticles with plants involve uptake, translocation, accumulation and phytotoxicity mechanisms at whole plant and cellular level.

22.4.1 Nanoparticle Uptake by Plants

Research on uptake mechanism of nanoparticles by plants is yet limited (Nevius et al. 2012; Tani and Barrington 2005). Nanoparticles generally adhere to the roots and eventually either follow physical or chemical paths of entry in the plant system (Hartley and Lepp 2008). However, there are many studies focusing on how the nanoparticles interact with the plants (Besson-Bard et al. 2009). The uptake and accumulation of nanoparticles varies amongst different plant species and also depend upon the shape, form and size of nanoparticle (John et al. 1972). The uptake mechanism of nanoparticles by plants is generally studied on stock solutions (Hauck et al. 2008; Smirnova et al. 2012) at germination stage or in cell culture. In case of carbon nanotubes, the nanoparticles enter the plant system through aquaporins, ion channels or sometimes endocytosis. They bind to a carrier protein or organic chemical as a part of their uptake mechanism (Kurepa et al. 2010; Maine et al. 2001; Smirnova et al. 2012). Higher reactivities are observed in nanoparticle due to greater surface area to mass ratio in comparison to their metal counter-parts (Larue et al. 2012). Eventually, just prior to the transportation process in plants, nanoparticles tend to form complexes with root exudates and membrane transporters (Tani and Barrington 2005). The nanoparticles are transported from one cell to another via plasmodesmata either following apoplastic or symplastic mode of transportation (Hauck et al. 2008). As stated earlier, the efficiency of nanoparticle uptake depends upon the particle size. Hence, nanoparticle uptake steadily increase with declining granular size (Larue et al. 2012). For instance, 1.5 mm ZnO nanoparticles are more efficiently taken up by the plants than 2.0 or 2.5 mm ZnO nanoparticles. This results in better distribution of Zn and eventually better uptake and utilization (Wang et al. 2004).

22.4.2 Nanoparticle Translocation in Plants

The translocation of nanoparticles is dependent upon the concentration of nanoparticles and species under investigation (Yang and Ma 2010). The nanoparticles are translocated from leaves to roots, stems or other parts of the plant system and similarly from roots to leaves, developing grains or other parts of the plant system. The penetration of nanoparticles in the cell membrane or cell wall of root cells initiates the translocation mechanism. Xylem acts as the central passage for translocation of nanoparticles to the shoot (Birbaum et al. 2010; Miralles et al. 2012). Definitely for this process, the pore size of the cell wall and cell membranes needs to be larger (3–8 nm) than the size of the nanoparticle. *Allium porrum* (Leek) was studied to analyze the penetration rate and translocation of nanoparticles. It was concluded by the study that the nanoparticles follows a pathway for translocation from leaf to stomata (Birbaum et al. 2010).

22.4.3 Nanoparticle Transmission in Plants

Understanding the transmission of nanoparticles can help decipher the underlining advantages of nanotechnology application to agriculture. The plant cell wall acts as a blockade to the entry of all external particles including nanoparticles. Hence, only nanoparticles smaller than the pore size of the cell wall (pore size generally: 5–20 nm) can enter the plant cell (Lin et al. 2009b; Zhang et al. 2008). However, nanoparticles can also very well penetrate into the leaf cuticle and cell cytoplasm (Sharif et al. 2013; Zhang et al. 2008). Nanoparticles when injected can accumulate near the point of application or can flux from one cell to the other (Lin et al. 2009b). Nanoparticles are transmitted to the environment sometimes by plants and can further bioaccumulate in the food chain (Sharif et al. 2013). It is also plausible that the pore size of the cell wall increases upon contact with nanoparticles, which eventually facilitates its transmission (Nair et al. 2012). Additionally, nanoparticles can enter plant cell via ion channels or through carrier proteins. Upon entering the cytoplasm of the plant cell, nanoparticles can also interact or bind with different organelles. This interaction can lead to altered metabolic processes inside the cell (Zhang et al. 2008). They can even enter through stomata or trichomes when applied on leaf surface and can then translocate to other plant organs and tissues.

The aggregation of nanoparticles induces foliar heating on photosynthetic surfaces. This causes variations in gaseous exchange through stomata and hence induces alterations in cellular and physiological machinery of the cell. With the help of microscopy tools and techniques it has become possible to trail down the transmission and accumulation of nanoparticles inside the plant cell (Abdi et al. 2008). Sometimes the nanoparticles are coupled to agrochemicals or various compounds. This tagging reduces the damage to plants as well as helps in biodegradation of nanoparticles in the environment; however, certain damage by nanoparticles to the plants and the environment still is unavoidable (Ma et al. 2010). For instance, Fe_3O_4 nanoparticles are coated with carbon to provide larger adsorption area and biocompatibility. The kind of administration of nanoparticles was made on *Cucurbita pepo* (Hartley and Lepp 2008). In similar lines, Ag nanoparticles of less than 20 nm are transported through the plasmodesmata inside the plant cell (Cairns et al. 1975; Unrine et al. 2012; Wang 1986). *Brassica juncea* and *Medicago sativa* are reported as hyperaccumulators of Ag nanoparticles. *Medicago sativa* accumulated more Ag nanoparticles than *Brassica juncea* with increase in exposure time and metal concentration (Gardea-Torresdey et al. 2000).

22.4.4 *Nanoparticle Phytotoxicity in Plants*

Phytotoxicity studies are important for deeper comprehension of toxicity induced by nanoparticles. Plethora of information regarding adverse, advantageous and insignificant effects of nanoparticles on crop plants already exists (Bystrzejewska-Piotrowska et al. 2009; Ghodake et al. 2010; Muller et al. 2005; Sohaebuddin et al. 2010; Stampoulis et al. 2009). For instance, germination rate for rice seeds significantly enhanced upon CNTs treatment (Smirnova et al. 2012; Tan and Fugetsu 2007; Wang et al. 2012). The CNT treated seeds also withheld more water content in comparison to their untreated counter-parts. Thus, CNTs were reported as plausible enhancers in regulating rice seedling growth. Additionally, silica nanoparticle labelled with fluorescein isothiocyanate and photo-stable Cadmium-Selenide quantum dots are verified as bio-labels. They have been reported to enhance seed germination (Currie and Perry 2007; Fenoglio et al. 2000; Torney et al. 2007). On contrary, Al_2O_3 nanoparticles inhibited root growth in a variety of plants including soybeans, corn, cabbage, cucumber and carrot (Kollmeier et al. 2000; Ryan et al. 1992; Tian et al. 2007; Yamamoto et al. 2001). Similarly, ZnO was also found to be extremely toxic for root growth (Huang et al. 2002; Ma et al. 2009; Wong et al. 2010). Effect of Al_2O_3, SiO_2, ZnO, and Fe_3O_4 nanoparticles on seed germination were studied on model plant *Arabidopsis thaliana*. It was found that ZnO nanomaterials at 400 mgL^{-1} suppressed seed germination (bin Hussein et al. 2002; Wang et al. 2004). On the other hand, these nanoparticles also enhanced seedling growth and seed germination *Brassica juncea* and *Phaseolus mungo* (black gram) (Ghodake et al. 2010). However, in contrast to other nanoparticles, ZnO is extra toxic towards seed germinations, root elongations, and the count of leaves (Sharma et al. 2009).

Particle size and surface area are two important parameters to study the phytotoxological effects of nanoparticles. The surface area of the nanoparticles increases with the decrease in size. This exposes larger percentage of atoms and molecules on the surface comparative to its interior (Begum et al. 2011; Fugetsu and Parvin 2011). The small size of nanoparticles renders alteration in their physiochemical and structural properties. This results in multiple interactions of the nanoparticles with other substances, eventually leading to toxicological effects (Lee et al. 2012). This is evident by the study which indicated that 21 nm TiO_2 nanoparticles induced 43 $\times$ more inflammation as compared to 250 nm TiO_2 nanoparticle (Castiglione et al. 2011; Feizi et al. 2012, 2013b; Qiu et al. 2013; Song et al. 2013). The enhanced inflammation is attributed to the larger surface area of small nanoparticles for equivalent masses. It is very evident and proved experimentally that micro-particles are less harmful to plants than nano sized particles (Currie and Perry 2007).

22.5 Conclusions and Prospects

In conclusion, nanoparticles depending upon their size, form, chemical characteristics and species exerts physical or chemical toxicity on plants. The current understanding of the phytotoxicity initiated by nanoparticles is limited and the knowledge of adverse effect generated due to unique characteristics of the nanoparticles is still preliminary. Thus, it is imperative to study the permissible range of nanoparticles that can be applied to the plant system without imposing any stress. Furthermore, extensive research is intended to study the uptake kinetics of nanoparticles by plants as well as their interaction mechanisms within the plant cell. Proper engineering of nanoparticles, their delivery, accumulation within the plant system, biodegradation and eventually its disposal needs to be carried out with great precision to reduce the harmful effects on agricultural and environmental systems. Nonetheless, upon combatting the negative effects associated with nanotechnology, it will facilitate agriculture in the near future by precision farming techniques, regulating the capability of plants to absorb nutrients, targeted use of inputs, disease detection and disease control.

References

Abdi G, Salehi H, Khosh-Khui M (2008) Nano silver: a novel nanomaterial for removal of bacterial contaminants in valerian (*Valeriana officinalis* L.) tissue culture. Acta Physiol Plant 30(5):709–714

Adamson I, Prieditis H, Hedgecock C, Vincent R (2000) Zinc is the toxic factor in the lung response to an atmospheric particulate sample. Toxicol Appl Pharmacol 166(2):111–119

Adeleye AS, Conway JR, Garner K et al (2016) Engineered nanomaterials for water treatment and remediation: costs, benefits, and applicability. Chem Eng J 286:640–662

Akhavan O, Ghaderi E (2010) Toxicity of graphene and graphene oxide nanowalls against bacteria. ACS Nano 4(10):5731–5736

Al-Ghamdi AA, Gupta R, Kahol P et al (2014) Improved solar efficiency by introducing graphene oxide in purple cabbage dye sensitized TiO_2 based solar cell. Solid State Comm 183:56–59

Al-Halafi AM (2014) Nanocarriers of nanotechnology in retinal diseases. Saudi J Ophthal 28(4):304–309

Anjum NA, Singh N, Singh MK et al (2013) Single-bilayer graphene oxide sheet tolerance and glutathione redox system significance assessment in faba bean (*Vicia faba* L.). J Nano Res 15(7):1770

Anjum NA, Singh N, Singh MK et al (2014) Single-bilayer graphene oxide sheet impacts and underlying potential mechanism assessment in germinating faba bean (*Vicia faba* L.). Sci Tot Env 472:834–841

Arora S, Sharma P, Kumar S et al (2012) Gold-nanoparticle induced enhancement in growth and seed yield of *Brassica juncea*. Plant Growth Reg 66(3):303–310

Astruc D (2012) Electron-transfer processes in dendrimers and their implication in biology, catalysis, sensing and nanotechnology. Nature Chem 4(4):255

Bagheri H, Ayazi Z, Es'haghi A, Aghakhani A (2012) Reinforced polydiphenylamine nanocomposite for microextraction in packed syringe of various pesticides. J Chroma A 1222:13–21

Bai W, Zhang Z, Tian W et al (2010) Toxicity of zinc oxide nanoparticles to zebrafish embryo: a physicochemical study of toxicity mechanism. J Nano Res 12(5):1645–1654

Bandyopadhyay S, Peralta-Videa JR, Gardea-Torresdey JL (2013) Advanced analytical techniques for the measurement of nanomaterials in food and agricultural samples: a review. Env Eng Sci 30(3):118–125

Baughman RH, Zakhidov AA, De Heer WA (2002) Carbon nanotubes—the route toward applications. Sci 297(5582):787–792

Becana M, Moran J, Iturbe-Ormaetxe I (1998) Iron-dependent oxygen free radical generation in plants subjected to environmental stress: toxicity and antioxidant protection. Plant Soil 201(1):137–147

Beebe DU, Turgeon R (1992) Localization of galactinol, raffinose, and stachyose synthesis in Cucurbita pepo leaves. Planta 188(3):354–361

Begum P, Fugetsu B (2013) Induction of cell death by graphene in *Arabidopsis thaliana* (Columbia ecotype) T87 cell suspensions. J Haz Mat 260:1032–1041

Begum P, Ikhtiari R, Fugetsu B (2011) Graphene phytotoxicity in the seedling stage of cabbage, tomato, red spinach, and lettuce. Carbon 49(12):3907–3919

Besson-Bard A, Gravot A, Richaud P et al (2009) Nitric oxide contributes to cadmium toxicity in *Arabidopsis* by promoting cadmium accumulation in roots and by up-regulating genes related to iron uptake. Plant Physiol 149(3):1302–1315

Bianchini A, Wood CM (2003) Mechanism of acute silver toxicity in *Daphnia magna*. Env Toxicol Chem: An Int J 22(6):1361–1367

bin Hussein MZ, Zainal Z, Yahaya AH, Foo DWV (2002) Controlled release of a plant growth regulator, α-naphthaleneacetate from the lamella of Zn–Al-layered double hydroxide nanocomposite. J Cont Rel 82(2–3):417–427

Birbaum K, Brogioli R, Schellenberg M et al (2010) No evidence for cerium dioxide nanoparticle translocation in maize plants. Env Sci Tech 44(22):8718–8723

Blamey F, Edwards D, Asher C (1983) Effects of aluminum, oh: al and pal molar ratios, and ionic strength on soybean root elongation in solution culture. Soil Sci 136(4):197–207

Boisselier E, Astruc D (2009) Gold nanoparticles in nanomedicine: preparations, imaging, diagnostics, therapies and toxicity. Chem Soci Rev 38(6):1759–1782

Brock DA, Douglas TE, Queller DC, Strassmann JE (2011) Primitive agriculture in a social amoeba. Nat 469(7330):393–396

Bystrzejewska-Piotrowska G, Golimowski J, Urban PL (2009) Nanoparticles: their potential toxicity, waste and environmental management. Was Manage 29(9):2587–2595

Cairns J, Heath AG, Parker BC (1975) The effects of temperature upon the toxicity of chemicals to aquatic organisms. Hydrobiol 47(1):135–171

Canas JE, Long M, Nations S et al (2008) Effects of functionalized and nonfunctionalized single-walled carbon nanotubes on root elongation of select crop species. Env Toxicol Chem: An Int J 27(9):1922–1931

Castiglione MR, Giorgetti L, Geri C, Cremonini R (2011) The effects of nano-TiO_2 on seed germination, development and mitosis of root tip cells of *Vicia narbonensis* L. and *Zea mays* L. J Nano Res 13(6):2443–2449

Chai M, Shi F, Li R et al (2013) Interactive effects of cadmium and carbon nanotubes on the growth and metal accumulation in a halophyte *Spartina alterniflora* (Poaceae). Plant Grow Reg 71(2):171–179

Chekin F, Bagheri S, Arof AK, Hamid SBA (2012) Preparation and characterization of Ni (II)/polyacrylonitrile and carbon nanotube composite modified electrode and application for carbohydrates electrocatalytic oxidation. J Solid State Electrochem 16(10):3245–3251

Cheng Y, Yin L, Lin S et al (2011) Toxicity reduction of polymer-stabilized silver nanoparticles by sunlight. The J Phy Chem C 115(11):4425–4432

Cho U-H, Seo N-H (2005) Oxidative stress in Arabidopsis thaliana exposed to cadmium is due to hydrogen peroxide accumulation. Plant Sci 168(1):113–120

Clarke B, Brennan E (1989) Differential cadmium accumulation and phytotoxicity in sixteen tobacco cultivars. Japca 39(10):1319–1322

Collins PG, Bradley K, Ishigami M, Zettl dA (2000) Extreme oxygen sensitivity of electronic properties of carbon nanotubes. Sci 287(5459):1801–1804

Cui D, Ruan H, Zhang X et al (2012) Effects of single walled carbon nanotubes on Arabidopsis mesophyll cells. ECS Trans 41(40):43–48

Cumming JR, Cumming AB, Taylor GJ (1992) Patterns of root respiration associated with the induction of aluminium tolerance in *Phaseolus vulgaris* L. J Exp Bot 43(8):1075–1081

Currie HA, Perry CC (2007) Silica in plants: biological, biochemical and chemical studies. Ann Bot 100(7):1383–1389

Dehkourdi EH, Mosavi M (2013) Effect of anatase nanoparticles (TiO_2) on parsley seed germination (*Petroselinum crispum)* in vitro. Biol Trace Elem Res 155(2):283–286

De La Torre-Roche R, Hawthorne J, Deng Y et al (2013) Multiwalled carbon nanotubes and C60 fullerenes differentially impact the accumulation of weathered pesticides in four agricultural plants. Env Sci Tech 47(21):12539–12547

Delgado-Ramos GC (2014) Nanotechnology in Mexico: global trends and national implications for policy and regulatory issues. Tech Soci 37:4–15

Diao M, Yao M (2009) Use of zero-valent iron nanoparticles in inactivating microbes. Water Res 43(20):5243–5251

Dimkpa CO, McLean JE, Martineau N et al (2013) Silver nanoparticles disrupt wheat (*Triticum aestivum* L.) growth in a sand matrix. Env Sci Tech 47(2):1082–1090

Dutschk V, Karapantsios T, Liggieri L et al (2014) Smart and green interfaces: from single bubbles/drops to industrial environmental and biomedical applications. Adv Coll Inter Sci 209:109–126

Elghniji K, Hentati O, Mlaik N et al (2012) Photocatalytic degradation of 4-chlorophenol under P-modified TiO_2/UV system: Kinetics, intermediates, phytotoxicity and acute toxicity. J Env Sci 24(3):479–487

Elliott KC (2011) Nanomaterials and the precautionary principle. Env Hea Perspect 119(6):A240–A240

Feizi H, Amirmoradi S, Abdollahi F, Pour SJ (2013a) Comparative Effects of nanosized and bulk titanium dioxide concentrations on medicinal plant *Salvia officinalis* L. Ann Res Rev Bio 814–824

Feizi H, Kamali M, Jafari L, Moghaddam PR (2013b) Phytotoxicity and stimulatory impacts of nanosized and bulk titanium dioxide on fennel (*Foeniculum vulgare* Mill). Chemo 91(4):506–511

Feizi H, Moghaddam PR, Shahtahmassebi N, Fotovat A (2012) Impact of bulk and nanosized titanium dioxide (TiO_2) on wheat seed germination and seedling growth. Biol Trace Elem Res 146(1):101–106

Fenoglio I, Croce A, Di Renzo F et al (2000) Pure-silica zeolites (Porosils) as model solids for the evaluation of the physicochemical features determining silica toxicity to macrophages. Chem Res Toxicol 13(6):489–500

Feretti D, Zerbini I, Zani C et al (2007) *Allium cepa* chromosome aberration and micronucleus tests applied to study genotoxicity of extracts from pesticide-treated vegetables and grapes. Food Addi Contam 24(6):561–572

Ferguson EA, Hogstrand C (1998) Acute silver toxicity to seawater-acclimated rainbow trout: influence of salinity on toxicity and silver speciation. Env Toxicol Chem: An Int J 17(4):589–593

Fincheira P, Tortella G, Duran N et al (2020) Current applications of nanotechnology to develop plant growth inducer agents as an innovation strategy. Criti Rev Biotech 40(1):15–30

Fiskesjó G (1993) Allium test I: a 2–3 day plant test for toxicity assessment by measuring the mean root growth of onions (*Allium cepa* L.). Env Toxicol Water Qua 8(4):461–470

Flores D, Chaves JS, Chacón R, Schmidt A (2013) A novel technique using SWCNTs to enhanced development and root growth of fig plants (*Ficus carica*). In: Technical proceedings of the NSTI nanotechnology conference and expo (NSTI-Nanotech'13), pp 167–170

Foltête A-S, Masfaraud J-F, Bigorgne E et al (2011) Environmental impact of sunscreen nanomaterials: ecotoxicity and genotoxicity of altered TiO_2 nanocomposites on *Vicia faba*. Env Poll 159(10):2515–2522

Franke ME, Koplin TJ, Simon U (2006) Metal and metal oxide nanoparticles in chemiresistors: does the nanoscale matter? Small 2(1):36–50

Fugetsu B, Parvin B (2011) Graphene phytotoxicity in the seedling stage of cabbage, tomato, red spinach, and lettuce. CC BY-NC-SA. http://doi.org/10.5772/18393

Galvez F, Wood CM (1997) The relative importance of water hardness and chloride levels in modifying the acute toxicity of silver to rainbow trout (*Oncorhynchus mykiss*). Env Toxicol Chem 16(11):2363–2368

Gao F, Liu C, Qu C et al (2008) Was improvement of spinach growth by nano-TiO_2 treatment related to the changes of *Rubisco activase*? Biometals 21(2):211–217

Gao J, Wang Y, Folta KM et al (2011) Polyhydroxy fullerenes (fullerols or fullerenols): beneficial effects on growth and lifespan in diverse biological models. PLoS One 6(5):

Gardea-Torresdey J, Tiemann K, Gamez G et al (2000) Reduction and accumulation of gold (III) by Medicago sativa alfalfa biomass: X-ray absorption spectroscopy, pH, and temperature dependence. Env Sci Tech 34(20):4392–4396

Geremias R, Fattorini D, Fávere VT, Pedrosa RC (2010) Bioaccumulation and toxic effects of copper in common onion *Allium cepa* L. Chem and Ecolo 26(1):19–26

Ghodake G, Seo YD, Park D, Lee DS (2010) Phytotoxicity of carbon nanotubes assessed by *Brassica juncea* and *Phaseolus mungo*. J Nanoelectro and Optoelectro 5(2):157–160

Ghormade V, Deshpande MV, Paknikar KM (2011) Perspectives for nano-biotechnology enabled protection and nutrition of plants. Biotech Adv 29(6):792–803

Goodman CM, McCusker CD, Yilmaz T, Rotello VM (2004) Toxicity of gold nanoparticles functionalized with cationic and anionic side chains. Bioconj Chem 15(4):897–900

Green S, Renault S (2008) Influence of papermill sludge on growth of *Medicago sativa, Festuca rubra* and *Agropyron trachycaulum* in gold mine tailings: a greenhouse study. Env Poll 151(3):524–531

Greipsson S, Crowder A (1992) Amelioration of copper and nickel toxicity by iron plaque on roots of rice (*Oryza sativa*). Can J Bot 70(4):824–830

Griffitt RJ, Hyndman K, Denslow ND, Barber DS (2009) Comparison of molecular and histological changes in zebrafish gills exposed to metallic nanoparticles. Toxicolo Sci 107(2):404–415

Groppa M, Rosales E, Iannone M, Benavides M (2008) Nitric oxide, polyamines and Cd-induced phytotoxicity in wheat roots. Phytochem 69(14):2609–2615

Hansen S, Baun A, Michelson E et al (2009) Nanomaterials in consumer products, nanomaterials: risks and benefits, edited by I. Linkov and J. Steevens. Springer, The Netherlands, NAPSC; pp 359–367

Hansen SF (2010) A global view of regulations affecting nanomaterials. Wiley Interdiscip Rev: Nanomed and Nanobiotech 2(5):441–449

Hao Y, Yang X, Shi Y et al (2012) FITC delivery into plant cells using magnetic single-walled carbon nanotubes. J Nanosci and Nanotech 12(8):6287–6293

Hartley W, Lepp NW (2008) Remediation of arsenic contaminated soils by iron-oxide application, evaluated in terms of plant productivity, arsenic and phytotoxic metal uptake. Sci Tot Env 390(1):35–44

Hasan F, Parviz R, Nasser S, Amir F (2012) Effects of concentrations of bulk and nanosized titanium dioxide (TiO_2) on germination of wheat seed. In: Proceedings of the international congress on nanoscience and nanotechnology (ICNN'12), pp 1–10

Hata K, Futaba DN, Mizuno K et al (2004) Water-assisted highly efficient synthesis of impurity-free single-walled carbon nanotubes. Science 306(5700):1362–1364

Hauck TS, Ghazani AA, Chan WC (2008) Assessing the effect of surface chemistry on gold nanorod uptake, toxicity, and gene expression in mammalian cells. Small 4(1):153–159

Hazarika A, Maji TK (2014) Strain sensing behavior and dynamic mechanical properties of carbon nanotubes/nanoclay reinforced wood polymer nanocomposite. Chem Eng J 247:33–41

Hogstrand C, Galvez F, Wood CM (1996) Toxicity, silver accumulation and metallothionein induction in freshwater rainbow trout during exposure to different silver salts. Env Toxicol and Chem: An Intl J 15(7):1102–1108

Howe G, Merchant S (1992) Heavy metal-activated synthesis of peptides in *Chlamydomonas reinhardtii*. Plant Physiol 98(1):127–136
Huang FM, Tai KW, Chou MY, Chang YC (2002) Cytotoxicity of resin-, zinc oxide–eugenol-, and calcium hydroxide-based root canal sealers on human periodontal ligament cells and permanent V79 cells. Int Endo J 35(2):153–158
Huang Y-W, Cambre M, Lee H-J (2017) The toxicity of nanoparticles depends on multiple molecular and physicochemical mechanisms. Int J Mol Sci 18(12):2702
Hund-Rinke K, Simon M (2006) Ecotoxic effect of photocatalytic active nanoparticles (TiO_2) on algae and daphnids (8 pp). Env Sci Poll Res 13(4):225–232
Inokuchi R, Itagaki T, Wiskich JT et al (1997) An NADP-glutamate dehydrogenase from the green alga Bryopsis maxima: purification and properties. Plant and Cell Physiol 38(3):327–335
Jiang X, Luo Y, Zhao Q et al (2003) Soil Cd availability to Indian mustard and environmental risk following EDTA addition to Cd-contaminated soil. Chemosp 50(6):813–818
John MK, Van Laerhoven C (1976) Differential effects of cadmium on lettuce varieties. Env Poll (1970) 10(3):163–173
John MK, VanLaerhoven CJ, Chuah HH (1972) Factors affecting plant uptake and phytotoxicity of cadmium added to soils. Env Sci Tech 6(12):1005–1009
Johnston HJ, Hutchison G, Christensen FM et al (2010) A review of the in vivo and in vitro toxicity of silver and gold particulates: particle attributes and biological mechanisms responsible for the observed toxicity. Crit Rev Toxicol 40(4):328–346
Justin S, Armstrong W (1991) Evidence for the involvement of ethene in aerenchyma formation in adventitious roots of rice (*Oryza sativa* L.). New Phytol 118(1):49–62
Kairong W (1996) Comparative study on Cd phytotoxicity to different genes of rice. Rural Eco Env 12:18–23
Karamushka VI, Gadd GM (1999) Interaction of Saccharomyces cerevisiae with gold: toxicity and accumulation. BioMet 12(4):289–294
Kashem MA, Kawai S (2007) Alleviation of cadmium phytotoxicity by magnesium in Japanese mustard spinach. Soil Sci Plant Nut 53(3):246–251
Katsoyiannis IA, Zouboulis AI (2002) Removal of arsenic from contaminated water sources by sorption onto iron-oxide-coated polymeric materials. Water Res 36(20):5141–5155
Ke PC, Lamm MH (2011) A biophysical perspective of understanding nanoparticles at large. Phy Chem Chem Phy 13(16):7273–7283
Keltjens W, Van Beusichem M (1998) Phytochelatins as biomarkers for heavy metal stress in maize (*Zea mays* L.) and wheat (*Triticum aestivum* L.): combined effects of copper and cadmium. Plant and Soil 203(1):119–126
Khataee A, Fathinia M, Naseri A et al (2014) Modeling and optimization of simultaneous photocatalysis of three dyes on ceramic-coated TiO_2 nanoparticles using chemometrics methods: phytotoxicological assessment during degradation process. Res Chem Inter 40(3):1283–1302
Khiew P, Chiu W, Tan T et al (2011) Capping effect of palm-oil based organometallic ligand towards the production of highly monodispersed nanostructured material. Palm oil: nutrition, uses and impacts. Nova Sci:189–219
Khlebtsov N, Dykman L (2011) Biodistribution and toxicity of engineered gold nanoparticles: a review of in vitro and in vivo studies. Chem Soc Rev 40(3):1647–1671
Khodakovskaya M, Dervishi E, Mahmood M et al (2009) Carbon nanotubes are able to penetrate plant seed coat and dramatically affect seed germination and plant growth. ACS Nano 3(10):3221–3227
Khodakovskaya MV, Kim BS, Kim JN et al (2013) Carbon nanotubes as plant growth regulators: effects on tomato growth, reproductive system, and soil microbial community. Small 9(1):115–123
Kim BS (2013) Biological synthesis of nanomaterials using plant leaf extracts. In: 13th IEEE international conference on nanotechnology (IEEE-NANO 2013) IEEE 1204–1207
Kinraide TB, Ryan PR, Kochian LV (1992) Interactive effects of Al^{3+}, H^+, and other cations on root elongation considered in terms of cell-surface electrical potential. Plant Physiol 99(4):1461–1468

Kole C, Kole P, Randunu KM et al (2013) Nanobiotechnology can boost crop production and quality: first evidence from increased plant biomass, fruit yield and phytomedicine content in bitter melon (*Momordica charantia*). Bmc Biotech 13(1):37

Kollmeier M, Felle HH, Horst WJ (2000) Genotypical differences in aluminum resistance of maize are expressed in the distal part of the transition zone: is reduced basipetal auxin flow involved in inhibition of root elongation by aluminum? Plant Physiol 122(3):945–956

Kolmakov A, Moskovits M (2004) Chemical sensing and catalysis by one-dimensional metal-oxide nanostructures. Annu Rev Mater Res 34:151–180

Kong H, Gao C, Yan D (2004) Controlled functionalization of multiwalled carbon nanotubes by in situ atom transfer radical polymerization. J Am Chem Soc 126(2):412–413

Kraemer SM (2004) Iron oxide dissolution and solubility in the presence of siderophores. Aq Sci 66(1):3–18

Kumar V, Sharma M, Khare T, Wani SH (2018) Impact of nanoparticles on oxidative stress and responsive antioxidative defense in plants. Nanomaterials in Plants, Algae, and Microorganisms. Concepts and controversies, vol 1. Academic Press/Elsevier, London, San Diego, Cambridge, Oxford, pp 393–406

Kumari M, Mukherjee A, Chandrasekaran N (2009) Genotoxicity of silver nanoparticles in Allium cepa. Sci Tot Env 407(19):5243–5246

Kurepa J, Paunesku T, Vogt S et al (2010) Uptake and distribution of ultrasmall anatase TiO_2 Alizarin red S nanoconjugates in *Arabidopsis thaliana*. Nano Lett 10(7):2296–2302

Lai F, Wissing SA, Müller RH, Fadda AM (2006) Artemisia arborescens L essential oil-loaded solid lipid nanoparticles for potential agricultural application: preparation and characterization. Aaps Pharm Sci Tech 7(1):E10

Lang X, Hirata A, Fujita T, Chen M (2011) Nanoporous metal/oxide hybrid electrodes for electrochemical supercapacitors. Nat Nanotech 6(4):232–236

Larue C, Veronesi G, Flank A-M et al (2012) Comparative uptake and impact of TiO_2 nanoparticles in wheat and rapeseed. J Toxicol Env Health, Part A 75(13–15):722–734

Lazof DB, Holland MJ (1999) Evaluation of the aluminium-induced root growth inhibition in isolation from low pH effects in Glycine max, *Pisum sativum* and *Phaseolus vulgaris*. Fun Plant Bio 26(2):147–157

Lee D-Y, Fortin C, Campbell PG (2005) Contrasting effects of chloride on the toxicity of silver to two green algae, *Pseudokirchneriella subcapitata* and *Chlamydomonas reinhardtii*. Aq Toxicol 75(2):127–135

Lee G, Kim BS (2014) Biological reduction of graphene oxide using plant leaf extracts. Biotech Prog 30(2):463–469

Lee S, Kim S, Kim S, Lee I (2013) Assessment of phytotoxicity of ZnO NPs on a medicinal plant, *Fagopyrum esculentum*. Env Sci Poll Res 20(2):848–854

Lee W-M, Kwak JI, An Y-J (2012) Effect of silver nanoparticles in crop plants *Phaseolus radiatus* and *Sorghum bicolor*: media effect on phytotoxicity. Chemos 86(5):491–499

Lee WM, An YJ, Yoon H, Kweon HS (2008) Toxicity and bioavailability of copper nanoparticles to the terrestrial plants mung bean (Phaseolus radiatus) and wheat (Triticum aestivum): plant agar test for water-insoluble nanoparticles. Env Toxicol Chem: An Int J 27(9):1915–1921

Li HJ, Lu W, Li J, Bai X, Gu C (2005) Multichannel ballistic transport in multiwall carbon nanotubes. Phy Rev Lett 95(8):086601

Li W, Xie S, Qian LX et al (1996) Large-scale synthesis of aligned carbon nanotubes. Science 274(5293):1701–1703

Liden G (2011) The European commission tries to define nanomaterials. Ann Occup Hyg 55(1):1–5

Lin C, Fugetsu B, Su Y, Watari F (2009a) Studies on toxicity of multi-walled carbon nanotubes on Arabidopsis T87 suspension cells. J Haz Mat 170(2–3):578–583

Lin S, Reppert J, Hu Q, Hudson JS, Reid ML, Ratnikova TA, Rao AM, Luo H, Ke PC (2009b) Uptake, translocation, and transmission of carbon nanomaterials in rice plants. Small 5(10):1128–1132

Lin Y, Rao AM, Sadanadan B et al (2002) Functionalizing multiple-walled carbon nanotubes with aminopolymers. The J Phy Chem B 106(6):1294–1298

Liu D, Jiang W, Wang W et al (1994) Effects of lead on root growth, cell division, and nucleolus of *Allium cepa*. Env Poll 86(1):1–4

Liu D, Zhang Y, Sun X, Chang PR (2014) Recent advances in bio-sourced polymeric carbohydrate/nanotube composites. J App Poly Sci 131(12). https://doi.org/10.1002/app.40359

Liu Q, Zhang X, Zhao Y et al (2013) Fullerene-induced increase of glycosyl residue on living plant cell wall. Env Sci Tech 47(13):7490–7498

Liu Q, Zhao Y, Wan Y et al (2010) Study of the inhibitory effect of water-soluble fullerenes on plant growth at the cellular level. ACS Nano 4(10):5743–5748

Liu R, Lal R (2015) Potentials of engineered nanoparticles as fertilizers for increasing agronomic productions. Sci Tot Env 514:131–139

Ljubimova JY, Holler E (2012) Biocompatible nanopolymers: the next generation of breast cancer treatment? Nanomed 7(10):1467–1470

Lombardi L, Sebastiani L (2005) Copper toxicity in Prunus cerasifera: growth and antioxidant enzymes responses of in vitro grown plants. Plant Sci 168(3):797–802

Lombi E, Nowack B, Baun A, McGrath SP (2012) Evidence for effects of manufactured nanomaterials on crops is inconclusive. Proc Nat Acad Sci 109(49):E3336–E3336

Lombi E, Wenzel WW, Sletten RS (1999) Arsenic adsorption by soils and iron-oxide-coated sand: kinetics and reversibility. J Plant Nut Soil Sci 162(4):451–456

Long Z, Ji J, Yang K et al (2012) Systematic and quantitative investigation of the mechanism of carbon nanotubes' toxicity toward algae. Env Sci Tech 46(15):8458–8466

Lou JC, Jung MJ, Yang HW et al (2011) Removal of dissolved organic matter (DOM) from raw water by single-walled carbon nanotubes (SWCNTs). J Env Sci Health, Part A 46(12):1357–1365

Ma H, Bertsch PM, Glenn TC et al (2009) Toxicity of manufactured zinc oxide nanoparticles in the nematode *Caenorhabditis elegans*. Env Toxicol Chem: An Int J 28(6):1324–1330

Ma X, Geiser-Lee J, Deng Y, Kolmakov A (2010) Interactions between engineered nanoparticles (ENPs) and plants: phytotoxicity, uptake and accumulation. Sci Tot Env 408(16):3053–3061

Mahmoodzadeh H, Aghili R (2014) Effect on germination and early growth characteristics in wheat plants (*Triticum aestivum* L.) seeds exposed to TiO_2 nanoparticles. J Chem Health Risks. https://doi.org/10.22034/jchr.2018.544054

Mahmoodzadeh H, Aghili R, Nabavi M (2013) Physiological effects of TiO_2 nanoparticles on wheat (*Triticum aestivum*). Tech J Eng Appl Sci 3:1365–1370

Mahmoudi M, Shokrgozar MA, Simchi A et al (2009) Multiphysics flow modeling and in vitro toxicity of iron oxide nanoparticles coated with poly (vinyl alcohol). The J Phy Chem C 113(6):2322–2331

Maine MaA, Duarte MaV, Suñé NL (2001) Cadmium uptake by floating macrophytes. Water Res 35(11):2629–2634

Mani MK, Viola G, Reece MJ et al (2014) Fabrication of carbon nanotube reinforced iron based magnetic alloy composites by spark plasma sintering. J Allo Comp 601:146–153

Mao C, Yi K, Yang L et al (2004) Identification of aluminium-regulated genes by cDNA-AFLP in rice (*Oryza sativa* L.): aluminium-regulated genes for the metabolism of cell wall components. J Exp Bot 55(394):137–143

McGeer JC, Playle RC, Wood CM, Galvez F (2000) A physiologically based biotic ligand model for predicting the acute toxicity of waterborne silver to rainbow trout in freshwaters. Env Sci Tech 34(19):4199–4207

Mehta RJ, Zhang Y, Karthik C et al (2012) A new class of doped nanobulk high-figure-of-merit thermoelectrics by scalable bottom-up assembly. Nat Mat 11(3):233–240

Miralles P, Church TL, Harris AT (2012) Toxicity, uptake, and translocation of engineered nanomaterials in vascular plants. Env Sci Tech 46(17):9224–9239

Mocquot B, Vangronsveld J, Clijsters H, Mench M (1996) Copper toxicity in young maize (*Zea mays* L.) plants: effects on growth, mineral and chlorophyll contents, and enzyme activities. Plant and Soil 182(2):287–300

Mogharabi M, Abdollahi M, Faramarzi MA (2014) Safety concerns to application of graphene compounds in pharmacy and medicine. DARU J Pharm Sci 22:23. https://doi.org/doi.org/10.1186/2008-2231-22-23

Mohamed EF (2017) Nanotechnology: future of environmental air pollution control. Env Manag Sust Devp 6(2). https://doi.org/10.5296/emsd.v6i2.12047

Moisala A, Nasibulin AG, Kauppinen EI (2003) The role of metal nanoparticles in the catalytic production of single-walled carbon nanotubes—a review. J Phy: Con Matt 15(42):S3011

Molina-Barahona L, Vega-Loyo L, Guerrero M et al (2005) Ecotoxicological evaluation of diesel-contaminated soil before and after a bioremediation process. Env Toxicol: An Int J 20(1):100–109

Muller J, Huaux F, Moreau N et al (2005) Respiratory toxicity of multi-wall carbon nanotubes. Toxicol App Pharmacol 207(3):221–231

Munzuroglu O, Geckil H (2002) Effects of metals on seed germination, root elongation, and coleoptile and hypocotyl growth in *Triticum aestivum* and *Cucumis sativus*. Arch Env Cont Toxicol 43(2):203–213

Murphy CJ, Gole AM, Stone JW et al (2008) Gold nanoparticles in biology: beyond toxicity to cellular imaging. Acc Chem Res 41(12):1721–1730

Musante C, White JC (2012) Toxicity of silver and copper to Cucurbita pepo: differential effects of nano and bulk-size particles. Env Toxicol 27(9):510–517

Musee N (2011) Nanowastes and the environment: potential new waste management paradigm. Env Int 37(1):112–128

Nadiminti PP, Dong YD, Sayer C et al (2013) Nanostructured liquid crystalline particles as an alternative delivery vehicle for plant agrochemicals. ACS Appl mat Interf 5(5):1818–1826

Nair R, Mohamed MS, Gao W et al (2012) Effect of carbon nanomaterials on the germination and growth of rice plants. J Nanosci Nanotech 12(3):2212–2220

Nair R, Varghese SH, Nair BG et al (2010) Nanoparticulate material delivery to plants. Plant Sci 179(3):154–163

Neto OPV (2014) Intelligent computational nanotechnology: the role of computational intelligence in the development of nanoscience and nanotechnology. J Comput Theo Nanosci 11(4):928–944

Nevius BA, Chen YP, Ferry JL, Decho AW (2012) Surface-functionalization effects on uptake of fluorescent polystyrene nanoparticles by model biofilms. Ecotoxicol 21(8):2205–2213

Nguyen BD, Brar DS, Bui BC (2003) Identification and mapping of the QTL for aluminum tolerance introgressed from the new source, *Oryza rufipogon* Griff., into indica rice (*Oryza sativa* L.). Theor Appl Gene 106(4):583–593

Niederberger M (2007) Nonaqueous sol–gel routes to metal oxide nanoparticles. Acc Chem Res 40(9):793–800

Niederberger M, Garnweitner G, Buha J et al (2006) Nonaqueous synthesis of metal oxide nanoparticles: review and indium oxide as case study for the dependence of particle morphology on precursors and solvents. J Sol-Gel Sci Tech 40(2–3):259–266

Ocsoy I, Paret ML, Ocsoy MA et al (2013) Nanotechnology in plant disease management: DNA-directed silver nanoparticles on graphene oxide as an antibacterial against *Xanthomonas perforans*. ACS Nano 7(10):8972–8980

Ocsoy I, Tasdemir D, Mazicioglu S, Tan W (2018) Nanotechnology in plants. In: Plant genetics and molecular biology. Springer, pp 263–275

Ouda SM (2014) Antifungal activity of silver and copper nanoparticles on two plant pathogens, *Alternaria alternata* and *Botrytis cinerea*. Res J Microbio 9(1):34–42

Paret ML, Vallad GE, Averett DR et al (2013) Photocatalysis: effect of light-activated nanoscale formulations of TiO_2 on Xanthomonas perforans and control of bacterial spot of tomato. Phytopathol 103(3):228–236

Parise A, Thakor H, Zhang X (2014) Activity inhibition on municipal activated sludge by single-walled carbon nanotubes. J Nanopart Res 16(1):2159

Patel KD, Singh RK, Kim H-W (2019) Carbon-based nanomaterials as an emerging platform for theranostics. Mat Horiz 6(3):434–469

Pérez-de-Luque A (2017) Interaction of nanomaterials with plants: what do we need for real applications in agriculture? Front Env Sci 5:12

Perreault F, Bogdan N, Morin M et al (2012) Interaction of gold nanoglycodendrimers with algal cells (*Chlamydomonas reinhardtii*) and their effect on physiological processes. Nanotoxicol 6(2):109–120

Piccapietra F, Allué CG, Sigg L, Behra R (2012) Intracellular silver accumulation in *Chlamydomonas reinhardtii* upon exposure to carbonate coated silver nanoparticles and silver nitrate. Env Sci Technol 46(13):7390–7397

Podsiadlo P, Kaushik AK, Arruda EM et al (2007) Ultrastrong and stiff layered polymer nanocomposites. Science 318(5847):80–83

Potera C (2010) Nanomaterials: transformation of silver nanoparticles in sewage sludge. Environ Health Perspect 118(12):A526–A527

Prasad R, Bhattacharyya A, Nguyen QD (2017) Nanotechnology in sustainable agriculture: recent developments, challenges, and perspectives. Fronti Microbiol 8:1014

Qi M, Liu Y, Li T (2013) Nano-TiO_2 improve the photosynthesis of tomato leaves under mild heat stress. Biologic Trace Elem Res 156(1–3):323–328

Qiu Z, Yang Q, Liu W (2013) Photocatalytic degradation of phytotoxic substances in waste nutrient solution by various immobilized levels of nano-TiO_2. Water Air and Soil Poll 224(3):1461

Ratte HT (1999) Bioaccumulation and toxicity of silver compounds: a review. Env Toxicol Chem: An Int J 18(1):89–108

Reddy KM, Feris K, Bell J et al (2007) Selective toxicity of zinc oxide nanoparticles to prokaryotic and eukaryotic systems. App Phy Lett 90(21):

Ren D, Guo Z, Du F et al (2007) A nanohybrid material of SWNTs covalently functionalized with porphyrin for light harvesting antenna: synthesis and photophysical properties. J Nanosci Nanotech 7(4–5):1539–1545

Rizzello L, Pompa PP (2014) Nanosilver-based antibacterial drugs and devices: mechanisms, methodological drawbacks, and guidelines. Chem Soci Rev 43(5):1501–1518

Roessler PG, Lien S (1984) Purification of hydrogenase from *Chlamydomonas reinhardtii*. Plant Physiol 75(3):705–709

Roh JY, Sim SJ, Yi J et al (2009) Ecotoxicity of silver nanoparticles on the soil nematode Caenorhabditis elegans using functional ecotoxicogenomics. Env Sci Tech 43(10):3933–3940

Ryan PR, Shaff JE, Kochian LV (1992) Aluminum toxicity in roots: correlation among ionic currents, ion fluxes, and root elongation in aluminum-sensitive and aluminum-tolerant wheat cultivars. Plant Physiol 99(3):1193–1200

Safari J, Zarnegar Z (2014) Advanced drug delivery systems: nanotechnology of health design a review. J Saudi Chem Soci 18(2):85–99

Safiuddin M, Gonzalez M, Cao J, Tighe SL (2014) State-of-the-art report on use of nano-materials in concrete. Int J Pavem Eng 15(10):940–949

Saison C, Perreault F, Daigle JC et al (2010) Effect of core–shell copper oxide nanoparticles on cell culture morphology and photosynthesis (photosystem II energy distribution) in the green alga, *Chlamydomonas reinhardtii*. Aq Toxicol 96(2):109–114

Santos SM, Dinis AM, Rodrigues DM et al (2013) Studies on the toxicity of an aqueous suspension of C60 nanoparticles using a bacterium (gen. Bacillus) and an aquatic plant (*Lemna gibba*) as in vitro model systems. Aq Toxicol 142:347–354

Sanzari I, Leone A, Ambrosone A (2019) Nanotechnology in plant science: to make a long story short. Front Bioeng Biotech 7:120

Sartain J, Kamprath E (1978) Aluminum tolerance of soybean cultivars based on root elongation in solution culture compared with growth in acid soil 1. Agron J 70(1):17–20

Sasidharan A, Panchakarla L, Chandran P et al (2011) Differential nano-bio interactions and toxicity effects of pristine versus functionalized graphene. Nanoscale 3(6):2461–2464

Saxena A, Tripathi R, Singh R (2010) Biological synthesis of silver nanoparticles by using onion (*Allium cepa*) extract and their antibacterial activity. Dig J Nanomater Bios 5(2):427–432

Schwerdtfeger P, Ori O, Wirz L, Avery J (2015) Playing with hexagons and pentagons: topological and graph theoretical aspects of fullerenes. In: Proceedings of the 14th international conference on computational and mathematical methods in science and engineering (CMMSE), pp 1–3

Shang Y, Hasan M, Ahammed GJ et al (2019) Applications of nanotechnology in plant growth and crop protection: a review. Mol 24(14):2558

Sharif F, Westerhoff P, Herckes P (2013) Sorption of trace organics and engineered nanomaterials onto wetland plant material. Env Sci: Process Imp 15(1):267–274

Sharma V, Shukla RK, Saxena N et al (2009) DNA damaging potential of zinc oxide nanoparticles in human epidermal cells. Toxicol Lett 185(3):211–218

Shen CX, Zhang QF, Li J et al (2010) Induction of programmed cell death in Arabidopsis and rice by single-wall carbon nanotubes. Am J Bot 97(10):1602–1609

Shoults-Wilson WA, Reinsch BC, Tsyusko OV et al (2011) Role of particle size and soil type in toxicity of silver nanoparticles to earthworms. Soil Sci Soci Am J 75(2):365–377

Siddiqi KS, Husen A (2016) Engineered gold nanoparticles and plant adaptation potential. Nanoscl Res Lett 11(1):400

Siddiqi KS, Husen A (2017) Plant response to engineered metal oxide nanoparticles. Nanoscl Res Lett 12(1):92

Slade S, Pegg G (1993) The effect of silver and other metal ions on the in vitro growth of root-rotting Phytophthora and other fungal species. Annl App Bio 122(2):233–251

Smirnova E, Gusev A, Zaytseva O et al (2012) Uptake and accumulation of multiwalled carbon nanotubes change the morphometric and biochemical characteristics of Onobrychis arenaria seedlings. Front Chem Sci Engin 6(2):132–138

Sohaebuddin SK, Thevenot PT, Baker D et al (2010) Nanomaterial cytotoxicity is composition, size, and cell type dependent. Parti Fibre Toxicol 7(1):22

Song G, Gao Y, Wu H et al (2012) Physiological effect of anatase TiO_2 nanoparticles on Lemna minor. Env Toxicol Chem 31(9):2147–2152

Song U, Shin M, Lee G et al (2013) Functional analysis of TiO_2 nanoparticle toxicity in three plant species. Biolo Trace Elem Res 155(1):93–103

Stampoulis D, Sinha SK, White JC (2009) Assay-dependent phytotoxicity of nanoparticles to plants. Env Sci Tech 43(24):9473–9479

Stoimenov PK, Klinger RL, Marchin GL, Klabunde KJ (2002) Metal oxide nanoparticles as bactericidal agents. Langmuir 18(17):6679–6686

Tan X-m, Fugetsu B (2007) Multi-walled carbon nanotubes interact with cultured rice cells: evidence of a self-defense response. J Biomed Nanotech 3(3):285–288

Tani F, Barrington S (2005) Zinc and copper uptake by plants under two transpiration rates. Part I. Wheat (*Triticum aestivum* L.). Env Poll 138(3):538–547

Tarafdar J, Sharma S, Raliya R (2013) Nanotechnology: interdisciplinary science of applications. Af J Biotech 12(3). https://doi.org/10.5897/ajb12.2481

Tian QY, Sun DH, Zhao MG, Zhang WH (2007) Inhibition of nitric oxide synthase (NOS) underlies aluminum-induced inhibition of root elongation in Hibiscus moscheutos. New Phytol 174(2):322–331

Torney F, Trewyn BG, Lin VS-Y, Wang K (2007) Mesoporous silica nanoparticles deliver DNA and chemicals into plants. Nat Nanotech 2(5):295–300

Unrine JM, Colman BP, Bone AJ et al (2012) Biotic and abiotic interactions in aquatic microcosms determine fate and toxicity of Ag nanoparticles. Part 1. Aggregation and dissolution. Env Sci Technol 46(13):6915–6924

Vasquez Y, Henkes AE, Bauer JC, Schaak RE (2008) Nanocrystal conversion chemistry: a unified and materials-general strategy for the template-based synthesis of nanocrystalline solids. J Solid State Chem 181(7):1509–1523

Wagner PA, Hoekstra WG, Ganther HE (1975) Alleviation of silver toxicity by selenite in the rat in relation to tissue glutathione peroxidase. Pro Soci Exper Bio Med 148(4):1106–1110

Wang J, Chen S, Quan X et al (2007) Enhanced photodegradation of PNP on soil surface under UV irradiation with TiO_2. Soil and Sedi Contamin: An Int J 16(4):413–421

Wang M, Zhou Q (2005) Single and joint toxicity of chlorimuron-ethyl, cadmium, and copper acting on wheat *Triticum aestivum*. Ecotoxicol Environ Saf 60(2):169–175
Wang S, Kurepa J, Smalle JA (2011) Ultra-small TiO_2 nanoparticles disrupt microtubular networks in Arabidopsis thaliana. Plant, Cell Env 34(5):811–820
Wang W (1986) Toxicity tests of aquatic pollutants by using common duckweed. Env Poll (B Chem Phys) 12(1):1–14
Wang X, Han H, Liu X et al (2012) Multi-walled carbon nanotubes can enhance root elongation of wheat (*Triticum aestivum*) plants. J Nanoparti Res 14(6):841
Wang X, Summers CJ, Wang ZL (2004) Large-scale hexagonal-patterned growth of aligned ZnO nanorods for nano-optoelectronics and nanosensor arrays. Nano Lett 4(3):423–426
Warheit DB, Laurence BR, Reed KL et al (2004) Comparative pulmonary toxicity assessment of single-wall carbon nanotubes in rats. Toxicol Sci 77(1):117–125
Wong SW, Leung PT, Djurišić A, Leung KM (2010) Toxicities of nano zinc oxide to five marine organisms: influences of aggregate size and ion solubility. Analy Bioanaly Chem 396(2):609–618
Wood CM, Hogstrand C, Galvez F, Munger R (1996) The physiology of waterborne silver toxicity in freshwater rainbow trout (*Oncorhynchus mykiss*) 1. The effects of ionic Ag^+. Aq Toxicol 35(2):93–109
Wu S, Huang L, Head J et al (2012) Phytotoxicity of metal oxide nanoparticles is related to both dissolved metals ions and adsorption of particles on seed surfaces. J Pet Environ Biotechnol 3(4):126
Yamamoto Y, Kobayashi Y, Matsumoto H (2001) Lipid peroxidation is an early symptom triggered by aluminum, but not the primary cause of elongation inhibition in pea roots. Plant Physiol 125(1):199–208
Yan S, Zhao L, Li H et al (2013) Single-walled carbon nanotubes selectively influence maize root tissue development accompanied by the change in the related gene expression. J Haza Mat 246:110–118
Yang F, Hong F, You W et al (2006) Influence of nano-anatase TiO_2 on the nitrogen metabolism of growing spinach. Bio Trace Elem Res 110(2):179–190
Yang K, Ma Y-Q (2010) Computer simulation of the translocation of nanoparticles with different shapes across a lipid bilayer. Nat Nanotech 5(8):579
Ye L, Yong KT, Liu L et al (2012a) A pilot study in non-human primates shows no adverse response to intravenous injection of quantum dots. Nat Nanotech 7(7):453
Ye X, Gu Y, Wang C (2012b) Fabrication of the Cu_2O/polyvinyl pyrrolidone-graphene modified glassy carbon-rotating disk electrode and its application for sensitive detection of herbicide paraquat. Sens and Actuators B: Chem 173:530–539
Yi H, Nisar S, Lee SY et al (2005) Patterned assembly of genetically modified viral nanotemplates via nucleic acid hybridization. Nano Lett 5(10):1931–1936
Yin L, Cheng Y, Espinasse B et al (2011) More than the ions: the effects of silver nanoparticles on Lolium multiflorum. Env Sci Technol 45(6):2360–2367
Yin L, Colman BP, McGill BM et al (2012) Effects of silver nanoparticle exposure on germination and early growth of eleven wetland plants. PLoS One 7(10)
Yuan H, Hu S, Huang P et al (2011) Single walled carbon nanotubes exhibit dual-phase regulation to exposed Arabidopsis mesophyll cells. Nanoscale Res Lett 6(1):44
Zhai G, Walters KS, Peate DW et al (2014) Transport of gold nanoparticles through plasmodesmata and precipitation of gold ions in woody poplar. Env Sci Technol Lett 1(2):146–151
Zhang L, Feng C, Chen Z et al (2008) Superaligned carbon nanotube grid for high resolution transmission electron microscopy of nanomaterials. Nano Lett 8(8):2564–2569
Zhang N, Ping Q, Huang G et al (2006) Transport characteristics of wheat germ agglutinin-modified insulin-liposomes and solid lipid nanoparticles in a perfused rat intestinal model. J Nanosci Nanotech 6(9–10):2959–2966
Zheng L, Hong F, Lu S, Liu C (2005) Effect of nano-TiO_2 on strength of naturally aged seeds and growth of spinach. Biolog Trace Elem Res 104(1):83–91

Zhou X, Torabi M, Lu J et al (2014) Nanostructured energetic composites: synthesis, ignition/combustion modeling, and applications. ACS App Mat Interf 6(5):3058–3074

Zhu H, Han J, Xiao JQ, Jin Y (2008) Uptake, translocation, and accumulation of manufactured iron oxide nanoparticles by pumpkin plants. J Env Monitor 10(6):713–717

Chapter 23
Use of Nanomaterials in Plants to Coup with Abiotic Stress Conditions

Neelu Raina, Preeti Sharma, Parvez Singh Slathia, Rimpi Thakur, Vinaik Raina, Garima Slathia, Nedhu Rani, Sakshi Sharma, Priyanka Bhagat, Rakshit Manhas, and Atin Kumar Pathak

Abstract Nanomaterials (NMs) have an important role to play in relation to mobility, fate and toxicity of soil pollutants and thus are pivotal to abiotic stress (AS) remediation strategies. AS is caused by abiotic (non-living) factors, e.g., salinity, temperature, floods, UV-B radiation and drought. Role of NMs in supporting extensive crop yields under challenging environmental conditions has facilitated use of sustainable agricultural practices. The prominent role of NMs in reduction of nutrient loss from the soil, i.e., nano remediation strategies for reducing the infertile lands, rendered unfit for cultivation, is one of the goals for maintaining soil structure and improving its fertility. Nanopesticides, nanofertilizers and nanobiosensors influence the suppression of soil as well as crop borne diseases and, thereby enhance crop yields. Various case studies highlighting the use of nanobiosensors, nanofertilizers, nano-enabled remediation strategies for contaminated soils and nanopesticides in the agricultural sector have been dealt in this chapter. NMs play an important role in principle events of plant growth including seedling vigor, seed germination, growth, photosynthesis, flowering and root initiation. Also, these NMs play an important role in plant protection under oxidative stress. NMs increase the buildup of reactive oxygen species (ROS) in plants and replicate the functions of some of the enzymes (anti-oxidative), i.e., catalase (CAT), peroxidase (POX) and superoxide dismutase (SOD). Therefore, it becomes necessary to decipher the cellular, biochemical and molecular mechanism of NMs in plants during AS conditions. Future research directions have been discussed to meet challenging environmental conditions.

Keywords Abiotic stress · Nanomaterials · Nanopesticide · Nanotechnology · Sustainable agriculture

---

N. Raina (✉) · P. Sharma · P. S. Slathia · R. Thakur · V. Raina · G. Slathia · N. Rani · S. Sharma · P. Bhagat · R. Manhas
School of Biotechnology, Shri Mata Vaishno Devi University, Katra J&K 182320, India

A. K. Pathak
School of Energy Management, Faculty of Engineering, Shri Mata Vaishno Devi University, Katra J&K 182320, India
e-mail: 15dem001@smvdu.ac.in

J. M. Al-Khayri et al. (eds.), *Nanobiotechnology*,
https://doi.org/10.1007/978-3-030-73606-4_23

23.1 Introduction

The world population, per capita income and consequently the demand for food continues to rise (Fig. 23.1). This is evident from the fact that annually, the global production of crops is more than 3 billion mega tonnes which entails 4 million mega tonnes pesticides, 187 million mega tonnes fertilizer, 2.7 trillion m^3 water and approximately 2 quadrillion BTU (British thermal units) energy (Kah et al. 2019; Usman et al. 2020). In order to feed a constantly growing population, which will reach around 10 billion by the year 2050, the key challenge faced by the research fraternity is to enhance the worldwide production of food crops by more than 2/3rd towards sustainable world agriculture (Duque-Acevedo et al. 2020). In addition, one of the sustainable development goals of United Nations is to achieve "zero hunger", which can be attained only through sustainable agricultural practices. Therefore, in such an evolving world environment it has become impertinent to identify novel sectors and rigorous research and development is carried out in order to surpass challenges in developing a reliable technology in addressing the blockades in agricultural production for developing sustainable agriculture.

Abiotic stress (AS) is a foremost limitation which negatively affects plant growth and crop productivity on a global scale. Rise in the prevalence of a large number of abiotic stresses as a result of natural and man-made activities have become a major

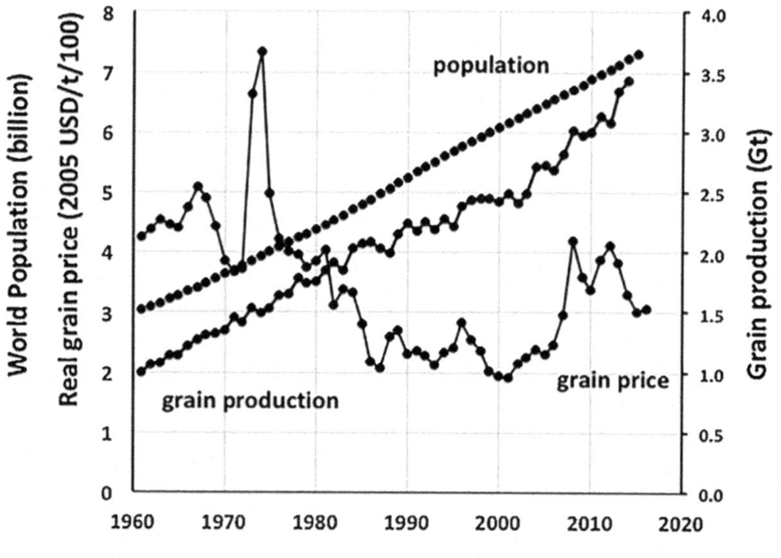

Fig. 23.1 World population, grain crop production and prices for grains between 1961 and 2016 (Fischer and Connor 2018) (*Sources* World population statistics from United Nations Department of Economic and Social Affairs [UN 2017], grain production numbers have been adopted from FAOSTAT [Shenashen et al. 2017], including all types of pulses, cereals and more than a dozen different kinds of oilseed crops; export prices comprise mean prices of soybean, rice, wheat and maize [World Bank 2017])

concern for mankind and there is an urgent need to mitigate their effects in order to increase yield of crops. Amongst different kinds of AS such as drought, salinity, temperature, metal, UV radiation and post-harvest stress; drought and salinity are the most widespread and common types of stress. In the recent decades, the field of nanotechnology (NT) has been gaining momentum in sustainable agricultural sector and thus occupied a promising position in mitigating the impact of AS by using NMs. NT has not only shown promising results in the field of plant growth and crop improvement but also proved to be an effective tool against plant AS.

Since materials tend to have diverse properties (chemical, physical, and biological) at the nanoscale than they do at their larger dimensions therefore, they acquire some properties different from their bulk counterparts due to re-organization of matter. Thus, Paul Ehrlich once called materials at nanoscale dimension (NMs) as "Magic Bullets" (Kreuter 2007). Hence, they have been widely studied for their applications in various areas over a period of time by a number of workers. It was in the year 1959, when Richard Feynman first suggested that materials and objects could be put together to an atomic classification. He said, "The principles of physics, as far as I can see, do not speak against the possibility of maneuvering things atom by atom." NMs have since then become one of the leading researched materials of the century giving rise to a new offshoot of science called 'nanotechnology' (Khan et al. 2017). It is the study of two fields, i.e., engineering and chemistry and defined as the assembly, arrangement, and application of materials engineered at the nanoscale (10^{-9} m) at least in a single dimension (Raina et al. 2020). NT explores novel and wide range of diverse applications and its potential benefits could be exploited in the area of agricultural production as well. Its use in agricultural sector is said to have started when humans employed clay and porous soil substitutes for water filtration and decontamination purposes. NMs have the ability to improve plant growth and development by their use in the form of herbicides, nanopesticides and nanofertilizers having the ability of efficient content release in requisite amounts in order to target plant cellular organelles. Studies on targeted mustard (*Brassica* sp. L.) and non-targeted maize (*Zea mays* L.) organisms have shown that the nano herbicide atrazine formulation showed strong control of the targeted species (Acharya and Pal 2020). This chapter gives the insights of various NMs used in agriculture sector and their potential applications in overcoming stress caused due to various abiotic factors thus enhancing crop yield.

23.2 Use of Nanomaterials (NMs) in Sustainable Agriculture

Agriculture has been, since time immemorial, the life-force for driving the survival of human race and the only prime contributor of food for humans. It produces and provides the basic requirements for food and feed industries and acts as fuel for the developing world economies. Rapid growth in the world population, sudden explosion in the technological sector along with the materialistic drive of humans have made agricultural sector more vulnerable, eventually draining natural reserves in addition to bringing concurrent degradation of environment as well (Raina et al. 2020). In an effort to increase agricultural yield we have practiced the use of pesticides and fertilizers disproportionately. This on one hand has increased the nutrient content of food but on the other hand also increased the toxicity levels in the soil and the agricultural yield. Prolonged irrigation and drainage malpractices has beefed up the rate of weathering process of soil minerals, increasing soil acidity levels and eventually made farm lands barren. Unwanted soil acidic conditions severely impact soil nutrients resulting in plant nutrient deficiency thereby resulting in deviation from the common physiological abilities of plants like growth and development. Exhaustive use of pesticides, fertilizers and wrong irrigation techniques has also caused extensive harm to the soil carbon profile and contaminated the soil biome with harmful trace elements and pesticide residues eventually making the soil saline. The main effect of prolonged soil salinity is toxicity of plant cells along with disruption in the osmotic equilibrium. The combined result of disruption in ionic and osmotic equilibrium of plants is on the overall growth and development of the plant (Jalil and Ansari 2019). As a result it becomes almost impossible to refine these lands without relocating the human population. Therefore, in order to maximize the agricultural yield and protect the related biomes, a number of modern technologies have been put into place. Recent developments in NT have seen a global renewed interest in the agricultural arena. NMs are having diverse potential applications in agriculture spanning from nanofertilizers to genetic engineering. Although there are a large number of NMs, obtained from essential metals (Zn, Fe, Mn, Cu), and their oxides are suitable for use in agriculture (Pérez-Labrada et al. 2020). Some of the potential applications of NMs in agriculture have been illustrated in Fig. 23.2.

The area/s of application of NT in sustainable agriculture consists of nanobiotechnological manipulation of genes and proteins, increased soil hydrophilicity, improvement of productivity by nutrient uptake at nanoscale, waste management, targeted delivery of chemicals and many more exciting ones. This technology has resulted in an enhanced resource management in the agricultural field, improved modes of drug delivery in the plants and maintained soil fertility as well. It uses sensors and monitoring devices, thereby stimulating the global food production and bringing positive impact in the world ecosystem by decreasing the percentage of pesticides, water and fertilizers used. Various improvements have been made in the field of nano-enabled remediation of contaminated soils and soil nanonutrition such as nanopesticides, nanofertilizers and nanobiosensors. Different types of NMs have thus been studied having applications in mitigating AS in plants in relation to the overall sustainable improvement in crop production as shown in Fig. 23.3.

Developments made using nanofertilizers and nano-encapsulated nutrients are proving to be valuable instruments towards sustainable crop production using efficient nutrient release in the soil, thereby making them available to the plants. Use of nanofertilizers enhances the chances of availability of soil micronutrients to plants in

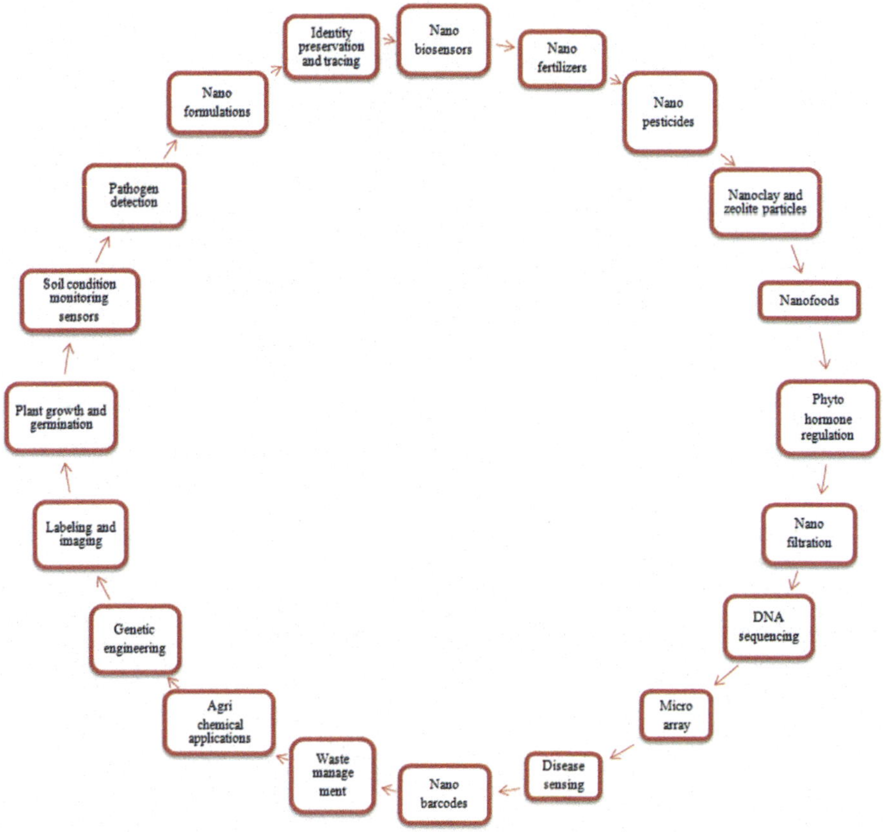

Fig. 23.2 Applications of nanomaterials in agriculture (Figure constructed by Neelu Raina)

addition to alleviating reduction in soil fertility due to buildup of chemical fertilizers and pesticides in the soil. The effect of NMs on plant growth and germination has been explored by various researchers which have led to their use in agriculture and food sector. A number of NMs produced using metals or metal oxides are being studied in order to estimate their possible benefits in the production and modulation of plant development and growth processes and their defense against various AS types. Several recent studies have been tabulated in order to evaluate the effect of different NMs against various AS types on a number of plant species (Table 23.1).

Plants are immobile; therefore they cannot escape the effect of any kind of environmental stress. They require optimal abiotic conditions for proper growth and development. Any alteration in the optimal physical and chemical environmental condition results in AS. AS such as drought, temperature (freezing, heat and cold), irradiation, heavy metals, water logging, nutrient deficiency, alkalinity of soils, heavy metal toxicity (Pasala et al. 2016) have a serious impact on the developmental processes of

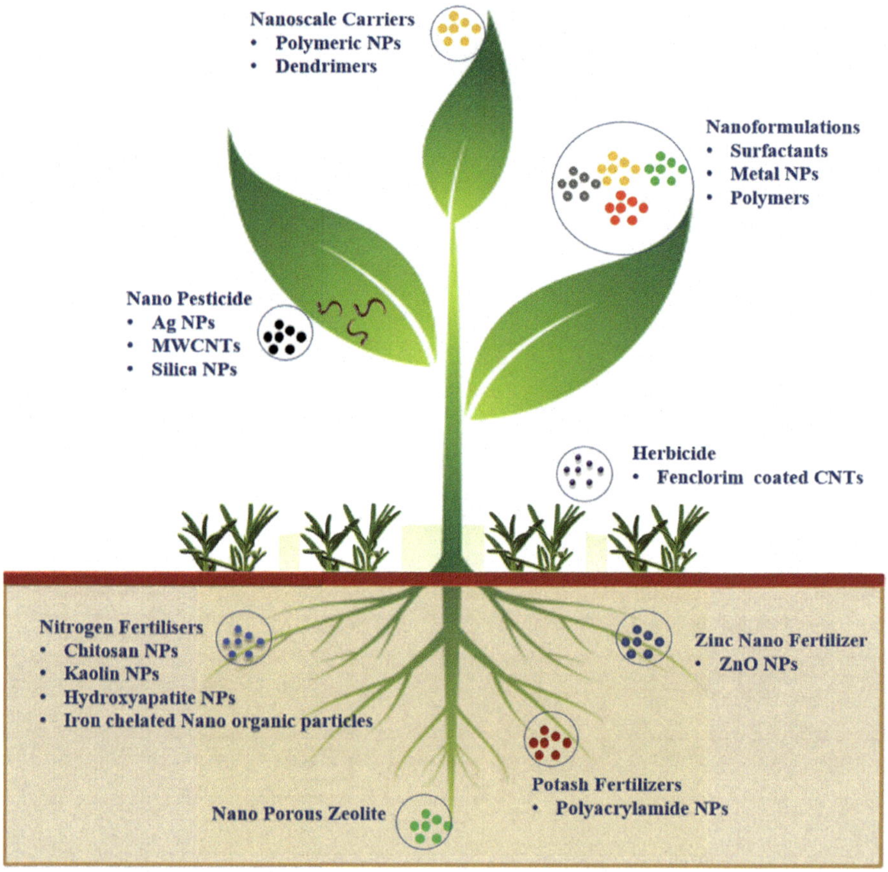

Fig. 23.3 Potential uses of nanomaterials in crop protection for sustainable agriculture development using materials at nanoscale dimension (*Source* Vishwakarma et al. 2018)

plants effecting the production of agricultural crops negatively (Emamverdian et al. 2015). It also impacts the growth, development and productivity of crop plants critically by approximately more than 50% (Jalil and Ansari 2019). Therefore, plants have a network of defense systems to counter any kind of stress. But, the precise identification, sequence and stimulation of these defense systems in reaction to the state of stress are crucial for the strengthening of plants defense against any kind of environmental stress.

Table 23.1 Effects of nanomaterials on various stress types in plants

Nano material	Plant species	Mode of application	Concentration Used	Duration of treatment	Abiotic stress	Effect	References
Ag	*Arabidopsis thaliana* (L.) Heynh.				Cold	Enhanced antioxidant activity of genes, Less than 40% of similar genes were affected by both Ag NMs and low temperature stress	Khan et al. (2017), Kohan-Baghkheirati and Geisler-Lee (2015)
Ag	*Carum copticum* (L.) Sprague ex Turrill				Drought	No considerable effect on the produce and efficacy in the use of water	Khan et al. (2017), Seghatoleslami et al. (2015)
Ag	*Chrysanthemum morifolium* L.				Post-harvest stress	Improved succulence and life span of cut flowers, reduced loss in fresh weight and in the colonies of stem bacteria	Kazemipour et al. (2013), Khan et al. (2017)
Ag	*Crocus sativus* L.				Flooding	Hindrance in ethylene signaling, increased root growth	Khan et al. (2017), Rezvani et al. (2012)
Ag	*Glycine max* (L.) Merr.				Flooding	Decrease in the production of cytotoxic by-products from glycolytic pathway, improved production of stress- related proteins, increased seedling growth	Khan et al. (2017), Mustafa et al. (2015a)
Ag NMs	*Triticum aestivum* L.	Mixed with pot soils	50 mg/L and 75 mg/L	Trifoliate stage		Enhanced tolerance to high temperature and increased plant growth	Iqbal et al. (2019), Shang et al. (2019)
Ag NMs	*Vigna sinensis* L.	Foliar application	50 mg/L	40 days		Rise in plant biomass by increased root nodulation and diversity of soil bacteria	Mehta et al. (2016), Shang et al. (2019)
Ag NMs	*Vigna unguiculata* (L.) Walp.	Foliar application	50–100 μg/mL	7 days		No signs of phytotoxicity, but may lead to decrease in the in vitro growth of some species of genus *Xanthomonas*	Shang et al. (2019), Vanti et al. (2019)
Ag	*Pelargonium zonale* (L.) L'Hér. ex Aiton				Dark stress	Enhanced activity of antioxidative enzymes, chlorophyll and carotenoid content of leaves, increased petal longevity and reduced lipid peroxidation	Hatami and Ghorbanpour (2014), Khan et al. (2017)
Al_2O_3	*Glycine max*				Flooding	Rise in the plant growth and regulation of energy metabolism including cell death	Khan et al. (2017), Mustafa et al. (2015b)

(continued)

Table 23.1 (continued)

Nano material	Plant species	Mode of application	Concentration Used	Duration of treatment	Abiotic stress	Effect	References
Al_2O_3	*Solanum lycopersicum* L.	Foliar application	400 mg/L	20 days		*Fusarium* root rot in Tomato is successfully inhibited	Shang et al. (2019), Shenashen et al. (2017)
Analcite	*Triticum aestivum* L., *Zea mays* L.				Drought	Increased buildup of flavonoids and carotenoids, reduced proline accumulation, improved germination of seeds, increased photosynthetic pigments and generation of plant biomass	Khan et al. (2017), Zaimenko et al. (2014)
Anatase-TiO_2	*Spinacia oleracea* L.				UV-B radiation	Reduced content of ROS and MDA species, improved antioxidative enzyme activity and higher oxygen generation	Khan et al. (2017), Lei et al. (2008)
Au^{3+}	*Vigna unguiculata* L.				Au^{3+}	Spike in the production of Au- NMs and reduced phenolics content, drop in Au^{3+} production	Khan et al. (2017), Shabnam et al. (2014)
Cd-Telluride Quantum Dots	*Triticum aestivum* L.				UV-B radiation	Restrained root and shoot growth, DNA laddering and programmed cell death	Chen et al. (2014), Khan et al. (2017)
CeO_2	*Chlorella vulgaris* Beijerinck				UV-B radiation	Better UV absorption by NMs and less damage caused by oxidative stress	Khan et al. (2017), Sicard et al. (2011)
CuO	*Elodea nuttallii* (Planch.) H.St. John				UV rays	Rise in the aggregation of Cu ions from CuO-NMs, decreased chlorophyll percentage and capacity to photosynthesize, increased activity of peroxidase enzyme	Khan et al. (2017), Regier et al. (2015)
CuO	*Solanum lycopersicum* L.	Foliar application	150–340 μg/mL	10–12 days		Improved control over late blight disease in plants	Giannousi et al. (2013), Shang et al. (2019)
CuO	*Spinacia oleracea* L.	Mixed with soils	200 mg/kg	55–65 days		Enhanced production of plant biomass and increased rate of photosynthesis	Shang et al. (2019), Wang et al. (2019)

(continued)

Table 23.1 (continued)

Nano material	Plant species	Mode of application	Concentration Used	Duration of treatment	Abiotic stress	Effect	References
Cu	*Petroselinum crispum* Mill.				Post-harvest stress	Decrease in lipid peroxidation regulating ascorbic acid concentration, thereby maintaining the quality of parsley	Khan et al. (2017)
FeS_2	*Sesamum indicum* L., *Cicer arietinum* L; *Daucus carota* L., *Brassica juncea* (L.) Czern.	Seed priming	90 µg/mL	13 h		Improved seed germination rate and thus increased crop yield	Das et al. (2016), Shang et al. (2019)
Fe/SiO_2	*Arachis hypogaea* L., *Zea mays* L.	In the form of fertilizers	14–16 mg/kg	4–5 days		Increased biomass accumulation and plant growth	Najafi Disfani et al. (2017), Shang et al. (2019)
Hydroxyapatite	*Brassica chinensis* L.				Cd	Improved biomass content, chlorophyll level and ascorbic acid, increased SOD, CAT, and POD activities, reduced MDA concentration	Khan et al. (2017), Siddiqui et al. (2014)
Kaolin, montmorillonite, hydroxyapatite, Fe_3O_4, α-Fe_2O_3 and γ-Fe_2O_3	*Daucus carota, Cucumis sativus* L., *Lycopersicon esculentum, Lactuca sativa* L.				Cd	Increased root growth in carrots by hydroxyapatite, reduced root growth in tomatoes by hydroxyapatite, and montmorillonite, improved root growth in lettuce by iron oxide particles	Khan et al. (2017)
Maghemite (γ-Fe_2O_3)	*Helianthus annuus* L.				Drought	Resistance to stress due to shortage of water with no effect on praline, accumulative amino acid content and movement of trace elements	Khan et al. (2017), Martínez-Fernández et al. (2015)
MgO	*Solanum lycopersicum*	Saturation	8–9 µg/mL	6–8 days		Suppressing the growth of wilt causing pathogen (*Ralstonia Solanacearum*)	Imada et al. (2016), Khan et al. (2017)

(continued)

Table 23.1 (continued)

Nano material	Plant species	Mode of application	Concentration Used	Duration of treatment	Abiotic stress	Effect	References
MWCNTs	*Arachis hypogaea* L., *Allium sativum* L., *Triticum aestivum*, *Zea mays* L.	Seed priming	45–55 μg/mL	Over night		Rapid rate of germination, enhanced accumulation of biomass and water absorbing potential of seeds	Shang et al. (2019)
Na_2SeO_4	*Lycopersicum esculentum* Mill.				High and low temperature	Enhanced growth in plants and chlorophyll content	Shang et al. (2019)
Na_2SiO_3	*Pisum sativum* L.				Cr(VI)	Increased activity of antioxidant defense system against Cr(VI) phytotoxicity	Shang et al. (2019), Tripathi et al. (2015)
Se	*Cucumis sativus* L.				Cold	The plant physiological response to extreme low temperatures is affected in addition to enhanced proline content in leaves and reduced lipid peroxidation in roots	Khan et al. (2017)
SiO_2	*Crataegus* sp.				Drought	Plant photosynthetic rate is significantly affected in addition to plant biomass and stomatal conductance	Ashkavand et al. (2015), Khan et al. (2017)
SiO_2	*Cucurbita pepo* L.				Salinity	Reduced melondialdehyde levels, chlorophyll degradation and oxidative damage, improved seed germination, enhanced antioxidant enzyme concentration and various photosynthetic parameters	Khan et al. (2017), Siddiqui et al. (2014)
SiO_2	*Lens culinaris* Medik.				Salinity	Enhanced seed germination and seedling growth	Khan et al. (2017)
SiO_2 NMs	*Oryza sativa* L.	Foliar application	2–3 mM/L	70 days		Increased heavy metal toxicity and increased growth by reducing bio-concentration and transport in plants	Shang et al. (2019), Wang et al. (2016a)
SiO_2	*Ocimum basilicum* L.				Salinity	Enhanced dry and fresh weight, chlorophyll and proline percentage	Khan et al. (2017)

(continued)

Table 23.1 (continued)

Nano material	Plant species	Mode of application	Concentration Used	Duration of treatment	Abiotic stress	Effect	References
SiO_2	*Solanum lycopersicum* L.				Salinity	Repressed the effect of salt stress on the rate of seed germination and length of root; Increased expression of four salt stress genes i.e. *AREB*, *TAS14*, *NCED3* and *CRK1*, whereas six genes i.e. *RBOH1*, *APX2*, *MAPK2*, *ERF5*, *MAPK3* and *DDF2* were down- regulated	Almutairi (2016), Khan et al. (2017)
TiO_2 and SiO_2	*Oryza sativa* L.	Foliar application	20 and 30 mg/L	53–57 days		Enhanced the plant growth by decreasing Cd toxicity and improving the antioxidant producing potential of plants	Rizwan et al. (2019), Shang et al. (2019)
TiO_2	*Cicer arietinum* L.				Cold	Improved the activity of phosphorenol pyruvate carboxylase increased Rubisco and chlorophyll binding protein gene expression	Hasanpour et al. (2015), Khan et al. (2017)
TiO_2	*Glycine max* L.				Cd	Increased uptake of Cd, decreased Cd toxicity, improved rate of photosynthesis and growth rate parameters, lowered lipid peroxidation and proline percentage	Khan et al. (2017), Singh and Lee (2016)
TiO_2	*Linum usitatissimum* L.				Drought	Rise in the level of chlorophyll and carotenoids, increased growth and yield parameters, lower H_2O_2 content	Aghdam et al. (2016), Khan et al. (2017)
TiO_2	*Spinacia oleracea*	Seed priming and foliar application	0.25% suspension	48 h and 35 days		Enlarged accumulation of biomass, nitrogen protein and chlorophyll content	Shang et al. (2019)
Zerovalent Fe	*Arabidopsis thaliana*				Drought	Activation of plasma membrane H^+-ATPase, enhanced chlorophyll percentage and concentration of plant biomass, increased CO_2 content	Khan et al. (2017), Kim et al. (2015)

(continued)

Table 23.1 (continued)

Nano material	Plant species	Mode of application	Concentration Used	Duration of treatment	Abiotic stress	Effect	References
ZnO and Fe_3O_4	*Moringa peregrina* (Forssk.) Fiori				Salinity	Decreased levels of Na^+ and Cl^- contents, enhanced N, P, K^+,Ca^{2+}, Mg^{2+}, Fe, Zn concentration, total chlorophyll, carotenoids, proline, carbohydrates and enzyme antioxidants	Khan et al. 2017), Soliman et al. (2015)
ZnO	*Coffea Arabica* L.	Foliar spray	8–12 mg/L	40–50 days		Rise in growth, biomass accumulation and net photosynthetic products	Rossi et al. (2019), Shang et al. (2019)
ZnO, CuO and Ag NMs	*Prunus domestica* L.	Fruit spray	100 and 1000 µg/mL	4 days		Suppressed grey mold symptoms caused by *B. cinerea* and soil borne diseases	Malandrakis et al. (2019), Shang et al. (2019)
ZnO	*Cyamopsis tetragonoloba* (L.) Taub.	Foliar spray	10 mg/L	6 weeks		Increased nutrient content, enhanced plant growth, accumulation of plant biomass and nutrient levels	Shang et al. (2019)
ZnO	*Helianthus annuus* L.				Salinity	Enhanced plant growth, net rate of CO_2 assimilation, concentration of sub- stomatal CO_2 and chlorophyll content, and decreased Na^+ ions in leaves	Khan et al. (2017), Torabian et al. (2016)
ZnO	*Nicotiana tabacum* L.	Hydroponics	0.2 µM and 1 µM	21 days		Enhanced physiological growth, enhanced metabolites, enzymatic activities of plants	Shang et al. (2019), Tirani et al. (2019)
ZnO	*Triticum aestivum* L.	Added to growth substrate	20 mg/L	Growth cycle		Improved grain yield in addition to rise in biomass accumulation	Du et al. (2019), Shang et al. (2019)

23.3 Plants and Abiotic Stress

All plants have an innate ability to adjust to unfavorable environment such as, drought, chilling, heat stress, salinity. Molecular and cellular responses of plants to all these types of abiotic stresses have been broadly studied (Duque et al. 2013). It has been shown that miRNAs are involved in plant response to AS (Frazier et al. 2014). The cell wall of plants restricts the affect of stress conditions and acts as a dynamic player in making the plants adapt to AS conditions. The initial reaction of plants against AS conditions consists of a transient rise in the cytoplasmic Ca^{2+} content, elevated secondary messengers (polyphosphate, inositol) in the intracellular space, abscisic acid, ROS and enhanced mitogen-activated protein kinase (MAPK) pathways (Baxter et al. 2014). The progressive stages of AS response by plants comprises of regulation of the expression of stress-specific genes and regulation of proteins involved in cellular damage protection. Secondary metabolites have an important role to play in plants in order to combat AS by maintaining cell structure, in signal transduction, biosynthesis of polyamines and defense of photosystem against ROS (Oh et al. 2009). Extracellular peroxidases modify cell wall; build up ROS and oxidative stress when encountering AS. Building up of oxidative stress activates production of ROS, increases phenylpropanoid content, regulates gene expression and biosynthesis of enzymes during plant defense response (Fig. 23.4) (Daudi et al. 2012).

Abiotic stress is the primary cause of generation of ROS in the plants cell organelles such as chloroplasts, mitochondria and peroxisomes. ROS have a key role to play as they act as indicators to any kind of stress and thus activate the defense system of plants as well as intensify damage to the cellular machinery (Jalil and Ansari 2019). Plants have an in-built antioxidant system to counter oxidative stress by scavenging ROS (Khan et al. 2017). Under normal conditions, the production of ROS and their elimination are regularly maintained in a balance mode. However, due to AS this equilibrium shifts more towards generation of ROS affecting severely the structure and function of proteins which then results in phytotoxicity. In mitochondria, over reduction of electron transport chain (ETC) produces hydrogen peroxide and oxygen. However, most of the generation of O_2 and H_2O_2 takes place in the chloroplasts (Davletova et al. 2005; Jalil and Ansari 2019). These super oxides are changed into hydrogen peroxide either spontaneously or by the help of enzyme named 'superoxide dismutase'. In the matrix of peroxisomes, the generation of O_2^- free radicals takes place in presence of xanthine oxidase enzyme by the oxidation of hypoxanthine and xanthine into uric acid. This event degrades/damages biomolecules present in the cell such as DNA, lipids, proteins and carbohydrates which results in cell death. All these stresses generate reactive oxygen species (ROS) inside plants and thus result in oxidative burst. Generation of ROS in huge number causes degradation of membrane lipids and various macromolecules (Jalil and Ansari 2019), induces cells toxicity (Yadav et al. 2014), in addition to decreased plant growth (Khan et al. 2016). Plants fight osmotic stress by enhancing the production of polyols, trehalose (glycerol, inositol, sorbitol) and amino acids (glycine, betaine, taurine and proline) that equilibrate the required osmotic level in plant cells. Plants, in response to heavy metal

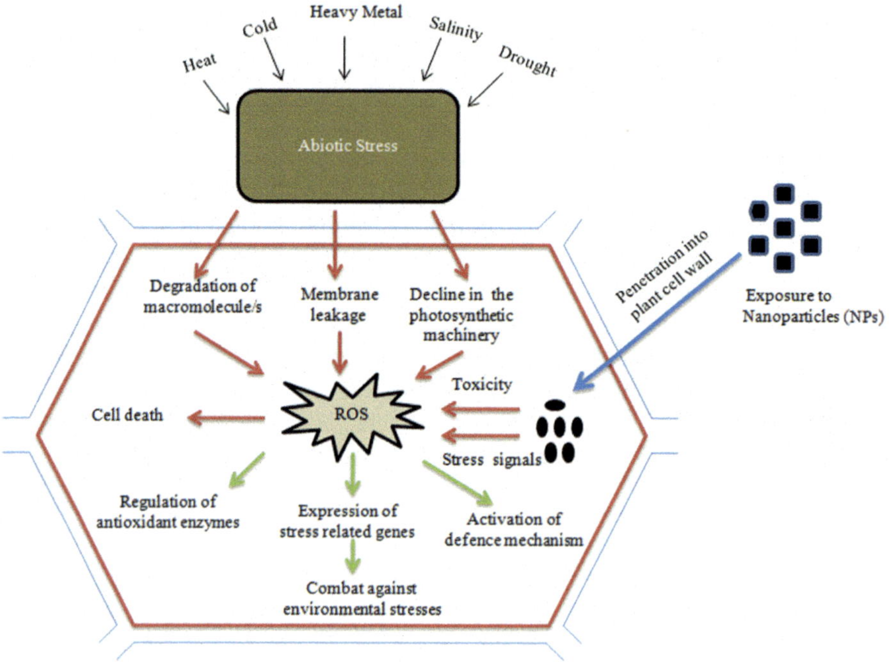

Fig. 23.4 Mechanism of action of nanomaterials under abiotic stress conditions in a particular plant cell (*Source* Jalil and Ansari [2019]. Arrows in red specify abiotic stress effect and the toxic effect of nanomaterials eventually leading to cell death. Arrows in green illustrate the affirmative role of nanomaterials by acting as stress signal molecules that initiate the defense machinery on order to mitigate abiotic stress conditions in plants [ROS = reactive oxygen species])

stress generate a large number of organic acids, polyphosphates and metal-chelates which result in accumulation of toxic metals in the plant cell plasma membrane.

23.4 Signaling Mechanism of Nanomaterials Under Abiotic Stress

NMs have a prime role in the growth and developmental processes of plants and protect them against AS (Khan et al. 2017). During AS the most common result of plant NM interaction is the production of ROS. NMs have the unique ability to imitate the activities of antioxidative enzymes and scavenge these ROS. NMs thus not only stimulate the production of ROS (Qi et al. 2013), but also replicate the activity of antioxidant enzymes to scavenge ROS (Fig. 23.4) (Wei and Wang 2013). The action of NMs can be viewed as a two way process where on one hand the ROS production is a trigger for the start of plant defense mechanism and can be scavenged by them but on the other hand also cause oxidative stress. In the plant photosynthetic process under

AS, NMs enhance the rate of photosynthesis by conquering osmotic and oxidative stress and thus protect the photosynthetic machinery (Fig. 23.4). TiO_2 NMs affect the plant photosynthetic process positively by increasing light absorbance, increasing transport and conversion of light energy, protecting chloroplasts against aging and increasing the photosynthetic time of chloroplasts by removing extreme light and enhancing antioxidant enzymatic activities (Ioannou et al. 2020). However, response of plants to NMs differs depending on the type of plant species and the type and concentration of NMs used. There are evidences also which have shown that NMs present many toxicity symptoms (Slomberg and Schoenfisch 2012). Decrease in shoot, root and germination rate of plants induce oxidative stress, decrease crop yield, rate of photosynthesis (Wang et al. 2016b) and nutritive value of crop plants (Peralta-Videa et al. 2014). Also, studies have shown that NMs interfere with gene expression involved in cell organization, cell biosynthesis, energy pathways and electron transport in AS response (Van Aken 2015). The properties of NMs such as small size and large surface area are readily available for binding to the toxic metals, therefore decreasing the toxicity and accessibility of heavy metals.

Nanomaterials enter the plant system via symplastic or apoplastic pathway depending upon a number of factors. The possible transport route of NMs through the plant system and its entry in a particular plant cell is shown diagrammatically in Fig. 23.5. Moreover, the transport of NMs takes place from the shoot to the root via phloem and from the root to the shoot via xylem. However, the path and the mechanism of entry are highly variable factors depending upon the concentration and the type of NM in use; in addition to the plant species, irrigation conditions and the soil composition.

A number of studies indicate that NMs mediated effect on growth and development of plants is concentration dependent and is also involved in the upregulation of activity of a number of antioxidant enzymes (Laware and Raskar 2014). Laware and Raskar (2014) performed a study on onion seedlings in presence of TiO_2 NMs; where it was observed that TiO_2 NMs promote the activity of superoxide dismutase enzyme which in turn improved the concentration of NMs. But, seedling growth and seed germination in onion enhanced at low TiO_2 NM concentration while the effect was suppressed at elevated concentrations. In addition to concentration dependent increase of superoxide dismutase, there was a considerable generation of amylase (hydrolytic enzyme); peroxidase and catalase enzyme activities, even though activity of TiO_2 was higher at lower concentration (~20 μg/mL) and low at a higher enzyme concentration (40 and 50 μg/mL) (Laware and Raskar 2014). A number of studies have shown TiO_2 and SiO_2NMsto have the ability to augment seed germination along with growth of soya bean (*Glycine max*) seeds (Jalil and Ansari 2019). Experiments done using Ag and $AgNO_3$ NMs with *Eruca sativa* L. Cav. has been reported to alter the proteins involved in redox reaction and sulfur metabolism in its roots (Vannini et al. 2013). Figure 23.6 shows the uptake route of AgNMs in major cell organelles of plants. Exposure of Ag and Ag^+NMs encased using polyvinylpyrrolidone (PVP) in the thale cress (*Arabidopsis thaliana*) plant regulated the expression of stress related genes (Kaveh et al. 2013). In tobacco plants, application of TiO_2 and Al_2O_3NMs upregulates expression of miRNAs in presence of metal stress, although

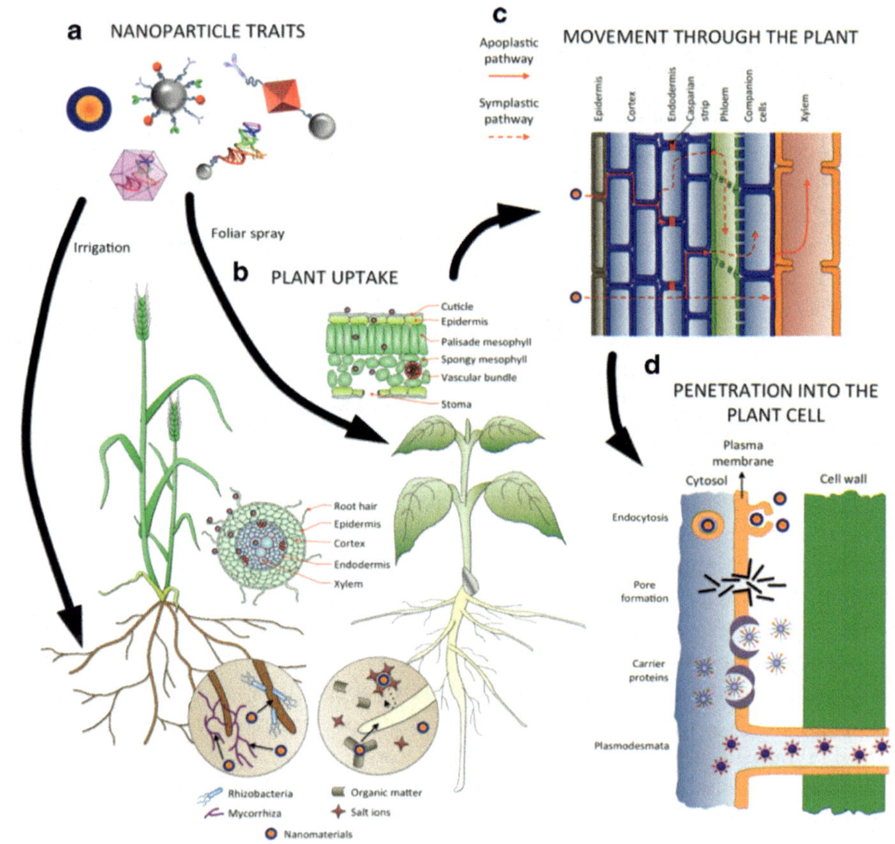

Fig. 23.5 Factors affecting uptake, movement and penetration of nanomaterials in plants systems. **a** Nanomaterial traits affect application, intake, transportation and translocation in the plant. **b** Interaction of nanomaterials in the soil system. **c** Nanomaterials can follow the apoplastic and/or the symplastic pathways or radial movement for travelling through the plant body. **d** Intake mechanisms of nanomaterials into plant cells (*Source* Pérez-de-Luque 2017)

increase in NMs concentration resulted in reduced biomass, wilting, decreased leaf size, root growth and reduction in leaf counts (Frazier et al. 2014). In another case, it has been found that the exposure of iron NMs (zerovalent) in *Arabidopsis* upregulated *AHA2* (involved in stomatal opening) gene expression enhanced tolerance to drought (Kim et al. 2015). Besides, under different stress conditions the treatment of Ag NMs TiO_2 and multiwall carbon nanotubes (CNTs), on *Arabidopsis thaliana* suppressed the expression of genes regulating root-development and phosphate starvation (García-Sánchez et al. 2015). A recent study, showing that SiO_2NMs play a protective role in barley in response to NiO NM stress, described that the application of NiO NMs on barley results in excessive production of ROS causing oxidative stress and enhanced lipid peroxidation. Interestingly, the application of SiO_2 NMs on plants already treated with NiO NMs enhanced rapid antioxidant response, decreased

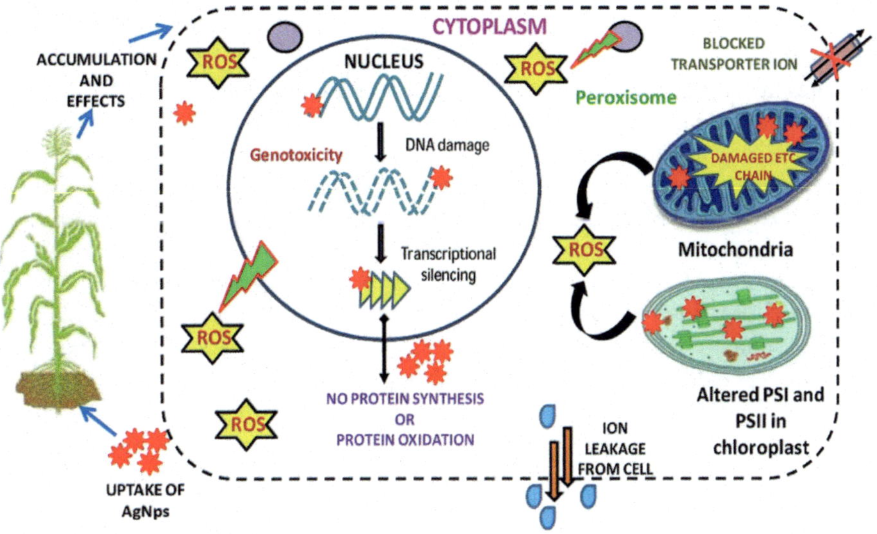

Fig. 23.6 Uptake route and metabolic pathway of AgNMs in major plant cell organelles (*Source* Tripathi et al. [2017]. This is an open-access article distributed under the terms of the Creative Commons Attribution License [CC BY])

lipid peroxidation levels and stimulated redox pathway thus mitigating phytotoxicity of NiO NMs (Soares et al. 2018). Several works have presented increases in plant vigor when applying NMs. In a study by Anandaraj and Natarajan (2017) increases in germination, bud and root length, and vigor index was observed when the seeds of *A. cepa* L. were coated with 1000 mg/kg of ZnO, Ag, CuO, and TiO_2 NMs. Another study reported that the application of ZnO NMs (0.75 g) enhanced germination of seeds, root, sprout and plant length (Pérez-Labrada et al. 2020).

Plants have an in-built defense mechanism that aids in combating AS (Yolcu et al. 2016). It leads to the reprogramming of metabolic processes in plant systems (Massad et al. 2012) and assists progressions in bio-physicochemical processes of AS (Jalil and Ansari 2019; Mickelbart et al. 2015). Plant defense system is activated by a signaling network which activates the molecular machinery in response to stress conditions. In signal transduction, calcium (Ca) ions act as second messengers. Stress signals via signaling network causes entry of Ca^{2+} ions into the cytosol using Ca^{2+} ion channels, which results into build-up of Ca^{2+} ions in the cytosol (Khan et al. 2014). Nitric oxide (NO) has been observed to increase the number of Ca^{2+} ions in the cytosol during various stress conditions (Khan et al. 2012) and therefore Ca^{2+} ions in turn leads to the synthesis of NO (Jalil and Ansari 2019). In the roots of *Oryza sativa* the application of Ag NMs result in formation of NM responsive proteins that are involved in the signaling and regulation of Ca^{2+}ions, apoptosis, protein degradation process, oxidative stress response pathway, synthesis of cell wall, transcription and cell division. Also, it is postulated that Ag NMs bind with

Ca^{2+} ion channels using Ca^{2+} ion receptors affecting cell metabolism (Mirzajani et al. 2014). It is observed that the interaction of C60 nanocrystals brings about functional changes in Ca^{2+}/calmodulin-dependent protein kinase II (Miao et al. 2014). The C60 molecule has a cage-like fused-ring structure which is formed when a single carbon atom is placed at each vertex and has all valences satisfied by the presence of two single bonds and a double bond. Also, in *Arabidopsis thaliana* the use of cadmium sulfide QDs resulted into overexpression of Ca-dependent protein kinase 23 and Ca-binding protein CML45 (Marmiroli et al. 2015). In plants, against various AS conditions these Ca-binding proteins have been known to influence stress response and overexpression resulting into improved resistance (Boudsocq and Sheen 2013).

NMs improved nitrate reductase enzyme activity in plants, increasing concentration of nitric oxide to modify immune response (Chandra et al. 2015). On the other hand, it has been shown that nitric oxide results into NMs-induced toxicity and stimulates the gene expression of antioxidant molecules in addition to suppressing the generation of ROS and lipid peroxidation (Chen et al. 2015). NMs mimic Ca^{2+}ions and thus bind to Ca-binding proteins accelerating the course of stress responsive genes (Mirzajani et al. 2014). In addition, the application of NMs increase the expression of genes related to cell division, cell elongation and stress response (Almutairi 2016). Evidences from studies in plant cells on NM-induced phytotoxicity have shown greater generation of ROS which act as toxic compounds and signaling molecules (Fig. 23.4). ROS play a number of roles governed by their production and scavenging activity. However, any kind of imbalance in any of these processes results in highly increased or decreased production and the accessibility of ROS, causing oxidative stress thus leading to signal disturbance. Alternatively, a balance is maintained by production and subsequent scavenging of ROS. Also, it has been observed that higher concentrations of NMs are lethal, while its lower concentrations are helpful or have no visible effect on plant's system. It is thus concluded that lower concentration of NMs maintain an active antioxidant defense system which in turn regulates the production of ROS into an accurate concentration sufficient for signaling system but insufficient to trigger damage (Syu et al. 2014).

23.5 Mode of Action of Nanomaterials Under Abiotic Stress

23.5.1 Drought Stress

Of all types of abiotic stresses, drought is the most prevalent type of stress and is the main cause for limiting crop production in the arid zones of the world. A number of studies have indicated that the use of micronutrients can be used to improve the effects of drought stress. Studies conducted on hawthorns (*Crataegus* sp.) have revealed that application of different concentrations of Silicon (Si) NMs result in increased tolerance of plants to water stress(Ashkavand et al. 2015). Physiological and biochemical responses in seedlings of hawthorn vary with concentration of Si NMs at varied levels of dehydration stress. These results indicated that by pretreatment of Si NMs there is a positive impact on photosynthetic parameters, water content, proline, ion leakage of membranes, leaf pigments, melondialdehyde (MDA) and carbohydrate content. Experiments have shown that the use of Si NMs on two sorghums (*Sorghum bicolor* (L.) Moench) cultivars having differential susceptibility to drought conditions have shown enhanced drought tolerance irrespective of their drought susceptibility by decreasing shoot to root ratio and the maintenance rate of photosynthesis. Similar conclusion could be drawn for drought tolerance in case of sorghum by increasing their water uptake efficiency (Jalil and Ansari 2019). SiNMs have possibly been employed to alleviate the effects of stress due to water shortage. Pei et al. (2010) have performed a study using wheat which points out that the exposure of plants to sodium silicate (1.0 mM) could reasonably lessen the detrimental effects of stress due to drought. Application of Si NMs have been known to improve shoot growth, maintains water potential of leaves and enhances their chlorophyll content. Also, in wheat it decreases membrane lipid peroxidation(Pei et al. 2010). In germinated seeds, Zn NMs can enhance growth of radicle and its high content in grains can amplify seed viability particularly in areas having Zn-deficiency. Sedghi et al. (2013) have verified through experimentation that ZnO NMs have the potential to increase rate and percentage of seed germination in soybean in comparison to those under water stress. They further suggested that under water stress conditions, the application of ZnO NMs enhances resistance to drought, reduces dry and fresh weight of seeds which illustrates that they are effective in growth and germination of seedlings (Sedghi et al. 2013). Another essential micronutrient is 'iron' which has an important role to play in the growth and development of plants. Its deficiency is related to crucial changes in metabolic machinery and causes chlorosis. Studies have revealed that under water stress the most crucial effects of iron (Fe) NMs is on plant features for, e.g., the number of bolls/branch, the number of seeds/boll, thousand-seed weight and the probable produce. Application of Fe NMs directly on the leaves of the safflower cultivars exhibiting drought conditions nullifies its effects on the biomass and its oil content. Treatment of Fe NMs also augments biomass at two phases of granulation and flowering, even though it was improved at the flowering period than at the seed formation stage in comparison to the plants under water stress without the application of Fe NMs (Zareii et al. 2014). The deficiency of the micronutrient iron leads to chlorosis and forms a major constituent of several enzymes. Foliar application of iron NMs on safflower has been reported by Zareii et al. (2014) to reduce the undesirable effects of water stress (Zareii et al. 2014). Foliar application of titanium NMs on wheat diminishes the adverse effects of water stress showing agreeable results on its gluten and starch content. Results have indicated that the use of 0.02% TiO_2NMs show improvement in terms of many agronomic characters such as ear number, ear weight, plant height, seed number, thousand-seed weight, final yield, biomass, starch content, harvest index including gluten, under drought stress (Jaberzadeh et al. 2013). AgNMs have shown appreciable results in decreasing the negative effects of water stress on lentils (*Lens culinaris* Medic). A combined study of application of AgNMs and Poly Ethylene Glycol (PEG) in lentil seeds has shown positive results in terms of germination percentage and rate, root freshness and length and seed dry weight. Also, AgNMs use could be ascribed to decreasing water stress as well as loss of plant yield and growth (Hojjat and Ganjali 2016; Jalil and Ansari 2019). Analcite [($AlSi_2O_6$)-H_2O], a naturally occurring mineral has been used in wheat and corn seedlings by Zaimenko et al. (2014) to study its effect in improving resistance to drought. By their experimental study they described that the use of analcite NMs appreciably improved water stress by increasing photosynthetic pigments and buildup of antioxidants that are protective in nature (Zaimenko et al. 2014).

The plant root cell walls is the main entry point through which Ag NMs enter in plant cells (Fig. 23.6). Seghatoleslami et al. (2015) found that under drought use of silver (Ag) NMs had no role in enhancing water use efficiency and yield of *Carum copticum* plant (Seghatoleslami et al. 2015). Also, the NMs have been shown to hold on to root cells and trigger physical injury, obstruct pores and decrease root hydraulic conductivity causing decreased water absorption and capacity to uptake nutrients (Khan et al. 2017; Martínez-Fernández et al. 2016). Mingyu et al. (2007) studied the effect of nano-anatase TiO_2 in spinach by analyzing oxygen evolution and energy transfer in photosystem II (PSII) protein complex. It was analyzed that the appropriate TiO_2 concentration applied increases the absorbance of visible light and improves energy transfer among amino acids within PSII complex and improves energy transport to chlorophyll a (Chl a) from the tyrosine residue and oxygen evolution in PSII complex (Khan et al. 2017; Mingyu et al. 2007). In contrast, Kiapour et al. (2015) have shown experimentally that the use of TiO_2 NMs could not improve drought stress but enhanced the harmful effects of drought stress (Kiapour et al. 2015). Physiological and biochemical performance of common flax or linseed (*Linum usitatissimum* L.) has been analyzed. Researchers studied the effect of different concentrations of nano-anatase TiO_2 (0, 10, 100, and 500 mg l^{-1}) under drought stress conditions. They concluded from this study that the application of lower dose of nano TiO_2on the external plant surface helped in alleviating the damage induced due to drought in comparison to that using higher dose (Aghdam et al. 2016; Khan et al. 2017).

23.5.2 Salinity Stress

Major crop plant species in the world belong to lycophytes, which are also susceptible to salt stress. Therefore, under critical environmental conditions such as salt stress crop productivity can dwindle (Munns and Tester 2008). It impacts many physiological and biochemical processes related to plant development and production. Most common consequences of salinity stress are nutritional imbalance, ionic toxicity (salt stress), lowering of osmotic potential. Some of the essential plant processes which are severely affected by salt stress are photosynthesis, protein synthesis and lipid metabolism. the effect of a hydrophilic polymer coating (polyacrylamide) and plant growth regulators i.e. gibberellic acid (GA3) and salicylic acid (SA) were evaluated on the performance of dill (*Anethum graveolens* L.) plants under salinity conditions (Ioannou et al. 2020). The application of nano-silicon enhanced tolerance to salt stress in cherry tomato (*Solanum lycopersicum* var. *cerasiforme* Dunal) by activating antioxidant defense system (Tanveer et al. 2020). Salinity poses a big challenge to the overall plant productivity as it has the capacity to limit productivity of food crops. Thus, salinity stress is a hurdle in the path of sustainable food production. In present times, NMs are used as a pivotal tool for enhancing growth and productivity of all crop plants under adverse salt stress conditions. Excessive accumulation of NaCl in the soil causes salinity stress which causes ionic and osmotic stress in the plants. The ability of the plants to take up minerals and water to its apex is hampered by osmotic stress, while ionic stress caused due to buildup of Na^+ions in the cytosolic space leads to lower K^+/Na^+ ratio (Khan et al. 2012) and generation of ROS in large amount results into acute damage to biological molecules thus resulting into leakage of electrolytes affecting metabolic activities of cytosol (Ismail et al. 2014). Silicon (Si), the second most abundant element present in soil, considerably decreases salt stress and improves the process of seed germination and activities related to antioxidative enzymes, rate of photosynthesis and water content present in the leaves (Qados and Moftah 2015).

The application of nanofertilizers is a prospective method to deal with issues such as soil toxicity. Plants take up a very less concentration of chemical pesticides and fertilizers than the quantity that is being added to soil. The left over chemicals therefore remain unutilized, collect in the soil strata and result in soil toxicity. The foliar application of Si NMs (5–15 nm) in peregrina (*Jatropha integerrima* Jacq.) under salt stress enhanced the thickness of palisade, mesophyll, and spongy parenchyma (Pérez-Labrada et al. 2020). Si NMs and Si based fertilizers have experimentally been used to treat basil (*Ocimum basilicum* L.) under salt stress by showing promising effects on its vegetative features and morphological characteristics such as considerable increase in chlorophyll content, proline level and other growth and development indices. It was inferred that this could be due to increased tolerance to salt stress in basil (Kalteh et al. 2018). Some other studies have shown that the application of SiO_2NMshave increased leaf dry weight, leaf fresh weight, chlorophyll content, upregulation of antioxidant enzyme activity and proline accumulation under salt stress. Such studies have substantiated the fact that use of Si NMs is related to increase in the AS tolerance in plants (Kalteh et al. 2018). Use of Si NMs on genotypes of lentil (*Lens culinaris* Medik.) under salinity stress has resulted into major rise in seed germination, early seedling growth and other related traits in comparison to non-treated plants under salt stress. Therefore, in plants SiO_2NMs improve diverse defense mechanisms against salinity stress (Sabaghnia and Janmohammadi 2015). Under salt stress, increase in Na^+ ion concentration takes place which leads to ionic toxicity in turn causing decreased crop growth and productivity. Studies performed in maize under salt stress indicate that after the application of SiO_2 nanomaterials, the fresh and dry weight of the plant shoot and its root increase (Jalil and Ansari 2019). Si NMs mitigate salt stress probably by decreasing the concentration of Na^+ ions by reducing their absorption by tissues of plants. Another experimental study in case of broccoli (*Brassica oleracea*) under salinity stress involves the use of multi-walled carbon nanotubes (CNTs) known to stimulate transportation of water by enhancing aquaporin transduction and the net assimilation of CO_2, changing the properties of salt stressed root plasma membrane slightly in order to increase growth (Martínez-Ballesta et al. 2016).

23.5.3 Temperature Stress

It is a prime cause affecting the growth and productivity of crop plants. It comprises- low temperature stress (cold/chilling stress) and high temperature stress. Chilling stress (0–15 °C) results from temperature low enough to cause injury in plant tissues without the formation of ice crystals, whereas high temperature stress is induced by the rise in temperature over a key threshold level for significant durationof time adequate enough to result in an irreversible damage to the plant processes.

23.5.3.1 Chilling Stress

Very low temperatures have been known to cause chilling stress capable of damaging plant cells and tissues (Hasanuzzaman et al. 2013). It causes distortion of cell wall permeability and ion leakage from the membrane, which impacts the plant negatively by reducing its growth and germination (Jalil and Ansari 2019). Plant species have the ability to adapt and respond to different kinds of chilling stress according to their tolerance ability (Heidarvand et al. 2011; Jalil and Ansari 2019). TiO_2NMs have got the capacity to decrease the negative effects of extreme low temperature stress by reducing the damage caused by ion leakage from the membrane (Mohammadi et al. 2013). Photosynthesis, an integral plant mechanism is vulnerable to chilling stress. Plant photosystems are affected by chilling stress in a number of ways such as reduction in their chlorophyll content, rate of transpiration, CO_2 assimilation and degradation of Rubisco (photosystem enzyme) (Liu et al. 2012). In the plant photosystem, NMs increase production of the enzyme Rubisco (Jalil and Ansari 2019), ability of chloroplasts to capture light (Ze et al. 2011) and inhibiting ROS production (Giraldo et al. 2014). In the presence of TiO_2NMs there is a spike in the production of chlorophyll binding protein gene expression and Rubisco (Hasanpour et al. 2015), antioxidant enzyme activity (Mohammadi et al. 2014), susceptibility to chilling stress and leaf pigments. Plants suffering with chilling stress have increased levels of dehydroascorbate reductase, glutathione reductase and monodehydroascorbate reductase activities that scavenge ROS and upregulated *MeCu/ZnSOD* and *MeAPX2* genes, which results in decrease in oxidative stress, i.e., chlorophyll degradation, lipid peroxidation, H_2O_2 generation and finally increased levels of stress tolerance (Xu et al. 2014). Though, exposure of plants to NMs along with extreme low temperatures have shown improved biochemical physiognomies and growth profile in presence of chilling stress (Haghighi et al. 2014; Kohan-Baghkheirati and Geisler-Lee 2015).

23.5.3.2 Heat Stress

Stress caused due to heat or high temperature at such a critical point for a long duration of time so as to result in an irreversible loss to the growth and developmental activities of plants is called heat stress. Extreme temperatures have the ability of triggering off the generation of ROS and production of oxidative stress, which causes lipid membrane degeneration and ion leakage from the membrane, preceded by protein degradation (Karuppanapandian et al. 2011) in addition to reduction in chlorophyll content and the rate of photosynthesis (Prasad et al. 2011). Se NMs when used in low concentrations lessen the effect of stress caused by heat by improving chlorophyll content of plants, hydration ability and plant development (Haghighi et al. 2014). At low concentrations, Se NMs exhibit antioxidative properties to plants; on the other hand their high concentrations induce oxidative stress (Hasanuzzaman et al. 2014). In the duration of heat stress, some molecular chaperones and heat shock proteins are synthesized by plants. Heat shock proteins are involved in heat stress resistance and aid other proteins in carrying on their constancy during the period of stress conditions. Studies have described that the multiwall CNTs play a role in the upregulation of gene expression of heat shock proteins, e.g., *HSP90* (Khodakovskaya et al. 2011). Also, in maize the exposure of CeO_2NMs results in disproportionate generation of H_2O_2 particles and thereafter the upregulation of *HSP70* (Zhao et al. 2012). In addition, the application of plants using TiO_2 NMs decreases the effect of heat stress by regulating the process of opening of stomata (Qi et al. 2013).

23.5.4 Heavy Metal Stress

Globally, heavy metal stress is a severe menace to crops plants due to increased levels of toxicity and reduced plants growth (Chibuike and Obiora 2014). Plant growth is hampered due to heavy metal toxicity in the soil since it results in deficiency of essential nutrients in the soil due to interruption of the process of uptake of essential minerals and suppression of enzyme activities (Capuana 2011). Addition of heavy metals enhances extensive production of ROS, thereby causing oxidative damage to the cell by changing cell structure, decreasing membrane permeability, and degrading proteins (Sharma et al. 2012). On the other hand, plants have developed their defense systems in order to combat stress due to heavy metals. Also, plants produce polyphosphates, organic acids and metal-chelates which minimize the intake of heavy metals, activate antioxidant enzymes that scavenge ROS and release metal ions. Activation of the plant defense systems is an important requisite for developing resistance against heavy metals. Experimental results have shown that NMs are extremely successful in decreasing phytotoxicity induced by heavy metal stress (Tripathi et al. 2015). NMs being very small in their size and having large surface area can easily penetrate into plant cell wall and also possess high affinity towards heavy metals. Studies have shown that introduction of quantum dots (QDs) decreases access of Pb and Cu (Worms et al. 2012). Though, the entry of heavy metals through the plant cell wall (crossing the biophysical barriers and entering into the plant cell) have the ability to respond to stress by activating antioxidant enzymes and building up nutrients and biomolecules in order to combat heavy metal stress. Studies have suggested that the exposure to TiO_2NMs decreases cadmium toxicity, enhances photosynthetic rate and plant growth (Singh and Lee 2016). In brown mustard (*Brassica juncea* L. Czern.), hydroxyapatite NMs treatment decreases cadmium toxicity (Siddiqui et al. 2014). In pea, supplementing Si NMs in plant growth media lessens the toxicity of chromium (Tripathi et al. 2015). In one such recent study, the application of Si NMs in plants has ameliorated the effects on Cr induced toxic effects (Tanveer et al. 2020). In addition, treating cowpeas using gold ions stimulates the decrease of Au^{3+} to non-toxic gold NMs by phenolic compounds in germinating seeds (Shabnam et al. 2014).

23.5.4.1 Nanomaterials and Phytoremediation Against Heavy Metal Stress

NMs in addition to reducing metal-induced toxicity in the plants, act as effective adsorbents of metal ions and are essential in phytoremediation process of metal ion removal from the contaminated zones (Singh and Lee 2016). Studies have shown that maghemite nanotubes (NTs) can be used for the extraction of metal ions from water by acting as a potential magnetic nano-adsorbent (Roy and Bhattacharya 2012). Nano-composites made up of silica/graphite oxide act as a useful adsorbent for removing metal ions, while graphite oxide removes nearly 90% metal ions (Sheet et al. 2014). There are reports that strong reductant ions such as zerovalent iron (ZVI) NMs have the ability to decompose halogenated hydrocarbons (HCs) in water and soil (Khan et al. 2017). Use of ZVI NMs in farmyard manure improved Cr(VI) to Cr(III) reduction by giving rise to microbial activity causing reduced bioavailability of Cr(VI) by *Brassica juncea* plant (Table 23.1). The response of plants to NMs depends on the NMs applied and the type of plant species. Zn and ZnO NMs for example, considerably slow down the process of root growth and seed germination of cucumber, corn, ryegrass, radish, rape and lettuce while AlNMs support the growth of rape and radish root (Khan et al. 2017). Also, in presence of TiO_2NMs,microalgae *Microcystis aeruginosa* (Kützing) showed enhanced toxicity symptoms due to Cd when graphene oxide was introduced into the medium (Tang et al. 2015), while Singh and Lee (2016) reported that Cd toxicity decreased (Table 23.1). Other literature studies pointing to the fact that the root growth response of different plants varies with varying NMs can be observed in Table 23.1.

23.5.5 Other Stresses

In addition to above described types of stresses, plants do encounter a number of other stress types like flooding stress, post-harvest stress and UV-B radiation stress. These stresses damage flowers, vegetables and plants in a number of diverse ways.

23.5.5.1 Flooding Stress

Flooding stress is caused by an inadequate supply (up to 10^4-fold slower rate of diffusion) of oxygen (hypoxia) in water than in the air. Under hypoxic conditions, plant experiences energy deficient conditions, up-regulation of ethylene synthesis-related genes and decreased respiration rate (Khan et al. 2017). Hypoxic conditions hamper the process of seed germination, hypocotyl pigmentation and root growth (Komatsu et al. 2012) and vegetative and reproductive growth (Khan et al. 2017). Flooded plants are devoid of energy supply due to reduction in the number of Adenosine Triphosphate (ATP) molecules formed under hypoxic conditions, therefore in order to sustain energy level to run cellular process plants are enforced to switch to fermentation from their carbohydrate metabolism (Banti et al. 2013) and pyruvate decarboxylase and alcohol dehydrogenase genes are up-regulated during flooding stress (Mustafa et al. 2015b). NMs nevertheless, reduce flooding stress and enhance plant growth by hindering ethylene biosynthesis (Syu et al. 2014). Ag NMs treated plants may possibly experience less oxygen-deprivation conditions therefore, these genes are down-regulated and glyoxalate II production is decreased at the transcriptional level (Mustafa et al. 2015b) causing better growth of treated plants under flooding stress (Rezvani et al. 2012). A gel-free proteomic technique in soybean under flooding stress showed that Al_2O_3NMs performed better than ZnO and Ag in improving plant growth by regulating the process of energy metabolism and cell death (Mustafa et al. 2015a). Based on the information available on use of NMs in plant interaction during flooding stress, NMs play a significant part in reducing hypoxic conditions under flooding stress by altering metabolism and gene expression thereby leading to improved plant performance under flooding stress (Table 23.1).

23.5.5.2 Post-Harvest Stress

Transport of horticultural items to long-distances and different handling issues related to their commercial movement result in a number of post-harvest stresses. Some of these is the generation of ROS, reduction in the production of plant chlorophyll (chl) content, enhanced ethylene production, membrane damage, suppression in photosynthetic and antioxidant enzyme activities (Khan et al. 2017). All these factors collectively lead to modifications in the cellular machinery thus leading to senescence and abscission altogether affecting the commercial worth by reducing the shelf life of plant and plant products. Horticultural items suffering from post-harvest stress is attributed by oxidative stress, lipid peroxidation, increased respiration and water loss which affect weight and nutritive value of the product (Ouzounidou and Gaitis 2011). Applications of NMs in plants alter their physiological and biochemical processes and help in prevention of loss of fruits, vegetables and ornaments. Silver (Ag) ions are studied to restrict abscission process of flower buds and flowers (Khan et al. 2017). Use of Ag NMs and copper (Cu) NMs improve longevity of chrysanthemum flowers and shelf life of parsley leaves respectively (Kazemipour et al. 2013). Increased petal longevity has been reported by use of Ag NMs on oxidative stress induced by dark stress. It was also found experimentally that the combined application of thidiazuron and Ag NMs during storage in the dark decreased petal abscission in four cultivars of geranium (Hatami and Ghorbanpour 2014).

23.5.5.3 UV-B Radiation

Ultraviolet-B (UV-B, 280–315 nm) radiations which are non-ionizing and non-photosynthetically active radiations promote production of ROS in plant cells (Khan et al. 2017) that damage several cellular processes (e.g., photosynthesis), chloroplast structure and DNA as well (Hideg et al. 2013). In order to combat harmful UV-radiations plants have adapted to enzymatic and non-enzymatic antioxidant defense system by accumulation of phenolic compounds (Khan et al. 2017). NMs protect plant photosynthetic systems against UV-B stress by increasing chlorophyll content, improving Rubisco enzyme activity, process of light absorption, transformation and transport of light energy, suppressing oxidative stress and absorbing harmful UV-radiations (Table 23.1). In contrast, presence of NMs in the plant growth medium may stimulate the harmful effects of UV-radiations. For example, the application of CuO NMs alone had no harmful effects; however their use in combination with UV-radiation caused major negative influence on many biochemical and physiological aspects of western waterweed, i.e., *Elodea nuttallii* (Planch.) H. St. John (Regier et al. 2015). Likewise, another study in wheat plants pointed that the combination of Cd telluride-QDs and UV-B radiation reduced antioxidant enzyme activities, chlorophyll content and enhanced DNA damage (Chen et al. 2014).

23.6 Conclusions and Prospects

Nanotechnology has a number of applications in the agricultural sector like nanobiosensors, nanopesticides, nanofertilizers or as agents for environmental remediation. The response of application of NMs varies according to the plant species and may help to improve the growth and yield of crop plants. Based on the data available, it is clear that NMs help in decreasing the damage caused by the stress induced by abiotic factors by activating plants' defense mechanism. NMs being small in size allow easy penetration and regulation of water channels which promotes plant growth and germination of seeds. Increase in surface area promotes enhanced adsorption and directed delivery of materials. On the other hand, NMs are known to activate ROS generation along with many other toxic effects in plants. Increase in level of ROS by NMs can be linked to increase in stress signals that might trigger defense system of plants in more effective way. However, with regard to the mechanism of action of NMs, the data currently available is emerging and incoherent to sum up the complete mode of action. Though, on the basis of some of the studies and findings it can be said that NMs either imitate Ca^{2+}ions or other signaling molecules present in cytosol or some other NMs-specific proteins. As a result, induction of a range of signaling molecules promotes gene expression resulting into improved stress tolerance. However, a definite understanding about nanomaterials' fate and environmental impact/s will facilitate complete picture about the behavior, role, fate and ecotoxicity of NMs. Effectiveness and mechanism of action of NMs can be optimized by regulating their properties. Hence, extensive progress in the development of improved methods of synthesis will be of immense use to improve their efficacy. Further, knowledge at the field level and molecular studies based on a varied class of NMs in different plant species would be extremely helpful for implementation of nano-based strategies at a large-scale. In order to tackle existing bottlenecks in the sustainable development of agriculture, incorporating nanotechnology in new and emerging technologies is urgently required. Developments in the near future include incorporating nanoparticulate formulations having better selectivity and higher efficacy in the agricultural sector. These should be made easily available to the farmers at a viable rate. Since not much is known about the fate and toxicity of nanomaterials in the environment this aspect needs to be further researched. Use of nanotechnology in a balanced way that does not harm the environment or its components is the key to developing sustainable agriculture.

References

Acharya A, Pal PK (2020) Agriculture nanotechnology: translating research outcome to field applications by influencing environmental sustainability. NanoImpact 100232

Aghdam MTB, Mohammadi H, Ghorbanpour M (2016) Effects of nanoparticulate anatase titanium dioxide on physiological and biochemical performance of *Linum usitatissimum* (Linaceae) under well-watered and drought stress conditions. Braz J Bot 39(1):139–146

Almutairi ZM (2016) Effect of nano-silicon application on the expression of salt tolerance genes in germinating tomato ('*Solanum lycopersicum*' L.) seedlings under salt stress. Plant Omics 9(1):106

Anandaraj K, Natarajan N (2017) Effects of nanomaterials for seed quality enhancement in onion (*Allium cepa* (Linn) cv. CO (On)]5. Int J Curr Microbiol App Sci 6(11):3714–3724

Ashkavand P, Tabari M, Zarafshar M et al (2015) Effect of SiO_2 nanoparticles on drought resistance in hawthorn seedlings. For Res Pap 76(4):350–359

Banti V, Giuntoli B, Gonzali S et al (2013) Low oxygen response mechanisms in green organisms. Int J Mol Sci 14(3):4734–4761

Baxter A, Mittler R, Suzuki N (2014) ROS as key players in plant stress signalling. J Exp Bot 65(5):1229–1240

Boudsocq M, Sheen J (2013) CDPKs in immune and stress signaling. Trends Plant Sci 18(1):30–40

Capuana M (2011) Heavy metals and woody plants-biotechnologies for phytoremediation. IForest 4(1):7–15

Chandra S, Chakraborty N, Dasgupta A et al (2015) Chitosan nanoparticles: a positive modulator of innate immune responses in plants. Sci Rep 5:15195

Chen H, Gong Y, Han R (2014) Cadmium telluride quantum dots (CdTe-QDs) and enhanced ultraviolet-B (UV-B) radiation trigger antioxidant enzyme metabolism and programmed cell death in wheat seedlings. PloS One 9(10):e 110400

Chen J, Liu X, Wang C et al (2015) Nitric oxide ameliorates zinc oxide nanoparticles-induced phytotoxicity in rice seedlings. J Hazard Mater 297:173–182

Chibuike GU, Obiora SC (2014) Heavy metal polluted soils: effect on plants and bioremediation methods. Appl Environ Soil Sci 2014:752708

Das CK, Srivastava G, Dubey A et al (2016) Nano-iron pyrite seed dressing: a sustainable intervention to reduce fertilizer consumption in vegetable (beetroot, carrot), spice (fenugreek), fodder (alfalfa), and oilseed (mustard, sesamum) crops. Nanotechnol Environ Eng 1(1):2

Daudi A, Cheng Z, O'Brien JA et al (2012) The apoplastic oxidative burst peroxidase in *Arabidopsis* is a major component of pattern-triggered immunity. Plant Cell 24(1):275–287

Davletova S, Rizhsky L, Liang H et al (2005) Cytosolic ascorbate peroxidase 1 is a central component of the reactive oxygen gene network of *Arabidopsis*. Plant Cell 17(1):268–281

Du W, Yang J, Peng Q et al (2019) Comparison study of zinc nanoparticles and zinc sulphate on wheat growth: from toxicity and zinc biofortification. Chemosphere 227:109–116

Duque-Acevedo M, Belmonte-Ureña LJ, Cortés-García FJ et al (2020) Agricultural waste: review of the evolution, approaches and perspectives on alternative uses. Glob Ecol Conserv e00902

Duque AS, de Almeida AM, da Silva AB et al (2013) Abiotic stress responses in plants: unraveling the complexity of genes and networks to survive. In: Vahdati K, Leslie C (eds) Abiotic stress-plant responses and applications in agriculture. InTech, Croatia, pp 49–101

Emamverdian A, Ding Y, Mokhberdoran F et al (2015) Heavy metal stress and some mechanisms of plant defense response. Sci World J 2015:756120

Fischer R, Connor D (2018) Issues for cropping and agricultural science in the next 20 years. Field Crops Res 222:121–142

Frazier TP, Burklew CE, Zhang B (2014) Titanium dioxide nanoparticles affect the growth and microRNA expression of tobacco (*Nicotiana tabacum*). Funct Integr Genomic 14(1):75–83

García-Sánchez S, Bernales I, Cristobal S (2015) Early response to nanoparticles in the *Arabidopsis* transcriptome compromises plant defence and root-hair development through salicylic acid signalling. BMC Genom 16(1):341

Giannousi K, Avramidis I, Dendrinou-Samara C (2013) Synthesis, characterization and evaluation of copper based nanoparticles as agrochemicals against *Phytophthora infestans*. RSC Adv 3(44):21743–21752

Giraldo JP, Landry MP, Faltermeier SM et al (2014) Plant nanobionics approach to augment photosynthesis and biochemical sensing. Nat Mater 13(4):400–408

Haghighi M, Abolghasemi R, da Silva JAT (2014) Low and high temperature stress affect the growth characteristics of tomato in hydroponic culture with Se and nano-Se amendment. Sci Hortic 178:231–240

Hasanpour H, Maali-Amir R, Zeinali H (2015) Effect of TiO_2 nanoparticles on metabolic limitations to photosynthesis under cold in chickpea. Russ J Plant Physiol 62(6):779–787

Hasanuzzaman M, Nahar K, Fujita M (2013) Extreme temperature responses, oxidative stress and antioxidant defense in plants. In: Vahdati K, Leslie C (eds) Abiotic stress-plant responses and applications in agriculture. InTech, Croatia, pp 169–205

Hasanuzzaman M, Nahar K, Fujita M (2014) Silicon and selenium: two vital trace elements that confer abiotic stress tolerance to plants. In: Emerging technologies and management of crop stress tolerance, Emerging technologies and management of crop stress tolerance, vol 1. Elsevier, USA, pp 377–422

Hatami M, Ghorbanpour M (2014) Defense enzyme activities and biochemical variations of *Pelargonium zonale* in response to nanosilver application and dark storage. Turk J Biol 38(1):130–139

Heidarvand L, Amiri RM, Naghavi M et al (2011) Physiological and morphological characteristics of chickpea accessions under low temperature stress. Russ J Plant Physiol 58(1):157–163

Hideg É, Jansen MA, Strid Å (2013) UV-B exposure, ROS, and stress: inseparable companions or loosely linked associates? Trends Plant Sci 18(2):107–115

Hojjat SS, Ganjali A (2016) The effect of silver nanoparticle on lentil seed germination under drought stress. Int J Farm Allied Sci 5(3):208–212

Imada K, Sakai S, Kajihara H et al (2016) Magnesium oxide nanoparticles induce systemic resistance in tomato against bacterial wilt disease. Plant Pathol 65(4):551–560

Ioannou A, Gohari G, Papaphilippou P et al (2020) Advanced nanomaterials in agriculture under a changing climate: the way to the future? Environ Exp Bot 104048

Iqbal M, Raja NI, Hussain M et al (2019) Effect of silver nanoparticles on growth of wheat under heat stress. Iran J Sci Technol A Sci 43(2):387–395

Ismail A, Takeda S, Nick P (2014) Life and death under salt stress: same players, different timing? J Exp Bot 65(12):2963–2979

Jaberzadeh A, Moaveni P, Moghadam HRT et al (2013) Influence of bulk and nanoparticles titanium foliar application on some agronomic traits, seed gluten and starch contents of wheat subjected to water deficit stress. Not Bot Horti Agrobo 41(1):201–207

Jalil SU, Ansari MI (2019) Nanoparticles and abiotic stress tolerance in plants: synthesis, action, and signaling mechanisms. In: Khan M, Iqbal R, Reddy PS, Ferrante A, Khan NA (eds) Plant signaling Molecules, plant signaling molecules. Elsevier, Woodhead Publishing, United Kingdom, pp 549–561

Kah M, Tufenkji N, White JC (2019) Nano-enabled strategies to enhance crop nutrition and protection. Nat Nanotechnol 14(6):532–540

Kalteh M, Alipour ZT, Ashraf S et al (2018) Effect of silica nanoparticles on basil (*Ocimum basilicum*) under salinity stress. J Chem Health Risks 4(3):49–55

Karuppanapandian T, Wang HW, Prabakaran N et al (2011) 2, 4-dichlorophenoxyacetic acid-induced leaf senescence in mung bean (*Vigna radiata* L. Wilczek) and senescence inhibition by co-treatment with silver nanoparticles. Plant Physiol Biochem 49(2):168–177

Kaveh R, Li Y-S, Ranjbar S et al (2013) Changes in *Arabidopsis thaliana* gene expression in response to silver nanoparticles and silver ions. Environ Sci Technol 47(18):10637–10644

Kazemipour S, Hashemabadi D, Kaviani B (2013) Effect of silver nanoparticles on the vase life and quality of cut chrysanthemum (*Chrysanthemum morifolium* L.) flower. Eur J Exp Bot 3(6):298–302

Khan M, Khan NA, Masood A et al (2016) Hydrogen peroxide alleviates nickel-inhibited photosynthetic responses through increase in use-efficiency of nitrogen and sulfur, and glutathione production in mustard. Front Plant Sci 7:44

Khan MIR, Asgher M, Khan NA (2014) Alleviation of salt-induced photosynthesis and growth inhibition by salicylic acid involves glycinebetaine and ethylene in mungbean (*Vigna radiata* L.). Plant Physiol Biochem 80:67–74

Khan MN, Mobin M, Abbas ZK et al (2017) Role of nanomaterials in plants under challenging environments. Plant Physiol Biochem 110:194–209

Khan MN, Siddiqui MH, Mohammad F et al (2012) Interactive role of nitric oxide and calcium chloride in enhancing tolerance to salt stress. Nitric Oxide 27(4):210–218

Khodakovskaya MV, de Silva K, Nedosekin DA et al (2011) Complex genetic, photothermal, and photoacoustic analysis of nanoparticle-plant interactions. Proc Natl Acad Sci 108(3):1028–1033

Kiapour H, Moaveni P, Habibi D (2015) Evaluation of the application of gibbrellic acid and titanium dioxide nanoparticles under drought stress on some traits of basil (*Ocimum basilicum* L.). Int J Agro Agric Res 6:138–150

Kim J-H, Oh Y, Yoon H et al (2015) Iron nanoparticle-induced activation of plasma membrane H+-ATPase promotes stomatal opening in *Arabidopsis thaliana*. Environ Sci Technol 49(2):1113–1119

Kohan-Baghkheirati E, Geisler-Lee J (2015) Gene expression, protein function and pathways of Arabidopsis thaliana responding to silver nanoparticles in comparison to silver ions, cold, salt, drought, and heat. Nanomater 5(2):436–467

Komatsu S, Hiraga S, Yanagawa Y (2012) Proteomics techniques for the development of flood tolerant crops. J Proteome Res 11(1):68–78

Kreuter J (2007) Nanoparticles—a historical perspective. Int J Pharm 331(1):1–10

Laware S, Raskar S (2014) Effect of titanium dioxide nanoparticles on hydrolytic and antioxidant enzymes during seed germination in onion. Int J Curr Microbiol App Sci 3(7):749–760

Lei Z, Mingyu S, Xiao W et al (2008) Antioxidant stress is promoted by nano-anatase in spinach chloroplasts under UV-B radiation. Biol Trace Elem Res 121(1):69–79

Liu Y, Qi M, Li T (2012) Photosynthesis, photoinhibition, and antioxidant system in tomato leaves stressed by low night temperature and their subsequent recovery. Plant Sci 196:817

Malandrakis AA, Kavroulakis N, Chrysikopoulos CV (2019) Use of copper, silver and zinc nanoparticles against foliar and soil-borne plant pathogens. Sci Total Environ 670:292–299
Marmiroli M, Imperiale D, Pagano L et al (2015) The proteomic response of *Arabidopsis thaliana* to cadmium sulfide quantum dots, and its correlation with the transcriptomic response. Front Plant Sci 6:1104
Martínez-Ballesta MC, Zapata L, Chalbi N et al (2016) Multiwalled carbon nanotubes enter broccoli cells enhancing growth and water uptake of plants exposed to salinity. J Nanobiotechnol 14(1):42
Martínez-Fernández D, Barroso D, Komárek M (2016) Root water transport of *Helianthus annuus* L. under iron oxide nanoparticle exposure. Environ Sci Pollut Res 23(2):1732–1741
Martínez-Fernández D, Vítková M, Bernal MP et al (2015) Effects of nano-maghemite on trace element accumulation and drought response of *Helianthus annuus* L. in a contaminated mine soil. Water Air Soil Pollut 226(4):101
Massad TJ, Dyer LA, Vega G (2012) Costs of defense and a test of the carbon-nutrient balance and growth-differentiation balance hypotheses for two co-occurring classes of plant defense. PLoS ONE 7(10):e47554
Mehta C, Srivastava R, Arora S et al (2016) Impact assessment of silver nanoparticles on plant growth and soil bacterial diversity. 3 Biotech 6(2):254
Miao Y, Xu J, Shen Y et al (2014) Nanoparticle as signaling protein mimic: robust structural and functional modulation of CaMKII upon specific binding to fullerene C60 nanocrystals. ACS Nano 8(6):6131–6144
Mickelbart MV, Hasegawa PM, Bailey-Serres J (2015) Genetic mechanisms of abiotic stress tolerance that translate to crop yield stability. Nat Rev Genet 16(4):237–251
Mingyu S, Xiao W, Chao L et al (2007) Promotion of energy transfer and oxygen evolution in spinach photosystem II by nano-anatase TiO_2. Biol Trace Elem Res 119(2):183–192
Mirzajani F, Askari H, Hamzelou S et al (2014) Proteomics study of silver nanoparticles toxicity on Oryza sativa L. Ecotoxicol Environ Saf 108:335–339
Mohammadi R, Maali-Amiri R, Abbasi A (2013) Effect of TiO_2 nanoparticles on chickpea response to cold stress. Biol Trace Elem Res 152(3):403–410
Mohammadi R, Maali-Amiri R, Mantri N (2014) Effect of TiO_2 nanoparticles on oxidative damage and antioxidant defense systems in chickpea seedlings during cold stress. Russ J Plant Physiol 61(6):768–775
Munns R, Tester M (2008) Mechanisms of salinity tolerance. Annu Rev Plant Biol 59651–59681
Mustafa G, Sakata K, Hossain Z et al (2015a) Proteomic study on the effects of silver nanoparticles on soybean under flooding stress. J Proteomics 122:100–118
Mustafa G, Sakata K, Komatsu S (2015b) Proteomic analysis of flooded soybean root exposed to aluminum oxide nanoparticles. J Proteomics 128:280–297
Najafi Disfani M, Mikhak A, Kassaee MZ et al (2017) Effects of nano Fe/SiO_2 fertilizers on germination and growth of barley and maize. Arch Agron Soil Sci 63(6):817–826
Oh M-M, Trick HN, Rajashekar C (2009) Secondary metabolism and antioxidants are involved in environmental adaptation and stress tolerance in lettuce. J Plant Physiol 166(2):180–191
Ouzounidou G, Gaitis F (2011) The use of nano-technology in shelf life extension of green vegetables. J Innov Econ Manag 2:163–171
Pasala RK, Khan MIR, Minhas PS et al (2016) Can plant bio-regulators minimize crop productivity losses caused by drought, heat and salinity stress? An integrated review. J Appl Bot Food Qual 89:113–125
Pei Z, Ming D, Liu D et al (2010) Silicon improves the tolerance to water-deficit stress induced by polyethylene glycol in wheat (*Triticum aestivum* L.) seedlings. J Plant Growth Regul 29(1):106–115
Peralta-Videa JR, Hernandez-Viezcas JA, Zhao L et al (2014) Cerium dioxide and zinc oxide nanoparticles alter the nutritional value of soil cultivated soybean plants. Plant Physiol Biochem 80:128–135
Pérez-de-Luque A (2017) Interaction of nanomaterials with plants: what do we need for real applications in agriculture? Front Environ Sci 5:12

Pérez-Labrada F, Hernández-Hernández H, López-Pérez MC et al (2020) Nanoparticles in plants: morphophysiological, biochemical, and molecular responses. In: Tripathi DK, Singh P, Chauhan V, Kumar D, Sharma S, Prasad SM, Dubey NK, Ramawat N (eds) Plant life under changing environment. Elsevier, India, pp 289–322

Prasad P, Pisipati S, Momčilović I et al (2011) Independent and combined effects of high temperature and drought stress during grain filling on plant yield and chloroplast EF-Tu expression in spring wheat. J Agron Crop Sci 197(6):430–441

Qados AMA, Moftah AE (2015) Influence of silicon and nano-silicon on germination, growth and yield of faba bean (*Vicia faba* L.) under salt stress conditions. J Exp Agric Int 509–524

Qi M, Liu Y, Li T (2013) Nano-TiO_2 improve the photosynthesis of tomato leaves under mild heat stress. Biol Trace Elem Res 156(1–3):323–328

Raina N, Sharma P, Slathia PS et al (2020) Efficiency enhancement of renewable energy systems using nanotechnology. In: Bhushan I, Singh V, Tripathi D (eds) Nanomaterials and environmental biotechnology. Springer, Cham, pp 271–297

Regier N, Cosio C, Von Moos N et al (2015) Effects of copper-oxide nanoparticles, dissolved copper and ultraviolet radiation on copper bioaccumulation, photosynthesis and oxidative stress in the aquatic macrophyte *Elodea nuttallii*. Chemosphere 128:56–61

Rezvani N, Sorooshzadeh A, Farhadi N (2012) Effect of nano-silver on growth of saffron in flooding stress. World Acad Sci Eng Technol 6(1):517–522

Rizwan M, Ali S, Rehman MZ et al (2019) Effect of foliar applications of silicon and titanium dioxide nanoparticles on growth, oxidative stress, and cadmium accumulation by rice (*Oryza sativa*). Acta Physiol Plant 41(3):35

Rossi L, Fedenia LN, Sharifan H et al (2019) Effects of foliar application of zinc sulfate and zinc nanoparticles in coffee (*Coffea arabica* L.) plants. Plant Physiol Biochem 135:160–166

Roy A, Bhattacharya J (2012) Removal of Cu (II), Zn (II) and Pb (II) from water using microwave-assisted synthesized maghemite nanotubes. Chem Eng J 211:493–500

Sabaghnia N, Janmohammadi M (2015) Effect of nano-silicon particles application on salinity tolerance in early growth of some lentil genotypes/Wpływ nanocząstek krzemionki na tolerancję zasolenia we wczesnym rozwoju niektórych genotypów soczewicy. Annales UMCS, Biologia 69(2):39–55

Sedghi M, Hadi M, Toluie SG (2013) Effect of nano zinc oxide on the germination parameters of soybean seeds under drought stress. Ann West Univ Timisoara. Ser Biol 16(2):73

Seghatoleslami M, Feizi H, Mousavi G et al (2015) Effect of magnetic field and silver nanoparticles on yield and water use efficiency of *Carum copticum* under water stress conditions. Pol J Chem Technol 17(1):110–114

Shabnam N, Pardha-Saradhi P, Sharmila P (2014) Phenolics impart Au^{3+}-stress tolerance to cowpea by generating nanoparticles. PLoS ONE 9(1):e85242

Shang Y, Hasan M, Ahammed GJ et al (2019) Applications of nanotechnology in plant growth and crop protection: a review. Molecules 24(14):2558

Sharma P, Jha AB, Dubey RS et al (2012) Reactive oxygen species, oxidative damage, and antioxidative defense mechanism in plants under stressful conditions. J Bot 2012:217037

Sheet I, Kabbani A, Holail H (2014) Removal of heavy metals using nanostructured graphite oxide, silica nanoparticles and silica/graphite oxide composite. Energy Procedia 50(5):130–138

Shenashen M, Derbalah A, Hamza A et al (2017) Antifungal activity of fabricated mesoporous alumina nanoparticles against root rot disease of tomato caused by *Fusarium oxysporium*. Pest Manag Sci 73(6):1121–1126

Sicard C, Perullini M, Spedalieri C et al (2011) CeO_2 nanoparticles for the protection of photosynthetic organisms immobilized in silica gels. Chem Mater 23(6):1374–1378

Siddiqui MH, Al-Whaibi MH, Faisal M et al (2014) Nano-silicon dioxide mitigates the adverse effects of salt stress on Cucurbita pepo L. Environ Toxicol Chem 33(11):2429–2437

Singh J, Lee B-K (2016) Influence of nano-TiO_2 particles on the bioaccumulation of Cd in soybean plants (*Glycine max*): A possible mechanism for the removal of Cd from the contaminated soil. J Environ Manag 170:88–96

Slomberg DL, Schoenfisch MH (2012) Silica nanoparticle phytotoxicity to *Arabidopsis thaliana*. Environ Sci Technol 46(18):10247–10254
Soares C, Branco-Neves S, de Sousa A et al (2018) SiO_2 nanomaterial as a tool to improve *Hordeum vulgare* L. tolerance to nano-NiO stress. Sci Total Environ 622:517–525
Soliman AS, El-feky SA, Darwish E (2015) Alleviation of salt stress on *Moringa peregrina* using foliar application of nanofertilizers. J Hortic For 7(2):36–47
Syu Y-y, Hung J-H, Chen J-C et al (2014) Impacts of size and shape of silver nanoparticles on *Arabidopsis* plant growth and gene expression. Plant Physiol Biochem 83:57–64
Tang Y, Tian J, Li S et al (2015) Combined effects of graphene oxide and Cd on the photosynthetic capacity and survival of *Microcystis aeruginosa*. Sci Total Environ 532154–161
Tanveer M, Shahzad B, Ashraf U (2020) Nanoparticle application and abiotic-stress tolerance in plants. In: Durgesh KT, Vijay PS, Devendra KC, Shivesh S, Sheo MP, Nawal KD, Naleeni R (eds) Plant life under changing environment. Elsevier, pp 627–641
Tirani MM, Haghjou MM, Ismaili A (2019) Hydroponic grown tobacco plants respond to zinc oxide nanoparticles and bulk exposures by morphological, physiological and anatomical adjustments. Funct Plant Biol 46(4):360–375
Torabian S, Zahedi M, Khoshgoftar AH (2016) Effects of foliar spray of two kinds of zinc oxide on the growth and ion concentration of sunflower cultivars under salt stress. J Plant Nutr 9(2):172–180
Tripathi DK, Singh VP, Prasad SM et al (2015) Silicon nanoparticles (SiNp) alleviate chromium (VI) phytotoxicity in *Pisum sativum* (L.) seedlings. Plant Physiol Biochem 96:189–198
Tripathi DK, Tripathi A, Singh S et al (2017) Uptake, accumulation and toxicity of silver nanoparticle in autotrophic plants, and heterotrophic microbes: a concentric review. Front Microbiol 8:7
UN (2017) United Nations department of economics and social affairs, Population Division
Usman M, Farooq M, Wakeel A et al (2020) Nanotechnology in agriculture: current status, challenges and future opportunities. Sci Total Environ 721:137778
Van Aken B (2015) Gene expression changes in plants and microorganisms exposed to nanomaterials. Curr Opin Biotech 33:206–219
Vannini C, Domingo G, Onelli E et al (2013) Morphological and proteomic responses of Eruca sativa exposed to silver nanoparticles or silver nitrate. PLoS ONE 8(7):e68752
Vanti GL, Nargund VB, Vanarchi R et al (2019) Synthesis of Gossypium hirsutum-derived silver nanoparticles and their antibacterial efficacy against plant pathogens. Appl Organomet Chem 33(1):e4630
Vishwakarma K, Upadhyay N, Kumar N et al (2018) Potential applications and avenues of nanotechnology in sustainable agriculture. In Nanomaterials in plants, algae, and microorganisms. Elsevier, pp 473–500
Wang S, Wang F, Gao S et al (2016a) Heavy metal accumulation in different rice cultivars as influenced by foliar application of nano-silicon. Water Air Soil Pollut 227(7):228
Wang X, Yang X, Chen S et al (2016b) Zinc oxide nanoparticles affect biomass accumulation and photosynthesis in *Arabidopsis*. Front Plant Sci 6:1243
Wang Y, Lin Y, Xu Y et al (2019) Divergence in response of lettuce (var. ramosa Hort.) to copper oxide nanoparticles/microparticles as potential agricultural fertilizer. Environ Pollut Bioavail 31(1):80–84
Wei H, Wang E (2013) Nanomaterials with enzyme-like characteristics (nanozymes): next-generation artificial enzymes. Chem Soc Rev 42(14):6060–6093
World Bank, 2017. Global Economic Monitor (commodities)
Worms IA, Boltzman J, Garcia M et al (2012) Cell-wall-dependent effect of carboxyl-CdSe/ZnS quantum dots on lead and copper availability to green microalgae. Environ Pollut 167:27–33
Xu J, Yang J, Duan X et al (2014) Increased expression of native cytosolic Cu/Zn superoxide dismutase and ascorbate peroxidase improves tolerance to oxidative and chilling stresses in cassava (*Manihot esculenta* Crantz). BMC Plant Biol 14(1):208
Yadav T, Mungray AA, Mungray AK (2014) Fabricated nanoparticles: current status and potential phytotoxic threats. In: Whitacre D (ed) Reviews of Environmental Contamination and Toxicology

volume, Reviews of Environmental Contamination and Toxicology, vol 230. Springer, Cham, pp 83–110

Yolcu S, Ozdemir F, Güler A et al (2016) Histone acetylation influences the transcriptional activation of POX in *Beta vulgaris* L. and *Beta maritima* L. under salt stress. Plant Physiol Biochem 100:37–46

Zaimenko NV, Didyk NP, Dzyuba OI et al (2014) Enhancement of drought resistance in wheat and corn by nanoparticles of natural mineral analcite. Ecol Balk 6(1):1–10

Zareii DF, Arash R, Amir H (2014) Evaluation the effect of water stress and foliar application of Fe nanoparticles on yield, yield components and oil percentage of safflower (*Carthamus tinctorious* L.) Int J Adv Biol Biomed Res 2(4):1150–1159

Ze Y, Liu C, Wang L et al (2011) The regulation of TiO_2 nanoparticles on the expression of light-harvesting complex II and photosynthesis of chloroplasts of *Arabidopsis thaliana*. Biol Trace Elem Res 143(2):1131–1141

Zhao L, Peng B, Hernandez-Viezcas JA et al (2012) Stress response and tolerance of *Zea mays* to CeO_2 nanoparticles: cross talk among H_2O_2, heat shock protein, and lipid peroxidation. ACS Nano 6(11):9615–9622

Chapter 24
Nanomaterials in Combating Plant Stress: An Approach for Future Applications

Zahra Iqbal, Mohammad Israil Ansari, Anjuman Gul Memon, Garima Gupta, and Mohammed Shariq Iqbal

Abstract In a present scenario, the global agriculture system has been adversely affected by climatic changes. Apart from this, natural as well as human activities have added various xenobiotics type nanomaterials to our environment, leading to deprived crop productivity and food security. In order to overcome such problem, nano-engineering has been emerged as a noble tool for improving production of crops and food sustainability. However, at present nano-engineering techniques has been employed in the crop fields and gaining more and more consideration, but plant- nanomaterial interaction is still in early stages of study. Apart from this, the fortune and transformation of the nanomaterials within plants system is still the subject of consideration. As a result, many investigations have been carried out concerning nanomaterials and their applications in diverse fields together with nano-agricultural products like nano-fertilizers, nano-herbicides, and nano-pesticides. In addition, these nanomaterials can be used as a source of nano-remediation in the field of agriculture having various environmental issues such as sustainability in agriculture, management of plant diseases with crop protection, management of wastewater, limiting nutrient loss and most importantly, enhancing plant's ability towards abiotic stresses. However, nanomaterials when used at higher concentrations may cause toxic effects due to the production of free radical species, thereby affecting plant growth as well as development. Consequently, an extensive research is required at

Z. Iqbal
Department of Biochemistry, Faculty of Science, Chulalongkorn University, Bangkok 10330, Thailand

M. I. Ansari
Department of Botany, University of Lucknow, Lucknow 226 007, India

A. G. Memon
Department of Biochemistry, College of Medicine, Qassim University, Buraydah, Kingdom of Saudi Arabia

G. Gupta · M. S. Iqbal (✉)
Amity Institute of Biotechnology, Amity University Uttar Pradesh, Lucknow campus, Lucknow 226 028, India

G. Gupta
e-mail: msiqbal@lko.amity.edu

J. M. Al-Khayri et al. (eds.), *Nanobiotechnology*,
https://doi.org/10.1007/978-3-030-73606-4_24

cellular, sub-cellular and molecular level so as to define the performance of nanomaterials in restraining the plants against various stress factors. Thus, the present chapter provides an overview on the role of nanotechnology in sustainable agricultural practices with an understanding of plant-nanomaterials interaction, in order to better understand their implications with respect to stress physiology of plants for their future applications in plant science.

Keywords Abiotic stresses · Agriculture · Nanomaterials · Nanotechnology · Plant stress

24.1 Introduction

It has been observed that high yield production from crop plants primarily depends upon the soil quality, water availability and various other environmental conditions like light, temperature and so on (Hakeem 2015; Meena et al. 2017; van der Laan et al. 2017). Any fluctuations in the aforesaid environmental conditions may affect the yield of plant, leading to diminished productivity (Daryanto et al. 2017; Donfouet et al. 2017; He et al. 2017). However, climatic variations together with human involvements resulted in severe declined in soil fertility, thereby causing decrease in crop productivity (Chen and Lackner 2017; Khan and Akhtar 2015). Apart from this, due to the practice of outdated techniques for fertilization as well as irrigation including other methods of agriculture has been led to escalate this issue (Manyi-Loh et al. 2018; van der Laan et al. 2017). The remedy for such a problem is the usage of novel technologies in fertilization and irrigation with identifying plant nutrition, which is less toxic to food crops (Bargaz et al. 2018). Apart from this, the advancement of new strains of plants species that can tolerate biotic and abiotic stresses are needed (Rejeb et al. 2014). As abiotic factors are concern, nanoparticles that are naturally synthesized are common element of biological systems that have various structures and broad range of biological roles (Hedayati et al. 2016). Amongst novel technologies, nanotechnology is a developing and fast emerging field of science that can assist in improving these stress parameters by various methods such as development of competent antioxidant system and fertilizers that are more competent and less harmful (Zuverza-Mena et al. 2017). Abiotic and biotic stress factors responsible for plant stress have be reported by various researchers (Abiri et al. 2017; Calanca 2017; Pandey et al. 2017; Wani et al. 2016). The abiotic stresses that involve salinity, flooding, drought, ultraviolet radiation, freezing and chilling including various other factors leads to worldwide damage in crop production (He et al. 2018; Li et al. 2017; Wani et al. 2016). Several researchers have reported plant abiotic stress, including other stresses as a result of adverse environmental conditions, such as temperature fluctuations, high intensity UV rays, drought conditions, freezing, salinity and heavy metal concentrations and hypoxia (Ahmad and Prasad 2012; Hirt and Shinozaki 2004; Pandey et al. 2015; Waqas et al. 2019). Moreover, recent research on plant abiotic stress at cellular, sub-cellular and molecular level have also been reported (Asensi-Fabado et al. 2017; Iqbal et al. 2020; Mudalkar et al. 2017; Rossini et al.

2016; Wang et al. 2017). Due to cumulative chances of climate variation around the world and numerous abiotic stresses, pressing need is for adaptation and modification of plant abiotic stress factors (Grover et al. 2011; Raza et al. 2019; Wani et al. 2016). Thus, an intense exploration on plant growth and developmental stages due to diverse effect of abiotic stresses needs to be identified at morpho-physiological, biochemical and molecular levels (Hasanuzzaman et al. 2013; Wani et al. 2016). Diverse plant mechanisms in response of various abiotic stresses together with safeguarding their growth and development would be of remarkable importance (Zhu 2016). Though, the usage of nanomaterials is evolving, but there is need to come up with significant solutions for plants response towards abiotic stresses with respect to previously accepted solutions of plant abiotic stresses (de la Rosa et al. 2017; Hatami et al. 2016; Iqbal et al. 2020; Reddy et al. 2016) (Fig. 24.1).

Owing to their distinctive physiognomies, nanomaterials have been already used in numerous applications comprising of industrial, medicinal and agricultural divisions (Jeevanandam et al. 2018; Khan et al. 2017; Patra et al. 2018; Servin and White 2016). Regarding the agricultural usage of nanomaterials, it has been observed that an upsurge and indefinite use of nanomaterial was found; such as in nanopesticides (Chhipa and Joshi 2016; Kah and Hofmann 2013), nanofertilizers (Ahmed et al. 2019; Chhipa and Joshi 2016; Tarafdar et al. 2014), nanosensors for nano-farming (Bogue 2009; Chhipa and Joshi 2016; Yılmaz et al. 2017), soil nano-reclaimants

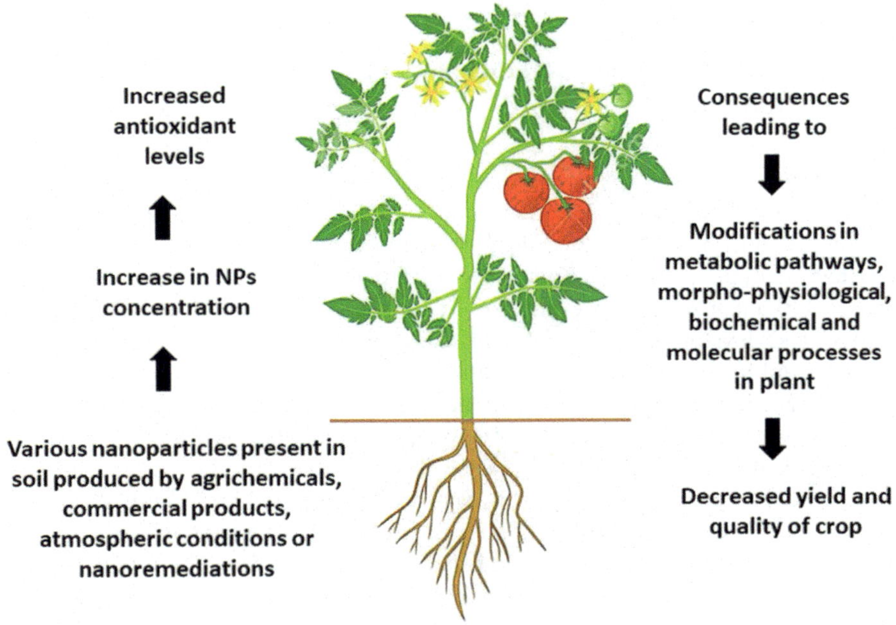

Fig. 24.1 Probable sources of nanoparticles (NPs) and its effect on plant with respect to its yield and productivity. (Figure constructed by Mohammed Shariq Iqbal)

(Floris et al. 2017; Patra et al. 2016), soil and water nanoremediators (Gomes et al. 2016; Gil-Díaz et al. 2017; Paul et al. 2018). The usage of such nanomaterials might assist in limiting the utilization of various agro-chemicals such as fertilizers and pesticides, which may lead to curtail environmental pollution thereby, helping in sustainable agriculture (Panpatte et al. 2016; Shang et al. 2019). Concerning the connection between plant stress and nanomaterials, various investigations have been reported elucidating more information about their relationship (Kole and Banerjee 2016; Khan et al. 2017; Zaytseva and Neumann 2016). The foremost notable outcome of plant abiotic stress is oxidative stress, which is in general first mode of action of defense by the plants (Servin and White 2016). Due to oxidative stress, nanomaterials could assist plants under stress in improving their defense system that comprises of enzymatic antioxidants such as catalase (CAT), superoxide dismutase (SOD), and peroxidase (POX) (Patra et al. 2016). However, when these nanomaterials used in higher concentration may leads to oxidative stress in plants, which is due to the production of reactive oxygen or nitrogen species (ROS or RNS) that leads to damage the cell or its organelles (Iqbal et al. 2018; Khan et al. 2017; Zaytseva and Neumann 2016). Thus, plants and nanomaterials interactions under abiotic stress must be explored under various morpho-physiological, biochemical and molecular levels. Therefore, the objective of the present chapter is to provide better understanding of nanoparticles-plant interactions with plant's stress physiology, which could deliver the future application of nanoparticles in sustainable agricultural practices.

24.2 Plants Uptake, Translocation and Biological Effects Nanoparticle

Nanotechnology implicational strategies in plants requires a precautionary but exact assessment of interactions between plant and nanoparticles (Shang et al. 2019). It could deliver the understanding for the mechanisms of nanoparticles uptake, its translocation within the plant body and finally its accumulation (Fig. 24.2). It also further provides the evaluation of probable confrontational effects on the growth and development of plant (Behzadi et al. 2017). Nanoparticles uptake by plants is unforeseeable, as various factors are responsible for it such as net charge, surface functionalization, chemical composition, and shape and size of nanoparticles (Gupta and Xie 2018). Apart from this, several other factors are also responsible for the effect of nanoparticles on plants, like application routes, interactions with components in environment like soil type, water accessibility and microorganism present in the soil (Remedios et al. 2012). However, it was reported that plant's anatomy, morphology physiology and individual plant species differently interact with nanoparticles (Remedios et al. 2012). It was reported in previous studies that easy tracking and detection using microscopy techniques facilitated us to understand the uptake and interaction of metal oxide nanoparticles and metal ions in plants (Plascencia-Villa et al. 2012). However, the interaction of nanoparticles with

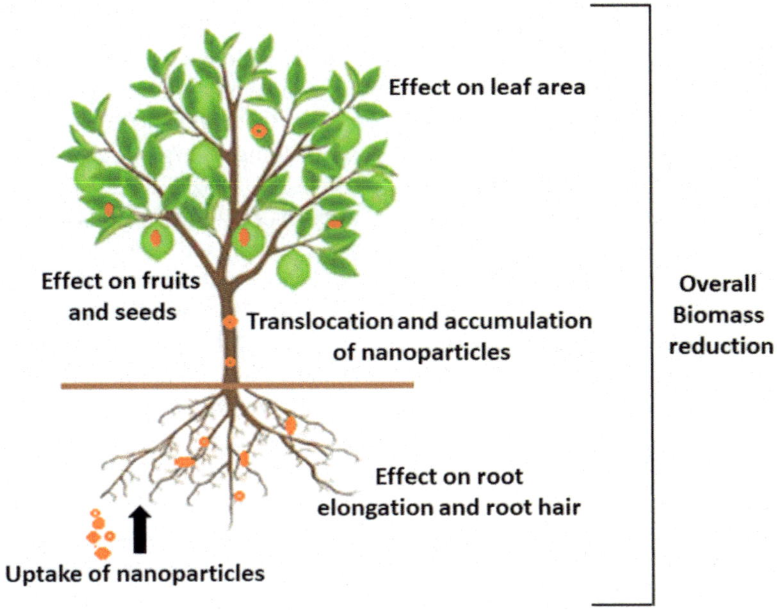

Fig. 24.2 Nanoparticles uptake, translocation and accumulation in plants and its biological effects (Figure reconstructed by Mohammed Shariq Iqbal based on Yan and Chen 2019)

plants a handful of information is available, but only limited information is available about the types of nanoparticles, i.e., shape, size, charge and coating interaction with plants is concerned (García-Gómez et al. 2018; Moon et al. 2016; Song et al. 2013; Vidyalakshmi et al. 2017; Zhu et al. 2012).

24.3 Application of Nanomaterials in Agroecosystems

The physicochemical characteristic of nanomateriales or nanoparticles like particle morphology, surface area, pore size, and high reactivity, led the pavement to nanotechnology for its novel application in agro based i ndustry (Ahmad et al. 2020; Gatoo et al. 2014; Jeevanandam et al. 2018; Mani and Mondal 2016; Panpatte et al. 2016; Servin and White 2016). The application of nanotechnology, with respect to agricultural sectors can be useful in various ways, which includes plant protection (Li et al. 2017), remediation of terrestrial environments (Gomes et al. 2016; Gil-Díaz et al. 2017; Guerra et al. 2018; Khan et al. 2014; Patra et al. 2016;), fertilization sector (Chhipa and Joshi 2016; Derosa et al. 2010; Dubey and Mailapalli 2016; Mani and Mondal 2016; Panpatte et al. 2016; Shang et al. 2019), food sector (Ghanbarzadeh et al. 2016; Li et al. 2017; Rashidi and Khosravi 2011). The nanomaterials or nanoparticles could improve the capability of up-taking nutrients by the plants, thereby proliferating the fertility of the soil along with enhanced crop production. Additionally, soil affected due to increased concentration of salts could be managed by the use these nanomaterials, which could bring better crop productions (Patra et al. 2016). The environments for agroecosystem comprises of water, soil, air, plants, sediments, microbes and various other components, which are generally affected by various biotic or abiotic stresses. As a result of these stresses, agricultural production is affected and gets deteriorated. Considering this, various studies have been carried out on these stress factors such as marginal lands, soils affected due to high salt concentrations, wetlands, water and soils pollution, agroforests, changes in climatic conditions, and bushfires (Amini et al. 2016; Smith et al. 2016). The afore-mentioned agroecosystem has numerous stresses like soils affected due to high salt concentrations because of drought stress and salinity stress, land pollutions leading to oxidative stress, changes in climatic conditions because of flooding or drought and salinity (Gupta et al. 2020; Omena et al. 2019; Rizwan et al. 2017; Venkat-achalam et al. 2016). One of the serious and extreme environmental challenges is salinity, which causes extreme decline in crop productions. However, it is considered as salinity to be the major factor in limiting crop production in dry lands or semiarid lands around the globe (Ashraf et al. 2009). It is very well recognized that, a great danger to land efficiency could be due to the presence of salts that are soluble in water as well as surface water and underground water. The poor agricultural productivity was also observed due to high concentration of salt content in plants supplied from soil or water, thereby limiting the uptake of essential nutrients in plants. Additionally, soil affected due to high salts concentration leads to various problems comprising of high Na^+ concentration, nutrients loss, poor porosity, waterlogging and water limitation in plants. Various techniques have been applied to manage lands that are salt affected by the process of chemical reclamation and as a modern approach by nano-materials (Patra et al. 2016). Thus, nanomaterials based reclamation of soil is now a fast developing approach for soils that are affected with high salt concentrations. The nanomaterials-based reclamation of soil by the use of nano-calcium, nano gypsum is more effective and easily producible, which can improve hydraulic property of plants and better stability of soil (Mukhopadhyay and Kaur 2016). It has been reported by Patra et al. (2016) that by using nanomaterial could help in soil reclamation that are affected by high salt concentrations. Hence, it could be elucidated that, nano-materials generally have an influence on various agroecosystems having negative as well as positive properties. With respect to negative properties, nanomaterials might affect plant with toxic effect, also affect microbes present in soil, which are useful for soil as well as for plants and other organisms found in water. When, positive factors are observed nanomaterials can be used for the remediation of soil and water pollution, nano-nutrients uptake by the plants, and limiting the adverse effect on plants due to abiotic stresses. Therefore, a prudent and maintainable approach in sustainable agroecosystems must be maintained by the use of nanomaterials. The futuristic approaches of nanotechnology in combating against plant stress condition is detailed in Fig.24.3.

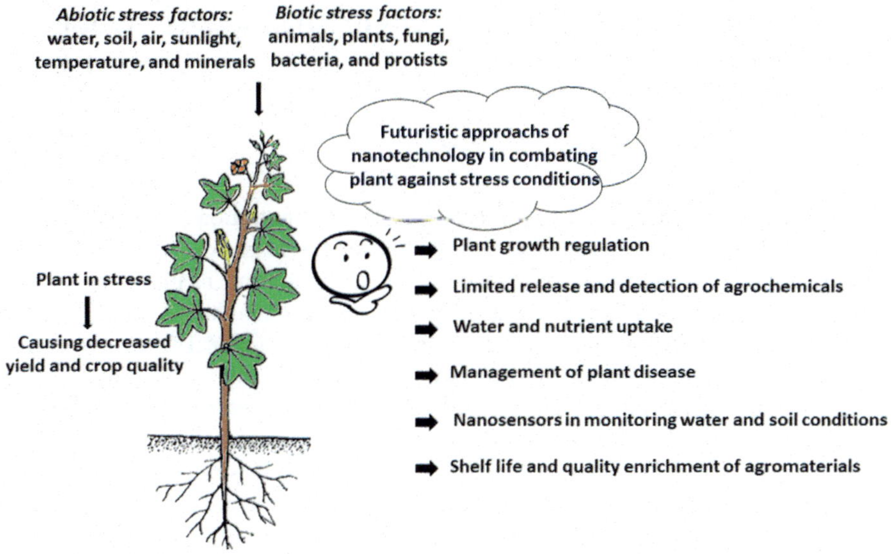

Fig. 24.3 Various approaches of nanotechnology in stress management of plants in near future (Figure constructed by Mohammed Shariq Iqbal)

24.4 Limiting Effect of Nanomaterials in Plant Stress

It is being observed in various reports that nanomaterials could be employed for sustainable crop production by limiting the loss of nutrients, and inhibiting plant diseases, thereby improving yields of crop plants (Khan et al. 2017). It was reported that nanomaterials hold numerous benefits including improvement in plant growth and developmental stages, initializing from germination of seeds, seedling formation, root initiation and growth, and photosynthesis up to flowering stage, under low concentrations of nanomaterials used (Kole and Banerjee 2016). Regarding the defense of plants, against oxidative stress, the nanomaterials may well perform or simulate the role of enzymatic antioxidants like peroxidase (POX), superoxide dismutase (SOD) and catalase (CAT) (Zaytseva and Neumann 2016). The use of nanomaterials in higher concentrations might be phyto-toxic, due to the formation of reactive oxygen species (ROS) and its accumulation might damage the cell with respect to cell membrane, nucleic acids and proteins (Khan et al. 2017).

24.5 Exploitation of Nanomaterials in Limiting the Abiotic Stress

24.5.1 Silica Nanomaterial (SiO_2)

Plants generally possess silicon (Si) in substantial concentrations (approx. 1 to 10% of the dry matter), or sometimes even in higher amounts in different plant species. The Si uptake capability of roots leads to its accumulation in plants at different levels causing variation in Si amounts (Parveen and Hussain 2008). Silicon, being a constructive element for plants, which delivers substantial benefits in several ionic conformations. Silicon nanomaterials plays a significant role in alleviation of salt stress. Several reports about the capability of silicon nanomaterials to counteract the adverse effects of salt stress on plant growth and development rates were documented (Wang et al. 2010). Therefore, the usage of nano-silica could more effective than normal silica, which enables nano-silica advantageous and to be used by root system of plant. The leaves relative water content can be enhanced by the use of nano-silica as it helps in retaining xylem humidity, translocation of water and maintains turgor pressure by using water efficiently.

24.5.2 Silver Nanomaterials (Ag^+)

The prospect of the use of silver nanoparticles for enhancement of crop plants productivity was evaluated by several researchers. It was analyzed for growth of plant and improving photosynthesis by enhancing chlorophyll content (Hatami and Ghorbanpour 2013; Shang et al. 2019; Shelar and Chavan 2015; Vannini et al. 2013). The antimicrobial activity of silver nanoparticles is been used to manage various plant diseases, caused by microbes (Lamsal et al. 2011). It was studied that silver nanoparticles have been used quiet efficiently in enhancing the germination rate against salinity stress of cumin and fennel (Ekhtiyari and Moraghebi 2011). It was further reported by Almutairi (2016) that by the use of silver nanoparticles enhanced the resistance of tomato seeds under salt stress (NaCl stress), which boosted the germination rate as well as fresh weight and dry weight of seedling and root length.

24.5.3 Zinc Oxide Nanomaterials (ZnO)

One of the important micronutrients for plants, needed for favorable growth and development is zinc (Zn) that helps in directing vital metabolic reactions and stimulates better growth and development for plant. Apart from this role of zinc in plants growth and development, it plays a significant role in limiting the heavy metal toxicity in plants, thus inhibiting plant from cadmium-heavy metal toxicity (Baybordi 2005).

It was reported by Cakmak (2008), that zinc plays a vital role in plants endurance against environmental stress, where it enhances the plant's resistance to tolerate against drought stress. As stated previously, micro element in minute size, i.e., in nano size may enables plants for enhanced uptake and as well as required in lesser quantity. Therefore, the role of Zn can be more efficiently accomplished by the use of Zn in nano size. It was reported by Seghatoleslami and Forutani (2015) that, the ability of the sunflower plant in using water under water stress was shown better in nano-ZnO, rather than with bulk ZnO, thereby increasing its yield. The results pointed out that the complete irrigation management led to the maximum biomass and seeds with bulk ZnO treatments, however under the water stress conditions the maximum biomass and seeds were found to be with ZnO nanoparticle treatment. Thus, it could be assumed that by the use of ZnO nanoparticle, the usage of water by the plant can be improved efficiently, and furthermore, yield of seeds can also be enhanced.

24.5.4 Titanium Oxide Nanomaterials (TiO_2)

Biologically plants are readily affected by titanium, which is favorable at low concentrations, however toxic at higher levels. Pesticides photo-catalytic degradation using TiO_2 as well as other catalysts had revealed a potential technique for water remediation (Lee et al. 2003). In a report by Akbari et al. (2014), titanium oxide nanomaterials can increase photosynthetic apparatus in plants, thereby enhancing the plant's capability to capture the sunlight, which could affect the production of pigments as well as the alteration of the light energy to energize electrons within the cell and, thus chemical activity, hence improving the photosynthetic efficiency in maize plant, specially under drought stress condition. TiO_2 nanomaterials was also detected in stimulating the developments in spinach plant, by increasing the photosynthetic rate and metabolism of nitrogen in the plant (Yang et al. 2006). It was reported by Mahmoodzadeh et al. (2013) in canola plant and again by Jaberzadeh et al. (2013) in wheat plant, that TiO_2 nanomaterial could improve nitrogen content and water uptake and furthermore it can excite enzymatic antioxidants like catalase (CAT), superoxide dismutase (SOD), and peroxidase (POX) within the plant. In another interesting study by Shallan et al. (2016) explored the effects of silica nanomaterial and titanium oxide nanomaterial on drought stressed cotton plants by estimating yield and chemical components of the plant The effects revealed that pretreatment with silica nanomaterial and TiO_2 nanomaterial on cotton plants subjected to drought stress, resulted in improved pigment content, sugar content, total phenolics content, soluble protein content, free amino acids content, proline, reducing power, antiradical ability and enzymatic antioxidant activity as well as improvement in yields. To conclude, it can be assumed that the foliar usage of silica nanomaterial and titanium oxide nanomaterial may perhaps enhance tolerance towards drought stress in cotton plants. However, the usage of TiO_2 nanomaterial in soybean plants can contest with cadmium (Cd) induced stress in plant which may be due to the establishment of new bonds between cadmium (Cd) and TiO_2 nanomaterial within the plant (Singh and Lee 2016). Thereby, to understand the role of nanomaterials in lower concentrations in improving the plant system against abiotic stress by initiating the signals to the cell of the plant through the formation of reactive nitrogen species (RNS) or/and reactive oxygen species (ROS) or by stimulating the enzymatic and non-enzymatic antioxidants (plant defense system), or by storing osomolytes, free amino acids and other nutrients required for plants (Khan et al. 2017). However, subjected to high concentrations, these nanomaterials might cause toxicological effect on the plants (Husen and Siddiqui 2014). It is noteworthy to mention that, usage of nanomaterials for the plants under stress condition, subjected to environmental factors, might boost the formation procedure of various free radicals i.e. RNS and ROS, thereby causing cell injury due to the oxidative stress so produced (Chichiriccò and Poma 2015; Khan et al. 2017).

24.6 Conclusion and Prospects

The agricultural plant production is facing extreme problems around the world, due to climatic changes, exhaustion of land and water resources, biotic and abiotic stresses and depletion of energy. Therefore, the appropriate way in encountering these issues must be made with consideration to be eco-friendly as well as to be more sustainable. Now-a-days, one of the most imperative and favorable techniques used to overcome these issues is nanotechnology. Through various studies, it has been elucidated that nanomaterials could be used in lessening the damage caused due to various abiotic stresses, as nanomaterials helps plant by activating their defense system in order to combat any stress conditions. It is very well implicit that because of the small size of the nanomaterials, it can readily infiltrate through plant tissues and can bind to the active site, where it is required within the cell. The nanomaterials, increases the surface area, thereby enhancing the abilities to be more operative in the process of adsorption as well as in the delivery targeted constituents by the cells and its organelles. It is also been reported that nanomaterials can regulate the uptake of water by the plant vascular tissue, thus enhancing the germination of seeds and improving plant growth and development. With consideration to the characteristic of nanomaterials in resisting the abiotic stresses of plant leading to various damages such as hindering growth and development of plant and/or soil and water toxicity, more research is needed at cellular and molecular levels to fully elucidate the effect of nanomaterials in combating against stress by the plants. Furthermore, as a future perspective, plant-nanoparticles interactions could deliver a base for manipulating desired genetic characters in plants. However, the uptake of nanoparticles, its translocation and accumulation is still completely unrevealed, thereby leading to ethical issues and safety perspectives. Thus, nanoparticles field application needs an improvement through further extensive research. It is important to reveal biomolecular interaction with nanoparticles, which causes regulation in expression of plant genes. Therefore, further research is desirable to fully elucidate the plant-nanoparticles interactions at molecular and sub-molecular levels which could be useful for a perky future of agro based industries.

References

Amini S, Ghadiri H, Chen C, Marschner P (2016) Salt-affected soils, reclamation, carbon dynamics, and biochar: a review. J Soils Sediments 16:939–953. https://doi.org/10.1007/s11368-015-1293-1

Ahmed S, Hassan AA, Salama DM et al (2019) Green synthesis of nanofertilizers and their application as a foliar for *Cucurbita pepo* L. J Nano. https://doi.org/10.1155/2019/3476347

Abiri R, Shaharuddin NA, Maziah M et al (2017) Role of ethylene and the APETALA 2/ethylene response factor superfamily in rice under various abiotic and biotic stress conditions. Env Exp Bot 134:33–44

Ahmad A, Ghouri M, Khan Z et al (2020) Nanotechnology: Transformation of agriculture and food security. Am J Biotech Biosci 3;19. https://doi.org/10.28933/ajbb

Akbari GA, Morteza E, Moaveni P et al (2014) Pigments apparatus and anthocyanins reactions of borage to irrigation, methylalchol and titanium dioxide. Int J Biosci 4(7):192–208

Almutairi Z M (2016) Influence of silver nanoparticles on the salt resistance of tomato (*Solanum lycopersicum* L.) during germination. Int J Agric Biol 18(2):449–457 https://doi.org/10.17957/ijab/15.0114

Ashraf M, Ozturk M, Athar H (2009) Salinity and water stress: improving crop efficiency. Springer, Netherlands ISBN: 1402090641, https://doi.org/10.1007/978-1-4020-9065-3

Asensi-Fabado MA, Amtmann A and Perrella G (2017) Plant responses to abiotic stress: The chromatin context of transcriptional regulation. Biochimica Biophysica Acta (BBA) - Gene Reg Mechan 1860(1):106–122

Ahmad P, Prasad MNV (2012) Abiotic stress responses in plants: metabolism, productivity and sustainability. Springer, New York, ISBN 978-1-4939-4192-6 doi:https://doi.org/10.1007/978-1-4614-0634-1

Bogue R (2009) Nanosensors: a review of recent research. Sensor Review - SENS REV. https://doi.org/29.10.1108/02602280910986539

Behzadi S, Serpooshan V, Tao W et al (2017) Cellular uptake of nanoparticles: journey inside the cell. Chem Soci Rev 46(14):4218–4244. https://doi.org/10.1039/c6cs00636a

Baybordi A (2005) Effect of zinc, iron, manganese and copper on wheat quality under salt stress conditions. J Water Soil 140:150–170

Bargaz A, Lyamlouli K, Chtouki M et al (2018) Soil microbial resources for impsroving fertilizers efficiency in an integrated plant nutrient management system. Front Microbiol 9:1606. https://doi.org/10.3389/fmicb.2018.01606

Calanca PP (2017) Effects of abiotic Stress in Crop Production. In: Ahmed M, Stockle CO (eds) Quantification of climate variability, adaptation and mitigation for agricultural sustainability. Springer International Publishing Switzerland, pp 165–180. https://doi.org/10.1007/978-3-319-32059-5_8

Cakmak I (2008) Enrichment of cereal grains with zinc: agronomic or genetic biofortification? Plant Soil 302(1–2):1–17

Chen WYTS, Lackner M (2017) Handbook of climate change mitigation and adaptation. Springer International Publishing Switzerland, ISBN 978-3-319-14408-5 https://doi.org/10.1007/978-3-319-14409-2

Chhipa H, Joshi P (2016) Nanofertilisers, nanopesticides and nanosensors in agriculture. In: Ranjan S, Dasgupta N, Lichtfouse E (eds) Nanoscience in food and agriculture 1, sustainable agriculture reviews, 20. Springer International Publishing, Switzerland pp 247–282. https://doi.org/10.1007/978-3-319-39303-2

Chichiriccò G, Poma A (2015) Penetration and toxicity of nanomaterials in higher plants. Nanomater 5:851–873
Derosa M, Monreal C, Schnitzer M et al (2010) Nanotechnology in fertilizers. Nat Nanotechnol 5:91. https://doi.org/10.1038/nnano.2010.2
Dubey A, Mailapalli DR (2016) Nanofertilisers, nanopesticides, nanosensors of pest and nanotoxicity in agriculture. In: Lichtfouse E (ed) Sustainable agriculture reviews. Dimension publications 19:307–330 https://doi.org/10.1007/978-3-319-26777-7_7
Donfouet HPP, Barczak A, Détang-Dessendre C, Maigné E (2017) Crop production and crop diversity in France: a spatial analysis. Ecol Eco 134:29–39. https://doi.org/10.1016/j.ecolecon.2016.11.016
Daryanto S, Wang L, Jacinthe PA (2017) Global synthesis of drought effects on cereal, legume, tuber and root crops production: a review. Agri Water Manag 179:18–33
de la Rosa G, García-Castañeda C, Vázquez-Núñez E et al (2017) Physiological and biochemical response of plants to engineered NMs: implications on future design. Plant Physiol Biochem 110:226–235
Ekhtiyari R, Moraghebi F (2011) The study of the effects of nano silver technology on salinity tolerance of cumin seed (*Cuminum cyminum* L). Plant and Ecosys 7(25):99–107
Gupta A, Singh SK, Singh MK et al (2020) Plant growth–promoting rhizobacteria and their functional role in salinity stress management. Abatement Env Poll 151–160 https://doi.org/10.1016/B978-0-12-818095-2.00007-2
Gupta R, Xie H (2018) Nanoparticles in daily life: applications, toxicity and regulations. J env patholo, toxicolo oncolo: official organ of the international society for environmental toxicology and cancer 37(3):209–230. https://doi.org/10.1615/JEnvironPatholToxicolOncol.2018026009
Ghanbarzadeh B, Babazadeh A, Hamishehkar H (2016) Nano-phytosome as a potential food-grade delivery system. Food Biosci 15:126–135
Gomes HI, Fan G, Ottosen LM et al (2016) Nanoremediation Coupled to Electrokinetics for PCB Removal from Soil. In: Ribeiro AB, Mateus EP, Couto N (eds) Electrokinetics across disciplines and continents. Springer publication 331–350 https://doi.org/10.1007/978-3-319-20179-5_17
Grover M, Ali SZ, Sandhya V et al (2011) Role of microorganisms in adaptation of agriculture crops to abiotic stresses. World J Microbiol Biotechnol 27:1231–1240
Gatoo MA, Naseem S, Arfat MY et al (2014) Physicochemical properties of nanomaterials: implication in associated toxic manifestations. Biomed Res Int, Article ID 498420(2014):8. https://doi.org/10.1155/2014/498420
García-Gómez C, Obrador A, González D et al (2018) Comparative study of the phytotoxicity of ZnO nanoparticles and Zn accumulation in nine crops grown in a calcareous soil and an acidic soil. Sci Total Environ 644:770–780. https://doi.org/10.1016/j.scitotenv.2018.06.356
Guerra FD, Attia MF, Whitehead DC, Alexis F (2018) Nanotechnology for environmental remediation: materials and applications. Mol 23(7):1760. https://doi.org/10.3390/molecules23071760
Gil-Díaz M, Pinilla P, Alonso J, Lobo MC (2017) Viability of a nanoremediation process in single or multi-metal(loid) contaminated soils. J Haz Mat 321:812–819
Hedayati M, Sharma P, Katyal D, Fagerlund F (2016) Transport and retention of carbon-based engineered and natural nanoparticles through saturated porous media. J Nanopart Res 18:57. https://doi.org/10.1007/s11051-016-3365-6
He C, Liu Z, Xu M et al (2017) Urban expansion brought stress to food security in China: Evidence from decreased cropland net primary productivity. Sci Tot Env 576:660–670
Hirt H and Shinozaki K (2004) Plant responses to abiotic stress. Topics Curr Gene 4 doi:10.1007/b84369
Hakeem KR (2015) Crop production and global environmental issues. Springer International Publishing Switzerland, ISBN 978-3-319-23161-7 https://doi.org/10.1007/978-3319-23162-4
He M, He CQ, Ding NZ (2018) Abiotic stresses: general defenses of land plants and chances for engineering multistress tolerance. Front Plant Sci 9:1771. https://doi.org/10.3389/fpls.2018.01771

Hasanuzzaman M, Nahar K, Alam MM et al (2013) Physiological, biochemical, and molecular mechanisms of heat stress tolerance in plants. Int J Mol Sci 14(5):9643–9684. https://doi.org/10.3390/ijms14059643

Hatami M, Ghorbanpour M (2013) Effect of nanosilver on physiological performance of pelargonium plants exposed to dark storage. J Horti Res 21(1):15–20

Hatami M, Kariman K, Ghorbanpour M (2016) Engineered nanomaterial-mediated changes in the metabolism of terrestrial plants. Sci Total Environ 571:275–291

Husen A, Siddiqi KS (2014) Phytosynthesis of nanoparticles: Concept, controversy and application. Nanoscale Res Lett 9:229–252

Iqbal MS, Singh AK, Singh SP, Ansari MI (2020) Nanoparticles and plant interaction with respect to stress response. Bhushan I, Singh VK, Tripathi D (eds) Nanomaterials and environmental biotechnology, nanotechnology in the life sciences 1–15 doi.org/https://doi.org/10.1007/978-3-030-34544-01

Iqbal MS, Abid M, Ansari MI (2018) Agricultural crops contamination with xenobiotics and its impact on human health. Perspective in Ecotoxicology, ISBN: 978-93-85995 Ch 13:171–181

Jeevanandam J, Barhoum A, Chan YS et al (2018) Review on nanoparticles and nanostructured materials: history, sources, toxicity and regulations. Beilstein J Nanotech 9:1050–1074. https://doi.org/10.3762/bjnano.9.98

Jaberzadeh A, Moaveni P, Moghadam HRT, Zahedi H (2013) Influence of bulk and nanoparticles titanium foliar application on some agronomic traits, seed gluten and starch contents of wheat subjected to water deficit stress. Not Bot Horti Agrobo 41:201–207

Kole C, Banerjee J (2016) Plant nanotechnology: An overview on concepts, strategies, and tools. In: Kole C, Sakthi KD, Mariya VK (eds) Plant nanotechnology: principles and practices. Springer International Publishing, Switzerland 1–14 https://doi.org/10.1007/978-3-319-42154-4

Kah M, Hofmann T (2013) Nanopesticide research: Current trends and future priorities. Env Int 63C:224–235. https://doi.org/10.1016/j.envint.2013.11.015

Khan I, Mohd F, Pratichi S, Padma T (2014) Nanotechnology for environmental remediation. Res J Pharma Biol Chem Sci 5(3):1916

Khan MA and Akhtar M S (2015) Agricultural adaptation and climate change policy for crop production in Africa. In: Hakeem KR (ed) Crop production and global environmental. Springer publication Switzerland https://doi.org/10.1007/978-3-319-23162-4_18

Khan AL, Waqas M, Asaf S et al (2017) Plant growth-promoting endophyte *Sphingomonas* sp LK11 alleviates salinity stress in *Solanum pimpinellifolium.* Env Exp Bot 133:58–69

Lamsal K, Kim SW, Jung JH et al (2011) Application of silver nanoparticles for the control of *Colletotrichum* species *in vitro* and pepper anthracnose disease in field. Mycobiol 39(3):194–199

Lee DJ, Senseman SA, Sciumbato AS et al (2003) The effect of titanium dioxide alumina beads on the photocatalytic degradation of picloram in water. J Agri Food Chem 51(9):2659–2664

Li N, Qian WJ, Wang L et al (2017) Isolation and expression features of hexose kinase genes under various abiotic stresses in the tea plant (*Camellia sinensis*). J Plant Physiol 209:95–104

Mukhopadhyay SS, Kaur N (2016) Nanotechnology in Soil-Plant System. In: Kole C, Sakthi KD, Mariya VK (eds) Plant nanotechnology: principles and practices. Springer International Publishing, Switzerland 329–348 https://doi.org/10.1007/978-3-319-42154413

Mahmoodzadeh H, Nabavi M, Kashefi H (2013) Effect of nanoscale titanium dioxide particles on the germination and growth of canola (*Brassica napus*). J Ornamental Hortic Plants 3:25–32

Mudalkar S, Sreeharsha RV, Reddy AR (2017) Involvement of glyoxalases and glutathione reductase in conferring abiotic stress tolerance to *Jatropha curcas* L. Env Exp Bot 134:141–150

Moon JW, Phelps TJ, Fitzgerald CL et al (2016) Manufacturing demonstration of microbially mediated zinc sulfide nanoparticles in pilot-plant scale reactors. Appl Microbiol Biotechnol 100:7921–7931. https://doi.org/10.1007/s00253-016-7556-y

Manyi-Loh C, Mamphweli S, Meyer E, Okoh A (2018) Antibiotic use in agriculture and its consequential resistance in environmental sources: potential public health implications. Mol 23(4):795. https://doi.org/10.3390/molecules23040795

Mani PK, Mondal S (2016) Agri-nanotechniques for plant availability of nutrients. In: Kole C, Sakthi KD, Mariya VK (eds) Plant nanotechnology: principles and practices. Springer International Publishing, Switzerland, 263–303. https://doi.org/10.1007/978-3-319-42154-411

Meena RS, Gogoi N, Kumar S (2017) Alarming issues on agricultural crop production and environmental stresses. J Clean Prod 142(4):3357–3359

Omena BO, Olanrewaju OS, Babalola OO (2019) Plant growth promoting rhizobacterial mitigation of drought stress in crop plants: Implications for sustainable agriculture. Agron 9(11):712. https://doi.org/10.3390/agronomy9110712

Patra AK, Adhikari T, Bhardwaj AK (2016) Enhancing crop productivity in salt-affected environments by stimulating soil biological processes and remediation using nanotechnology. In: Dagar JC, Sharma PC, D.K. Sharma DK, Singh AK (eds) Innovative saline agriculture, Springer India, pp 83–103 https://doi.org/10.1007/978-81-322-2770-04

Parveen A, Hussain F (2008) Salinity tolerance of three range grasses at germination and early growth stages. Pak J Bot 40:2437–2441

Patra JK, Das G, Fraceto LF, Campos E et al (2018) Nano based drug delivery systems: recent developments and future prospects. J Nanobiotechnol 16(1):71. https://doi.org/10.1186/s12951-018-0392-8

Pandey P, Ramegowda V, Senthil-Kumar M (2015) Shared and unique responses of plants to multiple individual stresses and stress combinations: physiological and molecular mechanisms. Front Plant Sci 6:723. https://doi.org/10.3389/fpls.2015.00723

Paul B, Merly C, Kvapil P, Hans-Peter K (2018) Status of nanoremediation and its potential for future deployment: Risk-benefit and benchmarking appraisals. The J Env Cleanup Costs Technolo Tech, Remediation. https://doi.org/10.1002/rem.21559

Plascencia-Villa G, Starr CR, Armstrong LS et al (2012) Imaging interactions of metal oxide nanoparticles with macrophage cells by ultra-high resolution scanning electron microscopy techniques. Integrative biology: quantitative biosciences from nano to macro 4(11):1358–1366. https://doi.org/10.1039/c2ib20172k

Panpatte DG, Jhala YK, Shelat HN, Vyas RV (2016) Nanoparticles: the next generation technology for sustainable agriculture. In: Singh D, Singh H, Prabha R (eds) Microbial inoculants in sustainable agricultural productivity Springer publication India 289–300 https://doi.org/10.1007/978-81-322-2644-418

Pandey P, Irulappan V, Bagavathiannan MV, Senthil-Kumar M (2017) Impact of combined abiotic and biotic stresses on plant growth and avenues for crop improvement by exploiting physio-morphological traits. Front plant sci 8:537. https://doi.org/10.3389/fpls.2017.00537

Rossini MA, Maddonni GA, Otegui ME (2016) Multiple abiotic stresses on maize grain yield determination: additive vs. multiplicative effects. Field Crops Res 198:280–289

Remedios C, Rosario F, Bastos V (2012) Environmental nanoparticles interactions with plants: morphological, physiological and genotoxic aspects. J Bot Article ID 751686:8. https://doi.org/10.1155/2012/751686

Rashidi L, Khosravi K (2011) The Applications of Nanotechnology in Food Industry. Crit Rev Food Sci Nutr 51:723–730. https://doi.org/10.1080/10408391003785417

Rizwan M, Ali S, Qayyum MF et al (2017) Effect of metal and metal oxide nanoparticles on growth and physiology of globally important food crops: A critical review. J Hazard Mater 322(A):2–16

Rejeb IB, Pastor V, Mauch-Mani B (2014) Plant Responses to Simultaneous Biotic and Abiotic Stress: Molecular Mechanisms. Plants 3(4):458–475. https://doi.org/10.3390/plants3040458

Reddy PVL, Hernandez-Viezcas JA, Peralta-Videa JR, Gardea-Torresdey JL (2016) Lessons learned: are engineered nanomaterials toxic to terrestrial plants? Sci Tot Env 568:470–479

Raza A,.Ali R, Sundas SM et al (2019) Impact of climate change on crops adaptation and strategies to tackle its outcome: A review. plants 8(2):34 doi.org/https://doi.org/10.3390/plants8020034

Shallan MA, Hassan HM, Namich AAM, Ibrahim AA (2016) Biochemical and physiological effects of TiO_2 and SiO_2 nanoparticles on cotton plant under drought stress. Res J Pharma Biol Chem Sci 7(4):1541

Shang Y, Hasan MK, Ahammed GJ et al (2019) Applications of nanotechnology in plant growth and crop protection: A review. Mol 24(14):2558. https://doi.org/10.3390/molecules24142558

Servin AD, White JC (2016) Nanotechnology in agriculture: Next steps for understanding engineered nanoparticle exposure and risk. NanoImpact 1:9–12. https://doi.org/10.1016/j.impact.2015.12.002

Seghatoleslami M, Forutani R (2015) Yield and water use efficiency of sunflower as affected by nano ZnO and water stress. J Adv Agri Tech 2(1):34–37

Song U, Jun H, Waldman B et al (2013) Functional analyses of nanoparticle toxicity: a comparative study of the effects of TiO_2 and Ag on tomatoes (*Lycopersicon esculentum*). Ecotoxicol Env Saf 93:60–67. https://doi.org/10.1016/j.ecoenv.2013.03.033

Smith P, House J, Bustamante M et al (2016) Global change pressures on soils from land use and management. Global Change Bio 22:1008–1028. https://doi.org/10.1111/gcb.13068

Shelar GB, Chavan AM (2015) Mycosynthesis of silver nanoparticles from *Trichoderma harzianum* and its impact on germination status of oil seed. Biolife 3(1):109–113

Singh J, Lee BK (2016) Influence of nano TiO_2 particles on the bioaccumulation of Cd in soybean plants (*Glycine max*): A possible mechanism for the removal of Cd from the contaminated soil. J Env Manag 170:88–96

Tarafdar JC, Raliya R, Mahawar H, Rathore I (2014) Development of zinc nanofertilizer to enhance crop production in pearl millet (*Pennisetum americanum*). Agric Res 3(3):257–262. https://doi.org/10.1007/s40003-014-0113-y

Vannini C, Domingo G, Onelli E et al (2013) Morphological and proteomic responses of *Eruca sativa* exposed to silver nanoparticles or silver nitrate. PLoS ONE 8(7):

Venkatachalam P, Jayaraj M, Manikandan R et al (2016) Zinc oxide nanoparticles (ZnO NPs) alleviate heavy metal-induced toxicity in *Leucaena leucocephala* seedlings: a physiochemical analysis. Plant Physiol Biochem 110:59–69. https://doi.org/10.1016/j.plaphy.2016.08.022

Vidyalakshmi N, Thomas R, Aswani R et al (2017) Comparative analysis of the effect of silver nanoparticle and silver nitrate on morphological and anatomical parameters of banana under in vitro conditions. Inorg Nano Metal Chem 47:1530–1536. https://doi.org/10.1080/24701556.2017.1357605

van der Laan M, Bristow KL, Stirzaker RJ, Annandale JG (2017) Towards ecologically sustainable crop production: A South African perspective. Agri Ecosys Env 236:108–119

Wang XD, Ou-yang C, Fan ZR et al (2010) Effects of exogenous silicon on seed germination and antioxidant enzyme activities of *Momordica charantia* under salt stress. J Anim Plant Sci 6:700–708

Wang P, Sun X, Jia X, Ma F (2017) Apple autophagy-related protein MdATG3s afford tolerance to multiple abiotic stresses. Plant Sci 256:53–64

Wani SH, Kumar V, Shriram V, Sah SK (2016) Phytohormones and their metabolic engineering for abiotic stress tolerance in crop plants. The Crop J 4(3):162–176

Waqas MA, Kaya C, Riaz A et al (2019) Potential mechanisms of abiotic stress tolerance in crop plants induced by thiourea. Front Plant Sci. https://doi.org/10.3389/fpls.2019.01336

Yılmaz EE, Özgür N, Bereli D et al (2017) Plastic antibody based surface plasmon resonance nanosensors for selective atrazine detection. Mat Sci Engine: C 73:603–610

Yang F, Hong F, You W et al (2006) Influence of nano-anatase TiO_2 on the nitrogen metabolism of growing spinach. Biolo Trace Elem Res 110(2):179–190

Yan A, Chen Z (2019) Impacts of silver nanoparticles on plants: a focus on the phytotoxicity and underlying mechanism. Int J Mol Sci 20(5):1003. https://doi.org/doi.org/10.3390/ijms20051003

Zaytseva O, Neumann G (2016) Carbon nanomaterials: production, impact on plant development, agricultural and environmental applications. Chem Biol Technol Agric 3:17. https://doi.org/10.1186/s40538-016-0070-8

Zuverza-Mena N, Martínez-Fernández D, Du W et al (2017) Exposure of engineered nanomaterials to plants: Insights into the physiological and biochemical responses-A review. Plant Physiol et Biochem 110:236–264. https://doi.org/10.1016/j.plaphy.2016.05.037

Zhu ZJ, Wang H, Yan B et al (2012) Effect of surface charge on the uptake and distribution of gold nanoparticles in four plant species. Environ Sci Technol 46:12391–12398. https://doi.org/10.1021/es301977w

Zhu JK (2016) Abiotic stress signaling and responses in plants. Cell 167(2):313–324. https://doi.org/10.1016/j.cell.2016.08.029

Index

J. M. Al-Khayri et al. (eds.), *Nanobiotechnology*,
https://doi.org/10.1007/978-3-030-73606-4

B

V

W

X

Y

Z

Printed by Printforce, the Netherlands